American Institute of Chemical Engineers

Sustainable Engineering Forum Conference

Topical Conference at the
2008 AIChE Spring National Meeting

April 6-10, 2008
New Orleans, Louisiana, USA

Volume 1 of 2

Printed from e-media with permission by:

Curran Associates, Inc.
57 Morehouse Lane
Red Hook, NY 12571
www.proceedings.com

ISBN: 978-1-60560-226-4

ISBN: 978-1-60560-226-4

American Institute of Chemical Engineers
Proceedings
Three Park Avenue
New York, NY 10016-5991
Phone: 212-591-8100

www.aiche.org

American Institute of Chemical Engineers

Sustainable Engineering Forum Conference
2008

TABLE OF CONTENTS

Volume 1

Volume 2

36c Advances in Molecular Modeling of Bulk and Interfacial Phase Behavior and Microstructure in Copolymers and Associating Fluids

Walter G. Chapman, Shekhar Jain, Adam Bymaster, Chris Emborsky, and Kenneth R. Cox. Chemical and Biomolecular Engineering Department, MS-362, Rice University, 6100 Main St., Houston, TX 77005

Many modern applications of macromolecular fluids take advantage of their novel microstructure and phase behavior. The challenge in modeling nanostructured fluids lies in understanding the new physics that emerges from finite-size effects, varying dimensionality, surface forces and interplay of multiple length and time scales. Over the past twenty years, significant advances in statistical mechanics have provided engineers with accurate equations of state for bulk phase behavior of mixtures containing associating and polymeric components. More recently, developments in statistical mechanics are providing scientists and engineers with valuable tools to correlate and predict bulk and interfacial phase behavior and microstructure in these complex systems. Insights will be provided in the development of these theories. Opportunities for applications and extension of these models also will be discussed along with future challenges.

36d New Picture of First-Order Phase Transitions in Metastable Fluids: Homogeneous Nucleation and Growth

David S Corti, School of Chemical Engineering, Purdue University, 480 Stadium Mall Drive, West Lafayette, IN 47907-2100

Homogeneous nucleation is the activated process by which the new phase (vapor or liquid) is formed from a bulk metastable fluid (superheated liquid or supercooled vapor, respectively) in the absence of impurities or solid surfaces. Recently, we showed that bubble nucleation is more appropriately described by an "activated instability". As the free energy barrier is surmounted, further growth of the post-critical bubbles must necessarily proceed via a mechanism appropriate for an unstable system. Ongoing work also indicates that limits of stability also appear for droplet nucleation, though important differences between droplet and bubble formation still remain. We speculate on what the appropriate signature is needed to identify these unstable growth stages. We discuss how some current expressions for the rate of nucleation need to be modified to incorporate the different properties of the free energy surface that emerge from our studies.

111e Nonequilibrium Dynamics and the Electrostatic Charging of Granular Systems

Daniel J. Lacks, Department of Chemical Engineering, Case Western Reserve University, Cleveland, OH 44106

As granular materials flow, the particle contacts can cause electrostatic charges to develop on the particles. Despite the widespread significance of this phenomenon, the origins of the electrostatic charging are unclear: e.g., when all particles are chemically identical, this would seem to preclude any driving force for charge transfer. We suggest that this electrostatic charging is a special case of a more general class of non-equilibrium phenomena. For species that cannot equilibrate between states on a particle surface (e.g., electrons in insulators), only particle collisions can facilitate transitions to different states. The non-equilibrium dynamics that follow from collision-induced transitions generate species accumulation on a particle-size-dependent subset of the system. These ideas rationalize experiments on granular systems which show that smaller particles charge negatively, and that the charging magnitude increases with the breadth of the particle size distribution.

36a Simulation Methods for the Thermodynamics and Kinetics of Complex Systems

Fernando A. Escobedo, School of Chemical and Biomolecular Engineering, Cornell University, 120 Olin Hall, Cornell University, Ithaca, NY 14850

This talk will review of some of the recent developments on the methodology for simulating of both equilibrium and kinetic properties. For equilibrium systems, a framework for the simulation of both isothermal and non-isothermal ensembles will be described to show how it can be used to generalize methods to simulate density of states of free-energy functions for a wide range of conditions. Optimized expanded ensembles for simulations involving molecular insertion and deletions will also be described. For kinetic properties, the focus will be on methods to sample rare events, in particular, on a class of transition path sampling methods known as forward flux sampling. In this context, a technique will be described that allows a suitable selection of effective reaction coordinates. In all cases, sample applications will be presented to illustrate some of the strengths and limitations of the methods, and to help identify where future productive directions may lie.

36e Mainstreaming Molecular Simulation in Chemical Engineering Education and Application

J. Richard Elliott, Chemical Engineering Dept, University of Akron, Akron, OH 44325-3906

Molecular simulation has made rapid progress in recent years, especially as a research tool. For example, recent results from the Industrial Fluid Properties Simulation Collaborative (IFPSC) have shown that simulation results can rival experimental measurements in accuracy, exposing flawed data and extending limited data. Demonstrations of fundamental phenomenology abound, promising future applications to self assembly, amyloid aggregation, entanglement, and transition states.

These developments suggest that molecular simulation is poised to supplant engineering methods like group contributions and corresponding states for engineering properties, at the least. More broadly, future chemical engineers may design molecular structures in the same way that civil engineers design bridges now.

Incorporating this rapid change into the classroom may seem daunting. The curriculum is already crowded. On the other hand, technology is changing the mode of presentation as well as its substance. ConcepTesting permits rapid introduction and assessment of qualitative concepts. Testing can take place in a computational laboratory. These observations suggest a trend with less emphasis on tedious hand calculations while placing greater emphasis on fundamental derivations and conceptual reasoning.

This presentation reviews these developments and describes practical ways of implementing them in the undergraduate classroom.

36b Molecular Simulation of Advanced Materials

Monica H. Lamm, Department of Chemical and Biological Engineering, Iowa State University, 2114 Sweeney Hall, Ames, IA 50011

New developments in synthetic and materials chemistry continue to produce a wide variety of building blocks for advanced materials that possess novel structure and function. This talk will describe our recent work aimed at connecting the molecular-level properties of nanoscale building blocks to the macroscopic properties of materials. Examples to be covered include block copolymer templating of nanoparticles, self-assembly of linear-dendritic polymers, and diffusivity of fractal aggregates.

166d Investigation of Fouling of Crudes with Low Asphaltenes

Parviz Rahimi[1], Irwin A. Wiehe[2], and Teclemariam Alem[1]. (1) National Centre for Upgrading Technology (NCUT), 1 Oil patch Drive, Suite A202, Devon, AB T9G 1A8, Canada, (2) Soluble Solutions, 3 Louise Lane, Gladstone, Gladstone, NJ 07934

The fouling of Light crudes was investigated using Hot Liquid Process Simulator (HLPS) with Carbon steel tubes. The fouling proensities of the crudes were correlated with the properties of the crudes.

166a Isokinetic Phenomena in Crude Oil Fouling

Christopher A. Bennett, Heat Transfer Research, Inc, 150 Venture Drive, College Station, TX 77845 and Barry D. Crittenden, Department of Chemical Engineering, University of Bath, Bath, BA2 7AY, United Kingdom.

Our studies of crude oil fouling have revealed two manifestations of the isokinetic phenomenon. First, initial fouling rates of four proprietary crude oils clearly exhibit an isokinetic temperature in the middle of their Arrhenius plot. A plot of natural logarithms of pre-exponential factors versus apparent activation energies is linear, confirming the validity of the isokinetic temperature and indicating the presence of the compensation effect. Second, the initial fouling rates of Maya crude oil and styrene fouling systems become velocity-independent, thereby creating another isokinetic condition, at temperatures of 274 °C and 77 °C, respectively. The compensation effect is also observed in the Maya crude oil and styrene systems but here the linearity in the logarithms of pre-exponential factors with apparent activation energies is due to changes in velocity instead of changes in fluid type. This paper will discuss the background of the compensation effect and its manifestation in the isokinetic phenomenon.

166b Design Approach for Mitigating Fouling of Reboilers and F/E Exchangers

C.B. Panchal, Argonne National Laboratory, Chicago, IL 60439 and Blazo Ljubicic, Koch Heat Transfer Company, 12606 FM 529, Houston, TX 77041.

Significant progress has been made in understanding the underlying mechanisms of hydrocarbon fouling and characterizing the effects of physical and chemical parameters. However, the research and development focuses on single-phase flows. Industrial heat exchangers in refining and petrochemical plants operate under multi-phase flow conditions and fouling and localized corrosion is a common problem in such heat exchangers, particularly in reboilers. This is because design guidelines are inadequate to minimize localized process conditions that induce severe fouling, which could also lead to localized corrosion. In a new design approach conditions under different operating conditions of reboilers and feed/effluent exchangers are evaluated and the propensity of high fouling rates are detected on the basis of fluid and thermal condiitons that may trigger chemical mechnisms of precursor formaiton. The design approach in this study is based on evaluating local threshold condiitons and designing reboilers and F/E exchanger to minimize propensity of precursor generation. Such a design may not be optimum from heat transfer and pressure drop for clean conditions, but it will minimize fouling and localized corrosion. The result is lower lifecycle ownership cost of the exchanger.

166c Heat Exchanger Performance Enhancement through the Use of Tube Inserts

Artur W. Krueger, Thermal Enhancement Systems, PETROVAL - HOUSTON, P.O. BOX 19686, Houston, TX 77224-9686

Heat exchanger tube inserts have been used for many years as practical methods for fouling control and heat transfer enhancement, and their benefits are clearly established in numerous industrial installations. Through a combination of increased turbulence and mechanical movements, a reduction of the tube side fouling layer and an improvement of the inner heat transfer rate are achieved.

Several application examples will be presented, showing comparisons between equipped and unequipped exchangers in refinery crude units, in different flow conditions, such as

- crude vs. residue

- crude vs. reflux

- reboilers

Also, the combination of tube inserts with antifoulant additives and chemical cleaning will be discussed.

Benefits derived from the utilization of tube inserts include energy savings, enhanced heat transfer levels, improved online availability of equipment, extended run lengths of exchangers, reduction in maintenance costs, production enhancements through avoidance of down-times, de-bottlenecking, and the possibility of plant emission reductions.

Thermodynamic Model of Sediment Deposition in the LC-FINING® Process

K.M.Sundaram, U.Mukherjee* and M.Baldassari*
Lummus Technology, Bloomfield, NJ 07003.
* Chevron Lummus Global

Abstract

The LC-FINING® ebullated bed hydrocracking process is used to hydrocrack residuum. In this process, sediment deposition in the downstream equipment sometimes affects the overall economics by limiting the operating conversion, even though the upstream reactor system is capable of achieving much higher conversion levels. The sediment is measured by SHFT value (Shell Hot Filtration Test method), which mainly measures that portion of the asphaltene that is insoluble in heavy oil at specific laboratory test conditions. A solubility-theory-based thermodynamic model is developed to predict the SHFT value. According to this model, asphaltenes are in equilibrium (in solution) with the surrounding fluid. The model parameters are asphaltene molecular weight and its solubility parameter. Both are related. Solubility parameter is modeled as a function of the hydrogen-to-carbon (H/C) atomic ratio of the heavy oil. The molecular weights are estimated from pilot plant data by minimizing the sum of squares of the residuals between the observed SHFT value and the calculated value for all runs. Asphaltene molecular weight decreases with increasing process severity, indicating that the asphaltene becomes more aromatic by losing side chains. The simple model is able to explain the observed trends with respect to different LC-FINING catalysts, feed sources, and diluent effects. Commercial data is compared with the pilot plant data and the model prediction validates the model.

Introduction

The search for Best Available Technology ("BAT") has intensified over the last few years because of diminishing supplies of sweet crudes and incremental supplies coming predominantly from heavy sour crudes and synthetic crudes. Various catalytic residue-upgrading technologies are available from Chevron Lummus Global ("CLG") including residue desulfurization technologies (ARDS, VRDS, UFR, OCR) and the LC-FINING® process (Dahlberg et.al, 2007; Mukherjee et. al, 2007). The LC-FINING process combined with integrated ISOCRACKING (the CLG hydrocracking Process) offers a proven high conversion option (Spieler et. al, 2006). The combined process is especially attractive in situations requiring high conversion of residuum with high metals content and where diesel demand is higher than gasoline demand. In the LC-FINING process a series of ebullated bed reactors are used. A suitable catalyst such as Ni/Mo extrudate catalyst particles, suspended in the liquid and in the presence of high hydrogen partial pressure (100-180 bar) and at moderate temperature (400-440°C) removes the heteroatoms and the metals while converting 60 to 80% of the residue. Subsequently, further upgrading can be done with integrated hydrocracking. Typically anywhere from one to three reactors are used for the LC-FINING process and one or two additional fixed bed reactors for the integrated hydrocracking segment. To keep the catalyst activity high at all times, a small amount of catalyst is removed from the reactors daily and an equivalent quantity of fresh catalyst is added. The feed usually contains the residuum and a small quantity of an aromatic diluent. Some internal recycle streams are used, as well.

In its natural state, residua have no or very low SHFT value. However, it is the conversion process that creates the sediments by disturbing the SARA (saturates, aromatics, resins, and asphaltenes) balance. The ebullated bed reaction system emulates a CSTR. The nature of the feedstock is such that high operating temperatures are required for conversion and a significant amount of thermal cracking occurs in addition to normal hydroconversion. Hydrogen partial pressure and residence time are major variables in aiding hydrogenation and ensuring that the conversion of residuum occurs in an environment where the resin-to-asphaltene ratio remains in a domain where sedimentation does not occur.

After the reaction section, the products are recovered in the separation section. Considerable research and development, backed by commercial feedback, has been spent in understanding the variables within the reactor system. This paper focuses on the other parts of the process that are impacted by sediment deposits. Unless addressed by process parameters and equipment design, the sediment issue can reduce on-stream factor and conversion targets significantly.

The sections most often impacted by sediment deposits are the segments from the bottom of the atmospheric and vacuum towers and in the vacuum tower bottoms product rundown circuit. Unlike the reactor system where sediment formation can be addressed by increased catalyst addition, sediment formation in equipment such as the vacuum bottoms product rundown exchangers cannot be addressed or even predicted, easily.

In the commercial application sediment deposition affects heat transfer and pressure drop. Unfortunately, the physical and chemical properties of the deposits are rarely analyzed. Sediment comes in different forms. It can be hard like coke, or in gelatinous, gummy form. In commercial units CLG has observed different types of sediments at different points in the residue rundown circuit, with the nature of the deposit being a strong function of operating temperature. It has also observed that these deposits can be rendered back into solution at reactor operating temperatures. Currently, the prevalent mechanism for removing sediment deposits from exchanger circuits is by mechanical cleaning or hydrojetting.

To design the LC-FINING unit for a specific feed, CLG performs detailed pilot plant work and collects necessary design data. Unlike the commercial unit, there is no continuous withdrawal or addition of catalyst in the pilot plant. After a specified decline in activity, a new batch of catalyst is used. A detailed product analysis of the gas and liquid samples was obtained for each run. This includes measuring the Shell hot filtration test (SHFT) value for the liquid products. The SHFT value approximately measures that portion of the asphaltenes that are insoluble in the heavy oil along with some inorganic materials. A portion of the asphaltenes deposits in the separation section and causes the observed increase in pressure drop in the equipment. Hence, in this paper, SHFT value is modeled and it is used to represent the asphaltene content of the sediment.

Solubility Model

Asphaltene is not a simple molecule like benzene or naphthalene. Because of its complexity it is defined by its solubility. It is measured by a SARA analysis, which measures the residue into saturates, aromatics, resins and asphaltenes. Therefore, the physical properties like molecular weight or hydrogen content of asphaltene can vary from feed to feed. Even for the same feed, these physical properties may vary when they are subjected to reaction conditions (Sheu&Mullins, 1995; Wiehe, 2007). Asphaltene in its neat form is a solid. In their classical paper on asphaltene deposition in crude oil processing Hirschberg et.al (1984) considered a modified Flory-Huggins polymer solution theory where asphaltene is treated like a polymer. The proposed solubility model is based on thermodynamic approach and it has two parameters related to asphaltene (solubility parameter and molecular weight). All other parameters are related to the surrounding fluid. For the proposed model, asphaltene (solute) is in equilibrium with the heavy oil (solvent) and the following equilibrium can be applied.

$$K_i = X_s/X_l \qquad (1)$$

where X_s and X_l are the mole fraction of asphaltene in the solid and liquid phases respectively and K_i is the distribution coefficient (or equilibrium constant) for i-th phase. Assuming the asphaltene concentration is low and the entropy of mixing can be adequately modeled by Flory-Huggins entropy of mixing, the following equation can be obtained.

$$\ln(K_s)= 1- V_s/V_l+ \ln(V_s/V_l)+V_s/(RT)*(\delta_l-\delta_s)^2 \qquad (2)$$

For the liquid phase, or the solid phase the solubility parameter (δ) is calculated by Hildebrand's definition. Both heat of fusion and heat of vaporization required for solubility parameter are difficult to measure for asphaltene and hence the solubility parameter of asphaltene is treated as an optimization parameter in the current model. Assuming solute (asphaltene) activity coefficient is unity, and only pure asphaltene is present in the asphaltene phase, Eq.(2) predicts the maximum amount of asphaltene deposited. In Eq. (2) all parameters except the molar volume (V_s) and the solubility parameter (δ_s) of asphaltene are known (measured) or calculable from an equation of state. Hence the two unknown parameters are the molecular weight and the solubility parameter of asphaltene.

Rogel (1998) measured and calculated many physical properties of oils, aromatics, resins and asphaltenes from various sources and processes and concluded that the solubility parameter can be correlated to H/C atomic ratio. The following equation is obtained from Rogel's data, which is also used in this work.

$$\delta = 35.87 - 10.477* (H/C) \qquad (3)$$

Estimation of Asphaltene Parameters

The physical properties of some of the undiluted feedstocks processed in different pilot plant campaigns are summarized in Table 1. The properties of some diluents are also included. Assuming each SHFT value measured represents the maximum asphaltene deposition at the corresponding operating conditions, Eq. (2) can be used to estimate the molecular weight or solubility parameter of asphaltene when the other parameter is known. When both parameters are independently known then the calculated asphaltene concentration can be compared with the experimental value. Initially the weighted sum of squares between the calculated and the measured SHFT values was minimized with one set of values for the asphaltene molecular weight and the solubility parameter for each feedstock: Within the experimental accuracy, it was also found here that the asphaltene solubility parameter was related to the solubility parameter of the surrounding heavy oil. The following equation was derived from a best fit all experimental data:

$\delta_{asph} = 1.33 * \delta_{HYO}$ (4)

Table.1 Feed Properties and Diluents used in Pilot plant Runs

	Vacuum Residues					Diluents and Additives		
Feed Source	Arabian Heavy	Cold Lake	Bitumen	900F+Athabasca Bitumen	Ural	HCO	LCO	MVGO
Specific gravity	1.0389	1.0602	1.0608	1.0452	1.012	1.1106	0.9279	0.971
Carbon, wt%	83.67	84.13	83.50	82.61	86.28	91.59	88.48	85.47
Hydrogen, wt%	10.17	9.51	9.52	9.98	10.12	6.73	10.22	10.84
Sulfur, wt%	5.41	5.08	5.6	5.89	2.76	0.63	0.52	3.43
Nitrogen, wt%	0.6	0.68	0.78	0.44	0.85	0.59	0.11	0.24
Oxygen, wt%(est)	0.15		0.6	1			0.67	
Ash			0.035	0.08	0.08			
Nickel, ppm	67	98	85	90	49	1.9		<0.01
Vanadium, ppm	156	255	263	276	140	0.4		<0.01
MCRT,wt%	22.25	20.79	21.57	20.15	17.46			
Pentane Insol., w%		21.53	26.19		10.74			
Heptane Insol., wt%	12.56	19.7	19.0		6.73			
Toluene Insol., wt%		0.13	0.12		0.03	0.03		
SHFT, wt%	0.03							
524C+Res. Vol%	92.3	78.5	89.265	84.5	95.1	13		12

As a result for a perfect fit of the measured SHFT value, the molecular weight of asphaltene can be calculated from Eq.(2). The calculated molecular weights are shown in Fig. 1.

Mannistu et al. (1997) showed that the asphaltene molecular weight was inversely proportional to its square of the solubility parameter. Therefore, in Fig.1 square root of molecular weight instead of the molecular weight of asphaltene is shown against its solubility parameter. From this figure, it is clear that asphaltene molecular weight spans a narrow band and often a single straight line can be drawn for each feedstock. Therefore, for each feed the following equation was used.

$M_a = (b_o + b_1 * \delta_{asph})^2$ (5)

where M_a is the molecular weight of asphaltene and b_0 and b_1 are constants for a given feed.

With the optimized values for the constants b_o and b_1, the calculated sediment values are compared with the reported values for all runs in Fig.2. The parity line is also shown in Fig.2. As discussed earlier, within the accuracy of SHFT value the model predicts the reported values satisfactorily. The simple model is able to explain some of the observed behaviors in the pilot units.

Fig.1 Estimated Asphaltene Molecular weight for Perfect Fit of Sediment Value

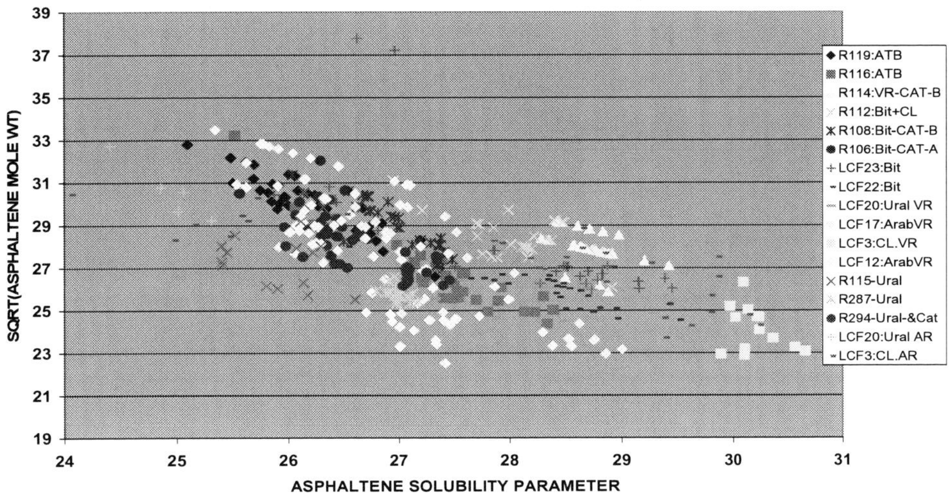

Fig.2 Calculated by Solubility Model and Reported Sediment Values in Heavy Oil

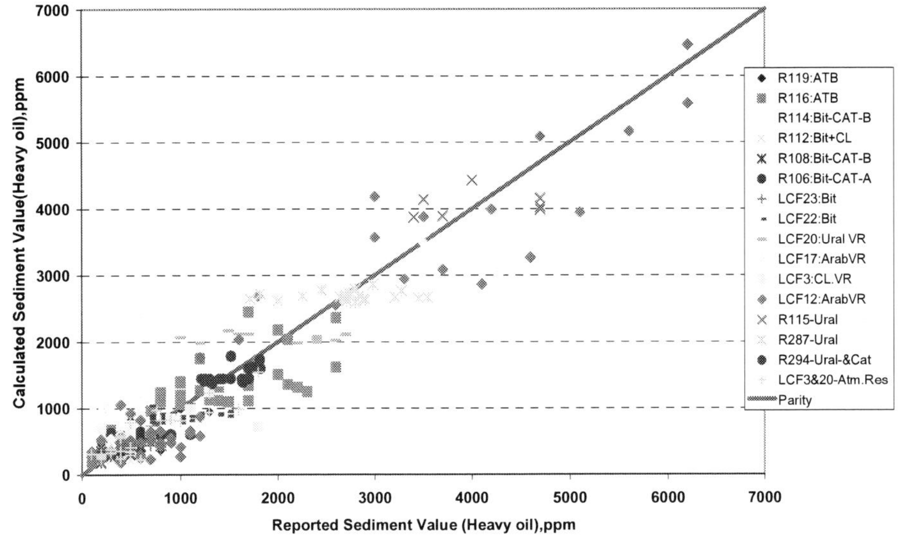

Catalyst Effect

Athabasca bitumen residue was processed with two different commercial catalysts. Run #106 was operated with Catalyst-A and Run #108 was operated with Catalyst-B.

Figure 3 shows the calculated asphaltene molecular weight versus asphaltene solubility parameter for the above experiments. It appears that Catalyst-A gave a low molecular weight at a given solubility parameter. Since the feedstock is the same for both catalysts, low asphaltene molecular weight represents the loss of possible side chains. At the operating temperatures, thermal cracking in liquid phase is significant. The products of thermal cracking are mainly methane and light hydrocarbon components. A plot of sum of C_1 to C_5 components, which are mainly formed by thermal cracking as a function of severity confirms this theory. Within the accuracy of experimental data, only qualitative trends can be established. It is possible that the Catalyst-A promotes cracking, but that can be confirmed by analyzing the asphaltene itself in detail.

Fig.3 Calculated Asphaltene Molecular Weight for different Catalysts for Processing 525C+ Bitumen

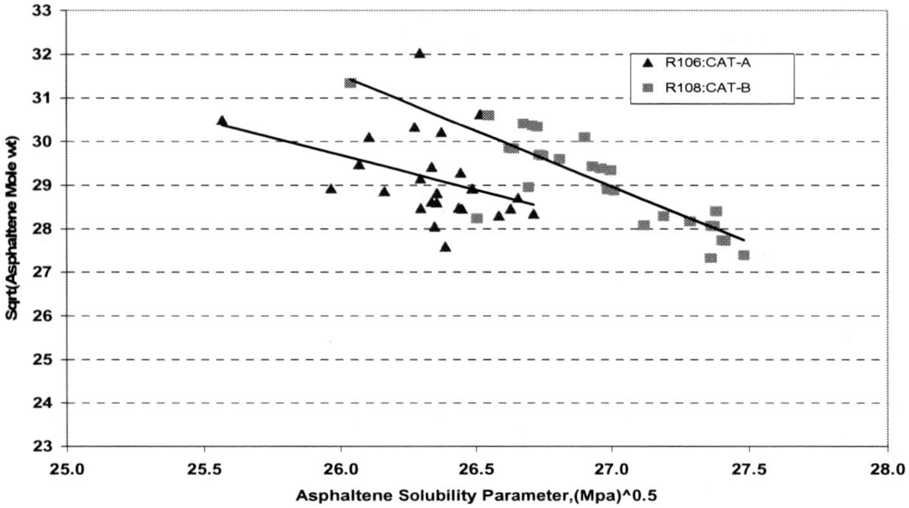

Effect of Diluents

Arabian heavy vacuum residue was processed with two different diluents: straight run VGO and heavy cycle oil. Table 2 shows some of the characteristics of the blends. Blend#2 and #4 are essentially identical: The differences simply reflect different drums used for blending and the accuracy of measurement.

Table 2 Feed Characteristics for different blends

Feed Blend#	1	2	3	4
Residue,wt%	92.8	87	100	87
VGO,wt%	7.2	0	0	0
HCO,wt%	0	13	0	13
Total	100	100	100	100

Measured sediment values in the heavy oil product are shown in Fig.4. Blend #3, which consists of vacuum residue plus internal VGO exhibits the highest sediment value. Blend #4, which is same as blend #2 was processed at high severity. Calculated asphaltene molecular weights are shown in Fig.5. Again a plot of sum of C_1 to C_5 component yields confirms the trend giving some validity to the hypothesis that the molecular weight of asphaltene changes with severity by losing its side chains.

Generalized correlation

Asphaltene content in the feed is an inherent property of the crude, and even for the same crude, it varies over time when the crude is pumped from the reservoir. It is similar to obtaining a correlation for the nitrogen content or sulfur content of crude: there are no direct correlations for estimating these properties, but general trends that have been observed. Asphaltene molecular weight is no exception to this trend. The following correlation was obtained as a function of feed specific gravity for a typical value of δ_{asph}= 26.8.

$M_a = 737.99(SG)^{2.4675}$

Generally heavy crudes have higher asphaltene molecular weight.

Fig.4 Effect of Adding Different Diluents on Sediment Value for Processing Arabian Heavy Vacuum Residue

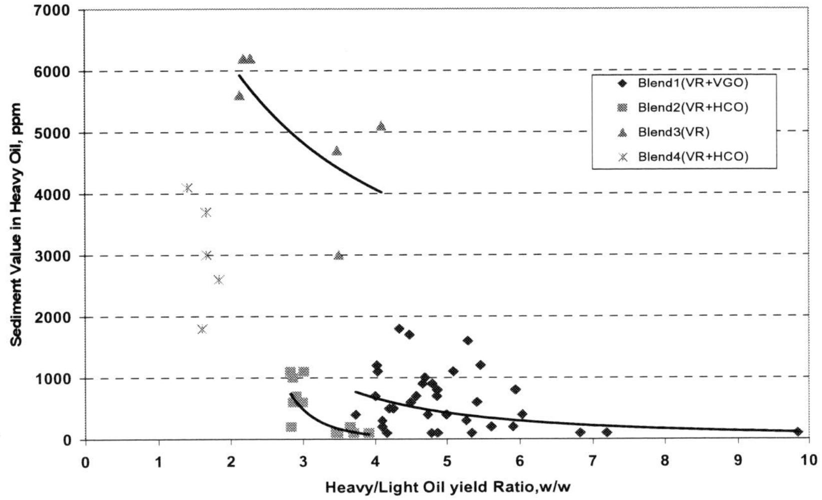

Fig.5 Calculated Asphaltene Molecular Weights for Mixing Different Diluents for Arabian Vacuum Residue

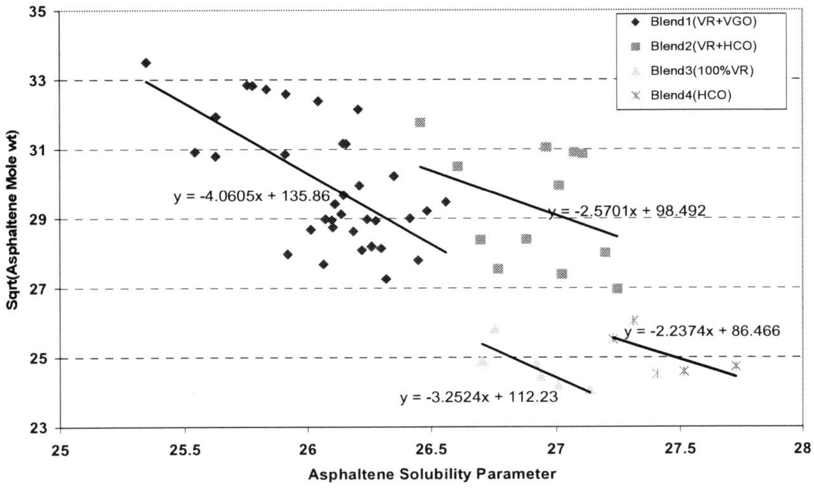

Summary and Conclusions

Commercial experience indicates that with certain feeds, conversion of residuum in residue hydrocracking processes such as the LC-FINING process can be restricted by rapid deposition of sediments in equipment, in particular is exchangers in the fractionation section. Over the last two decades, Lummus and Chevron Lummus Global have carried out many pilot plant experiments with various feeds to extract hydroprocessing kinetics and to obtain product distribution for scale-up. The potential for sediment deposits is measured by Shell hot filtration test (SHFT) method. A solubility-theory-based model was developed to predict the SHFT value measured in the pilot plants. The key parameters are the molecular weight and the solubility parameter of asphaltene, which are estimated from the data by an optimization technique that minimized the sum of squares of the residuals between the measured and the calculated sediment values.

The asphaltene solubility parameter was correlated as a function of the H/C atomic ratio of heavy oil. The progression of asphaltene molecular weight with severity was measured by the heavy oil/light oil product ratio: increasing severity showed a decrease in molecular weight. This is attributed to the loss of side chains in the asphaltene molecule. Thus it became more aromatic compared to the surrounding oil. When the right balance was not maintained, the asphaltene separated out and deposited in the equipment as soon as the temperature fell below certain levels. The reduction in molecular weight is also supported by the increased lighter component yields (C_1 to C_5), which are mainly produced by thermal cracking in LC-FINING process. This supports the theory that possibly more side chains in the asphaltene are lost, causing a lower molecular weight. This simple thermodynamic model explains qualitatively observed trends with respect to differences in catalysts and diluents.

Acknowledgements: We thank Lummus Technology and Chevron Lummus Global for granting permission to publish this paper. We thank our colleagues who reviewed the manuscript and gave suggestions and who collected the pilot plant data for many years.

References

Dahlberg, A., Mukherjee, U.; Olsen, C.W.; "Consider using integrated hydroprocessing methods for processing clean fuels", Hydrocarbon Process., p111, Sep. 2007.

Sheu E.Y. Mullins, O.C.(Editors), " Asphaltenes :Fundamentals and applications", Plenum Press, NY 1995. (for all papers in that book).

Hirschberg, A.; deJong, L.N.J.; Schipper, B.A.; Meijer, J.G.; "Influence of temperature and pressure on asphaltene flocculation", Soc.Petr.Engr.J. 283-293, June 1984.

Mannistu K.D.; Yarranton H.W.; Masliyah J.H.; "Solubility modeling of asphaltenes in organic solvents", Energy & Fuels, **11**, 615-622 (1997).

Mukherjee,U.; Reynolds, B.; Thiel, P; "Maximizing high quality diesel from LC-FINING residue hydrocracking", 1st Asia Bottom of the Barrel Technology Coneference and Exhibition, Kuala Lumpur, May 22-23, 2007.

Rogel E., "Theoretical approach to the stability of visbroken residues", Energy & Fuels, **12**, 875-880 (1998).

Spieler, S.; Mukherjee,U.; Dahlberg, A.; "Upgrading residuum to finished products in integrated hydroprocessing platforms- solutions and challenges", paper presented at NPRA meeting, Salt Lake City, Utah, March 19-21, 2006.

Wiehe, I.; "Tutorial on the phase behavior of asphaltenes and heavy oils", 2007. www.solublesolutions.com/petmacTutorial.html

Reduced Order Modeling of Rate-based Distillation Columns

M.L.S. Welz[1], K. Möller and A.C. Chakraborty

Sasol Technology, Process Development, Andries Brink Building,1 Klasie Havenga Road, Sasolburg, 1947, South Africa

Abstract

This work presents an application of orthogonal collocation to a rigorous, nonequilibrium rate-based packed column. The low order model is constructed with a focus on the feed discontinuity, which is treated as a separate adiabatic flash.
An objective function is defined as the model solution to a combination of degrees of freedom. This definition allows for multiple solutions to a single separation specification to be found. The optimal solution can be subsequently found using an optimization routine. The construction of the objective function ensures that there is always a solution to the design problem.
The resulting low order model has potential use in for the design of distillation trains. If a full-order model is used, numerical difficulties ensue due to the resulting high degree of problem stiffness.

Key words: Collocation Nonequilibrium Rate-based distillation

1 Introduction

1.1 Background

Collocation has been used extensively to model staged columns. An early work by Wong and Luus (1980) developed a simple model for a dynamic staged absorber. A few years later, a staged distillation model was developed by Cho and Joseph (1983b), with single collocation sections above and below the feed. This was extended to include multiple feeds and side draws using OCFE by Srivastava and Joseph (1986, 1985b). Later models have been developed by Seferlis and Hrymak (1994) using OCFE to simulate existing columns, as well as multiphase distillation by Swartz and Stewart (1987).

[1] *To whom correspondence should be addressed. Tel.: +27 (16)-940-2651. Fax:+27 (11) 219 4452. E-mail: Martina.Welz@sasol.com*

The majority of applications of collocation have been to staged columns. The difference equations are converted into differential equations through a transformation before collocation can be applied. Therefore the equations are no longer representative of a staged system, but rather of a rate based model. For a packed column, collocation is directly applied to the model equations. The additional equation conversion step in staged columns is undesirable and introduces error.

In an "equilibrium" staged column, equilibrium only occurs at discrete points i.e. the stages. There is no continuum. When applying collocation to the staged system, the equilibrium definition is changed to a continuum. Although applied by Srivastava and Joseph (1986, 1985a,b), this is fundamentally incorrect. The authors Stewart et al. (1985) have the correct approach as collocation points were chosen such that the error between the staged and collocated approach was a minimum. The stages were approximated by collocation equations inclusive of the equilibrium and collocation is *not* applied at the equilibrium points.

The incorrect assumption (within the context of a *staged* column) of a continuous equilibrium function to which collocation can be applied, was initially justified by Wong and Luus (1980). An analogy to a continuous stirred tank reactor (CSTR) was used. A system of CSTR's in series can be represented by difference equations which leads to the Kremser equations. The CSTR model solution asymptotically tends toward the solution to a plug flow reactor (PFR) as the number of CSTR's become large.

Analogously, the argument follows that a staged column can be represented by a single PDE. However this is incorrect as the staged operating lines do not have analogous behavior to that of the infinite number of CSTR's; namely as the number of stages increases, the operating lines overlap. Therefore no asymptotic analogy exists for an *equilibrium* column. For a *rate-based* approach this analogy is correct i.e. ChemSep rate-based packed column methodology[2] as described by Taylor and Kooijman (2000) and packed columns.
To the authors knowledge, this has not been addressed in the literature. Therefore it is proposed that the more correct use of collocation is for that of packed columns.

There have been few attempts at modelling packed columns on a continuous basis. An early study by Srivastava and Joseph (1984) modelled three simple packed-bed separation processes. Resistance to mass transfer was assumed to occur only in the vapor phase. The models were limited to a single feed and a single column section. This simplifies the collocation problem as there are no discontinuities. A batch distillation model using a differential contactor with mass transfer effects was developed by Wajge et al. (1997).

[2]ChemSep has two approaches; a rate based staged approach and a rate based packed approach. The packed approach fixes the height and the subsequently the separation. As the "stages" are increased, the operating line approaches a continuum.

A later work by Karacan et al. (1998) investigated a dynamic model of a packed binary distillation column. The system was more complex than the previous one by Srivastava and Joseph (1984), as axial dispersion was considered.

Integrating the packed column feed into the collocation structure is non-trivial. Two literature references mentioned this problem in reasonable detail for staged columns (Seferlis and Hrymak (1994) and Stewart et al. (1985)). There are no references to packed columns. The central idea is that of smoothing out the feed tray discontinuity through various interpolation schemes. For example, Cho and Joseph (1983a) used a single continuous polynomial to approximate the entire profile. While this reduces the model order, it was found by Srivastava and Joseph (1986) that the column profiles incur error, which is insensitive to the model order. The authors suggested an alternative spline technique approach. Pinto and Biscaia (1988) continued this approach using various spline interpolation schemes. However it is noted that feed discontinuities are an intrinsic to packed columns and therefore should not necessarily be smoothed out.

The natural progression after developing a single collocation model, is to develop an optimization routine. Initially an optimization routine was developed by Seferlis and Hrymak (1994) for a staged distillation model. This was extended to include reactive distillation by Seferlis and Grievink (2001). Temperature collocation was used by Zhang and Linninger (2004) to develop an algorithm for assessing separation feasibility. A separation sequencing superstructure was developed in a further study by Zhang and Linninger (2006). However, the concept was limited to non-azeotropic and quaternary mixtures, as well as constant molar overflow assumptions. More recently, a flowsheet for the synthesis of reactive absorption column was developed by Algusane et al. (2006).

1.2 Purpose of Study

The purpose of this study is twofold. Firstly, it aims to show the successful application of collocation to a rigorous packed nonequilibrium rate-based column. Packed columns are more amenable to the application of collocation than their staged counterpart.
Secondly, the study purports to highlight the advantages of an accurate, low order model by providing a methodology for assessing the separation feasibility. The ease and rapidity of data generation recommends future use of the low order model within a flowsheet for distillation train design and optimization. The resulting model has a fixed size irrespective of the column height or number of stages. The problem size does not increase in the vicinity of pinches.

2 Methodology

Figure 1 illustrates the collocation structure and the mass and energy transfer across the two-phase film, which is representative of a nonequilbrium rate-based packed distillation column. The following model assumptions were made:

- Ideal VLE described using the Antoine correlation
- Maxwell-Stefan mass transfer with no interaction effects in the matrix of diffusion coefficients
- Adiabatic and isobaric column operation
- Plug flow without backmixing
- Structured packing well represented by the Bravo correlation (Taylor and Krishna (1993))

2.1 Equation Development

The nonequilibrium rate-based packed distillation column can be described by the MERSQ equations as detailed in the work of Taylor and Krishna (1993). These equations apply staged equations to a rate based system. Mass transfer was assumed to be from the vapor to the liquid phase. The following overall and component mass balances were used:

$$\frac{dL}{dz} + \sum_{k=1}^{k=C} N_k = 0 \tag{1}$$

$$\frac{dV}{dz} + \sum_{k=1}^{k=C} N_k = 0 \tag{2}$$

$$\frac{d(L \cdot x_k)}{dz} + N_k = 0 \qquad k = 1, ..., C \tag{3}$$

$$\frac{d(V \cdot y_k)}{dz} + N_k = 0 \qquad k = 1, ..., C \tag{4}$$

The flux term in the above equations was described by:

$$N_k = \kappa_y \cdot a \cdot (y_k - y_k^i) = \kappa_x \cdot a \cdot (x_k^i - x_k) \qquad k = 1, ..., C \tag{5}$$

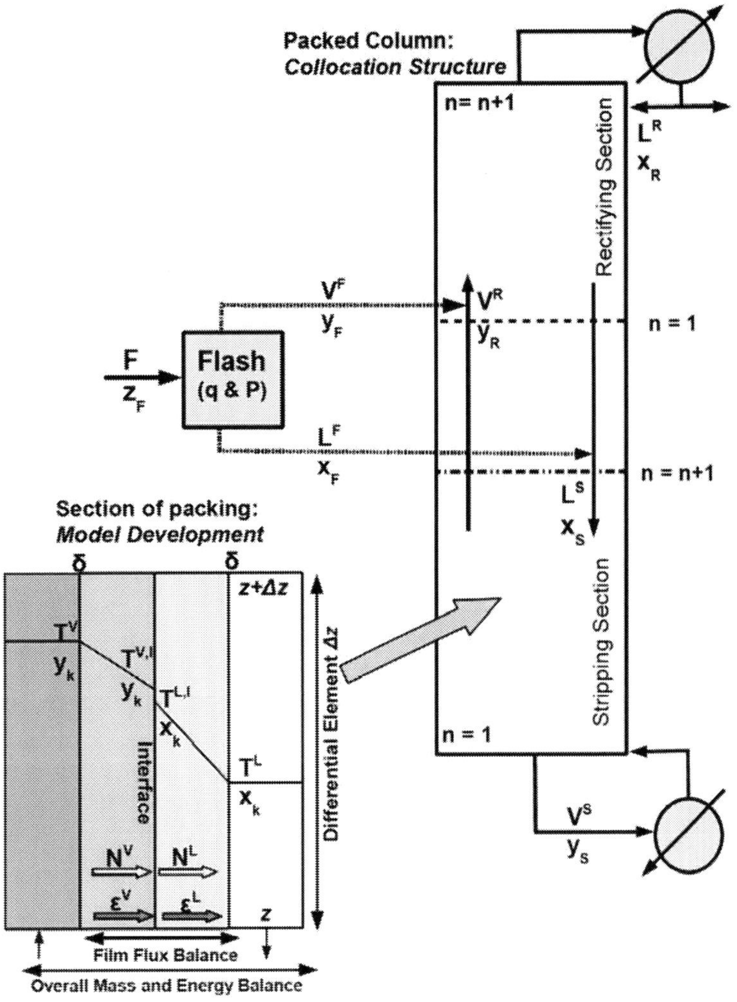

Figure 1: Collocation structure of the distillation column with flashed feed with detailed mass and energy balance across an infinitesimally small section of packing

An energy balance across each phase was included:

$$\frac{d(V \cdot H^V)}{dz} + \epsilon^V = 0 \tag{6}$$

$$\frac{d(L \cdot H^L)}{dz} + \epsilon^L = 0 \tag{7}$$

The energy flux terms in the above two equations are described below where the flow energy is neglected as no dispersion was considered:

$$\epsilon^V = h_v \cdot a \cdot (T^V - T^{V,i}) + \sum_{k=1}^{C} N_k \cdot H_k^V \tag{8}$$

$$\epsilon^L = h_l \cdot a \cdot (T^{L,i} - T^L) + \sum_{k=1}^{C} N_k \cdot H_k^L \tag{9}$$

The bootstrap condition was the adiabatic condition of the column described by:

$$\epsilon^L - \epsilon^V = 0 \tag{10}$$

Unlike ChemSep, all three temperatures at a single lateral point in the film were determined. There are $5C + 5$ variables at each collocation point n. These are ordered into a vector (g_n) according to the inner film balance and the outer bulk balance:

$$(g_n)^T = ((g_{film,n})^T, (g_{bulk,n})^T) \tag{11}$$

$$(g_{film,n})^T = (x_{1,n}^i, x_{2,n}^i, \ldots, x_{C,n}^i, y_{1,n}^i, y_{2,n}^i, \ldots, y_{C,n}^i, T_n^i, N_{1,n}, N_{2,n}, \ldots, N_{C,n}) \tag{12}$$

$$(g_{bulk,n})^T = (x_{1,n}, x_{2,n}, \ldots, x_{C,n}, L_n, T_n^L, y_{1,n}, y_{2,n}, \ldots, y_{C,n}, V_n, T_n^V) \tag{13}$$

The model differential equations are subsequently converted into algebraic equations through the use of orthogonal collocation. Evenly weighted Jacobi polynomials were used as the roots of the collocation polynomials. The collocation boundary conditions for each column section are listed in Table 1.

Table 1: Explanation of Boundary Conditions used

Rectifying Section		
Profile	*Position*	*Boundary condition*
x_k	$n+2$	$x\|_{k,n+2}^{R} = y\|_{k,n+2}^{R}$
y_k	1	$y\|_{k,1}^{R} \cdot V\|_{1}^{R} = y_k^F \cdot V^F + y\|_{k,n+2}^{S} \cdot V\|_{n+2}^{S}$
L	$n+2$	$L\|_{n+2}^{R} \cdot (1+Re) = V\|_{n+2}^{R}$
V	1	$V\|_{1}^{R} = V^F + V\|_{n+2}^{S}$
Stripping Section		
Profile	*Position*	Boundary condition
x_k	$n+2$	$x\|_{k,n+2}^{S} \cdot L\|_{n+2}^{S} = L^F \cdot x_k^F + x\|_{k,1}^{R} \cdot L\|_{1}^{R}$
y_k	1	$x\|_{k,1}^{S} = y\|_{k,1}^{S}$
L	$n+2$	$L\|_{n+2}^{S} = L^F + L\|_{1}^{R}$
V	1	$V\|_{1}^{S} \cdot (1+Rb) = Rb \cdot L\|_{1}^{S}$

2.2 Feed Discontinuity

The column was split into two collocation sections (rectifying and stripping), as shown in Figure 1. These are not finite elements as neither the value of the composition variable nor the value of the first derivative, is continuous across the interface i.e $x_1^R \neq x_{n+2}^S$ and $\frac{dx_1^R}{dz} \neq \frac{dx_{n+2}^S}{dz}$. This is due to the discontinuities introduced by the feed. The feed was treated as an adiabatic mixing stage as suggested by Stewart et al. (1985). It was flashed at the column pressure, with the condition specified. The resulting vapor stream was mixed with the vapor leaving the stripping section and vice versa for the liquid flows, as shown in Figure 1. This approach is only strictly true for adiabatic columns with constant molar overflow. But provided the temperature of the flashed feed is similar to the column temperature at that point and the feed position is near optimal, there is little reason to dispute this approach. The similar resulting compositions will result in a negligible ΔH^{vap}.

For a binary system with an optimal position, there should be no feed discontinuity. This is shown in Figure 2, which illustrates the effect of flashing the feed. Points A and C are subsequently defined. Point C represents the compositions on the feed tray assuming that is it well mixed. However neither point provide information as how to find the column composition, as indicated by point B. It must be emphasized that point B does not physically exist in a staged system *unless* the stage meets exactly at the point. Even if the stage does meet exactly at that point, it represents the passing streams and thus has no physical location. Therefore points on the operating line refer to passing streams *between* the stages. Point B always exists in a packed column as the discontinuity between the two column sections.

If the feed is optimally placed as is shown in Figure 2, 'stepping over' from the rectifying to stripping operating line occurs exactly at the flashed feed composition (point C). This results in a smooth composition profile in both phases. For a ternary or multicomponent system, the best continuity possible using this approach will be a smooth composition profile for *either* the light *or* heavy key.

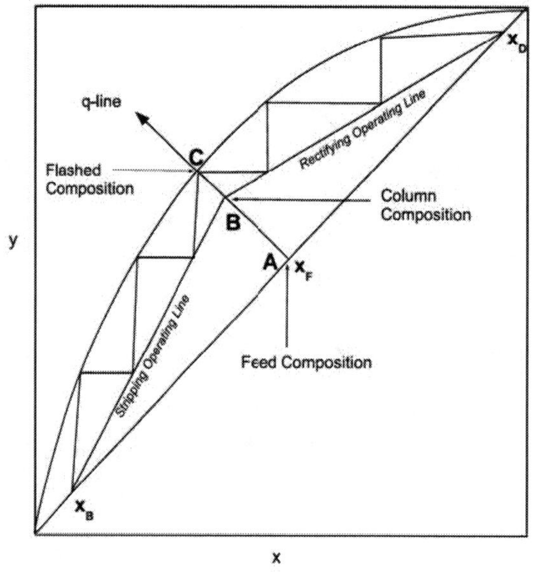

Figure 2: McCabe-Thiele representation for optimal feed positioning

2.3 Assessing Separation Feasibility using an Objective function

If the feed composition and quality are defined, there are four remaining degrees of freedom, which are independent of the number of components (Doherty and Malone (2001)). The reflux and reboil ratio and section heights were used as inputs to the low order model. This is shown in equation 14. The combination resulted in a robust problem which always had a solution. The more common and less stable alternative, is to specify a separation as an input parameter instead.

$$f_{Objective} = f(Re, Rb, Height^R, Height^S) \tag{14}$$

Table 2: Specifications used to fully define the feed

	Line to Column	*Flashed on Entry*	
Variables	*Feed*	*Liquid*	*Vapor*
T^V (K)	-	371	371
T^L (K)	368	371	371
Pressure (kPa)	300	200	200
V ($\frac{kmol}{min}$)	-	-	45
L ($\frac{kmol}{min}$)	100	55	-
Quality q	1	1	0
$y_{ethanol}$	-	-	0.3222
y_{water}	-	-	0.1789
$y_{methanol}$	-	-	0.4987
$x_{ethanol}$	0.3200	0.3170	-
y_{water}	0.2700	0.3810	-
$x_{methanol}$	0.4100	0.3014	-
Total flowrate ($\frac{kmol}{min}$)	100	45	55

When the model is solved, there are numerous variables which can be tracked or optimized. To find a feasible region for a specific separation, an objective function is defined as a single variable representing the required separation i.e. a distillate mole fraction. This does not affect the four available degrees of freedom. Out of these four, one is arbitrarily is fixed at a constant value. This leaves three degrees of freedom which can be varied. This is a necessary simplification for representation on a three-dimensional surface map.

To find the feasible region graphically, the objective function is plotted. A plane perpendicular to the objective function axis at the value of the desired specification is included. The section of the surface which is isolated by this plane represents the feasible region. The graphical approach has been used for explanatory purposes, but it is simpler to perform a data search and extract the information. The resulting matrix will be in four dimensions, i.e. combinations of height, reflux and reboil ratio which result in the desired separation specification. This data is then ready for NLP optimization.

3 Case Study

Table 2 lists the data used. Table 3 lists the degrees of freedom and associated parameters.

Table 3: Degrees of Freedom

Parameter	*Value*
Reflux Ratio	1.0
Reboil Ratio	2.4
Height	10.0 m
Feed Position from reboiler	5.0 m

3.1 Feed discontinuity

A ternary system was chosen such that ternary diagrams could be considered in the analysis i.e. Figure 3 where the points represent collocation points and *not* stages. Ternary diagrams are useful for identifying pinches. For more than three components, they start to loose application. In Figure 3, the overall mass balance is shown as a straight line intersecting the end-points of the rectifying and stripping profiles. The composition of the flashed feed has also been included. The end points of the stripping and rectifying operating lines, as well as the flashed feed composition, *must be collinear*. This is a necessary condition as a result of the adiabatic mixing feed stage boundary condition.

The corresponding pseudo McCabe-Thiele diagram is shown in Figure 4 where the points represent collocation points and *not* stages. A total reboiler was used, which is why there is no associated stage. The light key is normalized with respect to the heavy key. The position of the q-line is fixed as the feed composition and quality is both known. The feed location is optimal. This translates into a continuous profile for the heavy key.

3.2 Separation Feasibility

The recovery of water to the distillate is defined as the objective variable requiring optimization. The desired recovery is 5.0%. The feed position is arbitrarily fixed at a constant value of 0.5. The resulting surface map is shown in Figure 5. A horizontal plane is drawn perpendicular to the objective function axis at a value of $x_{water,D} = 0.05$. The surface below the plane represents the feasible region.
This region consists of various combinations of the degrees of freedom which result in the desired separation. In this example, as the height increases, the separation improves. There is an obvious trade-off between the relative costs of an increase in height against a purer product. These and other considerations can be taken into account when designing a NLP flowsheet optimizer. An important observation is that the collocation solution identified multiple minima. This solution would *not* be easily found using a NLP code. The collocation solutions can be used as the first step in a global optimization routine, as the peak information is readily supplied.

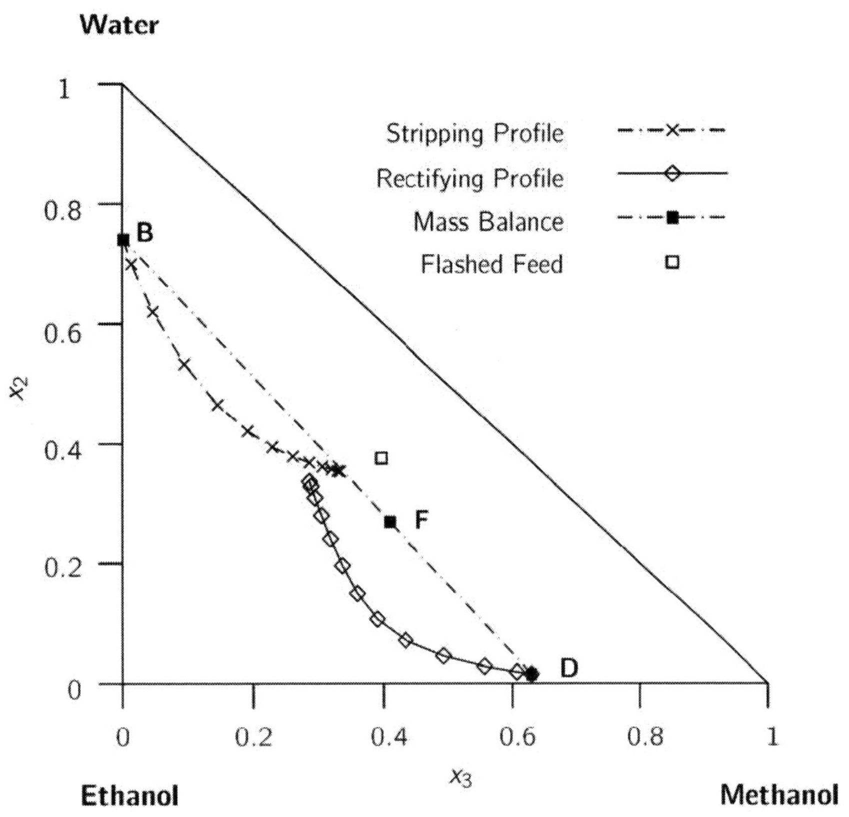

Figure 3: Liquid phase ternary plot of Water versus Methanol

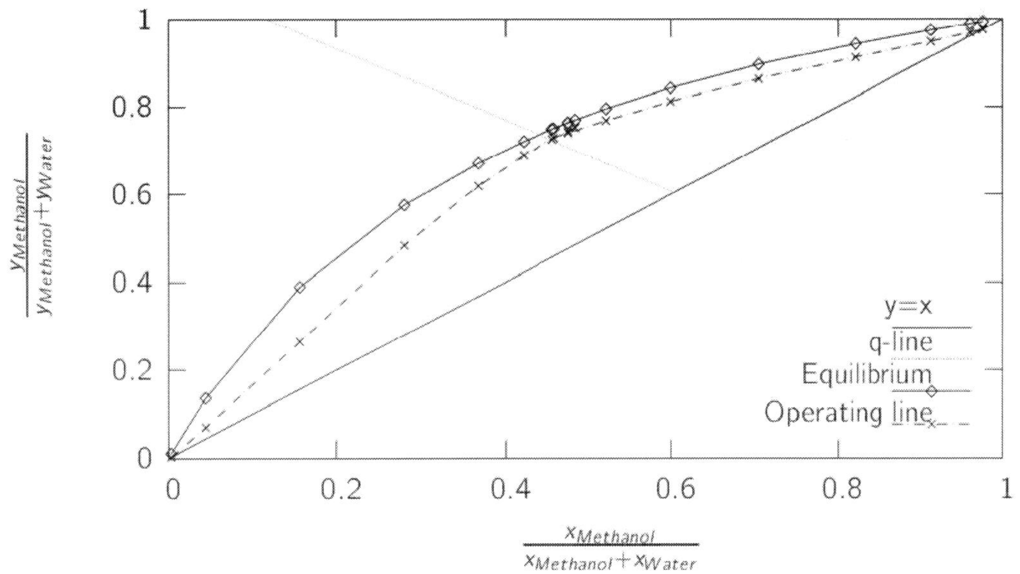

Figure 4: Pseudo McCabe-Thiele plot of Methanol with respect to Water

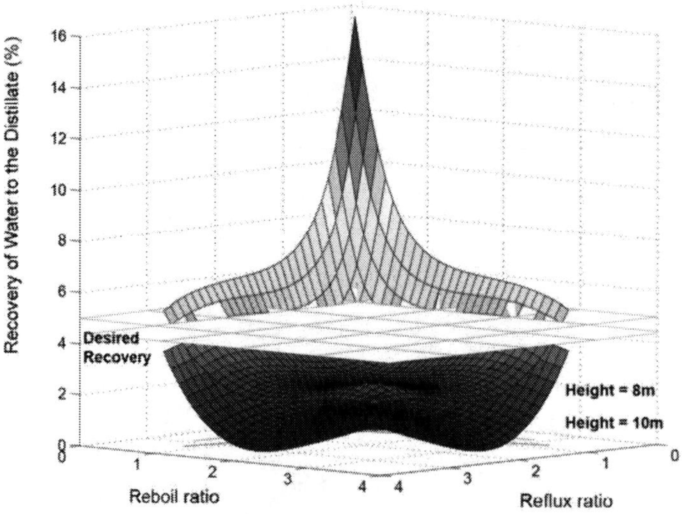

Figure 5: Separation Feasibility: Recovery of water to the distillate as a function of the *reflux ratio*, *reboil ratio* and the *height*. The feed position is 0.5 and the recovery $< 5\%$

The key idea is that the collocation code remains the same size (i.e. the order is unchanged), despite a changing height. This is particularly useful for pinches; if a standard integrator was used, pinches constantly reduce the step size, thereby increasing the code run-time and problem order. There is also the likelihood of a standard integrator getting 'trapped' at one of the local minima or maxima, which is not a concern with the robust collocation approach.

4 Conclusions and Significance

A low order model of a rigorous nonequilibrium rate-based distillation column was successfully developed using orthogonal collocation. An objective function was defined such that a solution was always found. This allows for the future inclusion of the low order model within an NLP flowsheet superstructure. The model is particularly suited to optimization as there are no integer decision variables due to the use of collocation.

5 Future Research Directions

The model is limited by the assumptions of ideal thermodynamics and negligible interaction parameters. This can easily be included in a future version. Future work will involve the development of the NLP flowsheet superstructure in which the low order model will be included.

Acknowledgements

The authors gratefully acknowledge the support of the University of Cape Town, SASOL and the South African National Research Foundation.

Nomenclature

a Interfacial area

h Heat transfer coefficient

H Enthalpy

K Equilibrium coefficient

L Liquid

n Collocation point

N Rate of mass transfer

P Pressure

T Temperature

V Vapor

x Liquid mole fraction

y Vapor mole fraction

Greek:

δ Film thickness

ϵ Rate of energy transfer

κ Mass transfer coefficient

Superscripts:

i Interface

R Rectifying

S Stripping

Subscripts:

C Number of components

Abbreviations:

CSTR Continuous Stirred Tank Reactor

DE Differential equation

MERQS Mass Energy Rate Summation Equilibrium

NLP Non-Linear Programming

OCFE Orthogonal Collocation on Finite Elements

PFR Plug Flow Reactor

Re Reflux ratio

Rb Reboil ratio

VLE Vapor Liquid Equilibrium

vap Vaporization

References

Algusane, T. Y., Proios, P., G., M. C., and Pistikopoulos, E. N. (2006). A framework for the synthesis of reactive absorption columns. *Chem. Eng. Proc.*, 45(4):276–290.

Cho, Y. and Joseph, B. (1983a). Reduced-order models for separation columns-III. Application to columns with multiple feeds and sidestreams. *Comput. Chem. Engng.*, 8:81–90.

Cho, Y. and Joseph, B. (1983b). Reduced-order steady-state and dynamic models for separation processes. Part 1. Development of model reduction procedure. *AIChE*, 29:261–269.

Doherty, M. and Malone (2001). *Conceptual Design of Distillation Systems*. McGraw-Hill Inc. New York.

Karacan, S., Cabbar, Y., Alpbaz, M., and Hapoglu, H. (1998). The steady-state and dynamic analysis of packed distillation column based on partial differential approach. *Chem. Eng. Proc.*, 37(5):379–388.

Pinto, J. C. and Biscaia, E. C. (1988). Order reduction strategies for models of staged separation systems. *Comput. Chem. Engng.*, 12(8):821–831.

Seferlis, P. and Grievink, J. (2001). Optimal design and sensitivity analysis of reactive distillation units using collocation models. *Ind. Eng. Chem. Res.*, 40:1673–1685.

Seferlis, P. and Hrymak, A. N. (1994). Optimization of Distillation Units Using Collocation Models. *AIChE*, 40:813–825.

Srivastava, R. and Joseph, B. (1984). Simulation of packed-bed separation processes using orthogonal collocation. *Comput. Chem. Engng.*, 8:43–50.

Srivastava, R. and Joseph, B. (1985a). Reduced-order models for separation columns-V. Selection of collocation points. *Comput. Chem. Engng.*, 6:601–613.

Srivastava, R. and Joseph, B. (1985b). Reduced-order models for separation columns-VI. Columns with steep and flat composition profiles. *Comput. Chem. Engng.*, 11:165–176.

Srivastava, R. and Joseph, B. (1986). Reduced-order models for separation columns-IV. Treatment of columns with multiple feeds and sidestreams via spline fitting. *Comput. Chem. Engng.*, 11:1519–164.

Stewart, W., Levien, K. L., and Morari, M. (1985). Simulation of fractionation by orthogonal collocation. *Chem. Eng. Sci.*, 40(3):409–421.

Swartz, C. and Stewart, W. (1987). Finite-element steady state simulation of multiphase distillation. *AIChE*, 33:1977–1987.

Taylor, R. and Kooijman, H. (2000). *The ChemSep Book*. Books on Demand.

Taylor, R. and Krishna, R. (1993). *Multicomponent Mass Transfer*. John Wiley and Sons.

Wajge, R., Wilson, J., Pekny, J., and Reklaitis, G. (1997). Investigation of numerical solution approaches to multicomponent batch distillation in packed beds. *Ind. Eng. Chem. Res*, 36:1738–1746.

Wong, K. and Luus, R. (1980). Model Reduction of High-Order Multistage Systems by the Method of Orthogonal Collocation. *Can. J. Chem. Eng.*, 58:382–388.

Zhang, L. and Linninger, A. A. (2004). Temperature collocation algorithm for fast and robust distillation design. *Ind. Eng. Chem. Res.*, 43:3163–3182.

Zhang, L. and Linninger, A. A. (2006). Towards Computer-Aided Separation Synthesis. *AIChE*, 52:1392–1406.

Investigation of Hydrodynamics, Heat and Mass Transfer Simultaneously in Liquid Film Flow on Inclined Plates Using CFD

M.R. Ehsani ,
M.R. Khosravi Nikou[1]
Chemical Engineering Department, Isfahan University of Technology (IUT), 84155 Isfahan, Iran

AIChE Spring Meeting, April 2008
8th Topical Conference on Natural Gas Utilization
New Orleans, LA, April 6-10, 2008

Abstract
A two-phase flow CFD model using volume of fluid (VOF) method is presented for predicting the hydrodynamics, heat transfer and mass transfer of falling film flow on inclined plates, corresponding to the surface of texture of structured packing. Using the proposed CFD model the influence of solid surface microstructure, liquid properties and gas flow rate on flow behavior was investigated.

1 Introduction

Liquid film flow on the surface of inclined corrugated plates corresponding to structured packing has drawn great attention, and some experimental and theoretical research work has been carried out in recent years, but regarding heat and mass transfer there is no specific investigation on research papers. Among the hydrodynamics research, the group of Cerro et al.(1992, 1993, 1995, 1997, 1998) carried out a number of experimental and theoretical studies into liquid film flow on solid surfaces with micro and micro-texture structures. In their studies a film evolution equation with an approximation of the viscose long-wave for the Navier-Stockes equations was presented. However, the inertia term was neglected, which is important to determine the stability of liquid film. The studies by Ataki and Bart (2002) described rivulet flow on an inclined plate using experimental and CFD methods as a basis to understand the film flow on the surface of structured packing.

Soare et al. (2001) proposed a simplified mathematical model for liquid flow on corrugated metal sheets in which the effect of surface tension was included by assuming that the force generated is uniformly distributed over entire flowing liquid film. Using analytical solutions the liquid film thickness, velocity and liquid flow rate can be calculated. Using FLUENT software package, Szulczewska et al. (2003) simulated two-phase counter-current flow in a plate-type structured packing. The effect of the liquid flow rates on the interfacial area formed on the surface of the structured packing was investigated, but the effective wetted model can be unreliable.

The objective of the present work in to construct a two-phase CFD model using the VOF (volume of fluid) approach, which enables us to investigate the influence of the solid surface characteristics, liquid properties and gas flow rates on falling film flow, heat and mass transfer coefficients on inclined plates.

2 Theoretical and Numerical Methodologies

Since the gas and liquid film on the plate do not inter-penetrate each other, a calculation using VOF approach was chosen. This method is a surface-tracking technique applied to a fixed Eulerian mesh, which enables the computation of two-phase flow under the condition that phases do not mix, i.e., the gas liquid interface is clearly identified. The VOF allows the construction of the interface to become part of the solution based on the same grid system, which is a flexible and efficient method to describe the changes in topology of a gas-liquid interface (Ataki and Bart (2002), Szulczewska et.al.(2003), Hirt and Nichols (1981), Raynal and Harter (2001), Taha and Cui (2002), Li et.al.(2000), Li et.al.(2001), Brackbill et.al.(1992)).

2.1 Governing Equation

The mass, momentum and energy conservation equations for two-phase flow throughout the domain are:

$$\frac{\partial}{\partial t}(\rho)+\nabla.(\rho\vec{v})=0 \tag{1}$$

$$\frac{\partial}{\partial t}(\rho\vec{v})+\nabla.(\rho\vec{v}\vec{v})=-\nabla P+\nabla.\left[\mu\left(\nabla\vec{v}+\vec{v}^{T}\right)\right]+\rho\vec{g}+\vec{F} \tag{2}$$

$$\frac{\partial}{\partial t}(\rho h)+\nabla.(\rho\vec{v}h)=\nabla.(k\nabla T)+\frac{Dp}{Dt}-\left[\mu\left(\nabla\vec{v}+\nabla\vec{v}^{T}\right):\nabla\vec{v}\right]-\nabla.\left(\sum_{k}h_{k}j_{k}\right) \tag{3}$$

The equations given above are dependent on the volume fraction of all phases, expressed implicitly by the properties ρ and μ. In a two-phase system, for example, if the phases are represented by the subscripts 1 and 2 and the volume fraction of phase 2 is known, the density and viscosity in each cell are given by equations (4) and (5).

$$\rho=\alpha_2\rho_2+(1-\alpha_2)\rho_1 \tag{4}$$

$$\mu=\alpha_2\mu_2+(1-\alpha_2)\mu_1 \tag{5}$$

In the VOF model the movement of the phase interface is affected by the distribution of a_q, the volume fraction of qth phase (secondary phase) in a computational cell. When in a particular computational cell $a_q=0$ the cell is empty of qth fluid, and when $a_q=1$ the cell is full of qth fluid. When $0< a_q<1$ the cell contains the interface between the qth fluid and one or more other fluids (Hirt and Nichols (1981). Thus, the interface between two phases can be traced by solving the continuity equation for the volume fraction function:

$$\frac{\partial a_q}{\partial t}+\vec{v}.\nabla a_q=0 \tag{6}$$

The volume fraction for primary-phase in Eq. (6) will be obtained from the following equation:

$$\sum_{q=1}^{n}a_q=1 \tag{7}$$

For thin liquid film flow, surface tension plays an important role. The surface tension model in CFX 11 is the continuum surface force model proposed by Brackbill et al. (1992).

2.2 Model Geometry and Boundary Conditions

All the model geometry and dimensions are given in figure 1 and table 1.

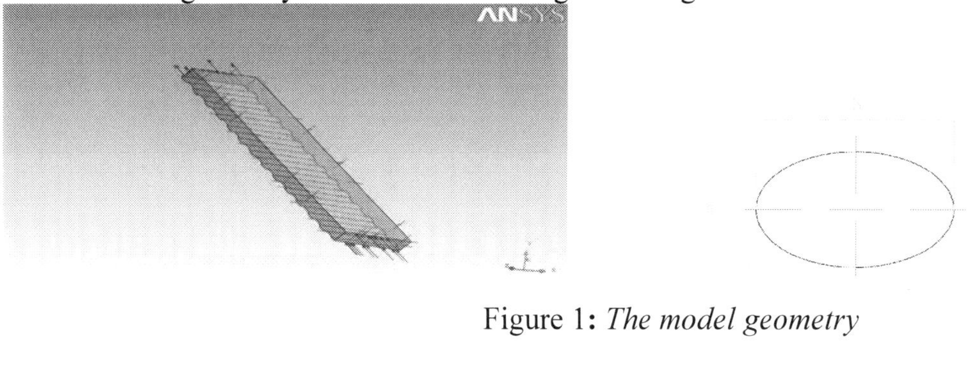

Figure 1: *The model geometry*

Table 1: *Structural parameters of various geometries*

Type	Design parameters		Plate dimensions	Plate inclination angle	Mesh statistics
	a(mm)	b(mm)	H×L×t (mm)	degree	-
Flat plate			53×20×3	45	3051
Corrugated plate1	0.25	4	53×20×3	45	1067375
Corrugated plate2	0.5	8	53×20×3	45	618278
Corrugated plate3	0.5	4	53×20×3	45	679716
Corrugated plate4	1	8	53×20×3	45	30664

Three-dimensional coordinate system is used for computation. No slip boundary condition is provided at the wall. All the boundary conditions are specified in table 2.

Table 2: *Boundary conditions*

Location	Boundary condition types
Inlet	Velocity, Volume fractions, mass fraction of methanol, Temperature
Outlet	Velocity
Wall	No-slip, Temperature

2.3 Simulation Scheme

The CFD software, ANSYS-CFX version 11, is used to simulate the motion of gas-liquid two-phase flow on structured packing. A high resolution was chosen as the solution of the momentum and energy equations. The SIMPLEC algorithm is used to solve the pressure field.

3 Results and Discussion

3.1 Influence of Plate Structure

Influence of corrugated plate structure on hydrodynamic, heat and mass transfer of two-phase flow of Methanol-Isopropanol mixture on smooth and four types of plates, whose structural parameters are listed in table 1, has been referred in figures 2-4.

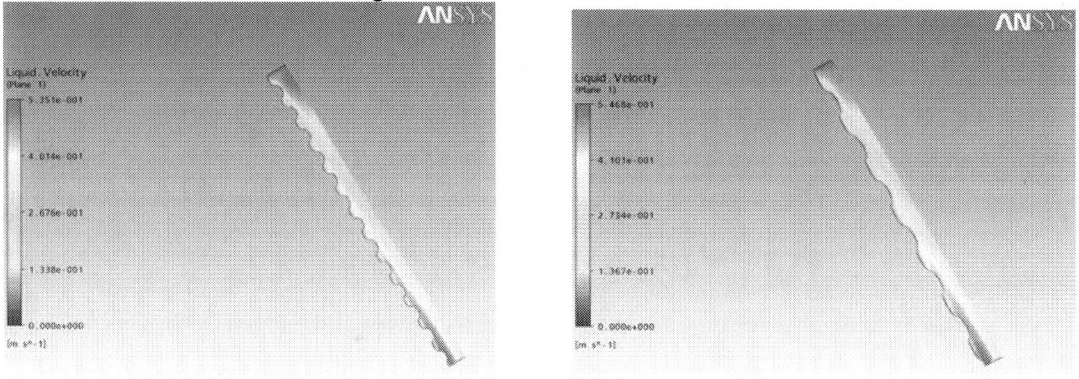

corrugated plate 1 corrugated plate 3

Figure 2: *Liquid velocity profile on corrugated plates no. 1 and 3(Re_l= 835, Re_g= 1685)*

As liquid moves over the solid surface the film breaks randomly, and breaking points of film occurs at the neighboring regions of the transition points from convexities to concavities. When changing the texture or structural parameters of the wavy plates the flow pattern of liquid films will change accordingly. It is found that for wavy plates with small waves the liquid film breaks up at many points and no continuous film forms. By contrast, continuous films can form easily on the plates with large waves.

It is expected that when there is continuous film over structured packing, heat and mass transfer enhanced by increasing turbulence on the film. From Fig.3 it can be distinguished that, when flat plate structure becomes wavy, methanol concentration gradient between inlet and outlet increased. It means interfacial area will be increased by waviness. Heat and mass transfer enhancement are increased by increasing waviness on plate till to reach to maximum and after that starts to decrease. By increasing too much waviness amplitude on corrugated plate, continuous film is broken and results to decrease transport phenomena drastically. It can be explained by this reason, increasing the value of waviness amplitude will increase the amount of stagnant points in liquid concaves. If the eddies inside stagnates could be replaced by passing the flow over structured packing, the amount of decrease is not significant (figure 4) but when stagnant pool is big enough, it effects on transfer process too much.

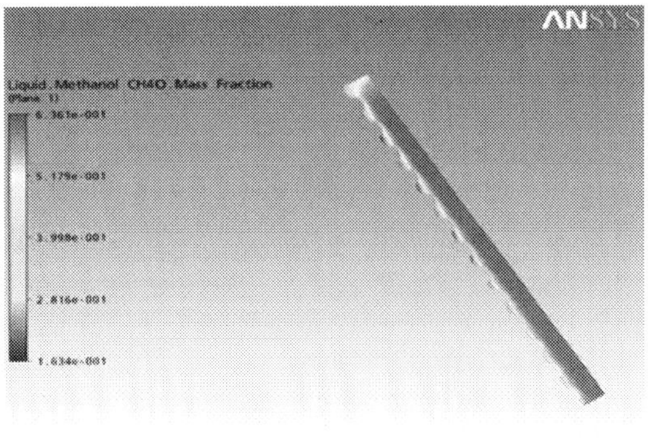

corrugated plate 1

corrugated plate 4

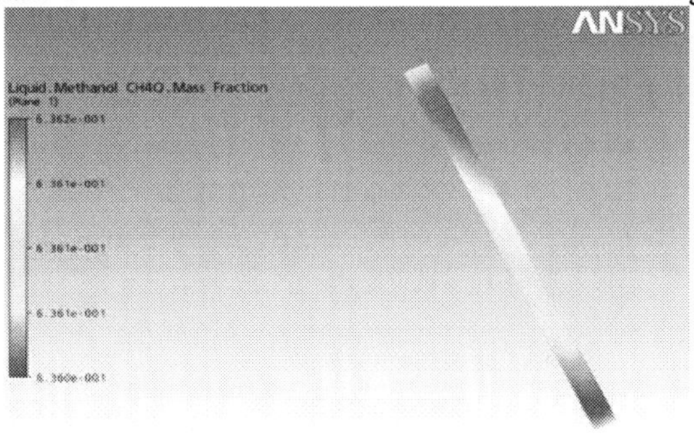

flat plate

Figure 3: *Methanol concentration profile in liquid phase over corrugated plates no. 1 and 4 and flat plate (Re_l= 835, Re_g= 1685)*

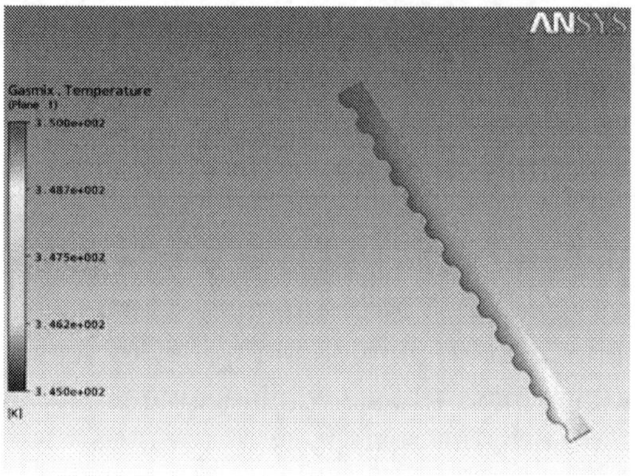

corrugated plate 1

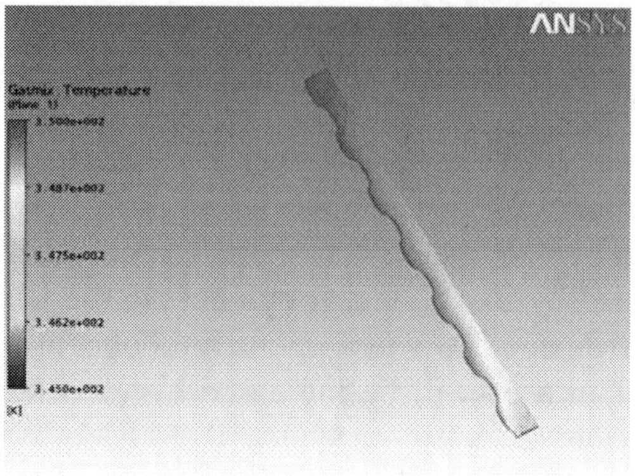

corrugated plate 3

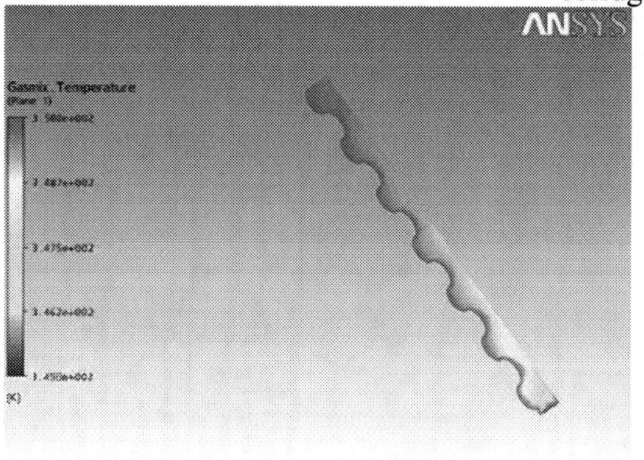

corrugated plate 4

Figure 4: *Temperature profile in gas phase over corrugated plates no. 1, 3 and 4 (Re_l= 835, Re_g= 1685)*

3.2 Influence of Liquid Properties

To investigate the effect of liquid properties on the flow behavior, Water-Air mixture, Glycerol-Air mixture and Methanol-Isopropanol mixture (properties are listed in table 3) were selected to cover a range of surface tension, viscosity, and contact angles etc. Figures 5, 6 show liquid profile and liquid temperature for theses two mixture. It was found that by increasing liquid viscosity, film thickness increased and film will be continuous.

Besides liquid viscosity, surface tension plays a very important role in liquid stability. In general, low surface tension systems tend to yield low contact angles which favor complete wetting of surface structure. This means that, for a given solid surface, one may expect better wetting for lower surface tension liquids. To investigate the influence surface tension on the film flow, systems of methanol-Isopropanol mixture was simulated.

Table 3: *Mixture properties*

Mixture	**ρ (kg/m^3)**	**μ (kg/m s)**	**σ (N/m)**	**θ (degree)**
Water-Air	998.2	0.001003	0.0722	57
Glycerol-Air	1262	1.495	0.0727	56
Methanol-Isopropanol	776.2	0.00776	0.02776	40

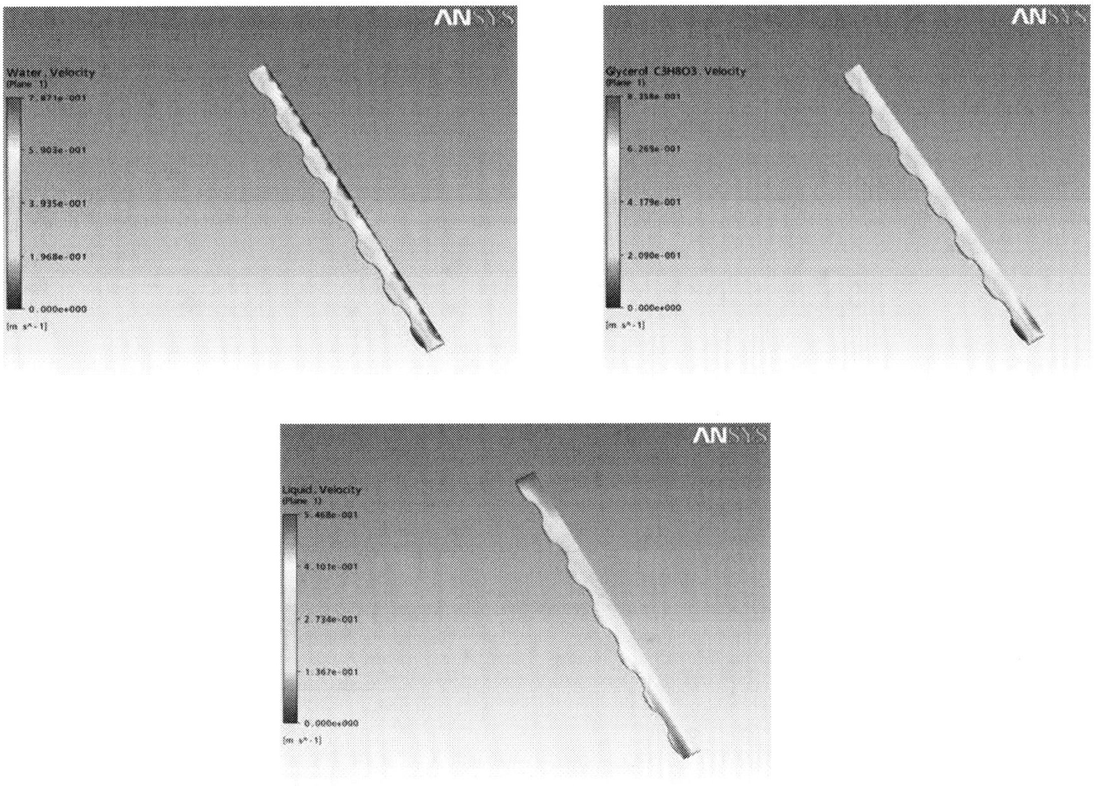

Figure 5: *Liquid velocity profile for Water-Air, Glycerol-Air and Methanol-Isopropanol mixture over corrugated plate 2*

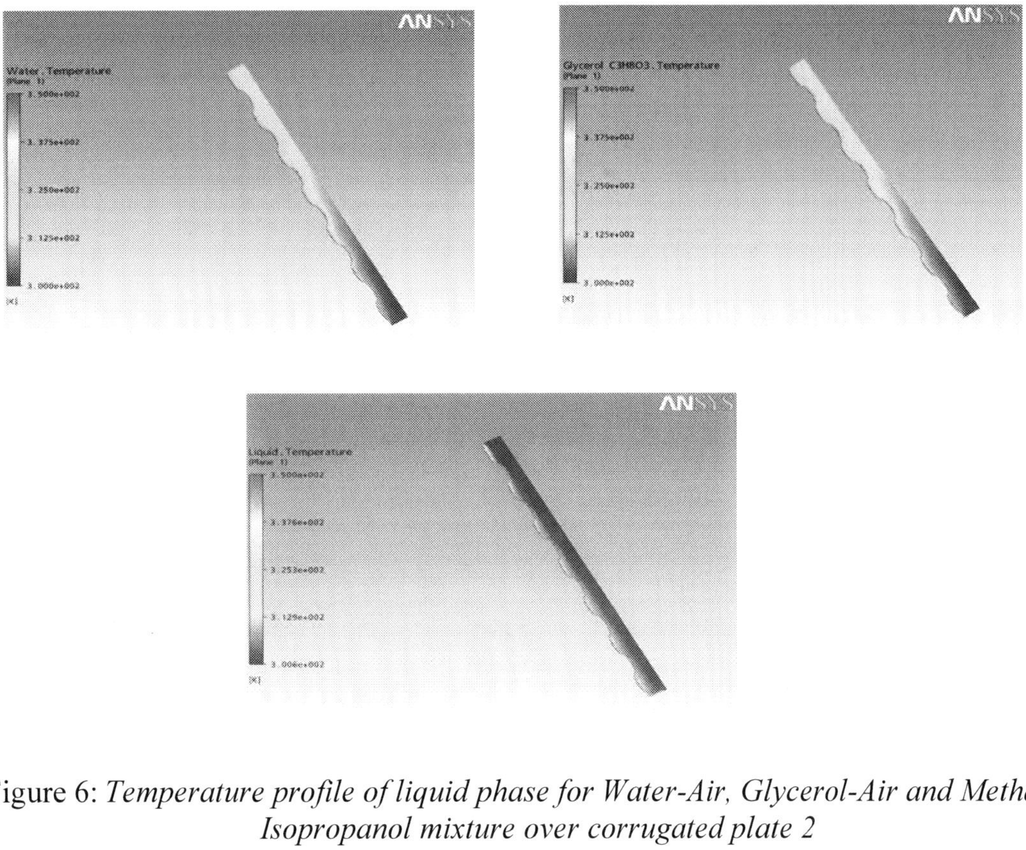

Figure 6: *Temperature profile of liquid phase for Water-Air, Glycerol-Air and Methanol-Isopropanol mixture over corrugated plate 2*

3.3 Influence of Gas Flow

The counter-current flow behavior and heat and mass transfer changes were simulated, and are shown in figure 7.

case 1

case 2

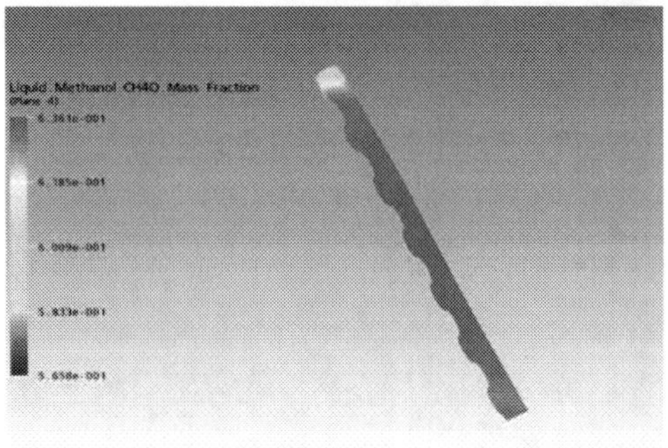

case 1

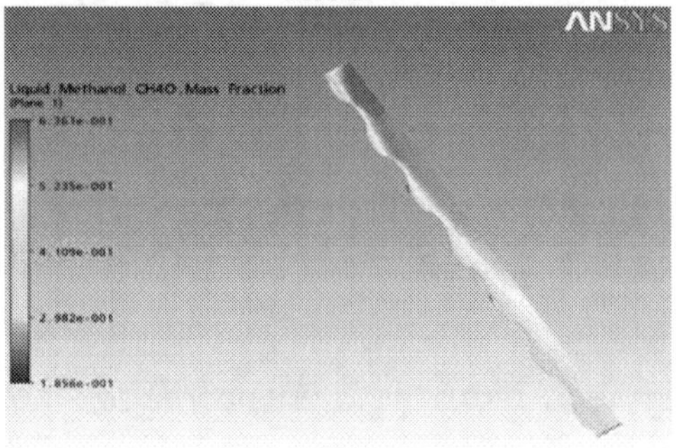

case 2

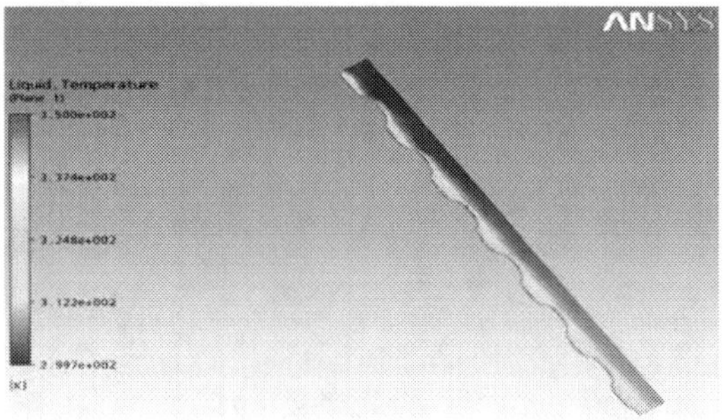

case 2

Figure 7: *Gas flow rate effects on pressure drop, methanol concentration and temperature profile of liquid phase for Methanol-Isopropanol mixture over corrugated plate 2 (Case1: u_g=0.36m/s and Case2: u_g=0.72m/s)*

The simulation results in this figure demonstrate that the flow behavior of two-phase flow and transfer processes on structured packing is greatly influenced by film thickness. Thus, all effects increasing the film thickness will be of advantage in forming a continuous film. Once the film breaks down a number of dry patches come into bring, which are extremely disadvantageous to heat and mass transfer processes.

4 Conclusion

A numerical method for simulation of coupling of hydrodynamic and heat and mass transfer for two-phase flow inside smooth and four types of geometry of structured packing has been presented.
It has been shown that adding waviness structure to smooth plates could increase heat and mass transfer phenomena but when concavity and convexity increase too much, because of generating stagnant point, this enhancement will be reversed and decrease transfer drastically.
The fluid properties, especially the liquid surface tension, play an important role in determining the flow hydrodynamic. The simulated results indicate that under counter-current two-phase flow, measures to increase continuity of film to reduce occurrence of dry patches.
Gas flow has positive effects because of turbulence but when discontinuity is generated on structured packing, transport phenomena decreased rapidly.

5 References

1. Zhan, L. and Cerro, R.L, 1992, Int. J. Multiphase Flow, 18, 495.
2. Shetty, S. and Cerro, R.L, 1993, Int. J. Multiphase Flow, 19, 1013.
3. Shetty, S. and Cerro, R.L, 1995, Ind. Eng. Chem. Res., 34, 4078.
4. Shetty, S. and Cerro, R.L, 1997, Ind. Eng. Chem. Res., 36, 771.
5. Shetty, S. and Cerro, R.L, 1998, Ind. Eng. Chem. Res., 37, 626.
6. Ataki, A. and Bart, J.H., 2002, Int. Conf. on Distillation & Absorption, BadenBaden, Germany.
7. Soare, G. and Piesu,V., 2001, 12th Romania Int. Conf. on Chem. & Chem. Engineering, Bucharest.
8. Szulczewska, B. and Zbicinski, I. and Gorak, A., 2003, Chem. Eng. Technol., 26, 580.
9. Hirt, C.W. and Nichols B.D., 1981, J. Comput. Phys. , 39, 201.
10. Raynal, I. and Harter, I., 2001, Chem. Eng. Sci., 56, 6385.
11. Taha, T. and Cui, Z.E., 2002, J. Membrane Sci., 210, 13.
12. Li, Y. and Zhang, J.P. and Fan, L.S., 2000, Chem. Eng. Sci., 55, 4597.
13. Li, Y. and Yang, G.Q. and Zhang, J.P. and Fan, L.S., 2001, Powder Technol., 116, 246.
14. Brackbill, J.U. and Kothe, D.B. and Zemach, 1992, J. Comput. Phys. , 100, 335.

Mechanistic Study of the Fischer-Tropsch Synthesis Using Transient Kinetics

Nilenindran Sundra Govender[1], Mart HJM. de Croon[2], and Jaap C. Schouten[2].
(1) Laboratory of Chemical Reactor Engineering, Eindhoven University of Technology, Den Dolech 2, Eindhoven, 5611cx, Netherlands, (2) Laboratory of Chemical Reactor Engineering, Department of Chemical Engineering and Chemistry, Eindhoven University of Technology, P.O. Box 513, Eindhoven, 5600 MB, Netherlands

The Fischer-Tropsch reaction mechanism over an Iron based catalyst is investigated with the Steady State Isotopic Transient Kinetic Analysis (SSITKA) technique using a plug-flow fixed-bed reactor. This reactor is operated under differential and intrinsic conditions in the absence of severe pressure effects related to the abrupt change in isotopic feed composition. Emphasis is put on evolution of the labeled carbon 13C in the C1 to C5 hydrocarbon products, leading to a kinetic analysis of the Fischer-Tropsch mechanism. A plug-flow reactor model is used to describe the experimental results. Data is obtained at 330 °C, H2/CO ratios from 4 to 20, and a pressure of 1.2 bar. The results for the methanation reaction, which includes a reaction mechanism and the relevant kinetic parameters, will be presented. The extension of this mechanism to include higher hydrocarbon formation will also be presented.

Promoted ZnO Sorbents for Low Temperature H_2S Removal: Influence of Temperature, Support Surface Area, Pore Size and Pore Volume

Priyanka Dhage, Vivekanand Gaur, Bruce J. Tatarchuk*
Center for Microfibrous Materials Manufacturing, Department of Chemical Engineering
Auburn University, Auburn, AL, USA

AIChE Spring Meeting, April 2008
8th Topical Conference on Natural Gas Utilization
New Orleans, LA, April 6-10, 2008

Abstract

Promoted ZnO sorbents with formulation $Zn_{0.95}M_{0.05}O$ (M = Mn, Fe, Co, Ni, Cu) were supported on various silica gels, MCM-41 and MCM-48. H_2S adsorption was conducted from room temperature to 400 °C in flowing 1%H_2S/H_2. Cu-doped ZnO (15wt %) /MCM-41 provided the highest sulfur capacity at room temperature [0.9 mmol S/mmol (Zn+Cu)]. H_2S adsorption at elevated temperatures (ca. 400 °C), resulted in near total sulfidization of the available reactant regardless of dopant. At intermediate temperatures, the utilization of the reactant (Zn+M) toward H_2S removal depended on the support employed in the order MCM-41 > MCM-48 > silica gels. This dependence was investigated in terms of the support: surface area, pore volume, and pore size; using adsorption-desorption isotherms. At equivalent levels of support surface area and pore size; higher pore volumes provided greater low temperature H_2S capacities presumably as a result of the lower Zn $(NO_3)_2.6H_2O$ concentrations used during impregnation/drying and a diminution in $Zn_{0.95}M_{0.05}O$ crystallite size. Estimates of crystallite size were assessed from XRD line broadening (where feasible). The sulfided sorbents were regenerated under air flow for repeated use in multiple adsorption-desorption cycles.

Keywords: H_2S removal; ZnO catalyst; Mesoporous silica; MCM-41; Breakthrough analysis

1. Introduction

Hydrogen sulfide and other sulfur-containing compounds are present in the fossil fuel, natural gas and other liquid hydrocarbon fuel. It is a worldwide challenge to develop efficient and economic ways for removing such pollutants. These impurities need to be reduced to less than 1 ppm level since they are extremely toxic and poisonous to many industrial catalysts during various applications [1]. Furthermore, H_2S in the excess of 3 ppmv causes pipeline corrosion [2]. Fuel cells have very stringent conditions for sulfur impurities [3]. For PEM fuel cell application, the sulfur level in the hydrogen-rich fuel needs to be less than 100 ppb [4].

Metal oxides, in particular ZnO, are widely applied for gas desulfurization processes [5-7]. Westmoreland and Harrison [8] have shown that the oxides of Fe, Mn, Zn, Ca, V, Cu, Co and W are the most suitable sorbents at temperatures above 300 °C. Among various metal oxides, the ZnO based sorbents have shown advantages of high sulfur capacity and favorable sulfidation thermodynamics at moderate temperatures [9].Silica supported on ZnO and/or doped with Cu is widely used catalyst for desulfurization [10]. In combination with indispensable high-temperature stability, low-temperature activity is highly desirable for a catalyst. The removal of H_2S at high temperatures (ca. 350 °C) has been extensively studied but little work has been reported in the literature for the development of low-temperature H_2S adsorbents [11-12]. Concerns about the removal of H_2S at low temperatures are growing for fuel cell application and several other processes, including natural gas sweetening and Claus process, wherein H_2S is by-product at low temperatures.

The most extensively used porous adsorbent materials for desulfurization include activated carbon, γ-alumina, modified zeolites, etc [13-14]. The discovery of the highly ordered mesoporous materials has received considerable attention in the heterogeneous catalysis [15]. These materials have much promise for the development of novel solid catalyst due to their structural characteristics such as ordered pore structure, high specific surface area (1000-2000 m^2/g), uniform pore size distribution (varying from 1.5-10 nm), and high specific pore volume (1.0-2.0 cc/g) [16-17]. Such mesoporous silica has been recently used as a support for metal catalysts, resulting in several cases in significant improvements when compared to commercial and conventional amorphous silica-alumina catalysts. For example, Corma et al. [18] have reported a superior hydrogenation activity and sulfur tolerance for Pt/MCM-41 in comparison to Pt/zeolite, Pt/SiO_2, and Pt/Al_2O_3. Song and Reddy [19] reported that Co-Mo/MCM-41 showed higher hydrogenation and hydrocracking activities than conventional Co-Mo/γ-Al_2O_3. Fe_2O_3/MCM-41 was found to exhibit a superior performance for the conversion of SO_2 into SO_3 compared to Fe_2O_3/silica [20]. However, reviewing literature, it is worth to mention that so far little information is reported on the MCM-41 and MCM-48 application for H_2S uptake. Therefore, well-ordered, thermally stable mesoporous silica, MCM-41 and 48 were selected as H_2S sorbent and compared with conventional silica in this study.

It is well-known that the surface area, porosity, and chemical nature of the oxide support can affect supported metal-catalyzed reactions [21]. Therefore, it is interesting to compare adsorption capacity of metal oxide supported on mesoporous silica and conventional silica. Since ZnO has high equilibrium constant for H_2S removal at ambient temperature, in the present work, it is attempted to impregnate different silica (MCM-41, MCM-48 and silica gel) supports with ZnO by incipient wetness method. Additionally, the ZnO supported sorbents were doped with Cu and examined for the desulfurization performance at room temperature. These materials were characterized by XRD, N_2 adsorption, and SEM to obtain detailed information in the development of new sorbents. After the adsorption tests, the catalysts were thermally regenerated for multiple adsorption-desorption cycles.

2. Experimental

2.1 Silica support

Different types of silica supports including MCM-41, MCM-48 silica and various other silica gels with different pore volumes were used in this study. The MCM-41 and silica gels, procured from Sigma Aldrich were used as received. The MCM-48 was prepared as described by Schumacher and co-workers elsewhere [22]. 10.4 g of cetyltrimethyl ammonium bromide (CTAB, Aldrich) was dissolved in 480 ml of water and 200 ml of absolute ethanol (99.5%, Aldrich). 48 ml of ammonia solution (32%, Aldrich) was added to the mixture and constantly stirred for 15 min. Then, 13.6 g of tetraethoxysilane (TEOS, 98%, Aldrich) was added, and the whole mixture was constantly stirred at room temperature for 10 h. The obtained white suspension was filtered, washed with hot distilled water and dried at 100 ^{o}C for 12 h. The white powder was then calcined at 550 ^{o}C in air for 10 h to obtain the MCM-48.

2.2 Sorbent impregnation

The un-promoted and promoted silica supported catalysts were prepared by incipient wetness method. Nitrate solutions were used as the metal precursors for impregnating the MCM (-41 and -48) and other silica supports with different loadings. The impregnated supports samples were dried overnight in room air and subsequently calcined in a tubular reactor at 350°C for 1 h under mild airflow.

2.3 Adsorption experiment

The adsorption experiments for desulfurization were carried out at ambient conditions (20 °C, 1 atm). A vertical quartz reactor (10 mm I.D. x 30 mm L) coaxially mounted in a 200 mm long tubular furnace. The temperature of the furnace during desorption experiments was controlled using a PID temperature controller. The gas flow rates were controlled by mass flow controllers (Omega FMA A2404). An inlet concentration of 1 % (v/v) H_2S in H_2 (ultra high purity grade; from Airgas South, Inc.) was used as sulfur source at a face velocity of 2.12 cm/s, corresponding to the volumetric gas flowrate of 0.1 slpm. H_2S uptakes during the dynamic adsorption experiments were measured using a gas chromatography (GC) instrument (Varian CP3800) equipped with thermal conductivity detector (TCD) and pulse flame photometric detector (PFPD). The breakthrough time was defined as the time required for the outlet H_2S concentration to reach 100 ppm. All the tests are been repeated for three times for obtaining the consistency of the results. Using the saturation time and the metal wt % loading on the support, metal utilization is calculated as the ratio of moles of H_2S captured at saturation to total number of ZnO moles present.

2.4 Sorbent Characterization

Nitrogen adsorption/desorption isotherms at 77 K were measured using Quantachrome Instrument Corporation (Autosorb 1-C model, USA). Prior to measurement, all samples were degassed for 10h at 200 °C. Specific surface area, S_{BET} was calculated using the BET equation. Total pore volume, V_P was calculated at $P/P_0 = 0.99$. The pore width, P_w distribution over the range of ~(2–80 nm) was generated from the adsorption branches of the isotherms via BJH method. XRD patterns were obtained using a Rigaku Miniflex diffractometer at room temperature using CuK_α radiation. Diffraction patterns were obtained with Ni-filtered CuK_α radiation (λ = 0.15418 nm) using a scanning speed of 2°/min and an accelerating voltage of 30 kV. The surface morphology of the MCM samples before and after metal impregnation was investigated with Scanning Electron Microscopy (SEM).

3. Results and discussion

3.1 N_2 physisorption

Table 1 describes the specific surface area, total pore volumes, and average pore size of different silica samples before and after impregnation. As observed from Table 1, the BET area as well as the pore volume of all the silica samples decreases on impregnation with the metal oxide. Further increase in the metal loading, the surface area and pore volume decreased. On promoting the ZnO-based samples with 5 mol% CuO, the BET area and pore volume further decreased considerably. On the other hand, in Cu-promoted samples, the pore size was observed to increase significantly (in SiO_2) or marginally (in MCM). This indicates that impregnation of silica supports with metal oxides may result in the decrease in the microporosity and increase in the macro or mesoporosity. In other words, it may be concluded that the metal oxides are preferentially dispersed in the interior of the porous texture of silica, which results in the blocking of mainly the micropores and in the development of pore opening. Table 2 shows the pore volume, surface area and pore size of different commercial silica samples used in this study.

Table 1. Structural characteristics of Silica sorbents determined by N_2 adsorption

Sample	ZnO Loading (wt%)	S_g (m^2/g)	V_t (cc/g)	Avg. Pore Size (nm)	W_{avg} (nm)
Silica C	0	300	1.65	25.0	-
ZnO/SiO_2 C	15	218	0.68	6.5	2.36
$Cu/ZnO/SiO_2$ C	21	165	0.38	9.3	2.45
MCM-41	0	1260	1.30	4.1	2.45
ZnO/MCM-41	8	988	0.92		2.31
	15	850	0.74	3.5	2.58
	25	672	0.71		2.25
Cu/ZnO/MCM-41	15	524	0.52	3.6	2.51
MCM-48	0	1420	1.10	3.1	2.53
ZnO/MCM-48	15	631	0.47	3.0	2.26
	25	592	0.45		2.26
Cu/ZnO/MCM-48	15	303	0.32	4.2	2.58

3.2 Effect of Silica type

To determine the influence of support properties, in particular the structural properties, on H_2S adsorption capacity, different types of silica materials including the conventional amorphous silica (SiO_2, Type C) and mesoporous silica (MCM-41 and MCM-48) were chosen as support for ZnO and tested for their desulfurization performance. Both, the conventional silica gel and highly ordered mesoporous silica were impregnated with ZnO at identical loading of 14% (w/w). The desulfurization activity of these samples was examined using breakthrough capacity measurements described in previous section. Figure 1 shows the breakthrough curves obtained from the adsorption experiments on these ZnO-based silica samples under identical conditions. The adsorption behavior is almost similar for all the three samples. The breakthrough of H_2S was observed to occur at 5, 17, and 20 minutes for the SiO_2, MCM-48 and MCM-41 samples, respectively. The longer breakthrough times indicate that MCM-sorbents were superior candidates for H_2S removal at ambient temperature. The saturation capacities were calculated using the $t_{1/2}$ values, the breakthrough curve is assumed to be symmetric over the entire range.

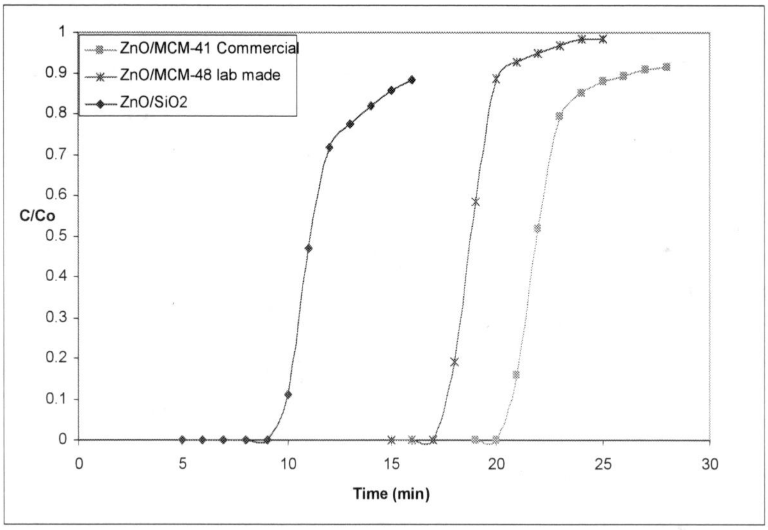

Figure 1. Breakthrough curves for different types of silica impregnated with ZnO
T = 20 °C, Q = 100 ccpm, C_o = 1vol% H_2S/H_2, ZnO Loading = 14% (w/w)

This enhanced adsorption capacity of MCM samples may be explained on the basis of their high surface area and well-structured pores. The BET surface area and total pore volume of MCM samples are significantly larger that those of SiO_2 (Type C) as shown in Table 1. Even, after the loading of ZnO, MCM-41 and MCM-48 samples found to have approximately three fold larger surface areas as compared with that of the corresponding ZnO/SiO_2 sample. Metal utilization values were calculated as described in previous section. ZnO/MCM-41 showed 98% of metal utilization at room temperature for H_2S as against 52% for ZnO/SiO_2 at similar metal oxide loading. This suggests that though the amount of metal oxide loaded is same in MCM and silica, the higher surface area and unique structural integrity of the MCM samples enables them adsorb more H_2S per gram of the sorbent.

3.3 Effect of pore volume

Pore volume is one of the most important factors in catalyst preparation, especially in wet incipient impregnation method. In order to study the effect of pore volume of silica on adsorption capacity of H_2S, various silica samples with different pore volumes but almost similar surface area and pore size have been procured. Table 2 shows the pore volume, surface area and pore size of different silica samples.

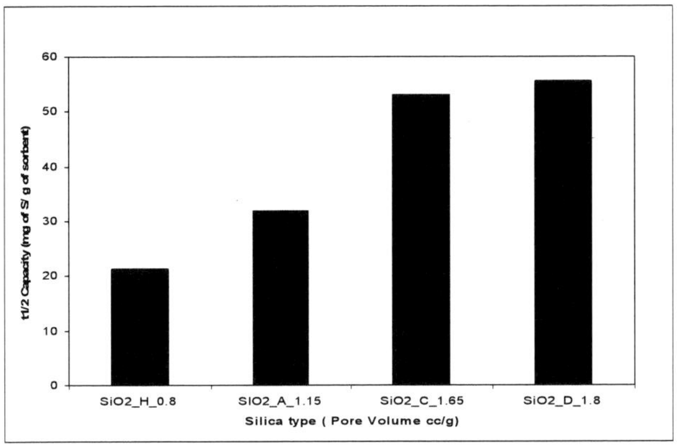

Figure 2. H_2S saturation capacity of ZnO-based SiO_2

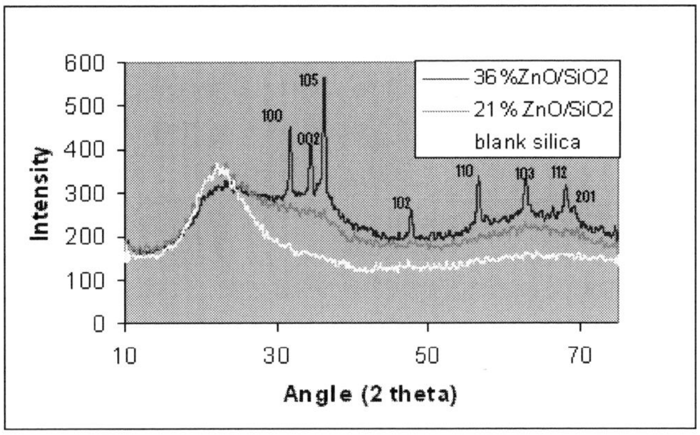

Figure 3. XRD patterns of ZnO/SiO_2

Higher loading of ZnO is possible on silica samples of larger pore volumes. As presented in Figure 2, the adsorption capacity increases significantly with the increase in pore volume and subsequently the ZnO loading. The increasing trend of the saturation capacity is directly attributed to the increasing content of active ZnO phase inside the pores of silica. XRD patterns, shown in Figure 3, indicate that there is almost no aggregation or ZnO crystallite formation taking place on these samples when the impregnation is done as per pore volume (21 wt%) whereas excess/higher loading (36 wt%) of metal oxide shows the development of larger crystallite. In the same line, the data obtained from N_2 adsorption isotherms also suggests that no blocking of pores takes place up to the concerned loading of ZnO.

3.4 Effect of temperature

To understand the effect of temperature on the adsorption of H_2S, ZnO supported Silica (Type C) was tested for its reactivity in the temperature range between 20 and 400 ^oC. Figure 4 shows the saturation capacity obtained from the adsorption of H_2S at different temperatures, when the inlet H_2S concentration (C_o) is 1% (v/v) in H_2 at a face velocity of 2.12 cm/s. The H_2S adsorption saturation capacity remained almost constant up to 200 ^oC and then increased significantly at 400 ^oC. The forward reaction of ZnO and H_2S is favored, leading to enhanced capacities at higher temperatures. The zinc utilization value is also observed to increase by 63% with increase in temperature from 20 to 400^oC. This suggests that the H_2S adsorption on

ZnO or the sufidation reaction is limited to surface at low temperatures, whereas, at high temperatures, the diffusion of H_2S and coverage onto ZnO is almost complete.

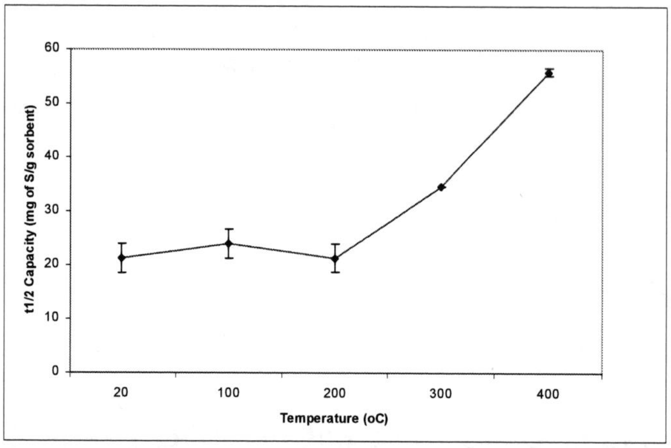

Figure 4. Saturation capacity of ZnO/SiO_2 at different temperatures Q = 100 cc/min, C_o= 1 vol % H_2S/H_2, Loading = 14 wt%, Calcination condition = 350°C/1h

3.5 Screening test for the metal oxide

To find an optimal sorbent for H_2S at ambient temperature, a number of adsorption experiments were carried out on silica dispersed with different metal (Zn, Cu, Mn, Fe, Co and Ni) oxides. Figure 5 (A) compares the breakthrough capacities H_2S for various supported metal oxides at room temperature. Different metal oxides supported on SiO_2 (Type C) at almost identical loading of 21% (w/w) showed distinct performances. ZnO/SiO_2 showed the highest (~48 mg sulfur/g sorbent) capacity. On the other hand, iron, cobalt and nickel oxides supported silica samples were not effective candidates because they showed almost no capacity under identical operating conditions. As a result, the H_2S adsorption performance of the supported metal oxides increased in the order: Fe $\cong$ Co $\cong$ Ni < Mn < Cu < Zn. This was also observed in a study [9] where ZnO based sorbents have shown advantages of high sulfur capacity and favorable sulfidation thermodynamics at moderate temperatures. Also silica supported with ZnO and/or doped with Cu is widely used catalyst for desulfurization [10]. To investigate the influence of the doped ZnO on the H_2S removal, a series of M-ZnO/SiO_2, with M/Zn atomic ratio of 5/100, were prepared by incipient wetness method. Here, M represents transition metal: Mn, Fe, Co, Ni, Cu. The total (Zn + M) metal oxide loading is kept at 21% (w/w). Figure 5 (B) shows the comparative performance of the promoted and un-promoted ZnO-Silica. The ZnO/SiO_2 sorbents doped with Cu and Fe showed significant increase in adsorption capacity. On the other hand, the performance remained almost unchanged for Ni, Co and Mn-doped samples. The Cu-doped ZnO showed the highest capacity of 71 mg S/g sorbent.

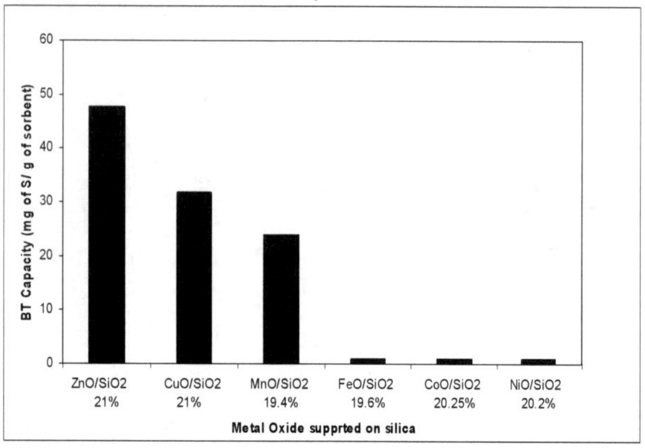

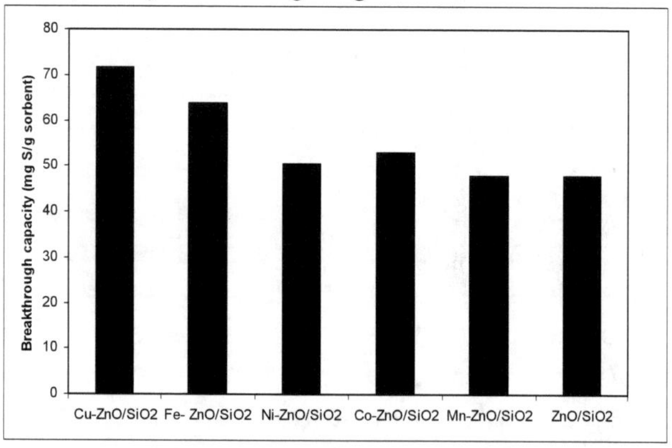

Figure 5. Breakthrough capacity values for (A) different metal oxides and (B) different promoters on ZnO/SiO_2 Q = 100cc/min C_o = 1 vol% H_2S/H_2. Calcination condition = 350°C/1h

To understand the role of ZnO on different sorbent desulfurization capacity, zinc content was varied ranging between 8 and 36% (w/w) and the corresponding sorbents were examined for their adsorption capacity under identical conditions. Table 3 compares the $t_{1/2}$ capacity values obtained for SiO_2 and MCM-41 impregnated with different ZnO loadings. For SiO_2, the desulfurization capacity increased significantly from 37 to 53 mg S/g sorbent with increasing ZnO loading from 12 to 21%, respectively than the increase in capacity from 21 to 29%. This is in line with the Zn utilization values calculated for the corresponding samples. It may be inferred that low ZnO loading results in low desulfurization efficiency. Similar trend in the results was obtained on ZnO/MCM samples. The MCM materials exhibited a superior affinity to H_2S and the desulfurization capacity is 58.53 mg S/g sorbent at ~15% ZnO loading. The Zn utilization of 96.48 % was observed for ZnO/MCM-41. This suggests that the excess of 15% (w/w) ZnO loading may result in the formation of relatively larger metal crystallites. Similar explanation is reported elsewhere [2] for the excess of ZnO loading. With time, during desulfurization reaction, the reaction-product may plug the pores and limit the gas diffusion to result the decrease in H_2S capture.

Table 3. Capacity values for varying loading of ZnO on silica and MCM

Support	Metal (wt% loading)	$t_{1/2}$ capacity (mg S/ g sorbent)	Zn Utilization (%)
SiO_2 (Type C)	12	37.18	78.43
	21	53.12	64.03
	29	55.72	48.68
MCM-41	8	29.28	91.12
	15	55.87	96.48
	25	58.53	59.67

3.6 Effect of calcination temperature

The ZnO/SiO_2 (Type C) sample was calcined at different temperatures ranging between 250 and 550 °C in order to investigate the effect of calcination temperature on the saturation capacity and the crystallite size. As can be seen in the Figure 7, the sample calcined at 250 °C showed a saturation capacity of 61.2 mg S/g sorbent, which increased to the highest level at 71 mg S/g sorbent for that calcined at 350 °C. Samples calcined at higher temperatures than 350 °C resulted in the decrease in the saturation capacity, suggesting that the calcination temperature has to be greater than 250°C and less than 450°C

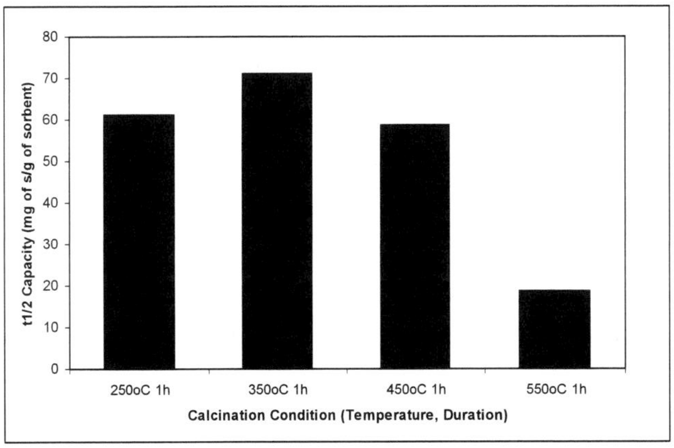

Figure 7. Saturation capacity of ZnO (21 wt%)/SiO_2 C at various calcination conditions

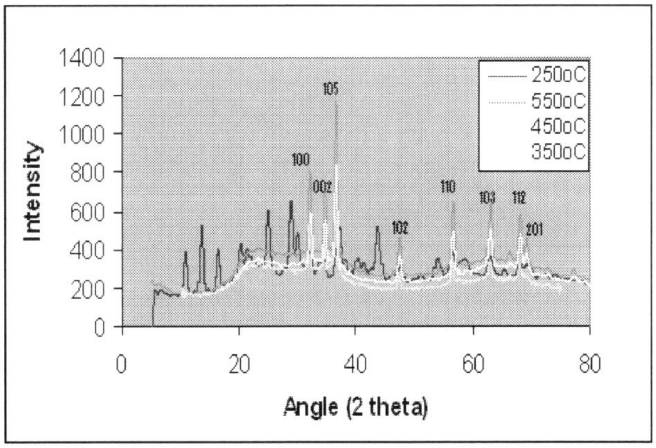

Figure 8. XRD graphs of ZnO(36wt%)/SiO_2 C at various calcination conditions

It may be attributed to the fact that with the increase in calcination temperature, the grain size of ZnO increases resulting in decrease in the total active surface area of the metal oxide crystallites. The XRD graphs also indicate the change in the crystallite size taking place as the calcination temperature changes. At higher calcination temperature larger crystallites are formed which explain the lower adsorption capacity. At 250°C, it can be seen in Figure 8 that several Zn metallic phases other than ZnO are present indicating the incomplete decomposition of nitrates into oxide. As described in Figure 3 in earlier section, ZnO (21wt %) /SiO_2 calcined at 350 °C showed no ZnO crystallites. Based on these results, the 350 °C was chosen as the most favorable calcination temperature and kept constant for the preparation of all sorbents throughout this study.

3.7 Regeneration studies

The sorbents were tested for their repeated use over number of adsorption-desorption cycles. The ZnO/SiO_2 samples were regenerated at 600 °C in air for 1h. As presented in Figure 9, after regeneration the adsorption capacity of the ZnO/SiO_2 samples dropped almost 74% and then remained constant for next cycles. On the other hand, ZnO/MCM-41 showed no regeneration under the same conditions. It was observed that the regeneration of these samples started at low temperature, ca. 400 °C, but the recovered capacity was low (11%). More efforts are required to investigate the favorable regeneration conditions for ZnO/MCM samples.

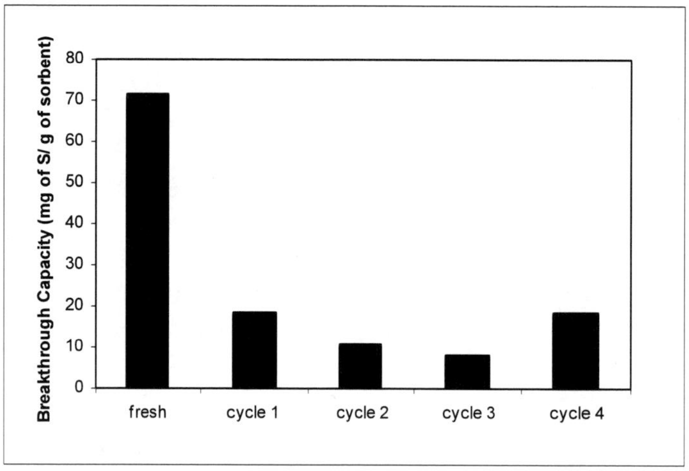

Figure 9. Breakthrough capacity of ZnO/SiO_2 regenerated at 600°C 1h. Q = 100cc/min C_o = 1 vol% H_2S/H_2. Calcination condition = 350°C/1h

4. Conclusions

The experimental results revealed that the support's surface characteristics like the BET area, pore volume and pore size play significant role in the adsorption behavior of the sorbent. Higher the pore volume, higher is the amount of active metal that can be loaded on the support and hence higher is the adsorption capacity. The relatively higher surface area and well-structured pores of MCM-41 and MCM-48 resulted in the promising adsorption capacity for H_2S at room temperature. Amongst all the metal oxides (Zn, Ni, Mn, Co. Fe and Mn), ZnO exhibited highest capacity. ZnO based sorbents showed decrease in their microporosity on increasing the metal loading and/or doping. Doping with Cu and Fe showed significant increase in adsorption capacity of ZnO. The Cu-promoted ZnO (15%)/MCM-41 were found to be the most effective sorbent for desulfurization. Calcination temperature should be low but above the nitrate decomposition temperature because higher calcination temperature leads to development of larger crystallite size, which causes lower capacity. As the temperature of the system is increased from room temperature to 400°C, the adsorption capacity rises by more than 50%. ZnO/MCM-41 shows 52% higher capacity than ZnO/SiO_2 with similar metal loading at room temperature.

References

1. Control of Air Pollution from New Motor Vehicles: Proposed Tier 2 Motor Vehicle Emissions Standards and Gasoline Sulfur Control Requirements. Environmental Protection Agency, 40CFR, Parts 80, 85 and 86, AMS-FRL:-6337-3, RIN 2060-A123; Federal Register/Vol. 64, No. 92/Thursday, 13 May 1999.
2. X. Wang, T. Sun, J. Yang, L. Zhao, J. Jia, Chem. Eng. J. in press (2007).
3. L. Li, D.L. King, Catalysis Today, 116 (2006) 537-541.
4. C.S. Song, catalysis Today 77 (2002) 17-49
5. T. Baird, K.C. Campbell, P.J. Holliman, R.W. Hoyle, M. Huxam, D. Stirling, B.P. Williams, M. Morris, J. Mater. Chem. 9 (1999) 599-605.
6. E. Laperdrix, G. Costentin, O. Saur, J.C. Lavalley, C. Nedez, S. Savin-Poncet, J. Nougayrede, J. Catal. 189 (2000) 63-69.
7. K. Polychronopoulou, J.L.G. Fierro, A.M. Efstathiou, Appl. Catal B: Env. 57 (2005) 125-137.
8. P.R. Westmoreland, D.P. Harrison, Environ. Sci. Tech. 10 (1976) 659-661.
9. E. Sasaoka, Energy and Fuels 8 (1994) 1100-1105.
10. H. Yang, Y. Lu, B.J. Tatarchuk, J. Power Sources 174 (2007) 302-311.
11. K. Polychronopoulou, J.L.G. Fierro, A.M. Efstathiou, App. Catal. B: Environ. 57 (2005) 125-137.
12. M. Xue, R. Chitrakar, K. Sakane, K. Ooi, Green Chemistry 5 (2003) 529-534.
13. M.A. Daley, C.L. Mangun, J.A.D. Barr, S. Riha, A. Lizzio, G. Donnals, J. Economy, Carbon, 35 (1997) 411-417.
14. M.N. Bae, M.K. Song, Y. Kim, K. Seff, Micropor. Mesopor. Mater. 63 (2003) 21-31.
15. J.S. Beck, J.C. Vartuli, W.J. Roth, M.E. Leonowicz, C.T. Kresge, K.D. Schmitt, C.T-W. Chu, D.H. Olson, E.W. Sheppard, S.B. McCullen, J.B. Higgins, J.L. Schlenker, J. Am. Chem. Soc. 114 (1992) 10834-10843.
16. A.B. Bourlinos, M.A. Karakassides, D. Petrdis, J. Phys. Chem. B. 104 (2000) 4375-4380.
17. X-Y. Hao, Y-Q. Zhang, J-W. Wang, W. Zhou, C. Zhang, S. Liu, Micropor. Mesopor. Mater. 88 (2006) 38-47.
18. A. Corma, A. Martinez, V. Soria, L. Catal. 169 (1997) 480.
19. C. Song, K.M. Reddy, Appl. Catal. A 176 (1999) 1.
20. F. Schuth, A. Wingen, J. Sauer, Micropor. Mesopor. Mater. 44-45 (2001) 465.
21. S.J. Gregg and K. S. W. Sing, (1982), *Adsorption, surface area and porosity, 2ed*, Academic Press, London.
22. K. Schumacher, P.I. Ravikovitch, A.D. Chesne, A.V. Neimark, K.K. Unger, Langmuir 16 (2000) 4648-4654.

Natural Gas Desulfurization for Fuel Cell Applications by Adsorption

Gokhan O. Alptekin, Ambal Jayaraman, Margarita Dubovik, Matthew Schaefer, Robert Amalfitano, Michael Ware, and Kristin Bradley
TDA Research, Inc, 12345 W. 52nd. Avenue, Wheat Ridge, CO 80033

The effective utilization of natural gas in fuel cell applications requires removal of odorants (sulfur compounds) to below 0.2 ppm. Adsorption is a promising technology for accomplishing such low levels of sulfur in natural gas. TDA has developed the SulfaTrap™ series of sorbents that are tailored for natural gas desulfurization, and has been carrying out field demonstrations with fuel cell developers. It is important to realize that pipeline natural gas usually has few ppm to few hundred ppm of moisture, and this has an impact on adsorptive desulfurization. Water either competes for adsorption sites or may induce side reactions depending on the odorant. Removal of various sulfur species from natural gas by TDA's SulfaTrap™ will be discussed in this paper.

Hydrogen Generation Process

Zhijun Jia, Larry Stryker, Doug Decker, and Mark Wood
Chart Energy & Chemicals, 2502 Losey Blvd, La Crosse, WI 54601

A new and patented process is described for hydrogen production from liquid fuel using supercritical water. In this process, hydrogen is produced in a single step , even from various fuel sources. The single step process reaches almost perfect conversion and more than 90% yield. In addition, one catalyst appears effective for various liquid fuels. Meanwhile the small amount of sulfur contamination appears not to affect the catalyst activity during the period of operation. In the described process, hydrogen is generated not only from liquid hydrocarbons but a greater quantity is generated from the water. As a consequence, CO2 emission is greatly reduced for the same amount of hydrogen production from conventional hydrogen generation processes. This new process for hydrogen production has a great potential to be used in various scenarios, such as fuel cells for stationary, automotive or portable applications and also large scale hydrogen production.

Low-Cost High-Efficiency Distributed Hydrogen Production

Kandaswamy DuraiSwamy[1], Josh L Mauzey[1], Donovan A. Pena[2], and Anand S. Chellappa[2].

(1) Intelligent Energy, Inc, 2955 Redondo Ave, Long Beach, CA 90806, (2) Intelligent Energy, 9019 A Washington NE, Albuquerque, NM 87113

Intelligent Energy is developing a new-generation 1-10 kWe CHP unit that achieves the high efficiency, high durability and low cost targets simultaneously includes a bold optimization and integration of existing IE technology platforms. The CHP unit will be based on IE's open architecture integration philosophy that maintains a high purity hydrogen interface between the hydrogen generation and fuel cell subsystems. The fuel cell subsystem will be derived from IE's 2kWe CHP platform and its advanced 10kWe auxiliary power unit platform that achieves 60% efficiency on pure hydrogen. An innovative hydrogen generation subsystem will be developed to support the aggressive costs and performance targets, but will leverage IE's experience from two validated technology platforms.: - IE's 100 to 500We membrane reformer that achieves 99.9+% purity on seven different fuel types, and its 10kWe steam reformer integrated with a fast cycle pressure swing adsorption hydrogen purification system. Both of these H2 generation platforms achieve only efficiencies in the range of 60-65% currently. IE plans to investigate significant improvements in these technologies to increase the H2 generation efficiency over 75%.

The greatest challenge of the development will be to achieve an optimized balance between increased stack performance (high cell voltage at high current densities), low cost cell components, increased hydrogen generation efficiency (high fuel conversion, lower steam/carbon ratios, maximum recuperation of heat and water vapor, and high hydrogen recovery factors), low parasitic power components and efficient grid connected inverter, and least cost balance of plant in a fully integrated system.

A Preliminary Physical-Chemical Analysis of Steam Reforming of Biogas from a Small Wastewater Treatment System of a University Campus

Antonio Carlos Caetano Souza, José Luz Silveira Sr., Marcio Evaristo Silva, and Manuela Corghi

Energy, UNESP (São Paulo State University), Av Ariberto Pereira da Cunha, 333, Guaratinguetá, 12514-410, Brazil

The need to diminish environmental impact caused by energy generation and consumption has leaded researches concerning renewable resources such as biogas.

A small wastewater treatment system with biogas production was developed in a campus of São Paulo State University, campus of Guaratinguetá.

Recently, a project was approved under support by FAPESP, a state-owned company responsible by financial support of projects performed in the Brazilian state of São Paulo. Firstly, a theoretical analysis was developed to evaluate the content of hydrogen to be produced through steam reforming. The content of filtered and dried biogas produced in this wastewater treatment system was 80% CH_4 and 20% CO_2. At 1 atm, the predicted content of the gases in the syngas was 57% H_2 at 300°C and 58% H_2 at 800°C, where the reaction could be terminated. At this temperature, 27% CO_2, 9% CO, 6% solid carbon, and small traces of ethylene and acetylene could be encountered.

A Study of Steam Reforming of Ethanol for Hydrogen Production: Technical and Experimental Analysis Utilizing NiCu/Al2O3

José Luz Silveira Sr.[1], Antonio Carlos Caetano Souza[1], Marcio Evaristo Silva[1], Maria Isabel Sosa Sr.[2], Manuela Corghi[1], and Vivian Soliani[1].
(1) Energy, UNESP (São Paulo State University), Av Ariberto Pereira da Cunha, 333, Guaratinguetá, 12514-410, Brazil, (2) Mechanics, UNLP - La Plata National University, Calle 48 y 118, La Plata, 1900, Argentina

Various researches have been developed utilizing Nickel (Ni), Copper (Cu), and ã-Al2O3 as support for biohydrogen production purpose due to its high efficiency of conversion of ethanol and due to its low cost. In this work, results of an theoretical and experimental study concerning steam reforming of ethanol utilizing both materials have been evaluated and compared with results extracted from other works.

The experimental results cited in this work were similar to theoretical studies evaluated in this and other works. However, due to acid effect of Al2O3, this process contributed to dehydration process of ethanol, thus producing ethylene and water. This fact could be diminished introducing Potassium in the composition of studied catalyst. The dehydrogenation occurs specially due to the presence of Cu, and CH4 and CO production occurs mainly due to C-C bonds rupture due to presence of Ni.

Biomass to Biofuels: Catalytic Conversion of Biomass-Derived Synthesis Gas to Oxygenates

Mouzhgun Anjom
Materials Science & Engineering Department, SUNY at Stony Brook, Stony Brook, NY 11794
Sheena Joseph
Chemical and Molecular Engineering, Stony Brook University, 210 Old Engineering, Stony Brook, NY 11794-2275, and Devinder Mahajan, Energy Science and Technology Department, Brookhaven National Laboratory, Upton, NY 11973.

Biomass can be converted to useful fuels and chemicals by two methods: biological (bio) and thermochemical. The bio route is mediated by microorganisms that can process simple sugars as a source of carbon. An example of bio route is the well-established fermentation pathway to ethanol. The thermochemical route is a two-step process and involves: 1) biomass gasification to yield synthesis gas (or syngas), a mixture of CO and H2, followed by catalytic conversion of syngas into oxygenates. For example, conversion of biomass to ethanol via the thermochemical pathway requires development of highly efficient catalysts to achieve total carbon utility and produce ethanol economically.

Since methanol can be produced selectively and in high yields by catalyzed reaction of syngas at low temperatures, we are investigating catalysts that can efficiently convert methanol into ethanol. Mechanistically, the reaction must undergo C-C coupling (homologation). First, Rh-based catalysts are being used because Rh is known to promote methanol homologation to ethanol at 200OC. The reaction is conducted in a 300 mL Parr batch reactor for fast catalyst screening. The goal is to select a non-Rh (less expensive) catalyst that operates in aqueous phase but maintains efficient ethanol conversion under moderate conditions. Ultimately, alcohols can be stored as liquids for hydrogen production. Recent results from our laboratory will be discussed.

1-butene alkylation reaction on $PtSO_xZr/TiZrO_y$ catalyst

Roberto Galiasso Tailleur[1] and Jose V. Andretti Salva
Department of Thermodynamics, Simon Bolivar University,
Baruta, Sartenejas, MirandaVenezuela

Abstract

The apparent chemistry in solid acid alkylation of 1-butene with isobutane was studied at different temperatures and concentrations in a continuous slurry reactor (CSTR) at a total pressure of 1.2 MPa. Different operating variables were explored to obtain 230 experimental points. Samples of liquid and gas phases were taken at different contact times and analyzed by GC-MS techniques using a PONA-type column. A novel $PtSO_xZr/TiZrO_y$ catalyst was used to study the reaction network in the presence of hydrogen that, for this process, is made up of thousands of carbenium ion elementary steps. The complexity of the corresponding reaction network was overcome using an apparent heterogeneous kinetic model to represent the alkylation, hydrogenation, and polimerization of 1-butene. The catalyst deactivation was determined by running the reactor at constant operating conditions for two days with the same batch of catalyst. The procedure was repeated at two different operating conditions. The catalyst was characterized before and after reaction. It was found that in presence of Pt, at low hydrogen partial pressure and low concentration of olefins, the mechanism of alkylation seems to be controlled primary by on-surface rate of reaction, following an order one respect to the hydrocarbons reactants, and secondarily by significantly adsorption of C_{10}^{+}. Noteworthy is the fact that this catalyst, with Pt incorporated, suffered 10 times lower deactivation in presence than in absence of hydrogen.

Keywords: alkylation, spouted bed reactor, solid super acid catalyst, mass transfer.

INTRODUCTION

The development of an alkylation process that uses a solid catalyst is one of the most challenging activities in efforts to improve the production process of a "green" gasoline component. Currently, commercial alkylation of olefins is accomplished using strong acids (HF and H_2SO_4) with good selectivity and activity but which present strong safety limitations (HF) or generate a toxic sludge (H_2SO_4). The development of an active and stable solid super acid alkylation catalyst will introduce interesting technical and economical possibilities, but presents a big challenge. Solid acid catalysts have been developed and tested in very different conditions, some of which are difficult to compare in the presence of rapid catalyst deactivation, and whose mechanism of reactions is difficult to understand. Weitkamp *et al.* (1997), [1] after reviewing the state of the art in solid alkylation, noted that the problem remains unsolved despite numerous contributions in this area. In research presented in this paper, we used different catalyst formulations based on a wide porous Ti-based mesoporous material in which Pt and ZrSOx species were present on the surface. As early as 1975, Yang [2] from Union Carbide discovered that spent zeolitic alkylation catalysts can be regenerated under relatively mild conditions and under hot hydrogen atmosphere if small amounts of Pt are present on the surface. We extend this concept and perform the reaction itself in the presence of the metal and under

[1] Correspondent author. Present address Department of Chemical Engingeering, Texas A&M University, College Station, Texas 77843, galiassa@cantv.net

small partial pressure of hydrogen to control polymerization. To perform the isobutane/alkene alkylation on solid catalysts, a reactor with an efficient back-mixing characteristic and facilities to regenerate the catalyst was selected. In this particular alkylation process (Figure 1) [3], the feeds entering at the bottom of the reactor comprise a liquid phase that contains olefin, isobutane, and a large proportion of recycled isobutane or a C_{10}^+ paraffinic cut, and a hydrogen-rich gas stream (Figure 1). Gas and liquid phases are internally contacted with additional solid acid catalyst coming down from the regenerator. All components entering the riser are propelled upward by the energy of the gas. At a level about 4/5 of the height of the reactor, the gas disengages (or is stripped) from the reacting mixture, is cooled down, and the carried-over liquid is eliminated in a gas-liquid pressure separator. Gas (mainly H_2) is recycled to the inlet of the reactor, and the condensed liquid sent to the fractionation section. The slurry phase leaves laterally (carrying over some gas) near the top of the reactor and is sent to the hydrocyclon. There, liquid and gas are recovered from the top and fractionated, while the wet solid is collected at the bottom of the hydrocyclon, regenerated by hot hydrogen, and sent back to the bottom of the reactor. The high proportion of isobutene or C_{10}^+ recirculated from the fractionation to the inlet of the reactor guarantees the necessary dilution of the olefins in a liquid phase to control the rate of polymerization. Within the reactor, other secondary reactions occur, such as the hydrogenation of olefins into paraffins and their polymerization into a larger molecule (C_{10}^+: soft coke). This soft coke remains adsorbed at the acid sites of the catalyst, generating an important catalyst deactivation. Other details discussed in refs. [4-5] are the fluid dynamics of the continuous slurry in the hydrocyclon. For purposes of designing this system, the kinetic and deactivation rate expressions and fluid dynamic information are needed. To start, a preliminary kinetics and deactivation model to complement the previous results must to be developed. The present experimental effort, therefore, focuses on developing a preliminary apparent kinetics model tested under conditions in which deactivation can be controlled.

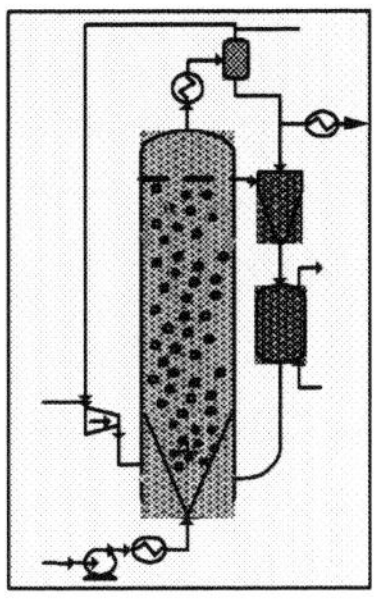
Figure 1. Process.

EXPERIMENTS AND MODEL EQUATIONS

Catalyst. A novel catalyst was prepared by a gel co-precipitation of beta-diketone (2,4-pentane-dione) tetra-ethyl-orthotitanate and a tetra-ethyl-orthozirconate (Aldrich) in ethanol. The solid was stirred for 48 hours in the presence of an ammoniac buffer (pH 8.9). The precipitate was then washed with ammoniac solution at 25°C and dried at 120°C for 8 hours. If this precipitate is calcinated at 500°C, it shows an incipient-structure-type beta zeolite at DRX. The hydrogel was treated in a centrifugal drier in the presence of steam and air at 300°C to promote the migration of Zr to the surface and to destroy micro-porosity. The solid had a particle size between 50 to 100 microns and presents a completely amorphous (DRX) structure. This solid is then treated in a batch reactor at 1.0 MPa in the presence of oleum for 4 hours at 40°C, filtered, washed with methanol, and dried. Then the solid was impregnated with an aqueous solution of diamine platinum dichloride, dried at 120°C, and calcinated in air at 450°C for 12 hours. The catalyst was analyzed using DRX and XPS to check for crystal structure and metal dispersion; and by IR spectroscopy of the pyridine remaining at 300°C to determine its acidity relative to the known SO_4Ti/SiO_2 catalyst.

Kinetic of reactions. The underlying chemistry in the solid acid alkylation of isoalkanes with alkenes follows these elementary steps: alkene protonation, intramolecular hydride shift, methyl shift, protonated cyclopropane (PCP) branching, oligomerization, beta-scission, and intermolecular hydride transfer [6]. The reaction network for this process is made up of thousands of these elementary steps. The complexity of the corresponding reaction network makes the formulation of kinetic models a difficult task, since it requires hundreds of rate coefficients. The alkylation reactions are very complex, and according to the traditional interpretation, they proceed via π bounding of alkenes into a Brönsted site followed by protonation to generate an alkyl carbenium ion. Alkanes, probably adsorbed in a basic site via penta-coordinate carboniun ions, participate through hydride transfer [7]. It is believed that the reaction follows a cycle in which the hydride transfer is the controlling step. In our case, in the presence of Pt and with high amounts of hydrogen and low concentrations of olefins on surface, the mechanism could be different because the product formed seems different. For the simplified analysis of the results, the following three lumps of apparent reaction for 1-butene disappearance were used together with an empirical evaluation of the selectivity.

Table 1. Lump of reactions.

Alkylation (1)	Polymerization "C_{10}^{+}"(2)	Hydrogenation (3)
1-butene + isob. → alkylate	3(1-butene) → C_{10}^{+}	1-butene + H_2 → butane

Deactivation of the catalyst has been studied by several authors [2,8-9] and modeled in different ways. Here, a Levenspiel type-model for two sites active centers was used [3].

Kinetic and deactivation experiments. Extensive experimental studies have been performed with the above-mentioned catalyst using different concentrations of 1-butene and isobutane (olefins/isobutane (O/I) ratio: 0.05:0.2), all diluted in n-decane (olefins/n-decane (O/D) ratio: 0.04-0.08), and treated in the presence of hydrogen partial pressure (H_2/HC ratio: 0:6 by mol). In this preparation, a 50-cm^3 (CSTR) continuous stirred tank reactor (Figure 2) was employed operating at a total pressure of 1.2 MPa, space velocities from 0.8 to 1.2 h^{-1} (respect to the olefin throughput) and temperatures between 333 to 370K (230 experimental points). Duplicate samples of liquid and gas phases were taken at different contact times, and full analysis of the product was performed by GC-MS techniques using a PONA-type column. A reconciliation procedure followed by a mass balance allowed a check of the operational stability. The previously obtained experimental information [5] was used to program a test to verify the kinetics for the alkylation of 1-butene and catalyst deactivation. Here, a simple 5^3 experimental design was programmed to develop a model for olefins disappearance. In addition, three tests were performed at constant temperature for three weeks with all other operating conditions kept constant (residence time 1 h, T: 350 K, O/I: 0.1, O/D:0.1). The conversion and products

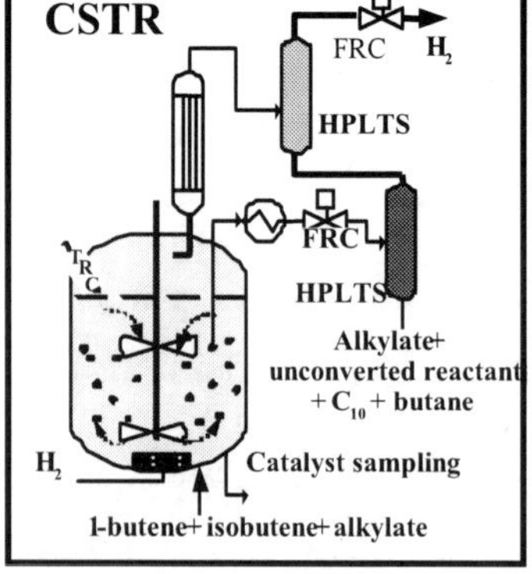

Figure 2. CSTR set-up.

qualities were followed up by daily analysis. This equipment allowed sampling of catalysts for C/H and ^{13}CNMR analysis.

Model development and genetic algorithm. A genetic algorithm (GA) optimization tool was used to determine the apparent reaction rate constants of the three parallel pseudo-heterogeneous rate expressions (equation (4-6)).

$$-R^{i}_{O-Alk} = \frac{k_{o1}K_{i}K_{O}.\exp\left(\frac{-E_{1}}{R.T}\right)*CiC_{O}}{\left[1+K_{C10+}\sum C_{C10+}+K_{H2}C_{H2}\right]} \quad (4), \quad -R^{i}_{O-C^{+}_{10}} = \frac{k_{o2}K_{O}.\exp\left(\frac{-E_{2}}{R.T}\right)*C_{O}^{no}}{\left[1+K_{C^{+}_{10}}\sum C_{C^{+}_{10}}+K_{H2}C_{H2}\right]} \quad (5)$$

$$-R^{i}_{O-P} = \frac{k_{o3}K_{H2}K_{O}.\exp\left(\frac{-Ei}{R.T}\right)*CiC_{H2}^{nH2}}{\left[1+K_{C10+}\sum C_{C10+}+K_{H2}C_{H2}\right]} \quad (6); \quad \tau = \frac{Ci}{-\eta_{i}.R_{i}.a_{p}} \quad (7), \quad a_{m} = e^{k_{dm}t} \quad (8) \quad a_{a} = e^{k_{da}t} \quad (9)$$

$$k_{lO}a_{s}(C_{lO}-C_{O}) = \sum R_{O} \quad (10); \quad k_{li}a_{s}(C_{li}-C_{i}) = R_{O-A} \quad (11); \quad k_{lH2}a_{s}(C_{lH2}-C_{O}) = R_{O-P} \quad (12)$$

Where R_i is the reaction rate (gmol/liter h), k_{oi} is the pre-exponential factor, E_i is the activation energy; K_O, K_i,, and K_{H2} (liter/mol) are the constants for olefins (o), isobutane (i) and H_2 equilibrium of adsorption on catalytic surface (K_i=K_{oi}*exp (λ/RT, being λ the adsorption energy (cal/mol)). Reaction 1 and 2 take places on acid site and 3 in metal site.

To be able to predict the conversions, the three reactions rate equations (mol/liter h) will be calculated using equation (7), where C_O and C_{lO}, C_i, and C_{li}, C_{H2}, and C_{lH2} are the concentrations in solid and in liquid phase (mol/liter). The apparent order for the olefins is n_O(-), and that of the hydrogen, n_{H2} (-); τ is the residence time in hours, η_i is the efficiency factor for the reactions in pores (-) that depends on the Thiele module ($\varphi=3/d_p\sqrt{Ri/D_{eff}C_i}$), a_m a_a (-) is the activity factor for acid and metal sites, and $k_{dp}=k_{dop}\exp(-E_{dp}/RT)$ (1/day). There, k_{dop} (1/day) is the pre-exponential factor for deactivation; E_{dp} (cal/mol) is the deactivation energy; and t (day) the time on stream. A computer program in Visual Basic 6 calculates the rate constants of equations (4-6) using the mass balance for a CSTR equation (7), the mass transfer rate for the reactants equations (8-10), the data obtained at the reactor outlet in 240 independent experimental measurements, and a Carol-type genetic algorithm. The GA adapted the "population" (κ) of constants seeded with the pass of "generations" (ĭ) to fulfill the objective function. This objective function was defined in equation (12). The program requires: (a) thermodynamic information, (b) operating variables and initial concentrations, (c) a seed of constants, and (d) the genetic parameters (population size, seed, number of generations, and mutation probability). The program iterates until it fulfills the objective function, and reports the best set of constants.

$$\text{Objective function} = Abs\left(C_{O}^{Exp} - C_{O}^{\Pr ed}\right) (12)$$

To evaluate the reactivity of n-decane, two tests were performed, with and without olefins, in the presence of the hydrogen and catalyst. The GA also determined the constants (a-e) of an empirical equation (13) that predicts the octane number in products as a function of conversions (X_i is the conversion of 1-butene in wt %).

$$C_8 \text{ in alkylate} = aX_1 + bX_2 + cX_3 + dX_1X_2 + eX_1X_3 \qquad (13)$$

RESULTS AND DISCUSSION

Catalyst. The catalyst contains 0.1 wt% of Pt well dispersed (XPS, TEM) on the surface, and decorating the [SO_4Zr]Pt and [SO_4Ti]Pt clusters (XPS). The catalyst has a superficial concentration of 7.4% on Zr, 0.34% on Pt and 3.4% (mol) on S (XPS). The catalyst shows a 2.5 Brönsted/Lewis ratio (measured by IR of remaining adsorbed pyridine at 573 K) and a total acidity of 2.3 mmol pyridine/g of catalyst. Acidity values twice as high as those of a USY50, and 1.5 times higher than those of SO_4Ti/SiO_2 catalysts were measured. The solid has a surface area of 220 m^2/g, a pore volume of 0.45 gr/cm^3, and an average pore diameter of 14 nanometers. The average particle diameter used was 88 microns. Other detail can be seen in ref. [6].

Products. Alkylated products cannot be described adequately by a simple stoichiometric factor, as alkylate is a complex mixture of isoalkanes. With this catalyst, the major constituents observed (Table 2) in the products were the tribranched isooctanes: 2,2,4-, 2,3,4-trimethyl pentane, followed distantly by the 2,2,3- and 2,3,3-trimethyl pentane, a result different from the distribution observed, for example, by Taylor and Sherwood [8]. These two are desired products, due to their octane numbers in the vicinity of 100. Besides, alkylate is also composed of other isoalkanes, especially of dibranched isooctanes (2,3- and 2,4-dimethylpentanes and dymethylhexane). Small amounts of monobranched isooctanes (methylheptanes and 3-ethylhexane) occurred. Note in Table 2 the presence of cycloparaffins $C_{7\text{-}9}$ never reported before with other catalysts. There are also i-C_{10} (isomers, not shown) from the isomerization of the solvent. The test of n-decane plus olefins indicated that there is, in the main, no conversion into products other than i-C_{10} isomers.

Effect of temperature and residence time. As an example, six products are reported in Table 1, obtained at 333, 343, and 353 K for 0.8 and 1.2 h of residence time (t = 3 h).

Table 2. Alkylate composition at three temperatures and two contact time.
(P_T:1.2 MPa, H_2/O: 4 O/D: 0.07, O/I:0.15, t=3 h)

Residence time τ (h)		**0.8**			**1.2**	
Component $C_{5\text{-}9}$ % mol / Temp K	**333**	**343**	**353**	**333**	**343**	**353**
Isopentanes	1	0.8	0.65	0.5	1	1.4
Dimethylbutane	0.3	0.45	0.5	0.8	1	1.2
Methylpentane	0	2.2	3.5	2.8	3.2	4.1
2,3 dimethylpentane	5.4	6.4	6.7	6.3	8.1	9.5
2,4 dimethylpentane	10.5	15.2	19	0	0.5	1.6
Methylhexane	0	0.1	0.1	0	1	2
2,2,4-trimethylpentane	32.3	30.2	29.5	35	31.4	30.1
2,2,3-trimethylpentane	15	14.1	12.1	13	12.5	9
2,3,4- trimethylpentane	7.4	6.2	5.4	12.8	11.2	10.3
2,3,3-trimethylpentane	8	5	4.2	9.3	10.5	11.1
Dimethylhexane	13.2	13.1	13	10.1	9.2	8.7
Cycloparafins	3	2	1	2.22	1.7	1.1
C_9	3.7	3.9	4.1	5.7	5.8	6
C_{10}^+	0.3	0.5	0.9	2	3	4
Olefins to alkylate wt% of feed	55	61	68	60	68	71
Olefins to C_{10}^+ wt% of feed	7	8	5	11	12	13
Olefins to $C_{1\text{-}4}$ wt% of feed	4	8	13	6	8	14

The 1-butene is primarily converted to alkylate and. to a lesser extent, converted to C_{10}^+ and paraffins, as shown in Table 2, rightmost three columns. With increasing temperature, there is an increase in the cracking and the production of C_8^- and C_{10}^+ compounds.When the residence time is duplicated, the olefin conversion increases but the production of C_8

is reduced, to the detriment of the other products. It seems that the cracking and hydrogenation have higher activation energies than the alkylation reactions. Notice that by changing the temperature and residence time, the distribution of C_8 isomer changes, indicating that the mechanism of reactions is quite complex.

Catalyst deactivation. Taylor and Sherwood [8] reported that after a certain time on stream, the butene conversion begins to drop and finally stabilizes at a low level. They noticed that when the 2-butene conversion begins to drop, the C_8 production was dramatically reduced. The new catalyst behaved differently for 1-butene and H_2 because the activity drops slowly and continuously as a function of time (Figure 3, right side), while the carbon distribution remain quasi constant (Table 3). The C_8 composition varies as a function of time. Figure 3a shows the effect of temperature deactivation. The higher the temperature, the faster the catalyst deactivation, with similar carbon distribution in the products. It seems more a decrease in number of active sites on catalyst than a selective elimination of them as a function of time. In the absence of hydrogen, the catalyst has deactivation activity ten times faster (Figure 3a) for the same other operating conditions than in the presence of hydrogen.

Table 3. Alkylate composition as a function of time on stream.
(4.0% olefin, 1.2 MPa, 1.0 h , H_2/O: 4, O/D: 0.07, O/I: 0.15), * 8% olefin

$C_{5\text{-}9}$ (% mol) *t* (h)	**1**	**14**	**24**	**48**	**48**	**24***
Temperature K		**343**			**353**	**343**
$C_{5\text{-}7}$	25	24.1	24.5	24.95	25	23
2,2,4-trimethylpentane	30.2	30.2	32	33	30	22
2,2,3-trimethylpentane	14.1	15	16	15	14	17
2,3,4- trimethylpentane	6.2	6	4.4	3.8	6	4
2,3,3-trimethylpentane	5	7.0	8.1	9.1	5	6
Dimethylhexane	13.1	10	7.3	6.2	13	20
Cycloparafins	2	1.7	0.9	0.55	2	2
C_9	3.9	5.0	6.0	7.1	4	5
C_{10}^+	0.5	1.0	0.9	0.5	1	2

Effect of olefin concentration. The effect of olefin concentration at constant O/I ratio was studied, keeping all other parameters constant. Figure 3b shows that doubling the olefin concentration in the feed dramatically decreased the conversion to alkylate and increased the production of nC_4 and C_{10}^+, reducing the benefit of the process (less yield and octanes)

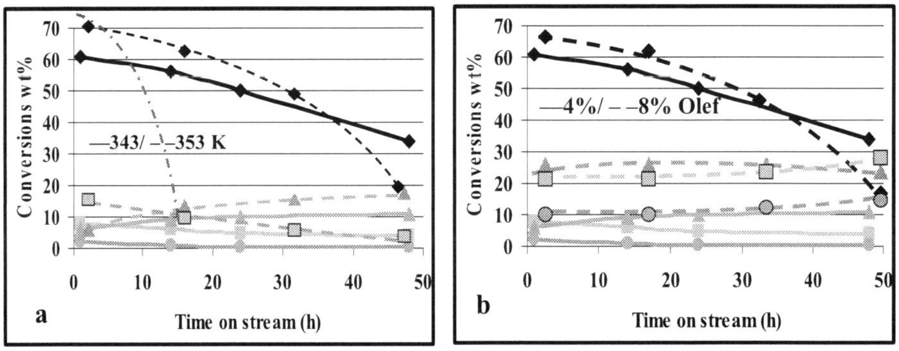

Figure 3. a) Effect of temperature – 343 K, --353 K and ... without H_2 (343 K); b): effect of olefins concentration on conversions (♦ O→Alk, ■ O→-C_{10}^+, ● O→$C_{1\text{-}3}$, ▲O→nC_4).

In addition, the deactivation of the catalyst is faster with higher olefin concentrations due to the polymerization increase; thus, the catalyst required more frequent regeneration. If hydrogen is shut off, the catalyst lost its activity with 4% O in 10 hours.

Products distribution and catalyst deactivation. Table 3, rightmost column, shows the product distribution after 24 hours on stream operating at 343K with 8% of olefins in the feed. Comparing the C_8 distribution with the 4% olefins at 343 K, it can be seen that the deactivation is faster in conditions with 8% olefins in feed, and the C_8 distribution changed (there is more 2,2,4- and less 2,2,3-trimethylpentane, among other changes). The conventional mechanism in the literature [7-9] failed to explain the formation of 2,2,4-, 2,3,3-, 2,3,4-trimethylpentane, cycloparaffins, and dibranched and monobranched isooctanes. A new mechanism is needed to explain the results with the present catalyst. We suspect that cyclo-intermediaries are formed and then transformed into these isomers. In addition, the increase in oligomerization activity with increasing olefin at constant ratio O/I content in the feed is believed to be responsible for a faster deactivation of the acid site, other results that need a deeper analysis outside the scope of this paper. The catalyst is re-generated in hot hydrogen (373 K) in a continuous process that allows operating at constant acceptable activity, in agreement with most of the actual technologies under development.

Effect of isobutane and hydrogen partial pressure. The analysis of the results indicated that the main effect of hydrogen is in reducing the deactivation (not shown) at the expense of increasing the hydrogenation of the olefins. The rate of reactions increased with a very low reaction order: 0.3. The effect of isobutane at this high O/I ratio was small, and an increase of alkylation with the O/I ratio was observed.

Kinetics and deactivation model development. To simulate the continuous slurry hydrocyclon process for an economical feasibility study, a simplified model for the main conversion and one empirical octane equation, as well as a more elaborate kinetics, were developed. Table 4 depicts the value of the constant found for the three lumped reactions and for the deactivation equations. and Figure 5 presents the agreement between the model and the 40 experimental points (25% of total data). The model determined that for the highly diluted olefin feed, the mass transfer rate of reactants does not limit the rate of reactions. The small catalyst particle diameter also reduces the effect of internal diffusion control for the reactions, and the efficacy factor approached 1. Thus, constants (Table 4) were found by the GA with ±1.5% error in prediction for a population size of 4 and after 12 generations, disregarding the effect of mass transfer along the cycle length. Clearly, the results arrived at using data obtained at different times on stream indicated a higher hydrogenation and polymerization sensitivity to the temperature and olefins than the alkylation reaction. - Deactivation of the metal site is three times slower than that of the acid site.

Table 4. Kinetic, deactivation and adsorption constants for the simplified model.

k_{o1} L*min/mol	E_1 cal/mol	k_{o2} l*min/mol	E_2 cal/mol	k_{o3} l*min/mol	E_3 cal/mol	k_{doa} 1/day
6,5x10^{8}	9500	5.4x10^{13}	21000	2.0x10^{12}	17500	2.4x10^{16}
E_{da} cal/mol	k_{dm} 1/day	E_{dm} cal/mol	K_{oC10+} l/mol	λ_{C10+} cal/mol	n_O (-)	n_{H2} (-)
32000	1.1x10^{13}	22000	2	1200	1.7	0.3

Results verified the need to keep both variables as low as possible (an economical compromise due to that increase the costs of reactor and fractionation). The C_{10}^{+} effect of activity is clearly indicated by its adsorption coefficient. The adsorption of hydrogen is not as significant, and its role might be only to control the olefins concentration on surface. Figure 4 shows how well the prediction fits the measured values for a selected set of experimental points. For all data, the error in predicted conversion for reaction 1 is ±1.5%; for reaction 2, ±2.5%; and for reaction 3, ±3.0%.

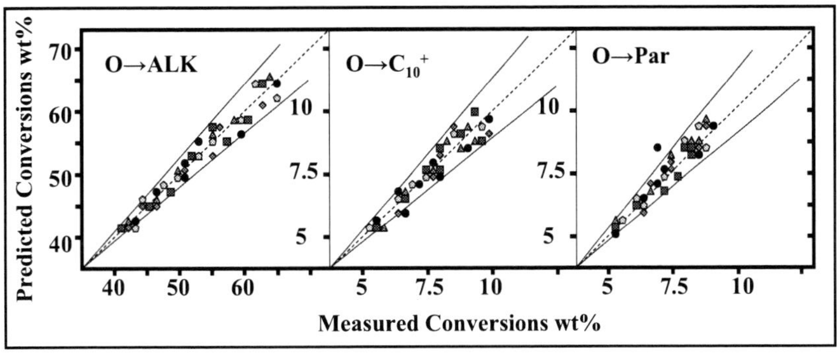

Figure 4. Conversion predicted vs. conversions measured: (temperature (K) - % olefin in feed - time on stream (h)): ■ (343-2-2), ▲ (353-2-2), ♦ (343-4-2), ● (343-2-4), ● (343-2-10).

CONCLUSION

A model that predicts the main conversion and the catalyst deactivation was developed based on extensive experimental work developed in an isothermal CSTR. The results indicate the necessity of working at low temperature and low olefins concentration to maximize the selectivity. The model is based on a lump of three reactions and a Levenspiel type – two sites – catalyst deactivation. The new catalyst shows higher stability in presence of hydrogen than the previous one tested by us without hydrogen or Pt on the catalyst. The metal sites deactivated three times more slowly than acid sites.

References

[1] J Weitkamp J., and Y.Traa, *Catalyst Today*, 49 (1999), 193-199.

[2] C.L. Yang, *US Patent* 3893942, July (1975)

[3] R Galiasso Tailleur, *US patent* pending (2007)

[4] R Galiasso Tailleur, "Spouted bed simulation" *Preprint of AICHE Meeting*, San Francisco Nov. 14th, (2006).

[5] F Gonzalez, M. Rodrigez, and R Galiasso Tailleur, "Simulacion de la alquilación con catalizador sólido usando un reactor slurry" *Preprint of the XX SICAT Congress (Brasil)*, October 17th (2006).

[6] R Galiasso Tailleur, G Andretti Salva "Cold flow studies of a slurry transport reactor-hydrocyclon system" sent to Chem. Eng. Sc. (2008).

[7] A. Feller and J.A. Lercher, Adv. Catal., 48, (2004)229-295

[8] R. Taylor and D. Sherwood Jr., *Appl. Catal. A Gen.*, 157, (1997)195-217

[9] J. Martinis, G.F. Froment, *Ind. Eng. Chem. Res.* 45, (2006), 940-953.

[10] C.A. Querini, E Roa, *App. Catal. A: Gen.* 163, (1997), 199-215

66c Surface Chemistry of Carbon on Ni and Ni-Alloys: Promotion of the Long-Term Stability of Reforming Ni Catalysts by Surface Alloying

Suljo Linic and Eranda Nikolla. Department of Chemical Engineering, University of Michigan, Ann Arbor, MI 48109

Reforming of hydrocarbon fuels is one of the most important heterogeneous catalytic processes. Current reforming catalysts such as Ni, however, suffer from two major problems: (i) carbon-induced deactivation and (ii) sulfur poisoning of the catalyst.

We have utilized Density Functional Theory (DFT) calculations to study the surface chemistry of carbon atoms and carbon fragments on various Ni facets. These studies demonstrated that there is a strong thermodynamic driving force to form extended carbon networks which diminish the long-term activity of Ni. We have further utilized DFT to identify Ni-containing alloys that are more carbon-tolerant than monometallic Ni, i.e., that have a lower propensity to form the carbon networks. Subsequent steady-state reactor studies validated the predictions of the DFT calculations. Aside from the steady-state reactor tests, the catalysts were characterized with multiple spectroscopic and microscopic tools.

We will provide a very general and physically intuitive model that can explain the enhanced performance of the Ni alloys.2

1 E.Nikolla, A. Holewinski, J. Schwank, S. Linic; "Controlling Carbon Surface Chemistry by Alloying: Carbon Tolerant Reforming Catalyst", Journal of the American Chemical Society, 2006; 128(35); 11354-11355.

2 E.Nikolla, A. Holewinski, J. Schwank, S. Linic "Promotion of the long-term stability of reforming Ni catalysts by surface alloying", Journal of Catalysis, 250(1), 85-93, 2007

66b Isomerization of Hexane on Pt-Au Nanoparticles Supported on Zeolites

J. Fraissard Sr.[1], Vasilyi Gerda[2], Kazymyr I. Patrylak[3], and Yu. G. Voloshyna[3]. (1) Laboratoire de Physique Quantique. ESPCI, University of Pierre and Marie Curie, 10 rue Vauquelin, 75005 Paris, France, (2) Department of Chemistry, Taras Shevchenko National University, 64 Volodymyrskaya str., Kiev, Ukraine, Kiev, 01033, Ukraine, (3) Institute of Biorganic Chemistry and Petrochemistry, Kiev, 02660, Ukraine

It is planned to eliminate benzene and its derivatives from petrol and to replace them by methylated hydrocarbons, whence the importance of the isomerization of n-alkanes, which is currently performed in liquid superacid media, since current solid superacids are very rapidly poisoned. We show that some Pt-Au based zeolite catalysts can be useful in this reaction.

Samples were prepared according to reference [1] with 1% w/w of Pt and a ratio Au/Pt=20. UV-Visible, microtomy-EM and EDX spectroscopies show that the particles are essentially in the pores and are bimetallic.

The hydroisomerization of hexane was carried out at 200-350°C in a flow-type fixed-bed reactor. HY, Pt–HY and PtAu-HY are of no interest for this reaction. The efficiency of PtAu-HZSM5 is noteworthy; it is remarkably stable over one week with a selectivity and an activity for methylpentane of about 80% at 260°C. Moreover, cracking is negligeable, the other reaction products being i-C5 and i-C4.

[1] Riahi G., Guillemot D., Polisset-Thfoin M., Khodadafi A., Fraissard J., J. Catal. Today, 72, 115 (2002)

66f Whole-Cell Biocatalysis in Biphasic Media for Oxidation of Naphthalene

S. V. B. Janardhan Garikipati, University of Iowa, Iowa City, IA 52242 and Tonya L. Peeples, Department of Chemical and Biochemical Engineering, The University of Iowa, Iowa City, IA 52242.

Biocatalysis involves the use of enzymes to perform stereo- and enatio-specific reactions producing valuable products in an environmentally benign way. However, the hydrophobic nature of most of the substrates affects the productivity due to poor solubility in aqueous solutions and high toxicity to cells expressing the biocatalyst. These drawbacks can be eliminated by using biphasic media, where a second phase of organic solvent serves as a reservoir of substrate and helps in extracting the products in-situ. The use of biocatalysis in biphasic media is applied to oxidation of napthalane to produce either 1-naphthol or 1,2-naphthalene dihydrodiol.

Simulation Process of Biodiesel Production Over Heterogeneous Catalysts

Biodiesel is a promising nontoxic and biodegradable fuel comprised of alkyl esters of long chain fatty acids, which are derived from renewable biological sources such as vegetable oils and animal fats[1]. It has similar physical properties to petrochemical diesel and is considered the best alternative fuel candidate for use in diesel engines[2]. Biodiesel was firstly tested as engine fuel by Rudolf Diesel one hundred years ago[3]. However, it was only used in emergency situations in past few decades. Recently, due to the increases in crude oil prices, limited resources of fossil oil and the environmental concerns, biodiesel is becoming a developing area of high concern. Compared to diesel fuel, one of the advantages of biodiesel is that the raw materials used to produce it are natural and renewable, and therefore biodiesel has low emission profiles, less toxic, and is more environmentally beneficial.

Different studies have been carried out extensively using different ways of biodiesel production, with different oils as raw materials[4]. The most commonly used method is by transesterification of triglycerides, the main constituent of vegetable oils and animal fats, with an alcohol to form esters and glycerol. Transesterification (also called alcoholysis) is the reaction of a fat or oil with an alcohol to form esters and glycerol. Methanol and ethanol are most frequently used alcohols in the transesterification processes, especially methanol, because of its low cost and its physical and chemical advantages[5]. Because the reaction is reversible, excess alcohol is used to shift the equilibrium to the products side. Different kinds of catalysts, homogeneous ones such as sodium hydroxide, potassium hydroxide, sulfuric acid and supercritical fluids, and heterogeneous ones such as lipases, can be used to improve the reaction rate and yield. Currently, homogeneous alkaline catalysts, such as *K* or *Na* alkoxides or hydroxides are most often used commercially for the transesterification of biodiesel production[6]. However, removal of the soluble base after reaction is a major problem, because aqueous quenching results in the formation of stable emulsions and saponification, rendering separation and purification of the methyl ester difficult. As a result, biodiesel production by these routes is still not cost-competitive with petrochemical diesel fuel[7]. These disadvantages could, however, be removed by use of a heterogeneous catalyst, which can eliminate of a quenching step (and associated contaminated water waste) to isolate the products and operate a continuous process[8]. Heterogeneous catalysts, including zeolites[9], alkali earth oxides[10] and hydrotalcites[11,12], have been investigated in transesterification reactions.

Heteropolyacids (HPAs) are an interesting class of solid acid catalysts, which can be used for diverse reactions that require strong acidity. Although the acid forms are themselves useful solid catalysts, they are highly soluble in, and difficult to separate from polar media. This high solubility makes such solid acid catalysts unsuitable for catalyzing polar chemistry[13]. Recently, Narasimharoa et al.[6] introduced a series of

Cs-doped HPAs (e.g., $Cs_xH_{3-x}PW_{12}O_{40}(x=1-3)$), which exhibit dramatic increases in surface area and profound changes in solubility over the parent HPA. Partial substitution of protons by Cs^+ also changes the number of available surface acid sites with significantly higher activity than the parent acid[14]. In their study[6], this series of insoluble HPA salts of general formula $Cs_xH_{3-x}PW_{12}O_{40}(x=1-3)$ was characterized using a range of bulk and surface sensitive probes, and investigated for solid acid transesterification reactions. However, few studies concerning the kinetic mechanism of this Cs-doped heteropolyacid catalyzed transesterification have been proposed. Thus, in our recent work, a heuristic diffusion-reaction kinetic model has been established to simulate the transesterification of alkyl ester with methanol over a series of heterogeneous Cs-doped heteropolyacid catalysts.

Firstly, we applied the Eley-Rideal type of mechanism[15] to the simulation of the heterogeneous transesterification of biodiesel production over Cs-doped heteropolyacid catalysts. The transesterification of tributyrin by methanol consists of the following three consecutive transesterification reactions:

$$T + MeOH \Leftrightarrow D + MeBu$$

$$D + MeOH \Leftrightarrow M + MeBu$$

$$M + MeOH \Leftrightarrow G + MeBu$$

The triglyceride, tributyrin (T), firstly reacts with methanol, to produce a diglyceride, dibutyrin(D), and the fatty acid ester, in this case methylbutanoate (MeBu).Then the diglyceride formed reacted further with the alcohol to form another molecule of the fatty acid ester and a monoglyceride, monobutyrin (M). This then reacts with the alcohol to obtain glycerine (G) and the fatty acid ester.

The elementary steps and the corresponding reaction mechanism for each transesterification reaction are shown in Table 1. The three parameters of the model are the activation energy $E_{a,MeOH}$, the pre-exponential factor A of the rate-determining step, and the adsorption equilibrium coefficient K_A. The rate equations for the three glyceride transesterification are presented as follows

$$r_1 = \frac{k_{MeOH}(a_{MeOH} - (1/K_{eq})(a_{MeBu}a_D / a_T))}{1 + (K_A / K_{eq})(a_{MeBu}a_D / a_T) + K_A a_D + K_A a_M + K_A a_G} \quad (1)$$

$$r_2 = \frac{k_{MeOH}(a_{MeOH} - (1/K_{eq})(a_{MeBu}a_M / a_D))}{1 + (K_A / K_{eq})(a_{MeBu}a_M / a_D) + K_A a_D + K_A a_M + K_A a_G} \quad (2)$$

$$r_3 = \frac{k_{MeOH}(a_{MeOH} - (1/K_{eq})(a_{MeBu}a_G / a_M))}{1 + (K_A / K_{eq})(a_{MeBu}a_G / a_M) + K_A a_D + K_A a_M + K_A a_G} \quad (3)$$

And

$$E_{a,MeOH} = -RTLn(\frac{k_{MeOH}}{A}) \quad (4)$$

Where k_{MeOH} is the methanol adsorption rate constant equal for all three reactions, and a_i is the activity of each reaction component *i*. K_{eq} is the equilibrium coefficient of the overall transesterification reaction.

Table 1 Elementary steps and reaction mechanism for the biodiesel transesterification reactions

Elementary reactions	Reaction mechanism		
	T-D	D-M	M-G
$CH_3OH + * \Leftrightarrow CH_3OH*$	1	1	1
$CH_3OH*+T \Leftrightarrow D*+MeBu$	1		
$CH_3OH*+D \Leftrightarrow M*+MeBu$		1	
$CH_3OH*+M \Leftrightarrow G*+MeBu$			1
$D* \Leftrightarrow D+*$	1		
$M* \Leftrightarrow M+*$		1	
$G* \Leftrightarrow G+*$			1

Since it was observed that the reaction mixture was not ideal, activities, rather than concentrations should be used in the rate expression. The activity a_i for each component i is calculated from the component concentration C_i and the activity coefficient γ_i (Eq. (5)). The UNIFAC contribution method was used to calculate activity coefficients[16].

$$a_i = C_i \times \gamma_i \tag{5}$$

Thus for a batch reactor with the assumption that no by-product is observed, differential equations of reaction components are expressed as:

$$\frac{dn_T}{dt} = -r_1 W \tag{6}$$

$$\frac{dn_D}{dt} = r_1 W - r_2 W \tag{7}$$

$$\frac{dn_M}{dt} = r_2 W - r_3 W \tag{8}$$

Where n_T is the number of mole of triolein, n_D is the number of mole of diolein, and n_M is the number of mole of monoolein. W is the mass of the catalyst and

r_1, r_2, r_3 are the reaction rate derived from Eqs. (1)-(3). And the estimation of the kinetic parameters was performed by minimization of the objective function S:

$$S = \sum_{k=1}^{N} (n_{T,k} - \hat{n}_{T,k})^2 + \sum_{k=1}^{N} (n_{D,k} - \hat{n}_{D,k})^2 + \sum_{k=1}^{N} (n_{M,k} - \hat{n}_{M,k})^2 \qquad (9)$$

Where $\hat{n}_T$, $\hat{n}_D$, $\hat{n}_M$ are the *k*th experimental number of moles of triglyceride, diglyceride and monoglyceride, and n_T, n_D, n_M are the corresponding value calculated by the model. *N* is the total number of experimental values. The minimisation is achieved by a Genetic algorithm developed by MATLAB.

For the kinetics model using Eley–Rideal type of mechanism we proposed above, it is assumed that there is no mass transfer limitation at the surface of particles and hence the concentration at the surface of a particle is equal to the bulk concentration. Another important assumption in this model is the presence of high mass transport limitations inside the catalyst particles as a result of the relatively low diffusivity of triglycereide. Thus, no diffusion of reactants inside the pellet has been considered. However, for a given fixed bed reactor, the effect of mass transfer inside the homogeneous catalyst bed should be taken into account[17]. Thus, a diffusion-reaction bulk scale model for a fixed bed batch reactor is developed. The mass balance for each bulk specie *i* can be expressed as a function of diffusion transfer (effective diffusivity $D_{L,i}$), convection (axial velocity u), and rates of reaction ($r_{cat,j}$) (Eq. (10)).

$$\frac{\partial C_i}{\partial t} = \frac{D_{L,i}}{\varepsilon}[\frac{\partial^2 (C_i)}{\partial z^2}] - \frac{u}{\varepsilon}\frac{\partial C_i}{\partial z} + \frac{(1-\varepsilon)}{\varepsilon}\rho_{cat}\sum_{j=1}^{nrxn} v_{ij} r_j \qquad (10)$$

Where C_i represents the bulk concentration of specie *i*, *z* is the bed axial length, and ε is the bed void fraction. ρ_{cat} is the density of catalyst, v_{ij} represents the stoichiometric coefficient for specie *i* in reaction *j*, and r_j is the rate of jth reaction (derived from Eqs. (1)-(3)).

The novelty of this framework lies in detailed modeling of surface reacting kinetic phenomena and integrating that with particle-level transport phenomena all the way through to process design and optimisation, which has been done for biodiesel production process for the first time. This multi-disciplinary research offers better insights into catalyst design and process intensification for the industrial application of Cs-doped heteropolyacid catalysts for biodiesel production. A case study of the transesterification of tributyrin with methanol has been demonstrated to establish the effectiveness of this methodology.

References

1. Krawczyk T. Biodiesel - Alternative fuel makes inroads but hurdles remain. INFORM 7, 1996; 801-829.
2. Romano S. Vegetable Oils Fuels, in: Proc. Int. Conf. on Plant and Vegetable Oils as Fuels, ASAE, MI, USA, 1982; p. 106.
3. Shay E. G. Diesel fuel from vegetable oils: status and opportunities. Biomass and Bioenergy. 1993; 4, 227-242.
4. Marchetti J.M., Miguel V.U., and Errazu A.F. Possible methods for biodiesel production. Renewable and Sustainable Energy Reviews. 2005; 6:1300-1311.
5. Ma F., Hanna M. A. Biodiesel production: a review. Biosource technology. 1999; 70:1-15.
6. Narasimharao K., Brown D.R., Lee A.F., Newman A.D., Siril P.F., Tavener S.J., Wilsona K. Structure–activity relations in Cs-doped heteropolyacid catalysts for biodiesel production. Journal of Catalysis. 2007; 248: 226–234.
7. Freedman B., Pryde E.H., Mounts T.L., Am J. Oil Chem. 1984; 61: 1638.
8. Ono Y., Baba T.. Catal. 1997; 38: 321.
9. Leclercq E., Finiels A., Moreau C, Am J.. Oil Chem. 2001; 78: 1161.
10. Gryglewicz S.. Bioresour. Technol. 1999; 70: 249.
11. Cantrell D.G., Gillie L.J., Lee A.F., Wilson K., Appl. Catal. A Gen. 2005; 287: 183
12. Di Serio M., Ledda M., Cozzolino M., Minutillo G., Tesser R., Santacesaria E., Ind. Eng. Chem. Res. 2006; 45: 3009.
13. Okuhara T., Mizuno N., Misono M.. Adv. Catal. 1996;41: 113.
14. Mizuno N., Misono M.. Chem. Lett. 1987; 967.
15. Dossin T.F., Reyniers M. F., Marin G.B.M., Appl. Catal. B Environ. 2006; 61: 35-45.
16. Reid C., Prausnitz J.M., Poling D.E.. The Properties of Gases and Liquids, fourth and fifth ed., McGraw-Hill, New York, 1987.
17. Ding Y.. Alpay E. Adsorption-enhanced steam-methane reforming. Chemical Engineering Science. 2000; 55:3929-3940.

66d Biocatalytic Systems for Aromatic Oxidations

Angela M. McIver and Tonya L. Peeples. Department of Chemical and Biochemical Engineering, The University of Iowa, 4133 Seamans Center, Iowa City, IA 52242

The purpose of this research is to engineer a biocatalytic system to facilitate the production of oxidation products in an economic and environmentally benign fashion. Oxygenases are powerful stereoselective and regioselective catalysts that are useful in the preparation of valuable pharmaceutical and specialty chemical intermediates. The need for cofactor regeneration necessitates the use of whole-cells in such bioprocesses. We have selected to work with various organisms carrying mono- and dioxygenases. Immobilization of these microorganisms, including solvent tolerant host organisms, will result in more stable biocatalysts that will be more amenable to meet process requirements. Specifically, retention of the immobilized catalyst will facilitate isolation of valuable products. This effort is critical to producing environmentally beneficial biotransformation systems by providing an environmentally benign oxidation process with reaction and separation of products. This work highlights some results comparing the effectiveness of naphthalene and toluene dioxygenases expressed in Escherichia coli for aromatic oxidations. The use of biphasic reaction media is used to increase productivity and enhance the reaction. We will show methods of creating an environmentally benign system.

Detection and Measurement of Fugitive Methane Gas Emission Using Infrared Imaging Camera

Anisa Safitri and Dr. Sam M. Mannan
Mary Kay O'Connor Process Safety Center
Artie McFerrin Department of Chemical Engineering
Texas A&M University

AIChE Spring Meeting, April 2008
8th Topical Conference on Natural Gas Utilization
New Orleans, Louisiana, April 6-10, 2008

Introduction

Infrared thermography is a modern gas detecting method that provides a means of temperature measurement of the gas cloud. Thermography determines the surface temperatures of the gas using a special-purpose camera that captures matters that human cannot see. The infrared camera consists of permeable lens, a transmission line and a sensitive detector. The detector captures the infrared radiation which converted into electric signals. The signals are then transformed into pixels so that a thermogram, the figure of the temperature profile of the gas, appears on the screen.

Fundamental of Infrared Technology

Infrared detection technique using infrared thermography is developed based of the theory of radiation by Kirchhoff, Planck's law of radiation, Wien's displacement law and Stefan-Boltzmann law. All objects at non-zero temperature radiate electromagnetic energy. Kirchhoff"s law of thermal radiation states that at thermal equilibrium, the emissivity of a body (surface) is equal to its absorptivity. A perfect black body absorbs all light that strikes it, no electromagnetic radiation passes through it and none is reflected.[1,2]

Real objects never behave as a perfect black body; instead the emitted radiation is less that the ideal emission. This usually recognized as 'grey body'. The ratio of emissivity E of a real body to the emissivity E_s of a black body is given as a non-dimensional emission coefficient, ε:[1,2]

$$\varepsilon = \frac{E}{E_S}$$

The emission coefficient depends on the wavelength, temperature and surface texture of the body. This concept is the basis used in thermography technique. The spectral of electromagnetic radiation from a back body at specific temperature at all wavelengths is described by Planck's law:[1,2]

$$I(\upsilon,T)=\frac{2h\upsilon^3}{c^2}\frac{1}{e^{\frac{h\upsilon}{kT}}-1} \quad [radiance/unit\ frequency\ \text{int}erval]$$

Plank's law can also be written as a function of wavelength, λ:

$$M(\lambda,T)=\frac{2hc^2}{\lambda^5}\frac{1}{e^{\frac{h\upsilon}{\lambda kT}}-1} \quad [radiance/unit\ wavelength\ \text{int}erval]$$

where λ is wavelength, T is the absolute temperature, h is the Planck constant and c is the speed of light. Wien's displacement law describes the relationship between the temperature of a blackbody and the wavelength at which it produces the maximum intensity of radiation.[1,2]

$$T\lambda_{\max}=2.898\times10^{6} \quad nm\,K$$

Stefan-Boltzman's law discovered that total energy radiated per unit are of a black body resulted from integrating the Planck's law over all wavelengths (j*) is proportional to the fourth power of its temperature.[1,2]

$$j^*=\sigma T^4$$

where σ is the Stefan-Boltzman's constant:

$$\sigma=\frac{2\pi^5k^4}{15c^2h^3}=5.670400\times10^{-8}\ Js^{-1}m^{-2}K^{-4}$$

For a grey body, the Stefan-Boltzman's law is also a function of emissivity:

$$j^*=\varepsilon\sigma T^4$$

The absolute power of energy radiated for an object is obtained by multiplying the energy radiant flux with the surface area of the object:

$$P=Aj^*=A\varepsilon\sigma T^4$$

Conceptual design

This research project will focus on the use of the infrared camera to detect a gas and/or leak and measure the temperature profile. The final goal is to estimate the emissions of the methane gas and therefore a vapor dispersion model will be developed for different cases in order to convert the temperature profile into a concentration gradient.

The temperature profile of the methane gas plume will be shown from the infrared image. The hypothetical result is illustrated in **1**. The plume will disperse principally into the direction of the wind, and slightly in the vertical and crosswind direction.

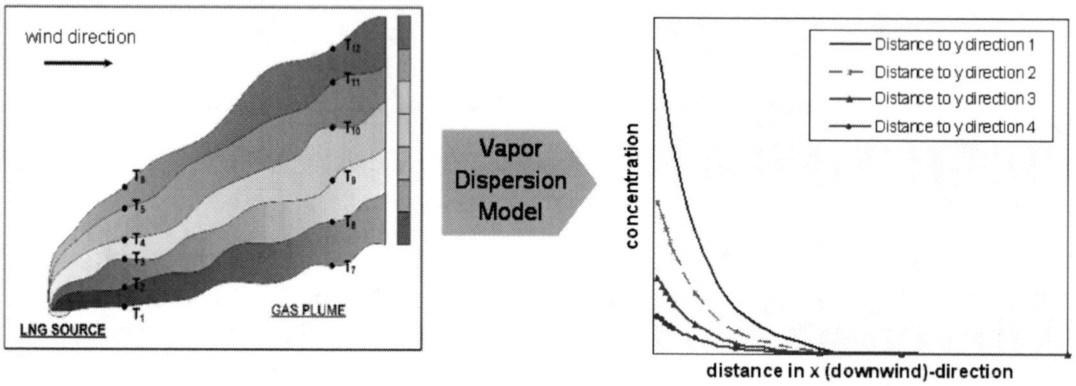

Figure 1. *The hypothetical 2-dimensional plume temperature profile and concentration gradient for the dispersion taking place under the actual environmental and meteorological conditions.*

Methodology

This research project is going to be accomplished in three phases as shown in **Figure** . The first phase is the laboratory work, the second phase will be the field tests and the third phase will be the measurements in the natural gas industry, i.e., plant and pipelines. For each phase, a vapor dispersion model will be developed.

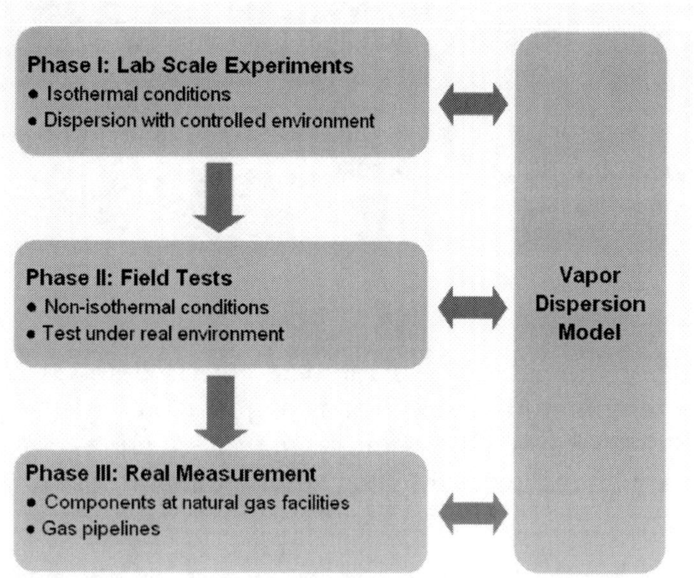

Figure 2. *The research plan: phase I is the laboratory experiments, phase II is the field tests and phase III is the real natural gas industry application. Vapor dispersion model is developed for each measurement.*

Phase I: Laboratory-scale experimentation

The laboratory experiments will be employed to accomplish the following objectives:

- To identify the correlation between the temperature and concentration at a certain volumetric flow rate for pure methane gas and mixtures of methane and air.
- To test the ability of the infrared camera in detecting and monitoring the leak.
- To capture the plume temperature profile using the infrared camera under the simple (controlled) environmental conditions.

Two experimental setups will be built to accommodate these objectives. The first setup consists of a methane cylinder, a cylinder containing air, pressure gauge, volumetric flow meter, a tube with attached thermocouples, and a series of pipelines. Pure methane gas will be introduced into the tube at a certain (volumetric) flow rate. The thermocouple will detect the temperature of the gas inside the tube. Assuming ideal gas, the concentration of the methane gas can be determined. The same experiments will be done for mixture of methane and air at different compositions. The result of these experiments will provide the data of temperature and concentration for both pure methane gas and the mixture of methane and air at different flow rates, such that their correlations can be established.

The configuration of the second setup is similar to the first setup. The only differences are the tube design and that it only uses pure methane gas. An adjustable hole is prepared on the tube to allow the gas to seep out of the tube. The gas escaping from the hole is going to be detected by the infrared camera. In order to know the maximum ability of the infrared camera in identifying the leak, the hole size is reduced until it reaches the minimum size in which the leak can still be spotted by the IR camera. This experiment is set in order to give an illustration of detection of a fugitive leak since most of the methane gas emissions in the natural gas industry come from fugitive leaks. Therefore, understanding the capability of the camera will help to specify the lowest methane concentration that can be measured in the natural gas industry.

The outcome from these laboratory experiments is the image of the temperature profile of the plume. However, the emission rate requires concentration of the methane gas and therefore converting the temperature profile into the concentration is important. This goal is achieved by developing a vapor dispersion model of the methane gas based on the conservation of mass using the continuity equation.

Phase II: Field tests

The second phase of this research project will be the field tests. These tests need to be accomplished in order to perform the measurements under the real environmental and meteorological conditions that come close to the conditions in the natural gas industry and test the accuracy, stability and operative performance of the infrared camera. The tests will be performed by taking into account the effect of wind velocity, humidity, moisture, and daylight or night conditions. The concept of the field tests will be similar to that of the second laboratory setup; the only difference is that the system will be placed outdoor where a variety of parameters will affect the atmospheric dispersion of the methane gas. The gas flow meter and vapor analyzer will also be required for measuring the leak flow rate and the concentration at the leak source as means to validate the measurements.

Prior to performing the measurements, the environmental and meteorological conditions including wind, humidity, atmospheric stability and ground condition has to be identified. The atmospheric stability affects the vertical mixing of the air. Their classes depend on the wind speed and the quantity of the sunlight. The increased wind speed will increase the atmospheric stability during the day whereas reverse consequence will occur at night time. The atmospheric stability and the distance downwind from the release will dictate the dispersion coefficients in the downwind, crosswind and vertical directions. Ground conditions will affect the mechanical mixing at the surface and the wind profile. For a flat open area the mixing in the vertical direction will be minimal. Understanding these parameters will also help to determine the placements of the infrared camera in order to gain high-quality data. The infrared camera should be placed perpendicular to the wind direction in order to visualize the plume shape in the downwind direction because the methane dispersion will be dominant in that direction.

The physical features that need to be observed from the measurements are: above-ground point-source leak conditions, leak rate, pipe internal pressure, above-ground plume conditions to be controlled or measured, plume thickness, exit plume velocity, distance of the measurement, depth, area extent, measurements of temperature as a function of surface position and height, and measurements of surface temperature as a function of distance from the source. From the measurement of those parameters, a vapor dispersion model can be established.

Phase III: Measurement in industry

The use of the infrared camera will be tested in the actual natural gas industry, mainly to detect the fugitive gases. Some measurements will be performed in different equipment including valve fittings, connections, flanges, compressors, pumps and pipelines. The same procedure as the field tests will be carried out. However, since this technique is novel in the natural gas industry, the conventional method using a vapor analyzer and a mass flow meter will support and complete the measurement as well as validate if the new proposed technology is reliable for estimating the methane emissions.

Vapor Dispersion

Dispersion model can be used to illustrate the transport phenomena of gas or vapor released from its source to the atmosphere. The vapor concentration will be at the highest concentration at its release point and will continuously decrease as it is carried downwind. The atmospheric dispersion is calculated by solving the conservation equations of mass, energy and momentum. [3]

The equation of continuity (conservation of mass) is given as:

$$\partial\rho/\partial t + (\nabla \cdot \rho v) = 0$$

For three dimensional Cartesian coordinate the equation becomes

$$\frac{\partial p}{\partial t} + \frac{\partial}{\partial x}(\rho v_x) + \frac{\partial}{\partial y}(\rho v_y) + \frac{\partial}{\partial z}(\rho v_z) = 0$$

The expression of the equation of energy is given as follow:

$$\frac{\partial}{\partial t}\left(\frac{1}{2}\rho v^2 + \rho\hat{U}\right) = -\nabla \cdot \left(\left(\frac{1}{2}\rho v^2 + \rho\hat{U}\right)v\right) - (\nabla \cdot q) - (\nabla \cdot pv) - (\nabla \cdot [\tau . v]) + \rho(v.g)$$

And the equation of motion (conservation of momentum) is described as follow:

$$\frac{\partial}{\partial t}\rho v = -[\nabla \cdot \rho vv] - \nabla p - [\nabla \cdot \tau] + \rho g$$

For non-isothermal flow, the fluid density and viscosity depend on temperature and pressure. This variation in the density is important and can not be neglected because it determines the buoyant force which results in convection of the fluid.

Prior to doing the dispersion model, one has to identify the type of the release incident, develop a source model to describe how the material is being released as well as the rate of the release and the downwind concentration of the vapor can be determined thereafter. The factors which can affect the atmospheric dispersion are:[4]

1) Wind speed
2) Atmospheric stability
3) Ground conditions
4) Height of release above the ground
5) Momentum and buoyancy of the initial material released

Depending on the nature of the natural gas that being observed (light or dense gas), different dispersion modeling will be employed. Methane gas at atmospheric temperature and pressure has a characteristic of light gas. Therefore, Gaussian dispersion model is applied to the system.

For Liquefied Natural Gas (LNG), vapor dispersion of dense gas release will be pertinent since at very low temperature, natural gas is considered heavier than air. Three categories of dense gas dispersion model of LNG developed by Ermak will be utilized, i.e.: Germales-Drake (GD) modified Gaussian model, SLAB and FEM3. These three level approaches of vapor dispersion model are used to simulate the atmospheric dispersion of a cold, denser-than-air gas release. The GD model is based on the single of species equation and neglects the momentum and energy effects to determine the vapor cloud dimensions. The SLAB model includes the conservation of mass, momentum and energy in addition to the species equation, but it is time-averaged and the neglects some parameters such as the crosswind plane. In this case all properties of the vapor cloud only vary in the downwind direction only. FEM3 is the most complete model among the three; it includes the conservation laws by treating explicitly in three dimensions.[3]

Conclusion and Future Work

This research focuses on the use of a new technology, the infrared camera, to detect, monitor and measure the methane gas release. This allows for the estimation of the methane emissions from natural gas industry due to the camera's ability to visualize, localize and provide the image of the temperature distribution of the methane in a wide area of coverage.

The problem related to the current passive infrared camera is that it only provides the temperature profile even though the emission can be estimated from the concentration of the methane gas only. This research bridges the gap by performing experiments, tests and vapor dispersion modeling to obtain the correlation between the temperature profiles obtained from IR camera measurements and the concentration distribution calculated by the vapor dispersion model. The proposed work plan for this research is given in **Figure** .

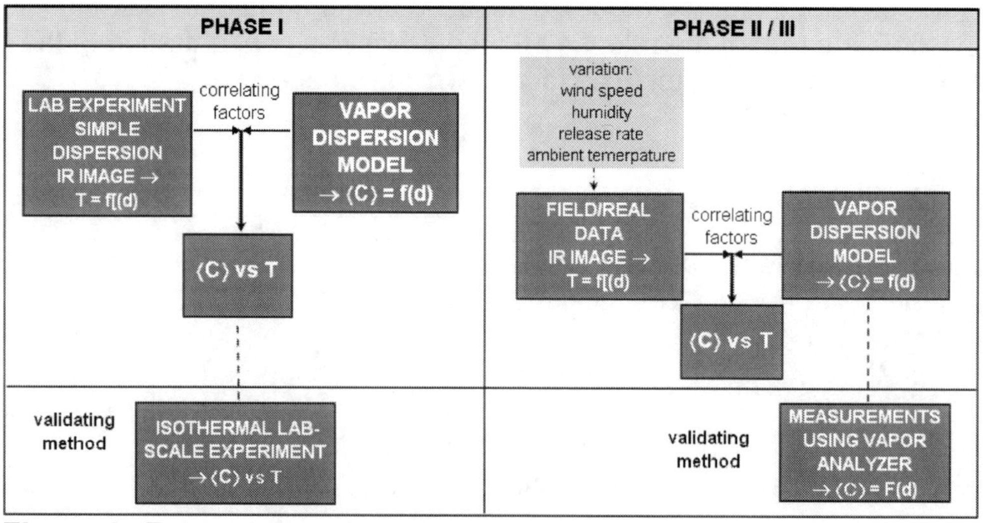

Figure 3. *Future research work*

Vapor dispersion modeling will confirm the concentration distribution of the gas release from the source to the surroundings and the result will be validated from the experimental data. Dispersion under controlled and real environmental and meteorological conditions will be studied. From the experiments and the modeling, the correlation between temperature and concentration can be established.

References

1. Lifshitz, Evgeny and L.P. Pitaevskii, *Statistical Physics:Part 2*, 3rd edition, Elsevier, 1980
2. Reif, F., *Fundamentals of Statistical and Thermal Physics*, McCgraw-Hill, Boston, 1965
3. Ermak, Chan, Morgan, Morris, *A Comparison of Dense Gas Dispersion Model Simulations with Burro Series LNG Spill Test Result*, Journal of Hazardous Materials, 6:129-160, 1982
4. Crowl, Louvar, *Chemical Process Safety: Fundamentals and Applications*, 2nd edition, 2002

Anisa Safitri

Research Assistant
Mary Kay O'Connor Process Safety Center, Texas A&M University
Cell: 979-204-0967, Office: 979-862-1310
Email: anisa.safitri@chemail.tamu.edu
Web: psc.tamu.edu

Anisa Safitri is a PhD candidate at Artie McFerrin Department of Chemical Engineering at Texas A&M University. She obtained her bachelor degree in chemical engineering in October 2002 at Bandung Institute of Technology in Indonesia. She continued her study at Eindhoven University of Technology the Netherlands, specializing in environmental technology and graduated with Master of Science degree in chemical engineering in November 2005.

During her master program, she involved in several projects and internship. Several multidisciplinary projects that she worked on include: study on self-healing material, tar removal using plasma corona discharge, bio-gasification using fluidized bed gasifier. In 2004, she interned with Siemens AG Department of Research and Technology in Erlangen, Germany and her task is to conduct research on ozone generation using plasma discharge.

She joined Mary Kay O'Connor Process Safety Center in August 2006 to pursue a PhD degree under the guidance and supervision of Dr. Sam Mannan. Her current research topic is the detection and measurement of natural gas leak using infrared technology.

Improvement of Radiographic Test Method on 9% Ni Steel Welds for LNG Storage Tanks

Seung Rim Lee, Korea Gas Safety Corporation, Seoul, Korea
Ji Yoon Kim, Korea Gas Safety Corporation, Seoul, Korea
Jang Sik Park, Korea Gas Safety Corporation, Seoul, Korea
Young Soon Lee, Professor, Dept. of Safety Eng'g, SNUT, Korea

AIChE Spring Meeting, April 2008
8th Topical Conference on Natural Gas Utilization
New Orleans, Louisiana, April 6-10, 2008

The inner tanks of LNG storage tanks have been traditionally made of 9% Ni steel plates for their cryogenic applications. Great efforts have been made to develop heavier 9% Ni plates and their welding methods which can follow the trend toward the development of large-sized tanks. Since the tanks have recently become larger and been operated under severer conditions, the shell material of the tanks now in use is at least 30mm thick and the welded joints must have mechanical properties equivalent to those of the base metal.

In order to meet these requirements, radiographic test is adopted as one of test methods on these welding joints.

Electrodes of Inconel or Hasteloy series are generally used in 9% Ni steel welds that can be considered as dissimilar metal welds. It is similar to the weld of austenitic stainless steel; the properties of the weld metals are quite different from those of the base metal and the grain is coarsened.

Therefore, as a result of radiographic test, false indications are generated from X-ray diffraction and difference in absorption coefficients between the base metal and the weld metal. There have been controversies over whether these are real flaws or not at construction sites.

In order to solve this problem, we carried out mechanical tests on the welds having false indications to verify its integrity and established the standard for its assessment method by retests, changing radiographic conditions and comparing the radiographs with the original one. As the result of application of the new method, we could reduce unnecessary repairs and support the economical construction of LNG tanks by reducing the working periods.

1. Introduction

High Ni electrodes such as Inconel or Hasteloy series are used in 9% Ni steel welds, which can be considered as a type of dissimilar metal welds. These welds are similar to those of austenitic stainless steel; the properties of the weld metal are significantly different from those of the base metal and their grain is coarsened. Therefore, when they are examined with radiography, deflection mottles in the shapes of "bands" appears due to "X-ray diffraction indications" and significant difference in absorption coefficients between the base metal and the weld metal.

This paper aims at supporting the economical construction of LNG storage tanks by reducing unnecessary weld repairs and reasonably managing their construction schedules through the establishment of the standard for final interpretation on whether the defects are real or not by carrying out research on specifications and technical bases related to the establishment of the interpretation method for false defects (X-ray diffraction images and band images) spotted in the radiographs of the 9% Ni inner tank welds of LNG storage tanks.

2. Current Nondestructive Test Methods and Criteria for Defect Diagnosis

Radiographic tests (KS D 0237, Methods of Radiographic Test and Classification of Radiographs for Stainless Steel Welds) are conducted on the whole weld lengths in accordance with KGS PV015 (Standard for Full Containment LNG Storage Tanks), Article 8.5.2(Scope of Radiographic Tests), and the welds classified into Class 2 shall be accepted in accordance with the Standard for High-Pressure Gas Safety Control, Unified Notifications No 15-1-46 (Radiographic Test).

In the case of field tests, when X-ray diffraction indications are encountered, the welds are partially retested by radiography and finally classified by the film interpreter or inspector, while most band indications are classified into lack of fusion. In the case of controversy over the classification, the weld is classified by its retest.

Therefore, retests by radiography or repairs of rejected welds are repeated at the site without consistent standards or guidelines for these two types of false indications. Such economical loss and technical uncertainty related to this matter shall be eliminated.

3. Outline of Deflection mottles

In general, deflection mottles spotted in radiographs of 9% Ni steel storage tanks are caused by several factors. The images shown on the film are different from ordinary images shown on radiographs depending on their characteristics.

Such deflection mottles appear largely due to the following 3 factors:

(1) X-ray diffraction
(2) Difference in X ray absorption
(3) Application of mask

The characteristics of the above deflection mottles are as follows:

3.1 Deflection mottles caused by X-ray diffraction

The deflection mottles caused by X-ray diffraction are due to the dissimilar metal welds which are made by welding 9% Ni steel plates with high Ni electrodes such as Inconel or Hasteloy series electrodes. These welds are similar to those of austenitic stainless steel. Since their deposited metal is non-magnetic, their characteristics are significantly different from those of the base metal, and their grain is coarsened, X-ray diffraction is caused by these structures when they are tested by radiography.

In general, the shape of the deflection mottles spotted at this time is the one in which black and white bands are alternated in the blackened background, and the image is easily distinguished from those of volumetric defects (porosity, lack of fusion). In order to minimize deflection mottles caused by diffraction, high-energy radiographic sources may be used. However, high-energy sources are not applied because they deteriorate the image quality and weaken the capability to identify defects.

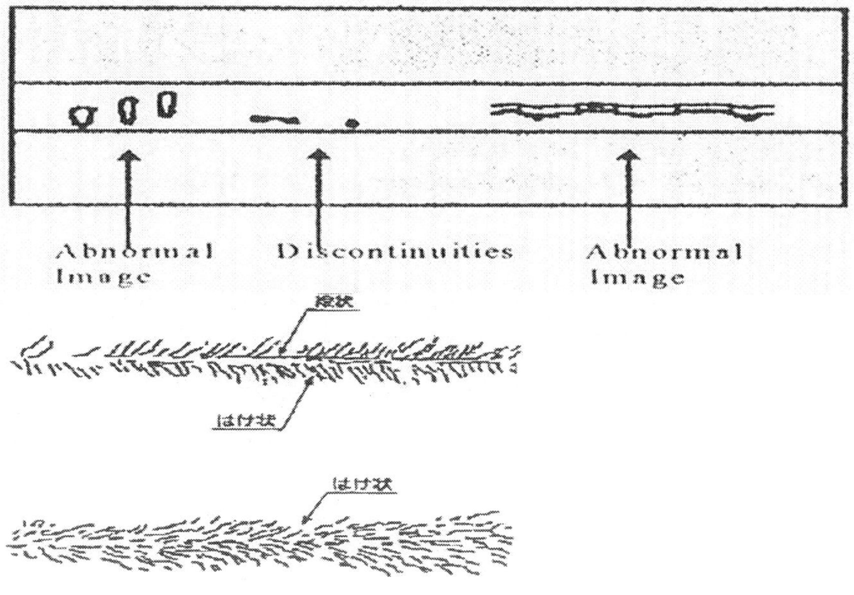

Fig. 1 Shapes of Deflection mottless Caused by Diffraction

3.2 Deflection mottles caused by difference in X-ray absorption

Deflection mottles such as Band–shaped images are generated at the interface between the base metal and the deposited metal because of the difference in their X-ray absorption coefficients. The shape of the deflection mottle is multi-layer liners (==), and many images are indicated remarkably in SAW welds. They can be identified by changing the incident angle of X-ray. In general, the mass absorption coefficients of the base metal and the deposited metal are as follows when the weld of 22 mm 9% Ni steel is tested at 200kVp:

(1) 9% Ni steel = 0.1470cm^2/g
(2) Deposited metal = 0.1558cm^2/g

Band-shaped deflection mottles appear at the interface where the difference in mass absorption occurs.

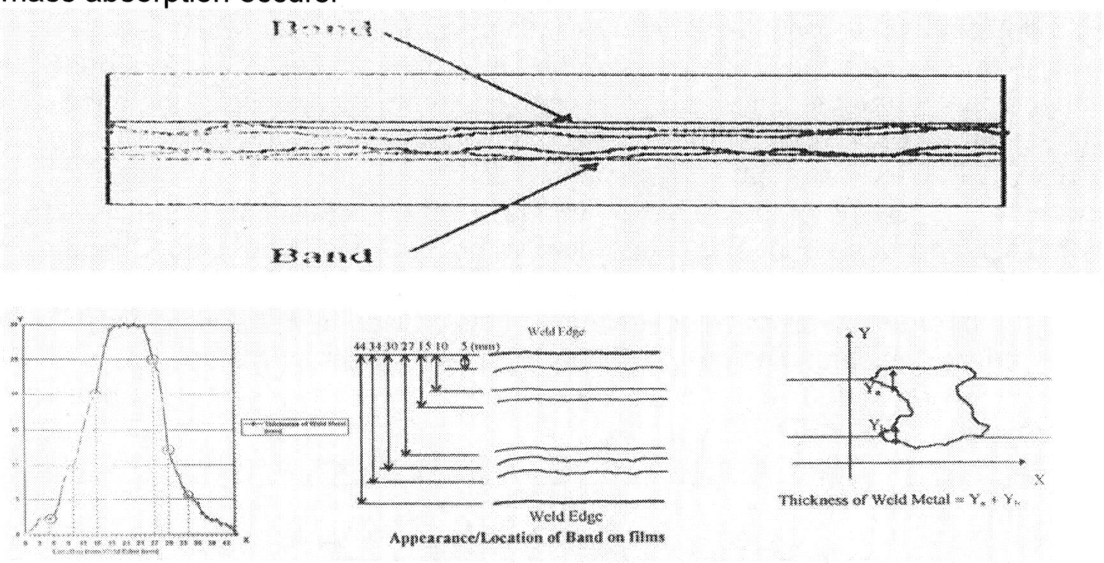

Fig. 2 Band-Shaped Deflection mottless Generated at the Interface Where Difference in Mass Absorption Occurs

3.3 Deflection Mottles Cause by Use of Mask

When the welds are tested by keeping the change in mass absorption coefficients to a minimum to distinguish deflection mottles from defect images, the weld is irradiated at different incident angles. The incident angle may be affected by the groove angle, but the band-shaped images disappear or diminish when the angle is tilted by 10 to 20 degrees. In the case of volumetric defects, their images may not disappear but become more distinctive when the radiation angle is tilted in the reverse direction.

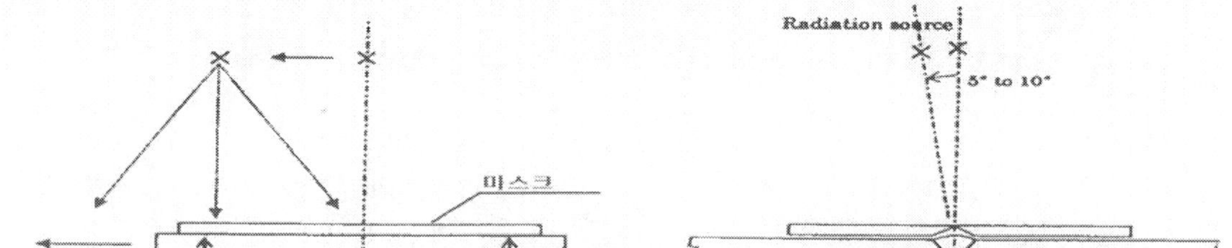

Fig. 3 Radiographic Test Using a Mask

In order to correct the difference in density coefficients of radiographs caused by the difference in the above-mentioned absorption coefficients, masks (material: barium, tungsten) may be used, but they cause linear deflection mottles along the groove interference.

The occurrence frequency of deflection mottles to be caused by the mask will be remarkably reduced if the groove edges are finished smooth during their edge preparation. The deflection mottles generated at this time will look like those of undercuts along the interference between the weld and the base metal. The deflection mottles can be identified by whether the band shapes and deflection mottles lines are continuous and straight or irregular and linear.

4. Technical Review and Improvement Methods

4.1 X-ray diffraction indications

1) Related basis data

JIS Z 3106, Appendix 5 (see the reference enclosed hereunder) stipulates the guideline for distinguishing between false defect images and real defect images because false defect images are generated from the diffraction images of the solidified metal structure which are irrelevant to any defects and they are not subject to the defect classification.

X-ray diffraction images are false defect images, which are irrelevant to the strength of welds, but may be confused with welding defects such as lack of radiation and blow holes. In order to avoid confusion, it is necessary to fully understand how to distinguish false defect images and the causes of their generation.

2) Method of defect interpretation

Defect images can be identified by comparing the spotted diffraction images with the reference radiograph of the test material of similar thickness prepared in similar welding conditions.

In the case of radiation technique method, the location of a diffraction image can be shifted or the diffraction may disappear depending on the radiation conditions of X-ray such as object-to- film distance, incident angle, and radiation energy. The radiation technique includes angle change method, diffusion method, grit method, radiation quality change method, mask method, and mesh utilization method.

There are certain diffraction elimination methods which may be practically applicable depending on the related data while other methods are impractically applicable only in laboratories. In other words, these methods are very limited in the terms of time and space in their field application. Therefore, it is impossible to eliminate diffraction images for practical purpose.

In case hair cracks are hidden in brush-shaped images, since the brush-shaped images have the characteristics of being formed continuously in one single direction as being brushed in that direction only, and hair cracks are formed in random directions different from the brushing direction, the directions of hair cracks are surely different from that of brush-shaped images and there will be distinctive difference in their brightness and darkness. Those parts shall be retested by using the angle change method.

4.2 Band-shaped images

1) Related basis data

In the actual interpretation of the radiographs of 9%Ni steel welds, experienced interpreters can do most of interpretation in consideration of welding deposit. In case the interpretation is difficult, the radiation method is changed, and the image can be interpreted as not being a defect, if the image has disappeared after the change.

In consideration of the fact that there is no big difference in the mass absorption coefficients between the base metal and the weld when X-ray passes through them, defect images and band images can be classified by changing the radiation method(incident angles). When X-ray is radiated, the band moves up and down as much as the changed incident angles; one part moves to the centerline of welding and the other part moves to the center of the weld but there is no defect in the band, and the band moves as much as the change in the incident angle.

When X-ray is radiated at inclined incident angles to a base metal and a weld having any defect, and if there is any hidden defect in an area where the band image is overlapped, the band image will move up and down or just disappear, while the defect hidden in the band image will move as much as the inclination. If there is no defect, the defect image will not appear when incident angles are changed.

2) Defect assessment method

In order to distinguish between “bands” and defects, X-ray may be radiated at changed incident angles in consideration of the fact there is no big difference in the mass absorption coefficients of the base metal and the weld.

Radiographs show that in some welds there are discontinuity images which may be considered as band images but have slightly different shapes. Currently, these images appear near band images on the film, are interpreted as lack of fusion, and are partially repaired. However, when the surface is examined after grinding or gouging, it is confirmed that there is no defect contrary to anticipation. Therefore, it is necessary to check whether there is any defect by taking samples and carrying out structural test and macro test in addition to RT or UT applied at the site.

5. Conclusion

It is confirmed by this research that since 9% Ni steel welds are made by using high Ni electrodes such as Inconel or Hasteloy series, this welding is a type of dissimilar metal welding similar to austenitic stainless steel welding, the characteristics of the deposited metal are significantly different from those of the base metal, and its crystal grain is coarsened, deflection mottles appear on radiographs, such as X-ray diffraction images caused by coarsened crystal grain and band images caused by big difference in absorption coefficients between base metal and deposited metal.

In Japan, X-ray diffraction images monitored in the vertical welds of 9% Ni steel plates have been reviewed from 40 years ago. JIS Z 3106(Methods of Radiographic Test and Classification of Radiographs for Stainless Steel Welds) stipulates that they are false defects and how to distinguish between false defects and real defects. As an effective interpretation method, the object material is retested by changing radiation techniques (angle change method, diffusion method, grit method, radiation quality change method, mask method, and mesh utilization method) to check whether there is any defect.

In addition, for band images appearing at the horizontal welds of 9% Ni steel because of the difference in the absorption coefficients between base metal and deposited metal, masks are used to correct the density difference of radiographs caused by the difference. In such a case, linear deflection mottles may appear at the interface between the band image and the groove of the mask. This is a false defect, which can be finally checked whether it is a real defect by retest by changing the radiation technique based on related engineering data.

By these methods, unnecessary welding repairs because of false defects may be reduced, and efficiency can be improved not only in terms of time but also in economical aspects.

REFERENCES

1. KGS PV015 "Code for Full-containment LNG Storage Tank", Korea Gas Safety Corp., 2002
2. KOGAS Technical Report, "False Indications on the Weld Joints of 9% Ni steel Plate for LNG Storage Tank", 2005
3. KOGAS, "Welding Procedure Qualification", December 2003
4. Korea Gas Technology Corp., "Specification for 9% Ni Steel Plate", 2003
5. Dealim Industrial Company Limited, "WPS/PQR(Pyeong-Taek Terminal #11-14)", 2005
6. KS B 0845, "Methods of Radiographic Examination for Welded Joints in Steel", 2005
7. JIS Z 3106, "Methods of Radiographic Examination for Welded Joints in Stainless Steel", 2001

8. KOTRIC, "Test Report(PQ Test, LNG-PQR-SA-003, A553 Type 1)", 2005
9. Mitsubishi Heavy Industry Technical Report, "Study on Welding of Heavy 9% Ni steel for LNG plant Pressure Vessel", vol. 24 No.3, May 1987
10. Jong-Bong Lee, etc, "Welding Characteristics of Recently Developed 9% Ni Steel", Korea Weld Society , Vol. 13 No.3, September 1995

Biography of Speaker

Institute of Gas Safety R&D,
Korea Gas Safety Corporation
332-1, Deaya-dong, Shheung-si, Gyeonggi-do, 429-712, Korea
Phone: +82.31.310.1532, Fax: +82.31.315.4363
E-mail: srlee@kgs.or.kr
Cell : +82.11.290.6668

Seung-Rim Lee is a Senior Researcher in the Institute of Gas Safety R&D at Korea Gas Safety Corporation. He has been working for Korea Gas Safety Corporation for twelve years. He has been involved research and development projects related with domestic standardization and risk assessment in LNG industry. Especially, he has been participated in development of Korean Standards and International Standards (ISO/TC67/WG10) related with LNG storage tank and LNG terminal from 1999. He obtained a Master of Science in field of Safety Engineering in 2001 and he is under the doctor´s course in field of Gas Engineering.

Design Considerations for Propane Condenser Approach Temperature in an Air Cooled LNG Plant

J.C. Kuo, P.E., Senior Advisor
Pankaj M Shah, P.E., Team Leader
Bobby Martinez, P.E.Team Leader

Chevron Energy Technology Company

AIChE Spring Meeting, April 2008
8th Topical Conference on Natural Gas Utilization
New Orleans, Louisiana, April 6-10 2008

1.1 INTRODUCTION

A majority of the existing base load LNG plants use air cooling technology to reject process heat required to liquefy the natural gas. In addition, an increasing number of LNG plants under development plan to use air cooling technology. A 5 MMTPA LNG plant will typically require an air cooler with approximately 75 bays with an equivalent heat duty of about 1000 Million BTU/Hr which equates to an estimated total installation cost of over $55 million dollars in today's dollars.

A question frequently asked during conceptual design phase is how one defines the optimal approach temperature for the propane condenser. In the past, approach temperatures ranging from 14-30 °C have been employed in LNG facilities. A systematic method to determine the optimal approach temperature of the propane condenser during the early conceptual design stage would be very useful in the design of LNG plants.

1.2 OBJECTIVE

The objective of this paper is to identify the factors contributing to an optimum propane condenser (air cooler) design. The following design parameters will be defined quantitatively suitable for economical analysis:

- Low temperature approach of propane condenser versus high temperature approach.
- Impact of ambient temperature on both instantaneous and annual LNG production and fuel consumption.
- Propane condenser surface area and number of bays.

This study includes the following:

- Develop a suitable design basis and methodology for this paper.

- Develop sensitivity analysis on LNG production rate and fuel consumption with four different propane condenser approach temperatures at three ambient environments. Both instantaneous as well as annual LNG production will be addressed as part of the analysis.

- Determine propane condenser surface area and number of bays for all study cases.

2.1 DESIGN BASIS

A typical LNG process is utilized as a basis for this work. LPG, condensate recovery and related storage are provided in the LNG plant.

Feed Gas to LNG Plant

The composition (mol %) of the feed as selected from the database is:

	Feed
N_2	0.62
CO2	1.49
CH_4	83.06
C_2H_6	7.17
C_3H_8	4.29
I- C_4H_{10}	0.69
N- C_4H_{10}	1.47
I- C_5H_{12}	0.36
N- C_5H_{12}	0.32
C_6+	0.53

LNG Product Specification

	Minimum	Maximum
HHV (High Heating Value) (Btu/Scf)	1,015	1,075

Average Ambient Temperature

Cases A, B and C represent cold, medium and warm ambient environments at different locations such as Asia, South America and Africa, respectively.

There are 20 °C for Case A, 25 °C for Case B and 30 °C for Case C

One (1) °C shall be added to all ambient temperatures in use to account for hot air recirculation.

Propane Condenser Temperature Approach

The following four approach temperatures are selected for Cases A, B and C.

There are 12 °C, 18 °C, 24 °C and 30 °C

Report of LNG/NGL Productions

LNG production in storage tank will be reported as an index, with 100.00% as the reference point for 25 °C (Case B) at 12 °C. NGL production will be proportional to the LNG production rate as the propane percent recovery is kept the same for all study cases.

2.2 ASSUMPTIONS

The assumptions used in this work are listed below:

- There is no market limitation on LNG production. The only limitation is the available power provided by all the turbine drivers and helper motors.

- The 36 °C spread of ambient temperature is based on 100% and 0% exceedance. It will be difficult to design compressor casings capable of handling this wide ambient temperature spread. However, it is assumed that all compressor casings are designed to accommodate the whole range of ambient conditions. There is no concern on getting into the surge at the highest ambient, nor any concern on hitting the stone-wall at the lowest ambient in condition. In actual design, the spread is based on 99% and 1% or even 98% and 2% exceedance. The spread will be 20 °C or lower.

2.3 METHODOLOGY

The following design practices and limitations are used in this work to minimize unnecessary complexities in simulations and simplify cost estimating requirements.

Use of Common Condenser Bay Design

A common condenser bay design is selected for all study cases instead of optimizing bay design case by case, in order to simplify future cost estimating. Design characteristics are listed below:

Size (WxL):	4.826 m x 18.288 m
Tube OD:	25.4 mm
Bare Surface:	630 m^2
No. of Tubes:	438
No. of Fans:	3
Fan Motor:	29.8 kw/Fan

Equipment Design Other Than Propane Condenser

The propane condenser is the only equipment that changes case by case. Plant design and all other equipment, except vessels and columns related to LNG liquefaction, remain the same in order to identify only the impact on the propane condenser and to simplify cost estimating. The potential differences of certain capacity-dependent units such as acid gas removal, dehydration, and fractionation including vessels and columns in LNG liquefaction, are estimated

Turbine Power Available for Design

The available power for refrigerant compressor turbines is a function of ambient temperature. No turbine is allowed to have a negative power for any case.

Compressor Efficiency Used in Design

The compressor efficiencies used in the design of the reference case (Case B with 12 °C condenser approach) are estimated for the average and two extreme ambient temperatures from past experience of similar work. For the purpose of comparison, all other cases use the same set of efficiencies for the corresponding average ambient and two extreme ambient temperatures, respectively.

LPG Recovery

The LPG recovery is maintained the same for all study cases. Therefore, NGL production will be proportional to the LNG production rate.

Design Margin

No design margin will be employed.

3.1 SUMMARY TABLE OF RESULTS

Table 3-1 summarizes the results for all three ambient cases, 20 °C, 25 °C, and 30 °C, each with four condenser approach temperatures, 12 °C, 18 °C, 24 °C, and 30 °C. LNG production is reported as an index number. The instantaneous production at 25 °C ambient for Case B with 12 °C approach is used as the reference point at 100.00%. The values for annual LNG production in storage and average fuel consumption are calculated through the cumulative distribution of ambient temperatures, and are discussed in Section 3.2.

Figure 3-1 illustrates the variation of annual LNG production in storage versus propane condenser temperature approach. As observed, annual LNG production increases as the condenser approach temperature decreases.

Figure 3-2 illustrates the variation of annual LNG production in storage versus average ambient temperature. As observed, annual LNG production increases as the ambient temperature decreases.

Figure 3-3 shows the variation of annual average fuel consumption versus propane condenser temperature Approach. Annual average fuel consumption increases as the condenser approach temperature decreases.

Figure 3-4 shows that the total number of propane condenser bays increases as the condenser temperature approach decreases.

Table 3-1: Summary Table of Results

Study Case	Case A				Case B				Case C			
Average Ambient Temperature, °C	20				25				30			
Propane Condenser Temp. Approach, °C	12	18	24	30	12	18	24	30	12	18	24	30
Min./Average/Max. Ambient Temp., °C	3/20/39				8/25/44				13/30/49			
LNG in Storage @ Average Ambient, %	104.55%	102.37%	100.16%	97.44%	**100.00%**	97.63%	95.28%	92.55%	94.98%	92.90%	90.55%	87.44%
LNG in Storage @ Maximum Ambient, %	85.36%	83.47%	81.18%	78.40%	80.31%	78.28%	75.56%	72.30%	75.08%	72.81%	69.87%	66.45%
LNG in Storage @ Minimum Ambient, %	117.21%	115.26%	113.29%	111.03%	112.90%	110.92%	108.58%	105.27%	108.34%	106.09%	103.74%	100.18%
Annual LNG Production in Storage, %	**105.38%**	**103.21%**	**101.02%**	**98.34%**	**100.92%**	**98.50%**	**96.15%**	**93.38%**	**95.85%**	**93.76%**	**91.41%**	**88.26%**
Ratio of Annual to Average Ambient Production	1.0079	1.0083	1.0086	1.0092	1.0092	1.0089	1.0091	1.0090	1.0092	1.0093	1.0095	1.0095
Fuel Consumption @ Avg. Ambient, GJ/Hr	3,821.2	3,780.5	3,760.5	3,737.1	3,755.5	3,715.5	3,697.5	3,667.3	3,688.5	3,649.2	3,632.9	3,600.2
Fuel Consumption @ Max. Ambient, GJ/Hr	3,576.0	3,539.8	3,513.6	3,480.2	3,508.1	3,472.5	3,437.2	3,400.6	3,437.1	3,400.3	3,358.4	3,319.5
Fuel Consumption @ Min. Ambient, GJ/Hr	4,052.0	4,002.6	3,976.3	3,951.2	3,976.0	3,930.6	3,908.4	3,873.3	3,903.4	3,862.8	3,842.4	3,802.7
Annual Average Fuel Consumption, GJ/Hr	**3,836.7**	**3,795.3**	**3,774.9**	**3,751.4**	**3,769.8**	**3,729.9**	**3,711.5**	**3,680.9**	**3,702.8**	**3,663.5**	**3,646.7**	**3,613.5**
Propane Condenser Bare Area, m^2	59,058	40,833	30,821	24,502	57,792	39,590	29,542	23,265	56,542	38,300	28,940	22,607
Propane Condenser Number of Bays	**94**	**65**	**49**	**39**	**92**	**63**	**47**	**37**	**90**	**61**	**46**	**36**
% Increase in Production Per Additional Bay, %	0.072%	0.104%	0.204%	0.273%	0.085%	0.119%	0.225%	0.297%	0.077%	0.124%	0.264%	0.357%

Figure 3-1: Annual LNG Production in Storage versus Propane Condenser Temperature Approach

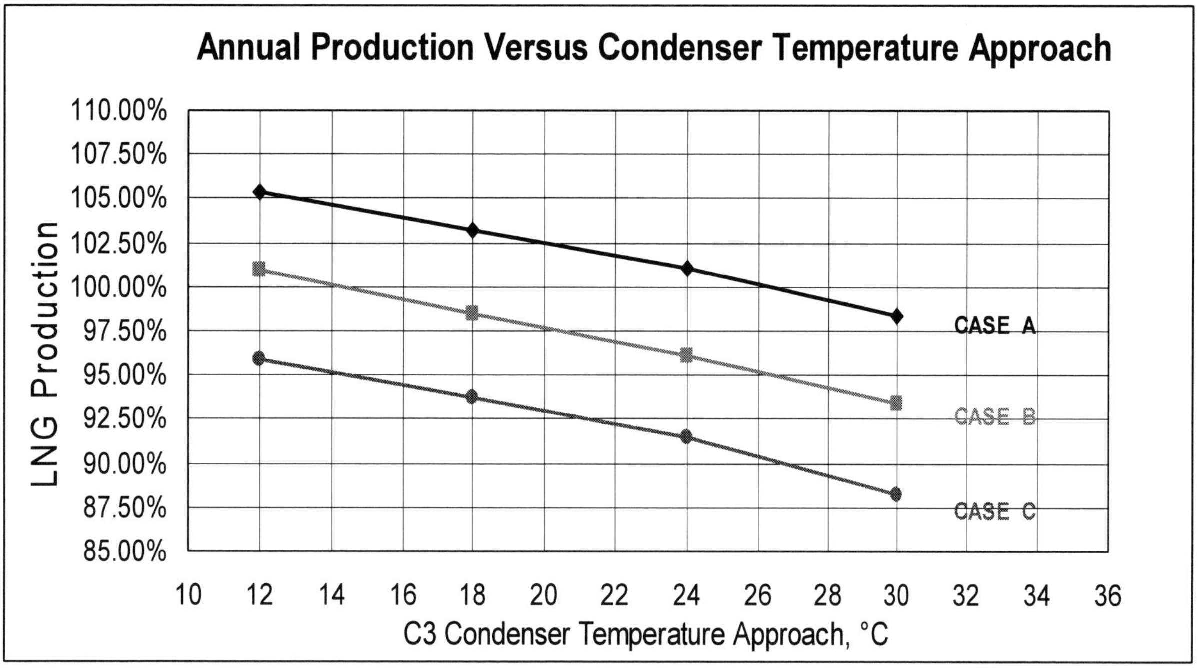

Figure 3-2: Annual LNG Production in Storage versus Average Ambient Temperature

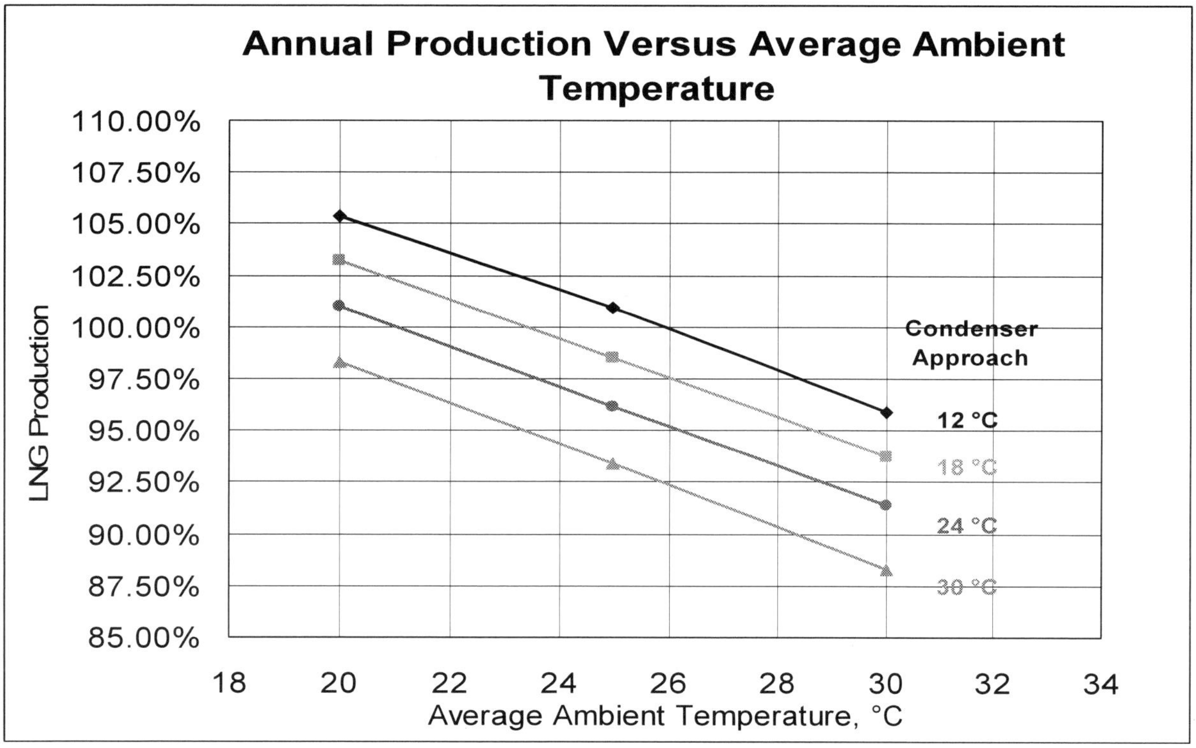

Figure 3-3: Annual Average Fuel Consumption versus Propane Condenser Temperature Approach

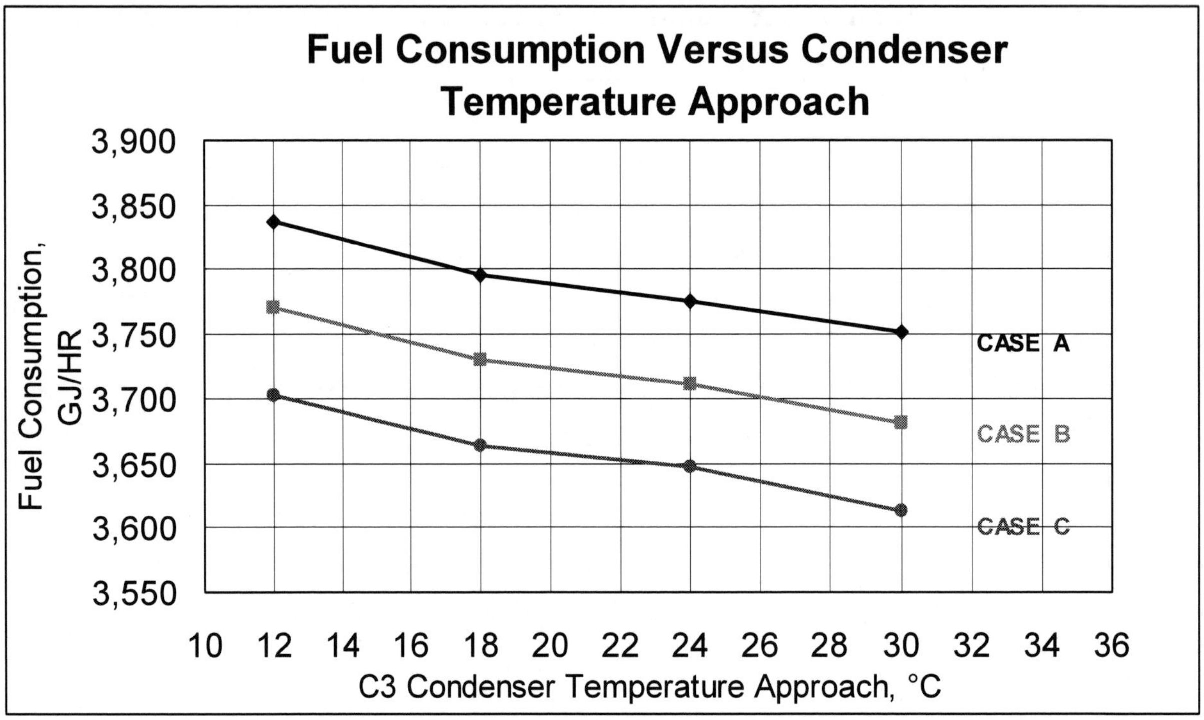

Figure 3-4: Number of C3 Condenser Bays versus Propane Condenser Temperature Approach

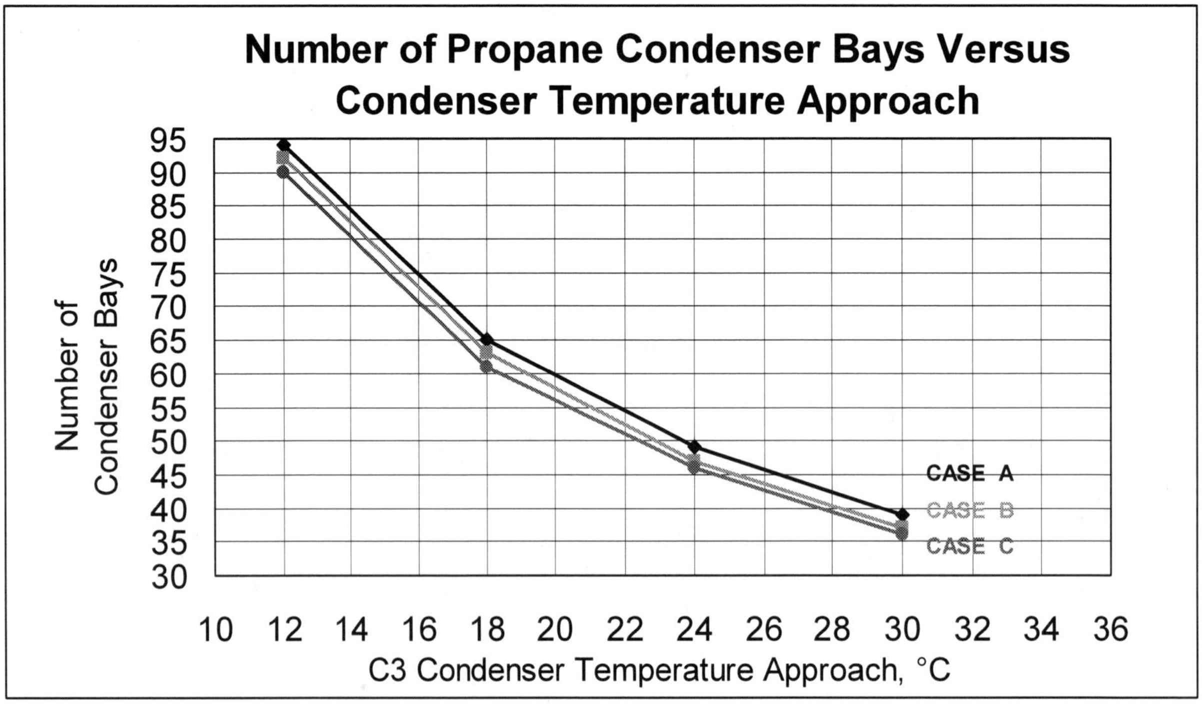

3.2 LNG PRODUCTION PROFILE

Table 3-1 shows that for each case LNG production is calculated at three ambient temperatures: the average, the maximum, and the minimum.

LNG Production Profile for Case A, B and C

A LNG production profile for each approach temperature of Case A, B and C are calculated through a second–order polynomial equation fitted with the three data points.
LNG production data at four approach temperatures for Case A, B and C are plotted in Figure 3-5,6and 7 as a function of ambient temperatures. Four nearly parallel curves show that a closer approach will produce more LNG.

Figure 3-5: Production Profile Curves for Case A

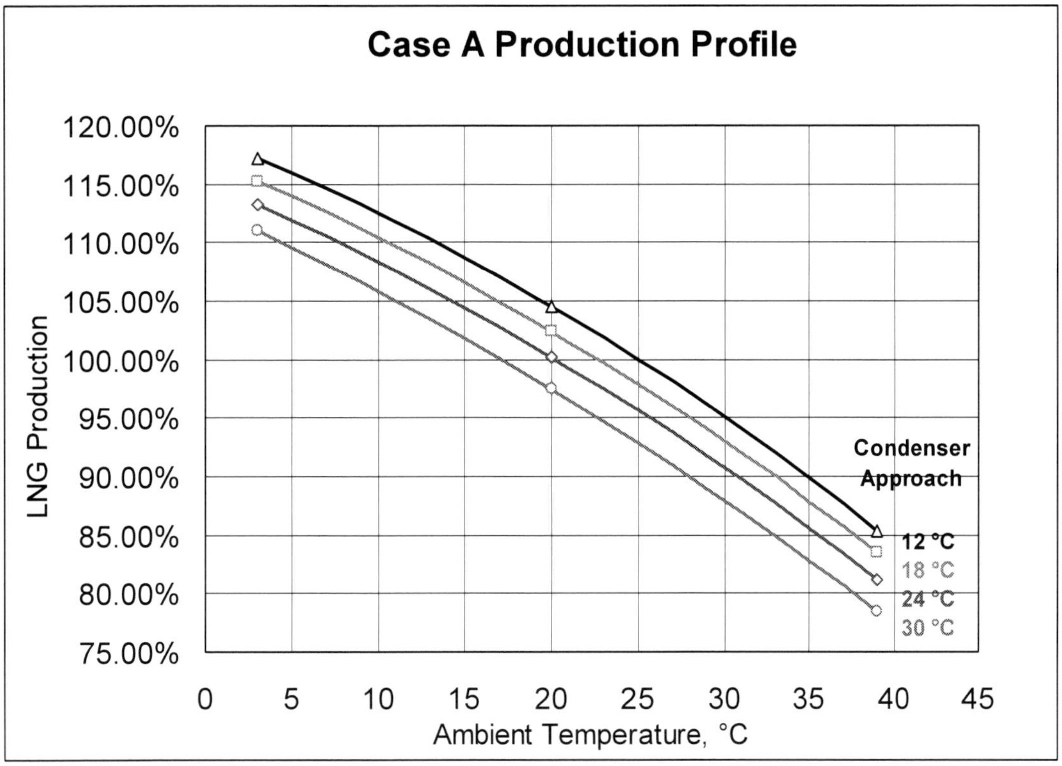

Figure 3-6: Production Profile Curves for Case B

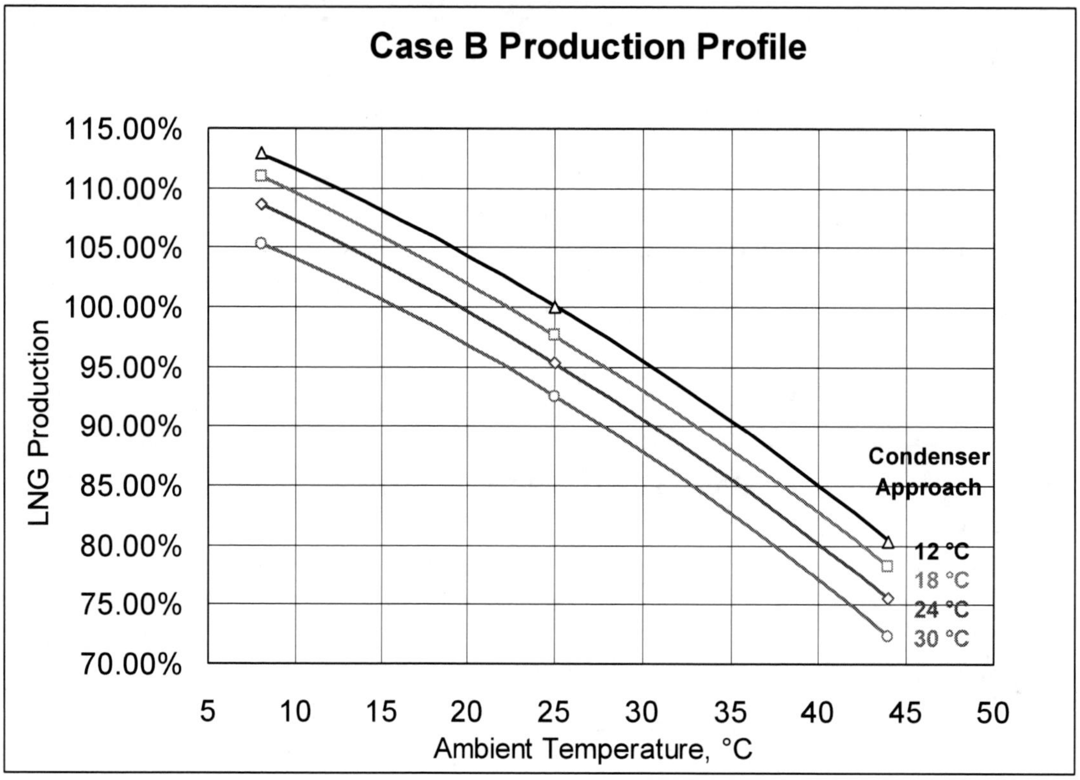

Figure 3-7: Production Profile Curves for Case C

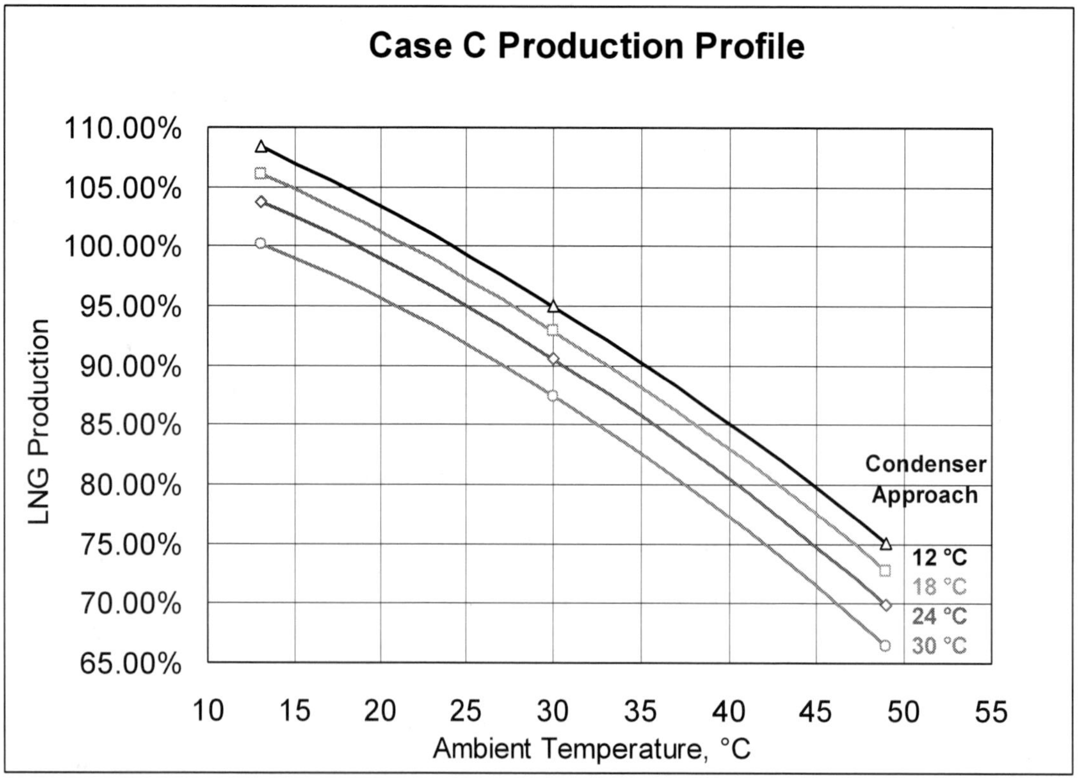

3.3 PRODUCTION VERSUS PROPANE CONDENSER BAYS

Figure 3-8 plots annual LNG production data against the required number of propane condenser bays.

Figure 3-8: Plot of Annual LNG Production against Number of Propane Condenser Bays

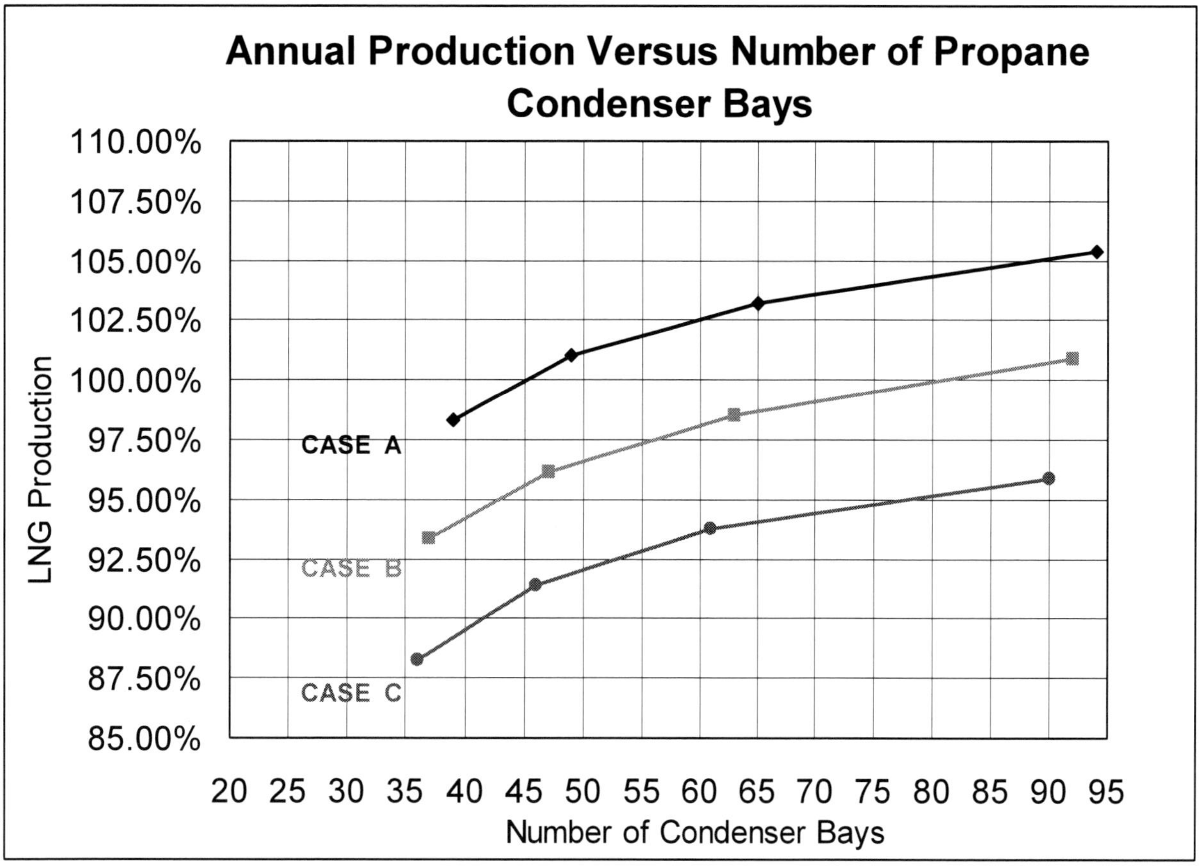

As observed, LNG production increases as the number of condenser bays increases, when the approach temperature is getting narrower. The three nearly parallel curves show that production is higher at a lower ambient temperature. The slope of these three curves, which represents the rate of increase in LNG production, becomes more and more flat. The potential of capacity increase is generally higher as indicated by the steeper slope when it is far away from the optimum. These curves in Figure 3-8 use straight lines to connect data points, which may not represent the actual situation at around 80-85 bays. Figure 3-9 re-draws the data and adds a trend line fitted with a second order polynomial for each case.

Figure 3-9: Plot of Annual LNG Production against Number of Propane Condenser Bays – Trend-lines Added

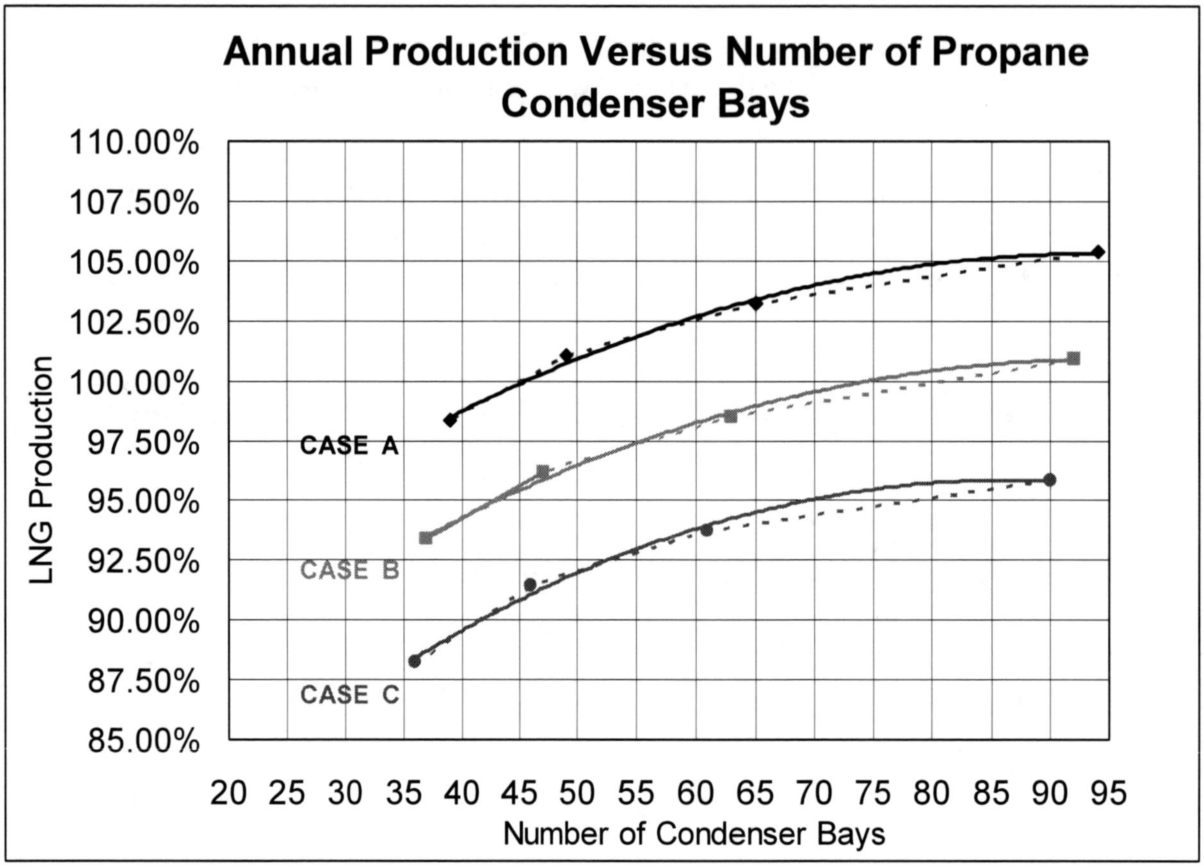

As observed from Figure 3-12, a plateau is reached or closely approached prior to hitting the end point, which represents the 12 °C approach temperature. It is more and more difficult to increase LNG production as the end point is approached.

3.4 PRODUCTION CHANGE WITH AMBIENT TEMPERATURE

During the design of the LNG plant, ambient temperature impacts on LNG production is normally of great interest for preliminary estimating. Table 3-2 lists this information. Production increases from 0.9% to 1.16% per °C drop of ambient temperature. The percentage of increase is higher at wider condenser approaches than at closer approaches. The design with a wider condenser approach is slightly more sensitive to changes in the ambient condition. The potential of capacity increase is generally higher when it is farther from the optimum.

Table 3-2: Percent Increase of LNG Production per °C of Ambient Temperature Drop

Study Case	**Average Ambient Temperature, °C**	**Percent Production Increase Per °C of Ambient Temperature Drop**			
Propane Condenser Temp. Approach, °C		12	18	24	30
Case A	20	0.88%	0.96%	1.01%	1.06%
Case B	25	0.97%	0.98%	1.03%	1.11%
Case C	30	1.06%	1.01%	1.04%	1.16%

4.1 CONCLUSIONS

Lower Practical Limit on Approach Temperature

It is believed that 12 °C is nearing the lower practical limit of propane condenser temperature approach. The plot space required for Case A with 12 °C approach is at least 230 meters long, which is based on 94 condenser bays in a double-bank arrangement. A further increase in the number of condenser bays may impact the natural arrangement of the plant, thus adding to construction costs. As observed from Figure 3-9, a plateau is reached prior to attaining the lower limit, represented by the 12 °C approach temperature. It becomes more and more difficult to increase LNG production as this lower limit is reached.

Optimal Approach Temperature

Figure 3-9 suggests that the optimal approach temperature may fall into a range between 12 °C and 18 °C. However, every project needs to do economical analyses on identifying the optimal approach temperature with additional project data points.

Closer Approach Has More Benefit in Warmer Environment

As observed from all figures the percent increase of production reaches the highest point at 30°C approach for Case C. Each propane condenser bay will contribute more to production increases in warmer environments as illustrated by Case C, until the approach temperature is closer than 15 °C. In other words, it is more beneficial to add condenser bays for warmer ambient environments.

Situations that Justify Wider Approach Temperatures

A design with higher approach temperature for the propane condenser, which represents a lower cost option, may be beneficial under the following conditions but not limit to:

- Feed gas is limited or the sales target of LNG is already fulfilled

- Compressor limitation due to the mechanical design such as capacity
- Excess power is available in propane refrigeration service.
- For colder environment, propane service has plenty of excess power available.
- Limited initial TIC (Total Investment Cost)

4.2 RECOMMENDATIONS FOR ECONOMICAL STUDIES

Economical Analysis

This paper provides quantitative values for all key design parameters. However, the increase of LNG production is at the expense of more propane condenser bays, larger inlet feed gas receiving, gas treating, LNG liquefaction, fractionation and related utilities. The optimal condenser temperature approach cannot be identified without a suitable model for economic analysis. This analysis should be designed to evaluate different values of product revenue, which may shift the optimum point.

More Data Points Needed

During the progress of economic analysis, we discover that more data points closer to the optimal approach temperature are needed to demonstrate the optimum more precisely.

5 .0 REFERENCE

1. 2004 GPSA Engineering Data Book, Vol. I, Gas Processors Suppliers Association, Section 14"Refrigeration", 2004
2. 2001 Ashrae Handbook Fundamentals, American Society of Heating, Refrigerating and Air-Conditioning Engineers, Inc., ©2001, Chpt. 1, pgs 12-13

BIOGRAPHY

J.C. Kuo is a veteran of the Gas Processing/LNG industry, with 34 years experience in the field. He is senior advisor for Chevron.LNG Systems. He has served as the Process Manager for many projects, including the Gorgon LNG, Delta Caribe LNG, Casotte Landing and Sabine Pass LNG terminal projects. He also served as the Technical Reviewer for Angola, Olokola, Algeria and Shtokman LNG projects. Before working with Chevron, he was the Technology Manager for IPSI/Bechtel and served as the Process Manager for the Pemex Catarell offshore project, the Egyptian LNG (Idku) train 1/2 and China Shell Nan Hai Projects. He has contributed to LNG and gas processing technology improvements through book and many papers on the subject. He has a BS from Chung Yuan Christian University, Taiwan and a MS from the University of Houston, both for Chemical Engineering. He also holds a MS in Environmental Engineering from Southern Illinois University. He is a registered Professional Engineer in Texas and a member of AIChE. He is the president of the 99 Power Qi Qong Texas divisions.

Pankaj Shah is Team Leader in the LNG Project Support Group, in Chevron Energy Technology Company. Pankaj has over 25 years of experience in gas processing, LNG, GTL and refining. He has extensive experience in various phases of projects as a technology manager, project engineer and process engineer. Pankaj graduated from the University of Oklahoma in Norman, Oklahoma, with a Masters degree in Chemical Engineering. He also has a Masters in Business Administration from the University of Houston. He is a registered Professional Engineer in the State of Texas.

Mr. Martinez has over 20 plus years in the oil and gas industry. He earned a bachelor degree in Chemical Engineering and Chemistry from New Mexico State University, and is a registered professional engineer of the state of Texas. His professional career began with Phillips Petroleum where he developed an expertise in the art of natural gas gathering and processing, and commercialization; offshore oil and gas process design including natural gas liquefaction; and LNG technology licensing. He is presently a LNG Systems Team Leader in the Chevron's Energy Technology Company.

Thermodynamics of Liquid-Vapour Two-Phase Draft Tubes for Liquefied Gas Expanders

Widodo Wahyu Purwanto
Head of Chemical Engineering Department
University of Indonesia, Faculty of Engineering
Depok, Jakarta, Indonesia
widodo@che.ui.edu

Chen-Hwa Chiu
Senior Technology Advisor
Chevron Energy Technology Company
Houston, Texas, USA
Chen-Hwa.Chiu@Chevron.com

Yoshitsugi Kikkawa
Senior Engineering Consultant
Chiyoda Corporation
Yokohama, Japan
ykikkawa@ykh.chiyoda.co.jp

Atsushi Mukaiboh
Rotating Machinery Unit Manager
Chiyoda Corporation
Yokohama, Japan
amukaiboh@ykh.chiyoda.co.jp

Hans E. Kimmel
Executive Director Research & Development
Ebara International Corporation
Sparks, Nevada, USA
hkimmel@ebaraintl.com

Eginhard Berger
Marketing & Sales
IMPaC Offshore Engineering
Hamburg, Germany
eginhard.berger@impac.de

AIChE Spring National Meeting, April 2008
8th Topical Conference on Natural Gas Utilization
New Orleans, Louisiana, April 6-10, 2008

Abstract

In modern refrigeration process plants the liquefaction of gases is achieved at high pressure to improve the overall efficiency of the cryogenic process. After the condensation of the refrigerated gas the pressurized liquefied gas is expanded across a liquid-vapour two-phase expander to a lower pressure suitable for storage and transportation of the liquefied gas. The expansion process generates a certain amount of vapour and the remaining liquid is cooled down.

The objective is to increase the amount of liquid and to decrease the amount of vapour at the outlet of the two-phase expander. By employing a two-phase draft tube at the exit of the two-phase expander an increase in the amount of liquid is accomplished. The fluid dynamic operation and the thermodynamic performance of two-phase draft tubes is presented and analyzed.

Two-Phase Expansion Process

By expanding saturated liquefied gas to a lower pressure, vapour is evaporating from the liquefied gas. Figure 1 shows the expansion process for saturated liquefied gases. At the pressure p_1 the point 1 is on the saturated curve. The mass ratio between vapour and liquid portion is defined as x. At the point 1 the value of x_1 is zero, because no vapour occurs at this point.

Using a Joule-Thomson valve to expand the saturated liquid from pressure p_1 to the lower pressure p_2 as the specified outlet pressure, the enthalpy h of the two-phase fluid remains constant, $h_1 = h_2$. The amount of vapour at the point 2 is given by x_2. The expansion across a Joule-Thomson valve is approximately isenthalpic.

Expanding the saturated liquid across a two-phase expander with power generation from p_1 to the lower specified outlet pressure p_2 to the point 3, the enthalpy of the fluid is reduced from h_2 to h_3 by the amount of the generated power (h_2 - h_3). The vapour-liquid ratio x_3 is smaller than the ratio x_2, hence producing a higher liquid portion than in the case of a Joule-Thomson expansion.

Two-phase expanders with two-phase draft tubes are of the type of radial outflow turbines or of similar designs and the expanded fluid exits with a certain amount of remaining rotational or translatory kinetic energy. By employing a two-phase draft tube at the exit of the two-phase expander the fluid is expanded to a pressure p_4 below the specified pressure p_2 and the power generation increases. The increased power generation reduces the enthalpy from h_2 to h_4 at the point 4. The vapour-liquid ratio x_4 is smaller than the ratio x_3 producing more liquid at point 4 than at point 3.

Because the specified outlet pressure for the expander is p_2 and not p_4 a certain portion of the remaining kinetic energy of the fluid has to be converted into static energy in form of pressure using a two-phase draft tube. The two-phase draft tube is a diffuser that increases the fluid pressure from p_4 to the specified pressure p_2 at the point 5 with a partial re-condensation of the vapour. The vapour portion x_5 at the point 5 is smaller than the vapour portion x_4 at the point 4. The enthalpy of the fluid across the draft tube remains constant with $h_4 = h_5$. The liquid portion at point 5 is larger than at point 4, and the liquid portion at point 4 is larger than at point 3.

Figure 1: Pressure – Enthalpy Diagram for a Two-Phase Draft Tube

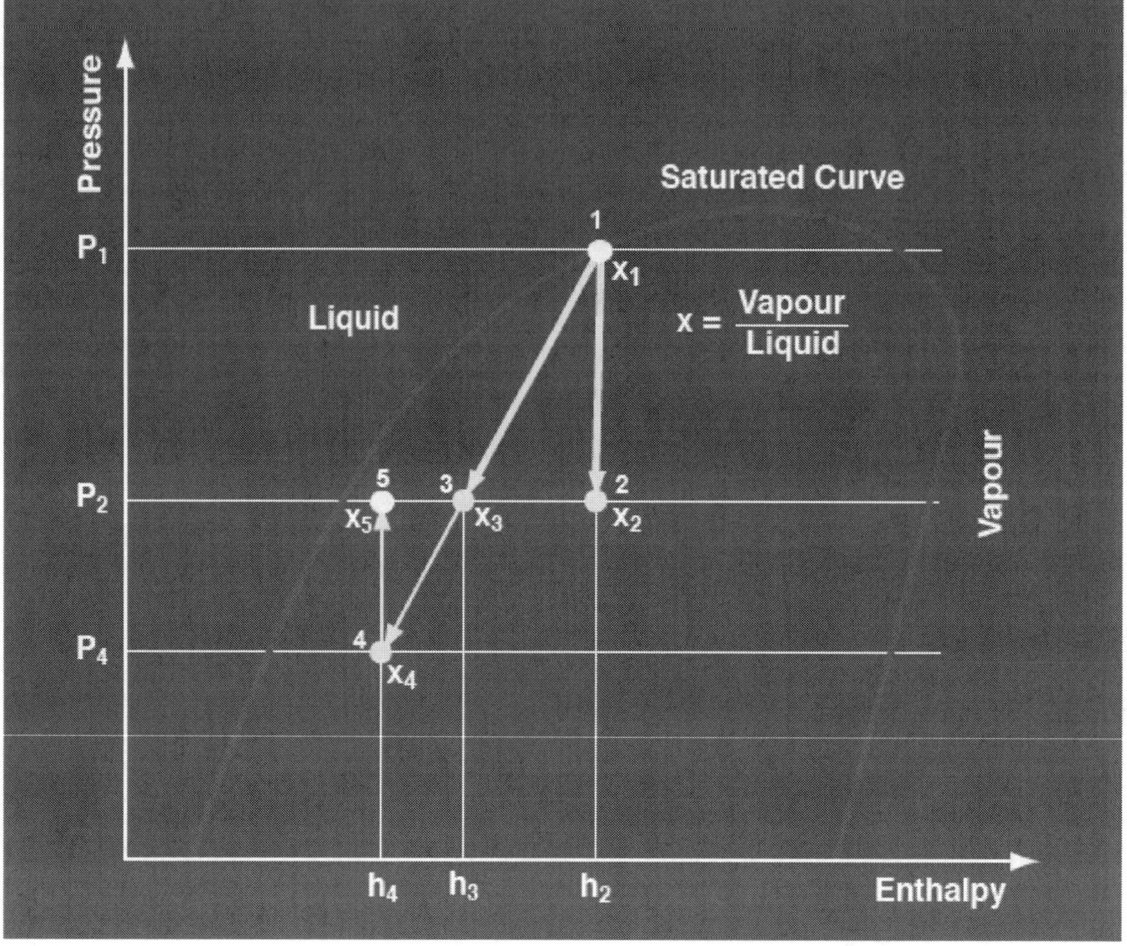

Conceptual Design of a Two-Phase Draft Tube

Figure 2 shows a conceptual design of a two-phase draft tube for radial outflow turbines. The draft tube consists of an inner converging cone and an outer diverging cone, increasing gradually the cross section of the fluid passage. At the inlet of the draft tube the cross section is shaped as a ring and at the outlet as a circle. The ring shaped inlet is designed to fit the ring shaped outlet flow of the radial outflow turbine. Between the inner and the outer cone are vanes shaped to convert the fluid rotation into translatory flow and reducing the kinetic energy of the fluid.

Summary

Using two-phase draft tubes at the outlet of two-phase expanders for the expansion of saturated liquefied gas increases substantially the amount of liquid and decreases the amount of vapour at the outlet of the draft tube.

Figure 2: Conceptual Design of a Two-Phase Draft Tube

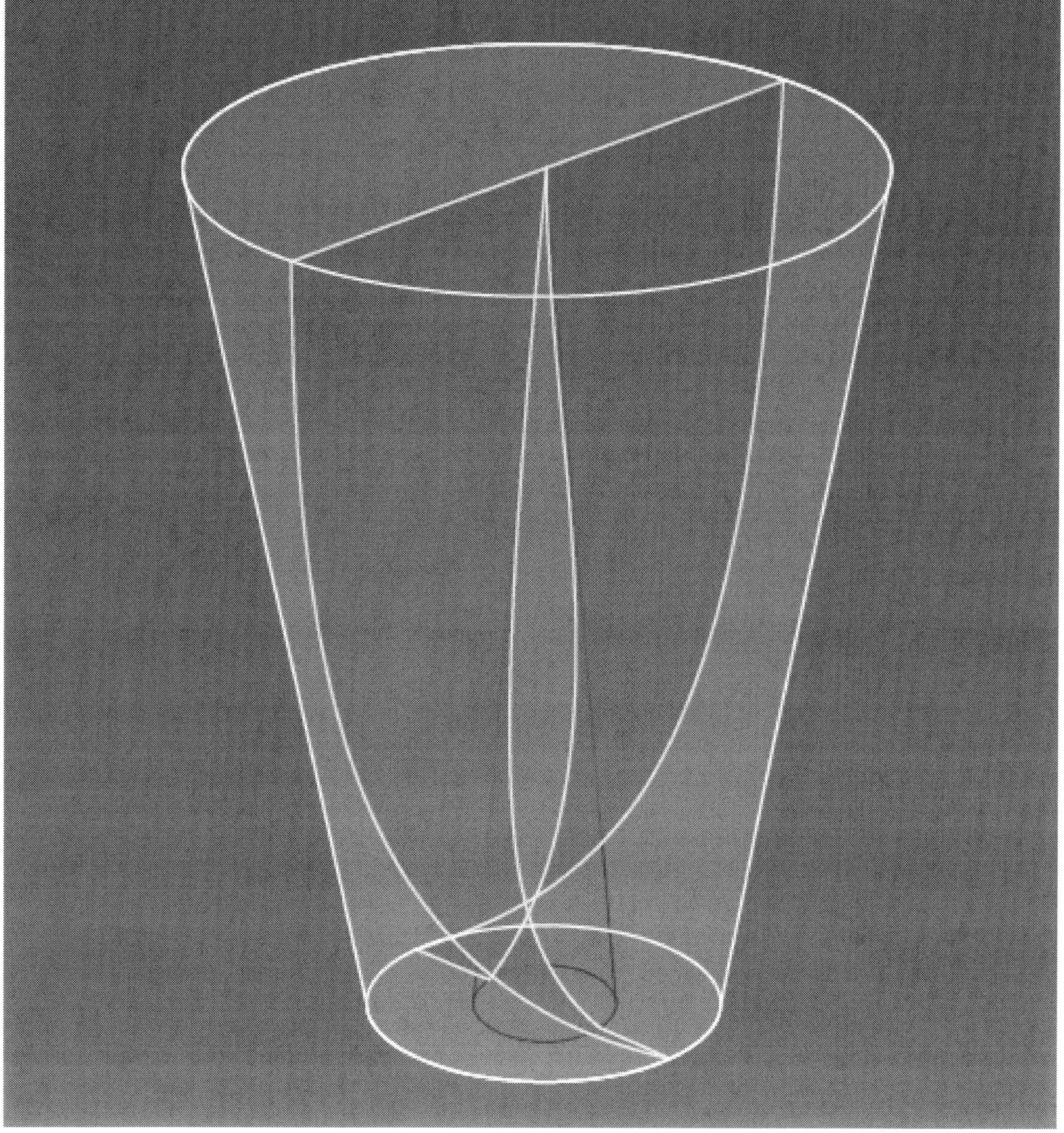

Recent Developments in the Prediction and Optimization of the Mixing of Different Qualities of LNG in a Storage Tank

Dominik Uznanski
Senior Project Manager

Fabienne Waz
Project Manager

Yacine Zellouf
Research Engineer

Research & Development Division
Gaz de France
Paris, France

AIChE Spring Meeting, April 2008
8th Topical Conference on Natural Gas Utilization
New Orleans, Louisiana, April 6-10, 2008

Introduction

The liberalization of gas markets worldwide, as well as the multiplication of LNG production sites resulting in an increase in LNG trade and the inherent increase of gas demand, has put increased pressure on gas companies to find innovative solutions in order to adapt to this changing environment. In particular, LNG import terminals are often faced today with the need to handle a variety of different qualities of LNGs in the same tanks, in order to reduce capital and operating costs. Also, this need is reinforced by the prospect of performing third party access to the gas network.

It is within this context and as a response to the above mentioned trends, that Gaz de France, in collaboration with Enagas and Osaka Gas, developed new tools, standards and procedures, which help their operators to be better prepared in dealing with the diversification of LNG supply sources. In particular, efficient mixing of various LNG qualities in a same tank was aimed for, while at the same time avoiding LNG stratification occurrence.

General description of the project

A three year collaborative project was carried out between 1999 and 2001, focusing on the characterization of commonly used industrial filling devices and featuring dedicated LNG storage tank filling trials carried out in a 500 m^3 pilot LNG storage tank.

The main objectives of the project were the following :

- to create an experimental database of trials characterizing LNG tank filling operations using various industrial filling devices,

- to develop a comprehensive predictive filling model for industrial LNG storage tanks validated with the trials database.

Within this project, a total of 40 trials were conducted over a 21 month period. These trials characterized 6 industrial filling devices (four bottom fill devices and two top fill devices) and were realized in the Nantes (France) 500 m^3 LNG storage tank.

The research methodology adopted in this project was based on an experimental and theoretical approach carried out in parallel. The experimental approach allowed the measurement of the main variables (BOG rate, vertical density and temperature profiles, operating pressure, ...) characterizing the mixing process of different LNGs in a same tank. The theoretical approach allowed the development of a predictive software tool capable of simulating filling operations carried out in large industrial tanks with commonly used filling devices.

Description of the testing station setup

The experimental trials were conducted at Gaz de France's Cryogenic Testing Station which was located in Nantes, France. This testing station included a 500 m^3 vertical cylindrical atmospheric LNG storage tank (8.5 m inner diameter), an auxiliary 250 m^3 vacuum insulated horizontal cylindrical LNG storage tank and had all the necessary equipment to reproduce LNG behavior under industrial operating conditions. It functioned much like a pilot scale methane terminal.

Figure 1 hereafter shows a schematic diagram of the experimental setup which was used during the trials. The LNG was supplied from the Montoir-de-Bretagne LNG terminal by 40 m^3 tanker trucks and stored in the 250 m^3 tank. The LNG was then transferred by the PO5 or PO7 pumps and filled into the 500 m^3 tank. The flowrate, density and temperature of the transferred LNG was monitored by a Coriolis flowmeter. A cryogenic LNG/LN_2 heat exchanger installed on the LNG transfer line between the 250 m^3 and 500 m^3 tanks allowed the precise control of the temperature of LNG injected into the 500 m^3 tank. LNG sampling devices connected to a chromatograph allowed the monitoring of the transferred and stored LNG's composition, as well as the gas phase composition in the 500 m^3 tank.

Vertical density and temperature profiles of the LNG stored in the 500 m^3 tank were obtained by a high precision Whessoe model 1146 traveling Level/Temperature/Density (LTD) gage. A vertical temperature mast equipped with 40 platinum (PT 100) temperature sensors allowed the continuous monitoring of the temperature at sensor locations and complemented the LTD gage's vertical temperature profiles. These two measuring devices, located at different positions in the tank, were useful for checking the axial symmetry of the tank's contents. Gas phase temperature and pressure were also measured, as well as the tank's roof and side-wall temperatures.

The BOG flowrate exiting the 500 m^3 tank was recorded with flow-meters located on the roof of the tank and downstream of the BOG gas heating elements used prior to BOG gas compression. The tank was equipped with a vent and two pilot controlled safety valves (6500 m^3(n)/h capacity each) to limit overpressures. Outflowing cold BOG was first heated using atmospheric or electric heaters. This BOG ran through a flow-meter and was then compressed to about 4 bars with up to 2 compressors and sent off to the local gas network. A site vent gave added safety for overpressure control.

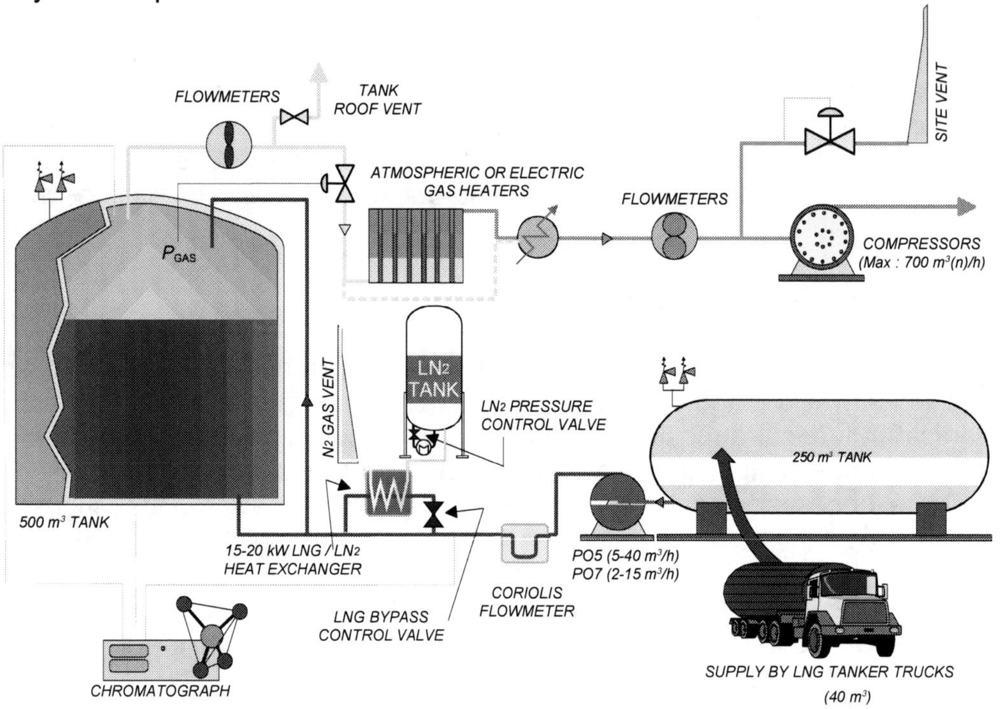

Figure 1 : Nantes testing station setup used for the tank filling trials

Characterization of LNG tank bottom filling devices

When unloading a ship's LNG cargo into a terminal's LNG storage tanks, the terminal operator needs to decide which filling device to use in the tank being filled to ensure that the two LNGs mix appropriately. Depending on the density difference between the cargo LNG and initial LNG heel contained in the storage tank, a choice needs to be made between top and bottom filling into the tank.

For the unloading of cargo LNG which has a lower density than the LNG heel in the storage tank, complete mixing between the two LNGs can be achieved by bottom filling the cargo LNG into the heel LNG. Also, flashing of the injected LNG can be avoided due to the hydrostatic pressure imposed by the initial LNG heel onto the incoming jet of LNG.

However, the unloading of cargo LNG which is heavier than the heel LNG is more difficult to handle. To achieve complete mixing of the LNGs and thus avoid stratification occurrence, top filling can be implemented. However, usually high vaporization (flashing) rates of the incoming LNG often accompany such an operation, since the LNG is injected directly into the vapor space of the tank and is often superheated with respect to the operating pressure of the tank.

In order to avoid this scenario, bottom filling can be chosen. In this case, there is a risk of LNG stratification occurrence, which needs to be managed. Therefore, it is important for terminal operators to know under which conditions two LNGs will completely mix when bottom filling an LNG heavier than the stored LNG. Stratification prevention standards under a variety of unloading conditions and for given bottom filling devices need to be developed, in order to ensure the feasibility of managing different qualities of LNG in a same tank.

Four bottom filling devices were characterized experimentally: a horizontal radial single nozzle, a simple T-shaped nozzle, a T-shaped nozzle with a gas phase funnel for BOG entrainment and a multiple slot device with an inverted cone in the lead pipe, also for gas phase BOG entrainment. These four devices are represented in figures 1 to 4 hereafter.

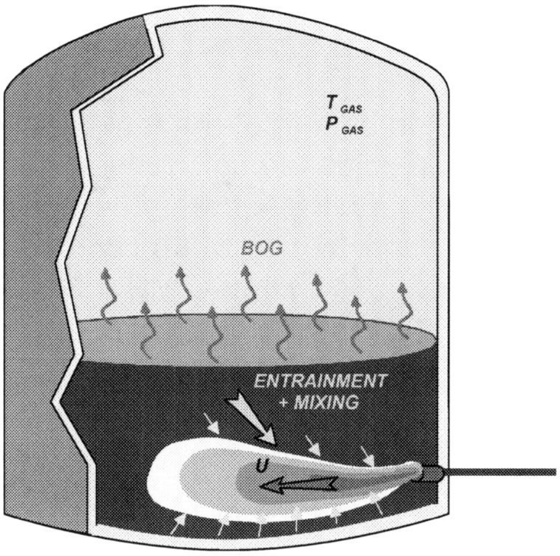

Figure 2 : horizontal radial nozzle

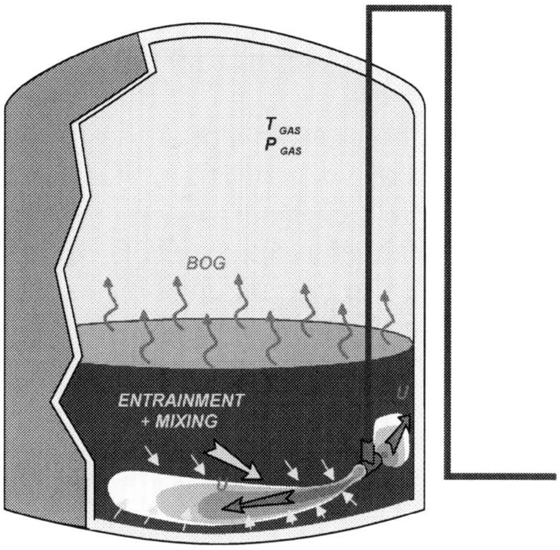

Figure 3 : T-shaped nozzle

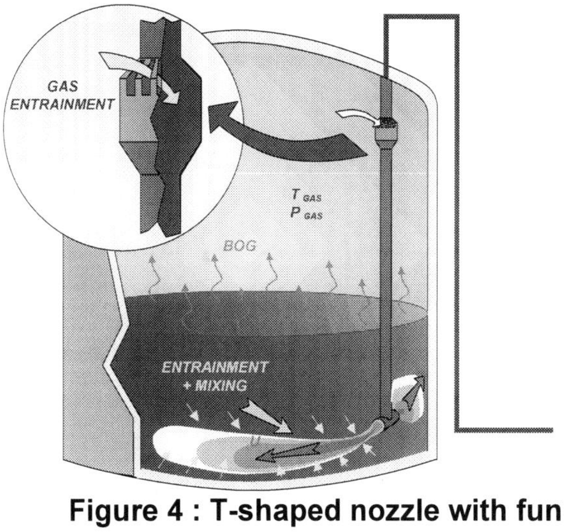

Figure 4 : T-shaped nozzle with funnel

Figure 5 : multiple slot device

Two top feed devices were also characterized : a T-shaped nozzle and a ring-shaped device. For these top feed devices, focus was placed on characterizing the production of BOG during the unloading of heavier LNG cargo into lighter LNG heel, resulting in the complete mixing of the two LNGs. For the remainder of this paper, only the bottom filling devices are addressed.

The horizontal radial nozzle is located near the bottom and side-wall of the tank and injects LNG towards the center of the tank. It is usually installed in older generation single containment type tanks. The T-shaped nozzle (with or without the gas phase funnel) is also located near the bottom and side-wall of the tank and injects LNG parallel to the side-wall of the tank in two opposite directions. It is usually installed in pre-stressed concrete (PC) type tanks.

The funnel located in the vapor space of the tank allows reincorporation of BOG into the injected LNG stream in the lead pipe. Thus, the total BOG production during a filling operation can be reduced. Also, the LNG injected at the bottom of the tank can contain entrained vapor bubbles, which contribute to a better mixing with stored heel LNG.

Finally, the multiple slot device is also located at the bottom of the tank near the side-wall and is installed in PC type tanks. LNG is injected through the vertical slots in multiple directions away from the side-wall in a hemispherical way. This device also benefits from a BOG reincorporation capacity by use of an inverted cone attached to the entrance of the lead pipe. The LNG discharged by the feed pipe is spread by the inverted cone onto the lead pipe inner wall.

For both the T-shaped device with funnel and the multiple slot device, BOG in the gas phase is entrained into the LNG stream via the Venturi effect.

Governing parameters for bottom filling of heavier cargo LNG into lighter stored LNG

The governing parameters for the mixing of different qualities of LNGs using bottom feed nozzles are the incoming jet's Froude number ($Fr = \mathrm{u}_0 \Big/ \sqrt{\mathrm{gd}_0 \frac{|\Delta\rho|}{\rho_s}}$), the jet's Reynolds number ($\mathrm{Re} = \mathrm{u}_0\mathrm{d}_0/\nu$) and the Lewis number ($Le = \mathrm{k}_{\mathrm{th}}/\mathrm{k}_{\mathrm{m}}$). The symbol u_0 represents the speed of injected LNG at the nozzle outlet ($u_0 = 4\dot{q}/(\pi d_0^{\,2})$ for the horizontal radial device, $u_0 = 2\dot{q}/(\pi d_0^{\,2})$ for the T-shaped nozzle and $u_0 = \dot{q}/S_{slot}$ for the multiple slot device), where $\dot{q}$ is the total LNG filling flowrate into the tank, S_{slot} is the total cross-sectional area of the slot openings, d_0 represents the inner diameter of the injection nozzle outlet(s), ν is the kinematic viscosity of LNG, $\Delta\rho$ is the density difference between the stored and injected LNGs, ρ_s is the density of stored LNG, k_{th} is the thermal molecular diffusivity, k_m the species molecular diffusivity and g is the gravitational acceleration. In the Froude number formulation above for the multiple slot device,

d_0 is replaced by the hydraulic diameter D_{hyd} of the slots($D_{hyd} = 4S_{slot} / P$, where P is the total wetted perimeter of the slots).

The Froude number represents the ratio of the incoming jet's inertial force and the buoyancy force generated by the density difference between the injected and stored LNGs. The Reynolds number represents the ratio of the jet's inertial force and the viscous (shear) force it generates. Finally, the Lewis number represents the double-diffusive nature of LNG behavior, since heat and LNG species are diffused at very different rates (for LNG, the Lewis number is about 50).

It was assumed here that the main non-dimensional parameter governing the mixing process is the Froude number, as was postulated in several other studies [1,2]. Thus, the operating conditions at terminal scale to be studied were scaled down at 500 m^3 scale by conserving the Froude number. Reynolds number similarity cannot be maintained simultaneously with the Froude number, however if the scaled down Reynolds number is in the turbulent range, similarity will hold. Finally, the Lewis number similarity is naturally conserved by using LNG in the trials.

Results for bottom filling devices

A total of 26 trials which involved the injection of LNG heavier than stored LNG were carried out in the 500 m^3 LNG tank using a horizontal radial nozzle (d_0=0.0507 m), T-shaped devices with and without a gas phase funnel (d_0=0.066 m) and a multiple slot device (1/10 scale). The range of conditions under which these trials were conducted at 500 m^3 scale cover the range of tank filling conditions usually encountered in LNG terminal operations. The Reynolds number for these trials was always well within the turbulent range.

Three types of mixing behaviors were observed during the tank filling process :

- the immediate segregation of the two LNGs into a density stratification (figure 6),
- the complete mixing of the two LNGs, followed by the occurrence of density stratification (figure 8),
- the complete mixing of the two LNGs throughout the entire filling operation.

The behaviors corresponding to stratification occurrence are depicted in figures 6 and 8 below, which show the evolution of vertical density profiles measured by the LTD gage as a function of time.

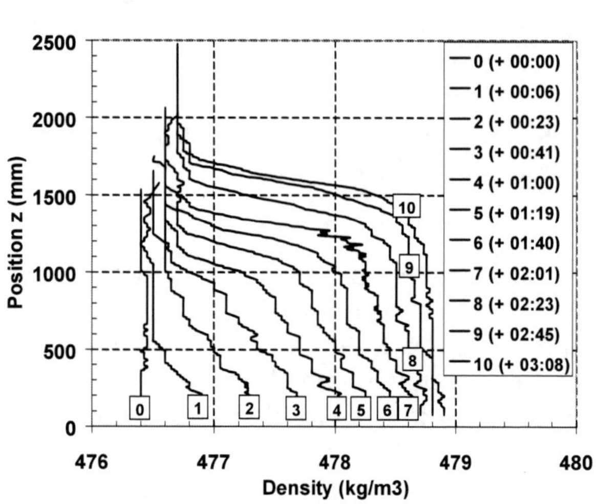

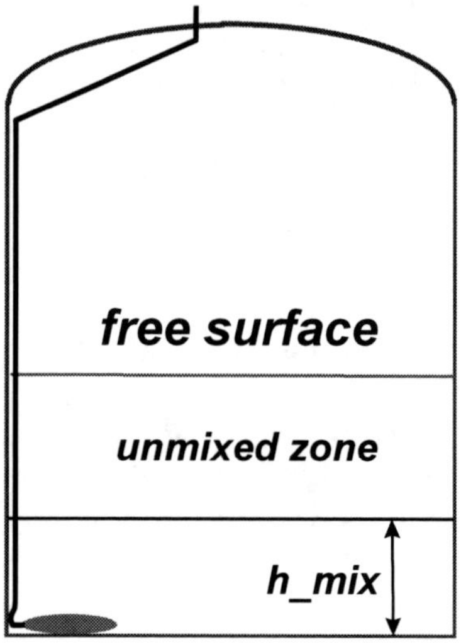

Figure 6 : vertical density profiles

Figure 7 : incomplete mixing

Figure 6 above shows that right from the start of the filling operation, a separation between the heavier injected LNG and the lighter stored LNG takes places. This means that the incoming LNG jet does not travel all the way up to the free surface of the stored LNG and only mixes with part of the heel LNG, as shown in figure 7 above. Once this density stratification is created, the incoming LNG continues to mix with the stratification's lower layer, while the upper layer remains unaffected. In this type of behavior, the jet's inertial forces are weak compared to the buoyancy forces generated by the density difference between the two LNGs, thus preventing their complete mixing.

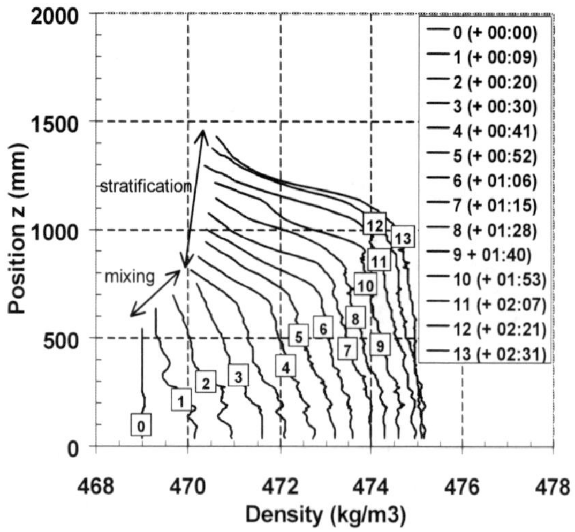

Figure 8 : vertical density profiles

Figure 9 : complete/incomplete mixing

Figure 8 above shows that during the first part of the filling process, the injected LNG completely mixes with the stored LNG : the incoming jet has sufficient inertia to reach the free surface of the stored LNG. However, as the filling process unfolds, a point is reached where the total LNG height is sufficiently large to prevent the incoming LNG jet from reaching the free surface, thus creating a density stratification, as depicted in figure 9. In this type of behavior, the jet's inertial forces are comparable to the buoyancy forces generated by the density difference between the two LNGs, thus allowing their initial mixing.

From these results, stratification prevention standards were developed for all four of the bottom filling devices, based on the Froude number of the injected LNG jet and on a reduced mixing height (h_mix/d_0), as was done in a previous study [3]. Figure 10 below shows the form of the stratification standards obtained for the horizontal radial pipe and the T-shaped device without funnel. As can be seen, a linear relationship was found between the reduced mixing height and the jet's Froude number. In figure 10, the region above the stratification standard line represents the stratification creation zone and the region under the line represents the stratification prevention zone.

The field trials also showed that the bottom fill T-shaped device with funnel and the multiple slot device were much more efficient in mixing heavier cargo LNG with lighter heel LNG, when compared to the horizontal radial nozzle and T-shaped device without funnel (all other conditions being equal). If you add on top of that the capacity of these BOG entrainment devices to limit the BOG production during a tank filling operation, it gives these devices a double advantage : increase the flexibility of mixing different grades of LNG in a same tank and minimize BOG production during tank filling.

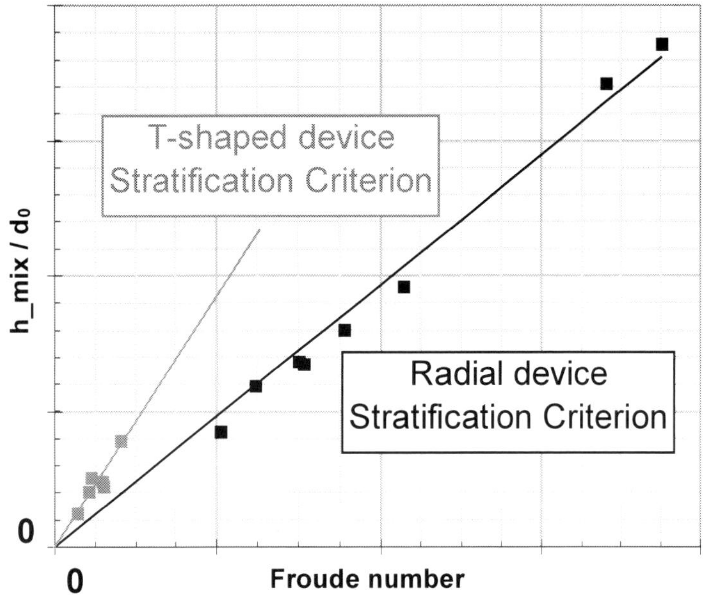

Figure 10 : generalized stratification standard for the horizontal radial nozzle and the T-shaped device without gas phase funnel

From these results, a generalized law was formulated for each bottom filling device, expressing the mixing height induced by the incoming jet of LNG as a function of its Froude number :

$$h_{mix} / d_0 = C \times Fr$$

where C is a constant which depends on the geometry of the filling device and its position in the tank. For the multiple slot device, d_0 is replaced by D_{hyd}.

An application of such stratification standards using different filling flowrates is shown in figure 11 below. It can be seen that the higher the filling flowrate, the less likely stratification appearance will be. Conversely, the higher the density ratio between the LNGs and/or the higher the height of stored LNG, the higher the chance of creating a stratification will be. This highlights the interaction between the jet's inertial forces and the buoyancy forces generated by the density difference between the LNGs. Figure 11 also shows that for the same filling conditions, the horizontal radial device is more efficient in mixing two LNGs than the T-shaped device without funnel. This is certainly due to the fact that the T-shaped nozzle splits the incoming LNG flow into two streams, thus reducing the inertial forces of the incoming jets.

However, to determine beforehand whether or not stratification will occur during the course of a filling operation, it is not sufficient to establish stratification standards as those depicted above, since the determination of whether or not complete mixing will occur between cargo and heel LNGs is necessarily based on the initial conditions of the LNGs. As was seen above, in some cases complete mixing occurs in the first stages of the filling operation and subsequently stratification occurs when the total height of LNG in the tank exceeds the mixing height induced by the incoming jet of LNG.

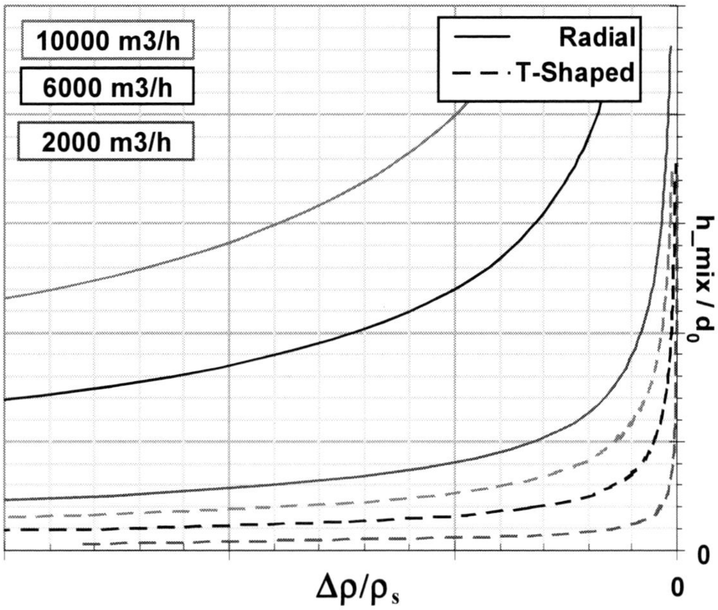

Figure 11 : stratification standard for different tank filling flowrates

Therefore, in order to predict the consequences of such a filling operation, it is necessary to have at one's disposal a dynamic tank filling model.

Dynamic tank filling model

A predictive dynamic tank filling model was developed by Gaz de France, based on the experimental databases which have been developed. This model uses a "global" mixing approach based on the empirical mixing height law exposed above, for bottom filling operations with heavier cargo LNG being introduced into lighter heel LNG. For bottom filling operations with cargo LNG lighter than heel LNG, it is assumed the two LNGs mix completely, due to the buoyancy effect.

At each time step of the calculation, the mixing height is calculated and compared to the height of the layer of LNG in contact with the bottom wall of the tank (for an LNG stratification, this is the lower layer). If the calculated mixing height is greater than the bottom layer height, then the incoming LNG volume filled during the elapsed time step will mix completely with the bottom layer LNG and eventually with the layers above, depending on the filling conditions. If conversely the mixing height is smaller than the bottom layer height, then the incoming LNG volume will only mix with the bottom layer volume comprised within the calculated mixing height. The remaining unaffected volume of this bottom layer then becomes part of the upper layers of the stratification (interface layers and upper layer). By repeating this process during the entire filling operation, a stratification composed of a lower layer, an upper layer and interfacial layers can be constructed.

To illustrate this, the results obtained with the dynamic tank filling model for the two mixing modes presented above (incomplete mixing and complete/incomplete mixing) are compared to field data in figures 12 and 13 below.

As can be seen, the model results compare well with the field data for both mixing modes, thus highlighting the pertinence of the modeling used. In figures 12 and 13 below, the field data vertical density profiles for the lower layer are slightly inclined. This is normal, as the traveling LTD gage takes a certain time to complete an up and down cycle from the tank bottom to the free surface. In this time, the lower part of the lower layer increases in density, as filling of heavier LNG continues near the bottom wall.

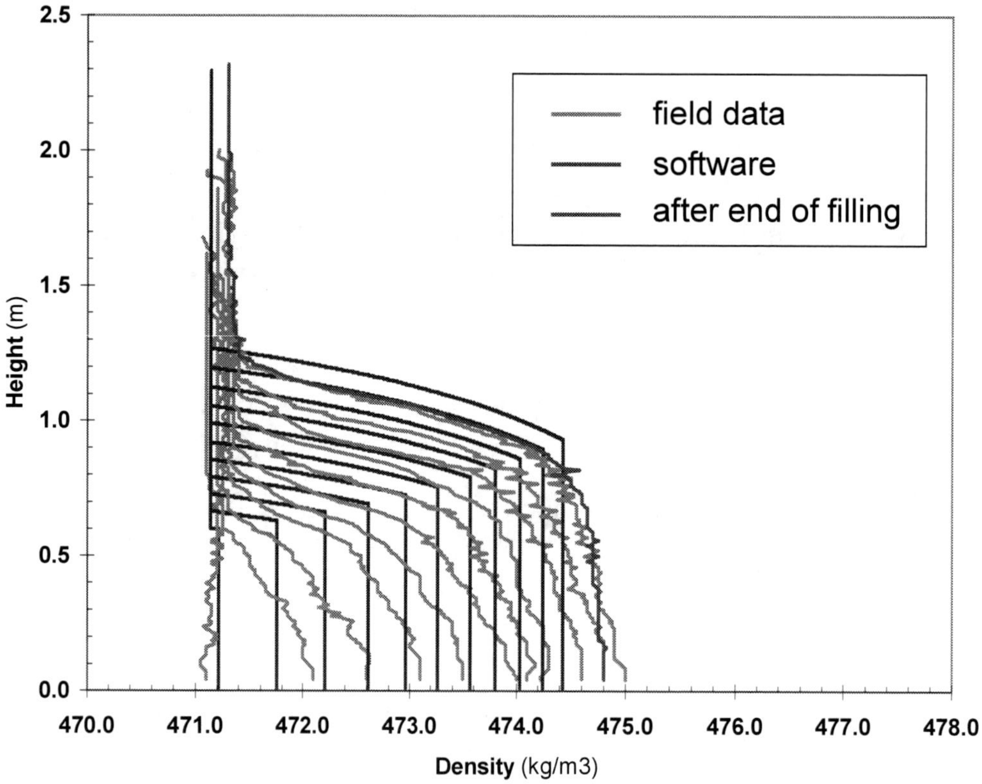

Figure 12 : incomplete mixing : filling model vs. field data

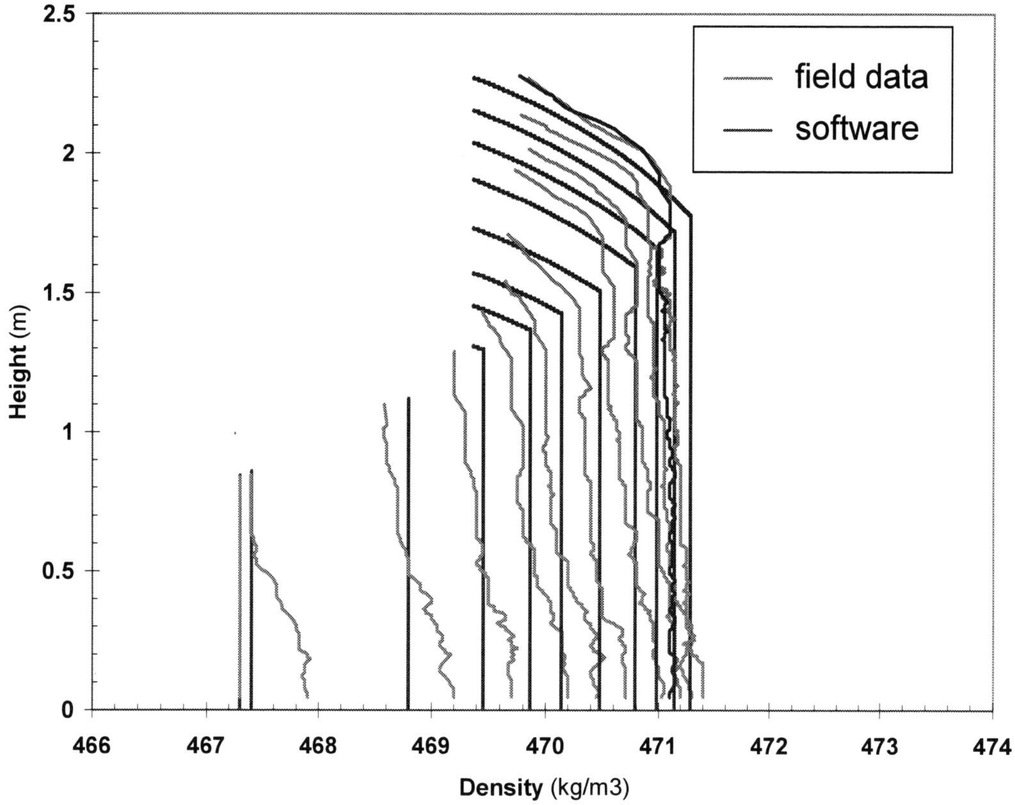

Figure 13 : complete/incomplete mixing : filling model vs. field data

This dynamic tank filling model will be part of a future version of Gaz de France's LNG tank management software named LNG MASTER®.

Impact and benefits of results on terminal operations

The results obtained in this project will benefit terminal operators in several ways. First of all, by developing comprehensive stratification prevention standards for a variety of filling devices commonly used in LNG storage tanks, operators can determine the operating conditions under which they will be able to completely mix LNGs having different densities, in particular for injected LNG which is heavier than stored LNG. In this way, instead of having to almost entirely empty a storage tank before introducing heavier LNG into it, operational flexibility can be gained by only partially emptying the storage tank before unloading new LNG into it, in accordance with the filling conditions (filling flowrate, stored LNG height and density difference of the LNGs).

Secondly, by developing a predictive tank filling model, validated with the dedicated experimental trials database, operators will have at their disposal a tool which will quantify the dynamic behavior of LNG during tank filling operations, both in terms of the mixing degree of different types of LNGs and BOG production. In this way, the consequences of tank filling operations can be quantified, thus allowing the operator to make the appropriate choices for optimizing the inventory management of his LNG storage tanks.

Conclusion

The work which was carried out in the collaborative project undertaken by Enagas, Osaka Gas and Gaz de France has yielded some important results concerning inventory management during LNG tank filling operations.

Different types of behaviors which occur when introducing heavy LNG cargo into light LNG heel by bottom filling have been identified. Stratification prevention standards have been developed for four commonly used bottom filling devices.

A predictive dynamic tank filling model has been developed which takes into account all of the four bottom filling devices which were characterized. The model results compare well with the trials carried out for the bottom filling devices.

It is hoped that the extensive database which has been established and the dynamic tank filling model which has been developed will bring increased operating flexibility and enhanced safety in dealing with the diversification of LNG supply sources in today's LNG industry.

Acknowledgments

The authors would like to gratefully acknowledge the contribution of Osaka Gas and Enagas to the success of this work. The authors would also like to thank Mr. Olivier Gorieu of the R&D Division of Gaz de France for his invaluable contribution in successfully carrying out the field trials and in developing the experimental database and dynamic tank filling model resulting from this work.

References

1. Tamura M. et al., "Prevention of LNG roll-over in an LNG tank", Poster session A, Paper A.2, LNG 12 Conference Proceedings, Perth, Australia, May 4-7, 1998

2. Uchida H. et al., "Study on mixing different density LNG during bottom-filling accompanied with jet mixing", IHI Technical Review, Vol. 35, N° 4, pp. 243-346, July 1995

3. Suzuki O., Takao S., "The formation of a non-linear density gradient in a tank", Authority of Mtg. A.S.M.E., New York, U.S.A., pp 1-8, 1983

4. Homan K.O., Soo S.L., "Model of the transient stratified flow into a chilled-water storage tank", Int. J. Heat Mass Transfer, Vol. 40, No. 18, pp. 4367-4377, 1997

5. Mahoney L.A. et al., "Modeling and analysis of ORNL horizontal storage tank mobilization and mixing", Pacific Northwest Laboratory, Richland, Washington, 99352, report PNL-9961, June 1994

6. Meyer P.A., "Computer modeling of jet mixing in INEL Waste Tanks", Pacific Northwest Laboratory, Richland, Washington, 99352, report PNL-9063, January 1994

7. Spall R.E., "A numerical study of transient mixed convection in cylindrical thermal storage tanks", Int. J. Heat Mass Transfer, Vol. 41, No. 13, pp. 2003-2011, 1998

Biography

361 avenue du Président Wilson, BP 33, 93211 Saint Denis la Plaine Cedex, France
Phone : (33) 1 49 22 52 91, Fax : (33) 1 49 22 50 40
E-mail : dominik.uznanski@gazdefrance.com

Mr. Dominik Uznanski joined Gaz de France's Research and Development Division in 1993 as a research engineer at the Cryogenic Studies Section. Since then, he has been involved in numerous studies on the behavior of LNG in storage tanks (including research on stratification behavior, weathering and tank filling operations), on LNG vapor heavy gas dispersion and on water curtain mitigation methods and has gained extensive experience in these fields. He is also involved in the development of Gaz de France's LNG tank management software named LNG MASTER and its delivery to end users, including Gaz de France's two LNG receiving terminals and external clients. He has managed multiple projects dealing with LNG Technologies and Operations of LNG storage tanks. In this respect, he has lead projects involving studies on LNG behavior, evaluation of the aging of LNG installations and their expected lifetime and technological advances in the LNG field.

Mr. Uznanski holds two Bachelor of Science degrees in chemical engineering (specialization in process control and optimization and numerical methods) and pure mathematics from the University of Maryland, U.S.A. (1986) and a Master of Science degree in fluid dynamics and heat transfer from the Ecole Centrale Nantes engineering school (1991).

28d Understanding Asphaltenes Conversion in Fixed Bed Conditions

João Marques[1], Arnaud Baudot[1], Isabelle Merdrignac[2], Denis Guillaume[1], Didier Espinat[2], and Sylvette Brunet[3]. (1) Catalysis and Separation Division, IFP Lyon, PO Box 3, 69390 Vernaison, France, (2) Physics and Analysis Division, Institut Français du Pétrole, PO Box 3, 69390 Vernaison, France, (3) UMR 6503 Catalyse en Chimie Organique, University of Poitiers, 40 avenue du Recteur Pineau, Poitiers, 86022, France

The nature of the catalyst carrier may affect residues hydrotreating performance. The association of an acid function with a strong hydrogenating phase can increase the hydrodeasphalting activity without substantial coking increment. Within this framework, catalysts doped with different acidity promoters at different levels were evaluated for a middle east residue hydrotreatment in a batch reactor. The catalysts carriers were carefully prepared in order to obtain the same porous distribution. Moreover, catalytic tests using model molecules were performed to screen catalysts hydrogenating and acid functions. For a better understanding of asphaltenes conversion, a strategy has been applied to reduce their polydispersity prior to characterization and catalytic tests. Therefore, a membrane separation process was used to obtain distinct asphaltene fractions with a narrow size polydispersity. This work presents the results of catalytic tests performed with reconstituted feeds made of narrow-sized asphaltenes fractions dispersed in the maltenes from initial residue.

28b Heavy Oil Tutorial: Polynuclear Aromatics, the Barrier to Conversion

Irwin A. Wiehe, Soluble Solutions, 3 Louise Lane, Gladstone, Gladstone, NJ 07934

The basic problem of the conversion of petroleum resids and heavy oils are the presence of polynuclear aromatics (PNS's) greater than four rings while the only aromatics in transportation fuels that are desirable are alkylated single rings in gasoline. It is these polynuclear aromatics that provide a thermal reaction limit and insolubility, resulting in fouling and coking. Although analytical tools cannot determine the full PNA distribution, HPLC provides data on the heptane soluble fraction for four rings and less and even hydrogen content provides insight. It is the objective of primary upgrading to produce the maximum feed for secondary upgrading that is relatively free of catalyst poisons, including PNA's greater than four rings, while rejecting a low valued byproduct. Fortunately, even with the limited understanding of the chemical structure of petroleum macromolecules, the path is clear on how to greatly improve such primary upgrading processes as coking and hydroconversion.

Improved Simulation Design Factors for Unconventional Crude Vacuum Units:

Cracked Gas Make

Stripping Section Performance

Darius Remesat, PhD, MSC, MBA, P.Eng.
Koch-Glitsch Canada LP
1829 Ranchlands Blvd NW
Calgary, AB, T3L3E5

Prepared for Presentation at

The AIChE 2008 Spring National Meeting – New Orleans

April 6- 8th, 2008

Session #28 - Heavy Oil Chemistry I

April 2008

" Submitted to Various Journals for Publication"

Currently "UNPUBLISHED"

Introduction:

The refining industry has been gaining experience in operating vacuum units processing unconventional heavy oil/bitumen derived crudes. New vacuum units are being planned and the knowledge gained from existing operations would improve the economics of these projects by incorporating the nuances of processing these crudes to create robust designs that represent the actual operation.

This paper focuses on sharing data specifically from unconventional heavy crude vacuum units operating in "wet" mode (stripping and velocity steam)[1] to provide insight into using a proven simulation topology[2] with heavy oil/bitumen Upgrader feeds to Vacuum distillation units.

When simulating/designing a heavy oil/bitumen Upgrader vacuum column the following design factors need to be incorporated, in order of importance:

1. Crude Characterization[3,4]
2. Thermodynamics choice[3,4]
3. Simulation topology (non-equilibrium)[2]

Each subsequent design factor cannot compensate for a poor crude characterization or thermodynamics choices. Assuming the heavy oil/bitumen crude has been characterized appropriately for the higher boiling point components (ex. using both ASTM 2887 and HTSD)[5], and the representative thermodynamic package is chosen (ex. variation of Peng-Robinson, Grayson-Streed, Esso Tubular, BK10), then the focus for the vacuum column design shifts to the simulation development . Figure 1 shows a proven vacuum unit simulation topology[2] that addresses the realities of the physical system built. The bottoms of the atmospheric crude column is heated (partially vaporized) in a vacuum heater to provide necessary energy to separate the higher boiling components at a reduced pressure in the vacuum column. In the simulation, the column is divided into two main sections (stripping section and above the flash zone). The transfer line is "partitioned" into separate vapour and liquid streams to address the non-equilibrium situation and various flashes are provided to address entrainment and the specifics of the flash zone and slop wax collector tray. It has been shown that the HVGO yield and liquid rate to the top of the wash section predictions from the figure 1 topology match well with plant data[2,3,6].

From reviewing operating data for various units, the Figure 1 simulation topology has been used to enhance the representation of bitumen based feed vacuum unit by capturing the thermal and variable stability characteristics of the crude. The amount of cracked gas generated, entrainment formed, and the stripping section performance should be critically evaluated to ensure these factors represent the bitumen characteristics.

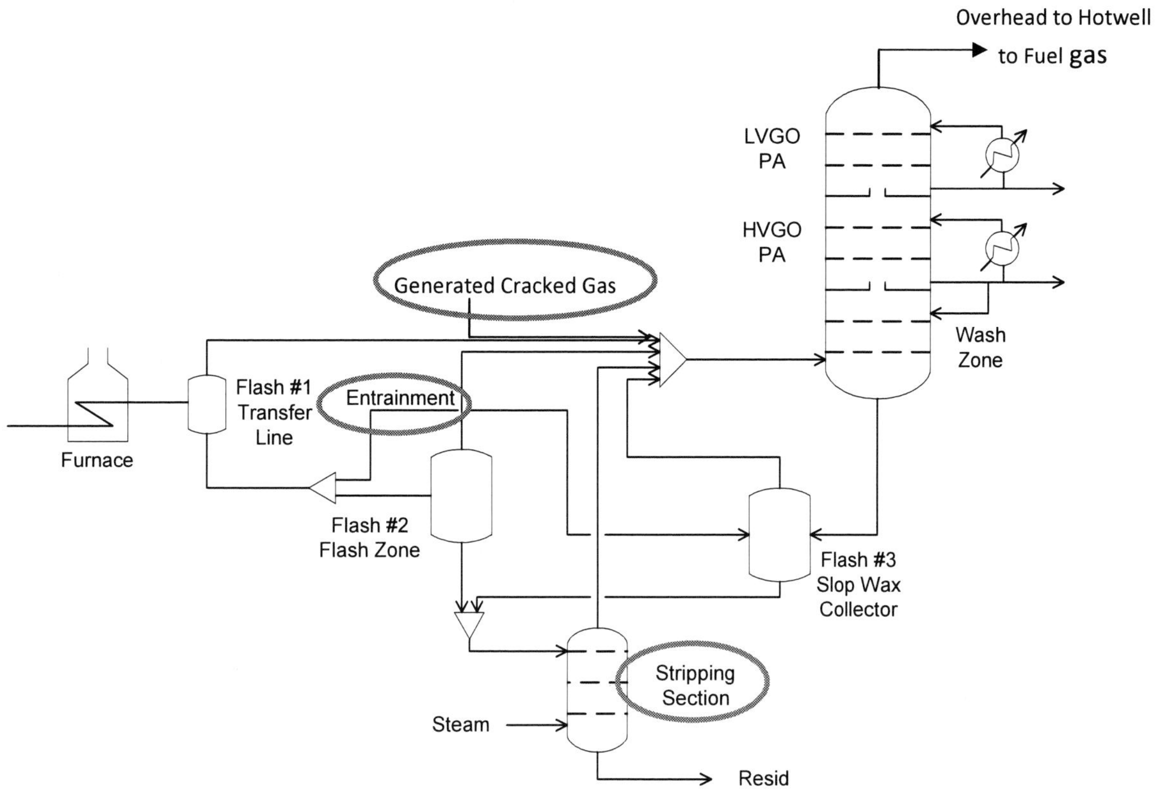

Figure 1 – Proven Vacuum Unit Simulation topology to represent non-equilibrium tendency in feed to vacuum column[2]. Experience "factors" for bitumen derived crude for cracked gas, entrainment and the stripping section should be provided to adequately represent actual operation.

Cracked Gas Make:

The operational (and thus economic) impact of underestimating crude cracking in the vacuum heater is large due to increased light hydrocarbon molecules that can reduce the effectiveness of the vacuum ejector system and increased coke particle generation that both can reduce the run length and/or throughput capacity of the column.

There are many factors influencing the generation of cracked gas including crude type, and heater design. The heater can be designed to reduce the tendency to crack by maintaining a well-distributed even heat flux (dual fired vs single fired)[7], reduced maximum film temperature and a low residence. These heater design guidelines are a function of the crude being processed, and typically the impact of the heater design is not fully included into the simulation.

Figure 2 shows a published relationship (Lieberman) that uses readily available measured plant data that relates transfer line temperature to the amount of offgas (SCFD) leaving the vacuum unit[8] (assumed downstream of the vacuum hot well) as a ratio to the crude flow (BPD) to the unit. Very simplistically, the trend is that as heater temperature increases, the quantity of offgas per crude feed increases as well. This relationship is only as good as the field measurements but provides a tangible appreciation for

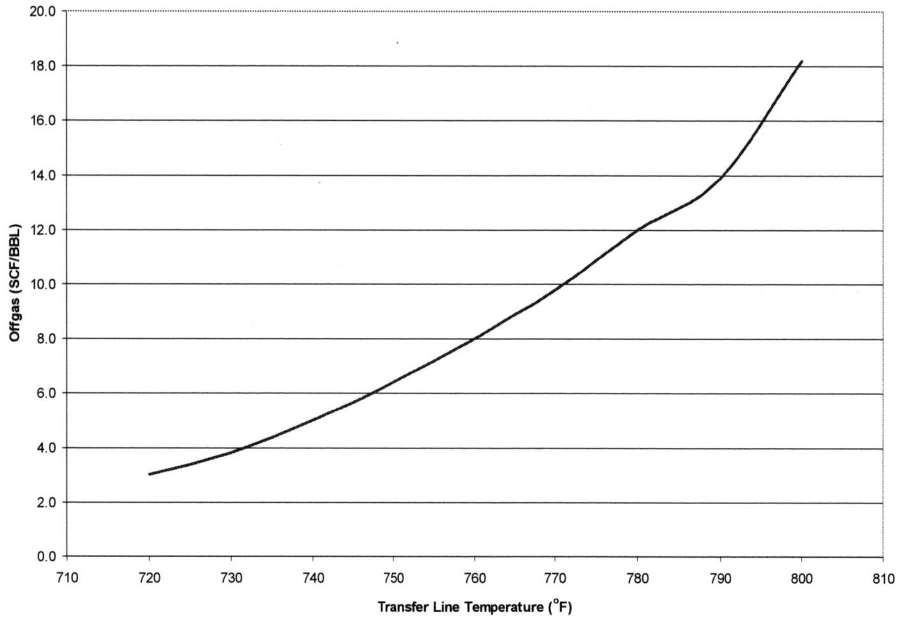

Figure 2 – Graphical Representation of Lieberman relationship; measured plant variables, transfer line temperature and offgas to crude flow ratio.

the degree of cracked gas generated from an unknown heater design. This offgas value is lumped consisting of air ingress, dissolved hydrocarbons in the feed, residual water and cracked gas. For the data gathered, a reasonable assertion appeared to be that the cracked gas comprised the majority of the offgas and was the prominent contributor to the increase in offgas flow when the heater temperature increased.

Figure 3 shows two data sets taken from vacuum units processing crudes with vacuum feed API's of 13 and 16. Both data sets support the Lieberman relationship trend that as the transfer line temperature increases, the offgas flow to crude feed ratio increases as well. The API 13 data set seems to match the Lieberman relationship well, while the API 16 data set seems to generate far less cracked gas. Trend lines were for each data set to reflect the Leiberman relationship representation.

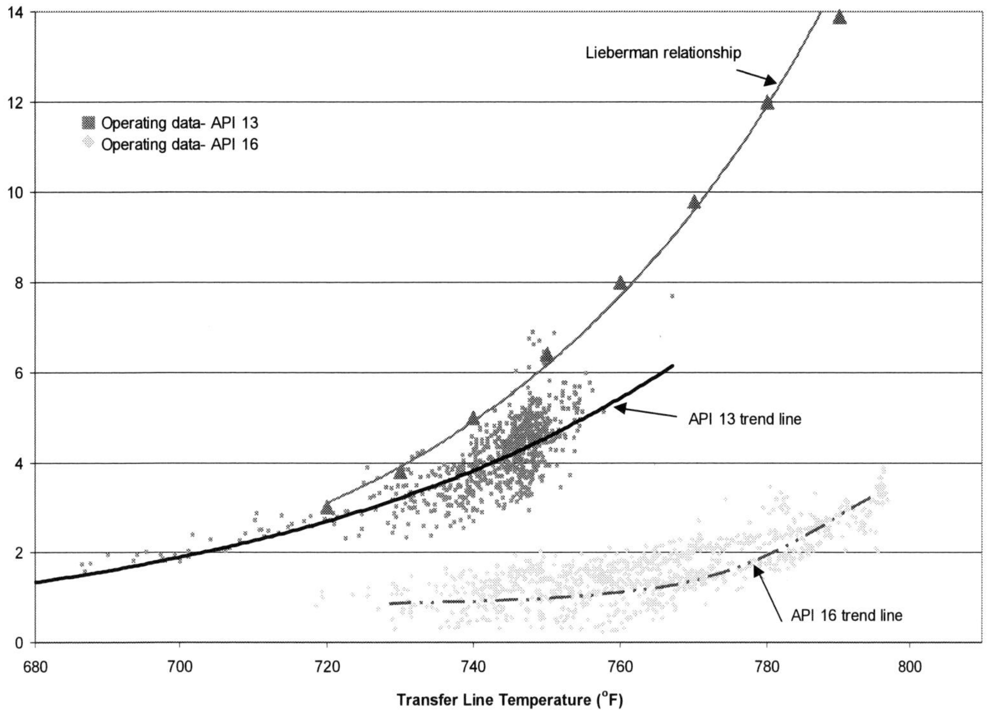

Figure 3- Addition of API 13 and 16 data sets to the transfer line temperature versus offgas to crude flow ratio relationship. Trend lines are provided for each data set.

Figure 4 includes trended data from vacuum units processing more heavy crude. Again, the relationship that as the transfer line temperature increases (from increased heater firing), the amount of offgas flow increases is reflected in the data sets. In general, vacuum units processing heavier crudes will operate at lower heater outlet temperatures than lighter crudes primarily due to the tendency for the heavier crudes to be more unstable at higher temperatures. The instability translates to increased coke generation and light hydrocarbon formation. Also, obviously as the API decreases, the degree of offgas increases. However, many vacuum unit designs do not reflect this situation. For example, at a temperature of 700oF, the API 13 crude tends to exhibit offgas to crude flow of 2-3 scf/bbl, while for the API 10 crude (ex. Cold Lake), the scf/bbl would be 6-7 (over double).

This type of over-simplified relationship between API (primarily bitumen derived) and off gas flow (cracked gas make) has implications for the design of the vacuum column unit. Assuming the desire to operate the vacuum unit at a pressure as low as possible (to maximize HVGO product yield), the overhead vacuum ejector design/size and secondly, the heat transfer capabilities of the quench/pumparound zones will be different if the data in figure 4 are used than what is available in literature[8].

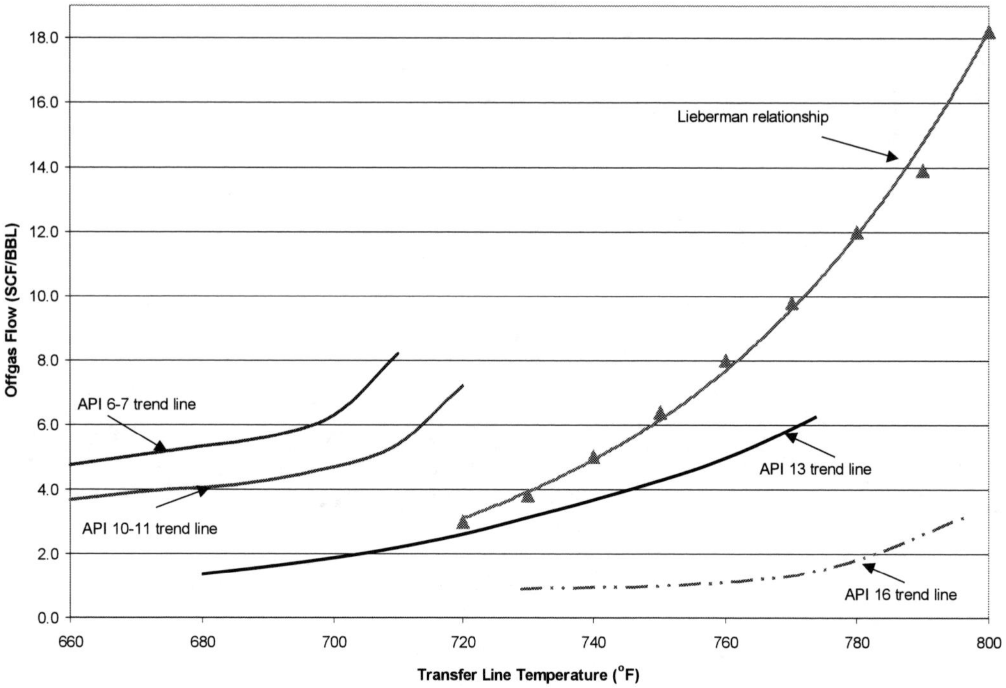

Figure 4 – Addition of heavier crude data sets to the transfer line to offgas flow to crude flow ratio relationship. From the data illustrated, as the API decreases (heavier crude), the degree of cracked gas (translated to offgas) increases.

Figure 5 shows one operating example and one re-design example (not in operation yet). Point A, for an unconventional crude with a feed API 10 represents a design point case where the actual offgas flow was significantly more than what was expected for designed heater outlet temperature and HVGO yield. The vacuum ejector system was overwhelmed, and the heater outlet temperature was reduced by approximately 35°F (point B), reducing HVGO yield significantly. It appeared that the design point A for the cracked gas make was based on the cracking characteristics of more conventional crude. Point B' is a suggested design point, using the benefits of an operating experience data base. Understanding and applying the specific crude characteristics (crude characterization and validating with similar data points) is crucial if a design is to operate as intended.

The second case illustrated in figure 5 is still in design mode with point C representing the initial design case. After sharing the operating experience with the process designer, point C was modified to point D. Basically, the amount of cracked gas generated was increased based on an extrapolation of the shared operating data. The concern with this approach is the high heater outer temperature still desired. The chance is high of accelerated coking in the heater at these high temperatures for unconventional crudes. Point D' is suggested that incorporates an increased cracked gas and a reduced outlet heater temperature.

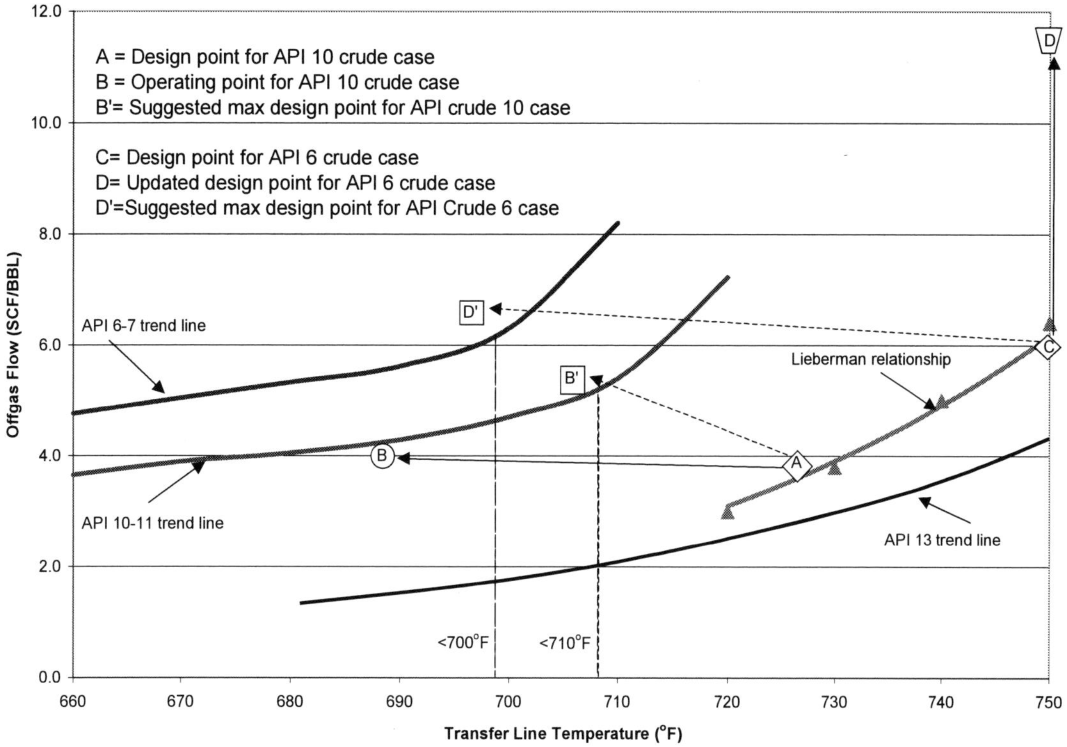

Figure 5 – *Operating Case* - Point A was a design point for an API 10 unconventional crude vacuum unit. The vacuum unit operated at point B, a significantly lower heater outlet/transfer line temperature. Point B' is a possible design point for the crude in question. *Design Case* - Point C was a design point for an API 6 unconventional crude vacuum unit. The vacuum unit is now desired for point D while point D' is suggested as a design point (reduced Heater design temperature).

A few generalizations can be considered from inspection of figure 5. For both the API 6 and 10 crude trend lines, there is a temperature (approx. 710°F for API 10, and 700°F for API 6) at which the offgas flow accelerates. Each crude should be inspected on a case by case basis, but from the data in Figure 5 and similar comments from experts, heater outlet temperatures for heavy crude with instability tendencies and asphaltenes (eg. Canadian Bitumen) should be designed not to exceed a maximum of 710-715°F[9].

Stripping Section (Residue Stripping):

The increased recovery of VGO due to stripping steam is related to the operating pressure in this section. Pressure drop is generated by the internals. Steam stripping lowers the hydrocarbon partial pressure to allow the material to boil at a lower temperature. Typically, operating tray pressure drop (0.07-0.1 psi/tray) is a small percentage of operating pressure, but in vacuum service, the tray pressure drop constitutes a large portion for the overall section pressure drop. In the simulation, the appropriate tray efficiency and expected pressure drop generated is input into the simulation as per Figure 1 (a separate column) so that the rapid approach to equilibrium occurs within one to two stages[10]. Figure 6 provides data from over 10 operating units[3], illustrating the observed staged count from matching plant data to a representative simulation. A simple way to explain this observation is that the pressure drop added from the increase in trays offsets the added separation provided by the tray. The stripping factor (KV/L; K = equilibrium constant, V= Vapour flow, L= liquid flow) reflects this relationship, with the equilibrium constant represented as PvHC/PT (PvHC = hydrocarbon vapour pressure, PT= total pressure). So as PT increases, the stripping efficiency reduces. This relationship needs to be incorporated in to the vacuum column design.

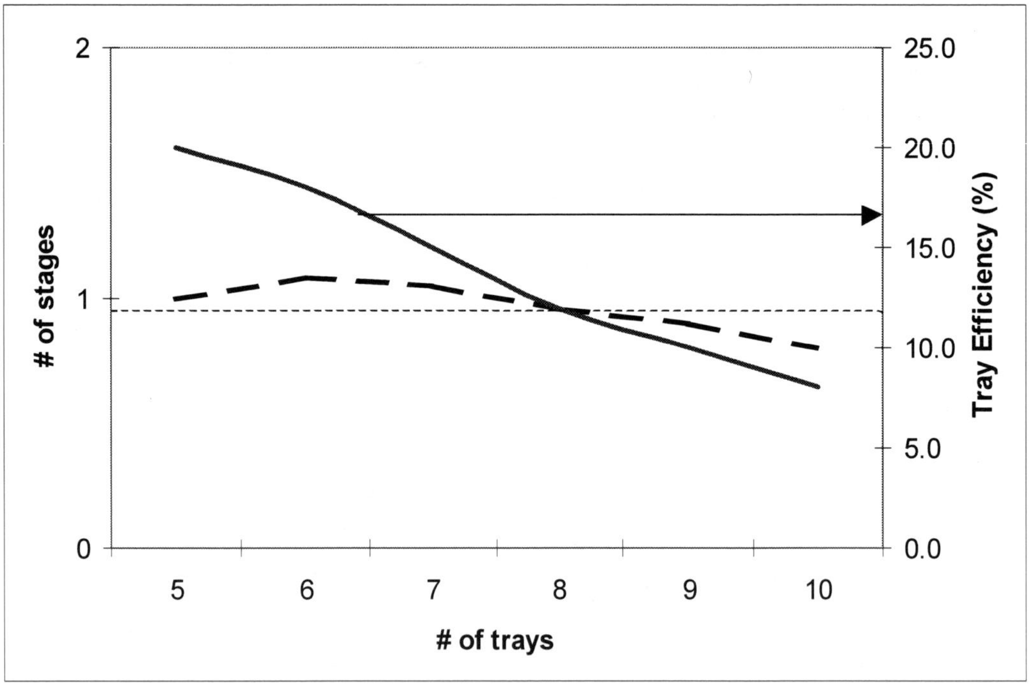

Figure 6 – An observed set of tray separation stages for different tray counts in operating vacuum units. The tray count appears to be insensitive to the actual number of stages, represented in a reduced tray efficiency.

Entrainment:

Entrainment in the flash zone (impacting HVGO quality) is a function of the geometry of the transfer line and performance of the flash zone (inlet nozzle and feed device). This paper focuses on sharing operating data to help with developing a representative simulation. Since there are no direct or indirect field measurements for entrainment (calculations can be performed with quench and slop wax flow measurements) entrainment numbers should include detailed hydrodynamic evaluation of the transfer line and flash zone geometry.

Conclusions:

Operating data has been gathered and shared to attempt to provide insight for specific design considerations for unconventional heavy oil/bitumen vacuum crude units, specifically data for cracked gas make and stripping section performance. The data provided in this paper can be used a validation point for designs with unconventional crudes.

References:

1. Habibullah, A., et al, Crude Vacuum Distillation: Wet, Dry or Damp, Institute of Chemical Engineers Conference, 1997, Maastricht, Netherlands.

2. Golden S., et al, Improved Flow Sheet Topology for Petroleum Refinery Crude Vacuum Distillation Simulation, 44th Annual CSChE Conference, 1994

3. Koch-Glitsch Canada Design Practices, Upgrader Vacuum Unit design, 2006, Confidential

4. Kaes, G. ,Refinery Process Modeling, A Practical Guide to Steady State Modeling of Petroleum Processes, Athens Printing Company, 2000, March

5. Golden S., et al, Feed Characterization and Deep Cut Vacuum Columns: Simulation and Design, AIChE 1994, Spring National Meeting

6. Remesat, D., Inside-Out Design Approach, Hydrocarbon Processing, 2006, August.

7. Golden, S., and Barletta, T., Designing Vacuum Units, Petroleum Technology Quarterly, 2006, Q2.

8. Lieberman, N. P., Process Design for Reliable Operations, Gulf Publishing Company, 1988, p.131.

9. Golden, S., Canadian Crude Processing Challenges, PTQ, 2008, Q1

10. Nelson, W. L., Petroleum Refinery Engineering, Mc-Graw-Hill Book Company,4th Edition

Depicting and Separating "Pro-Linking" Compounds to Enhance Upgrading Performance and Heavy Oils Processing

Pedro R. Pereira Almao
Department of Chemical and Petroleum Engineering, University of Calgary, Schulich School of Engineering, 2500 University Drive N.W, Calgary, AB T2N 1N4, Canada

Abstract
It is a common understanding that the asphaltene content of a crude oil alienates the performance of its upgrading and refining. Although not directly correlated in any case, they are held responsible for the limits of moderate thermal cracking; for the extent of coke production in more severe thermal cracking (delayed coking); for increased fouling during processing the vacuum bottoms of heavy and extra heavy oils; and are thought to affect the life and performance of hydroprocessing catalysts.

Own (and other's) experiences show that only a fraction of heavy molecules (most of them recognized as asphaltene but many within the range of resins), are really at the basis of this behavior. These molecules can be referred to as 'pro-linking' compounds based on the promotional effects they have on resins to grow into asphaltenes, and also on their interlinking, flocculating, depositing and promoting coking activity during vacuum residues processing. Identification and selective fast separation or transformation of these pro-linking compounds is the focus of this article. Their speciation and attempts of identification are suggested.

Understanding the specificities of pro-linking molecules will lead to smarter processing. Some process schemes tailored for dealing with these compounds are presented including their impact on both thermal cracking and hydroprocessing.

Introduction

Bitumen and extra heavy oils chemical upgrading is essentially about attempting to transform the most noxious substances there present, mostly of large molecular size, into harmless and if possible valuable products. Also parallel molecular weight reduction of numerous heavy compounds existing in large proportions within the residual fraction of these hydrocarbons takes place during upgrading. This parallel pattern would not necessarily generate significant constraints or limitations to the upgrading processes. However, this doesn't mean that some moieties of these latter compounds' transformation could not result making part of the undesirable patterns that the most noxious molecules open the way to.

Revising years of research by many groups allows finding a wide variety of models and mechanisms of reaction that support dozens of different heavy oil residue conversion processes. Most of them have in common the reference to molecules that are at the basis of undesirable patterns, molecules that constitute the essential bricks from which these reaction patterns develop. The single characteristic that makes these molecules so determinant is their capability of fast agglomerating or liking with other heavy molecules (or free radicals from) to quickly progress toward solid deposits or coke. In this paper these molecules will be referred to as Pro-Linking Molecules (PLM), thus avoiding the use of categories that are either too wide or imprecise for an adequate chemical understanding of their transformation.
Since the late 80's the author initiated research procuring a comprehensive approach and to develop processes and conditions, all targeting the specific transformation of PLM within different environments. Several parallel research activities were started to

progress on the understanding of the selective transformation of these substances; first under an industrial research environment and now within a more academic one. A revision of learning and experience on the way these essential target molecules are processed during upgrading using different schemes is exposed below, accompanied with a process upgrading proposition that is currently making part of further development.

Thermal cracking (TC)

TC deals with the PLM by partly transforming them into alkyl reduced large molecules with increased aromatic domains. The alkyl branches detached from the pattern molecules end into the distillate and gas fractions are produced. The PLM converted contain within difficult to extract heteroatoms, thiophenic and quinoleinic aromatic forms of sulphur and nitrogen compounds respectively.

Depending on thermal cracking severity, typically achieved by increasing residence time or temperature or both, this transformation pathway will transform these molecules either by leaving them precariously suspended in the liquid making part of a residual fuel or forming a solid carbonaceous material (coke). The former characterizes the mild visbreaking process; the latter is at the essence of coking processes.

During thermal cracking the PLM become the elemental bricks of intermolecular strong interactions that progressively form agglomerates and finally coke. Additives have been used to alter this process; some are used during visbreaking conditions either to accelerate or to initiate the formation of free radicals and to obtain similar conversion levels at slightly lower temperatures. Others are used to retard the PLM agglomeration process in an attempt to produce higher conversions (basically via dealkylation) from the larger molecules before they become agglomerated. This is the case of the aquaconversion process, designed to interrupt the normal pattern of PLM during TC, as shown in figure 1.

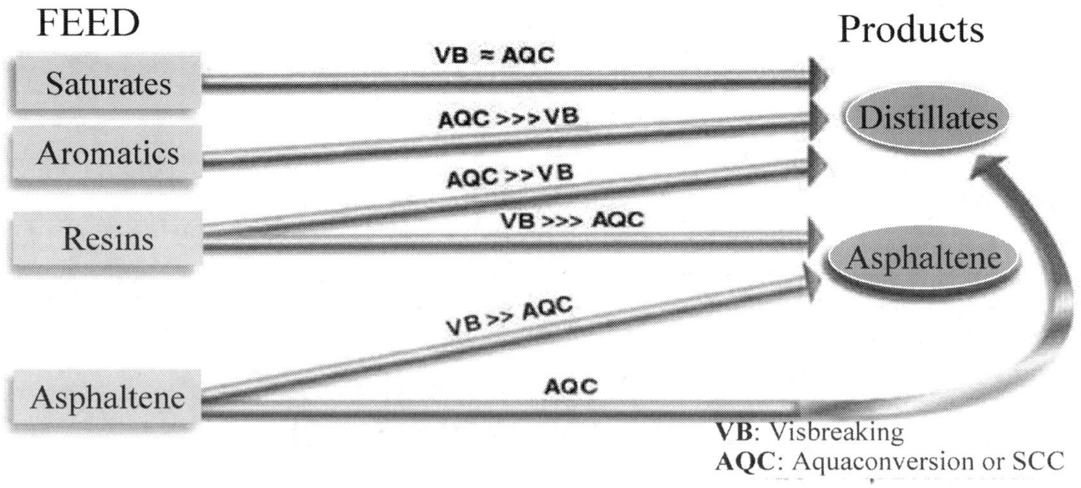

Figure 1. Reaction scheme proposed for the steam catalytic intervention of TC processing derived from kinetic studies with Vacuum Residues of naphtenic oils [7]

Catalytic Steam Cracking (Aquaconversion, AQC)
An interesting proposition that ended in a residual mild conversion process developed during the early 90s was to use a water dissociation catalytic pattern in parallel with the mild thermal cracking or visbreaking. Based on previous catalytic research on steam dealkylation of alkyl aromatics (BTX) and steam gasification of chars from coal and petroleum coke, the catalysts for these processes [3-6] were converted into ultradispersed formulations and used to test the visbreaking process in their presence with the addition of a higher than usual (for VB conditions) proportions of steam. The essential of the "aquaconversion" proposition targeted the interference of the classical VB pattern in which the PLM agglomerate and serve as platforms for growing larger clusters that become flocculates; if further heated these evolve into coke.
By running parallel catalytic steam dissociation into hydrogen, hydroxyl and oxygen radicals, the chances of quickly saturating the stable PLM free radicals forming from thermal cracking could increase. It is well known that water dissociation can't be used at moderate temperatures to hydrogenate aromatics and olefins, but instead it can be used to saturate aromatic free radicals and the catalytic steam dealkylation of alkyl-aromatics. As confirmed from thermodynamic spontaneity calculations -free energy calculations indicate dealkylation of alkyl-aromatics is complete in the presence of steam with preservation of the aromatic nuclei at temperatures around 440ºC-. Figure 2 shows the progression of an average hypothetical asphaltene structure as deduced from collating analytical information including ^{1}H and ^{13}C NMR.

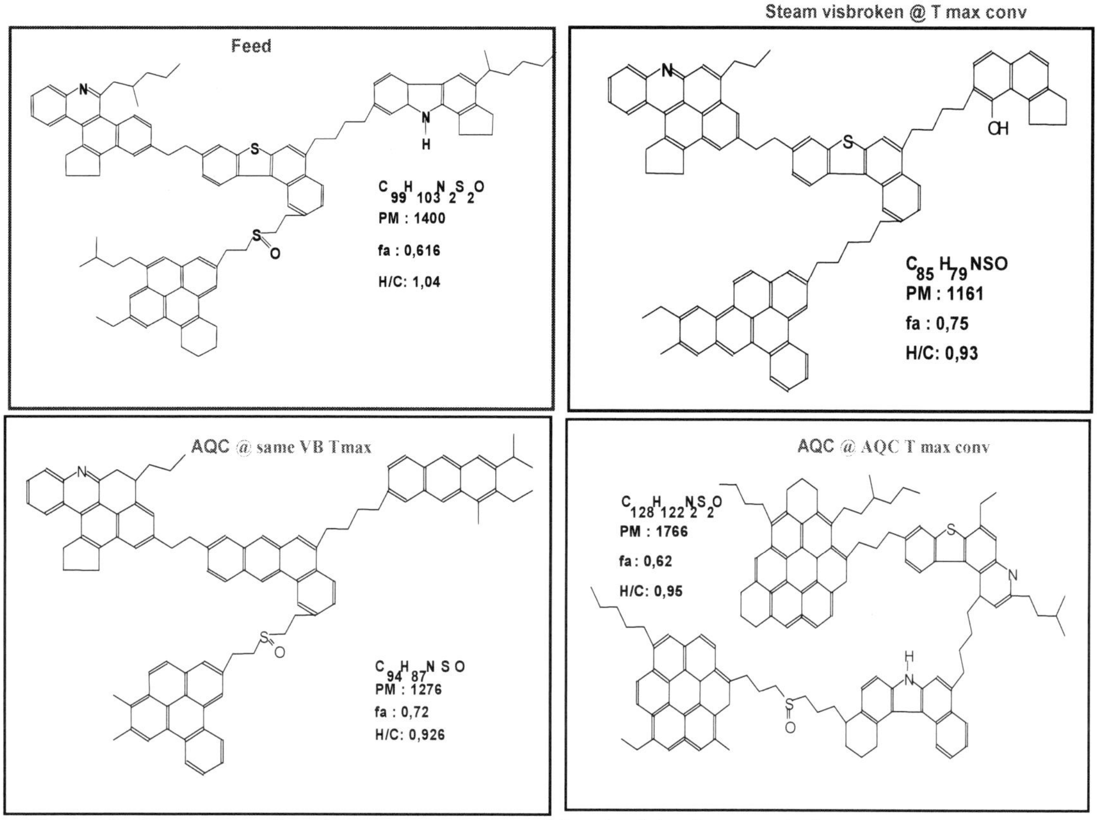

Figure 2. Average molecular representation of feed, visbroken asphaltene, and aqua-converted asphaltene at two different severities [8]

An example of reactivity tests performed with toluene steam dealkylation using a catalyst combination of a transition metal with alkali/ alkali-earth metals supported on alumina is shown in Table 1. Toluene was steam dealkylated following the reaction:

$$C_6H_5CH_3 + H_2O \rightarrow C_6H_6 + CO_2 + 3H_2$$

The steam dealkylation of toluene produces benzene, hydrogen, carbon oxides, and methane as main products, and traces of xylenes.

Table 1: Conversion and selectivity of Toluene with different catalysts, 440 °C, 1 atm. WHSV 1h^{-1}

Alumina supported catalysts	Cat A	Cat B	Cat C
Transition metal content/Alkali-earth alkali metal content, (%)	0.6/0	0.6/3	0.3/3
Toluene conversion, (%)	60	64	41
Benzene selectivity, (%)	70	70	80
Yield per TM unit mass, ($X \bullet S_B$/%TM)	70	75	109

[9]

It can be observed that the conversion of toluene steam dealkylation as well as the selectivity to the non degraded product (benzene) is enhanced by the addition of well known water dissociating catalytic phases [1-2]. It is to be observed that the temperature range of this reaction is similar to the typical temperature range used for visbreaking [3].

Table 2 shows the results obtained from processing Athabasca vacuum gas oil under thermal cracking and under steam thermo-catalytic cracking with an ultradispersed version of the catalyst presented in Table 1 for the steam dealkylation of toluene. Similar enhancements from the steam thermo-catalytic processing with respect to thermal processing were obtained in the past by processing vacuum residues and vacuum gas oil from Venezuelan extra heavy oils [11-14].

Table 2: Compared thermal and catalytic reactivity of Athabasca VGO in the presence of steam [10]

Athabasca VGO processing	**Steam Tests (5% steam/VGO)**		**UD Tests (5% steam/VGO)**			
			Cat A	Cat A	Cat B	Cat B
Temperature, C	430	430	430	430	430	440
VHSV, hr^{-1}	2	1	2	1	1	1
Conv 350+C	33	61.5	35	51	50.1	55
Yield 350+C to distillates	25	27	23	32	33	43
Relative µCarbon	1.0	4.6	1.0	1.9	2.2	2.3
Conv 350+C/Yield to distillates	1.3	2.3	1.5	1.6	1.5	1.3

The fact that initially the chemistry of this "Aquaconversion" process was at large scale evaluated and demonstrated for vacuum residues and that at the time asphaltene were widely attributed responsible for the undesirable molecular behaviour that restricts thermal cracking from higher conversion levels, made processing of vacuum gas oil (VGO) streams with steam and catalysts a conceptual challenge. VGO from bitumen and extra heavy oils, typically a 350-550°C cut, does not contain asphaltene. VGO is a vacuum distillate fraction rich in the same hydrocarbon chemical functionalities and of heteroatomic content similar to that of vacuum residues. Besides, the molecules in the VGO cut are smaller and expectably slightly more refractory to the same cracking pattern found for vacuum residuals. Nevertheless, Table 2 suggests that the catalyst action helps to reduce the relevance of the trend that forms solids, in this case reflected by the relative micro carbon values. It also shows a reduction of the tendency to produce gas and coke via addition of the steam dissociating catalyst as can be observed by the increased yield of distillates with increase in residence time (reduction of space velocity) and by the ratio conversion of VGO to liquid yield.

Catalytic Hydroprocessing

Another way of dealing with these large molecules transformation (by all means classical), is to add hydrogen early in the process of thermal cracking. High pressures of hydrogen, effective catalysts and relatively long residence times are needed to hydrogenate, hydro-treat and to produce massive hydrogenolysis of the large, highly aromatic molecules present in the residuals of bitumen and extra-heavy oils, especially including the so-called PLM.

Hydroprocessing of heavy oils residuals needs to produce the highest conversion possible of the feed. This is the best way of economically justifying the high investment required for this type of processing. But, as indicated, the PLM react fast to produce early agglomerates, they can also become active surfactant mid way to their final hydroprocessing and generate hard to break foam inside the hydroprocessing reactors. They can also transit through a high sticking tendency inside the reactor until they either form coke deposits or evolve into hydrogenated and crackable material. Due to these constraints, catalysts for hydroprocessing of residuals attempting high conversions must care about incorporating highly hydrogenating sulphur resistant catalytic functions, must provide incipient coke scavenging capabilities, and must have foam breaking solid characteristics. All this learning about PLM's chemical transitions led to developing a catalyst for hydroprocessing based in a combination of highly active ultra-dispersed transition metal sulphides close to nanoparticles range, combined with a larger particle size, fit-for-purpose carbonaceous macroporous solid, with affinity for the incipient PLM agglomerates. This large diameter solid fraction needed to be two orders of magnitude larger than the ultradispersed catalyst fraction. It was designed to incorporate a solid hold-up inside the residue hydroprocessing reactor capable of minimizing the formation of foam and efficiently evacuating from the reactor the potential solid build ups originated by PLM. Typical reactivity data obtained with the PLM-dealing approach to develop the catalyst formulation for a vacuum residue hydroprocessing is shown in Figure 3.

Converted PLM adsorption

A complementary and potentially inexpensive way of dealing with the presence of converted PLM in the heavy fractions being processed either by thermal or catalytic steam/mild hydroprocessing, has been recently proposed by the author [Energy and Fuels, March 2008]. It consists of separating the PLM from the reaction stream just before initiating the massive agglomeration path. During thermal processing (even in the presence of catalysts and steam or hydrogen) PLM loose their alkyl branches and their stabilizing media becomes less peptizing which leads to a higher tendency to interact with polar 'substrates'. These substrates can be of similar nature than PLM or can be other free radicals and 'surfaces' from large molecules in the surroundings of the PLM. They could be surfaces such as reactor walls and eventually catalysts or adsorbents.

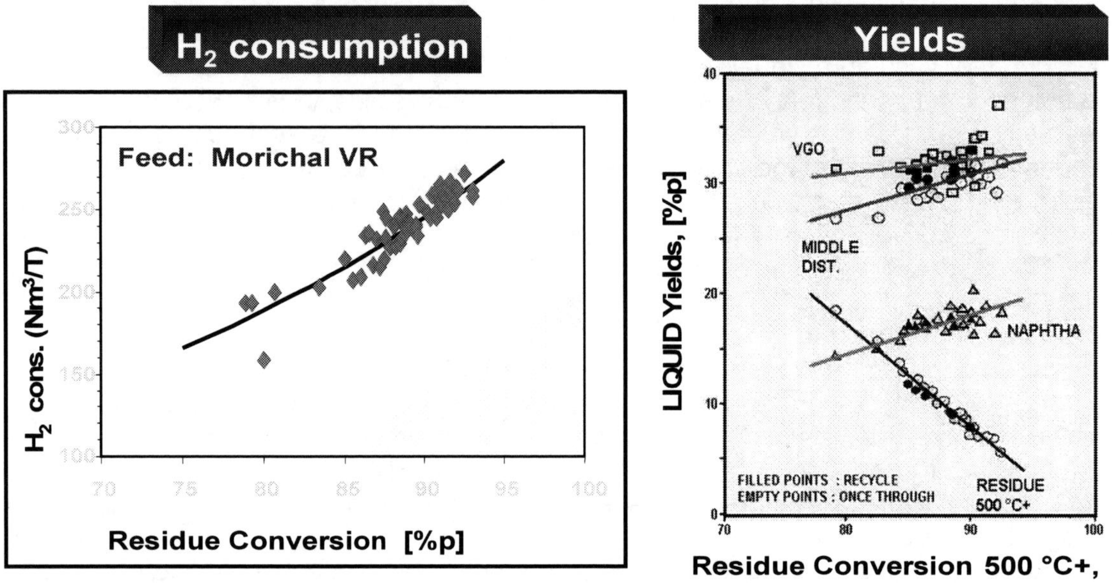

Figure 3. Typical hydroprocessing behaviour of a vacuum residue from Venezuelan extra-heavy oil using a bimodal catalytic formulation consisting of ultradispersed transition metal sulphides and a coarse carbonaceous macroporous solid fraction (P. Pereira et al, Heavy Oils International Seminars, Petrobras-Cenpes, 2002)

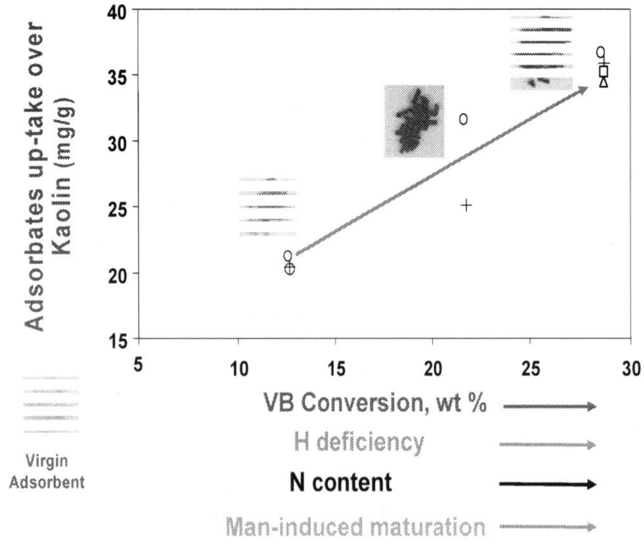

Micro Carbon Values

Description	Microcarbon determined (%)
Whole Bitumen	12.12±0.68
Vacuum residue	19.57±0.38
8.6 VB	21.44±0.30
13.3 VB	23.57±0.45
23.3 VB	27.61±0.61
28.5 VB	31.45±0.44
28.5 VB after adsorption on macroporous Kaolin	29.91±0.21

Figure 4, Table 3: Evolution of the adsorption of PLM with visbreaking and microcarbon of the visbroken samples after adsorption at different TC conversion levels (%conversion VB) respectively [.

At this stage, in the verge of instability and precipitation PLM can be selectively adsorbed on adequately designed surfaces, thus selectively extracted from the reaction stream. Deprived from the most instable or active PLM or reduced in their content, the VR residuals can be further converted beyond the conventional limits of thermal cracking. Figure 4 presents the adsorption of converted PLM close to their instability region vs. TC conversion of the 500C+ fraction. The PLM increase their stickiness to the adsorbent surface with progress in conversion, yielding after adsorption an enhanced material that with adequate further processing and iterated adsorption could yield much higher liquid product yields than conventional TC. Table 3 also shows how much the elimination of a relatively minor proportion of retained molecules contributes to a significant reduction of microcarbon, confirming that the molecules retained have a major impact on the formation of carbonaceous precipitates and solids, which justifies their denomination of Pro-Linking molecules.

The chemical nature of PLM close to their precipitation onset has been qualitatively partially identified via FTIR. These molecules show many chemical similarities to the ones detected on the material that has been found present on organo-clay particles present in the exploitation of bitumen by mining, see Figure 5.

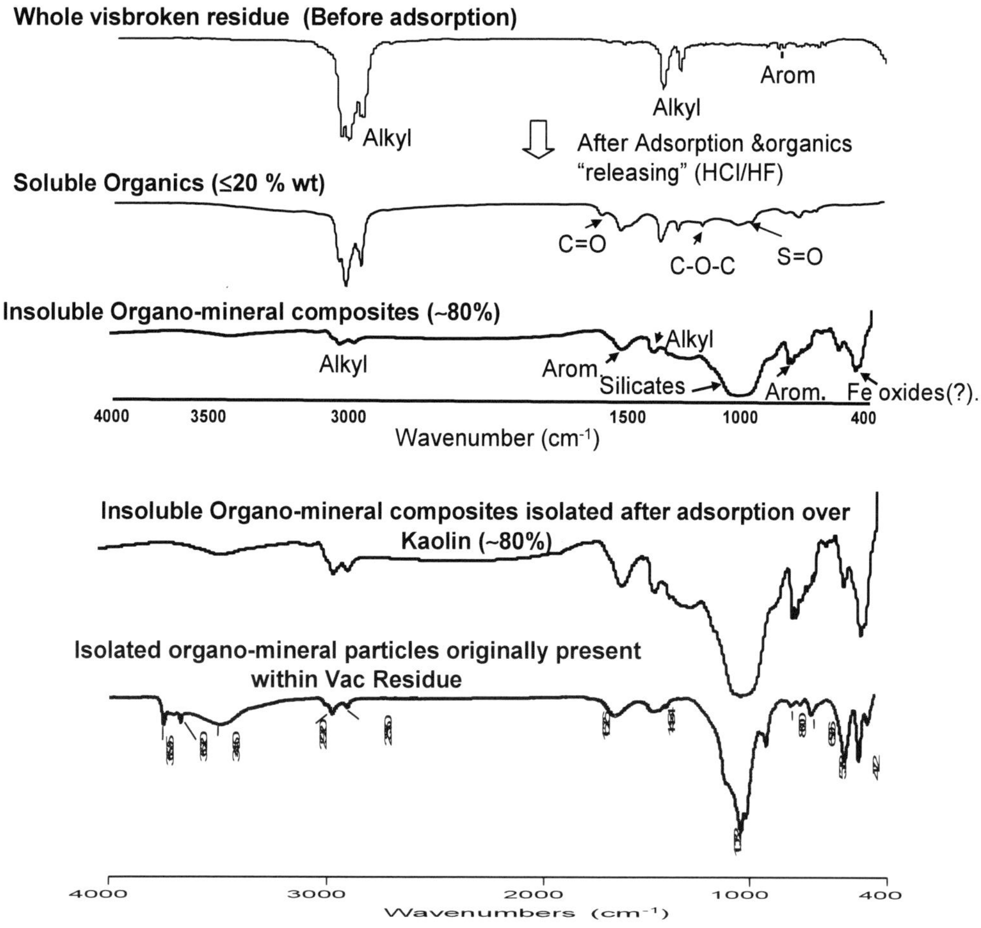

Figure 5. FTIR analysis revealing similar chemical nature of the polar compounds presents at the interlayer clays-hydrocarbon and the ones adsorbed on kaolin. Results by Lante Carbognani, Alberta Ingenuity Centre for In Situ Energy (AICISE). University of Calgary, 2007

Conclusions

Upgrading heavy hydrocarbon fractions essentially is regulated by and must be designed for coping with a relatively small number of molecules with particular chemical Pro-Linking functionalities. It was briefly illustrated how the recognition of the PLM has allowed understanding the real limits of thermal cracking of asphaltene containing and non-containing heavy fractions and to insert the catalysis of water dissociation known for partial reforming and steam gasification to modify the condensation trends that end producing instable molecules and coke. The concept of PLM also helped to develop a more suitable catalytic formula for high conversion hydroprocessing of extra heavy oils. Finally were introduced the ideas and results on selective adsorption of PLM, whose further attention and development may result in significant impact for heavy oil fractions processing.

References

1. Thiel P.A. and Madey T.E., *The dissociation of water over solid surfaces: Fundamental Aspects Surf. Sci. Rep* 7 (1987), p. 211.
2. Henderson M., *The dissociation of water over solid surfaces: Fundamental Aspects Revisited. Surf. Sci. Rep* 46, Issues 1-8, May 2002, Pages 1-308
3. Duprez, D., *Selective steam reforming of aromatic compounds on metal catalysts.* Applied Catalysis A: General, 1992. 82(2): p. 111-157.
4. Carrazza, J.T., W.T.; Heinemann, H. and Somorjai, G.A, *Gasification of graphite with steam below 900K catalyzed by a mixture of potassium hydroxide and transition metal oxide.* Journal of Catalysis, 1985(96): p. 234-241.
5. Pereira, P.Somorjai, G.;Heinemann, H., *Catalytic Steam Gasification of Coals.* Energy & Fuels, 1992(6): p. 407-410.
6. Pereira, P. Csentzis., R.;Somorjai,G.A., *Steam gasification of graphite and chars at temperatures less than 1000K over Potassium Calcium oxide catalysts.* Journal of Catalysis, 1989(123): p. 463-476.
7. Rosales N., *Chemical Engineering Undergraduate thesis*, Universidad de Carabobo, Valencia, Venezuela, 1999
8. Higuerey I. *PhD Thesis*, UCV, Caracas, Venezuela, 2000
9. Eldood A., *MSc Thesis*, Schulich School of engineering, University of Calgary, Canada, 2007
10. Trujillo G., *MSc Thesis* Schulich School of Engineering, University of Calgary, Calgary, Canada, 2008
11. Marzin R; McGrath M; Thompson G; Pereira P; Feintuch H; Houde E *A New Approach To Residue Processing...The Aquaconversion(TM) Process* AIChE 1999 Spring National Meeting (Houston 3/14-18/99) Preprint N.35e8P
12. Pereira P; Marzin R; Zacarias L; Lopez Trosell I; Hernndez F; Cordova J; Szeoke J; Flores C; Duque J; Solari R B *Aquaconversion(TM): A New Option For Residue Conversion And Heavy Oil Upgrading* Vision Tecnologica (ISSN 1315-0855) V6 N.1 5-14 (1998)
13. Pereira P. et al. Patent No US 6030522
14. Pereira, P. et al. Patent No US 5885441
15. Pereira P. *Pdvsa Strategic Technologies for Upgrading Venezuelan Heavy Crude Oils.* 1st CENPES Seminar on Heavy Crude Oils Upgrading Technologies. Rio de Janeiro, Brasil, October 2002
16. Gonzalez M., *MSc thesis*, Schulich School of Eng., University of Calgary, 2006
17. Sosa C., *MSc thesis*, Schulich School of Engineering, University of Calgary, 2007
18. Azfar Hassan, *unpublished results.* 2007
19. Lante Carbognani, *unpublished results.* 2007

Ionization reagents for enhanced speciation of petroleum compound classes by Electrospray Ionization FT-ICR Mass Spectrometry (ESI FT-ICR MS)

Priyanka Juyal, Ryan P. Rodgers, and Alan G. Marshall

Ion Cyclotron Resonance Program, National High Magnetic Field Laboratory, Florida State University, Tallahassee, FL, USA

Introduction

With global increase in consumption and thus depletion of the easily available resources of crude oil, the petroleum industry has moved toward production of extra heavy oils from "difficult" areas (e.g. deepwater production) and also to tap unconventional oil resources such as the "oil sands" (bitumen). However, the production, transportation and refining of these resources presents many problems and challenges for the oil industry.[1-3] The requisite advancement in technical knowledge for efficient generation of energy and for tailoring feasible solutions to overcome the problems associated with refining and utilization of these resources is imminent. From a health, safety and environment vantage, air pollution caused by emissions of SO_x and NO_x and soot / particulate matter from the combustion of transportation fuels is a matter of global concern. Intensive R&D efforts to achieve better catalysts, processes and solutions for rigorous refining operations are paramount in order to limit the amount of sulfur, nitrogen and aromatics to produce ultra-clean transportation fuels that are required to meet the current, stringent fuel regulations. The chemical structure of different heteroatom (S, N, and O) containing molecules determines their behavior in production and upgrading.[4-6] Therefore, efficient and economic exploitation of these resources hinges on the detailed understanding of the chemistry and structure of the constituents of this complex natural resource.

Advances in FT-ICR MS have made it possible to achieve ultrahigh resolving power and unprecedented mass accuracy required to resolve and provide detailed compositional information for complex mass spectra obtained from heavy oils and bitumen. Electrospray ionization Fourier transform ion cyclotron resonance mass spectrometry (ESI FT-ICR MS) efficiently provides critical information on the polar compound inventories of crude oils.[7-10] Ionization in this technique is achieved by generation of quasimolecular charged species of the type $(M+H)^+$ or $(M-H)^-$ by protonation or deprotonation effected by an acidic or basic reagent. Crude oil compound classes that are not sufficiently acidic or basic to yield simple protonation/deprotonation with dilute weak acids and bases are not efficiently ionized by conventional electrospray ionization. However, reagents/solvent systems that can react with nitrogen, sulfur, and oxygen in near-neutral species and hydrocarbons without major alteration of the molecular structure will enhance electrospray ion yield. Here, we report the use of modified solvent systems and reagents for the speciation of basic, acidic, and *nonpolar* crude oil molecules by positive and negative ion electrospray FT-ICR MS.

The ionization reagents employed in our study require brief reaction periods (< 60 s) and hold promise in highlighting both polar and nonpolar molecules for enhanced electrospray ionization. For positive ESI we use silver triflate as a complexing reagent that has a strong ability to bind to a source of electrons (n,π,σ) and offers a rapid method that highlights the basic crude oil matrix by generation of positively charged complexes. For negative ion ESI we employ tetramethylammonium hydroxide (TMAH) of pronounced basic strength that successfully highlights the near-neutral molecules as negative ions.

Experimental

All reagents and solvents were of highest purity grade and were procured from Fisher-Scientific and Sigma-Aldrich.

Positive ion ESI FT-ICR MS: One part of crude oil (1 mg/mL in 1:1 (v/v) toluene: methanol) was mixed with 3 parts of silver triflate and vortexed just before injection into the ESI source. The result was compared to that for conventional ESI with formic acid. Because the complexes form with electronegative atoms and the π-electron system of non-polar aromatic hydrocarbons, complex crude oils and bitumen offer many reactive sites for complexation.

Negative ion ESI FT-ICR MS: All crude oil samples were dissolved in a 2:1 (v/v) solution of toluene/methanol at 1 mg/mL concentration. TMAH (25% in methanol) and TMAH (25% in water) and ammonium hydroxide (28% of NH_3 in water) were spiked at a concentration of 10 μL. The reagents and sample solutions were vortexed for a few seconds before spraying.

Analysis: Samples were analyzed by ESI FT-ICR mass spectrometry with a custom-built 9.4T system equipped with a PREDATOR data station. Calibration of the instrument was achieved with Agilent (Palo Alto, CA) HP mix. Singly charged ions with a relative abundance greater than 6 times the standard deviation of the baseline noise were Kendrick-sorted and imported into Microsoft Excel for identification with a formula calculator as per previously reported procedures. The mass tolerance was set to ±1 ppm and Kendrick mass defect was used to identify homologous series and peak assignments.

Results and Discussion

Ag^+ESI FT-ICR MS: Silver adduction involves interaction of silver cation (Ag^+) with sources of n, π or σ electrons by cation-π interaction, host-guest, or donor-acceptor chemistry. Roussis and Proulx examined silver nitrate adducts with heavy aromatic petroleum fractions and non-boiling petroleum fractions to speciate aromatic hydrocarbons by ESI-MS.[11] Silver triflate is used in this study. Triflate anion or trifluoromethaneulfonate anion ($CF_3SO_3^-$) is resonance stabilized due to the dispersal of negative charge over three oxygens and the sulfur atom. An additional stabilization is achieved by the strong electron-withdrawing effect of the trifluoromethyl group, causing the triflate anion to be a very fast (one of the fastest) leaving group. The exemplary stability and non-nucleophilicity of the triflate anion enhances the electron-affinity of the silver cation and thus favors complex formation.

Figure 1 shows the isotopic distribution for ^{106}Ag and ^{108}Ag for each of the heteroatom classes S_1, S_2, HC and OS based on (+) ESI for several oils. The distinctive isotope distribution follows the natural relative abundance of the silver isotopes (^{106}Ag, 51.84% and ^{108}Ag, 48.16%) and allows for the verification of elemental composition assignments. The excellent nucleophilicity of sulfur, which is much greater than that for oxygen or nitrogen, allows for adduction with electrophilic Ag^+; thus the S classes are highly abundant. As expected, the nonpolar S classes are absent in conventional positive ion ESI mass spectra.

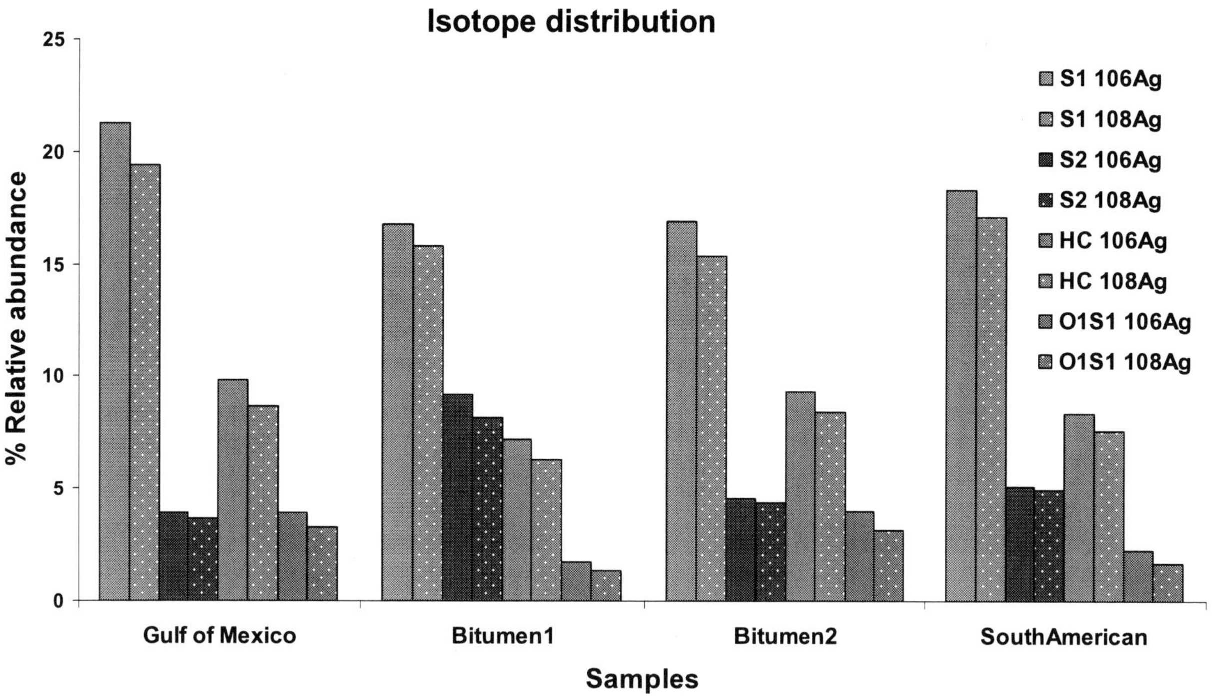

Figure1. Heteroatom class distribution ((+) ESI) for selected heteroatom classes for several oil and bitumen samples after silver adduction by (+) ESI. The Ag isotope distribution is unmistakable.

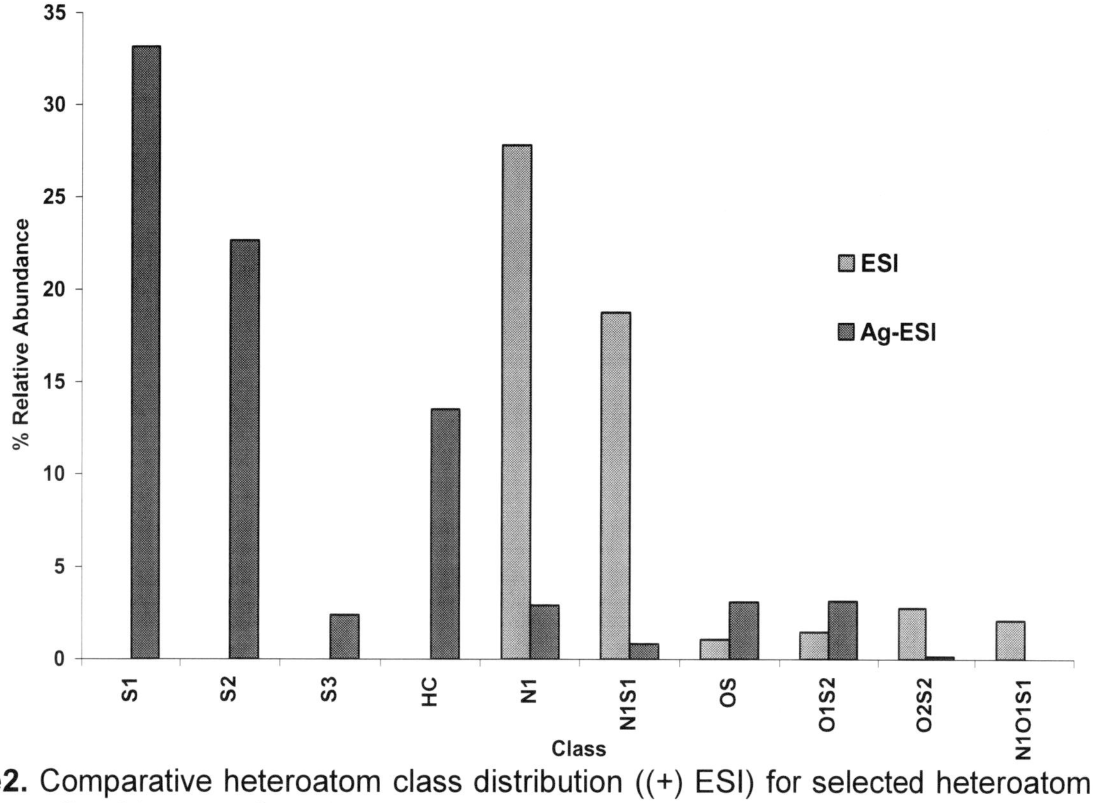

Figure2. Comparative heteroatom class distribution ((+) ESI) for selected heteroatom classes for a Canadian bitumen after silver adduction and conventional ESI.

Sulfur-containing classes included S_1, S_2, S_3, OS, and NS. The S_1, S_2, and S_3 classes are not detected by conventional positive-ion ESI mass spectra of oil samples spiked with formic acid (see **Figures 2 and 3a**). Further, the efficiency of Ag complexation in highlighting non-polar compound classes is evidenced from the speciation of hydrocarbon classes for all of the oil samples (see, e.g., **Figure 3b**).

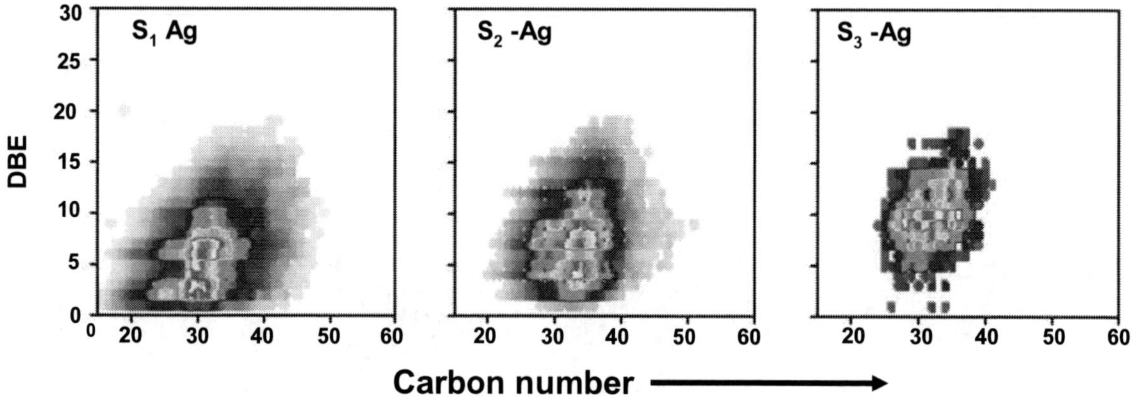

Figure 3(a). Isoabundance-contoured DBE (double bond equivalents = number of rings plus double bonds to carbon) versus carbon number images for S_1, S_2 and S_3 classes for a Canadian bitumen from Ag (+) ESI FT-ICR MS. Conventional ESI produced no spectral signals for these classes. Relative ion abundance within each class is color-coded on the z-axis.

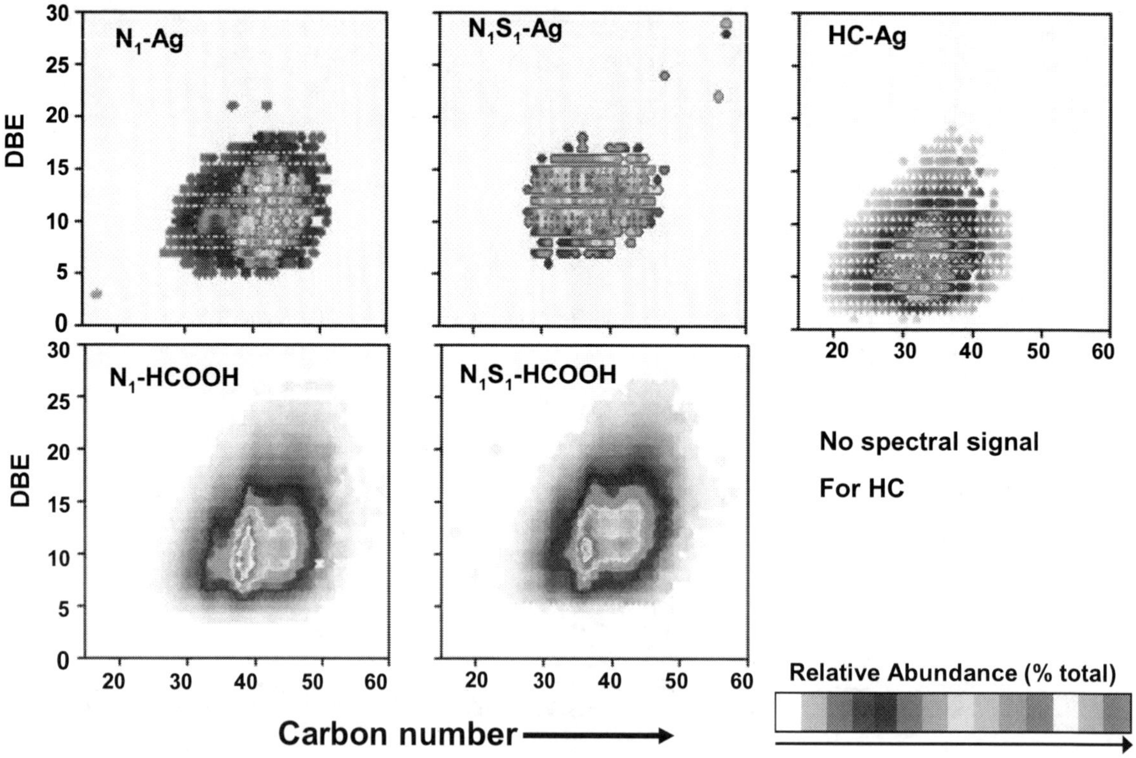

Figure 3(b) As in Fig. 3(a), but for N_1, N_1S_1 and HC classes for a Canadian bitumen based on Ag (+) ESI FT-ICR MS (top) and conventional (+) ESI FT-CR MS (bottom). Conventional ESI produced no spectral signal for the hydrocarbon class.

Other ionization methods for nonvolatile high-boiling petroleum materials, such as atmospheric pressure photoionization (APPI), field desorption (FD), matrix-assisted laser desorption/ionization (MALDI), and chemical ionization (CI), either have low throughput or involve thermal processes that may limit the observation of (or degrade) substrate molecules. Mix-and-spray Ag complexation avoids tedious separations and lengthy reaction periods without the need for thermal desorption. Hence, silver adduction overcomes the failure of conventional ESI-MS for rapid speciation of neutral heterocycles, specifically the sulfur-heterocycles, and hydrocarbons in complex crude oils without the thermal desorption limitations associated with APPI, APLI and CI.

Negative-ion ESI FT-ICR MS: Ammonium hydroxide used in conventional negative ion ESI provides low concentration hydroxide ions in solution. Thus, detection of weaker acidic species is limited by the weak basicity of ammonium hydroxide. Due to weaker ion pairing (Q^+X^-) than is possible with metal hydroxides (Na^+X^-) and excellent solubility in most organic solvents, tetramethylammonium hydroxide exhibits higher basicity and the ability to deprotonate molecules of varying acid strengths. Molecules other than carboxylic acids may be only weakly or moderately acidic with high pK_a values ranging from 9 to 34.

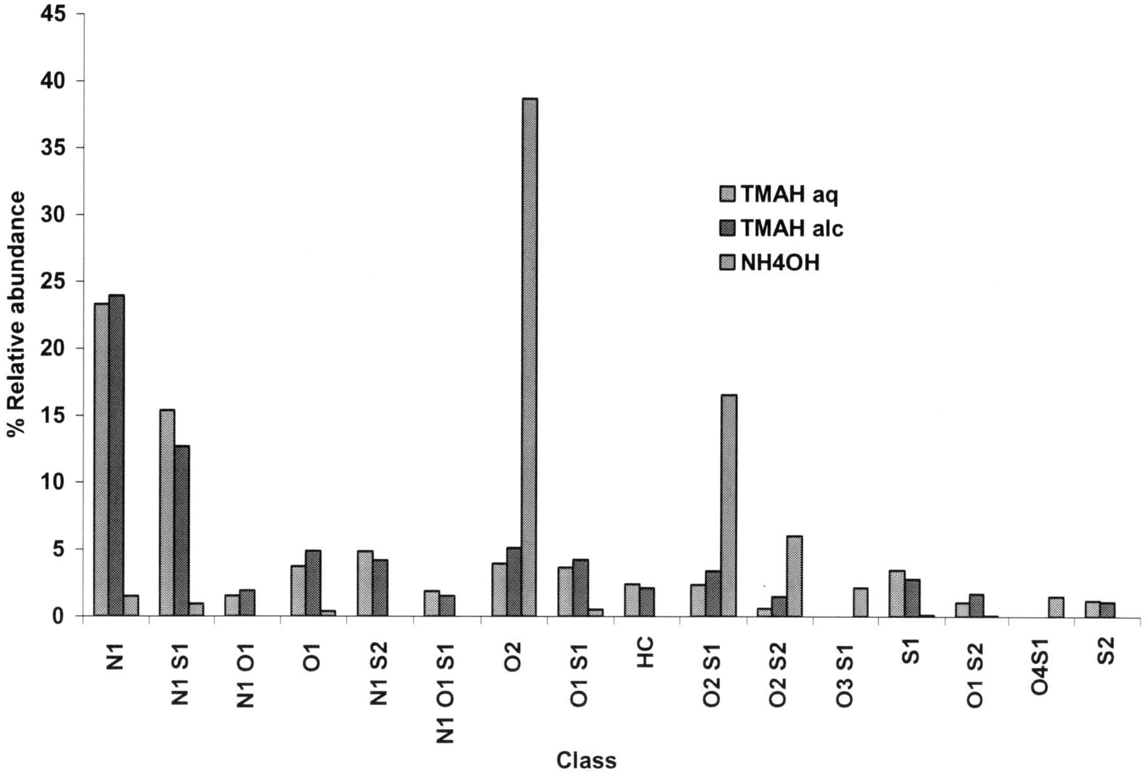

Figure 4. Comparative heteroatom class distribution for a Canadian bitumen. The most abundant classes for three different solvent conditions are represented. Although O_2 represents the most abundant class for ammonium hydroxide solution, N_1 is the most abundant for TMAH solution.

In all of the oil samples, O_2 is the most abundant class produced by ionization with ammonium hydroxide (see **Figure 4**). Carboxylic acids are more acidic (lower pK_a's) and readily deprotonate compared to the pyrrolic nitrogen species (high pK_a's). Hence, mass spectra generated with ammonium hydroxide display disproportionately large carboxylic acid species relative to other neutral and near-neutral species in crude oil. Deprotonation efficiency of neutral compounds is further diminished in the presence of carboxylic acids (matrix effect). Hence it is not easy to relate the observed ion relative abundances to those of neutral precursors in the original sample. So, even though O_2 class is most abundant for most crude oil and bitumen samples spiked with ammonium hydroxide in negative-ion ESI mass spectra, it may not be the most abundant in the source oil matrix.

Significantly, the N_1 class is most abundant in the TMAH spiked samples (**Figure 4**). Thus, the ionization of N_1 class is increased many-fold compared to that with ammonium hydroxide (**Figure 5**). Specifically, the classes with two or more oxygens (carboxylic acids) and sulfur oxy-acids show higher relative abundance with ammonium hydroxide whereas classes with one oxygen and nitrogen class, possibly phenols and pyrroles, exhibit higher relative abundance with TMAH. The effect can be ascribed to the ability of TMAH to extract a proton from relatively weak acids, so that the molecular classes that were insufficiently ionized with ammonium hydroxide (and hence exhibit low ion abundance) are now efficiently deprotonated.

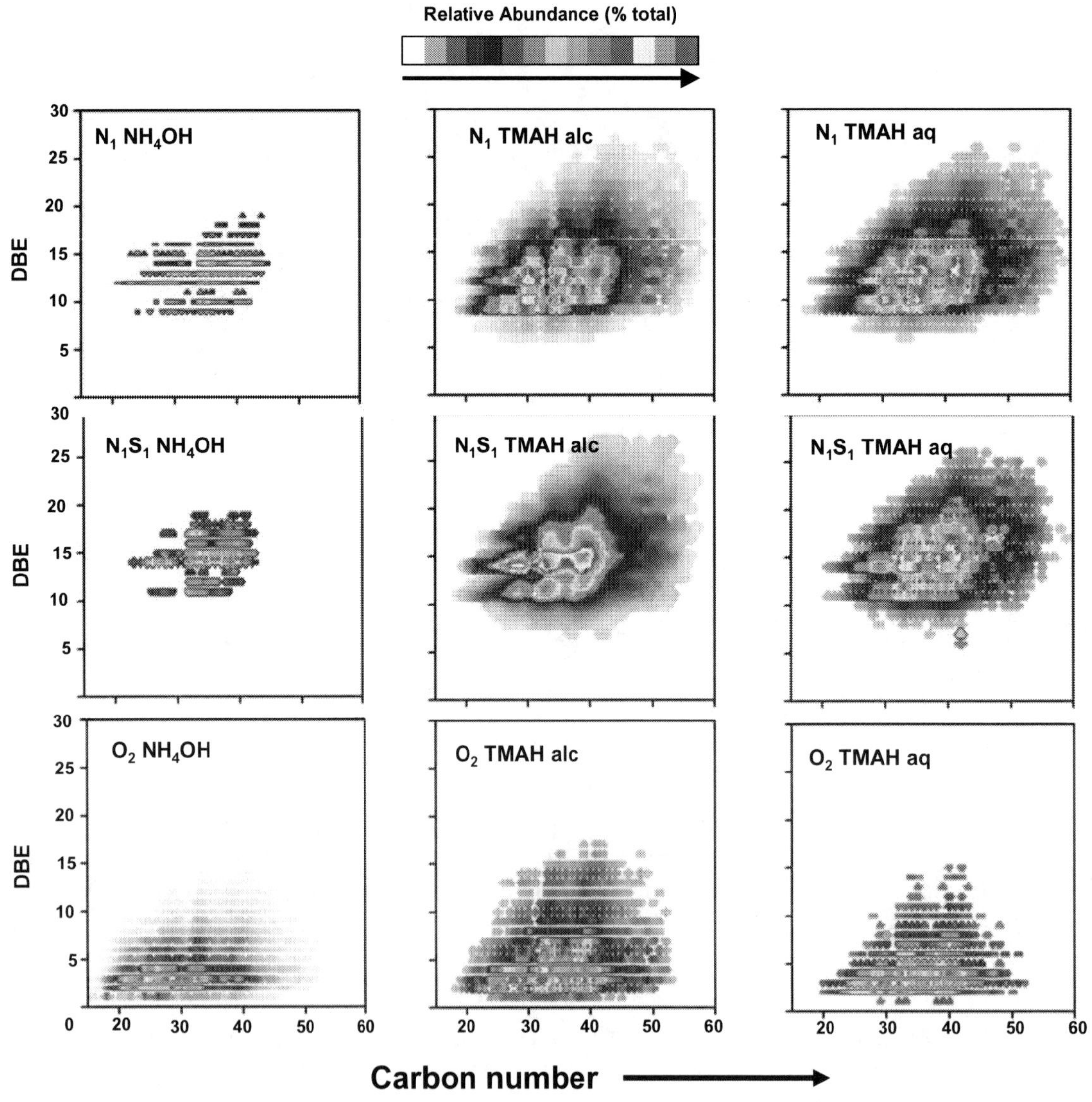

Figure 5. Isoabundance-contoured DBE versus carbon number images for heteroatom species for a Canadian bitumen analyzed with ammonium hydroxide (left), TMAH in organic solvent (middle) and TMAH aqueous (right).

ESI with TMAH also generates HC class ions that have never been observed as negative ions with ammonium hydroxide (**Figure 5**). It is well documented that aromatic hydrocarbons possessing benzylic hydrogen are weak acids and reported to deprotonate in the presence of strong bases. [12, 13] Sulfur classes S_1 and S_2, possibly thiophenols, are also detected with TMAH but not with ammonium hydroxide. For the Canadian bitumen (**Figure 6**), TMAH highlights S_1 and S_2 classes over a DBE and carbon number of approximately 10-25 and 20-50.

Thus we demonstrate that a slight modification in the basic strength of the solvent system for negative-ion ESI FT-ICR MS is extremely effective in generating comprehensive compositional profiles of crude oil acids over a wide DBE and carbon number range. TMAH allows for the efficient ionization and subsequent detection of stable species in much lower concentration in the sample matrix, and should thus more accurately represent the relative abundances of different acidic compound classes, in contrast to ammonium hydroxide, which over-emphasizes the carboxylic acids and sulfur oxy-acids.

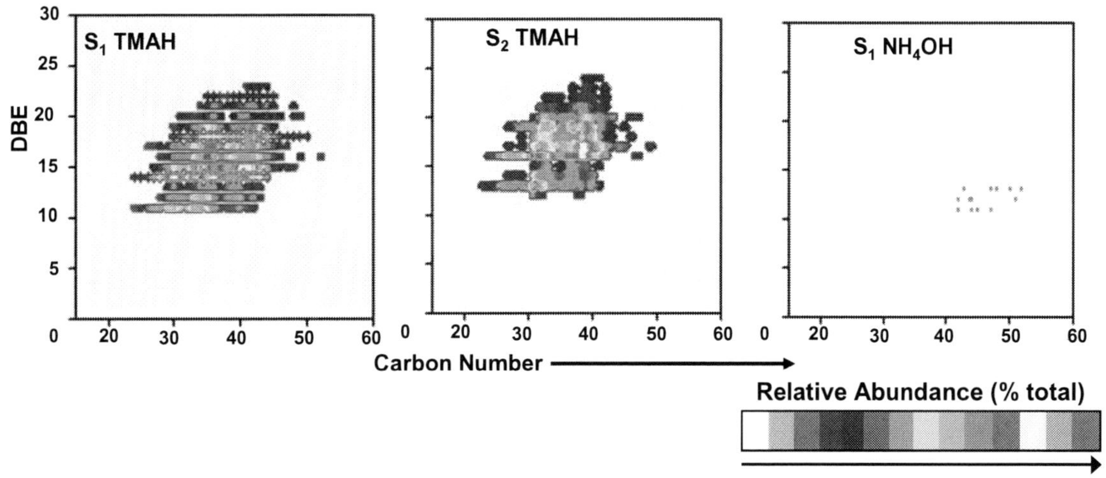

Figure 6: S_1 and S_2 class images from the TMAH generated mass spectrum (left and center) for a Canadian bitumen. S_1 for ammonium hydroxide is negligible (right) while S_2 is totally absent (not shown).

Acknowledgments

We acknowledge the support provided by the National Science Foundation (DMR 00-84173), Florida State University, and the National High Magnetic Field Laboratory in Tallahassee, FL.

References:

1. Algelt, K. H., and Boduszynski, M. M., *Composition and Analysis of Heavy Petroleum Fractions. In: Chem. Ind.* (Dekker), **1994**, *v* 54.
2. Speight, J. G., "The chemical and physical structure of petroleum: effects on recovery operations." *J. Pet. Sci. Eng.* **1999,** *22(1-3),* 3-15.

3. Joshi, N. B., et al. "Asphaltene Precipitation from Live Crude Oil." *Energy Fuels* **2001**, *15(4)*, 979-986.
4. Sullivan, R. F., et al. "Molecular transformations in hydrotreating and hydrocracking." *Energy Fuels* **1989,** *3*(5), 603-12.
5. Lee, K.-H., et al. "Catalytic cracking of vacuum gas oil on the dealuminated mordenites." *J. Catal.* **1998**, *178(1)*, 328-337.
6. Landau, M. V., et al. "Hydrocracking of Heavy Vacuum Gas Oil with a Pt/H-beta-Al2O3 Catalyst: Effect of Zeolite Crystal Size in the Nanoscale Range." *Ind. Eng. Chem. Res.* **2003**, *42(12)*, 2773-2782.
7. Zhan, D. and J. B. Fenn. "Electrospray mass spectrometry of fossil fuels." *Int. J. Mass Spectrom.* **2000, *194***(*2/3)*, 197-208.
8. Marshall, A. G. and R. P. Rodgers. "Petroleomics: The Next Grand Challenge for Chemical Analysis." *Acc. Chem. Res.* **2004**, *37(1)*, 53-59.
9. Marshall, A. G., et al. "Compositional characterization of petroleum by ultrahigh-resolution Fourier transform ion cyclotron resonance mass spectrometry with multiple ionization sources." *Abstracts of Papers, 232nd ACS National Meeting, San Francisco, CA, U.S.A., Sept. 10-14, 2006.*
10. Rodgers, R. P. and A. G. Marshall. Petroleomics: Advanced Characterization of Petroleum-Derived Materials by Fourier Transform ion Cyclotron Resonance Mass Spectrometry (FT-ICR MS). *Asphaltenes, Heavy oils, and Petroleomics*, ed. O. C. Mullins, E. Y. Sheu, A. Hammani and A. G. Marshall.*(Springer, New York, 2007),* pp. 63-89.
11. Roussis, S. G. and R. Proulx. "Probing the molecular weight distributions of non-boiling petroleum fractions by Ag+ electrospray ionization mass spectrometry." *Rapid Commun. Mass Spectrom.* **2004,** *18(15)*: 1761-1775.
12. Streitwieser, A., Jr., et al.. "Acidity of hydrocarbons. XXI. Equilibrium acidities of some hydrocarbons toward lithium cyclohexylamide in cyclohexylamine." *J. Am. Chem. Soc.* **1967**, *89(1)*: 59-62.
13. Starks, C. M., et al. *Phase Transfer Catalysis: Fundamentals, Applications, and Industrial Perspectives.* **1994** *New York, Chapman & Hall.*

Water (Resource) Conservation Using Closed-Loop, Evaporative Cooling Systems for Process, Power, and Refining Applications

Almost every large plant is evaluating how to use less (fresh) water or finding they are unable to acquire additional water for cooling. Many are also constrained by zero discharge restrictions. Economics of water use include the cost of purchase, treatment and disposal.

Consequently, water availability and water quality are becoming increasingly important considerations in facility design and operation. Closed-loop, evaporative cooling is a cost-effective heat transfer technology that also optimizes use of scarce water resources.

The usual "default" is to install dry, air-cooled units. Unfortunately, air coolers take up a large plot area, require a significant amount of parasitic HP, and cannot provide a cool enough process outlet temperature.

Systems such as Wet Surface Air Coolers are used in a wide variety of process, chemical and industrial facilities for fluid cooling and vapor condensing. These units combine a heat exchanger and cooling tower in a single structure, cooling the process stream by evaporative heat transfer.

These are several important features of the WSAC including providing lower outlet temperature and being able to use poor quality water as makeup. Water resources can be blowdown from other equipment, effluent from sewage treatment facilities, gray water, storm water collection, agricultural runoff and produced water from mining and drilling processes.

Units can be manufactured with a wide variety of materials depending on water quality, water treatment and cycles of concentration.

Makeup water can be from a variety of sources including blowdown, FGD water and seawater. The WSAC can also be used as a first stage evaporator upstream of ZLD systems.

The WSAC is also used for adding capacity in "thermally challenged" plants. This allows for additional direct cooling without having to purchase additional makeup water.

For water-limited applications (when not enough water is available for cooling the entire load), a hybrid unit incorporating a dry and wet section can be used.

Water (resource) issues will continue to have an increased impact on plant design and operation. Closed-loop, evaporative coolers can help maintain plant performance while utilizing water streams that are currently unusable with conventional towers and heat exchangers. This paper will also include a comparison to other alternative technologies, discuss materials of construction and specific provide case studies.

135f Efficient Degradation of Organic Pollutants Promoted by Hydrogen Peroxide Electrogenerated in Situ

Yanqing Cong, Department of Environmental Engineering, Zhejiang Gongshang University, Hangzhou 310035, China and Zucheng Wu, Department of Environmental Engineering, Zhejiang University, Hangzhou 310027, China.

An environmentally benign electrocatalytic system for organic pollutants degradation with hydrogen peroxide electrogenerated in situ was developed. Under the electrocatalysis, the stronger hydrogen peroxide could be generated in situ continuously by dioxygen activation and significantly improve the degradation of organic pollutants. The quantity of hydrogen peroxide electrogenerated was about 0.6 mmol L-1 at current density 3.5 mA cm-2, sparged dioxygen rate 0.05 m3 h-1. The removal rate of 4-chlorophenol could reach 90% under given operational condition. The generation of hydrogen peroxide was smartly controlled at relatively low concentration by electrochemical action. Thus, the potential hazards problem associated with the storage and transportation of concentrated hydrogen peroxide could be effectively avoided. For the reactions employed hydrogen peroxide as oxidants, our work may provide an attractive option both on environmental and economic basis.

Aqueous Reaction and Processing in Suzuki-Miyaura Couplings Using Immobilized Palladium Catalysts in a Phosphine-Free System

Peter Styring and Nam T.S. Phan, Department of Chemical & Process Engineering, The University of Sheffield, Mappin Street, Sheffield S1 3JD, UK.*
p.styring@sheffield.ac.uk

Abstract
The Suzuki-Miyaura reaction has been carried out under environmentally-friendly conditions in a process that is characterised by a reactive crystallisation of the product from the reaction solution. In particular, we have used a palladium(II) complex supported on an inert polymer as an 'androgynous' catalyst that combines the benefits of a homogeneous catalyst and a heterogeneous catalyst; the reaction is carried out in a high aqueous environment without the need to add a phase-transfer catalyst; no phosphines are present either in the catalyst or as additives; the reaction is quantitative. A process for the scale-up of the reaction has been developed that leads to 85% product recovery in 100% purity.

Introduction
The Suzuki-Miyaura reaction is one of the most versatile methods for the elaboration of the carbon framework in organic molecules.[1] The reaction is characterised by a cross-coupling of two aryl sub-units, one derived from an aryl boronic acid which possesses low toxicity and the other from an organohalide or triflate to give a biaryl motif.[2] Such structures are common in modern pharmaceutical and agricultural chemicals as well as technologically important molecules such as liquid crystals and polymers. The reaction is usually homogeneously catalysed using palladium phosphine complexes or palladocycles.[3] Typically, phosphines are added to stabilize the Pd(0) formed and to activate the catalytic species. The problem with this approach is that phosphines are expensive and toxic[4] and are not usually recovered to be re-used in further reactions. This therefore places an economic and environmental burden on the reaction making it less attractive for large scale industrial processes. Reports have been published at attempts to use heterogeneous catalysts and water as the reaction solvent. However there are various combinations of needs imposed on these reactions including the need to add phase transfer catalysts and the use of volatile organic solvents in the work-up and product isolation processes. Furthermore, the heterogeneous catalysts suffer from metal leaching.[5-7] We have previously reported the synthesis and catalytic behaviour of a number of homogeneous transition metal catalysts immobilized on to inert polymer supports,[8-9] including palladium complexes for the Heck and Suzuki-Miyaura reactions.[9-11] These catalysts have been called 'androgynous catalysts' as they combine the best features of both homogeneous (high activity and selectivity) and heterogeneous (ease of separation and recycling) catalysts. All have high efficiency but are characterised by the need to react in volatile organic solvents.

Here we report the use of a palladium(II) salen-type complex immobilized on Merrifield resin (**1**) in the Suzuki-Miyaura reaction using an essentially aqueous solvent system (3.3% toluene in water) under heterogeneous conditions with the addition of a mild base but without and phosphine ligands/additives or phase-transfer catalysts.[12] Furthermore, the reaction proceeds via a reactive crystallisation where the product crystallises immediately on formation, and where the work-up uses only aqueous solutions or pure water.

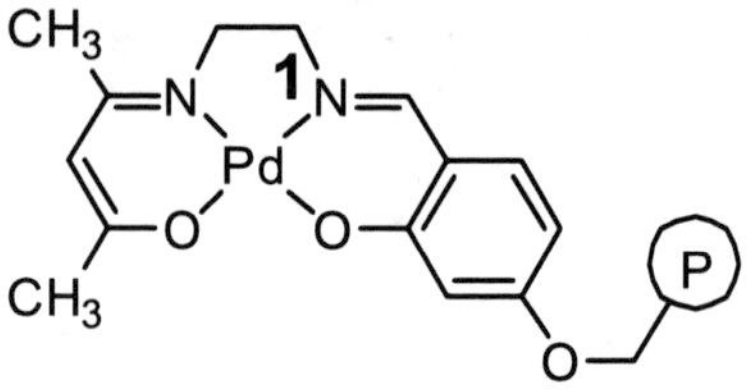

Experimental

The synthesis of the catalyst has been reported previously.[8] Catalytic reactions were performed using a 12-position Radley's Carousel Parallel Synthesiser fitted with a fuzzy logic controller unit giving temperature control to ±0.1 °C. Unless otherwise stated, a mixture of 4-bromoanisole (47.4 mg, 0.25 mmol), phenylboronic acid (46.2 mg, 0.38 mmol) and K_3PO_4 (0.17 g, 0.75 mmol) in water (3 ml) and the required amount of toluene was added to a Radley's Carousel reaction tube containing the required amount of palladium resin catalyst (6.0 mg, 0.5 mol% palladium) or palladium silica catalyst (12.0 mg, 0.5 mol% palladium). The volume of toluene was measured using a 0.25 ml Hamilton syringe. The mixture was heated at the required temperature for 24 h with magnetic stirring under a nitrogen atmosphere. To work-up the reaction, the mixture was allowed to cool to room temperature and saturated aqueous sodium chloride solution (3 ml) was added. The organic components were extracted into diethyl ether (3×3 ml) which was then dried over anhydrous $MgSO_4$ and the resulting solution analyzed by GC and GC-MS with reference to standard solution 4-methoxybiphenyl. The GC data showed conversion of 4-bromoanisole to 4-methoxybiphenyl as the sole product.

The reaction was also carried out at a ten-fold scale. After 24 h the crude reaction mixture was passed through a 125 μm nylon mesh to remove the catalyst while allowing the product to pass through. The catalyst beads were washed twice with aqueous sodium hydroxide solution (2M, 20 ml) and the filtrate added to the product suspension. The white solid product was isolated by filtration through a glass sinter. Two further washes were performed using pure water (2×20 ml) and the product air dried to yield 85% product with 100% purity by GC. No organic solvents were used in the work-up procedure. Microanalysis: Found C, 85.28; H, 6.19. $C_{14}H_{12}O$ requires C, 85.68; H, 6.16.

Results and Discussion

Catalytic studies were carried out initially on the reaction between phenylboronic acid and 4-bromoanisole (Scheme 1). Screening studies showed optimum conditions to be a 1.1 molar excess of the boronic acid in a mixture of water and toluene (3.3%) at 90 °C. The excess boronic acid pushed the reaction to completion as equimolar quantities gave 96% completion by GC. The amount of toluene was found to be critical to the efficiency of the reaction. More toluene was unnecessary as no improvement was observed while the conversion fell off noticeably if less than 3% toluene was used (90% at 2.5% toluene; 41% at 2%; 10% at 1.3%). Interestingly, 10% conversion was achieved in pure water.

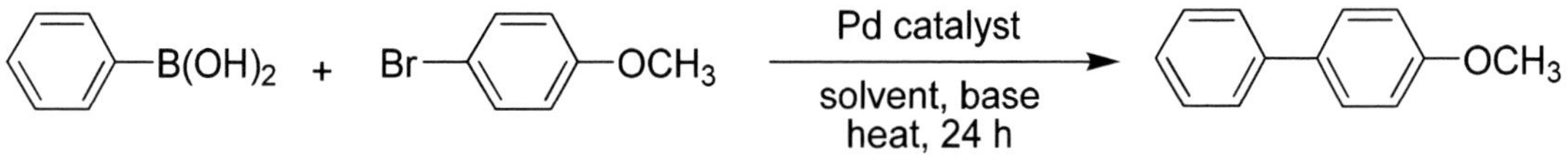

Scheme 1. The Suzuki reaction of 4-bromoanisole and phenylboronic acid

This is remarkable as the catalysts used was heterogeneous and no phosphines nor phase transfer catalysts needed to be added. The absence of phosphines in the system not only reduces process costs but also eliminates side reactions that may occur between the phosphines and phenylboronic acid.[13]

Kinetics studies (Figure 1) showed the reaction to be pseudo-first order overall with respect to either 4-bromoabisole, or the product 4-methoxybiphenyl. The calculated rate constant was 5×10^{-5} s^{-1} which is the same as that determined for the same catalyst under the same conditions but using DMF as the solvent.[11]

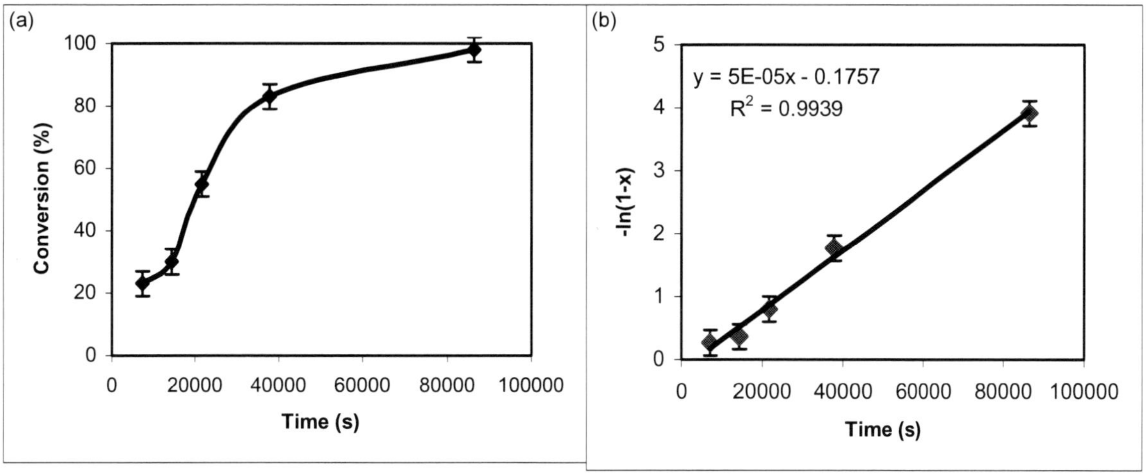

Figure 1. Kinetic data of the Suzuki reaction in aqueous media with 3.3% added toluene at 90 °C over 24 hours (raw data in a), showing in the log plot (b) an observed *pseudo* first order rate constant of $5 \times 10^{-5}\, s^{-1}$.

The key feature of the reaction that increases its 'Green Credentials' is the immediate precipitation of the product from the reaction solution under the reaction conditions. The small quantity of toluene in the reaction we believe solubilizes the reactants as well as causing some swelling of the catalyst support. However, the product is insoluble in the solvent so as it forms it immediately precipitates in a 'reactive crystallization' in the stirred tank reactor which is shown schematically in Figure 2. This has allowed us to design the complete synthetic process from reaction to isolation under increasingly green conditions. This was demonstrated in a ten-fold scale-up of the reaction. After completion of the reaction, aqueous sodium hydroxide solution was added through the port (1) to the stirred tank reactor to convert any phenylboronic acid (or anhydride) to the phenyl boronate salt. If necessary, any toluene could be recovered through distillation, the vapor passing through valve 4 to the condenser (5). Filtration through a 125 μm sieve (7) by opening valve 8 retained the catalyst in the reactor for further use while the product slurry passed in to the filter plate (9) where it was collected. The filtrate contained unreacted sodium phenyl boronate which could then be recycled in subsequent batches. Valve 8 was closed and pure water was added to wash any traces of product from the catalyst and reactor. Valve 8 was again opened and the suspension passed through the tilter plate. The latter process was repeated again before the white crystalline product was collected from the filter plate and air dried. An isolated yield of 85% was achieved and analysis of the product by GC, NMR and elemental analysis showed it to be 100% pure and free from any toluene. The catalyst could be used over repeated cycles although some deterioration of performance (70% completion) was observed in the fifth cycle. As a remedial step, the catalyst could be washed in situ using as small quantity of THF to regain full catalytic activity. The deactivation has been attributed simply to an accumulation of small quantities of product on the surface and in the pores of the catalysts support.

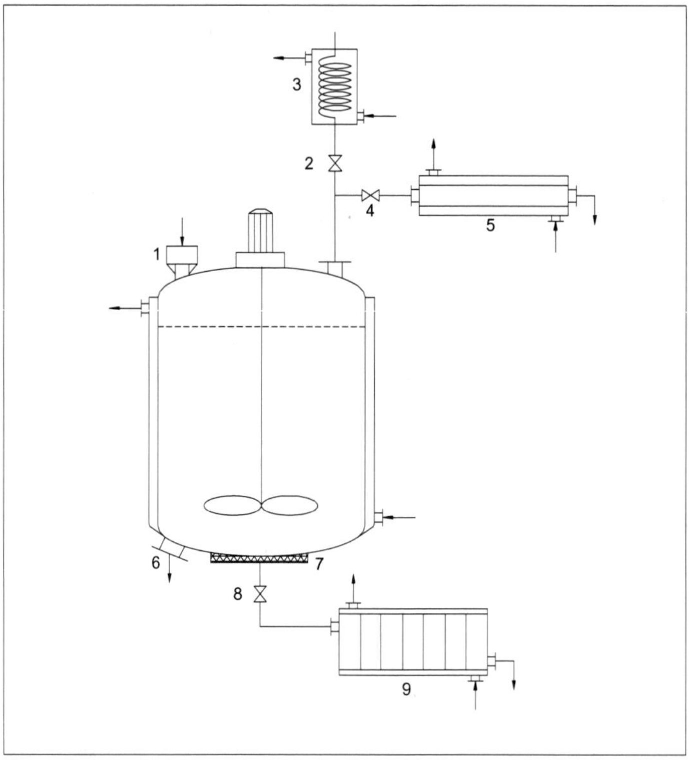

Figure 2. Proposed process for the scale-up of the aqueous Suzuki reaction using the palladium resin catalyst.

NMR analysis (Bruker AC250 spectrometer (^{1}H, 250.1) of the product showed resonances consistent with the literature data for 4-methoxybiphenyl.[14] δ_H ($CDCl_3$; 250.1 MHz, TMS) 7.95 (d, 2H), 7.62-7.50 (m, 4H), 7.40-7.25 (m, 3H), 2.55 (s, 3H); δ_C ($CDCl_3$; 62.9 MHz) 197.8, 145.8, 139.9, 135.9, 129.0, 128.9, 128.3, 127.3, 127.2, 26.7.

When the whole process is considered, the amount of toluene used is only 1.1%. Furthermore, it can be retained in the system on subsequent cycles so represents a very clean approach to the Suzuki-Miyaura reaction. ICP-EAS (Spectro Ciros ^{CDD}) showed it to be free from palladium contamination as any palladium detected was at the detection limit of the instrument and NMR showed the product to be free from toluene contamination.

The discovery scale process was repeated using a diverse library of organobromides to test the general applicability of the reaction. Analysis by GC against standards of the products yielded the data shown in Table 1. In each case a 1.5 times molar excess of the organobromide was used.
Most organobromides gave good to excellent yields in many cases at or approaching complete conversion. However, 4-bromoaniline and 4-bromonitro benzene gave trace product or no conversion at all. This are unusual results as these groups usually show high activity in the reaction.[15]

Reactions involving ethyl 4-bromobenzoate could be controlled depending on the added base. The usual base, K_3PO_4,gave 4-biphenyl carboxylic acid (entry 6) in 100% conversion. Here the coupling reaction takes place while at the same time the ester group is saponified to give the free carboxylic acid. Therefore we can perform the coupling reaction concurrently with a deprotection step which allows the molecular framework to be further functionalized. However, if the ester group is required to be left intact then the base is changed to the weaker K_2CO_3. Ethyl (4-bipnenyl carboxylate) is then formed in 96% conversion as the sole product (entry 7).

Conclusion

The supported palladium catalysts exhibit a high activity towards the Suzuki-Miyaura cross-coupling reaction without any added phosphine ligands. Using the silica-supported palladium catalyst requires a catalytic amount of PTC but this is not required in the case of the Merrifield resin. The fact that a water/toluene mixture with more than 95% (v/v) water can be used instead of common organic solvents is obviously of great benefit. The product is recovered without the need for organic solvent extraction as it is immediately precipitated from solution on formation through a reactive crystallization. The work-up procedure involves several aqueous base and pure water washes and hence the relative amount of toluene used in the complete process becomes increasingly small. The catalysts can be easily separated from the reaction mixture by simple filtration and reused after washing without a significant degradation in activity. Very little palladium if any leaches into solution from the catalyst during the course of the reaction. Therefore, this represents a green process that should have advantages for industrial applications.

Acknowledgement

We thank the Vietnamese Government for funding (NTSP) and Analytical Services in the Department of Chemistry at the University of Sheffield.

Table 1. The Suzuki reactions of substituted bromobenzenes using 0.5 mol% palladium resin catalyst, 1.5 eq. $PhB(OH)_2$, 3 eq. K_3PO_4 in water/toluene (30:1) at 90 °C for 24 hours under a nitrogen atmosphere (based on GC-MS data).

Entry	Substrate	Product	Conversion (%)
1	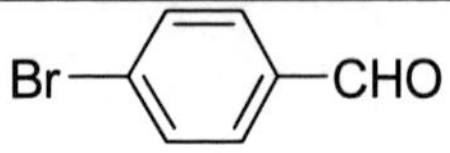	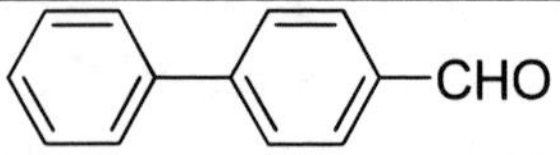	100
2	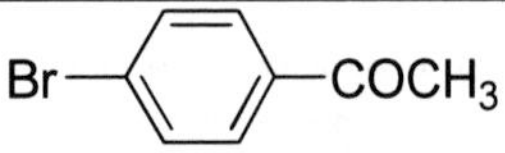	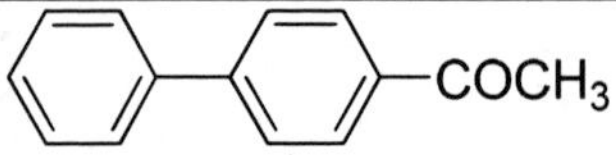	100
3		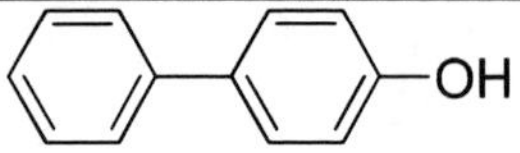	91
4	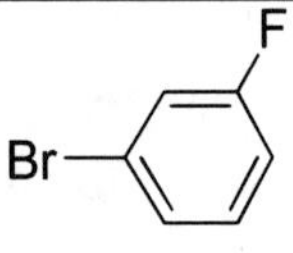	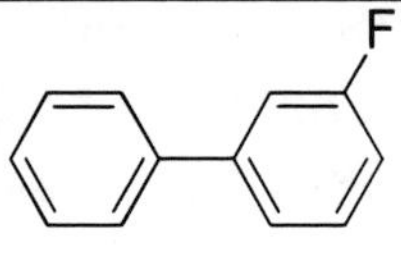	92
5	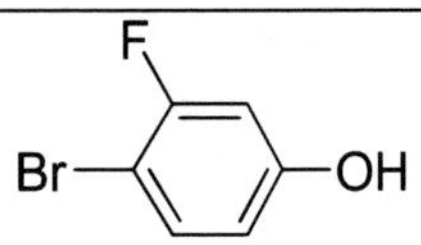	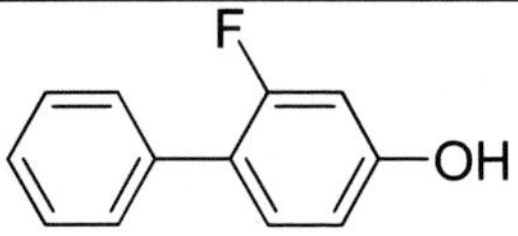	79
6	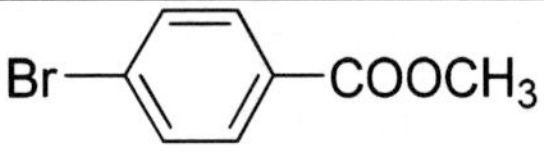	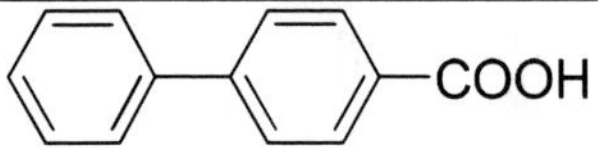	100
7	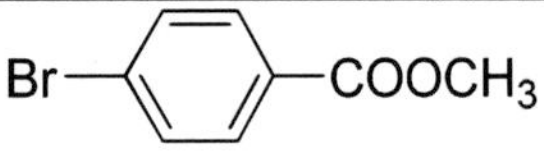	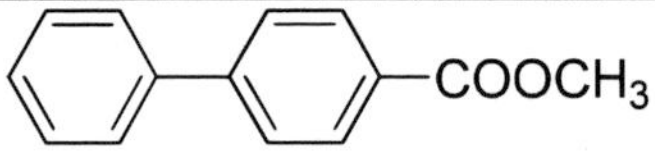	96*
8	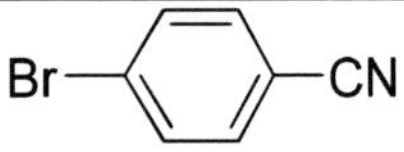	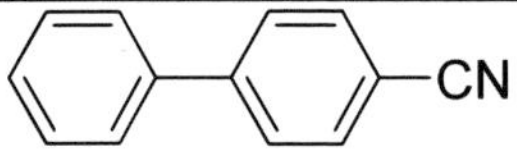	100*
9	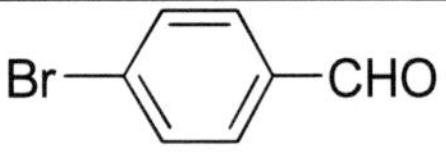	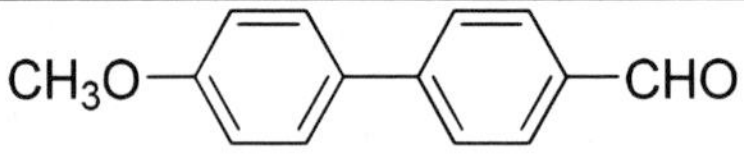	100**
10	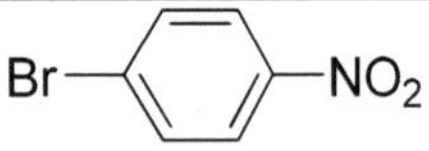	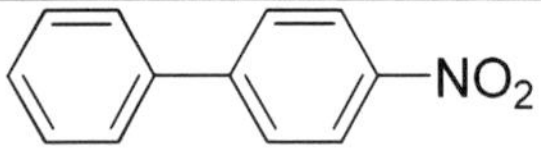	trace
11	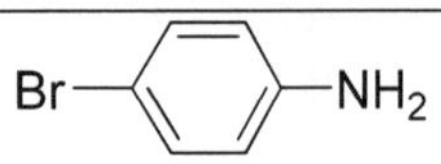	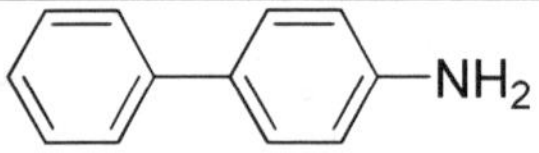	0

* K_2CO_3 was used instead of K_3PO_4

** 4-methoxybenzeneboronic acid was used instead of benzeneboronic acid

References

1. Paul, S.; Clark, J. H. *Green Chem.* **2003**, *5*, 635.
2. Miyaura, N.; Suzuki, A.*Chem. Rev.* **1995**, *95*, 2457.
3. Mubofu, E. B.; Clark, J. H.; Macquarrie, D. J. *Green Chem.* **2001**, *3*, 23.
4. Beletskaya, I. P.; Cheprakov, A. V. *Chem. Rev.* **2000**, *100*, 3009.
5. Cornils, B.; Herrmann, W. A.; Panster, P.; Wieland, S. In *Applied Homogeneous Catalysis with Organometallic Compounds*; Cornils B., Herrmann, W. A., Eds.; VCH: Weinheim 1996, vol. 2, p. 576.
6. Sakurai, H.; Tsukuda, T.; Hirao, T. *J. Org. Chem.* **2002**, *67*, 2721.
7. Uozumi, Y.; Danjo, H.; Hayashi, T. *J. Org. Chem.* **1999**, *64*, 3384.
8. Styring, P.; Grindon, C.; Fisher, C. M. *Catal. Lett.* **2001**, *77*, 219.
9. Phan, N. T. S.; Brown D. H.; Adams, H.; Spey; S. E.; Styring, P. *Dalton Trans.* **2004**, 1348.
10. Phan, N. T. S.; Brown, D. H.; Styring, P. *Tetrahedron Lett.* **2004**, *45*, 7915.
11. Phan, N.T.S.; Khan, J.; Styring, P. *Tetrahedron* **2005**, *61*, 12065.
12. Phan, N.T.S.; Styring, P. *J. Am. Chem. Soc.* **2007**, submitted.
13. Kong, K. C.; Cheng, C. H. *J. Am. Chem. Soc.* **1991**, *113*, 6313.
14. Wolfe, J. P.; Buchwald, S. L. *Angew. Chem.* **1999**, *38*, 2413.
15. Grushin, V. V.; Alper, H. *Chem. Rev.* **1994**, *94*, 1047.

135d New Type of Organic-Clay Intercalative Materials for Gene Delivery System

Xiaoying Wang and **Yumin Du**. Department of Environmental Science, Wuhan University, 129 Luoyu Road, Wuhan, 430079, China

Nonviral vectors have gained increasing interest in gene therapy, but generally they can not obtain satisfying transfection efficiency. Here we develop a hybrid material as a novel nonviral vector, which can combine both advantages of biopolymer with clay in gene delivery system. Soluble quaternized chitosan was intercalated into the interlayer of rectorite to obtain a new intercalated nanocomposite. MTT assay and in vivo studies of acute toxicity revealed that the nanocomposites were biocompatible and non-toxic. They can adsorb pDNA effectively and stably. In vitro study of cell transfection indicated that GFP expression was intensively continuous for 120h. Furthermore, in vivo transfection study, the most prominent GPF expression was observed in the gastric and duodenum mucosa, and good transfection efficiency was also obtained when injected into the muscle. All the results suggest that the nanocomposite is a novel and potential non-viral gene carrier.

135g Removal of SO2 and NO Simultaneously from Flue Gas Based on TiO2 Photocatalytic Oxidation

Zhao Yi, School of Environmental Science and Engineering, North China Electric Power University, 619 North Yonghua Street, Baoding, 071003, China

A new mechanism and process of simultaneous desulfurization and denitrification from flue gas by using TiO_2 potocatalysis were proposed. The influence factors on the efficiencies of simultaneous desulfurization and denitrification were studied. After the optimal values of the concentration of O_2 in flue gas,the relative humidity of flue gas, and the irradiation time in the photocatalysis reactor were used, the efficiencies of removal for SO_2 and NO can be achieved 98% and 67% respectively. The mechanism of SO_2 and NO removal by TiO_2 photocatlysis is that SO_2 is partly oxidized to SO_3, most of NO is oxidized to NO_3 and NO_2,these oxidation products are removed by resorbing.

High Quality Base Oil Production via the HyLube™ Process

***T.N. Kalnes and Mark VanWees**, UOP LLC, 25 E. Algonquin Rd., Des Plaines, IL 60017*
***Andreas Schuppel**, Puralube GmbH, Hauptstrasse 30, Gebaude 27, Elsteraue / OT Alttröglitz, D-06729, Germany*

ABSTRACT

The successful commercialization of the HyLube™ process by Puralube GmbH, a subsidiary of Puralube Global was presented at the 2007 AICHE meeting[1]. This paper will expand upon the previous paper by providing more detailed analytical characterizations of the re-refined base oil products and a more in depth discussion of how the process manages the wide range of contaminants present in used motor oil.

Unlike most other re-refining processes, the HyLube process produces a very consistent high quality base oil product. In addition, byproducts are managed to minimize environmental impact while maximizing energy recovery. Key process features include (1) continuous processing of the entire used oil charge stock in a hydrogen-rich environment without use of intermediate oil storage, (2) efficient catalytic removal of contaminants such as chlorinated, sulfurous, and oxygenated organic compounds and polyaromatic hydrocarbons, (3) management of all sulfurous and odorous compounds to eliminate malodorous and toxic emissions and (4) production of deeply desulfurized and saturated base oil and distillate fuel products.

Conventional base oil production methods are energy intensive, consume a diminishing fossil fuel resource, and place a large burden on the environment. The current trend toward increasing percentages of ethylene-based synthetic base oils in the blended lubricant product further raises the overall life cycle burden of the finished product. Used oil containing high percentages of high viscosity index and low pour point base oil represents a valuable resource and its proper management should be given the most attention.

INTRODUCTION

Tougher base oil standards have been established to facilitate auto manufacturers' requirement to improve engine performance from an environmental impact perspective. For example, General Motors' 2005 model year vehicles equipped with Engine Oil Life Systems averaged 8,500 miles between oil changes, compared to the 5,000 mile drain interval typical of GM's competitors. However, creating these longer-drain engine oils and fill-for-life transmission fluids and axle lubes requires Group II, II-plus or Group III base oils.[2]

The trend toward smaller engines that operate at higher engine revolutions per minute dictates the need for better motor oil quality. These quality requirements have been met by a variety of investments to produce enhanced quality base oils but also by the judicious application of performance related additive packages. The recent quality transition from International Lubricant Standard Approval Committee (ILSAC) GF3 designation (in 2000) to ILSAC GF4 (in 2005) and ILSAC GF5 (expected in 2009-2010) is a good example of this quality trend. Among other factors, these trends have sought to increase the level of saturates and decrease the volatility (in the base oil) both of which lead to superior quality finished oils that support extended oil drain intervals.[3]

The growing need for higher quality base oils provides an opportunity for a process that can rejuvenate used oil to these higher product standards. An example of such a process is the HyLube process commercialized by Puralube GmbH. This paper provides detailed analytical characterizations of the re-refined base oil products produced in the commercial HyLube process. It also provides an in depth discussion of how the process manages the wide range of contaminants present in used motor oil.

HIGH QUALITY BASE OILS BY RE-REFINING

Conventional base oil production methods are energy intensive, consume a diminishing fossil fuel resource, and place a large burden on the environment. The current trend of including higher percentages of hydrocracked base oils and/or ethylene-based synthetics in the lubricant blend further increases the overall life cycle burden of the finished product. Spent lube oils containing large quantities of high viscosity index and low pour point base oil represent a valuable resource which should be properly managed. While simple blending of used oils with low quality fuels will recover the energetic value of this material, the latent value of an engineered material containing very low aromatics and waxes is lost.

Puralube is an environmental technology company engaged in the global commercialization of HyLube re-refining technology. The HyLube process uses a unique feed preparation concept known as direct contact hydrogenation, which eliminates the need for much of the equipment used in other lube re-refining processes. The used lube oil feed is heated to reaction temperature by direct contact with a hot circulating hydrogen stream. As a result, the feed stream is maintained in a hydrogen environment which inhibits the formation of the polymeric and carbonaceous by-products which can cause equipment and catalyst fouling.

Figure 1 shows a simplified block diagram of the process. The first part of the process involves separation of the lube range and lighter components of the feed from the non-distillable residue portion. The HyLube process reactions are carried out at elevated temperatures and pressures in a hydrogen atmosphere using UOP proprietary catalysts. Contaminants are removed and the quality of the lube base oil is rejuvenated and enhanced. In addition to converting hetero-atoms such as sulfur and nitrogen, the HyLube catalyst increase viscosity index via saturation of multi-ring aromatic compounds. After exiting the reactor, the purified products are separated from the reaction by-products and excess recycle gas. Acid gases such as H_2S and HCl are present in the reactor effluent. To prevent these components from becoming corrosive, mild neutralization chemicals are injected into the lube separator vapor. The hydrogen rich vapor from the cold separator is scrubbed, compressed, reheated and returned to the mixer.

The hydrocarbon liquids collected in the separators are sent to the product fractionation section where the products are separated into various cuts to meet the desired lube oil viscosity grades. Unlike other re-refining processes, the lighter co-products are also severely hydrotreated. A photo of the Puralube unit, highlighting the lube fractionation column, is shown as Figure 2. Reference 1 provides more detailed description of the HyLube process flow scheme and product yields.

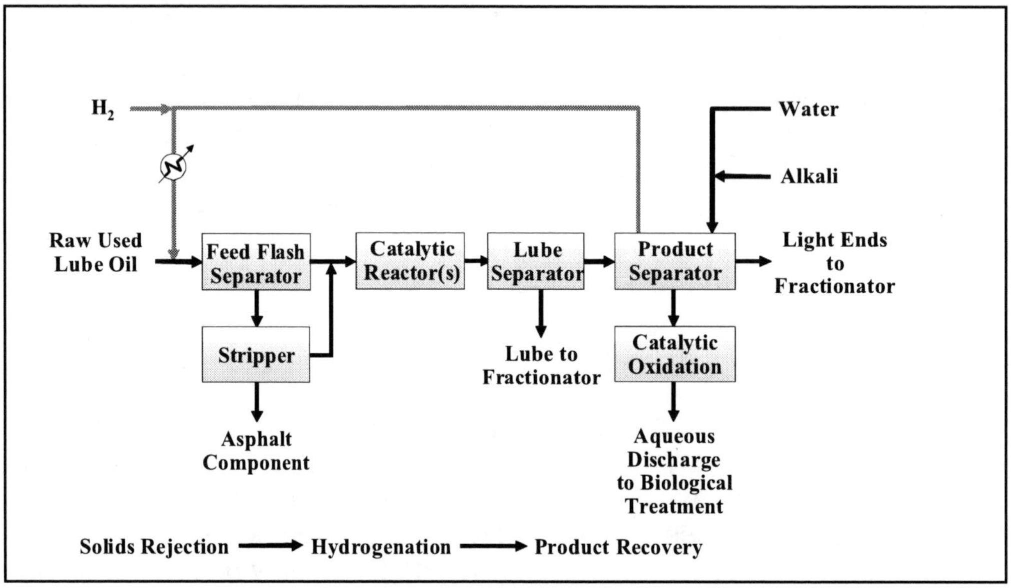

Figure 1: HyLube Block Flow Diagram

Figure 2: Puralube GmbH HyLube Unit Photo

Modern re-refined base oils meet the requirements of API Group II and are characterized by their high viscosity index, low sulfur content and water-white color.

Other characteristics are:

- ✓ Odourless
- ✓ High Oxidation stability
- ✓ Low evaporation loss
- ✓ Low content of aromatics
- ✓ Olefin free
- ✓ Halogen free
- ✓ No ash residues

The quality of the base oil products produced by Puralube is consistently very high (Group II, II-plus). This is due to both the improving quality of virgin base oils and to the attributes of the HyLube process which enables both rejuvenation and upgrading of the used oils. Tables 1-2 highlight the physical and chemical properties of the primary products; P-75, P-160, and P-300.

Properties	Standard	Typical Data			Test Method
		pur75	pur160	pur300	
Color		< L 0.5	< L 0.5	< L 1.5	DIN ISO 2049 / ASTM D-1500
Density @ 15 °C	kg / m³	850	855	860	DIN 51757 / ASTM D-1298
Flash point COC	°C	190	215	228	DIN ISO 2592 / ASTM D-92
Pour point	°C	- 12	- 12	- 12	DIN ISO 3016 / ASTM D-97
Noack evaporation loss	% wt	1.5 (150°C)	9	3	DIN 51581 / ASTM D-5800
Kin. Viscosity					
@ 40 °C	mm²/s	13.5	29.5	58	DIN 51562-1 / ASTM D-445
@ 100 °C	mm²/s	3.19	5.2	8.4	DIN 51562-1 / ASTM D-445
@ 100 °F	SUS	72	140	268	ASTM D-2161
@ 210 °F	SUS	35	43	53	ASTM D-2161
Viscosity index		100	115	116	DIN ISO 2909 / ASTM D-2270
Sulfur content	ppm	100	100	100	EN ISO 8754 / ASTM D-4294
Saturates	% wt	>> 90	>> 90	>> 90	ASTM D-2007

Table 1: Typical properties of Puralube base oil products

Properties	Standard	Typical Data			Test Method
		pur75	pur160	pur300	
Total Acid No.	mg KOH/g	< 0.03	< 0.02	< 0.03	ASTM D-974/DIN 51558-1,2,3
Copper Strip Corrosion		1a	1a	1a	ASTM D-130DIN ISO 2160
Aniline Point	°C	98.3	108	118	ASTM D-611
Refractive Index		1.4708	1.4725	1.4736	ASTM D-1218/DIN 51423-1
Sulfated Ash	wt.-%	<0.001	<0.001	<0.01	ASTM D-874/DIN 51575
Metals	ppm	< 1	< 1	< 1	ICP
Chlorine	ppm	0	0	0	DIN 51408 P1
Demulsibility	min	4	5	12	ASTM D-1401
Air Release	min	1	1	2	DIN 51381

Table 2: Additional properties of Puralube base oil products

To be successful in a competitive marketplace, it is imperative that quality of the base oil products be consistent day in and day out. This is especially true for re-refiners. Figure 3 provides a one year history of the key properties of the Puralube P-160 product (29.5 cSt @ 40^{0}C) demonstrating that Type II-plus quality can be consistently achieved. Because feed quality to a re-refinery can at times be difficult to control, the re-refining process needs to be robust in order to ensure product quality maintenance. Figure 4 illustrates the high conversion of aromatic oxygenates, sulfurs, and polyaromatic compounds achieved in the HyLube process as measured by high resolution mass spectrometry (HRMS) of feed and product.

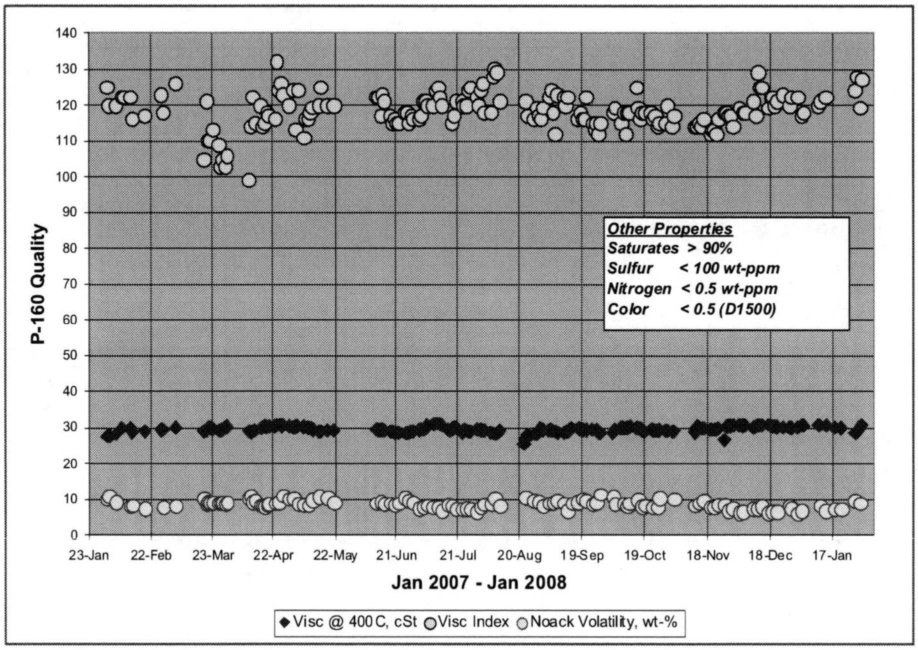

Figure 3: Consistent product quality demonstrated

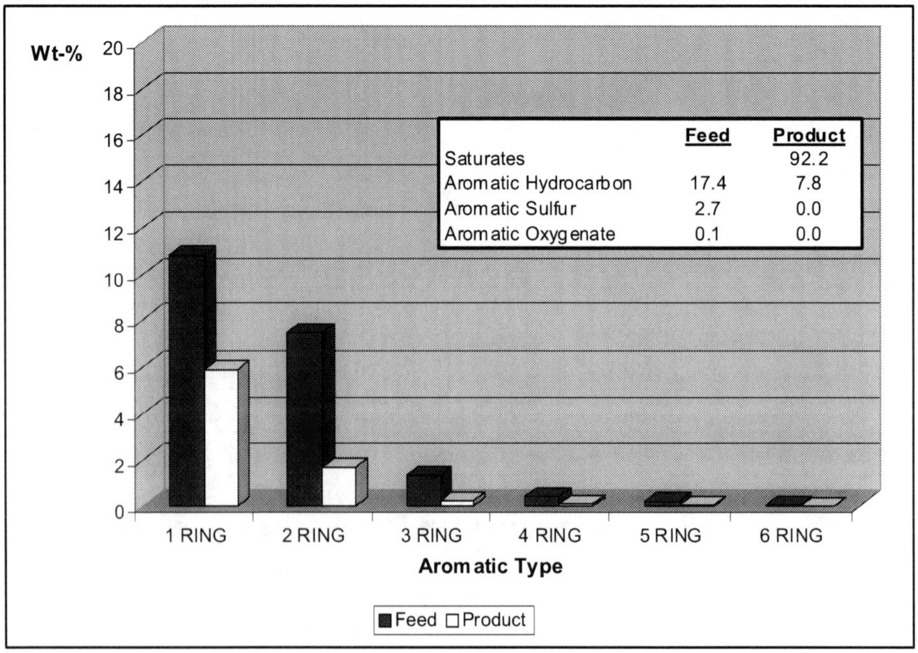

Figure 4: HRMS data showing aromatic conversion

As shown in Figure 5, the Puralube re-refined base oils passed all of the ecotoxicity tests with excellent grades. This is made possible by the high degree of polyaromatic saturation achieved in the HyLube process.

IP 346 DMSO Refractive Index Method

pur75 < 0.76 %
pur160 < 0.20 %
pur300 < 0.10 %

(no labelling < 3 wt%)

Modified Ames Test (ASTM E 1687)

MI

pur75 < 0.11
pur160 < 0
pur300 < 0

(MI < 1.0 oils are considered no mutagenic)

BUS Skin Irritation Test

pur75
pur160 << 3.0
pur300

(< 3.0 no skin irritation)

Benzo[a]pyrene:
Grimmer Method PAH – 0397
(industrial fluids should contain < 1 ppm benzo[a]pyrene)

SN 150	pur75	pur160	pur300	HC
< 0.1	< 0.01	< 0.2	< 0.02	< 0.01

[ppm]

Figure 5: Ecotoxicity data for Puralube base oils

Table 3 compares the Puralube base oils to key specifications for Group II and Group III oils. From this comparison it can be seen that this new generation of re-refined base oils produced in the HyLube process falls into a Group II-plus category similar to severely hydroprocessed base oils. Due to the high saturates, high VI and low volatility, Puralube base oils are able to outperform conventional Group I base oil products[4].

	Group I	Group II	puralube® pur160	Group III	Group IV	Group V
Saturates	< 90 %	> 90 %	> 95 %	> 90 %	PAOs	All base stocks not in Groups I, II, III and IV
	and/or	and	and	and		
Sulphur	> 0.03 %	< 0.03 %	< 0.01 %	< 0.03 %		
	and	and	and	and		
VI	≥80 - <120	≥80 - <120	115	>120		

Table 3: Puralube P-160 product exceeds Group II requirements

Significant advantages can be shown for lubricant formulations based on Group II or Group III base oils compared to Group I. These advantages include:

- the additive requirements to meet specifications such as ACEA E5-99 and API CH-4 are significantly lower for highly saturated, low sulfur base stocks
- Improved oxidation stability is observed in finished lubricants using Group II or III base stocks. Advantage can be taken of this by either the use of a more cost optimized inhibition system or by improved overall lubricant stability.
- Higher viscosity index base oils such as Group II+ or Group III (or Group IV) allow the blending of lower viscosity oils while continuing to meet viscometric and volatility requirements.

Recent literature indicates that severely hydroprocessed base oil can actually outperform a Group IV base oil (PAO) in several areas important to lubricants, such as additive solubility, lubricity and antiwear performance[5]. So even though there has been a worldwide excess base oil supply for some time, stocks in the Group II, Group II+ and Group III category should nevertheless be preferred for their advantages in engine oils, the biggest of all lubricant segments. This, plus the operating cost benefits of re-refining plants should position these stocks relatively well in an industry where worldwide future growth is expected to be modest.

CONTAMINANT MANAGEMENT USING THE HyLUBE PROCESS

Re-refining used oil is complicated by the fact that the oil is difficult to characterize. Contaminant levels are very dependent upon the source of the used oil, the types of used oil, how it was collected, and the intended end-use of the oil[6]. The highly variable quality of used oils drives the complexity of the facilities necessary to treat it for re-use. Used oil is made up of base oils and various additives that impart the desired quality characteristics to the finished product. High tech motor oil, for example, contains as much as 20–25 percent additive components in the finished product. This includes additives like viscosity improvers, demulsifiers, detergents, and antiwear additives, to name a few.

The additive content of lubricants and motor oils in particular, gives rise to the environmental concerns involved with combustion of used oils[3]. It is common that up to 20 to 25 percent of typical motor oil blends are made up of additives that are used to improve the quality, stability, and longevity of motor oils in combustion engine applications. Although industrial oils are also components of used oils, they do not contain the same level of additization and thus they tend to dilute the impact of motor additives. Table 4 details some of the common additives and their chemical makeup that are present in typical used oils.

High concentrations of a wide variety of synthetic additives also represent one of the major challenges to re-refiners that use hydroprocessing technology to rejuvenate used oils. Furthermore, contaminants such as soot, water, organic acids, and solvents that collect in the oil during normal service and collection make used oil a challenging feedstock that requires re-refining technology that is more robust than traditional oil refining processes.

Feed to the Puralube HyLube unit (Table 5) includes a mixture of recovered motor oils, hydraulic oils, gear oils, and other paraffin-rich industrial oils. In addition to lube base oil and assorted additives, the feed contaminants include gasoline, diesel, combustion byproducts, and water. Varying concentrations of particulate matter and corrosion metals are also present. Day-to-day variations in feed composition are buffered by blending in a primary feed tank which holds approximately seven days of inventory.

Antiwear	Zinc dithiophosphates, acid phosphates, organic sulfur and chlorine compounds. Sulphurized fats, sulfides and disulfides
Detergent	Metallo-organic compounds of sodium, Calcium and magnesium phenolates. Phosphonates and sulphonates
Anticorrosion	Zinc dithiophosphates, metal phenolates, fatty acids and amines
Dispersant	Alkylsuccinimides, alkylsuccinic esters
Friction Modifier	Organic fatty acids. Lard oil. Phosphorous based compounds
Pour point depressant	Alkylated naphthalene and phenolic polymers, polymethacrylates
Seal Swell Agent	Organic phosphates aromatic hydrocarbons
Viscosity Modifier	Polymers of olefins, methacrylates, di-enes or alkylated styrenes
Antifoamant	Silicone polymers, organic copolymers
Antioxidant	Zinc dithiophosphates, hindered phenols. Aromatic amines, sulphurized phenols
Metal Deactivator	Organic complexes containing nitrogen and sulfur amines, sulphides and phosphates

Table 4: Listing of Common Additives in Used Oils[7]

The patented HyLube approach to contaminant management is based on the continuous direct contact hydrogenation of the entire used oil feedstock. Volatile feedstock components are vaporized at reactor pressure by direct contact with a hot hydrogen rich recycle gas stream and then immediately hydrogenated without intermediate storage. Lighter than lube co-products are also severely hydrotreated. The non-volatile components and ash are concentrated and stabilized in a heavy asphalt product using a steam-stripped vacuum column.

Property	Units	Min	Max	Average
Viscosity @ 40C	cSt	46.8	70.6	56.2
Density	Kg/m3	881	907	888
Flash Point	°C	174	310	202
Total acid number	mg KOH/g	2.8	7.0	4.3
Water	wt%	3.1	10.7	6.0
Ash	wt%	0.33	1.11	0.72
Chloride	wt%	0.01	0.216	0.05
Sulfur	wt%	0.35	0.92	0.61
Nitrogen	wt%	0.00	1.60	0.12
Lead	wt ppm	6.9	72.7	29.4
Arsenic	wt ppm	0.02	3.8	0.81
Chromium	wt ppm	1	798	62
PCBs	wt ppm	0.1	1.2	0.3

Table-5: HyLube Process Unit Feed Characterization

The fate of the various feed contaminants is illustrated in Figure 6. Toxic organics are destroyed by severe catalytic hydrogenation which converts polyaromatic and heteroatomic compounds to useful products. Metals are stabilized in the asphalt product.

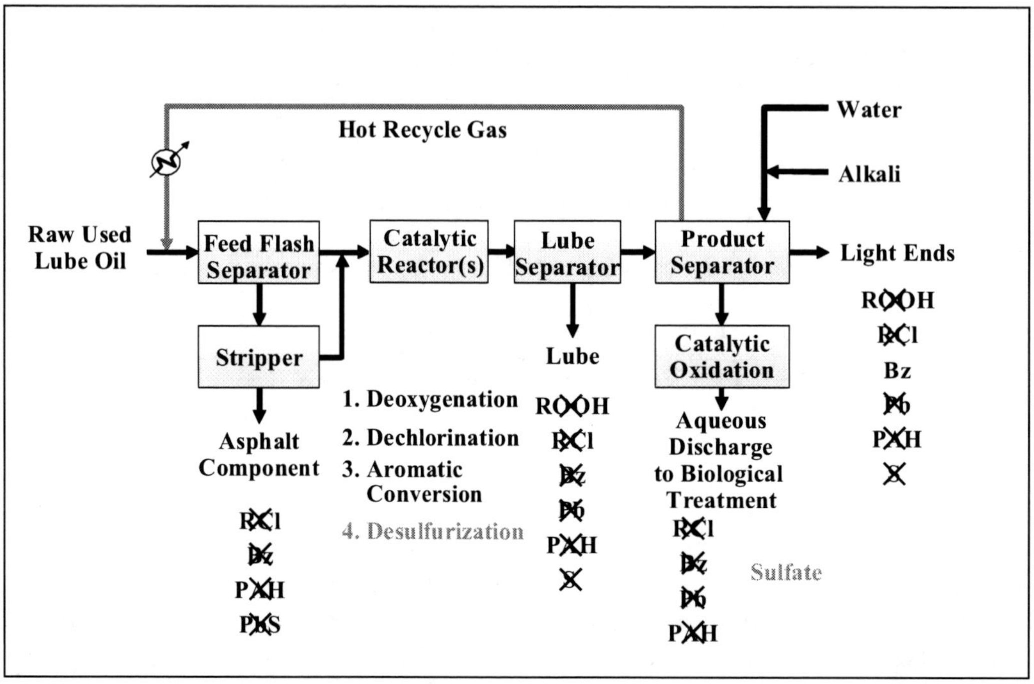

Figure 6: Fate of Contaminants in HyLube Process

Because the Puralube HyLube operation is a stand-alone facility, all of the sulfur, chloride, nitrogen, and oxygen removed from used oil feed must be managed in an environmentally acceptable manner. The byproduct of acid gas neutralization is an aqueous stream containing sulfide and chloride salts as well as small amounts of ammonia and water produced in hydrotreating reactions. All of the aqueous streams leaving the HyLube unit are collected in the main column overhead receiver and then routed to the Sulfide Oxidation section of the unit.

The Sulfide Oxidation process is a patented extension of the widely used UOP Merox™ process. Here the sulfides are catalytically oxidized to sulfate salts with a resultant reduction in chemical oxygen demand. After sulfide oxidation, the process water is sent to a two-stage biological treatment facility prior to discharge.

ENVIRONMENTAL CONSIDERATIONS

The waste management hierarchy detailed in Table 6 is expressed in the waste management policy sections of the Resource Conservation and Recovery Act and Pollution Prevention Act. The "Considered Action Step" is the logical application of this hierarchy to used oil from the most preferred environmental option to the least preferred environmental option[3].

Waste Management Ranking (from Environmental Perspective)	Option	Considered Action Step
Most Preferable	Prevent the waste in the first place	Source reduction (e.g. extended oil drain intervals)
	Re-use and reclaim the product	Re-refine used oil
	Recover energy by burning	Combust used oil for heating value recovery*
Least Preferable	Dispose of the waste by land filling or incineration	Recover and collect used oil for proper disposal

Table 6: RCRA Policy applied to Used Oil[3]

Several recent life cycle assessments (LCAs) have concluded that re-refining of used lube oils is a preferable form of recycling when compared to combustion for energy recovery from an environmental perspective [8, 9, 10, 11]. One of the most recent LCA studies modeled the effects of re-refining used oil generated in the European market in the HyLube process[12]. These effects were then compared to the impacts of burning the same oil in cement kilns, as a substitute for the primary energy sources of coal, heavy fuel oil (HFO) or natural gas (Nat Gas). ISO 14040 standards were followed in the study and SimaPro6.0 software was used for calculations.

Figure 7 presents a comparison of environmental impacts for the four alternatives. For the categories of acidification / eutrophication and fossil fuel consumption, re-refining of used oil (UO) is the best alternative[12]. The greatest benefit for climate change (primarily CO_2 emissions) is for the case of UO combustion with coal displacement. This is logical because coal has the highest emission of CO_2 per unit of fuel energy for any of the fossil fuels in this study, and therefore when coal is displaced by a less carbon intensive fuel, lower greenhouse emissions occur. The next best case is re-refining, followed by UO combustion with fuel oil displacement. The least desirable alternative for climate change is UO combustion with natural gas displacement. When UO displaces natural gas in cement kiln combustion, a higher rate of CO_2 release occurs, but this increase is almost exactly compensated for by avoiding emissions from natural gas production.

An analysis of the environmental fate of the key contaminants present in recycled lube oils was conducted by the DOE in 1995. It highlights the end disposition of the key contaminants depending on whether the used oil was combusted or re-refined. The key conclusion in the case of re-refining is that the noxious compounds are solidified and stabilized due to their disposition as a solid and pose minimal environmental risks. On the other hand, combustion of used oil results in air emissions, the magnitude dependent on the quality of the air pollution control equipment utilized.[3]

A peer reviewed study conducted by the California EPA and University of California (Berkeley) compared the life cycle environmental impacts of combusting used oil and re-refining used oils to produce base oils. The results showed that heavy metal emissions from uncontrolled combustion of used oil fuels could be hundreds of times higher than the comparable re-refining process or from the combustion of virgin fuel oil. The heavy metals emissions may lead to 150 times the ecotoxicity impacts compared to re-refining if air pollution control technology is not used.[8]

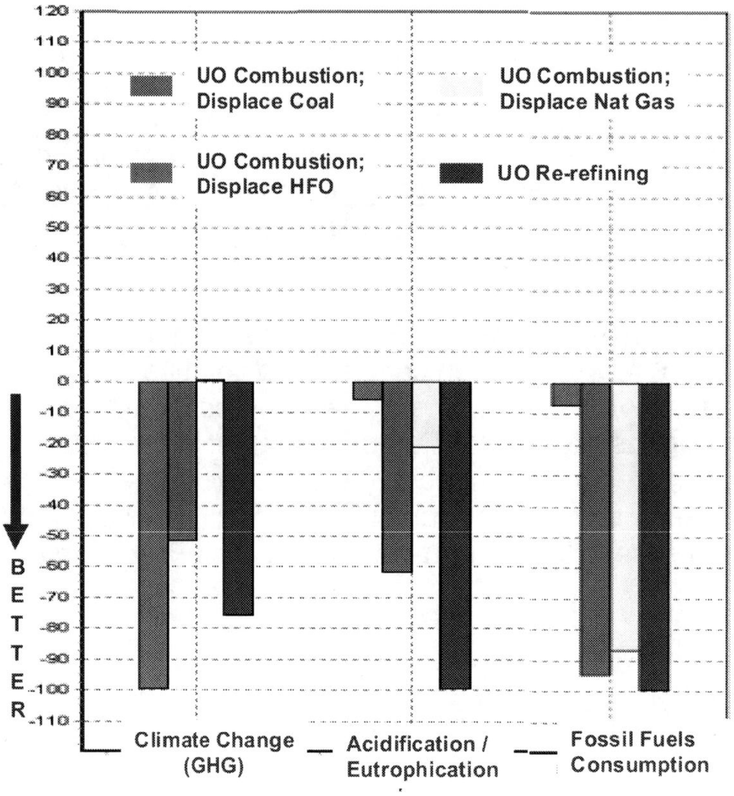

Figure 7: Environmental Impact Comparison (UO=Used Oil) [12]

SUMMARY

Spent lube oils containing large quantities of high viscosity index and low pour point base oil represent a valuable resource which should be properly managed. The HyLube approach to re-refining maximizes the recovery of high value lube products while destroying toxic organics and stabilizing metals. Sulfur is also managed in an environmentally friendly manner. Several recent life cycle assessments (LCAs) have concluded that re-refining of used lube oils is a preferable form of recycling when compared to combustion for energy recovery from an environmental perspective. As a result, RCRA guidelines for used oil management rank re-refining ahead of combustion for heating value recovery.

As automotive engine oil standards become more stringent and crude oil quality declines, production costs for base oil will increase. This trend coupled with more stringent environmental standards creates an opportunity to expand used oil recycling efforts worldwide. The successful commercialization of the first HyLube unit by Puralube GmbH represents an important step forward in the quest for a sustainable lube oil re-refining process.

The quality of the base oil products produced by Puralube is consistently very high (Group II, II-plus). Commercially re-refined base oils pass all of the ecotoxicity tests with excellent grades. This is made possible by the high degree of polyaromatic saturation achieved in the HyLube process. This new generation of re-refined base oils falls into a Group II plus category similar to severely hydroprocessed base oils. Based on high saturates content, high VI and low volatility these base oils are preferred for use in engine oils, the biggest of all lubricant segments. High quality coupled with the operating cost benefits associated with HyLube re-refining plants position these products relatively well in an industry where worldwide future growth is expected to be modest.

Puralube has moved forward with a project to build a second HyLube unit in Germany for startup in October 2008. Their longer term vision is to become the pre-eminent refiner of used oils in the world based on the UOP HyLube Process with a target to install 1.5 mm mta of processing capacity. Activities are already underway to expand the waste oil collection side of the business.

REFERENCES

1. Kalnes, T., Kadlec, L., Schuppel, A., "Hylube™ Process Commercialization: Recovering Value from Used Motor Oil", Spring 2007 AICHE Meeting, Houston, Texas

2. Lube Report (ISSN 1547-3392), "Lubes'n'Greases Magazine and Lubricants Industry Sourcebook"

3. "Used Oil Re-refining Study to Address Energy Policy Act of 2005 Section 1838", Office of Oil and Natural Gas Office of Fossil Energy, U.S. Department of Energy, July 2006

4. The 3rd Pan-American Base Oil & Lubricants Conference, "Re-refining Update, Puralube presentation", November 29, 2007, New York, C. Hartmann

5. David C. Kramer, Brent K. Lok, Russ R. Krug and John M. Rosenbaum, ChevronTexaco Global Lubricants, "Performance of Base Oils and Future Trends - The Evolution of Base Oil Technology - Part 3". *Machinery Lubrication* Magazine. September 2003

6. "End Uses for Used Oil, A Market Perspective", by Don Kress Par Excellence Developments Inc., Sudbury Canada, http://www.ped.vianet.ca/article1.pdf.

7. David Fitzsimons, Nicholas Morley, Peter Lee, Oakdene Hollins LTD, "UK Waste Oils Market 2001", page 8; http://www.defra.gov.uk/environment/waste/topics/oils/pdf/wasteoils.pdf

8. Bob Boughton, "Environmental Assessment of Used Oil Management Methods", Arpad Horvath Environmental Science and Technology, Vol. 38 No. 2, page 353, 2004, http://pubs.acs.org/cgibin/download.pl?es034236p/D2gn

9. Frankl, P., Fullana, P., Baitz, M., 2005, Europe Life Cycle Considerations on Waste Oils and Implications for Public Policy - Ecobilancio Italia, ESCI

10. Fehrenbach, H. 2005, Ecological and energetic assessment of re-refining used oils to base oils: Substitution of primarily produced base oils including semi-synthetic and synthetic compounds, Institut für Energie- und Umweltforschung GmbH (IFEU), a study commissioned by GEIR - Groupement Européen de l'Industrie de la Régénération.

11. Hartmann, C., 2005, Clean, Competitive, and Sustainable: Why the Recycling of Waste Oils Must Remain an EU Policy Priority, GEIR (Groupement Européen de l'Industrie de la Régénération), position paper, Square Marie Lousie 49 – 1000 Brussels High Performance Automotive Lubricants, 1995, Chem Systems, May 1995, 93S4

12. Kalnes, T., Shonnard D., Schuppel, A., 2006, LCA of a Spent Lube Oil Re-refining Process, 9th International Symposium on Process Systems Engineering, (Marquardt et al editors) © 2006 Elsevier B.V./Ltd.

Gold nanoparticle catalysts for the hydro-epoxidation of propene: an in-situ XANES and catalytic study into the adsorption of propene on gold

T. Alexander Nijhuis[a*], E. Sacaliuc[b], A.M. van der Eerden[b], B.M. Weckhuysen[b], J.C. Schouten[a]

[a]*Schuit Institute of Catalysis, Department for Chemical Reactor Technology, Eindhoven University of Technology,*
PO Box 513, 5600 MB, Eindhoven, The Netherlands
[b]*Department for Inorganic Chemistry and Catalyis, Utrecht University, Utrecht, The Netherlands*
Corresponding author: Tel: +31-40-2473671, Fax: +31-40-2446653,* ***T.A.Nijhuis@tue.nl*

Abstract: Gold nanoparticle catalysts for the direct epoxidation of propene are investigated. The adsorption of propene on the gold nanoparticles is studied using both catalytic experiments and by in-situ XANES analysis. The hydrogen oxidation over gold is used as a probe reaction for the catalytic activity of the gold nanoparticles. Both the hydrogen oxidation experiments and the XANES conclusively show that propene adsorbs strongly on gold under reaction conditions typical for the hydro-epoxidation of propene. It is currently being investigated how this adsorption fits into the epoxidation mechanism of the gold-titania based catalysts.

Keywords: propene, epoxidation, gold, titania.

1. Introduction

Propene oxide is a very important chemical intermediate, produced at about 6 million tons per year (2006) with demand still growing by 5 % annually [1]. One of the important new developments for the production of propene oxide are the gold titania based epoxidation catalysts[2, 3]. Gold-titania catalysts can very selectively epoxidize propene at mild conditions using molecular oxygen in the presence of hydrogen as sacrificial reductant. Because of the use of sacrificial hydrogen, this reaction is best called a hydro-oxidation rather than an oxidation reaction. The low hydrogen efficiency is the most important problem that needs to be solved before this catalyst system can be applied commercially. Since the mode of operation of these catalysts is still unclear. In this paper a kinetic study into the role of gold in the activation of propene is performed, since a kinetic understanding can be of great assistance in the development of improved catalysts. In this study, the main focus is on gold on silica catalysts, since it does not have any significant propene epoxidation activity, which makes it easier to distinguish between adsorbed propene and propene oxide. Furthermore, the silica support is inert under these reaction conditions, which makes it possible to attribute the catalytic and spectroscopic observations to exclusively the gold nanoparticles. Using in-situ XAFS experiments, the state of gold is probed and by using a delta-mu XANES analysis [4] changes in its state due to adsorbates are investigated. As a comparison, also the effect of ethene and propane are investigated.

2. Experimental

Gold catalysts were prepared using a deposition precipitation method at pH 9.5 using hydrogen tetrachloroaurate(III) solution aqueous ammonia [2]. Catalysts were prepared on silica, titania, titania on silica, and Ti-SBA-15. The catalysts were characterized using TEM, XRF, and UV-Vis. The catalytic performance was evaluated for the hydrogen oxidation (5 % H_2 and O_2 in He) at 353 K in a flow reactor system (GHSV 6000 h^{-1}). The effect of co-feeding hydrocarbons (propene, propane, ethene) during the hydrogen oxidation was determined, as a probe for their adsorption on gold. In the experiments first the hydrogen oxidation was measured until a stable activity was observer, after which 5 % of hydrocarbon was added to the feed. The analysis of the gases from the reactor was measured using an Interscience Compact GC-system using a Molsieve 5A column and a Porabond Q column with TCD detectors. The water concentration was used to determine the reaction rate.

XAFS measurements were performed on the working catalyst on the Au L3-edge and were carried out on station BM26A at the ESRF (Grenoble, France). In the experiments a 200 mg

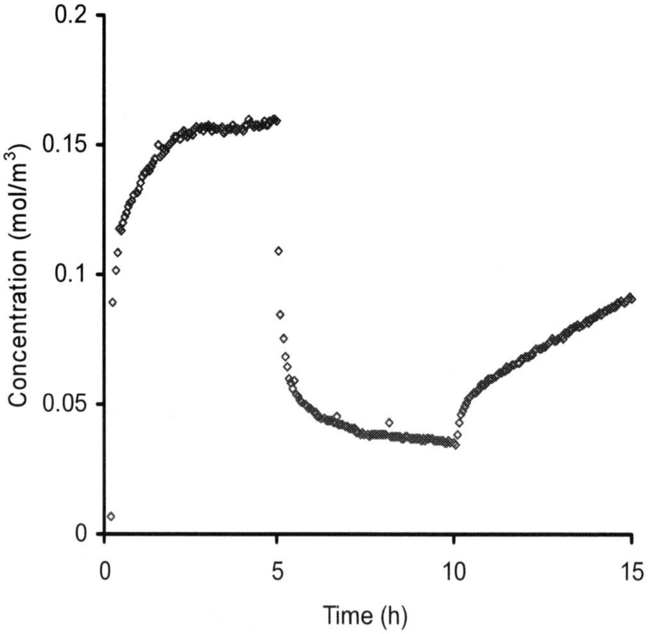

Figure 1. Propene co-feeding during hydrogen oxidation over Au/SiO_2 catalyst (352 K). The concentration of the water produced is shown as a function of time. During propene co-feeding (from 5-10 h), the water production rate drops dramatically.

sample was loosely pressed to form a self-supporting wafers/bed mounted in a specialized cell for recording in-situ XAFS data. The XAFS spectra were then recorded under conditions comparable to those used for the catalytic activity tests.

3. Results and discussion

The gold on silica catalysts prepared had a 1 wt% gold loading and a narrow gold particle size distribution as determined by TEM with an average of of 4.0 nm.

In the catalytic experiments, it was observed that propene (Figure 1) has a very strong rate inhibiting effect on the hydrogen oxidation, which can be explained by competitive adsorption of propene on the gold particles. A similar inhibiting effect was also observed for ethene, while propane only had a minor influence on the hydrogen oxidation rate. After removing the hydrocarbons, for both ethene and propene the activity only recovered slowly to its original value. No conversion of the hydrocarbons was observed.

In the delta-mu XANES analysis (Figure 2), it was observed that both ethene and propene have a significant peak at the white line position, while propane does not. The delta-mu XANES spectra were determined with the XANES spectrum of the catalyst in its steady state condition during the hydrogen oxidation as a reference. The appearance of the peak at the white line position was also observerd by Bus *et al.* [5] for the adsorption of ethene on platinum, which could be explained by di-sigma or pi-bonding of the alkene with the metal. In our study, this is able to explain the

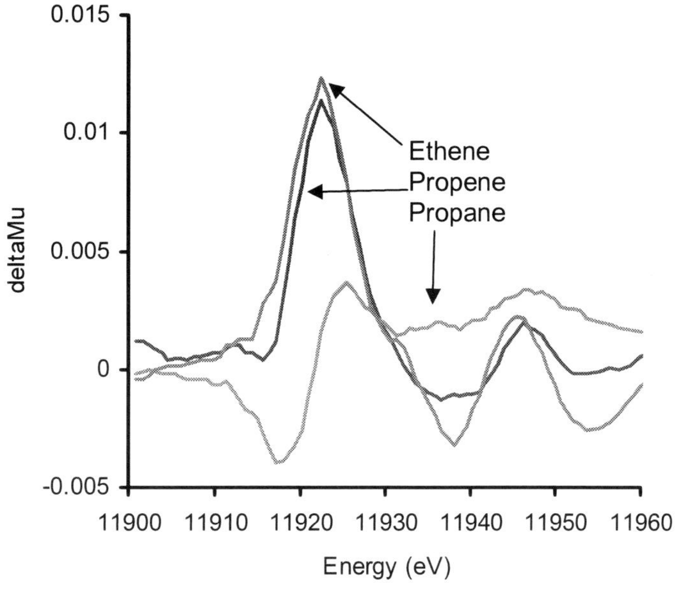

Figure 2. Delta-mu XANES spectra of gold during co-feeding of propene, propane, and ethene during the hydrogen oxidation over Au/SiO_2 (353 K).

inhibiting effect that both ethene and propene have on the hydrogen oxidation over gold, since the alkene adsorption on gold is blocking reaction sites for hydrogen/oxygen. The adsorption of propene on gold nanoparticles can explain the activating effect that Nijhuis reported previously for the reactive adsorption of propene on a gold-titania catalyst [6]. The exact implications for the mechanism of the propene epoxidation over gold-titania catalysts is currently being investigated.

4. Conclusions

From both the catalytic experiments and from the in-situ XANES experiments, it can be concluded that under propene hydro-epoxidation reaction conditions, propene is absorbing on gold nanoparticles. Propene adsorption decreases the hydrogen oxidation rate and interacts with surface gold atoms, causing a peak in the delta-mu XANES spectrum at the gold white line position. The relevance of this absorption is currently being investigated further.

Acknowledgement

STW/NWO is kindly acknowledged for the financial support for this research project (VIDI grant to T.A. Nijhuis). NWO is kindly acknowledged for providing XAFS beam time at the DUBBLE (BM26A) station at the ESRF in Grenoble.

References

1. T.A. Nijhuis, M. Makkee, J.A. Moulijn, B.M. Weckhuysen, The production of propene oxide: Catalytic processes and recent developments, *Ind. Eng. Chem. Res.* 45 (2006) 3447-3459.
2. T.A. Nijhuis, T. Visser, B.M. Weckhuysen, A mechanistic study into the direct epoxidation of propene over gold-titania catalysts, *J. Phys. Chem. B* 109 (2005) 19309-19319.
3. A.K. Sinha, S. Seelan, S. Tsubota, M. Haruta, Catalysis by gold nanoparticles: epoxidation of propene, *Top. Catal.* 29 (2004) 95-102.
4. A. Teliska, W.E. O'Grady, D.E. Ramaker, Determination of O and OH adsorption sites and coverage in situ on pt electrodes from Pt L-23 X-ray absorption spectroscopy, *J. Phys. Chem. B* 109 (2005) 8076-8084.
5. E. Bus, D.E. Ramaker, J.A. van Bokhoven, Structure of Ethene Adsorption Sites on Supported Metal Catalysts from in Situ XANES Analysis, *J. Am. Chem. Soc.* 129 (2007) 8094-8102.
6. T.A. Nijhuis, T. Visser, B.M. Weckhuysen, The role of gold in gold-titania epoxidation catalysts, *Angew. Chem. Int. Ed.* 44 (2005) 1115-1118.

107b Guanidine/alcohol and Amidine/alcohol Switchable Solvent Mixtures

Hillary A. Huttenhower[1], Lam Phan[2], Daniel Chiu[2], David J. Heldebrant[2], Ejae A. John[1], Xiaowang Li[2], Pamela Pollet[1], Charles A. Eckert[3], Charles L. Liotta[1], and Philip G. Jessop[2]. (1) School of Chemistry and Biochemistry, Georgia Institute of Technology, 311 Ferst Drive, NW, Atlanta, GA 30332-0400, (2) Department of Chemistry, Queen's University, Kingston, ON K7L 3N6, Canada, (3) School of Chemical and Biomolecular Engineering and Specialty Separations Center, Georgia Institute of Technology, 311 Ferst Drive, NW, Atlanta, GA 30332-0100

Organic solvents are a crucial part of chemistry on both a laboratory and an industrial scale. However, because solvents have very fixed properties, they often need to be removed and replaced between different steps of a reaction. This leads to an increase of economic cost and environmental impact for many processes. Developing a switchable solvent that can reversibly change its nature and properties at mild conditions would help solve this problem. Examples of solvents that consist of either a mixture of an amidine and alcohol or a guanidine and alcohol have been developed that form high polarity ionic liquids when exposed to CO2 at atmospheric pressure. Furthermore, these ionic liquids can be reversed back to a low polarity neutral form by exposure to nitrogen or mild heating. Such liquids have been used as a reaction medium utilizing their "built-in" facile separation.

Development and Characterization of Renewable Copolymers, Poly(glycerol-fumarate) and Poly(trimethylene-malonate) from By-Products of Biorefineries

Mathew D. Rowe and Keisha B. Walters, Dave C. Swalm School of Chemical Engineering, Bagley College of Engineering, Mississippi State University

Abstract

The goal of the research is the development of renewable copolymers from by-products of bio-refineries. The monomers, glycerol, 1,3-propane diol, malonic acid and fumaric acid, were selected based on availability and polymerization potential. Catalyst selection was based on the criteria of being environmentally benign, readily available and potentially biocompatible. The catalysts investigated in this study include zinc chloride, aluminum chloride, iron(III) chloride, and tin(II) chloride. The effects of catalyst, monomer ratio and temperature on copolymer yield and bulk properties were studied for these polycondensation reactions. Polymer synthesis was verified and the chemical composition was characterized using Fourier transform infrared spectroscopy (FTIR) and nuclear magnetic resonance (NMR). Gel permeation chromatography (GPC) determined average molecular weights and molecular weight distributions. Differential scanning calorimetry (DSC) was used to determine phase transitions in these new copolymers. Copolymers were produced with varying chemical composition, molecular weights, and bulk properties by controlling the monomer ratio.

Introduction

With increased concern for our environment and the increased cost of petroleum feedstocks, 'green' polymers are being sought from biomass-based resources as an alternative to petroleum based polymers. Products based on recycled and renewable resources are entering the market place as toys, packaging materials, clothing, and other products [1-4]. Polymerization of α-hydroxycarboxylic acids, glycerol (GLY), 1,3-propanediol (PDO), and amino acids, either individually or as copolymers, can generate commercially viable polymers for these applications [5]. Other researchers have shown that renewable-resource polymers, such as poly(lactic acid) (PLA) and others, can be viably produced and these materials are currently on the market for biomedical and packaging applications [4,5]. Our goal is to develop low-cost renewable copolymeres with comparable or even enhanced physical and chemical properties versus petroleum-based commodity plastics.

Poly(β-hydroxyalkanoate)s (PHA)s, Poly(glycolic acid) (PGA) and PLA are biocompatible plastics that have been shown to readily decompose [6-10]. Furthermore, the copolymer of lactic acid and glycolic acid, poly(lactic acid-co-glycolic acid) (PLG), is a widely used medical plastic [11]. In addition, NatureWorks™ family of PLA is being produced on industrial scales by Cargill Dow LLC [12]. DuPont and Shell Chemical Company have also formulated 'green' polymers under the names of Sorona® and Corterra® respectively. These new polymers are based on a PDO and diacid as the monomers [13, 14]. The polymerization of aliphatic dicarboxylic acids (with carbon chains lengths ranging from 4-14 carbons) with GLY has been successful with conversions ranging from 10 to 90% [15]. Moreover, diols and dicarboxylic acids can be used to produce biodegradable polyesters. PDO has been polymerized with terephthalic acid and sebacic acid to form a biodegradable polyester. The properties of these polyesters varied from brittle and opaque to flexible and transparent with varying degrees of biodegradability [16].

Production of polyesters is accomplished by polycondensation reactions. Reaction yields greater than 98% are necessary to produce high molecular weight polymers for step-growth polymerization [17]. Catalyst, monomer and temperature combinations that can achieve these high conversions are difficult to find, especially when the use of environmentally benign starting materials is also specified. The utilization of catalysts that are non-toxic or of low toxicity is necessary to achieve this goal. Lewis acids have been identified as a group of polymerization catalysts that have low or no toxicity when released into the environment [18]. The monomers need to be not only environmentally benign, but also low-cost and highly available [19]. The polymerization temperature is also a factor that needs to be addressed since there can be an increase in side reactions with increased temperature. Polyesterification with diols and dicarboxylic acids has been performed from 100-190°C [11,20]. By optimizing the reaction variables, including monomer, catalyst, and temperature, polymerization of biomass-derived polymers is possible.

The selection of PDO, GLY, fumaric acid (FA), and malonic acid (MA) for polymerization and copolymerization was based on functional groups, availability and cost. Polyester production can be achieved by using monomers with poly-acid or poly-ol functionalities. The monomers also need to be easily obtained from side products of biorefineries. The U.S. Department of Energy screened chemicals from biomass for building block potential based on their abilities to be transformed into a range of different starting materials, to replace a currently produced building block, and to become a 'super' commodity chemical in an effort to make biorefineries economical [21].

The catalysts used for renewable copolymer production need to be environmentally benign, biocompatible, easily handled and low cost. Aluminum, iron, zinc, and tin, for the most part, are environmentally benign, biocompatible, easily handled, and low cost [22-24]. While these metals may not perfectly fit the criteria outlined above, they are good starting catalyst components for renewable copolymer development. The reaction of these elements with hydrogen chloride leads to strong Lewis acids which have been shown to be good catalysts for the cationic polymerization of lactones and derivatives of glycerol [18, 25, 26]. The catalysts selected for this study include aluminum chloride, iron(III) chloride, tin(II) chloride, and zinc chloride.

Experimental Procedure

Reagents

Malonic acid (99%) and chloroform (98%) were used as received from VWR. 1,3-Propane diol (98%), $AlCl_3$ (98%), $ZnCl_2$ (98%), $FeCl_3$ (98%), $SnCl_2$ (98%), and diethyl ether (>99%) were used as received from Sigma-Aldrich. Glycerol (98%) and Fumaric acid (98%) were received from Fisher Scientific.

Polymerization

Copolymerization of PDO with MA and GLY with FA. The polymerizations were initially run in 10 mL test tubes at atmospheric pressure. PDO with MA and GLY with FA were fed in a 1:1 mole ratio with a 100:1 monomer to catalyst ratio and the test tubes were immediately placed into an oil bath and allowed to react for 24 h. These initial reactions were performed at temperatures from 125-175 ºC at 10 ºC intervals. The second set of experiments were done in 100 mL round bottom flask with varying molar ratio (0.1 to 0.9 alcohol fraction) with a 100:1 monomer to catalyst ratio. The system was at a pressure of 25 torr, 155 °C, and was stirred by a magnetic stir bar for four hrs.

Polymer separation. When the reactions were complete, excess monomer and catalyst were removed by dissolving the reaction product in chloroform. This chloroform solution is then poured into diethyl ether. The polymer precipitates out of solution and is removed by filtration through Whitman (grade 40) filter paper. This separation process is repeated until no additional polymer precipitates. The filtered polymer is then dried in a vacuum oven at 15 torr and 20 ºC for 24 hr and weighed.

Characterization

A Bruker AMX-300 NMR instrument was used to collect H^1 spectra at 300 MHz. The polymer was also characterized by Transmission FTIR spectroscopy using a Thermo Electron 6700 instrument and a transmission accessory. The FTIR system was purged with dry air. The product yields were calculated using the limiting monomer to determine theoretical yields based on complete conversion. The recovered polymer weight was divided by the theoretical weight and then converted to percent yield. Differential scanning calorimetry was done on a Setaram Instrumentation Technologies Model DSC 131 under a nitrogen purge of 30 mL/min and a heating rate of 5 ºC/min from -40 ºC to either 200 ºC or 400 ºC. A Water GPC was used with two 4.6 x 300 mm Styragel® columns, using THF as the effluent at 0.3 mL/min, and calibrated with polystyrene standards.

Results and Discussion

The initial experiments with PDO and MA where done in small volume and without vacuum or stirring, which limited the yields. All catalysts were initially used, but it was determined that the catalysts did not have a statistically significant affect on the yields. The yields for these first set of experiments ranged from 10% to 45% with the maximum being between 135 ºC and 155 ºC. For the later experiments that had vacuum and stirring, yields for PDO and MA reached approxemently 62% and for GLY and FA the yields reached approxemently 98%. Another important variable to be determined is the ratio of alcohol to acid, the ratio of PDO to MA and GLY to FA were examined. The yields from these experiments can be seen in Figure 1. The yield for PDO and MA is maximized at 0.75 molar fraction of PDO with a 62% yield. The maximum yield not being centered on 0.5 alcohol molar fraction is being furthered researched at this time. GLY and FA had maximum yields above 90% between 0.75 and 0.25 alcohol fraction and fell off at the extremes, which is expected as reactants become scare during polymerization.

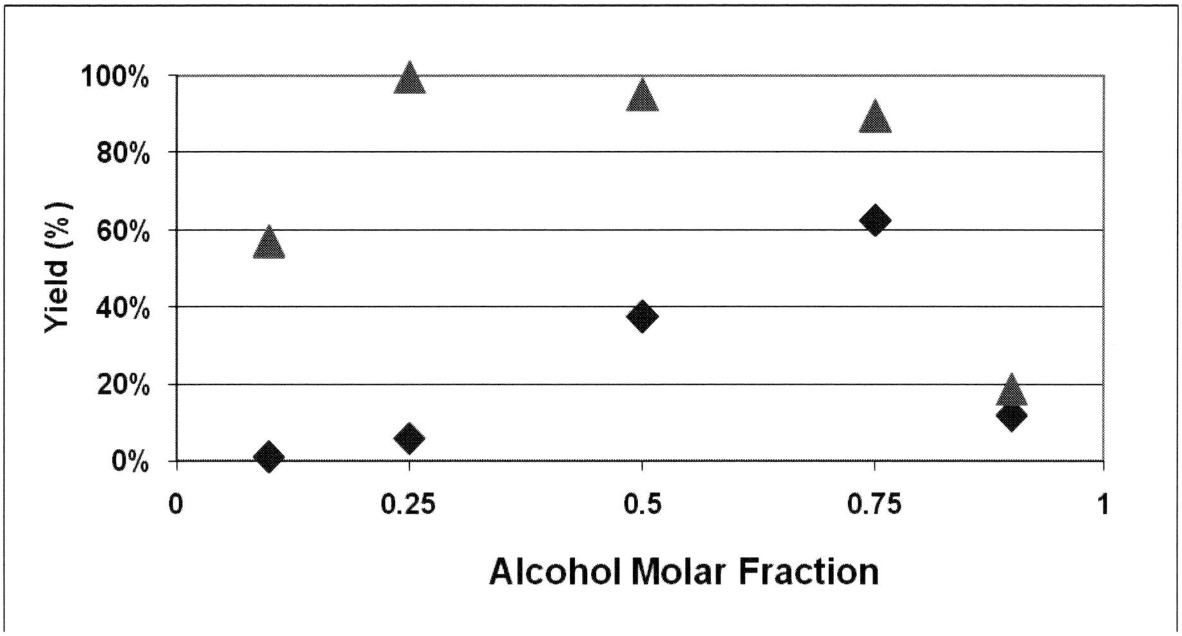

Figure 1. Yields compared against alcohol molar fraction for PDO and MA copolymer (blue diamond) and GLY and FA copolymer (red triangle).

Chemical characterization of the resultant polymers was performed using FTIR spectroscopy. Representative FTIR spectra of the polymers produced from PDO and MA with aluminum chloride as catalysts and GLY and FA with aluminum chloride as the catalysts are shown in Figure 2. The strong absorption peak at 1733 cm^{-1} corresponds to the carbonyl group of polyesters made from aliphatic diols that typically lies in range of 1715-1740 cm^{-1} [27]. In addition to the carbonyl of an ester, a peak at 1750 cm^{-1}, 1710 cm^{-1}, and 1040 cm^{-1} corresponds to a carbonyl of an anhydride. The series of peaks from approximately 2800 to 3000 cm^{-1} correspond to CH stretching vibration of alkane functionalities [28]. The broad absorption peak around 3500 cm^{-1} is due to hydrogen bonding and is more pronounced for the GLY and FA polymer than the PDO and MA polymer [29]. It was initially suspected that the difference in reactivity of the two functional groups would lead to a polymer with only ester bonds. However, the use of Lewis acid allowed for the chlorination of the carboxylic acid to form an acid chloride. The acid chloride does not preferentially react with the alcohols leading to the formation of both ester and anhydride linkage as shown by the FTIR spectrum.

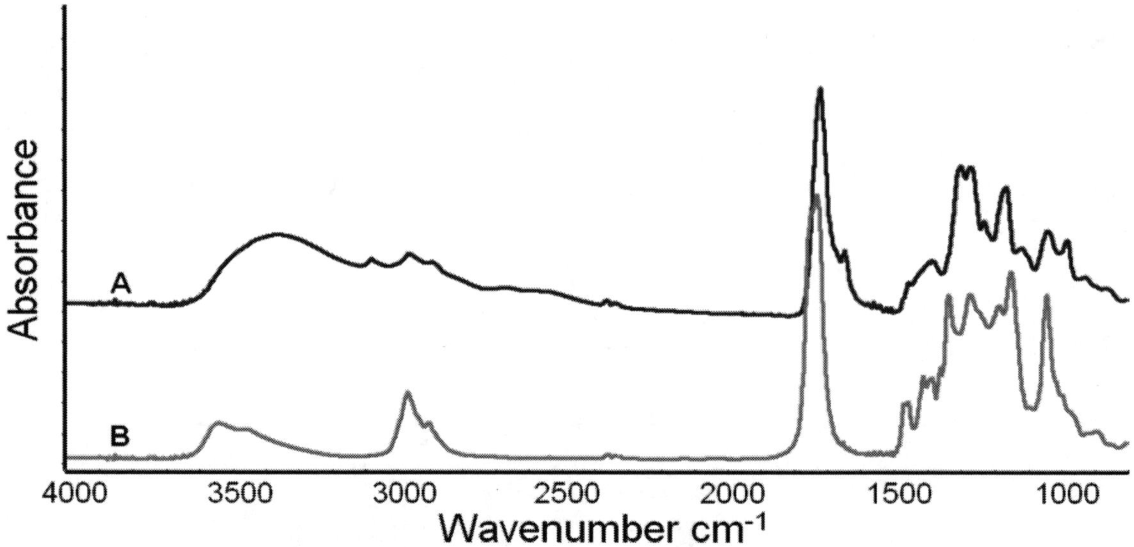

Figure 2. FTIR spectrum of GLY and FA copolymer (A) and PDO and FA copolymer (B).

The H[1] NMR spectrum (Figure 3) for PDO and MA copolymer produced using aluminum chloride as the catalyst had peaks at 2.28 and 4.25 ppm correspond to –CH_2-C(=O)OC(=O)- and –COO-CH_2. The peaks at 2.04 and 1.44 ppm correspond to the additional ethyl and propyl CH_2 groups from 1,3-propane diol. In addition, the peaks at 3.73 and 3.42 ppm correspond to the end group –CH_2OH and –CH_2- groups, respectively [27].

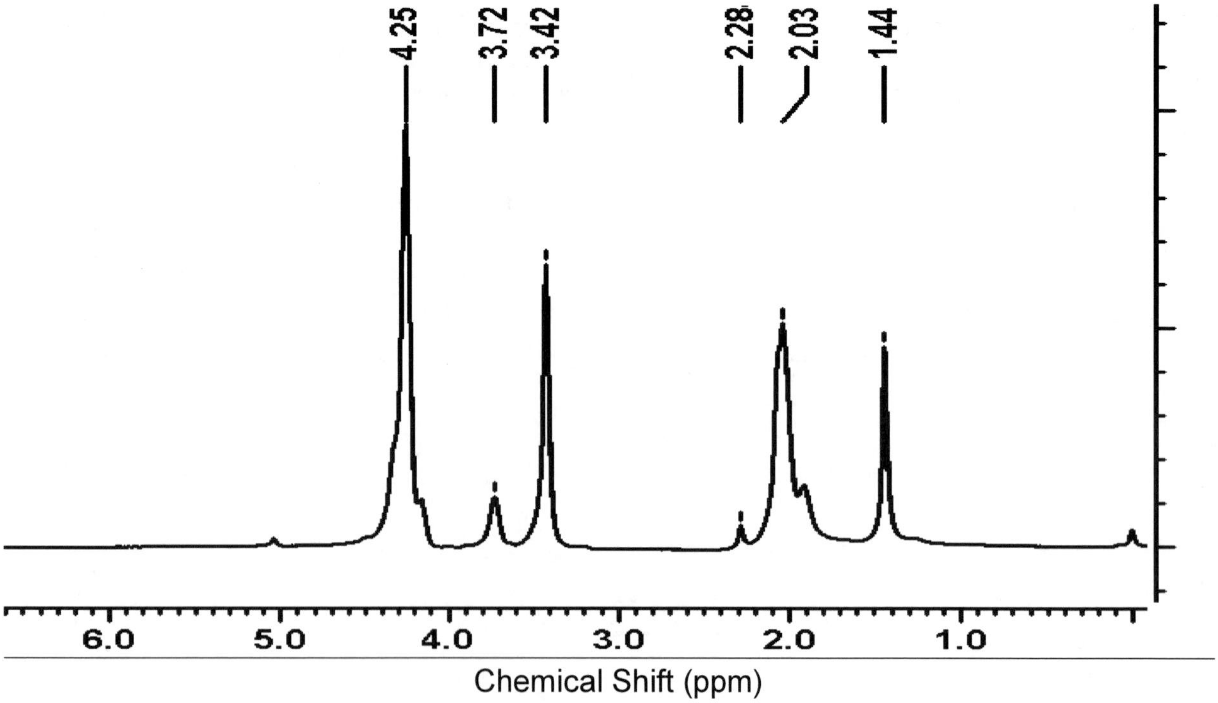

Figure 3. Representative H[1] NMR of PDO and MA copolymer.

Initial DSC characterization was performed to 400 °C to ensure that the glass transition and melting temperatures were captured and later experiments were performed only to 200 °C. For the samples made from PDO and MA, the glass transition temperature is preliminarily believe to lie between -40 °C and -20 °C, seen in Figure 4. The most notable feature of the experiments is the melting point seen between 40 °C and 50 °C. The melting point is 43.9 ±0.8 °C with a onset temperature of 37.9 ±2.2 °C. At approxementy 75 °C, there is a small, sharp peak indicating a change in crystallization, and between 110 °C and 130 °C, two broad peaks that are indicative of cross-linking or curing. GLY and FA were polymerized and examined via DSC under the same conditions as the PDO and MA samples. The thermograms of GLY and FA polymerization showed no apparent glass transition or melting point.

The effect of alcohol molar fraction on the molecular weight (Mw) of the synthesized polymers was examined. For the PDO and MA copolymers, there was no statistically significant effect on Mw with the Mw's ranging from 25 kDa to 50 kDa for all alcohol molar fractions (Figure 5). The consistent Mw of the PDO and MA copolymers was unexpected and is being further investigated. GLY and FA polymers had increasing Mw with increasing alcohol fraction, which was expected.

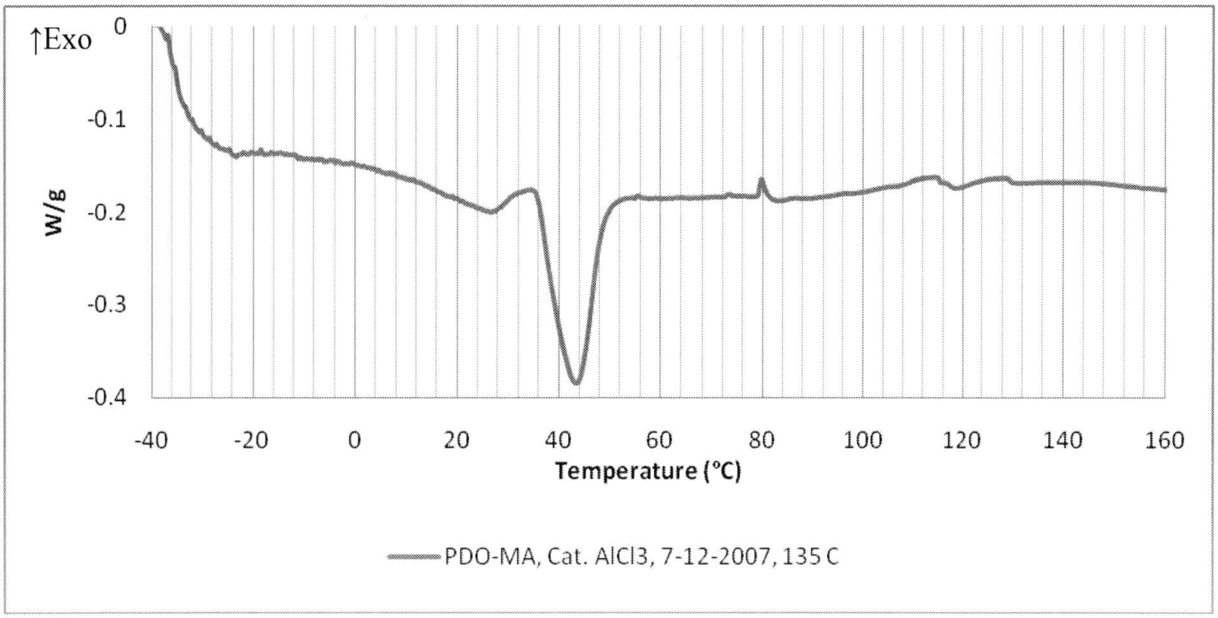

Figure 4. Representative thermogram of PDO and MA copolymer.

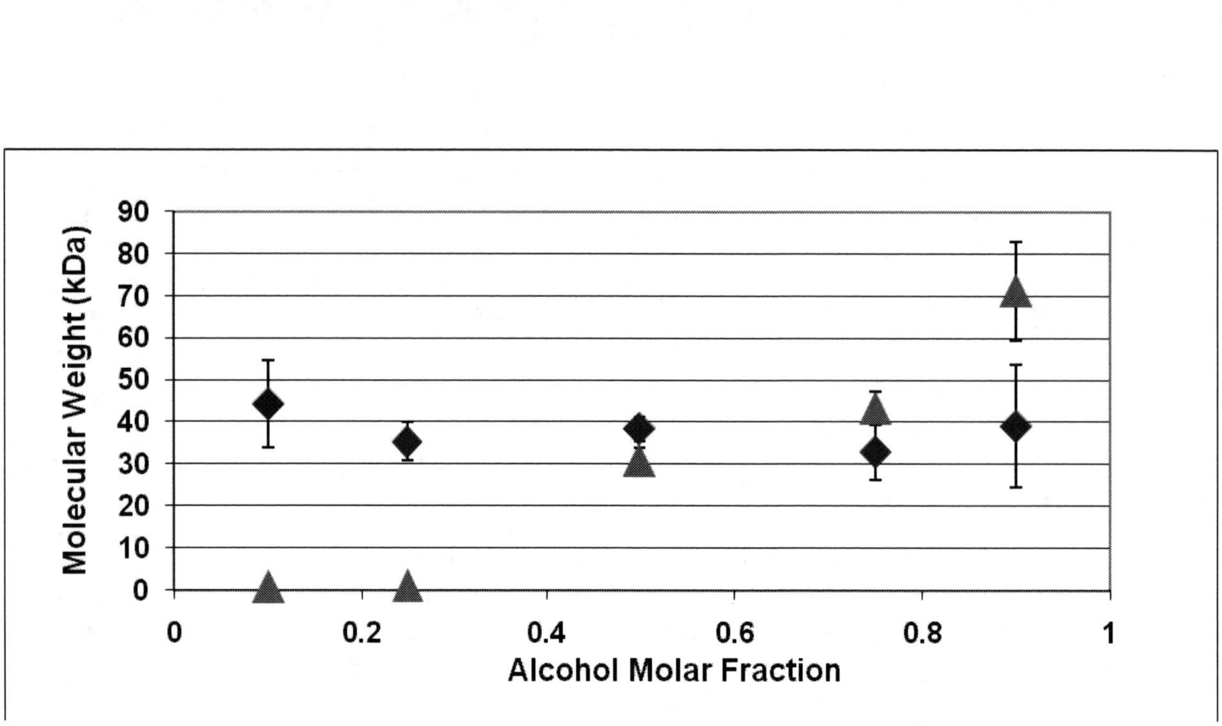

Figure 5. Mw compared against alcohol molar fraction for PDO and MA copolymer (blue diamond) and GLY and FA copolymer (red triangle).

Conclusions

Preliminary studies have shown that polyester copolymers can be produced using by-products of biorefineries as starting materials. A copolymer with both ester and anhydride bonds is formed with low to medium molecular weights. These reactions will be scaled-up and expanded to more complex monomers. In addition, continued physical and chemical property characterization of the resultant polymers is planned.

Acknowledgements

This work was supported by the Department of Energy through the Sustainable Energy Research Center (SERC) at Mississippi State University (MSU). Thanks to the Bagley College of Engineering at Mississippi State University (MSU) for their financial support through the Bagley Graduate Fellowship Program.

References

1. Mapleston, P. New Technologies for a Greener Industry, Plastic Engineering, 2008, 11-15.
2. Stewart, R. Going Green: Eco-Friendly Materials and Recycling on Growth Paths, Plastic Engineering. 2008. 17-23.
3. Robinson, G., Biofocals, Waste Management World, 2007, 77-83.
4. Tullo, A. H., World Chemical Outlook: United States, Chemical and Engineering News, 2008, 16-19.
5. Finkelstein, M., McMillan J. D., Davison, B. H., Varbara, E., Biotechnology for Fuels and Chemicals, Applied Biochemistry and Biotechnology, 2004, 113-116, 871-884.
6. Dechy-Cabaret, O., Martin-Vaca, B., Bourissou, D., Controlled Ring-Opening Polymerization of Lactide and Glycolide, Chemical Review, 2004, 104, 6147-6176.
7. Mohanty, A. K., Misra, M., Hinrichsen, G., Biofibres, Biodegradable Polymers and Biocomposites: An Overview, Macromolecular Materials and Engineering, 2000, 276-277, 1-24.
8. Okada, M., Chemical Synthesis of Biodegradable Polymers, Progress in Polymer Science, 2007, 27, 87-133

9. Middleton, J. C., Tipton, A. J., Synthetic Biodegradable Polymers as Orthopedic Devices, Biomaterials, 2000, 21, 2335-2346.

10. Li, S., Hydrolytic Degradation Characteristics of Aliphatic Polyesters Derived from Lactic and Glycolic Acids, Journal of Biomedical Materials Research (Applied Biomaterials) 1999. 48, 342-353.

11. Albertsson, A., Varma, I. K., Recent Developments in Ring Opening Polymerization of Lactones for Biomedical Applications, Biomacromolecules 2003, 4, 1466-1486.

12. Drumright, R. E., Gruber, P. R., Henton, D. E., Polylactic Acid Technology, Advanced Materials 2000, 12, 1841-1846.

13. Kurian, J. V., A New Polymer Platform for the future -- Sorona® from Corn Derived 1,3-Propanediol, Journal of Polymers and the Enviroment, 2005, 13, 159-167.

14. Werny, F., Chuah, H., Corterra PTT – A New Polymer for the Carpet Industry, Shell Chemicals, 1999.

15. Nagata, M., Kiyotsukuri, T., Ibuki, H., Tsutsumi, N., Sakai, W., Synthesis and Enzymatic Degradation of Regular Network Aliphatic Polyesters, Reactive & Functional Polymers, 1996, 30, 165-17.

16. Witt, U., Műller, R.-J., Augusta, J., Widdecke, H., Deckwer, W.-D., Synthesis, Properties and Biodegradability of Polyesters based on 1,3-Propanediol, Macromolecular Chemistry and Physics, 1994, 195, 793-802.

17. Odian, G., Principles of Polymerization, Fourth Ed., John Wiley & Sons, Inc. 2004, 39-41.

18. Abraham, G. A., Gallardo, A., Lozano, A. E., Roman, J. S., ε-Caprolactone/ZnCl2 Complex Formation: Characterization and Ring-Opening Polymerization Mechanism, Journal of Polymer Science, Part A: Polymer Chemistry, 2000, 38, 1355-1365.

19. Willke, T., Vorlop, K.-D., Industrial Bioconversion of Renewable Resources as an Alternative to Conventional Chemistry, Applied Microbiology Biotechnology 2004, 66, 131-142.

20. Lehtonen, J., Salmi, T., Immonen, K., Paatero, E., Nyholm, P., Kinetic Model for the Homogeneously Catalyzed Poly-esterification of Dicarboxylic Acids with Diols, Industrial and Engineering Chemical Research 1996, 35, 3951-3963.

21. Werpy, T., Peterson, G., Top Value Added Chemicals from Biomass: Volume I - Results of Screening for Potential Candidates from Sugars and Synthesis Gas, U.S. Department of Energy, 2004.

22. Dietary Reference Intake: Food and Nutrition Board, Institute of Medicine, National Academies, National Academy of Science, 2004.

23. Alzheimer's Disease, niehs.nih.gov/ external/faq/aluminum.htm (11/15/06).

24. Rouault, T. A., How Mammals Acquire and Distribute Iron Needed for Oxygen-Based Metabolism, PLoS Biology, 2003, 1, 326-328.

25. Kricheldorf, H. R., Mang, T., Jonte, J. M., Polylactones. 1. Copolymerization of Glycolide and e-Caprolactone, Macromolecules, 1984, 17, 2173-2181.

26. Moon, S, I,. Lee, C. W., Miyamtot, M., Kimura, Y. Melt Polycondensation of L-Lactic acid with Sn(II) Catalysts Activated by Various Proton Acids: A Direct Manufacturing Route to High Molecular Weigh Poly(L-lactic acid), Journal of Polymer Science, Part A: Polymer Chemistry, 2000, 38, 1673-1679.

27. Rogers, M. E., Long, T., Tuner, S. R., Synthetic Methods in Step-Growth Polymers, John Wiley & Sons, Inc. 2003, 91-93.

28. Silverstein, R. M., Bassler, G. C., Morrill, T. C. Spectrometric Identification of Organic Compounds, John Wiley & Sons, Inc., 1991, 105.

29. Kuptsov, A. H., Zhizhin, G. N., Handbook of Fourier Transform Raman and Infrared Spectra of Polymers, Elsevier, 1998, XXIV – XXVII.

107f Silica Surface Modifications for Bioapplications

Fei Shen[1], Jan Genzer[1], Orlando J. Rojas[2], Patrick V. Gurgel[3], and Ruben G Carbonell[1]. (1) Department of Chemical and Biomolecular Engineering, North Carolina State University, 911 partners way, EB1, Raleigh, NC 27695-7905, (2) Forest Biomaterials Science and Engineering, North Carolina State University, Campus Box 8005, Raleigh, NC 27695-8005, (3) Department of Chemical and Biomolecular Engineering, ProMetic Life Sciences and NC State University, 8168 chemin Montview, 911 Partners Way, Mont-Royal, QC H4P 2L7, Canada

Flat silica surfaces were modified for the immobilization of ligands or proteins to be used in a study of protein-ligand and protein-protein binding events on surfaces with low background from non-specific protein binding. A self-assembling monolayer (SAM) of 3-amino-propyl triethoxy silane (APTES) was formed on the silica surface as an initial anchor layer, then short chains of poly(ethylene glycol) with Fmoc protected amine group at one end and carboxyl group at other end were coupled to the silane SAM to reduce nonspecific binding, and finally the amine groups at the end were liberated by Fmoc deprotection, and these will be the active groups used for protein and ligand immobilization. The pK1/2 of the amine group on surface was determined by contact angle titration. The modified surface was also characterized by ellipsometry, X-ray photoelectron spectroscopy (XPS), and time of flight-second ion mass spectroscopy (ToF-SIMS). The amount of protein adsorption on the modified surface was determined by quartz crystal microbalance (QCM).

107g Green Catalytic Hydrogenolysis of Glycerol to Propylene Glycol Over a Promising Silica Supported Copper Catalyst

Zhiwei Huang, Fang Cui, Jing Chen, Xinzhi Zhang, **Zhongyi Ding**, and Chungu Xia. State Key Laboratory for Oxo Synthesis and Selective Oxidation, Lanzhou Institute of Chemical Physics,Chinese Academy of Sciences, Lanzhou, 730000, China

Glycerol is a main byproduct in the biodiesel production process, and the effective utilization of glycerol can contribute to the economy of the whole process. Catalytic hydrogenolysis of glycerol to propylene glycol was investigated over an environmentally friendly silica supported copper catalyst. This process provides a clean and economically competitive route for the production of commercially important 1,2-propanediol from renewable glycerol instead of from non-renewable petroleum. Propylene glycol was selectively produced (with a selectivity > 96%) in high yield (> 80%) under moderate conditions. Products with high purity were obtained easily without complicated separation processes. In addition, the catalyst showed excellent stability for glycerol reaction, no appreciable decrease in the activity was observed even after 1000 h of time-on-stream. The application of this copper based catalyst in glycerol hydrogenolysis is promising.

107e Environmentally-Friendly Glycerol Disposal Using Bacterial-Mediated Fermentation

Andrew T Ford[1], Bart HJ. van den Berg[2], Rafael Hernandez[3], William E. Holmes[4], and W. Todd French[3]. (1) School of Chemical Engineering, Mississippi State University, 408 11th St. S, Columbus, MS 39701, (2) Department of Basic Sciences, Mississippi State University, Box 6100, Mississippi State, MS 39762, (3) Dave C. Swalm School of Chemical Engineering, Mississippi State University, PO Box 9595, Mississippi State, MS 39762, (4) Mississippi State Chemical Lab, Mississippi State University, Starkville, MS 39762

In the search for alternative energy sources, biodiesel has become the most popular alternative fuel source. Glycerol is the by-product formed in current biodiesel production processes and has little market value. We aim to convert glycerol into other valuable industrial chemicals through bacterial-mediated fermentation. Under anaerobic conditions, and varying concentrations of bacterial cells, glycerol conversion was monitored using GC-FID. Reaction kinetics were examined based on the concentrations of formed products. Our primary results show that an initial glycerol concentration of 25g/L can be converted into a 2g/L concentration of acetic acid and a resulting glycerol concentration of 19g/L. Conversion of glycerol into butanoic and propanoic acid has also been observed. Determination of reaction kinetics would allow us to optimize this method for efficient glycerol conversion, which provides an environmentally friendly, cost efficient means of glycerol disposal, and could represent a valuable attribute to the biodiesel production pipeline.

107a Green Chemistry Considerations in the Design of Ionic Liquids

Richard E. Engler, Green Chemistry Program, US Environmental Protection Agency, 1200 Pennsylvania Ave. NW (7406M), Washington, DC 20460 and J. Vincent Nabholz, Office of Pollution Prevention & Toxics, U. S. EPA, Risk Assessment Division, Mail Code 7403, 1200 Pennsylvania Avenue, N. W, Washington, DC 20460.

Ionic liquids (ILs) are a still-growing field that may provide greener alternatives to traditional solvent systems. ILs are often touted as low volatility alternatives to traditional volatile organic solvents. However, low volatility is not sufficient to classify an IL as “green”. A wide variety of cations and anions can be combined to form ILs, allowing a wide range of properties and activity. The potential toxicity of anions and cations should be considered during the design of ILs. Human health and ecologic toxicology considerations of cations and anions will be discussed.

107c Screening Ionic Liquids for Green Chemistry

David A. Gallagher, CAChe Research, 13690 SW Otter Lane, Beaverton, OR 97008, Andreas Klamt, COSMOlogic GmbH&CoKG, Burscheider Str. 515, Leverkusen, 51381, Germany, and Michael Diedenhofen, COSMOlogic GmbH & Co.KG, Burscheider Str. 515, Leverkusen, D-51381, Germany.

Most methods for screening and predicting activity coefficients do not work well for solutes in ionic liquids. However, although originally developed for neutral solvents, the quantum-chemistry based statistical thermodynamics method, COSMO-RS [1], does appear to work equally well for ionic liquids [2], and without any modification.

The reason for the success of COSMO-RS with ionic liquids is that this method accounts for the strong short-range polar interactions in such systems, where long-range charge interactions are of negligible importance. The advantages of this method over the alternative simulation and prediction techniques with ionic systems, has led to its adoption in academia and industry for screening ionic liquids as green alternatives for certain separation processes.

In this presentation we shall discuss the current knowledge and experience on the applicability of the COSMO-RS method, as implemented in the software package COSMOtherm[TM], to ionic liquids screening with examples from recent publications.

[1] A. Klamt, COSMO-RS: From Quantum Chemistry to Fluid Phase Thermodynamics and Drug Design, Elsevier, Amsterdam 2005

[2] M. Diedenhofen, F. Eckert, and A. Klamt, Prediction of Infinite Dilution Activity Coefficients of Organic Compounds in Ionic Liquids Using COSMO-RS, J. Chem. Eng. Data, 48 (3), 475 -479, 2003

27c Proton - SO_3^- Interactions in Nafion Polymer Electrolyte Membrane Fuel Cells Gleaned from *Ab Initio* Molecular Dynamics Simulations of Mono-, Di-, and Tetra-Hydrate Trifluoromethanesulfonic Acids

Robin L Hayes and Mark E. Tuckerman. Department of Chemistry, New York University, 100 Washington Square East, New York, NY 10003

Polymer electrolyte membrane (PEM) fuel cells have the potential to replace petroleum-based fuels with more environmentally friendly energy sources in transportation and stationary residential applications. Nafion, the dominant PEM material, self-assembles into hydrophobic regions and hydrophilic channels decorated with SO_3^- that conduct protons via a water mediated mechanism. Structural proton diffusion is expected to dominate, but is hindered upon loss of water. Mono-, di-, and tetra- hydrate trifluoromethanesulfonic acids provide a model system to study proton transport as a function of hydration. *Ab initio* molecular dynamics reveal stable defect structures, SO_3^- - H^+ interactions, and proton conduction pathways. Results will be compared to path integral molecular dynamics calculations.

Fabricating Colloidal Crystals with Different Structures and Nanostructures Derived from Them

Zhiqiang Sun and Bai Yang*

State Key Lab of Supramolecular Structure and Materials, College of Chemistry, Jilin University, Changchun 130012, P. R. China. Email: byangchem@jlu.edu.cn

Colloidal crystals of ordered microspheres represent a new class of advanced materials. For example, they can be used as scaffolds of highly ordered macroporous materials and high-strength ceramics. Due to the novel light diffraction and photonic bandgap properties, colloidal crystals are also promising candidates for constructing devices such as optical filters and switches, chemical and biochemical sensors, and photonic chips. Various self-assembly techniques based on gravity sedimentation, electrostatic interaction, and capillary force have been developed to form colloidal crystals on different substrates, including the flow-cell methods, vertical deposition, micromolding in capillaries (MIMIC) and so on [1]. Although the existing methods allow fabrication of colloidal crystals with close-packed structures, efficient approaches to form high-stability and large scale colloidal crystals of different structures are still demanded. On the other hand, introducing ordered microstructures within colloidal crystals is of particular importance for preparation of optical devices.

Recently, we have developed a number of methods to organize polymeric, inorganic, even composite microspheres into various structures, generating various properties and functions.[2] Using these methods, various of colloidal crystals with different structures have been prepared, including colloidal crystals with ordered voids and two- or three-dimensional (2D or 3D) patterned arrays, colloidal crystals on nonplanar surfaces, heterogeneous colloidal crystals of different building blocks, colloidal crystals composed of non-spherical polyhedrons, and particularly colloidal crystals of non-close-packed colloidal microspheres. Application of these colloidal crystals as templates for different structures range from nanoscale to micron-scale has also been introduced [3].

1. Fabricating new colloidal crystals of different structures

Colloidal crystals have attracted extensive interest due to their potential applications in fields, such as optics, photonics, sensing, and surface patterning. Although a large number of methods have been developed to control the size, structure, and crystalline orientation of colloidal crystals, challenges still exist in introducing some specific microstructures into them for their promising device applications.

1. 1 Stable colloidal crystal chips and non-spherical colloidal crystals

Combining the vertical deposition with the micromolding in capillaries (MIMIC) method, we have demonstrated a versatile procedure of fabricating high-quality stable colloidal crystal chips. Figure 1 (I) shows typical scanning electron microscopy (SEM) image of the colloidal crystals obtained by two-substrate vertical deposition. The confinement of the two substrates renders colloidal crystal chips rather mechanically stable. Accordingly, we have developed an alternative approach towards non-spherical colloidal crystals (NSCCs) by thermal-pressing polymer colloidal crystals. Figure 2 (II) shows the section SEM image of the NSCCs.

Figure 1. SEM images of the colloidal crystals fabricated by two-substrate vertical deposition (I) and of the NSCCs (II).

1. 2 Patterned colloidal crystals

Applying lift-up soft lithography to colloidal crystallization, we have developed versatile approaches to patterned colloidal crystals of various structures. Figures 2 (I) shows a typical SEM image of parallel lines of 2D colloidal crystalline arrays fabricated by a one-step lift-up process. We also tried to apply this method to prepare microstructures of 3D colloidal crystalline arrays. Figure 2 (II) shows the resulting microstructures in a double-layered crystal film fabricated by a two-step lift-up process. Based on lift-up lithography and μCP, we have also succeeded in transferring colloidal crystals intentionally. Figure 3 (I) and (II) show the SEM images of the patterned 2D colloidal crystal transfered onto planar and nonplanar substrates, respectively.

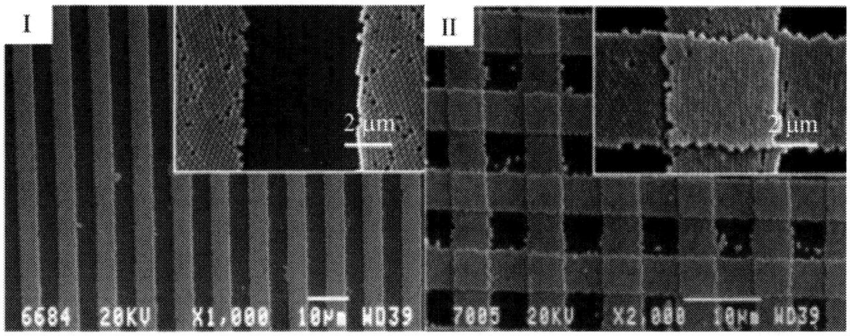

Figure 2. SEM images of 2D and 3D patterned colloidal crystals fabricated by lift-up process.

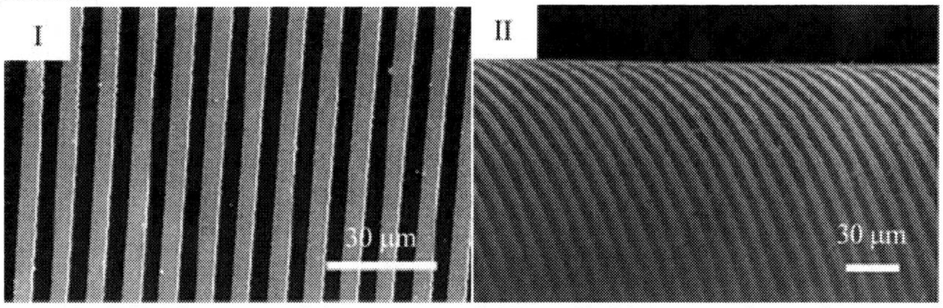

Figure 3. SEM images of the patterned 2D colloidal crystal on planar and nonplanar substrates.

1. 3 Non-close-packed arrays of colloidal microspheres

It is well-known that conventional self-assembly methods could be applied to produce close-packed 2D or 3D ordered arrays of colloidal microspheres. Using etching techniques or charged spheres, non-close-packed (ncp) arrays with controllable spacing between spheres can be created. Nonetheless, it is difficult to obtain lattice structures different from hexagonal packing. As we have demonstrated a lift-up soft lithography technique to form 2D hcp microsphere arrays on the surface of PDMS stamp. Based on the solvent-swelling and mechanical deformation behaviors of PDMS, we can adjust the lattice structures of these 2D microsphere arrays. Most importantly, the as-prepared 2D ncp arrays can be transferred onto the surfaces of solid substrates by using a modified μCP transfer technique. Figure 4 shows typical SEM images of the ordered silica microsphere arrays with different lattice structures.

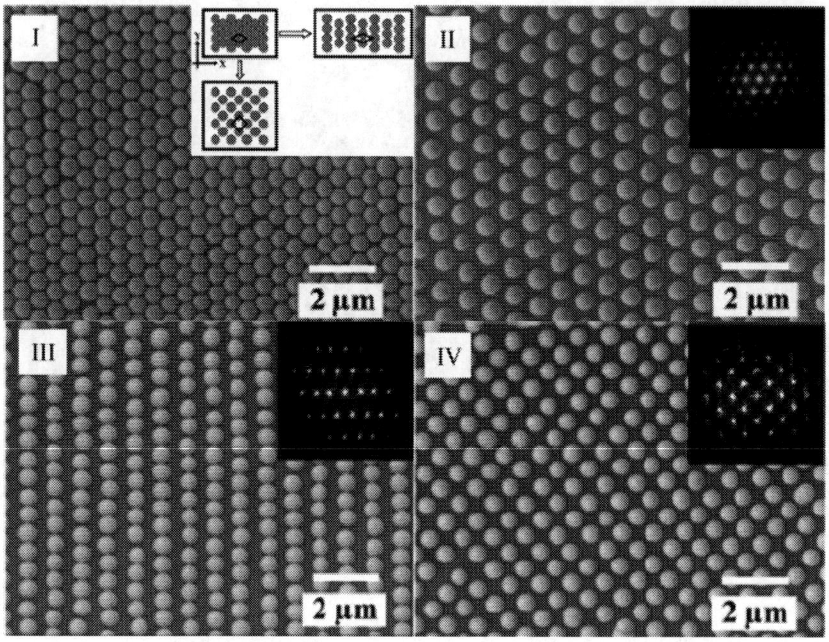

Figure 4. SEM images of the close-packed and non-close-packed array of silica microspheres

2. Application of colloidal crystals as templates for surface patterning

A number of approaches, involving μCP and soft lithography, self-assembly, and laser-assisted directed imprinting lithography, have been applied to pattern surfaces. However, to achieve 2D nanopatterned SAMs and desired morphologies on various substrates remains a challenge. Using the colloidal crystals as templates, we have developed a number of methods to generate surfaces patterned with different structures range from nanoscale to micron-scale. Particularly, we have developed colloidal-crystal-assisted-lithography (CCAL) techniques, in which we intentionally applied 3D colloidal crystals in preparing 2D nanostructures on various substrates. Figure 5 shows typical nanostructures fabricated by CCAL using 3D spherical and nonspherical colloidal crystals as templates.

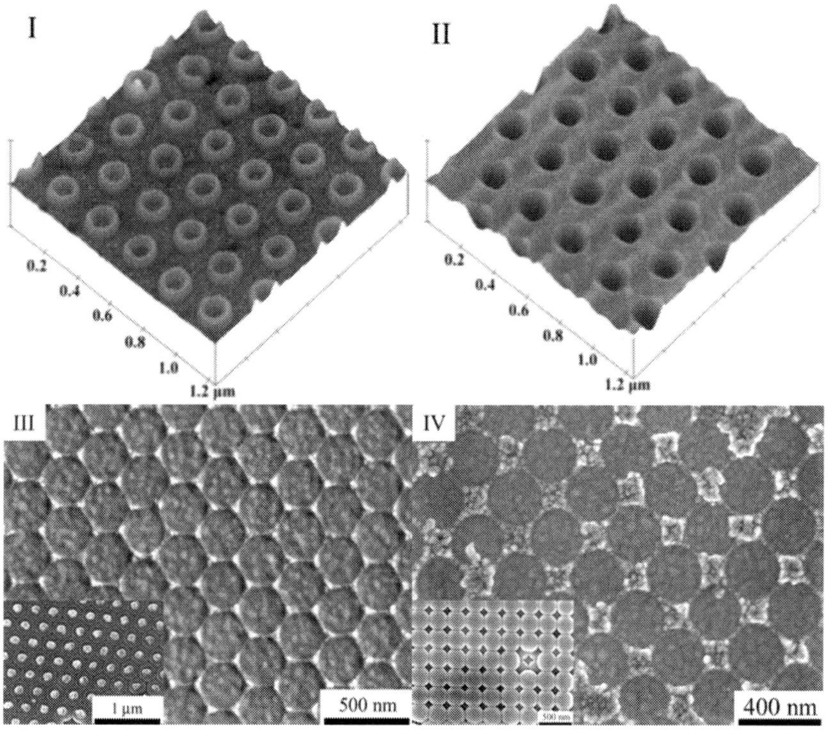

Figure 5. (I, II) Typical 3D AFM images of PVA nanostructures fabricated by CCACN. (III, IV) SEM images of silver nano-networks on gold substrates with different symmetries.

Conclusion:

In conclusion, we have demonstrated a set of approaches to fabricate new colloidal crystals with ordered voids, 2D or 3D patterned arrays, composed of non-spherical polyhedrons, patterned colloidal crystals on nonplanar surfaces, heterogeneous colloidal crystals of different building blocks, and particularly colloidal crystals of non-close-packed colloidal microspheres. These new colloidal crystals should be of importance in a wide range of applications, especially in photonics. Using various colloidal crystals obtained as templates, CCAL method has been established to generate surface patterns with different structures range from nanoscale to micron-scale. Therefore, our methods listed here should hold immersed promise in nanofabrication, nanosensing, microreactors, and control of colloidal crystallization.

Acknowledgment: This work was supported by National Basic Research Program of China (2007CB936402), the National Nature Science Foundation of China (Grant No. 90401020, 20534040) and the program for Changjiang Scholars and Innovative Research Team in University (No. IRT0422).

Reference

[1] S. H. Park, D. Qin, Y. Xia, *Adv. Mater.* **1998**, *10,* 1028; P. Jiang, J. F. Bertone, K. S. Hwang, V. L. Colvin, *Chem. Mater.* **1999**, *11*, 2132; Y. A. Vlasov, X. Bo, J. C. Sturm, D. J. Norris, *Nature* **2001**, *414*, 289; Z. Z. Gu, O. Sato, A. Fujishima, *Chem. Mater.* **2002**, *14*, 760; M. Trau, N. Yao, E. Kim, Y. Xia, G. M. Whitesides, I. A. Aksay, *Nature* **1997**, *390*, 674.

[2] G. Lu, X. Chen, J. M. Yao, W. Li, G. Zhang, D. Y. Zhao, B. Yang, J. C. Shen, *Adv. Mater.* **2002**, *14*, 1799; J. M. Yao, X. Yan, G. Lu, K. Zhang, X. Chen, L. Jiang, B. Yang, *Adv. Mater.* **2004**, *16*, 81; X. Yan, J. M. Yao,

G. Lu, X. Chen, K. Zhang, B. Yang, *J. Am. Chem. Soc.* **2004**, *126*, 10510; Z. Q. Sun, X. Chen, J. H. Zhang, Z. M. Chen, K. Zhang, X. Yan, Y. F. Wang, W. Z. Yu, B. Yang, *Langmuir* **2005**, *21*, 8987; X. Yan, J. M. Yao, G. Lu, X. Li, J. H. Zhang, K. Han, B. Yang, *J. Am. Chem. Soc.,* **2005**, *127*,7688;

[3] Z. Q. Sun, Y. F. Li, Y. F. Wang, X. Chen, J. H. Zhang, K. Zhang, Z. F. Wang, C. X. Bao, J. B. Zeng, B. Zhao, B. Yang, *Langmuir* **2007**, *23*, 10725; X. Chen, Z. M. Chen, N. Fu, G. Lu, B. Yang, *Adv. Mater.* **2003**, *15*, 1413; X. Chen, Z. Q. Sun, L. L. Zheng, Z. M. Chen, Y. Y. Wang, N. Fu, K. Zhang, X. Yan, H. Liu, L. Jiang, B. Yang, *Adv. Mater.* **2004**, *16*, 1632; Z. M. Chen, T. Gang, X. Yan, X. Li, J. H. Zhang, Y. F. Wang, X. Chen, Z. Q. Sun, K. Zhang, B. Zhao, B. Yang, *Adv. Mater.* **2006**, *18*, 924.

NANOTECHNOLOGY AND ITS APPLICATIONS TO ANTIMICROBIAL PESTICIDES: SCIENTIFIC AND REGULATORY ISSUES AND CONCERNS

Disclaimer: Viewed expressed in this presentation are those of the author and do not represent EPA's current thinking or Policy.

Background:

It is too early to assess full impact and importance of nanoscience and applications of nanotechnology in various fields of engineering, drug design and drug delivery for treating diseases, or manufacture of new products as industrial chemicals and pesticides, or daily-use products like cosmetics etc. Nanoscience is still in the developmental stage and a full risk assessment and benefits of nanoproducts can not be done at this time. Some characteristics of nanomaterials have been understood, some new characteristics are being discovered, but a wide variety of issues and concerns still remain unanswered. This has been demonstrated by EPA's White Paper on Nanotechnology[1], NSF's NSET Workshop Report[2], Maynard's work[3] on exposure issues, and this true for agricultural and food production science and technology [4, 5]

EPA as an Agency in general, and OPP in particular, deals with enormous amounts of scientific data and it has the burden of making regulatory decisions on industrial chemicals, agricultural and antimicrobials pesticides, and biopesticides. The new nanoscience and nanotechnology field presents an enormous challenge for the Agency in both domains.

In this presentation, we will focus on the exposure of food and water from antimicrobial products which are likely to be nanoproducts.

This presentation will also explore short-term benefits that can be achieved by using nanopesticides or a composite of an inert with a nanoppesticide. Long-term benefits are too early to assess.

EPA's Office of Pesticides Programs, mandated by FIFRA, is required to conduct risk assessment for exposure (workers, post-application, exposure into environmental media like air, soil, water, terrestrial and aquatic organisms and toxicity associated with these chemicals to humans and assess the hazards and risk associated for each environmental medium.

Uses, application rates, and application scenarios of agricultural pesticides are huge compared to the antimicrobial pesticides. Uses of antimicrobials, however, is closer to home and therefore are likely concerns for residential users primarily, outdoor users, in general. And this can be true for nanopesticides. Risk assessment of antimicrobials is conducted on twelve use patterns of these pesticides. These are: agricultural premises, and equipments, food handling/storage establishments and equipments, commercial, institutional and industrial premises and equipment, residential and public access premise, medical premises and equipment, human drinking water systems, material preservatives, industrial processes and water systems, antifoulant paints and ballast water treatments, wood preservatives, swimming pools, and aquatic areas. Of these use patterns, eight are indoor/residential and four are the so-called outdoor use patterns. Each use pattern triggers a set of data requirements.

Antimicrobials are used as: disinfectants, sanitizers, sterilants, antifoulants, materials preservatives and wood preservatives. As disinfectant/sanitizer/sterilant uses, antimicrobials are applied to drinking water systems, swimming pools, once-through and recirculating cooling towers, metalworking fluids, material preservatives like paints. These applications are likely to result in the presence of these antimicrobials into the wastewater streams and may end up in drinking waters.

As wood preservative uses (fungicide, insecticide), whether the wood is surface or pressure treated, it is likely to contaminate soils, and waterways when it enters into wastewater streams. Contaminated soils around the backyard decks, and children's play area have been shown to result in cancer risks for certain type of wood preservatives. As contaminants in water, antimicrobials may likely pose exposure and hazard concerns to humans, and aquatic organisms. As antifoulant use (molluscide etc.), contamination of water may result in drinking water problems and eco risks for the aquatic organisms. In addition, if the antifoulant reaches the sediments in lakes, rivers, estuaries, it is likely to adversely affect the bottom organisms.

The use of antimicrobials as disinfectant/sanitizer on hard surfaces such as counter tops, cutting boards, stove tops, microwave ovens, , the pesticide residues are likely to migrate into food indirectly and can a cause of concern for contamination of food. Similarly antimicrobials use in paper/pulp as disinfectant can cause an indirect migration of the residual pesticides into food by paper wrappers.

Antimicrobials use as disinfectant for drinking water, swimming pools, once-through and recirculating cooling towers are likely to contaminate the waterways and pose drinking water contamination concerns.

It should be pointed out the exposure and hazard concerns is not only by antimicrobials but also the degradates that these antimicrobials may form in various environmental media like water and soils, as well as compartments like fish tissues and humans when these are ingested by incident.

This paper shall address some of the enormous challenges that nanoproducts as pesticides may present.

Introduction:

A cursory glance in the research arena of nanoscience and nanotechnology can overwhelm both scientists and laymen. Similarly, searching through the number of books that have recently been published in the past six or seven years on nanoscience and nanotechnology is a phenomenon in itself. A manual count shows 120+ posters/papers/seminars were presented at the ACS Meeting in Boston (ACS Program Summary, C & E News, August, 2007) [*]

Environmental Protection Agency (EPA) like Food and Drug Administration (FDA) are unique Federal Agencies as these agencies pour over a tremendous amount of scientific data and conduct risk assessments. Office of Pesticides programs (OPP) within EPA is mandated under Federal Insecticide, Fungicide and Rodenticide Act (FIFRA) [6] to require scientific data on all pesticides submitted for registration. The submitted data undergoes a detailed and thorough review and risk assessment is done on each chemical. FIFRA. for safety, health, and environment. Code of Federal Regulations (40CFR) has undergone a couple of changes and more stringent guidance have been added to the act: Food Quality Protection Act (FQPA), August 1996 and more recently under PRIA II, of 2007 (Pesticide Registration Improvement Act II). FQPA addresses the improved scientific methods to assess risk assessments particularly for sensitive populations: infants and women. PRIA II addresses the regulatory method improvement for registering new or new uses of pesticides and a fee-based approach.

Office of Pesticides Programs has divided pesticides into to three broad categories: pesticides used in the agricultural arena (ag-pesticides); antimicrobials (disinfectants, sanitizers, sterilants, wood preservatives, and antifoulants); and biopesticides. All three categories require data submission from industry. It is some part of antimicrobial type pesticides like disinfectants, and sanitizers that data requirements and risk assessments of which overlap with FDA's Federal Food, Drug, and Cosmetic Act FFDCA (21CFR).[7]

Detailed Discussion:

For antimicrobials, it is important to know the use pattern which helps in determining in which environmental medium will the pesticide will likely deposit: air, water, soil, sediment, plants aquatic or terrestrial organisms or mammalian system. Traditionally, a few use patterns have been identified for antimicrobials. These are outlined in Table I.

Table I
Use Patterns for Antimicrobials

Use Pattern	Nature of Application	Application Method
Agricultural premises and equipment	Disinfectant/ Sanitizer	Formulations, Applied Externally on Surfaces
Food handling/storage establishment premises and equipment	Disinfectant/ sanitizer	Surface application
Commercial, institutional and industrial premises and equipment	Sanitizer/ disinfectant	Surface application
Residential and public access premises	Disinfectant/ sanitizer	Surface application
Medical premises and equipment	Sanitizer/ sterilant/ disinfectant	Surface application and internal application with water and other solvents as carriers
Human drinking water	Disinfectant	Added to water (mixing) or dissolved in water and then added to the water system
Material Preservatives	Disinfectant	Dissolved with or into matrices like paints
Industrial processes and water systems	Sanitizer/ disinfectant	Dissolved/mixed with aqueous systems in once-through cooling water, recirculating water systems
Antifoulant paints and ballast water treatments	Mollucide	Dissolved/ mixed with paints for application on boat/ship bottoms
Wood preservatives	Fungicide/ insecticides	Applied with pressure treatment or surface application when mixed with paint for antisapstain use
Swimming pools	Sanitizer/disinfectant	Added as powder or as a mixture liquid formulations
Aquatic areas	Sanitizer/disinfectant	Added as powder or as a mixture liquid formulations

With these use patterns, the possible formulations of pesticides can be: a) Antimicrobial pesticide is mixed in a powder; b) the antimicrobials is in solution mixture of water/other solvents; c) the antimicrobials is dissolved in a large water body for disinfection (drinking water or swimming pool); d) it is added into once through cooling and recirculating water or ballast water; e) it is applied as a spray to a hard surface; f) mixed with paint and used as antifoulant; g) it is applied as a fumigant; h) it is embedded into a hard surface, for example, a polymer (plastic).

Some typical ways antimicrobials are applied for various use sites are: Antimicrobials are mixed in a powder form; it is added in solution /mixture; added into large water body systems; as a spray to surfaces like counter tops or woods; as antifoulant paint on boat bottoms injected into wood under reduced pressure; applied as a mist; applied as a fumigant; is embedded into a hard surface, for example into a polymer (plastic).

OPP and Scientific and Regulatory Issues and Concerns for Nanomaterials As Nanopesticides

OPP in particular and EPA as an Agency in general has a double barrel of a problem if and when nanoproducts as pesticides are brought in for registration with the Agency. Nanotechnology, and therefore, manufacture and characterization of nanoproducts is a new field. Even when this stage becomes clearer in a short time, there are scientific uncertainties and possible regulatory tangles which have to be resolved before a nanopesticide can be registered. We briefly touch both aspects:

a. **Scientific Uncertainties**:

The ones identified below is by no means a complete list:

1. Even if NNI definition of a nanomaterial (1-100 nm) is accepted, the diversity of types of nanomaterials (metals, metal oxides, composites, SWCNT, MWCNT, fullerenes, shell, derivatives of carbon nanotubes, etc., databases for the basic characteristics of these nanonmaterials are at best ambiguous.
2. Purity of nanomaterials is a major issue and therefore data generated from these nanomaterials has uncertainties. A concise safety, health and environmental risk assessment can not be done.
3. Standards are not available for checking validity of data.
4. Nano scientists and engineers are not certain that if surface area a reliable parameter and be a required one or if surface activity is a better parameter scientifically.

5. Aggregation/agglomeration: Under what conditions it would or would not happen? And if it is not known how can a clean risk assessment be done?
6. Reliability of analytical instruments used to generate data.. like morphology of the manufactured nanomaterials

b. **Regulatory Uncertainties:**
A few easily recognizable ones are:

1. OPP is a data-rich organization and under FIFRA statuary it can and does require submission of scientific data form registrants. However, at this time the big question is if the existing guidelines are sufficient to ask for data for nanoproducts. If yes, what particular sections of existing guidelines would sufficiently address scientific data for nanoproducts. If not what kind of changes in the guidelines may be required. Any changing in the existing guidelines is a resource-intensive and will take a long time.
2. Should a nanoproduct be considered a new use or a new product. If we consider the nanoproduct as a new use, can the existing toxicity, eco effects, fate and transport of pesticides in air, water, soils, sediments, plants and aquatic organisms, and mammals, efficacy data suffice for registration. If a nanoproduct is considered a new product, do we require industry to generate new set of data on all disciplines mentioned above.
3. Should Agency require submission of selective data (Inhalation, dermal toxicity data, fate of nanomaterials in water at different pHs etc.)
4. Require submission of data on a nanoproduct for 1- 100nm material in at least one dimension or, require data submission for nanproducts between 1-10nm, 11-20 nm, 20-40 nm 40-80 nm 80-100 nm etc. as research in the nanoscience indicate that a number of nano-characteristics do change in small diameter range. Is a rationale for this type of data generation present in the existing guidelines?

ALTERNATE APPROACHES:

If the existing data requirements are not adequate, then what likely approach should EPA or any other Federal Agency take? A short list of possible approaches is presented:

1. Introduce a new statuary law and make comprehensive changes in the data requirements for nanoproducts.
2. Keep the existing statuary law and consider each nanoproduct for registration on a case-by-case basis and require product-specific data.
3. Ask industry to self certify their data based on the existing data requirements as UK has done[8]

4. Encourage academia, industry, federal and state labs to conduct research on nanomaterials and enrich, and populate the databases of nanomaterial characterizations as well as generate data on toxicity, eco effects, workers exposure, environmental fate and transport Some parts of federal organizations are already taking this approach; for example, EPA's Office and Research Development (ORD) is encouraging research on nanoproducts under their STAR Program for nanotechnology[9]
5. Encourage research and applications on green nanotechnology aspects only to introduce reduced risk products and also reduced contamination of the environmental media.

Green Chemistry and Green Nanotechnology:

Nanotechnology is unique in more than one way: It utilizes the concepts of all possible scientific disciplines to increase efficiency, cost and safety of the nanoproducts, and yet there is room for Green Chemistry contribution in the field of nanotechnology. In the scientific community there is a simultaneous push of using the Green Chemistry's Twelve Principles of: waste prevention, production of safer chemicals/products, deign less hazardous synthetic routes, use of reusable food stocks, preference of catalysts over stoichiometric reagents, avoiding use of derivatives, maximize atom economy, use of safer chemicals and reaction conditions, use energy efficiently, design chemicals and products which produce benign degradation products, conduct continuous analysis during the production of a chemical or product to minimize waste, or production of hazardous substances and finally minimize the possibility of chemical accidents, explosions, fires etc.[10] On similar grounds, engineers have defined a few principles for green engineering [11] Perhaps all this is a tall order, but is likely to be practical and useful for the long term goals. Efforts in producing green nanoproducts are already in the works, and a number of recent publications point to this trend.

Conclusions:

Perhaps a marriage of existing concepts, and listed approaches will produce a more defined field nanotechnology and speed up the development of nanoscience and nanotechnology field which is already moving at a fast pace. This will help all stakeholders of this arena: scientists of all scientific disciplines, engineers, industry, academia, and agencies like EPA and FDA to make sound scientific conclusions and conduct good risk assessments of all possible nano products whether these are simple food packaging, or disinfectants, or wood preservatives, or antifoulants or food additives, or drugs or drug delivery systems, or agricultural or biopesticides.

REFERENCES

1. US EPA's Nanotechnology White Paper, Science Policy Council, US EPA, Washington, DC 20460. EPA 100/B-07/001: http://es.epa.gov/ncer/nano/publications/white paper (accessed on Feb. 25, 2008)
2. National Science Foundation, Societal Implications of Nanoscience and Nanotechnology, Roco, C.M.; and Baindrige, W.S.; NSETs'Workshop Report, **2001**
3. a. Maynard, A.D.; Kuempel, E.D.; *J.Nantech.Res*, **2005**, 7(6), 587-624

c. Maynard, A.D.; Baron, P.A.; Foley, M.; Shvedova, A.A.; Kisin, E.R.; and Castranova, V. *J. Toxicol. Environ. Health, **2004,*** A 67: 87-107

d. Maynard, A.D.; *Phil. Trans. R. Soc. Lond.* **2000,** A 358: 2593-2610

4. Woodrow Wilson International Center for Scholars. *Project on Emerging Technologies.* Kuzuma, J. and VerHage, P.Nanotechnology in Agriculture and food Production: Anticipated Applications, **September 2006**
5. Down On The Farm, ETC Group. **2004. www.etc.group.org** (accessed on Feb.24, 2008)

6.Code of Federal Regulations, *40CFR, Parts 150 to 169*, Office of the Federal Register National Archives and Records, **July 2006**

7.Code of Federal Regulations, *21CFR, Parts 170 to 199,* Office of the Federal Register National Archives and Records, Revised in April **2000**

8.EU Approach: http://www.defra.gov.uk/environment/nanotech/policy/index.htm (Accessed on Feb. 25, 2008)

9.http://es.epa.gov/ncer/rfa.htlm. (accessed on Feb. 25, 2008)

10. Http://www.epa.gov/oppt/green engineering/pub/whats green engineering.htlm (accessed on Feb.25, 2008)

11. Anastas, P, and Warner J.C.; *Green Chemistry, Theory and Practice,* New York Oxford University Press, **1998**

27f Novel Biopolymer/rectorite Nanocomposite with Antimicrobial Activity

Xiaoying Wang and **Yumin Du**. Department of Environmental Science, Wuhan University, 129 Luoyu Road, Wuhan, 430079, China

Rectorite (REC) was used to prepare the intercalated nanocomposites with the biopolymer, N-(2-hydroxyl) propyl-3-trimethyl ammonium chitosan chloride (HTCC). Characterization by XRD and TEM revealed that HTCC were intercalated into the intergallery of REC. Moreover, it was confirmed from FT-IR, XRD and zeta potential analyses that interaction between HTCC and REC took place. In vitro antimicrobial assays indicated that all nanocomposites exhibited strong inhibition against Gram-positive bacteria, Gram-negative bacteria and Fungi under weak acid, water and weak basic condition, particularly against Gram-positive bacteria. The lowest minimum inhibition concentration values of the nanocomposites against Staphylococcus aureus and Bacillus subtilis were less than 0.00313% (w/v) in all media tested, and the killing rate on S. aureus reached more than 90% in 30min. The mechanism of the antimicrobial action was briefly discussed, suggesting that the fine dispersion and the interaction between HTCC and REC had a strong effect on the antimicrobial activity of the nanocomposites.

Synthesis and characterization of Fe doped mesoporous TiO_2 nanostructure using Supercritical CO_2

*Rahima A. Lucky, Paul A. Charpentier**
Department of Chemical and Biochemical Engineering
University of Western Ontario

Abstract: A series of Fe doped titanium nanomaterials were synthesized by an acid modified sol-gel technique using Ti(IV)-isopropoxide and $Fe(NO)_3$ in supercritical CO_2 ($scCO_2$). The precursors were mixed and subsequently hydrolyzed with water molecules generated by an *in situ* esterification reaction. The effects of Fe doping on the morphology, phase structure, mean crystallite size, specific surface area, and pore volume, surface acidity were investigated by using SEM, TEM, HRTEM, XRD, Raman, powder FTIR, EDX, BET nitrogen adsorption–desorption isotherm and TPD methods. SEM showed that pure titania nanofibers were formed whereas doped samples showed a flake type flat surface. The XRD results revealed that calcined undoped and doped samples contained anatase as the dominant phase. However, the crystallite size decrease from 13.8 nm in undoped samples to10 nm in the highest Fe doped sample. TEM images of the calcined materials reveled that the doped materials were more porous. Thermogravimetric analysis showed that the weight loss was accelerated and completed at lower temperatures as the relative concentration of iron in the Fe-TiO_2 samples increased.

Introduction. Supercritical carbon dioxide ($scCO_2$) is becoming an attractive replacement for organic solvents in a wide range of chemical processes because it is nontoxic, nonflammable, inexpensive, naturally abundant, and chemically inert. [1-3] Low viscosity, zero surface tension and the high diffusivity of $scCO_2$ are considered very favorable physical properties that can be exploited for synthesizing superior ultrafine and uniform nanomaterials. Moreover, these physical properties help to increases the reaction rates.[4, 5] In addition, the solvent density and solubility of the materials can be tuned by adjusting the operating temperature and pressure providing an easy tunable reaction medium. Moreover, when reactions are carried out in $scCO_2$, the reaction products can be easily recovered by simply venting the solvent, hence, no drying process is required and the porous structure can be maintained without collapsing the nanostructure. Furthermore, $scCO_2$ can be easily recycled after the pressure is diminished for potential scale-up for commercial employment. Therefore, exploring and utilizing this green solvent to synthesize advanced nanomaterials is of potential importance to many industrial and scientific areas.

Titania (TiO_2), a widely used photocatalyst, is receiving considerable attention due to titania's high activity, strong oxidation capability, and chemical stability. TiO_2 nanomaterials have many potential applications as in dye sensitized solar cells and lithium storage batteries, and photoconductive agents.[6-11] However, pure anatase TiO_2 has one well known significant disadvantage, i.e. a wide bandgap energy *ca.* 3.2 eV. Therefore, it requires UV light ($\lambda < 380$ nm) for its activation with only $\approx$5% of the total solar energy being utilized. Numerous studies have focused to shift TiO_2's bandgap to the visible spectral region, which would allow efficient utilization of solar energy and, hence, would greatly expand its application.[12, 13] Transition metal ion doping is one of the most promising approaches to shift the band gap by changing its electronic properties to the visible spectra for absorbing more light. This type of doping also alters the intrinsic properties of titania through the formation of ''shallow traps'' within the titania matrix and up to a certain concentration level it increase the photoinduced electron/hole charge recombination lifetimes by acting as an electron- hole trap site. Among the various metal ions, due to the unique electronic structure and size that closely matches that of titanium (IV), doping with iron(III) has been widely investigated.[14] However, no direct correlation between light absorption and photocatalytic activity could be found and the synthesis process is considered one of the most important factors controlling the reactivity of the iron doped titania.[15, 16] In addition, literature has shown that the performance of titania nanomaterials also strongly relies on their crystallinity, crystallite size, crystal structure, specific surface area, thermal stability and quantum efficiency.[17, 18] These properties significantly vary with the synthesis method, type and amount of iron loading and the subsequent heat treatment.[14, 16, 19, 20] Altering the textural properties of titania is another possible way to enhance catalytic performance of the nanomaterials. It has been suggested that a well ordered mesoporous titania material may show enhanced catalytic performance due to the large accessible surface area and uniform pore size. Previous research on mesoporous silica has shown that the structure of silica can be stabilized by incorporation of metal elements such as Fe(III), La(III) and Al(III) into its framework. Based on this concept, Wang et al synthesized mesoporous Fe-doped titania using surfactant by a hydrothermal process.[21] Different surfactants such as alkyl phosphate, block copolymers and cetyltrimethylammoniumbromide were used as templates to synthesize mesoporous titania. However, the template technique requires high calcination temperatures to remove the template and the hydrothermal process requires a large amount of inorganic solvents, which may lead to environmental pollution. These problems limit the practical usage of surfactant assisted synthesis process. Therefore, alternative processes using a green solvent such as $scCO_2$ is attractive and promising way to obtain mesoporous materials with enhanced surface area and pore volume of Fe doped TiO_2 nanomaterials. Fe-doped titania showed better performance not only by enhancing surface properties but also change band gap energy.

In this present research, we demonstrate a surfactant free one pot synthesize route to produce mesoporous Fe doped TiO_2 nanomaterials via an acid modified sol-gel process in $scCO_2$ which is scaleable, uses mild reaction conditions, provides a high yield and high quality nanomaterials. The effects of Fe doping on the morphology, crystal structure, crystal size, surface area, pore volume and pore size distribution and band gap energy of the synthesized nanomaterials are examined.

Experimental Section

Materials. Reagent grade titanium (IV) isopropoxide (TIP, 97%, Aldrich), acetic acid (99.7%, Aldrich), isopropanol(99.5%, Aldrich), Ferric nitrate ($Fe(NO)_3 . 9H_2O$, 99.7%, Aldrich), and (99.5%, Aldrich), and instrument grade carbon dioxide (99.99%, BOC) were used without further purification.

Synthesis. The experimental set up and procedure were explained in detail previously.[3] 1 wt % Ferric nitrate solution in iso-propanol was prepared and used as stoke solution. In a typical experiment, 2 ml titanium (IV) isopropoxide, appropriate amount of ferric nitrate solution, and 2 ml acetic acid were quickly placed in the view cell under stirring. After that CO_2 was added to the system. Final temperature and pressure was 60 °C and pressure was 5000 psig respectively. Based on the amount of stoke solution, colorless to reddish solution was observed. After the reaction mixture was stirred about 30 minutes to several hours, the fluids in the view cell became semitransparent then turn into white gel. After 5 days aging, the formed gel was washed continuously using $scCO_2$ (5000 psig, 60°C) at a controlled flow rate of 0.5 mL/min, followed by controlled venting. The resulting as-prepared material was amorphous. To get crystalline phase, synthesized materials were calcined in air at 500°C using a heating rate of 10 °C/min, the holding time was 2 h, and the cooling rate to room temperature was 0.5 °C/min. The preparation scheme of Fe doped TiO_2 nanomaterials is given in **Fig 1**.

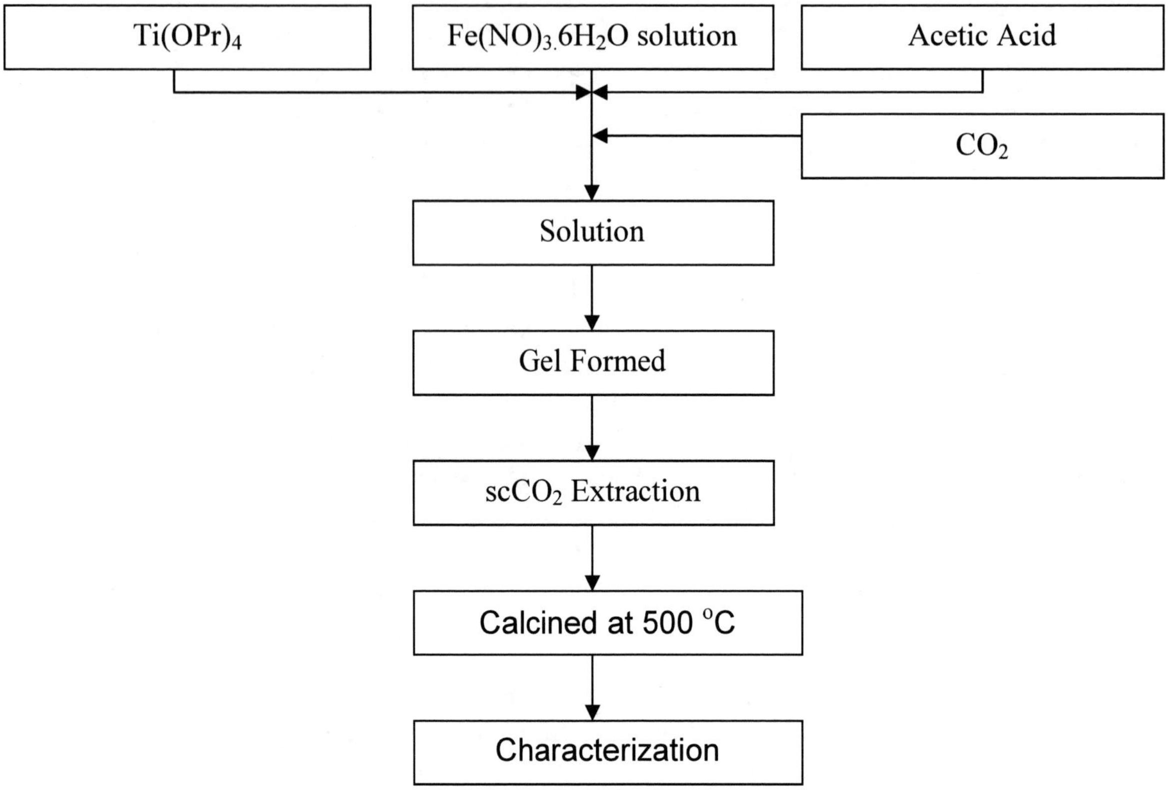

Figure1: The preparation scheme of Fe doped TiO_2 nanomaterials by sol-gel process in supercritical carbon dioxide.

Characterizations. The resulting nanomaterials were characterized by using SEM, TEM, TGA, EDX, XRD, HRTEM, Raman, N_2 isotherms, and UV-Vis.

Results and Discussion

Morphology. In this project, four samples with increasing amount of Fe stock solution were prepared and were characterized by SEM and TEM analysis. The SEM images, shown in Fig. 2, provide the general morphology and size of the as-prepared undoped and Fe-doped titania nanomaterials synthesized. From the image (Fig. 2(a)), it can be seen that a high yield of pure titania nanofibers having 20-50 nm dia. with lengths of up to tens of microns are formed whereas (Fig. 2(c)-(d)), the modified system is composed with mainly flake type structure. The widths of the flakes become larger with increasing concentration of Fe. To further confirm these results, TEM analysis was performed for the as prepared and calcined nanomaterials. TEM images (Fig. 3(a) &(b)) clearly demonstrate the morphological difference between the two systems. In addition, the calcined pure titania (Fig. 3(c)) formed larger crystallites whereas the doped nanomaterials(Fig. 3(d)) became porous consisted with smaller crystallites. This analysis confirms that Fe has a significant effect on the morphology and properties of the products.

Figure 2: SEM: Fe doped TiO_2 nanomaterials (a)S1- TiO_2, (b)S2-1m , (c)S3-1.5ml, and (d) S4-2 ml. *(All the samples were examined after platinum coating.)*

Figure 3: TEM: Fe doped TiO_2 nanomaterials:(a) S1- TiO_2, &(b)S4-2ml ,(c) S1-TiO_2, and (d) S4- 2ml.*(As-prepared (a)&(b), calcined (c)&(d), Bar represent 100 nm.)*

Composition

EDX mapping (recorded on a LEO 1530 operated at 20 kV) of the Fe-TiO_2 samples revealed that the iron and titanium elements are homogenously dispersed throughout the samples (Figure 4). However, the literature showed that iron has a very low solubility in anatase phase and only vary small amount can help to enhance titania's performance.[19, 22] Usually more than 1% Fe segregated during heat treatment and acts as recombination center. Therefore, very dilute iron stock solution was used to prepared doped titania. The surface iron amount is very small and XPS could not detect though the bulk amount of Fe/Ti ratio (0- 0.46 %) for different samples was determined by EDX.

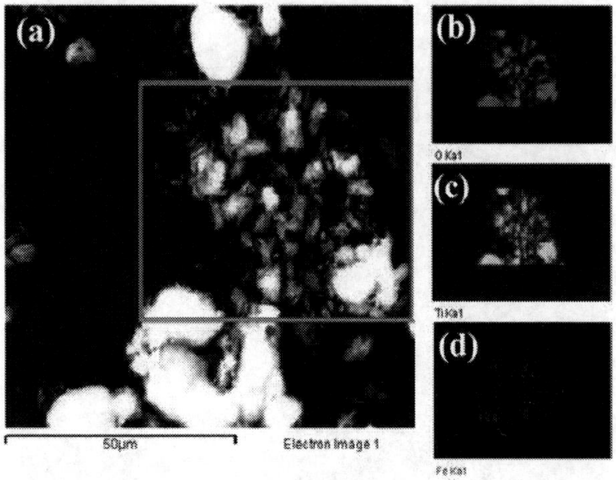

Figure 4: EDX O, Ti and Fe mapping for Fe doped TiO_2 nanomaterials.

Thermal analysis

TG–DTG analysis was carried out to study the thermal decomposition behavior of the synthesized materials. The results of TG and DTG for different samples are shown in Fig 5. There are three marked stage which are in range 25–125°C, 125–500°C and more than 500°C and The DTG curves present two main exothermic peaks. A small weight loss at first stage is caused by the desorption of physically adsorbed water. The second exothermic peak is sharp and narrow, and attributed to the burnout of the organic residue. This second strong exothermic peak in the DTG curve in the undoped TiO_2 sample shifted to the lower temperatures and became much more asymmetrical with increased iron doping. The weight loss over 500 °C is extremely small.

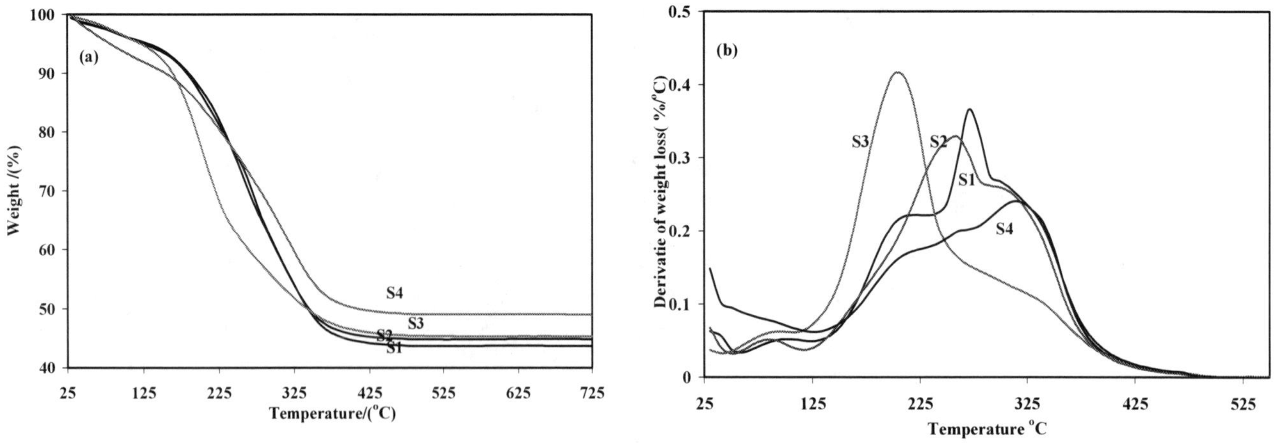

Figure 5: (a) TG and (b) DTG curves of Fe doped nanomaterials.

Phase structure

In order to examine the iron doping effects on the phase structure and crystallite size, XRD and HRTEM analysis were performed. The as-prepared materials were amorphous, while when increasing the calcination temperature up to 400 °C, the material reorganized itself, and the anatase particles began to grow, resulting in crystalline material. The XRD patterns (not shown) of all the calcined samples indicate that the Fe-TiO_2 consist of anatase crystal, with no rutile phase being present. Although all the modified samples consist of anatase crystal, the relative intensity of peaks at $2\theta = 25.2^{o}$ decreased significantly in doped TiO_2 with respect to the undoped TiO_2 nanocrystal. There was no distinct Fe peak indicating no phase separation, and the Fe was well integrated within the anatase crystal structure for this composition and calcination temperature. By increasing the calcination temperature from 400 to 600°C, the peak intensities increased as well as the width of the peaks became narrower, indicating an improvement of the anatase phase, and simultaneously the growth of anatase crystallites.

For the pure titania, the crystallite size was ca. 8.17nm calcined at 400°C, while with increasing calcination temperature, the crystallites became larger. The sizes were ca. 13.7nm and 15 nm at 500°C and 600°C respectively. Due to heat treatment, the crystallites also became larger for doped samples though the growth rate is lower resulting in smaller crystallite materials at the corresponding calcination temperature. For example, at 600°C doped samples had 14, 13.95 and 13.75 nm crystallites for samples S2, S3, and S4 respectively. This analysis along with literature confirmed that a small amount of Fe inhibited grain growth during heat treatment.

In addition to the XRD data, detailed information on the structural transformations and crystal growth can be obtained using HRTEM. The HRTEM micrographs of the lattice image of pure titania calcined at 500 °C is given in Fig.6(a), showing a grain size of ca.14 nm width with d-spacing 0.35 nm, very close to the lattice spacing of the (101) planes of the anatase phase. The lattice image of doped titania calcined at 500 °C (Fig.6 (b)) shows that these materials also composed with anatase phase. However, all grains were not the same in terms of size and shape. Some were long with significant lattice mismatch and grain boundaries. All these defects prevent rapid grain growth. The HRTEM image also revealed that doped materials are predominately mesoporous.

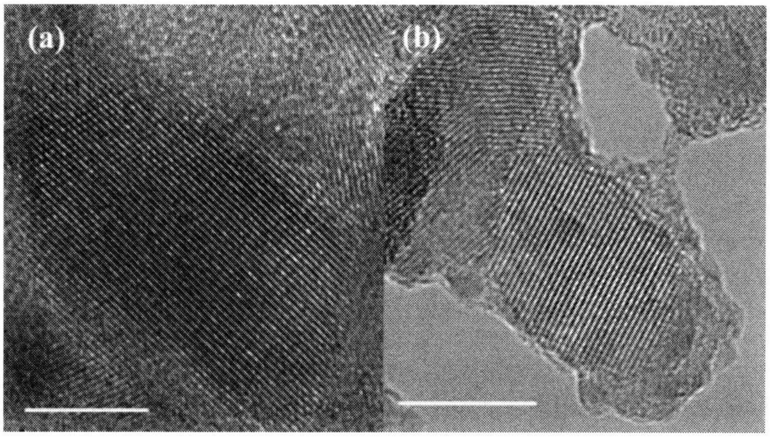

Figure 6: HRTEM: (a) S1- TiO_2, and (b) S4-Fe doped TiO_2 nanomaterials (*calcined at 500°C (a)&(b), Bar represent 5 nm.)*

Surface area and pore volume

The textural properties, i.e. the surface area, pore volume, and pore size distributions of the samples calcined at 400, 500 and 600 °C were characterized by nitrogen adsorption studies. Figure 7(a) shows the nitrogen adsorption isotherms for the as-prepared materials, which exhibit H3 hysteresis loops for all samples typical for mesoporous materials. To evaluate the pore size distribution, the calcined materials were plotted as shown in Figure 7(b). The average pore diameter for the pure titania and materials containing a small amount of iron is *ca.*11nm, whereas for sample S4 is 20nm. Upon calcination, the pore size became gradually larger and the pore size distribution shifted to forming larger pores in expense of the smaller ones. At 600 °C the pore size is *ca.* 15 nm for all samples except S4. The pore size became ca.24 nm for S4.

The surface properties of the Fe-TiO_2 nanomaterials calcined at different temperatures are summarized in Figure 11. The 400 °C calcined TiO_2 nanomaterials having a surface area of 69 m^2/g, gradually decrease through calcinations and became 43 m^2/g at 600 °C. Due to the sintering phenomena, the small pores collapse, reducing the pore volume and surface area. The transformations into anatase crystalline phases will also help to reduce the surface area. However, the presence of small amount of iron inhibits these transformations resulting doped nanomaterials showed higher surface area.

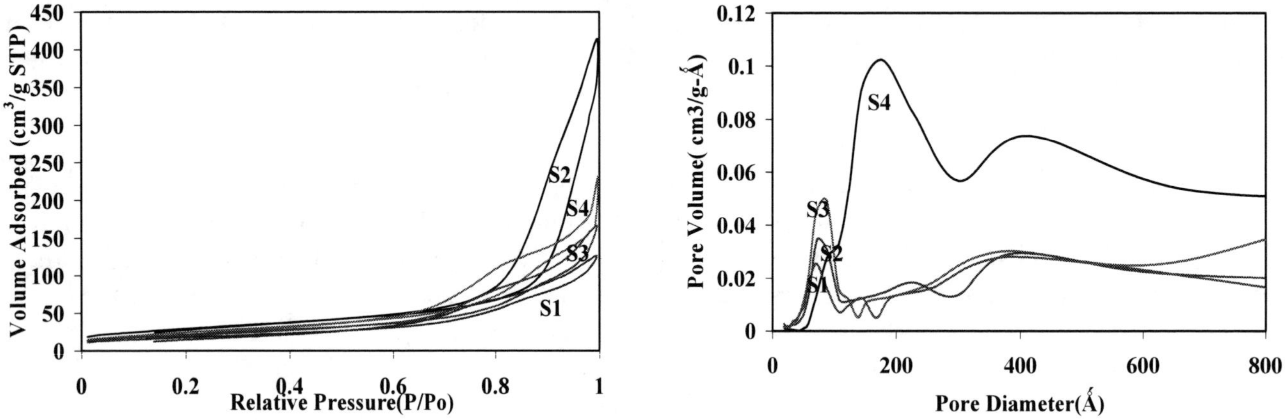

Figure 7: (a) N_2 adsorption/desorption isotherm, and (b) BJH pore size distribution of the Fe doped TiO_2 as-prepared nanomaterials.

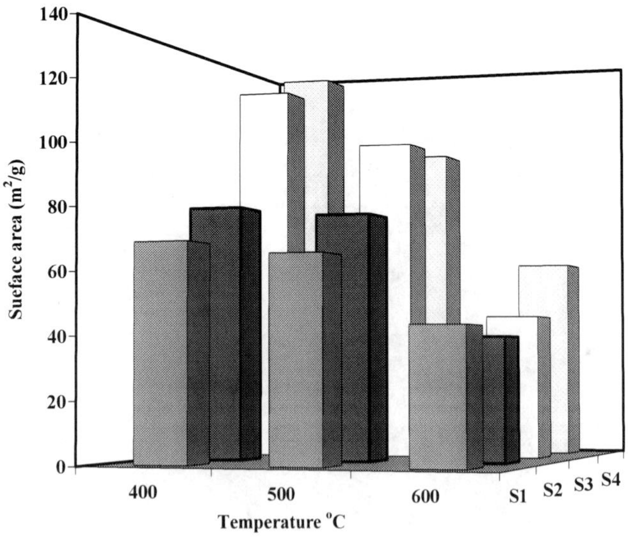

Figure 8: The BET surface area of the Fe doped TiO_2 nanomaterials calcined at different temperature.

IV. Conclusions

Fe doping was favorable for organizing a more ordered mesostructure and keeping the integrity of the pore structure at high temperature. The Fe-doped mesoporous TiO_2 showed a lager specific surface area and a more uniform pore distribution smaller crystallies compared with the undoped sample. It suggests that proper iron doping to the framework may become an effective way to improve the properties of mesoporous TiO_2 materials.

References

1. Clark, J. H.; Tavener, S. J., Alternative Solvents: Shades of Green. *Org. Process Res. Dev.* **2007,** 11, (1), 149-155.
2. Johnston, K. P.; Shah, P. S., Making Nanoscale Materials with Supercritical Fluids. *Science* **2004,** 303, (5657), 482-483.
3. Lucky, R. A.; Charpentier, P. A., A one-step approach to the synthesis of ZrO_2 modified TiO_2 nanotubes in supercritical carbon dioxide. *Advanced Materials.* (in press).
4. Smirnova, I.; Arlt, W., Synthesis of Silica Aerogels: Influence of the Supercritical CO2 on the Sol-Gel Process. *Journal of Sol-Gel Science and Technology* **2003,** 28, (2), 175-184.
5. Lee, M.-H.; Lin, H.-Y.; Thomas, J. L., A Novel Supercritical CO2 Synthesis of Amorphous Hydrous Zirconia Nanoparticles, and Their Calcination to Zirconia. *Journal of the American Ceramic Society* **2006,** 89, (12), 3624-3630.
6. Zhu, K.; Neale, N. R.; Miedaner, A.; Frank, A. J., Enhanced Charge-Collection Efficiencies and Light Scattering in Dye-Sensitized Solar Cells Using Oriented TiO2 Nanotubes Arrays. *Nano Lett.* **2007,** 7, (1), 69-74.
7. Sander, M. S.; Côté, M. J.; Gu, W.; Kile, B. M.; Tripp, C. P., Template-Assisted Fabrication of Dense, Aligned Arrays of Titania Nanotubes with Well-Controlled Dimensions on Substrates. *Advanced Materials* **2004,** 16, 2052-2057.
8. Chi, B.; Victorio, E. S.; Jin, T., Synthesis of Eu-doped photoluminescent titania nanotubes via a two-step hydrothermal treatment. *nanotechonology* **2006,** 17, 2234-2241.
9. Zhang, H.; Li, G. R.; An, L. P.; Yan, T. Y.; Gao, X. P.; Zhu, H. Y., Electrochemical Lithium Storage of Titanate and Titania Nanotubes and Nanorods. *J. Phys. Chem. C* **2007,** 111, (16), 6143-6148.
10. Kocher, M.; Daubler, T. K.; Harth, E.; Scherf, U.; Gugel, A.; Neher, D., Photoconductivity of an inorganic/organic composite containing dye-sensitized nanocrystalline titanium dioxide. *Applied Physics Letters* **1998,** 72, (6), 650-652.
11. Ramires, P. A.; Giuffrida, A.; Milella, E., Three-dimensional reconstruction of confocal laser microscopy images to study the behaviour of osteoblastic cells grown on biomaterials. *Biomaterials* **2002,** 23, (2), 397-406.
12. Kumbhar, A.; Chumanov, G., Synthesis of Iron(III)-Doped Titania Nanoparticles and its Application for Photodegradation of Sulforhodamine-B Pollutant. *Journal of Nanoparticle Research* **2005,** 7, (4-5), 489 - 498.
13. Chen, C.; Li, X.; Ma, W.; Zhao, J.; Hidaka, H.; Serpone, N., Effect of Transition Metal Ions on the TiO2-Assisted Photodegradation of Dyes under Visible Irradiation: A Probe for the Interfacial Electron Transfer Process and Reaction Mechanism. *J. Phys. Chem. B* **2002,** 106, (2), 318-324.
14. Ranjit, K. T.; Viswanathan, B., Synthesis, characterization and photocatalytic properties of iron-doped TiO2 catalysts. *Journal of Photochemistry and Photobiology A: Chemistry* **1997,** 108, (1), 79-84.
15. Maeda, M.; Yamada, T., Photocatalytic activity of metal-doped titanium oxide films prepared by sol-gel process. *Journal of Physics: Conference Series* **2007,** 61, 755–759.
16. Navio, J. A.; Colon, G.; Macias, M.; Real, C.; Litter, M. I., Iron-doped titania semiconductor powders prepared by a sol-gel method. Part I: synthesis and characterization. *Applied Catalysis A: General* **1999,** 177, 111-120.
17. Pal, M.; GarciaSerrano, J.; Santiago, P.; Pal, U., Size-Controlled Synthesis of Spherical TiO2 Nanoparticles: Morphology, Crystallization, and Phase Transition. *J. Phys. Chem. C* **2007,** 111, (1), 96-102.
18. Ohtani, B.; Ogawa, Y.; Nishimoto, S. i., Photocatalytic Activity of Amorphous-Anatase Mixture of Titanium(IV) Oxide Particles Suspended in Aqueous Solutions. *J. Phys. Chem. B* **1997,** 101, (19), 3746-3752.
19. Yu, J.; Zhou, M.; Yu, H.; Zhang, Q.; Yu, Y., Enhanced photoinduced super-hydrophilicity of the sol–gel-derived TiO2 thin films by Fe-doping. *Materials Chemistry & Physics* **2006,** 95, (2-3), 193-196.
20. Wang, C.-y.; Bo¨ttcher, C.; Bahnemannc, D. W.; Dohrmanna, J. r. K., A comparative study of nanometer sized Fe(III)-doped TiO_2 photocatalysts: synthesis, characterization and activity. *J. Mater. Chem.* **2003,** 13, 2322–2329.
21. Wang, Y.; Jiang, Z.-H.; Yang, F.-J., Effect of Fe-doping on the pore structure of mesoporous titania. *Materials Science and Engineering B* **2006,** 134, 76–79.

A one-pot route toward light selective quantum dot polymer nanofilms in $scCO_2$

William Z. Xu and Paul A. Charpentier, University of Western Ontario, London, ON, Canada

Abstract

The synthesis of polymer nanocomposites with incorporated quantum dots (QDs) has been attracting significant interest due to their unique applications in a variety of fields. Here we explored a novel one-pot synthesis for synthesizing QD-polymer nanocomposites for application as light selective nanofilms using the "green" solvent, supercritical carbon dioxide ($scCO_2$). CdS-ZnS core-shell QDs were first synthesized by means of pyrolysis of single molecular precursors at 180°C and then functionalized with a methoxysilane group, followed by the one-pot synthesis technique in $scCO_2$. FT-IR, EDX elemental analysis, NMR, solubility tests, photoluminescence and UV-Vis spectroscopy were used for characterization of the QDs, indicating successful synthesis and functionalization of the CdS-ZnS core-shell QDs. The synthesized QD-polymer nanocomposite films were characterized by TEM, EDX elemental mapping, GPC, photoluminescence and UV-Vis spectroscopy. It was found that the QDs were linked to the polymer chains and well distributed throughout the polymer matrix. With the QDs incorporated into the polymer matrix, the nanocomposites showed highly enhanced absorption in the UV and violet regions of the electromagnetic spectrum, and enhanced emission in the region from green to red light for potential applications for UV-protection nanofilms.

Introduction

Utilizing solar energy has been attracting considerable multidisciplinary research interest recently, as the Sun provides Earth with as much energy every hour as human civilization uses every year.[1] Although turning solar radiation into electricity is one option, another is finding ways to harness its heat and control the wavelengths of sunlight. Converting the undesirable higher-energy wavelengths, e.g., UV and violet, into desirable lower-energy lights will expand the applications of solar energy into new areas. Potentially, polymer films can be tailored for various applications such as greenhouses for controlling flowering or fruiting of plants and herbs or window films for sunlight control with heat gain/reduction.

Semiconductor nanocrystallites, known as quantum dots (QDs), are being widely explored in a variety of fields such as nanobiotechnology[2] and solar cell technology[3]. When the QDs are irradiated by light from an excitation source to reach their respective excited states, they emit photons corresponding to their respective energy band gaps. The band gaps of quantum dots can be tailored by changing the types of QDs, e.g., CdS, CdSe and CdTe, and/or the size, which are tunable by controlling the reaction conditions such as time and temperature.[4]

The main method for preparing QDs is classical colloidal chemistry[5] where various researchers have employed organometallic and/or metal organic compounds under anaerobic conditions to synthesize QDs. Several researchers used single-molecular precursors such as bis(diethyldithio-/diseleno-carbamato) cadmium (II) /zinc (II) for the preparation of QDs. [5-7] Single-molecular precursors have the significant advantages of being less toxic, insensitive to air and moisture, easier to purify, and provide highly monodisperse nanoparticles, with the synthesis generally undertaken at desirable lower temperatures.[5, 8]

In addition to the “bare” quantum dots such as CdSe and CdS, core-shell QDs have also been extensively studied due to their novel properties. Overcoating nanocrystallites with higher band gap inorganic materials has been shown to improve the photoluminescence quantum yields and photostability.[9, 10] Particles passivated with inorganic shell structures are more robust than organically passivated dots, and have shown greater tolerance to processing conditions necessary for incorporation into solid state structures.[11]

The colloidal QDs are surrounded by a layer of organic ligands in a “shell” which dictates the surface chemistry of the QDs, and therefore plays a significant role in the solubility or miscibility within a given environment.[12] As the high-temperature synthesis of QDs is not compatible with most organic functional groups, ligand exchange is often required for the introduction of new surface functionality to QDs. Compatible organic ligands with polymer matrices enable the applications of QDs in polymer composites, by using bifunctional molecules where the ligand contains a coordination “head” and a polymerizable “tail”.[13] Ligand exchange with various thiol ligands has been extensively studied.[14-16] Bewendi *et al* reported a family of oligomeric phosphine ligands allowing tunable compatibility in diverse environments and flexibility for further chemistry.[12]

In general, the major challenges for the successful synthesis of QD-polymer nanocomposites have been:[17, 18] (1) aggregation of QDs due to poor miscibility of the QDs with the polymer matrix, (2) migration of the incorporated QDs when the chemical environment is changed, and (3) a significant decrease in quantum efficiency of the QDs. Supercritical carbon dioxide ($scCO_2$) has been showing successful applications in fabrication of well-dispersed polymer nanocomposites[19]. The goal of this research was to explore a novel “green” one-pot technique for the synthesis of well-dispersed and chemically-bonded QD-PVAc nanocomposites, using our previously described technique.[20] In the present work, we synthesized CdS-ZnS core-shell QDs at relatively low temperature, functionalized the QDs with a methoxysilane group, and then incorporated the functionalized QDs into poly(vinyl acetate) matrix by means of parallel reactions of free radical polymerization and hydrolysis/condensation in the “green” solvent $scCO_2$.

Experimental

Materials

High purity CO_2 (BOC Gases, 99.99%, with dip-tube) and high purity N_2 (BOC, 99.99%) were further purified by passing through columns filled with 5 Å molecular sieves and reduced 20% copper oxide/Al_2O_3. Vinyl acetate (VAc) (>99%) purchased from Sigma-Aldrich Canada was further purified by passing through an inhibitor remover column (Sigma-Aldrich). (3-Mercaptopropyl)trimethoxysilane, vinyltrimethoxysilane (VTMO) (97%), pentane (≥99%), and cadmium chloride (tech grade, A.C.S. reagent), sodium diethyldithiocarbamate trihydrate (A.C.S. reagent), zinc diethyldithiocarbamate (98%), glacial acetic acid (≥99.99%), trioctylamine (TOA) (≥99.0%) (Fluka), trioctylphosphine (TOP) (≥90%) (Fluka), anhydrous methanol (99.8%), toluene (≥99.9%) (CHROMASOLV® Plus), poly(vinyl acetate) (PVAc) (Mw=113,000), and tetrahydrofuran (≥99%) were purchased from Sigma-Aldrich Canada and used as received. Anhydrous ethyl alcohol was purchased from Commercial Alcohols, Inc., Canada and used as received. The initiator diethyl peroxydicarbonate (DEPDC) was synthesized as previously described.[21] The single-molecular precursor $Cd[S_2CN(C_2H_5)_2]_2$ was synthesized with the method described by O’Brien and coworkers.[7]

Preparation of CdS-ZnS core-shell QDs

25mL TOA and a magnetic stirring bar were placed in a 100mL three-neck flask under nitrogen equipped with a reflux condenser, a thermometer, and a thermocouple for automatic temperature control. When the temperature was stable at 180°C, a solution of 0.5g $Cd[S_2CN(C_2H_5)_2]_2$ in 9mL TOP was rapidly injected into the flask. After 5 or 10min reaction time, a solution of 0.2g $Zn[S_2CN(C_2H_5)_2]_2$ in 3mL TOP was slowly added dropwise. 5min after the completion of addition of the zinc diethyldithiocarbamate solution, the heater was removed to cool down the reaction mixture. When the temperature dropped to approximately 75°C, a large excess of methanol/ethyl alcohol was added followed by separation of the quantum dots through centrifugation. The QDs were washed with methanol, and then dispersed into toluene. The mixture of QDs in toluene was further filtered under vacuum to remove any insoluble material.

Ligand exchange of the CdS-ZnS QDs

The CdS-ZnS core-shell QDs were subsequently reacted with 5mM (3-mercaptopropyl)trimethoxysilane with stirring at 70°C for 24 hours under a nitrogen flow. After the reaction, the reaction mixture was cooled to ambient temperature. Then a large excess of pentane was added into the mixture followed by separation of the QDs through centrifugation. The separated quantum dots were re-dispersed into 40mL of toluene, and then 5mM (3-mercaptopropyl)trimethoxysilane was added. The mixture was refluxed with stirring under a nitrogen flow at 70°C for 24 hours. This procedure was repeated several times. After centrifugation, the surface-displaced QDs were dispersed in toluene.

Polymerization procedure

Polymer nanocomposite synthesis was conducted in a 100-mL high-pressure stainless steel autoclave (Parr 4842) coupled with a digital pressure transducer. The designated chemicals were charged into the reactor, purging with a flow of nitrogen, then pumping CO_2 into the autoclave by means of a syringe pump (Isco 260D). After the reaction, CO_2 was carefully vented leaving the formed sample in the autoclave.

In situ Fourier transform infrared (FT-IR) monitoring of solution concentration in the stirred 100-mL high-pressure autoclave was performed using a high-pressure immersion probe (Sentinel-Mettler Toledo AutoChem).

Characterization

Melting point of the synthesized single-molecular precursor bis(diethyldithiocarbamato) cadmium (II) was measured on a Barnstead Electrothermal 9100. Photoluminescence (PL) spectra were recorded on a Photon Technology International QuantaMaster™. UV-Vis absorption spectra were measured on a Cary Varian UV-VIS spectrophotometer. Differential scanning calorimetric (DSC) analysis was performed using a Mettler Toledo DSC822e. Thermogravimetric analysis (TGA) analysis was performed using a TA Instruments TGA Q500. Gel Permeation Chromatography (GPC) was carried out using a Viscotek GPCmax (VE2001GPC Solvent/Sample Module) & Viscotek TDA302 (Triple Detector Array). Energy Dispersive X-ray (EDX) elemental analysis of the QDs was recorded using a Hitachi S4500 field emission scanning electron microscope. EDX elemental mapping of the QD-poly(vinyl acetate) nanocomposites was recorded using a Hitachi S-2600N scanning electron microscope. Transmission electron microscopy (TEM) images were recorded using a Philips CM10 transmission electron microscope. High resolution transmission electron microscopy (HRTEM) images were recorded using a JEOL 2010F FEG TEM/STEM. Nuclear magnetic resonance (NMR) spectra were recorded using a Varian Inova 600 or 400 NMR spectrometer.

Results and Discussion

CdS-ZnS core-shell QDs

The synthesized single-molecular precursor bis(diethyldithiocarbamato) cadmium (II) was characterized by means of measuring the melting point, FT-IR spectrum and NMR (^{1}H, ^{13}C, gCOSY, and gHSQC) spectra. All of these results indicate the successful synthesis of the single-molecular precursor bis(diethyldithiocarbamato) cadmium (II), as they are in good agreement with those reported in the literature.[22, 23] For the sake of ensuring that the ZnS shell was grown on the CdS core, all the bis(diethyldithiocarbamato) cadmium (II) solution was rapidly injected into the reaction flask to speed up nucleation of CdS nanocrystallites, while zinc diethyldithiocarbamate solution was slowly added dropwise to allow ZnS to grow on the CdS crystal nuclei, minimizing nucleation of ZnS crystallites.

The synthesized CdS-ZnS core-shell QDs were characterized with HRTEM, EDX, and photoluminescence. In order to compare the CdS-ZnS core-shell QDs with the "bare" CdS QDs, CdS QDs were also synthesized under similar conditions, except that a higher temperature (235°C) and longer reaction time (30 min) were used to grow dots of observable size to examine their crystal structure. From the HRTEM and the d-spacing images of the synthesized "bare" CdS and core-shell CdS-ZnS QDs, it was found that both the "bare" and core-shell QDs presented nanocrystal structure, with a clear difference between the two types of QDs. The nanocrystal structure of the CdS QDs appears to be more uniform and regular than that of the CdS-ZnS QDs. The d-spacing of the CdS QDs was measured to be approximately 4.5 Å while the CdS-ZnS QDs has a d-spacing value of approximately 3.5 Å. The slight "bending" of the lattice fringes of some CdS-ZnS QDs suggests some sort of strain exists in the core-shell structure, as also observed by Bawendi and coworkers[11] for CdSe-ZnS QDs. The CdS-ZnS QDs were also characterized with other methods as discussed below.

Ligand exchange of the CdS-ZnS QDs

In order to make the QDs compatible with the polymer matrix, the surrounding organic layer has to be modified. Previously we successfully synthesized silica-poly(vinyl acetate) nanocomposites in the "green" solvent $scCO_2$ by a one-pot technique where parallel reactions of free radical polymerization of vinyl acetate (VAc) and sol-gel hydrolysis/condensation of tetraethoxysilane (TEOS) or tetramethoxysilane (TMOS) took place simultaneously, while vinyltrimethoxysilane (VTMO) was used as a linker to participate in both reactions.[20] By functionalizing the organic shell of the QDs with such functional groups as Si-O-CH_3, we hypothesized that the QDs could be incorporated into a polymer matrix through hydrolysis/condensation. (3-Mercaptopropyl)trimethoxysilane was chosen as a candidate for the ligand exchange of the synthesized CdS-ZnS core-shell QDs because it has the required functionality. The thiol head can be coordinated with zinc or cadmium to stabilize the QDs, while the methoxysilane tail can be used for subsequent reaction to attach the QDs to polymer chains. **Scheme 1** displays a schematic diagram of the ligand exchange of the CdS-ZnS core-shell QDs, wherein outer layer TOA is replaced with (3-mercaptopropyl)trimethoxysilane.

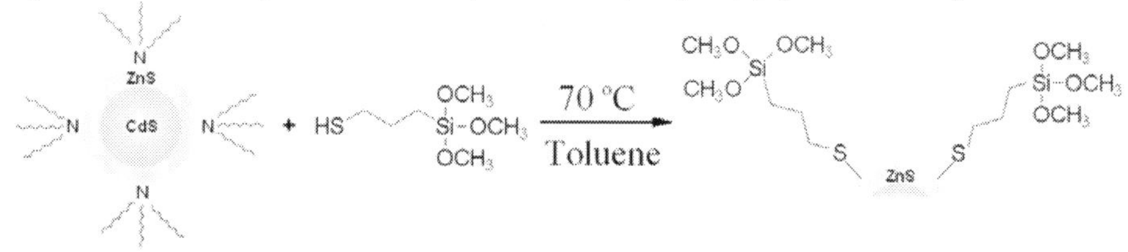

Scheme 1. Schematic diagram of the ligand exchange of CdS-ZnS core-shell QDs

In order to examine if the ligand exchange proceeded successfully, the QDs before and after ligand exchange were characterized with EDX elemental analysis, HRTEM, FT-IR, solubility testing, photoluminescence, and UV-Vis absorption spectroscopy. Due to the limitation of EDX elemental analysis for detection of hydrogen, the poor sensitivity to nitrogen, and the interference from the employed carbon supporting film, the absolute chemical composition cannot be obtained by means of this method. However, relative ratios of the composition are available and provide important information about the ligand exchange process. It was found that the ratios of Si/Cd and O/Cd increase significantly, while the ratio Zn/Cd drops after ligand exchange of the CdS-ZnS QDs. The increase in ratios of Si/Cd and O/Cd are consistent with (3-mercaptopropyl)trimethoxysilane being attached to the QDs. The decrease in the Zn/Cd ratio may be due to the loss of smaller zinc-enriched QDs during the repeated separation process through centrifugation. This was also evidenced by the photoluminescence spectra as discussed below. HRTEM images of the CdS-ZnS QDs after the ligand exchange show the well-preserved nanocrystallites. The FT-IR spectra of the CdS-ZnS QDs before and after the ligand exchange were collected. From the spectra, the dominant peaks from the TOA alkane groups observed at 2954 ($v_{as}CH_3$), 2923($v_{as}CH_2$), 2871(v_sCH_3), 2853(v_sCH_2), 1466(δ_sCH_2 & $\delta_{as}CH_3$), and 723 cm^{-1} (ρCH_2) disappear after ligand exchange, while new peaks are formed from the ligand exchange at 2926($v_{as}CH_2$ & $v_{as}CH_3$), 2842(v_sCH_2 & v_sCH_3), 1191(ρCH_3), 1087(v_{as}Si-O-C), and 813cm^{-1}(v_sSi-O-C). In addition, a special test was designed to examine the solubility difference of the QDs before and after the ligand exchange. In this test, the QDs were first dispersed in toluene, and then anhydrous methanol was added into the mixture followed by separation through centrifugation. It was found that the CdS-ZnS QDs before ligand exchange could be easily separated from the mixture, while the QDs after ligand exchange could hardly be separated, indicating the successful ligand exchange from the methanol-phobic TOA to methanol–philic (3-mercaptopropyl)trimethoxysilane. The photoluminescence and UV-Vis spectra of the CdS-ZnS QDs before and after the ligand exchange indicated that the spectral properties of the CdS-ZnS QDs after the ligand exchange were maintained, although a slight red shift was observed in all the excitation, emission, and UV-Vis spectra. This red shift may be due to the loss of smaller QDs during the repeated separation process through centrifugation as mentioned above from the EDX experiments.

From the above discussion, all of our results indicate successful synthesis of the CdS-ZnS core-shell QDs, and the subsequent ligand exchange with the Si-O-CH_3 functional group for subsequent incorporation into polymer matrix through hydrolysis/condensation.

Synthesis of QD-PVAc nanocomposites

The functionalized QDs can be incorporated into the polymer matrix by simply mixing the liquid QDs sample with polymer solution and then evaporating the solvent. Synthesis of QD-polymer composites by means of this procedure involves several steps. In addition to the poor interaction between the QDs and the polymer chains to cause aggregation and migration of QDs, each step consumes energy, labor, time and chemicals, and results in significant waste. Hence, it is desirable to make the nanocomposite with fewer steps, especially by using the "green" solvent, $scCO_2$. In this feasibility study, PVAc was chosen as the polymer matrix for its ability to form a clear film, it's relatively high solubility in $scCO_2$, and our interest in making light-emission nanofilm of the highly commercialized poly(ethylene-co-vinyl acetate).

For the synthesis of the CdS-ZnS QD-PVAc nanocomposites, the monomer vinyl acetate, initiator DEPDC, functionalized CdS-ZnS QDs, linker VTMO, hydrolysis agent acetic acid (HAc), and solvent CO_2 were added into the reactor. The temperature of the reactor was then raised and maintained at 60°C for 22 hours. The in-situ FT-IR measurement shows that the polymerization and the hydrolysis/condensation proceeded simultaneously.

The viscous liquid polymer product was collected from the reactor and subsequently dried under vacuum at 40°C overnight. The obtained solid product was then examined by TEM and EDX elemental mapping. The nanoparticles of approximately 100nm in diameter were well dispersed in the polymer matrix. These relatively large size nanoparticles are composites of silica and the CdS-ZnS QDs. EDX elemental mapping further proved the uniform distribution of the elements Cd, Si, Zn, and S.

In order to study the effect of the CdS-ZnS QDs on the synthesized QD-PVAc nanocomposites, a copolymer of vinyl acetate and VTMO and silica-PVAc composite were synthesized under similar conditions in $scCO_2$, respectively. Gel permeation chromatography (GPC) results showed that the virgin copolymer formed in the absence of acetic acid had a weight-average molecular weight of 27,000Da while the weight-average molecular weights of the PVAc composites with/without the QDs were at least one order of magnitude higher than that of the virgin copolymer. Further evidence of the crosslinks was obtained from the DSC and TGA experiments.

Photoluminescence and UV-Vis spectra of the synthesized QDs/SiO_2-PVAc nanocomposite were also measured and compared with the SiO_2-PVAc nanocomposite, as shown in **Figure 1a**. It is seen that by incorporating the functionalized CdS-ZnS QDs into the PVAc matrix, the nanocomposite film showed significant improvement of absorption at wavelengths from 300-500 nm, especially in the UV region (below 380 nm), and light emission at wavelengths from 500 to 700 nm. However, the previously observed emission peak of the CdS-ZnS QDs after ligand exchange at 620nm disappeared. In order to verify that the coating with silica was not a major factor in the disappearance of the QDs characteristic emission peak, commercial PVAc was purchased and mixed with the functionalized CdS-ZnS QDs in THF, then cast into a film. Photoluminescence and UV-Vis spectra of this mixed QD-PVAc composite was then measured and compared with virgin PVAc polymer, as shown in **Figure 1b**. Similarly, the mixed QD-PVAc composite showed higher absorption, especially in the UV region, and enhanced light emission from green to red light. Only a small emission peak could be observed at 601 nm. This indicates that the diminishment or disappearance of the characteristic emission peak of the QDs was not due to the chemical reaction of the silica coating during the one-pot synthesis. Similar results were also observed by Skaff et al[24] and Mattoussi et al[9] when they incorporated CdSe QDs into the conducting polymer poly(phenylene vinylene) (PPV). Improving interfacial contact by employing such methods as high quantum dot loading could facilitate energy transfer and produce QD-polymer composites with QDs dominant emission peaks.[24]

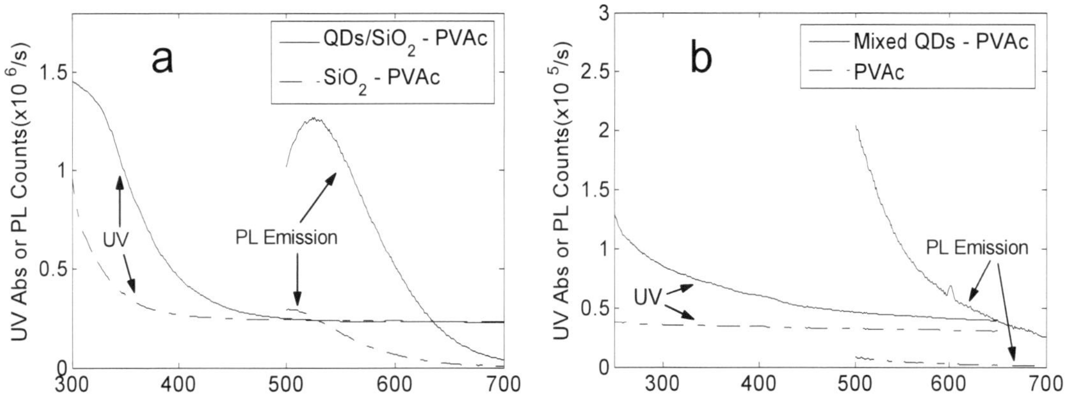

Figure 1. UV-Vis and photoluminescence spectra of the QDs/SiO_2-PVAc and SiO_2-PVAc composite (a) and the mixed QD-PVAc composite and PVAc (b)

Conclusions

A "green" chemical process containing parallel reactions of free radical polymerization and hydrolysis/condensation for the synthesis of QD-PVAc nanocomposites in $scCO_2$ was developed. CdS-ZnS core-shell QDs was synthesized by using single-molecular precursors at relatively low temperature, 180°C. The synthesized QDs were functionalized via ligand exchange with (3-mercaptopropyl)trimethoxysilane which enabled chemical bonding to link the QDs to PVAc chains. Highly crosslinked QD-PVAc nanocomposite was formed and the QDs were found to be well dispersed throughout the polymer matrix. The synthesized QD-PVAc nanocomposite showed attractive spectral properties such as strong absorption of short wavelength lights and emission of long wavelength lights. The one-pot synthesis technique provides advantages in energy-saving and waste-reduction by using $scCO_2$ as solvent. Future work will further synthesize QD-poly(ethylene-co-vinyl acetate) nanofilm with enhanced spectral properties via the one-pot "green" technique using $scCO_2$ to form desirable nanocomposites.

Acknowledgments

This work was financially supported by the Canadian Natural Science and Engineering Research Council (NSERC), the Ontario Centre's of Excellence (OCE) (through the EMK program), and the Canadian Foundation for Innovation (CFI). We thank Fred Pearson of Brockhouse Institute, McMaster University for his assistance on HRTEM, Ronald Smith of the Department of Biology, UWO, for TEM, Mark C. Biesinger of Surface Science Western, UWO for EDX elemental analysis, and Yun Yang and Dr. Robert H. E. Hudson of Chemistry Department, UWO for DSC and photoluminescence, respectively. Special thanks go to Ryan MacIver and Dr. Alex Henderson of AT Plastics for fruitful discussions.

References

1. Oliver, M. (2006). Solar energy: a new day dawning? Silicon Valley sunrise. Nature *443*, 19-22.
2. Gopalakrishnan, G., Danelon, C., Izewska, P., Prummer, M., Bolinger, P.-Y., Geissbuhler, I., Demurtas, D., Dubochet, J., and Vogel, H. (2006). Multifunctional lipid/quantum dot hybrid nanocontainers for controlled targeting of live cells. Angewandte Chemie, International Edition *45*, 5478-5483.
3. Kukimoto, H. (1990). Conductivity control of zinc selenide grown by MOVPE and its applications for blue electroluminescence. Journal of Crystal Growth *101*, 953-957.
4. Murray, C.B., Norris, D.J., and Bawendi, M.G. (1993). Synthesis and characterization of nearly monodisperse CdE (E = sulfur, selenium, tellurium) semiconductor nanocrystallites. J. Am. Chem. Soc. *115*, 8706-8715.
5. Pickett, N.L., and Brien, P.O. (2001). Syntheses of semiconductor nanoparticles using single-molecular precursors. The Chemical Record *1*, 467-479.
6. Barrelet, C.J., Wu, Y., Bell, D.C., and Lieber, C.M. (2003). Synthesis of CdS and ZnS Nanowires Using Single-Source Molecular Precursors. J. Am. Chem. Soc. *125*, 11498-11499.
7. Trindade, T., O'Brien, P., and Zhang, X.M. (1997). Synthesis of CdS and CdSe Nanocrystallites Using a Novel Single-Molecule Precursors Approach. Chem. Mater. *9*, 523-530.
8. Crouch, D.J., O'Brien, P., Malik, M.A., Skabara, P.J., and Wright, S.P. (2003). A one-step synthesis of cadmium selenide quantum dots from a novel single source precursor. Chemical Communications, 1454-1455.

9. Mattoussi, H., Radzilowski, L.H., Dabbousi, B.O., Fogg, D.E., Schrock, R.R., Thomas, E.L., Rubner, M.F., and Bawendi, M.G. (1999). Composite thin films of CdSe nanocrystals and a surface passivating/electron transporting block copolymer: Correlations between film microstructure by transmission electron microscopy and electroluminescence. Journal of Applied Physics *86*, 4390-4399.
10. Peng, X., Schlamp, M.C., Kadavanich, A.V., and Alivisatos, A.P. (1997). Epitaxial Growth of Highly Luminescent CdSe/CdS Core/Shell Nanocrystals with Photostability and Electronic Accessibility. J. Am. Chem. Soc. *119*, 7019-7029.
11. Dabbousi, B.O., Rodriguez-Viejo, J., Mikulec, F.V., Heine, J.R., Mattoussi, H., Ober, R., Jensen, K.F., and Bawendi, M.G. (1997). (CdSe)ZnS Core-Shell Quantum Dots: Synthesis and Characterization of a Size Series of Highly Luminescent Nanocrystallites. J. Phys. Chem. B *101*, 9463-9475.
12. Kim, S., and Bawendi, M.G. (2003). Oligomeric Ligands for Luminescent and Stable Nanocrystal Quantum Dots. J. Am. Chem. Soc. *125*, 14652-14653.
13. Skaff, H., Ilker, M.F., Coughlin, E.B., and Emrick, T. (2002). Preparation of Cadmium Selenide-Polyolefin Composites from Functional Phosphine Oxides and Ruthenium-Based Metathesis. J. Am. Chem. Soc. *124*, 5729-5733.
14. Chan, W.C.W., and Nie, S. (1998). Quantum Dot Bioconjugates for Ultrasensitive Nonisotopic Detection. Science *281*, 2016.
15. Mattoussi, H., Mauro, J.M., Goldman, E.R., Anderson, G.P., Sundar, V.C., Mikulec, F.V., and Bawendi, M.G. (2000). Self-Assembly of CdSe-ZnS Quantum Dot Bioconjugates Using an Engineered Recombinant Protein. J. Am. Chem. Soc. *122*, 12142-12150.
16. Wang, Y.A., Li, J.J., Chen, H., and Peng, X. (2002). Stabilization of Inorganic Nanocrystals by Organic Dendrons. J. Am. Chem. Soc. *124*, 2293-2298.
17. Sheng, W., Kim, S., Lee, J., Kim, S.W., Jensen, K., and Bawendi, M.G. (2006). In-Situ Encapsulation of Quantum Dots into Polymer Microspheres. Langmuir *22*, 3782-3790.
18. Tekin, E., Smith, P.J., Hoeppener, S., van den Berg, A.M.J., Susha, A.S., Rogach, A.L., Feldmann, J., and Schubert, U.S. (2007). Inkjet printing of luminescent CdTe nanocrystal-polymer composites. Advanced Functional Materials *17*, 23-28.
19. Yue, B., Yang, J., Huang, C.-Y., Dave, R., and Pfeffer, R. (2005). Synthesis of macroporous PMMA/silica nanocomposite monoliths in supercritical carbon dioxide. Macromolecular Rapid Communications *26*, 1406-1411.
20. Charpentier, P.A., Xu, W.Z., and Li, X. (2007). A novel approach to the synthesis of SiO2–PVAc nanocomposites using a one-pot synthesis in supercritical CO2. Green Chemistry *9*, 768-776.
21. Xu, W.Z., Li, X., and Charpentier, P.A. (2007). In situ ATR-FT-IR study of the thermal decomposition of diethyl peroxydicarbonate in supercritical carbon dioxide. Polymer *48*, 1219-1228.
22. Souza, A.G., Machado, M.C.N., Helker-Carvalho, L., and Trindade, M.F.S. (2000). Kinetic analysis of the decomposition of chelates of Di-alkyldithiocarbamate of Cd(II). Journal of Thermal Analysis and Calorimetry *59*, 633-642.
23. Hursthouse, M.B., Malik, M.A., Motevalli, M., and O'Brien, P. (1991). Mixed alkyl dialkylthiocarbamates of zinc and cadmium: potential precursors for II/VI materials. X-ray crystal structure of [MeZnS2CNEt2]2. Organometallics *10*, 730-732.
24. Skaff, H., Sill, K., and Emrick, T. (2004). Quantum Dots Tailored with Poly(para-phenylene vinylene). J. Am. Chem. Soc. *126*, 11322-11325.

136a Nitrone Cycloaddition Reactions in the Miceller System

Alaaeddin Alsbaiee, Department of chemistry, King Fahd University of Pet and Min, KFUPM, Dhahran, 31261, Saudi Arabia

I am a master student at King Fahd University for petroleum and minerals. My research work is being in the field of Organic synthesis in green solvents, which is a new trend universally. I am studying nitrone cycloaddition reactions in the miceller system in water as green system for synthesizing the isoxazolidine heterocycles which have very important applications in the natural products synthesis. The second importance of this study is the investigation of the miceller system as new method for the enantio- and diatereoselective nitrone cycloaddition reactions. We are studying asymmetric nitrone cycloaddition reactions of chiral cyclic nitrones with lipophilic parts in order to adjust the miceller system, we are studying the enantio-, diastereo-, and regio-selectivity of the reactions. So this system has two important advantages: 1) low cost, healthy, and environmentally clean system. 2) It can give better stereoseletivity (Enantio- and diatereoselectivity) than the traditional Organic synthesis in the Organic solvents.

136b Synthesis of New Types Gemini Ionic Liquids

Gangsen Li[1], Jun Wang[1], Xuzhao Yang Sr.[1], and Linxiu Zhao[2]. (1) School of material and chemical engineering, Zhengzhou University of Light Industry, No.5 of Dongfeng Road, Zhengzhou, China, (2) Chemical Industry and Ecology Institute, North University of China, Xueyuan road, Taiyuan, China

Two types of Gemini ionic liquids were synthesized with pyridine•AN-butylimidazole and 1,2-dibromoethane as raw materials. N-Butylimidazole reacted with 1,2-dibromoethane to form N-Butyl-N'-ethyldiimidazole(I), (I) reacted with Ammonium hexafluorophosphate to give N-Butyl-N'-ethyldiimidazole hexafluorophosphate. The monofactorial experiment and orthogonal experiment were applied to get the optimum processing conditions. The optimum processing conditions of Gemini ionic liquids were, with water as solvent, the appropriate proportion of pyridine and 1,2-dibromoalkane was 2.6:1, the temperature of reaction was 65°C, the time of reaction was 8h. Pyridine reacted with 1,2-dibromoethane to form N-bromoethyl pyridine, and then reacted with N-butylimidazole to give the asymmetric Gemini ionic liquids. The optimum processing conditions were, with ethyl acetate as solvent, the temperature of reaction was 55°C, the appropriate proportion of pyridine and 1,2-dibromoethane was 1:1.3, the temperature of reaction was 65°C, the time of reaction was 5h. The chemical composition of these production were identified by IR and NMR.

136c Synthesis of Aryl Esters and α,β-Alkynyl Ketones Via Palladium on Charcoal Catalyzed Carbonylation of Aryl Iodides

Jianhua Liu, **Jing Chen**, and **Chungu Xia**. State Key Laboratory for Oxo Synthesis and Selective Oxidation, Lanzhou Institute of Chemical Physics, Chinese Academy of Sciences, Lanzhou, 730000, China

Aryl esters and α,β-alkynyl ketones are highly attractive synthetic targets due to their versatility in organic synthesis as well as their presence in natural products with biological activity. Among various methods to these important intermediates, Pd-phosphine complexes catalyzed carbonylation of aryl halides with alcohol, phenol or alkyne as nucleophiles is drawing considerable interest. However, Pd-phosphine complexes are often used in the reaction under homogeneous conditions. So, it is urgent to find a phosphine-free and heterogeneous catalytic system for those interesting reactions. Herein, we report Pd/C-Et_3N catalyzed alkoxycarbonylation of aryl iodides (alcohols or phenols as nucleophiles) and carbonylative Sonogashira coupling reaction of aryl iodides with terminal alkynes to afford the corresponding aryl esters and α,β-alkynyl ketones in good to excellent yields under considerably mild conditions. It was found that both electron-withdrawing and electron-donating substituents on phenol enhance the phenoxycarbonylation of iodobenzene. In addition, the recyclability of Pd/C was also demonstrated.

136d Physical and Chemical Cellulose Surface Modification

Sergiu Coseri, "Petru Poni" Institute of Macromolecular Chemistry Iasi, 41A, Gr. Ghica Voda, Iasi, 700487, Romania

Autors:

S. Coseri1, G. Nistor1, S. Strnad2, V. Harabagiu1, A. Carpov.1

E-mail: coseris@icmpp.ro

Fax: +40 232 211299

1"Petru Poni" Institute of Macromolecular Chemistry Iasi, Romania

2 University of Maribor, Faculty of Mechanical Engineering, Slovenia

Members of the European Polysaccharide Network of Excellence (EPNOE)

ABSTRACT

The physical and chemical surface modifications of various cellulose types, including viscose fibers were carried out.

The physical activation was achieved by the exposure of the cellulose samples on cold plasma.

The chemical surface modification was performed by the use of two oxidative systems: i) a "classic" TEMPO/NaBr/NaClO system, see Scheme 1.

Scheme 1. TEMPO-mediated oxidation of cellulose primary hydroxyl groups to carbonxyl groups.

ii) A new and promising route involving different nitroxy radicals, which avoided the use of more polluting reagents. The modified samples were analyzed by means of FTIR, SEM, X-ray.

Both of these methods reveal only the mild modification on the cellulose's surface, without damages of the biomacromolecular chain.

Effects of ferrous ions on nitrobenzene reduction by zero-valent iron

YIN Jun[1,2], *LIU Lei*[1], *ZHANG Li-guo*[1], *LV Guo-xiao*[2]

1. School of Municipal and Environmental Engineering, Harbin Institute of Technology, Harbin, 150090, China; 2. Jilin Key Laboratory of Water pollution control and resources reuse, Jilin Architectural and Civil Engineering Institute, Changchun, 130021, China.

Introduction

As an industrial chemical, nitrobenzene is mainly used to produce a chemical called aniline, and it is also used to produce dyes, drugs, pesticides and synthetic rubber, etc [1]. The production and utilization of nitrobenzene has caused the pollution of groundwater.

Due to the innovation and low cost, the permeable reactive barriers (PRBs) have received a lot of attention during the past decade [2]. As a medium used in PRBs, zero-valent iron (ZVI) has been widely used to remove most of common environmental contaminants. The contaminants such as halogenated organic pollutants, pesticides and nitrate of water, can be effectively reduced by ZVI [3-5]. ZVI can be used in the in situ remediation of contaminated groundwater and soil as an electron donator [6]. Nitrobenzene, as a typical aromatic pollutant, had been examined in the ZVI system [7]. Nitro was recalcitrant to the biological degradation due to the lack of electrons, and it was reduced to amido by Fe^0. This reductive conversion could decrease the biological toxicity of nitro, accelerating the biological degradation.

However, most researchers focused on the degradation pathways and the reductive intermediate in the ZVI system. Not much information has been obtained upon the effect of environmetal factors. Therefore, considering the presence of Fe^{2+} in water, the effects of Fe^{2+} on the nitrobenzene reduction by Fe^0 were investigated using prepared samples containing Fe^{2+}, and the reaction kinetics of nitrobenzene reduction was analyzed with the presence of Fe^{2+}.

Materials and Methods

Materials and chemicals

Iron powder (A.R., Tianjin guangfu science and technology development Co. Ltd.) was used in this test without further pretreatment.The surface of iron grains retained a metallic glaze without apparent rust. The specific surface area (SSA) of iron powder was measured using BET method (NOVA1000, Quantachrome), and the SSA of iron powder was 3.5697 m^2/g. Nitrobenzene (A.R.) was obtained from Beijing chemical works. The Fe^{2+} solution was prepared using $FeSO_4 \cdot 7H_2O$ (A.R., Beijing chemical works). Acetonitrile (HPLC, Fisher scientific)。 The DI water had a resistivity of 18.0 MΩ, and it was obtained from a Millipore-Q system. The Millipore-Q water was sparged with N_2 for half an hour to ensure that the DO concentration was below 0.5 mg/L. The pH was adjusted with the acetate buffer solution.

Experimental methods

All experiments were conducted in the 40-ml borosilicate vials fitted with PTFE septum caps. The vials were filled with 0.5 g iron powder, 10-ml prepared nitrobenzene solution containing Fe^{2+}. The samples were sparged with N_2 for half an hour after the adjustment of pH. All the vials were shaken in a horizontal vibrator at 170 rpm and 30 ^{0}C. In addition, control sample, without the addition of iron powder and Fe^{2+}, was also tested. The experimental process lasted for 6 h. The pH and oxidation-reduction potential (ORP) were immediately measured after sampling. The solutions were then filtered through a 0.22 μm membrane filter, and the supernatant was used to measure nitrobenzene, aniline, Fe^{3+} and Fe^{2+}.

Analytical methods

The nitrobenzene and aniline concentrations were measured using an HPLC (SPD-10A_{VP}, Shimadzu Co.) with a UV detector and a C_{18} column (250mm×4.6mm, 5μm). The mobile phase was the acetonitrile/water mixtures in the proportion of 70/30 (v/v). The concentration of ferrous ions was measured using the phenanthroline method with a spectrophotometer (GBC, UV-911).

The reductive products of nitrobenzene were determined using an Agilent 6890 GC/MS system with an ECD detector and a DB-5 ms column (30m ×0.25mm×0.25μm). The flow rate of carrier gas (N_2) was 1 ml/min.

Results and Discussion

Effects of aqueous Fe^{2+} on nitrobenzene reduction by Fe^0

As compared to the system without the additon of Fe^{2+}, the different concentrations of Fe^{2+} samples were prepared. In addition, a control system without the additon of Fe^{2+} and Fe^0 was also prepared to examine the volatilization of nitrobenzene. As shown in Fig. 1, there existed only no more than 15% nitrobenzene loss in the control system, that may be attributable to the volatilization in headspace or the sorption of nitrobenzene on the septum. For the system without Fe^{2+}, no more than 90% of nitrobenzene was reduced in 6 h, and most of the reduction occurred in the first 4 h. While for the samples containing Fe^{2+} and Fe^0, nitrobenzene was fully reduced, and the reductive rate increased with the increasing of Fe^{2+} concentraiton. Nitrobenzene was completely reduced in the first 3 h when the Fe^{2+} concentration was up to 1000 mg/L. According to the analyses from GC/MS system, the reductive product of nitrobenzene was not changed with the addition of Fe^{2+}, and $ArNO_2$ was still reduced to aniline by Fe^0. That was in agreement with the conclusion of Agrawal et al [9]. Aniline, compared to nitrobenzene, was not recalcitrant to the biological degradation [10].

Variation of auqeous Fe^{2+} concentration at different dosage of Fe^{2+}

As shown in Fig. 2, the Fe^{2+} concentration was constant during the test for the system without the additon of Fe^{2+}. The initial pH was controlld at about 5.0 using the acetate buffer solution, therefore, the Fe^{2+} concentration did not increase. While for the samples containing Fe^{2+}, the Fe^{2+} concentration decreased slightly during the first 2 h, that may be attributable to the participate in reductive reaction. However, the Fe^{2+} loss was partially recovered by the end of test. There was a limited Fe^{2+} loss over the test period. An additional sample containing only 1000 mg/L of Fe^{2+} was tested to examine the function of Fe^{2+}. The results showed that there was a limited Fe^{2+} loss due to the volatilizaiton, and nitrobenzene was not reduced to aniline over 6 h, meaning that Fe^{2+} might not function as a major reductant.

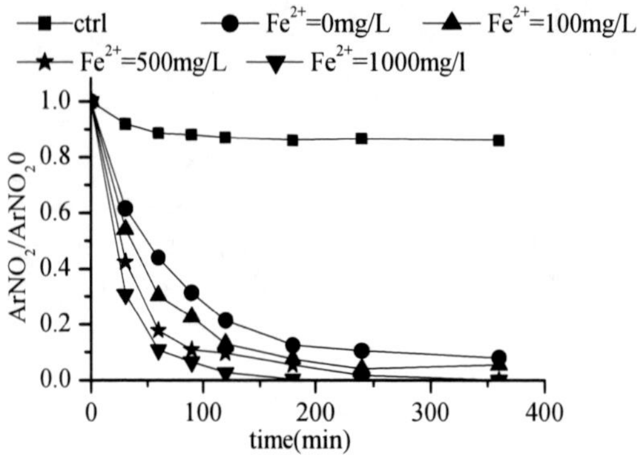

Figure 1. Nitrobenzene reduction at different initial concentrations of Fe^{2+}

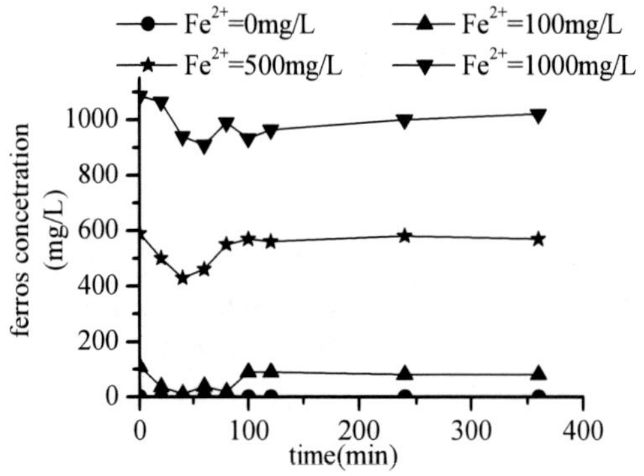

Figure 2. Time variation of Fe^{2+} concentration

Variations of pH and ORP

In the absence of acetate buffer solution, the pH of the system was dominated by the hydrolysis of Fe^{2+} species. The initial pH was controlled at about 5.0 to avoid the effect of $Fe(OH)_2$ that can be formed at a higher pH. As shown in Fig. 3, the pH was in a range of 5.0-6.0 for the samples containing Fe^{2+} species. However, the pH increased quickly for the samples without Fe^{2+} species. That was a result of the buffering caused by Fe^{2+}. Although the effect of Fe^{2+} was partially depended on the pH, the accelerating effect of pH on the reductive reaction was not ovserved in this test. Fig. 4 showed that the ORP value decreased from 133 to -273 mV for the system containing Fe^{2+} species, which suggested that the reductive reaction occurred in the liquid phase. The decrease of ORP was favourable to the reductive reaction.

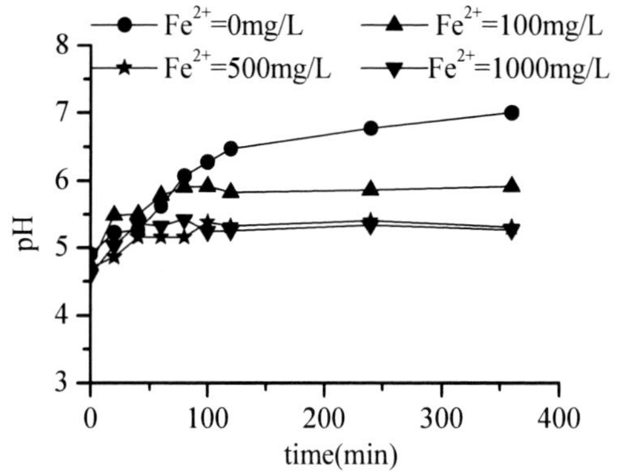

Figure 3. Time variations of pH

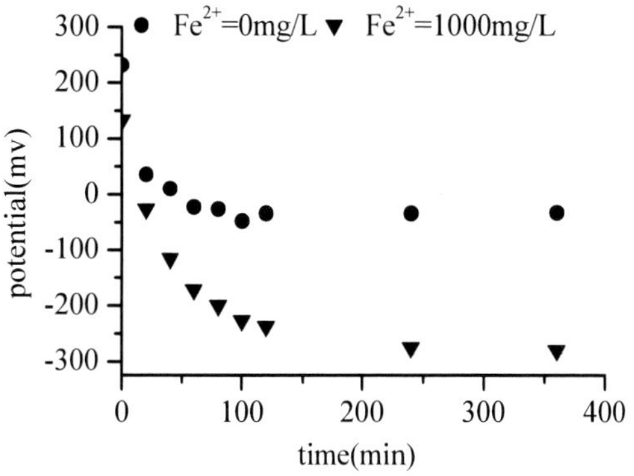

Figure 4. Time variations of ORP

Kinetic anlyses of reduction reaction

According to Mu et al. [11], the nitrobenzene reduction by Fe^0 followed a pseudo-first-order kinetics in terms of the concentration of nitrobenzene. The nitro reduction rate increased linearly with the concentration of nitrobenzene adsorbed on the iron surface, and the key step was the adsorption of nitrobenzene on iron surface [12]. However, the reaction kinetics of nitrobenzene reduction with the addition of Fe^{2+} were not examined. In the absence of Fe^{2+}, the reductive rate of nitrobenzene by Fe^0 was proportional to the Fe^{2+} concentration. The equation of pseudo-first-order kinetics as follows.

$$\frac{d[ArNO_2]}{dt} = -k_{obs}[ArNO_2] \quad (1)$$

The integral form of equation (1) was in the following form.

$$[ArNO_2] = [ArNO_2]_0 e^{-k_{obs}t} \quad (2)$$

Where, $K_{obs}(min^{-1})$ is the observed first-order rate constant, $[ArNO_2]_0$ is the initial nitrobenzene concentration and t is reaction time. The observed first-order rate constants can be calculated from the regression analysis of $\ln[ArNO_2]$ versus time.

The K_{obs} and correlation coefficients were calculated from Fig. 5 and shown in Table 1. The K_{obs} value increased with the increasing of Fe^{2+} concentration, indicating that the reductive rate was enhanced. The presence of Fe^{2+} in ZVI system was favorable to the nitrobenzene reduction.

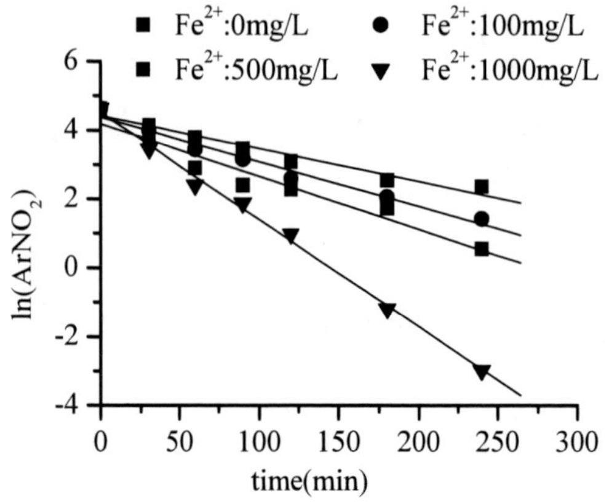

Figure 5. The relation of ln[$ArNO_2$] versus time

Table 1. Variation of K_{obs} with Fe^{2+} concentration in ZVI system

Fe^{2+} (mg/l)	K_{obs}(min^{-1})	R^2
0	0.0096	0.9573
100	0.013	0.976
500	0.0153	0.9443
1000	0.0311	0.9959

Brownish precipitate was observed during the tests except for the control sample, and the major component of precipitate may be maghemite. According to Farrell et al. [13], the passive layer on the surface of iron powder consisted of an inner layer (a semiconductor with good conductivity) and an outer layer (a semiconductor with poorer conductivity). The small band gap between its valence and the conduction bands provides it an electrical conductivity close to that of metal, enhancing the conductivity of the oxide layer and allowing electrons to be transferred from Fe^0 to the contaminant at the oxide-water interface. That was a result of that the addition of Fe^{2+} species accelerated the reductive reaction.

Conclusions

The reductive iron powder was used to investigate the nitrobenzene reduction in a ZVI system with the addition of Fe^{2+}. The results indicated that the reductive product of nitrobenzene was not changed with the addition of Fe^{2+}, and $ArNO_2$ was reduced to aniline by Fe^0. The reductive reaction rate of $ArNO_2$ was apparently enhanced with the addition of Fe^{2+}, however, Fe^{2+} was not the major reductant. According to the kinetics analysis of the reaction, the reductive reaction can be described using a pseudo-first-order kinetic model, and the observed pseudo-first-order rate constants (K_{obs}) increased with the increasing of Fe^{2+} concentration, indicating that the presence of Fe^{2+} accelerated the reduction of $ArNO_2$ by Fe^0.

References

[1] S.Contreras, M.Rodryguez, E.Chamarro, and S.Esplugas, "V-and UV/Fe(□)-enhanced ozonation of nitrobenzene in aqueous solution," Photochem. Photobiol, 2001, A142, 79-83.

[2] M.M.Schere, S.Richter, R.L.Valentine, P.J.Alvarez. "Chemistry and microbiology of permeable reactive barriers for in situ groudwarer cleanup,"Crit. Rev. Environ. Sci. Technol, 2000, l30, 363-411

[3] T.L.Johnson, M.M.Scherer, P.G.Tratnyek, "Kinetics of halogenated organic compound degradation by iron metal, "Environ.Sci.Technol, 1996,30, 2634-2640.

[4] G. Antoine, "Degradation of benomyl,picloram,and dicamba in a conical apparatus by zero-valent iron power, "Chemshere,2001,43, 1109-1117.

[5] C. P.Huang, H. W. Wang, P. C.Chiu, "Nitrate reduction by metallic iron， "Wat. Res,1998,32,2257-2264.

[6] A.Ghauch, C.Gallet, A.Charef, J. Rima, M.M. Bouyer, "Reductive degradation of carbary in water by zero-valent iron, "Chemosphere,2001, 42, 419-424.

[7] Yang Mu, Han-Qing Yu, Jia-Chuan Zheng, et al., "Reductive degradation of nitrobenzene in aqueous solution by zero-valent iron, " Chemosphere, 2004, 789-794.

[8] H.M.Hung, F.H.Ling, M.R.Hoffmann, " Kinetics and mechanism of the enhanced reductive degradation of nitrobenzene by elemental iron in the presence of ultrasound, "Environ.Sci.Technol, 2000, 34, 1758-1763.

[9] A.Agrawal, P.G.Tratnyek, " Reduction of nitro aromatic compounds by zero-valent iron metal, "Environ. Sci. Technol, 1996, 30, 153–160.

[10] L.S. Bell, J.F. Devlin, R.W. Gillham, P.J. Binning, "A sequential zero valent iron and aerobic biodegradation treatment system for nitrobenzene,"Journal of Contaminant Hydrology, 2003, 66 ,201– 217.

[11] Y.Mu, H.Q.Yu, J.C Zheng, S.J.Zhang, G.P.Sheng, "Reductive degradation of nitrobenzene in aqueous solution by zero-valent iron, "Chemosphere, 2004, 54, 789–794.

[12] J.F. Devlin, J. Klausen, R.P.Schwarzenbach, " Kinetics of nitroaromatic reduction on granular iron in recirculating batch experiments," Environ. Sci. Technol, 1998, 32, 1941–1947.

[13] J. Farrell, M. Kason, N. Melitas, T. Li, " Investigation of the long-term performance of zero-valent iron for reductive dechlorination of trichloroethylene, " Environ. Sci. Technol, 2000,34,514-521.

136f Coupling Multistage Distillation and Pervaporation

Pierre Schaetzel and Jean-Christophe Remigy. Chemical Engineering Department, Caen University, IUT Dpt Génie Chimique, Boulevard Marechal Juin, Caen cedex, 14032, France

To separate in a quantitative way water and IPA we have simulated a coupled pervaporation-multistage distillation process allowing treating a feed containing water soluble salts. We have performed the experimental study of aqueous solutions of IPA containing 1% to 15 % mass fraction of water at 70 °C, 80 °C, 95 °C and 105 °C with a pilot pervaporation test unit with 1m2 of flat sheet membrane surface area (Sulzer Chemtech). The test unit was equipped with a GFT 1001 PVA based membrane. We have modelized empirically the flux of IPA and water through the membrane vs. the temperature and the concentration of the feed. The water mass flux is proportional to the mass fraction of water in the feed. The IPA flux is linearly dependent on the water mass fraction in the feed. An Arrhenius like equation describes the dependency of both mass transfer coefficients on the absolute temperature.

136g Self-Diffusion of Monovalent Ions in Highly Concentrated Liquid and Solid Electrolytes

Pierre Schaetzel, Yongli Li, Quang Trong Nguyen, and Dominique Langevin. Chemical Engineering Department, Caen University, IUT Dpt Génie Chimique, Boulevard Marechal Juin, Caen cedex, 14032, France

We analyze data concerning tracer and self-diffusion of monovalent ions (K^+, Na^+, Li^+,H^+) in highly concentrated aqueous solutions of LiCl, NaCl and KCl ranging from 0.1 to 4 mol/l and self-diffusion coefficients of Na^+ in weak-acid ion exchanger rods (cross-linked polymethacrylic acid) in which the molarity of the counter-ion varied from 0.8 to 3.5 mol/l and that of water from 40 to 46 mol/l. To analyze the data we employ the generalized diffusion Nernst-Planck equation which was recently derived from the generalized Stefan-Maxwell equation. When the equation is applied to the solid electrolyte the polymer is modelled as a dusty gas, where the polymer species is immobile. The same statistical mechanics equation correlates all experimental data. We can conclude that solid and liquid electrolytes are solutions which only differ by the fact that the fixed charges in the solid ion-exchanger are immobile.

136h Pervaporation - Esterification Coupling

Pierre Schaetzel, Rachida Bouallouche, and Hamid Aït-Amar. Chemical Engineering Department, Caen University, IUT Dpt Génie Chimique, Boulevard Marechal Juin, Caen cedex, 14032, France

In this paper we describe the synthesis of a new polymeric membrane which is able to catalyze an esterification reaction and selectively extract water by pervaporation. The experimental example chosen here is the esterification of n-butanol by acetic acid at temperature ranging from 90°C to 110 °C. The membrane is a polymeric PVA(40%)-poly(styrenesulfonic acid-co-maleic acid)(60%) blended membrane. The membrane is stable and selective to water. With this process the water molar fraction in the reactor is low and the conversion rate is high.

136i Rheological Behavior and Drug-Controlled Release Property of Chitosan/rectorite Nanocomposites

Xiaoying Wang and **Yumin Du**. Department of Environmental Science, Wuhan University, 129 Luoyu Road, Wuhan, 430079, China

The rheological behaviour of chitosan/rectorite (CS/REC) nanocomposites was evaluated; the nanocomposites showed the low shear-rate non-Newtonian viscosity and rapid shear-thinning behaviour, zero shear rate viscosity and the storage and loss moduli for the nanocomposites displayed a monotonic increase with increasing REC content in comparison with pure CS. The results implied the interaction between CS and REC took place, the fact corroborated by the SEM images and FTIR analysis; the CS/REC nanocomposite beads prepared from the crosslink interaction of sodium tripolyphosphate (TPP) have compacter surface and internal structure than CS beads. The swelling degree of the nanocomposites in 0.9% (w/v) saline was only about 15% compared to CS beads of 30% swelling degree. The nanocomposite beads can release the drug continuously, in the given 6 days the minimum cumulative release only reached 51.9%, much lower than pure CS beads. CS/REC nanocomposites are considerably suitable to apply as mucoadhesive drug carriers.

136j Pulp Physical Property Improvement with LMS in TCF Bleaching of Masson Pine Kraft Pulp

Mengru Liu, Huaiyu Zhan Sr., **Shiyu Fu Sr.**, Hao Liu, and Hailong Li. Department of Resource Science and Papermaking Engineering, State Key Laboratory of Pulp and Paper Engineering, South China University of Technology, State Key Laboratory of Pulp and Paper Engineering, No. 381,Wushan road,, Guangdong Province,China, Guangzhou, 510640, China

Laccase from a white rot fungus Flammulina velutipes, in the presence of HBT (1-hydroxybenzotriazole), was applied to improve totally chlorine-free (TCF) bleaching of Masson pine kraft pulp. Different TCF bleaching sequences in combination with LMS treatment were evaluated. The results indicated the laccase from Flammìlina velitipes could improved the bleaching efficiency and pulp brightness. Kraft Masson pine pulps the can be bleached to 80%ISO using OLEQP and LEOQP bleaching sequence, which was 11.3%ISO higher and 10.7%ISO than that without biobleaching, respectively. In addition, with LMS (Laccase/Mediator System), the strength properties of the pulp increased significantly in comparison with TCF bleaching without LMS treatment. XRD was used to investigate the cellulose crystallinity change in the LMS treatment. FQA was also adopted to analyze the fiber properities. The results are expected to interpret the reason that the LMS treatment maintains the original fiber strength in TCF bleaching.

134c Monetizing Greenhouse Gas Emission Reductions: Opportunities in Carbon Trading

Scott Deatherage, Climate Change and Renewable Energy Practice Group, Thompson & Knight LLP, 1700 Pacific Avenue, Suite 3300, Dallas, TX 75201

The Kyoto Protocol and developing international programs that will follow, as well as exiting state and probable future US climate change legislation, now limit and will limit in the future greenhouse gas emissions (GHG) from industrial sources. Industries with significant GHG emissions will be able to use the "cap-and-trade" system existing at the international level and proposed under state and federal legislation to purchase emission offsets to meet at least in part their mandated emission reductions.

These "carbon credits" have become a $30 billion industry under Kyoto, and will balloon to much higher levels in the future. As a result, industries and other businesses and entrepreneurs may find investment in GHG reduction projects can generate valuable carbon credits to help offset their own GHG emission limitations or to sell in the international mandatory markets or the already existing voluntary markets in the US for significant profit.

April 8, 2008

Preparing for Greenhouse Gas Regulations

Greg L. Johnson
(504) 556-4115 Direct
gljohnson@Liskow.com

LISKOW&LEWIS

A Professional Law Corporation

New Orleans | Lafayette | Houston

Overview

- International Greenhouse Gas Initiatives
- U.S. Greenhouse Gas Issues
  - Litigation
  - EPA Developments
  - Regional Initiatives
  - State Level Developments
  - Proposed Legislation

International Greenhouse Gas Initiatives

LISKOW&LEWIS

A Professional Law Corporation

New Orleans | Lafayette | Houston

International Initiatives

- Kyoto Protocol in effect; 162 countries subject to the treaty.
- Australia, U.S. have not accepted the Protocol.
- Sets country-specific targets for reductions of 6 GHGs to 8% below 1990 levels.
  - CO_2, methane, nitrous oxide, hydrofluorocarbons, perfluorocarbons, and sulfur hexafluoride
- Initial targets are for 2008-2012.
- Expires in 2012.

International Initiatives

- On February 2, 2007, the United Nations Intergovernmental Panel on Climate Change issued *Climate Change 2007: The Physical Science Basis*.
- Report recognizes that an "unequivocal" warming of the climate system is taking place and suggests that it is "very likely" being caused by human activity.

International Initiatives

- G-8 Meeting – June 2007:
  - All nations agreed to "consider seriously" a target of cutting global carbon dioxide emissions in half by 2050.
- U.N. meeting in Bali Dec. 2007:
  - Goal: Negotiate framework for successor to Kyoto Protocol.
  - Set a 2009 deadline for drafting the new climate change strategy to replace Kyoto Protocol.
  - No specific emissions reduction targets yet set.

International Initiatives

- Major Economies Meeting on Energy Security & Climate Change
  - U.S. initiative to contribute to post-Kyoto framework.
  - 17 countries (including EU, Japan, China, India).
  - Each nation designs separate strategies to reflect each country's different energy resources, stages of development, and economic needs.

International Initiatives

- Major Economies Meeting on Energy Security & Climate Change
  - First meeting was September 2007.
  - U.S. Key Component: Advance clean energy technologies.
    - US will promote nuclear power plants.
    - US "Twenty in Ten" plan: reduce gasoline consumption by 20% in 10 years.
    - US will address unsustainable deforestation

International Initiatives

- Major Economies Meeting on Energy Security & Climate Change
  - Second meeting held in Honolulu in January 2008.
    - Discussion of sectoral approaches under Bali "road map."
    - U.S. giving "serious consideration" to proposals of halving 1990 level emissions by 2050.
    - No final agreements.
  - Third meeting set for France in April 2008.

U.S. Greenhouse Gas Issues

LISKOW & LEWIS

A Professional Law Corporation

New Orleans | Lafayette | Houston

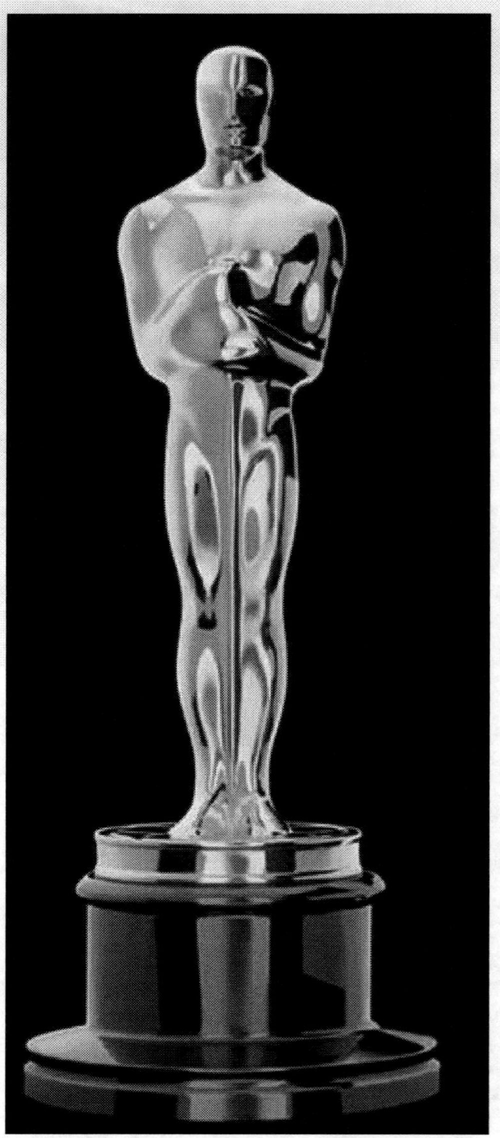

LISKOW&LEWIS

New Orleans | Lafayette | Houston

Greenhouse Gas Litigation

LISKOW&LEWIS

A Professional Law Corporation

New Orleans | Lafayette | Houston

Greenhouse Gas Litigation

- *Massachusetts v. EPA*, 127 S.Ct. 1438 (2007)
  - Background:
    - Organizations petitioned EPA to regulate GHGs in motor vehicle emissions under CAA § 202.
    - EPA determined that it lacked authority to regulate GHGs because not "pollutants."
    - Even if it had authority, EPA declined to regulate CO2 in relation to global warming for policy reasons.

Greenhouse Gas Litigation

- *Massachusetts v. EPA*, 127 S.Ct. 1438 (2007)
  - S. Ct. decision (5-4) issued April 2, 2007:
    - CO2 and other GHG are "pollutants" under CAA.
    - EPA's policy justifications for inaction not reasonable.
    - Remand back to EPA. EPA must ground its reasons for action or inaction in the statute.

LISKOW&LEWIS

New Orleans | Lafayette | Houston

Greenhouse Gas Litigation

- Implications of *Massachusetts v. EPA*
  - "Endangerment standard":
    - EPA must now determine whether GHGs "may reasonably be anticipated to endanger public health or welfare."
  - Definition of "pollutant" is applicable to entire CAA:
    - GHG regulation can apply to stationary sources.

Greenhouse Gas Litigation

- After *Massachusetts v. EPA*, EPA pledged to propose mobile source GHG regulations by the end of 2007.
- Congress passed energy legislation requiring an increase in CAFE standards and greater use of renewable fuels in auto fuels.
- After energy legislation, EPA indicated it may not need to conduct GHG rulemaking because the CAFE standards reduce GHG emissions.

Greenhouse Gas Litigation

- Massachusetts + 17 other states threatened suit if EPA fails to issue GHG rulemaking.
  - Demanded response from EPA by February 27, 2008, regarding intentions for rulemaking.

Greenhouse Gas Litigation

- *Green Mountain Chrysler-Plymouth-Dodge-Jeep v. Crombie*, D.Vt., No.05-cv-302, 9/12/07
  - Background:
    - Vermont adopted California's GHG regulations for new motor vehicles.
    - Auto industry claimed GHG regulations are actually fuel economy standards that conflict with federal law.
    - Auto industry claimed detrimental effect on industry.

Greenhouse Gas Litigation

- *Green Mountain Chrysler-Plymouth-Dodge-Jeep v. Crombie*
  - Federal Court Decision
    - Vermont GHG regulations sufficiently unrelated to fuel economy standards – NOT preempted by Energy Policy and Conservation At of 1975.
    - No catastrophic effect for auto industry.

Greenhouse Gas Litigation

- *Central Valley Chrysler Jeep v. Witherspoon* – E.D. Cal., No. 04-6663 (Dec. 12, 2007).
  - Court granted California summary judgment on EPCA and foreign policy preemption arguments.
  - If California Waiver granted by EPA, California GHG regulations are not preempted by federal law.

Greenhouse Gas Litigation

- Public Nuisance Claims
  - *New York v. TVA, et al.*, 05-5104, 2nd Cir.: Alleges companies creating public nuisance w/their GHG emissions. Dismissed by district court. Appealed to 2nd Circuit. Oral arguments held in June 2007.
  - *California v. GM*, 3:06-cv-5755, N.D. Cal.: State seeking financial payments from "Big Six" auto cos. Dismissed by district court. Appealed to 9th Circuit.

Greenhouse Gas Litigation

- Public Nuisance Claims
  - *Comer v. Murphy Oil*, 1:05-cv-436-LG-RHW, S.D. Miss.: Hurricane Katrina victims alleged GHG emissions strengthened storm. Dismissed by district court on Aug. 30, 2007.
  - *Korinsky v. EPA*, 1:05-cv-859-NRB, S.D. NY: Resident sued EPA, NY state and NYC for creating public nuisance with GHG emissions. Dismissed by district court for lack of standing. Appealed to 2nd Circuit.

Greenhouse Gas Litigation

- Other
  - *FOE v. Mosbacher*, 02-4106, N.D. Cal.: Alleges NEPA violations by agencies due to failure to assess GHG emissions in energy-intensive projects. Pending. Govt's motion to dismiss denied Feb. 2007.
  - *New York v. EPA*, 06-1148, D.C. Cir.: Alleges EPA failed to regulate CO2 from new coal-fired power plants and industrial boilers. Remand to EPA expected.

Greenhouse Gas Litigation

- Other
  - *Montana Envtl. Info. Center v. EPA*, 06-1059, D.C. Cir.: Settled. EPA repealed letter stating that power plant builders need not consider construction of a coal gasification plant.
  - *CBD v. Brennan*, 4:06-cv-7062-SBA, N.D. Cal.: Ct. ordered Govt to publish a new draft climate research plan by March 1, 2008 and a new scientific assessment by May 31, 2008.

EPA Greenhouse Gas Developments

LISKOW&LEWIS

A Professional Law Corporation

New Orleans | Lafayette | Houston

EPA Developments

- New Source Review for GHG Emissions
  - EPA considered proposing threshold for GHG emissions that will trigger NSR. (But not yet decided if NSR applies.)
  - To be issued with proposed GHG rule for mobile sources at end of 2007 – rules not yet proposed.
  - Theory: Once EPA regulates mobile source GHGs as pollutants, *any* emission of GHG from a stationary source could trigger NSR unless threshold level is set.

EPA Developments

- New Source Review for GHG Emissions
  - Problem: Even a 100 tpy or 250 tpy threshold would be burdensome on industry and agencies.
    - Typical threshold levels for increases: 25 tpy or 40 tpy.
    - Typical GHG emissions for a furnace or boiler in a commercial building: 250 tpy.

EPA Developments

- EPA GHG Emissions Reporting Rule
  - Mandated by the spending bill for fiscal year 2008.
  - Directs EPA to implement rules requiring GHG emissions reporting.
  - Appropriates $3.5 million to EPA for the rulemaking.
  - Requires "mandatory reporting of GHG emissions above appropriate thresholds in all sectors of the economy."
  - Includes "emissions resulting from upstream production and downstream sources."

EPA Developments

- EPA GHG Emissions Reporting Rule
  - Proposed rule – September 2008
  - Final rule – June 2009
  - EPA has discretion to use the existing reporting requirements for power plants, which has been in place since early 1990s.
  - Issues:
    - Emissions threshold for reporting?
    - Type of measurement tool?
    - Definition of upstream and downstream production?

EPA Developments

- EPA Region 9 comments on draft EIS for $1.2 billion coal-fired power plant in Nevada raised GHG emissions concerns.
- Recommendations:
  - Further analysis of alternative options for power generation, including wind and solar.
  - Discussion of carbon capture and sequestration.

EPA Developments

- EPA Region 9 EIS Recommendations (cont'd):
  - Comparison of annual GHG emissions from proposed project to existing and future projects and at a regional, national and global scale.
  - Discussion of anticipated effects of climate change.
- EPA Region 9 raised similar concerns in draft EIS for another proposed coal-fired power plant in Nevada.

Regional Greenhouse Gas Developments

LISKOW&LEWIS

A Professional Law Corporation

New Orleans | Lafayette | Houston

Regional Initiatives

- The Climate Registry
  - A common GHG reporting system.
  - Provides consistent, transparent, and verified set of emissions data for participating entities.
  - Majority of states are members.
  - Released draft *General Reporting Protocol*.

Regional Initiatives

- British Columbia
- Manitoba
- Québec
- Saskatchewan
- Ontario
- Prince Edward Island
- New Brunswick

- Sonora
- Tamaulipas

States, Provinces and Tribes that have joined *The Climate Registry* as of February 1, 2008

Regional Initiatives

- Western Climate Initiative (WCI)
  - Arizona, California, Oregon, New Mexico, Washington, Utah, and Canadian provinces of British Columbia and Manitoba.
  - Load-based, Cap and Trade program.
  - Goal: Reduce GHGs to 15% below 2005 levels by 2020.
  - Multi-sector action.
  - All participants are members of the Climate Registry.
  - Part of International Carbon Action Partnership (ICAP) (a forum to establish a global cap-and-trade market).

Regional Initiatives

- Northeast Regional Greenhouse Gas Initiative (RGGI)
  - Connecticut, Delaware, Maine, Maryland, New Hampshire, New Jersey, New York, and Vermont.
  - Initially, cap CO_2 emissions from power plants.
  - Second stage may include sources other than power plants.
  - Goal: an approximately 35 % reduction by 2020.
  - All participants are members of the Climate Registry.
  - Part of International Carbon Action Partnership (ICAP).

Regional Initiatives

- Midwestern GHG Reduction Accord
  - Minnesota, Wisconsin, Illinois, Indiana, Iowa, Michigan, Kansas, Ohio and South Dakota, and the Canadian Province of Manitoba.
  - Committed to establishing GHG reduction targets and developing cap-and-trade system.

State Greenhouse Gas Developments

LISKOW & LEWIS

A Professional Law Corporation

New Orleans | Lafayette | Houston

State Level Developments

- Kansas DHE denied an air permit for a proposed coal fired plant on basis that CO_2 emissions present "a substantial endangerment." (Oct. 2007).
- Kansas statute provides authority for DHE to take necessary action upon receipt of information that air pollution presents "substantial endangerment."
- DHE relied on *Massachusetts v. EPA* to establish CO_2 as a pollutant.

State Level Developments

- California Waiver
  - California law imposes more stringent GHG controls on motor vehicles.
    - Cut emissions from cars/light trucks by 25% and from SUVs by 18% starting with 2009 models.
  - Under CAA, a state may enact its own motor vehicle emissions regulations so long as it receives permission from EPA. *See* CAA § 209.
  - Other states can follow federal or waiver-approved California standards.

State Level Developments

- California Waiver
  - EPA is obligated to provide the waiver unless:
    - State determination is arbitrary and capricious;
    - State does not need the standards to meet extraordinary circumstances; or
    - State standards are not consistent with federal standards.
- Other states adopting California standards: VT, NY, NJ, MA, CT, ME, RI, PA, MD, WA, OR.

State Level Developments

- EPA denied waiver in December 2007 because:
  - GHGs need to be addressed nationally and globally.
  - No demonstration of "compelling and extraordinary need."
  - Federal energy legislation signed into law in December 2007 imposes national fuel economy standards with greater GHG reductions.
- 1st ever denial of waiver.

State Level Developments

- California + 15 other states filed suit January 2, 2008. *California v. EPA*, 9th Cir.).
  - EPA rationale for denial is flawed.
  - CA regs would cut GHGs by 17.2 million metric tons by 2016 and 33.4 million metric tons by 2020.
  - Fed stds will cut GhGs by 7.7 million metric tons by 2016 and 18.9 million metric tons by 2020.

State Level Developments

- Congressional reaction to EPA's denial:
  - Rep. Waxman (D.-Calif.), the chair of the House Oversight and Gov't Reform Committee, has opened an investigation into EPA's denial.
  - Sen. Boxer (D-Calif.) introduced legislation to overturn EPA's decision (S. 2555).

LISKOW&LEWIS New Orleans | Lafayette | Houston

Proposed Federal Legislation

LISKOW&LEWIS

A Professional Law Corporation

New Orleans | Lafayette | Houston

U.S. Proposed Legislation

- Global Warming Reduction Act (S. 485):
  - Kerry/Snowe
  - Cut GHG emissions 65% from 2000 levels by 2050.
  - Freeze GHG emissions in 2010.
  - 1.5% annual cut through 2020.
  - 2.5% annual cut through 2030.
  - 3.5% annual cut through 2050.

U.S. Proposed Legislation

- Climate Stewardship and Innovation Act (S. 280):
  - McCain/Lieberman
  - Reduce GHG emissions to 2004 levels by 2012.
  - Cap steadily declines until it is equal to one-third of the 2004 levels by 2050.
- Climate Stewardship Act (H.R.620):
  - Olver
  - Reduce GHG emissions to 2004 levels by 2012.
  - Cap steadily declines until it is equal to one-fourth of the 2004 levels by 2050.

U.S. Proposed Legislation

- Global Warming Pollution Reduction Act (S.309):
  - Boxer/Sanders
  - Cap GHG emissions economy-wide in 2010.
  - In 2020, cap national emissions at 1990 levels.
  - Reduce GHG emissions to 20% of 1990 levels by 2050.
- Safe Climate Act (H.R.1580):
  - Waxman
  - Reduce GHG emissions to 80% below current levels by 2050.

U.S. Proposed Legislation

- Bingaman-Specter: Low Carbon Economy Act (S. 1766):
  - Regulates: petroleum refineries, natural gas processing plants, LNG facilities, importers of liquid fossil fuels and non-CO2 GHGs, and coal-consuming facilities.
  - Reduce GHG emissions to 1990 levels by 2030.
  - Technology Accelerator Payment ("Safety Valve")
    - Cost cap on emissions allowances.
    - Initial carbon price of $12/metric ton (2012). Price rises 5% on top of the rate of inflation each successive year.
  - Directs money to support various industries in efforts to transition to a low-carbon economy.

U.S. Proposed Legislation

- Proposed Legislation Targeting Power Plants:
  - Electric Utility Cap-and-Trade Act (S. 317):
    - Feinstein/Carper
  - America's Climate Security Act (S. Draft):
    - Lieberman/Warner
  - Clean Air Planning Act (S. 1117):
    - Carper/Feinstein/Lieberman
  - Clean Air/Climate Change Act (S. 1168):
    - Alexander/Lieberman
  - Clean Power Act (S. 1201):
    - Sanders, Lieberman

U.S. Proposed Legislation

- Containing and Managing Climate Change Costs Efficiently Act:
  - Warner/Landrieu/Graham/Lincoln
  - To be incorporated as a cost-control mechanism into other cap-and-trade bill(s).
  - Allows firms to bank/borrow emissions credits.
  - Establishes a Carbon Market Efficiency Board (similar to Federal Reserve Board).

U.S. Proposed Legislation

- Save Our Climate Act (H.R.2069):
  - Stark/McDermott
  - Carbon Tax of $10/ton of carbon content.
  - Goal is to reduce GHG emissions until 80% below 1990 levels.
  - Tax raised $10 annually until goal attained.

U.S. Proposed Legislation

- Dingell (D-Mich.) Carbon Tax Draft:
  - Tax on carbon emissions of $50/ton for industry.
  - Add a 50-cent-per gallon tax on gasoline on top of the current 18.4-cents-per-gallon.
  - Phase out the mortgage intereste deductions for large homes (4200 square feet and greater).
  - Revenue to be used to expand Earned Income Tax Credit to help low-income families pay higher fuel bills; 60/40 split for highway trust fund and mass transit programs; and other social welfare programs.

U.S. Proposed Legislation

- America's Climate Security Act (S. 2191):
  - Lieberman/Warner
  - Cap-and-trade system administered by EPA.
  - Reduce GHG emissions to 70% below 2005 levels by 2005.
  - Regulates electrical utilities, industrial manufacturing, and petroleum refiners and importers.

U.S. Proposed Legislation

- America's Climate Security Act (S. 2191):
  - Bill approved by the Senate Environment and Public Works Subcommittee on Nov. 1, 2007.
  - Now before the full Senate environment committee.
  - This is the first time limits on GHGs have been approved in a congressional vote.
  - Likely to be scheduled for a Senate floor vote prior to Memorial Day.
  - Will likely undergo significant revisions to gain support.

LISKOW&LEWIS New Orleans | Lafayette | Houston

U.S. Proposed Legislation

- America's Climate Security Act (S. 2191):
  - All GHGs classified as CO_2 equivalents.
  - Right to emit 1 ton CO_2 = 1 allowance.
  - GHG data from 2004 to 2007 = "base year" emissions (i.e. emissions prior to reductions).
  - Free allowances given to entities each year. Other allowances auctioned by EPA. (# decreases every year until 2050, when capped at 30% of 2012 cap).
  - May purchase offset allowances to satisfy max. of 15% of compliance obligation.

U.S. Proposed Legislation

- America's Climate Security Act (S. 2191):
  - Benefits for "early action."
    - Actions "taken since January 1, 1994, that resulted in verified and credible reductions of [GHG] emissions."
  - Action must be registered before enactment of law.
    - Register in voluntary reduction program: EPA, DOE, state, regional, voluntary entity programs.
    - Possibly registration with Climate Registry, if reductions are result of a program.

U.S. Proposed Legislation

- America's Climate Security Act (S. 2191):
  - "Early Action" (cont'd)
    - No intent requirement (but House may add this).
    - 1 extra allowance for each ton of CO_2 not emitted.
    - 75% of available bonus allowances will be granted between 2012 and 2022.
  - Note: Duke University study estimates that one allowance will be worth $18 in 2015 and $101 in 2050.

	Safe Climate Act of 2007	Climate Stewardship and Innovation Act	Global Warming Pollution Reduction Act	Global Warming Reduction Act	Low Carbon Economy Act	Climate Stewardship Act
Bill Number	HR 1580	S 280	S 309	S 485	S 1766	HR 620
Sponsors	Waxman (D) + 141 cosponsors	Lieberman, McCain, Obama, Clinton + 7 others	Sanders (I), Boxer (D) + 17 others	Kerry (D), Snowe (R), and Kennedy (D)	Bingaman (D), Specter (R) + 5 others	129 cosponsors
Cap-and-Trade	Y	Y	Allowed	Y	Y	Y
Reduction Goal	80% below current levels by 2050	67% below 2004 levels by 2050	80% below 1990 levels by 2050	Reach 1990 levels by 2020; -2.5%/yr through 2030; -3.5%/yr through 2050	20% below 2006 levels by 2030	70% below 1990 levels by 2050
Regulated Industries	Sectors with "largest emissions" and "most cost-effective opportunities to reduce emissions"	Power plants; Commercial and Manufacturing Facilities; Transp.fuels at refinery or import terminal	Economy-wide	EPA discretion	Most sectors of economy	Power plants; Commercial and Manufacturing Facilities; Transp.fuels at refinery or import terminal
Allowances	Auction	Auction/ Giveaways	Allowed	Auction/ Giveaways	Auction/ Giveaways	Auction/Giveaways
Offsets	N/A	Y	Silent	Possible	Y	Y

	America's Climate Security Act	Clean Air Planning Act	Clean Air/Climate Change Act	Clean Power Act	Electric Utility Cap-and-Trade Act	Save Our Climate Act
Bill Number	S 2191	S 1177	S 1168	S 1201	S 317	HR 2069
Sponsors	Lieberman (I) and Warner (R)	Carper (D), Feinstein (D), and Lieberman (I)	Alexander (R) and Lieberman (I)	Sanders (I), Lieberman (I) + 2 others	Feinstein (D) and Carper (D)	Stark (D) and McDermott (D)
Cap-and-Trade	Y	Y	Y	Y	Y	N (Carbon Tax)
Reduction Goal	70% below 2005 levels by 2050	CO2 25% below 1990 levels by 2050. Other reductions for other GHGs.	CO2: Reach 2006 levels by 2011; -1.5 bill metric tons by 2025. Other reductions for other GHGs.	CO2: 17% below 1990 levels by 2025. Other reductions for other GHGs.	Reach 2006 levels by 2011	80% below 1990 levels; $10/ton carbon content, raised $10 annually, until goal reached
Regulated Industries	Elec utilities; industrial manf; petroleum refiners and importers	Power plants only	Power plants only	Power plants only	Power Plants only	N/A
Allowances	Auction/ Giveaways	Auction/ Giveaways	Giveaway 75%	Auction/ Giveaway	15% auctioned; 85% allocated based on output	N/A
Offsets	Possible	Y	Y	Possible	Y	N/A

U.S. Proposed Legislation

- Presidential frontrunners all support cap-and-trade for GHG emissions:
  - John McCain
    - Co-sponsored bill with Lieberman containing a GHG emissions trading program (S.280).
  - Hillary Clinton and Barack Obama
    - Co-sponsors of McCain-Lieberman containing a GHG emissions trading program (S.280).
    - Co-sponsors of Lieberman-Warner cap-and-trade bill, America's Climate Security Act (S. 2191).

Greg L. Johnson

(504) 556-4115 Direct

gljohnson@liskow.com

LISKOW&LEWIS

A Professional Law Corporation

New Orleans | Lafayette | Houston

134b Carbon Trading: Enhancing Business Value through Carbon Reductions I

Michael E. Heintz, Porter, Wright, Morris & Arthur LLP, 41 S. High Street, Columbus, OH 43215

Companies can generate carbon reductions in many ways. They may do so when they shut down a facility, change a production process, enhance energy efficiency, install pollution controls, or take any number of other actions. Which of these activities can generate carbon reduction credits? How much are these credits worth? Where can they be traded? This presentation will answer these questions. It will describe the principal carbon trading programs at both the international and domestic levels. These include the Kyoto Flexibility Mechanisms, the European Union Emissions Trading Scheme, the Regional Greenhouse Gas Initiative and the Chicago Climate Exchange. It will explain how each of these initiatives works – how each evaluates carbon reductions, assigns "credit" for these decreases, and creates a market in which participants can trade these credits. Having laid this foundation, it will explain how individual companies – including members of the chemical industry – can engage in these new markets and obtain value from their carbon reductions.

134d The Countdown to Greenhouse Gas Management I

Becky Keogh, Industrial Systems, CH2M HILL, 523 South Louisiana Street, Suite 304, Little Rock, AR 72201

This presentation will cover the current practice standards for development of a quality Greenhouse Gas Inventory program for an organization. It will feature key elements of the Greenhouse Gas inventory process, and tools and techniques used to collect data across an organization. Both current practices and possible evolution of the inventory management process will be discussed. The Climate Registry protocol guidelines will be compared to current voluntary registries established in the US, and case histories for various industry sectors will be discussed.

Floating Lng Liquefaction Facilities Using the Optimized Cascade Process

Michael A Wilkes
Global Gas LNG Licensing and Technology, ConocoPhillips, 3000 Post Oak Blvd, MS59, Houston, TX 77056

For several years, there has been a lot of discussion about the possibility of a Floating LNG Facility which in certain situations may provide a feasible economic option for the development of natural gas resources. In selecting the most economic configuration for a given development, several factors need to be considered. Important factors include cost, plant availability, safety, weight and space requirements, process efficiency, size and the impact of met ocean conditions on the facilities. ConocoPhillips has considered various configurations for Floating LNG and have evaluated a design based on the Optimized Cascade LNG Process. ConocoPhillips has determined that many of the same considerations that make the Optimized Cascade LNG Process a good choice for an onshore LNG facility also make it a good choice for Floating LNG. The Two-Trains-in-One configuration improves process efficiency and lends itself to the use of aero-derivative gas turbine drivers which are uniquely suited to the offshore environment. The low profile plate fin exchangers employed in the liquefaction system are well suited to the floating environment. The experience that ConocoPhillips has with the building and operating of their floating liquefied petroleum gas (LPG) processing facility at the Belanak field in Indonesia can also be applied to Floating LNG developments.

Concept Validation for a Very Large LNG Train

Heinz C. Bauer (speaker), VP Technology LNG & Gas Processing
heinz.bauer@linde-le.com
Linde Engineering, 82049 Pullach, Dr-Carl-von-Linde Str 6-14, Germany

Arne O. Fredheim, Sr Research Advisor
StatoilHydro ASA, 7005 Trondheim, Norway

Stefano De Simone, LNG Commercial Manager
GE Oil & Gas, 50127 Firenze, Italy

AIChE Spring Meeting, April 2008
8th Topical Conference on Natural Gas Utilization
New Orleans, Louisiana, April 6-10, 2008

Abstract

The capacity of Liquefied Natural Gas (LNG) trains remained stagnant at 4-5 million tons per annum (mtpa) throughout the late 90's. Six trains with liquefaction capacities of 7.8 mtpa, which are currently under construction in Qatar, have broken this trend. In addition to high performance main equipment, such as that supplied by GE Oil & Gas, new process technologies had to be developed and qualified to achieve a nameplate capacity of 7.8 mtpa. The Mixed Fluid Cascade (MFC®) process, owned by StatoilHydro and Linde, is such a technology.

The MFC® process is presently in the final phase of commercialization at Melkøya near Hammerfest, Norway. The MFC® process is perfectly suited for very large LNG trains, as the three underlying mixed refrigerant cycles can be arranged such that all three cycles have identical power consumption. Thus, the largest available gas turbine, GE's Frame 9, can be used as the driver for all three refrigerant compressor trains. This arrangement promises an unbeatable concept in terms of simplicity, reliability and cost effectiveness for reaching record capacities.

While StatoilHydro and Linde own technologies for the liquefaction process and for the spiral wound heat exchangers, a partner for the main rotating equipment was still required. GE's Oil & Gas business, with headquarters in Florence (Italy), as a world leader in advanced technology equipment and solutions for liquefaction trains was obviously the first choice. GE Oil & Gas agreed to participate in a concept validation program whereby the three parties defined a case for the validation work (a hypothetical LNG plant in a warm climate) and succeeded in optimizing the concept. Using only qualified and proven equipment such as Frame 9 gas turbines, large centrifugal compressors from the portfolio of GE Oil & Gas, and spiral wound heat exchangers manufactured by Linde AG, an LNG capacity in excess of 1,500 tons per hour was found to be possible (equivalent to more than 12 mtpa at 335 stream days). This result is especially remarkable because the total amount of equipment could be minimized i.e. there is no parallel rotating equipment and the number of refrigeration compressor casings is limited to five. This paper explains in detail the findings made during the optimization of the concept.

Introduction

StatoilHydro and Linde agreed to define and optimize a concept for large LNG trains. The main assumptions are listed below:

- Drivers: 3 x Fr9 + starter/helpers (fully utilized)
- Available power: 3 x 145 MW (including site rating and helpers)
- Cooling media: Air
- C3+ extraction: Upstream of liquefaction unit
- Reference temperature: 40 °C (defined as stream temperature after cooling)
- Feed pressure: 70 bar into liquefaction unit
- On-stream days: 330
- Heat exchanger configuration: all SWHE
- Flow rate: in the range of 10 mtpa (3x145 MW fully utilized)
- Feed gas composition:

N2	4.9000 vol%
CO2	0.0050 vol%
CH4	89.4849 vol%
C2H6	5.2550 vol%
C3H8	0.3480 vol%
i C4H10	0.0044 vol%
n C4H10	0.0027 vol%
total	100.0000 vol%

Process

The basic process scheme focused on simplicity and liquefaction, not on NGL extraction. Thus, NGL extraction was assumed to take place upstream of the liquefaction with subsequent recompression of the lean natural gas. This made it possible to select and optimize the liquefaction pressure within a certain range, which is mostly limited by mechanical constraints.

An all spiral wound heat exchanger arrangement was chosen based on the desire to qualify all relevant equipment, which is required for a single train concept without the use of parallel equipment, as is used in competing solutions[1]. Spiral wound heat exchangers are considered as monolithic equipment in contrast to plate-fin heat exchangers, which require many (several dozen) parallel and serial cores for large plant capacities. A plate-fin solution presents serious challenges to fluid distribution.

Figure 1 MFC Process with two pre-cooling levels

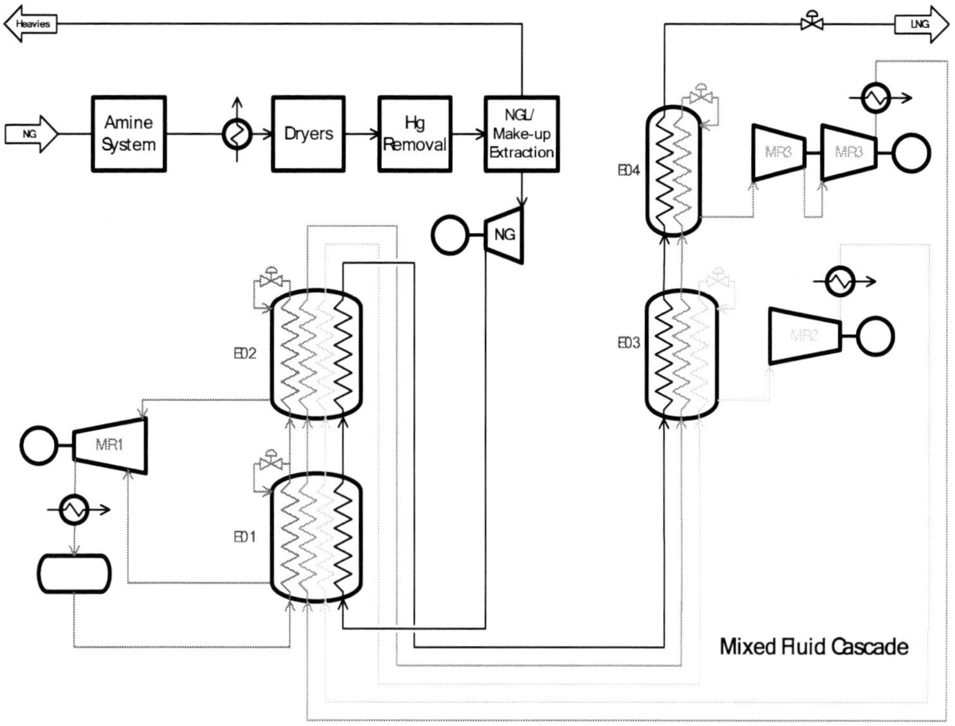

Figure 2 Composite Curves for two pre-cooling levels

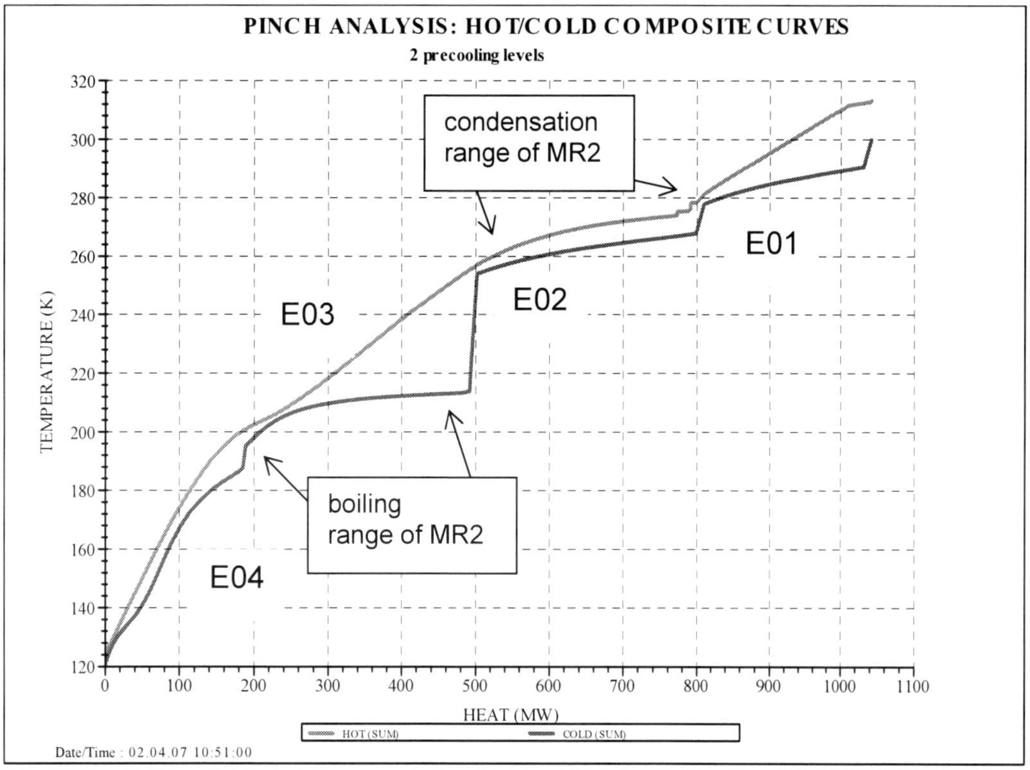

Pre-cooling of MR1 with two independent temperature ranges in E01 and E02, and consequently, two suction lines to the MR1 compressor, were adopted as an inherited best practice. During the first simulation runs, the power consumption of the three refrigerant compressors was not limited, i.e. there was no enforced load balancing. Only the total energy consumption was limited in the first place to 360 MW. The spread of individually optimized compressor power was about 3% for all three compressors. Thus, an enforced even load to the compressors was considered to have a minor effect on the specific energy consumption. In fact, a calculation check with exactly 3x120 MW shaft power showed only a slight increase in energy consumption of approximately 1.5%.

Optimizations

As the efforts to improve the thermodynamic efficiency reached their limits for the process shown in Figure 1, a more radical approach was necessary. According to the objective, the process flow scheme was still supposed to be built from a limited number of large building blocks. While MR2 and MR3 had already the essential goal of simplicity, MR1 still had room for improvement.

Figure 3 MFC Process with one pre-cooling level

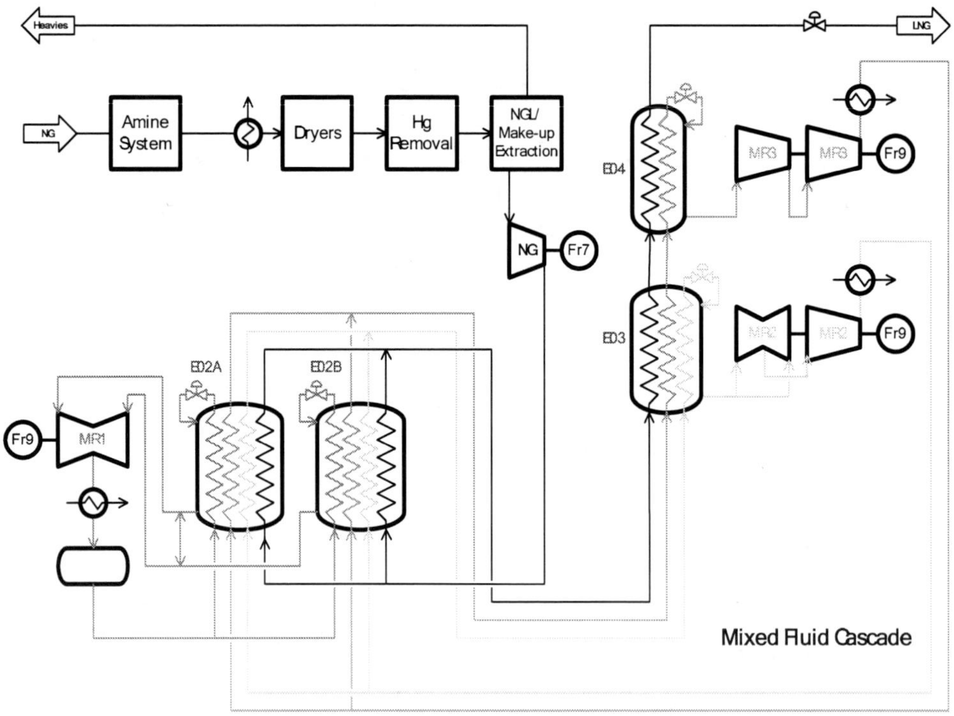

With the ambition to reduce the tube side pressure drop in the serial arrangement of E01 and E02, E01 was eliminated and E02 was doubled to E02 A/B (see Figure 3). In this way the equipment count and size could arguably be considered as unchanged. There were some reservations concerning the impact of this modification on overall efficiency, as the general opinion favored multi-staged pre-cooling to optimize the heat integration.

Figure 4 Composite Curves for one pre-cooling level

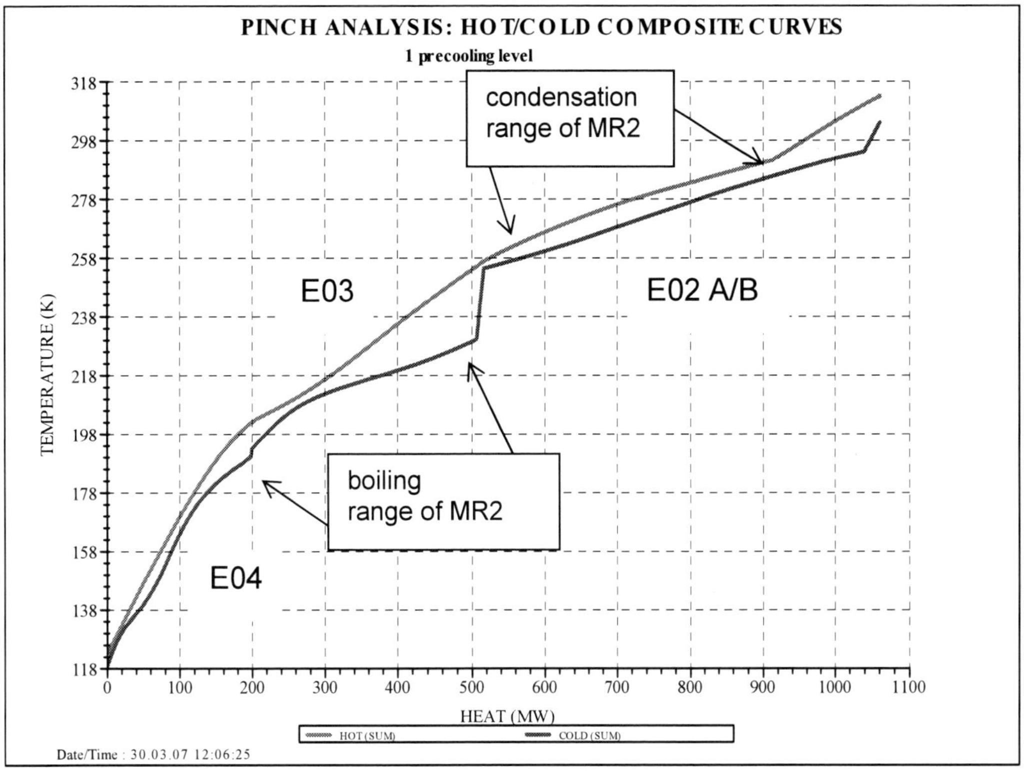

The improved temperature approach in both E02 and E03 (see Figure 4) is considered to be the major advantage of the single level pre-cooling concept in comparison to the two level concept. The intermediate refrigerant in MR2 is now a real three component mixture with a relevant concentration of propane. While 2-stage pre-cooling with MR1 causes a pronounced step between E01 and E02 in the vaporization curve, a 1-stage pre-cooling concept shows a step-less shape. The OPTISIM® process optimizer rewards the smooth refrigeration curve seen in E02 A/B with a wide boiling composition of MR2. A wide boiling range, coupled with the fact that E02 A/B has the largest heat load, contributes to the shape of the composite diagram during its condensation at the discharge pressure of the MR2 compressor. This effect is essential for the better temperature approach in E03 and again for the improved efficiency.

After successfully simplifying the process, the aim was to maximize the plant capacity without technological step-outs. One straight forward possibility is maximization of the feed gas pressure. As the net temperature level of gas liquefaction rises with increasing pressure, the required refrigeration can be supplied at a higher temperature level. According to Carnot, energy savings will result. Figure 5 shows that when going from 70 to 90 bar, the feed gas pressure will increase the liquefaction rate by nearly 10% at constant refrigeration power. The selected pressure of 90 bar corresponds to a yet reasonable mechanical design, especially for spiral wound heat exchangers with aluminum shells.

Figure 5 Increase of LNG production by feed gas compression

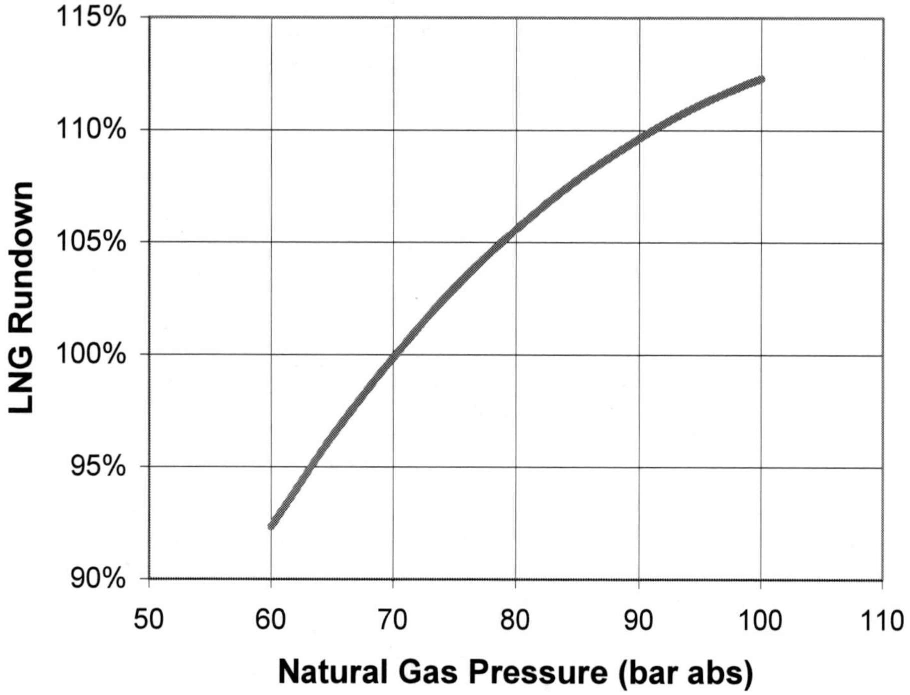

The final step in maximizing the plant capacity required maximizing the shaft power and the volumetric flow to the refrigerant compressors. A workshop in Florence showed that the most powerful gas turbine (GE Frame 9E), supported with a sizeable helper motor, is capable of driving the refrigerant compressors at 145 MW shaft power each.

All relevant equipment, especially compressors and spiral wound heat exchangers, did reach a reasonable upper limit of what is feasible today. Achieving a plant capacity of more than 12 mtpa at 8,040 annual operating hours (equivalent to a name plate capacity of 13.4 mtpa) is impressive and establishes the maximum level, which can be reached today with existing technology. Competing triple flow LNG liquefaction technologies[2,3] are either thermodynamically less efficient or not able to fully load all three main drivers.

Compressor Selection

Based on the optimized process with single level pre-cooling, natural gas compression at 90 bar and 145 MW power for each compressor train, GE Oil & Gas selected the following configuration:

Figure 6 Compressor Selection for a 12 mtpa LNG Plant

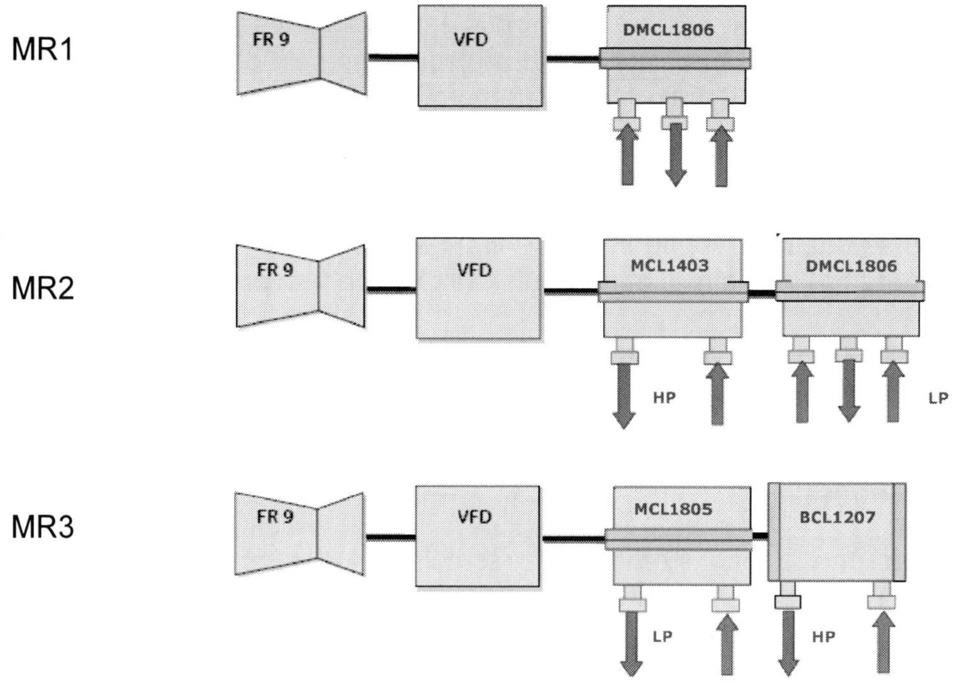

The suction flow rate in MR1 is far too large even for an impeller with a 1,800 mm diameter at 3,000 rpm. As a result, GE Oil & Gas selected a double flow configuration with two inlet nozzles and one common discharge nozzle in the middle of the casing. As the required head in MR1 is small, three impellers are sufficient. In the terminology of GE Oil & Gas, the MR1 compressor is called DMCL 1806. DMCL means double flow, horizontally split, 1800 mm nominal impeller diameter with 2x3=6 impellers.

The flow rate to the MR2 compressor is similar to the situation found with MR1. As a result, the same type of compressor was chosen for the LP portion. However, as the required head was greater than what could be provided by three impellers, an additional casing had to be considered. Furthermore, since the suction flow rate to the HP MR2 compressor was now below 200,000 m³/h, an arrangement with 3 impellers with 1400 mm diameter was sufficient. The name of the utilized HP MR2 compressor is MCL1403. The term MCL1403 means horizontally split, 1400 mm nominal impeller diameter with 3 impellers.

The MR3 compression system was the most challenging because it has highest compression ratio and the lowest molecular weight. Depending on the process parameters, between 12 and 15 impellers are required. The maximum number of impellers, which can be installed in a combination of a horizontally split compressor and a barrel type compressor is 12 or 13, depending on special case by case considerations. Any additional impellers beyond 13 would require a third casing, which is not desirable. The final design of MCL1805 + BCL1207, where BCL stands for barrel type compressor, was reached after adjusting the suction flow rate to the LP stage of MR3 to about 300,000 m^3/h and after removing the limitations to the discharge pressure of HP MR3.

This compression system solution is especially favorable as the only contribution to the outstanding LNG capacity of this project is the split of the MR2 compressor into a LP and a HP casing. All major components of the envisaged compressor concept have previously been supplied by GE Oil & Gas to the Qatargas[4] and Rasgas LNG projects in Qatar. Prior to shipment, full-load string tests were performed by GE Oil & Gas at its testing facility in Massa (Italy). The major compressor components can therefore be considered as qualified. As soon as the first large LNG plant with a capacity of 7.8 mtpa is up and running, all main rotating equipment will be proven.

The decision to delete the side stream to the MR1 compressor was a prerequisite to achieving the present elegant solution characterized by only one casing. Any other concept with more than one suction pressure in MR1 would require a second casing.

Heat Exchangers

Based on the results of the process calculation, a rigorous thermal and hydraulic design and a basic mechanical design were carried out. All spiral wound heat exchangers considered for the purposes of this study have heating surfaces of about 40,000 m^2 each and a shipping weight of about 250 tons. The design of the heat exchangers was reviewed and discussed by StatoilHydro and Linde experts during a workshop. With the exception of the increased tube wall thickness for natural gas, no changes to existing references were identified.

Under the assumption that the present mechanical design will be demonstrated successfully during operation, no critical observations or comments were made. The large dimensions and the heavy weight of the heat exchangers will not result in manufacturing or shipping problems.

The selected concept is only possible in conjunction with spiral wound heat exchangers. The temperature range of pre-cooling in E02 is arguably too large for the use of plate-fin heat exchangers. The MR1 refrigerant has a 36°C boiling range and the warm and cold ends of the heat exchanger have a 58°C temperature differential. These figures exceed the limits that experts consider to be safe for plate-fin heat exchangers service.

Conclusions

- 12 mtpa single train capacity can be reached with equipment, which is either already proven or about to be proven in an LNG plant already under construction.
- Mixed refrigerants require a wide boiling range to develop their potential for good heat integration. Multi-stage pre-cooling jeopardizes this effect. Therefore MR1 is designed as a single loop.
- The optimized concept requires spiral wound heat exchangers for pre-cooling due to the wide boiling range of the refrigerant.

References

---

[1] A novel Design for 10-12 mtpa LNG Trains, Sander Kaart, Wiveka Elion, Barend Pek, Rob Klein Nagelvoort, LNG 15, Barcelona, Spain, April 2007

[2] Aeroderivative Gas Turbine Drivers for the ConocoPhillips Optimized Cascade^{SM} LNG Process – World's first Application and future Potential, Cyrus B. Meher-Homji, Doug Yates, Hans P. Weyermann, Karl Masani, Weldon Ransbarger, Satish Gandhi, LNG 15, Barcelona, Spain, April 2007

[3] The C3MR Liquefaction Cycle: Versatility for a fast growing, ever changing LNG Industry, Mark Pillarella, Yu-Nan Liu, Joseph Petrowski, Ronald Bower, LNG 15, Barcelona, Spain, April 2007

[4] Design, Manufacture, and Test Campaign of the World's Largest LNG Refrigeration Compressor Strings, Roy Salisbury, Pete Rasmussen, Todd Griffith, Andrea Fibbi, LNG 15, Barcelona, Spain, April 2007

Biography

Dr. Heinz C. Bauer (heinz.bauer@linde-le.com) is Vice President of Technology for LNG and Gas Processing within Linde Engineering, Pullach, Germany. He has been with Linde for over 27 years and has been involved in many projects in the field of cryogenic gas processing. He graduated at the Technical University Munich in applied Physics.

Comparison of Present Day Peakshaving Liquefaction Technologies

Nancy C. Ballout, LNG – Process Engineer
Black & Veatch Corp.
Houston, TX

Brian C. Price, VP- LNG Technology
Black & Veatch Corp.
Overland Park, KS

AIChE Spring National Meeting, April 2008
8th Topical Conference on Natural Gas Utilization
New Orleans, Louisiana, April 6-10, 2008

ABSTRACT

While base load LNG plants provide a large quantity of the LNG market, there is an increasing need for small-scale plants to provide LNG for distributed markets. The small scale LNG industry in the US was initially developed in the 1960-70 time period and generally consisted of peak shaving facilities for winter gas supply. The numerous facilities developed during that time period are now reaching the end of their useful life and are being replaced with new, modern designs which are more efficient and easier to operate.

The selection of the appropriate LNG technology can be a difficult process. Although efficiency, ease of operation, and capital and maintenance costs are usually the main decisive factors, site specific requirements, such as pipeline available pressure and tail gas allocation, can also play a big role during the selection process. In addition, these main factors are sometimes sacrificed in favor of City and/or State regulations, or simply operator preference.

This paper covers a detailed comparison of three widely used peak shaving liquefaction technologies. For a small scale (8.5 MMSCFD) LNG production, three simulated cases were developed to show the differences between a Nitrogen Expander Cycle, an Open Expander Cycle, and a Single Mixed Refrigerant (SMR) Cycle.

US PEAK SHAVERS

The first LNG plant was built in West Virginia in 1912 and began operation in 1917. The first commercial liquefaction plant was built in Cleveland, Ohio, in 1941. The most common use of LNG in the U.S. is for "peakshaving." Small scale LNG facilities have been used for over 30 years in the US for gas supply "peakshaving" applications.

Peakshaving is a way local electric power and natural gas companies or utilities store gas for peak demand periods when their typical pipeline capacity is a limiting factor. In these facilities the LNG produced from the liquefier is stored in a large LNG tank for use in these peak periods. Peakshaving can occur during the winter heating season or when more natural gas is needed to generate electric power for air conditioning in the summer months. When needed, the LNG is pumped from storage, vaporized and sent to the consumers. These plants are set up to send out large volumes of gas at a moments notice. The utility companies liquefy natural gas when it is abundant and available at off-peak prices. The tank inventory is replenished during the off peak periods. Typical plants have liquefaction capacities of 5-20MMSCFD and are intended to fill a large LNG tank in 150-200 days. These facilities normally have one or more LNG tank(s) to provide 1-3 BCF of storage capacity. Facilities to vaporize and send out the stored gas during 10-20 days are also included.

There are currently 58 liquefaction facilities in the US. More than 80% of these are well over 20 years old. Most of these facilities are in need of process replacement due to aging equipment, and sometimes newer local and/or environmental regulations. Also, modern technology provides more efficient and lower cost liquefaction capacity. Operators are also striving to reduce operating cost by minimizing staff and optimizing running times to fill storage. Many of the older plants struggle with disposal of regeneration gas from gas treating system and tank vapor recovery. As plants age, the parameters of pipeline supply and take away can change dramatically.

LNG is also currently being used as an alternative transportation fuel in public transit and in vehicle fleets such as those operated by many local natural gas utilities companies for maintenance and emergencies.

TECHNOLOGY SELECTION

Existing LNG production facilities in the small scale area employ a variety of LNG production technologies. These older plants are either Cascade refrigeration, Open Expander, Single Mixed Refrigerant (SMR), Nitrogen Expander Refrigeration, or other multiple refrigerant schemes. After the initial developments in the 1960-70's, the predominant technology has been the mixed refrigerant system with over 60% of the installations. Modern LNG production facilities are either Open Expander, N_2 Expander Refrigeration or SMR designs. The cascade and multiple loop systems have proven to be too complex and costly to operate. Most of these multiple refrigerant systems as well as Expander plants with tail gas disposal limitations are currently being replaced by modern refrigeration units.

Pre-Treatment

All LNG processing units have some kind of feed gas pretreatment for removal of impurities, including aromatics and heavy hydrocarbons. At low temperatures these impurities solidify and plug the liquefaction equipment. For pipeline quality gas major impurities include water, carbon dioxide, benzene and heavy hydrocarbons. Due to size, a common technology used by peakshaver facilities for removal of impurities is molecular sieve adsorption. Any heavy hydrocarbons, which are not removed by adsorption are condensed and removed cryogenically.

Some facilities use two separate systems for the removal of water and CO_2. Other facilities combine the functionality of removing both water and CO_2 inside one single bed containing two different molecular sieve materials. In a single combined pre-treatment system there is a corresponding ease of operation plus a reduction in capital and maintenance costs. However, separate pre-treatment systems can give the operator more flexibility and are easier to troubleshoot in case of problems.

In this comparison document it is assumed that all discussed LNG processes use the same molecular sieve pre-treatment system.

Open Expander Cycle

The Open Expander process, sometimes referred to as a pressure letdown process, can only be used economically under certain conditions. In this process, the inlet feed gas is compressed to between 600 to 800 PSIG and then the majority of the gas is expanded and used as refrigeration. Since only 10% to 16% of the total feed gas flow to the plant is condensed to liquid, a very large tail gas stream of low pressure expanded gas is produced.

To make this process economical, plants are usually located where the natural gas is taken from a high pressure main line and is to be delivered to a low pressure local distribution system. These expander plants have been used frequently and are interesting where enough pressure drop is available. If the tail gas could not be re-introduced into a low pressure system, this type of process would not be considered.

The higher the feed gas inlet pressure, the more efficient the process since the liquefaction of that 10% to 16% of the feed gas can be accomplished with little or no compression or mechanical refrigeration. Some peak shaver facilities utilize a reduced energy process by taking advantage of the high natural gas pressure feeding the plant. These types of facilities use energy-free expander/compressor units and no feed gas booster compression (Figure 1).

Other peak shaver facilities with a lower pipeline available pressure need the extra boost to still utilize energy-free expander/compressor units of reasonable size (Figure 2). A higher feed gas pressure greatly simplifies the liquefaction cold box design and increases overall efficiencies. The choice of using a gas turbine versus an electric motor driven feed gas compressor is based on economics and operator preference. Gas turbine driven units generally result in lower operating costs because usage of natural gas as fuel as opposed to purchased electricity is more economical However, gas turbine driven compressors are generally more expensive to install, involve emissions permitting and have a higher maintenance costs. Gas turbines also produce waste heat that can be efficiently and economically recovered and used in the process to reduce the size and cost of supplemental fired heaters used for the regeneration of the pre-treatment vessels.

The Open Expander process is not only feed pressure sensitive but composition sensitive as well. Drastic changes in feed gas composition can considerably reduce the production capacity and cause operations problems. Black & Veatch recently designed and built a new LNG Open Expander plant which includes a demethanizer system for BTU control. Figure 3 shows this type of system for tailoring the LNG product from varying feed streams. This kind of design minimizes potential NGL issues and can be used in combination with any of the liquefaction technologies. In developing such a plant, a broad range of feed gas compositions must be considered in the design in order to specify equipment capable of handling all potential feed gases. Designing for a single feed is a risky decision.

Most open expander processes have very limited turn-down capabilities (about 90%). However, some expander/compressor units can be designed to operate at a 50% turn-down rate but the operation becomes less efficient and it creates extra wear and tear in the equipment. Turn-down capability allows the process to operate at lower production rates when tail gas allocation decreases. While operating at turn-down capacities, the feed gas pressure must be increased which is not always possible if the pressure is dependant on incoming pipeline pressure.

Recompression of the tail gas increases the LNG production but results in a process with the highest cost and lowest process efficiency of all processes. The addition of a recycle compressor would considerably increase LNG production. Unfortunately, the high capital and operating costs, substantial footprint, low investment return, and added complexity rapidly diminish the attractiveness of such a system. If it is necessary to recycle gas instead of releasing to a low pressure line, then a refrigerant process would be selected.

Nitrogen Expander Cycle

The Nitrogen Expander Refrigeration process (Figure 4) has been used on a limited basis for liquefaction, especially on smaller systems. The process is similar to the open expander process in using a vapor refrigeration stream with an additional large compressor for a closed loop nitrogen recirculation.

Since the process relies on nitrogen vapor for condensing, the refrigeration flow is quite large and the system will be significantly larger than a mixed refrigerant system. In most Nitrogen Expander cycles, a gas turbine or an electric motor driven nitrogen multi-stage compressor is used to re-pressurize the nitrogen once expanded to low pressure. In addition, dual expander/compressor units are used to attain the process temperatures needed for liquefaction and help on the energy efficiency of the process.

The power to run this system is much larger than other technologies making this cycle less efficient than an open expander or a mixed refrigerant cycle. However, the choice of selecting this cycle might come from the fact that there is no need to store flammable refrigerants on site.

Turn-down is achieved by running the cycle system with reduced inventory. Nitrogen is removed from the system through a vent valve. Operating pressures are then adjusted which allows the process to operate at peak efficiency at lower production rates. However, since the main compression is typically a centrifugal unit, the compressor loading will be the primary factor in limiting efficient turndown.

Mixed Refrigerant Cycle

The Single Mixed Refrigerant process (Figure 5) is normally the lowest cost design for small scale liquefaction systems. Also, unless free pressure drop is available for the expander process, the mixed refrigerant system is the most efficient process since it has the lowest power requirement. The PRICO® process offered by Black & Veatch is just such a mixed refrigerant process that has been applied in 20 facilities around the world. PRICO® uses a single closed loop mixed refrigerant process which uses a mixture of nitrogen and hydrocarbons normally found in the feed gas in all our facilities. The hydrocarbon constituents range from methane to isopentane. The design of the PRICO® system is such that only a small amount of refrigerant is required once the system is charged, so minimal makeup inventories are required.

This technology, in contrast to the others discussed, only has a single compression system for the refrigeration. The Refrigerant Compressor is typically a multi-stage centrifugal compressor driven by a turbine or an electric motor. There are no other expanders or compressors required in the process. The inherent simplicity of the single mixed refrigerant process leads to less piping, fewer controls, and a lower equipment count. Lower valve count and lack of expander/compressors results in lower fugitive emissions from such a facility as well.

Also, the main exchanger is a very simple plate-fin unit with a minimum number of connections which increases the ease of operation. A single plate fin can be used in systems up to about 20MMSCFD. For larger capacities, multiple plate fin cores can be used in the same cold box. The system is designed such that during a shutdown the refrigerant inventory is maintained in the system. No venting or pressure relieving is needed.

Improvements in the PRICO® process by B&V over the years have resulted in a 25-35% reduction in power versus older facilities. Generally, these processes require about 350-500 hp per MMSCFD of LNG capacity. The exact value depends on the system design parameters such as feed pressure, ambient conditions and process specifications. Besides the refrigerant compressor, the main exchanger is the key piece of equipment in the liquefaction system. Modern SMR plants have much more efficient main exchanger designs. Figure 6 shows a comparison of the modern plate fin box design versus a 1970 vintage unit. As can be seen from this photo, the PRICO® exchanger has all connection external to the cold box. There are no flanged connections, valves or instrumentation inside the box. This design avoids any leak potential inside the box from such connections. Exchangers such as this have been operated for 30 years without leaks or maintenance. Good examples of a modern installations are the Alagasco LNG plant (Figure 7) and the Brazilian White Martins plant (Figure 8) with liquefaction capacities of 12 & 14 MMSCFD respectively.

PROCESS COMPARISONS TABLES

Table 1 summarizes the main characteristics of all three processes.

The data logged in Table 2 was extracted from process designs developed by Black & Veatch for the three specific LNG processes. For all three processes, the production capacity is 8.5 MMSCFD.

Process Comparisons – Table 1

OPEN EXPANDER PROCESS	NITROGEN EXPANDER PROCESS	SMR PRICO PROCESS
Uses portion of feed gas as refrigerant	Expander process plus nitrogen refrigeration is required	Single refrigeration system
Needs HP feed gas availability and LP tail gas allocation	Single degree of freedom (flow only)	Mixed refrigerant gives more flexibility / another degree of freedom in operation for changing gas and conditions
Single or multiple expander/compressors units.	Utilizes two expander systems	Minimum utilization of rotating equipment
More complex heat exchanger design	Higher refrigeration power required	Refrigerant mix uses components already available in the gas
Feed gas pressure and composition sensitive	Sensitive to pressure/ approaches	On-site storage of refrigerants
LNG Temperature: -255°F	LNG Temperature: -280°F	LNG Temperature: -255°F
Rapid start-up and shutdown	Rapid start-up and shutdown	Rapid start-up and shutdown

Process Comparisons – Table 2

	OPEN EXPANDER PROCESS (with feed gas compression)	NITROGEN EXPANDER PROCESS	SMR PRICO PROCESS
Refrigeration Compressor Power (HP)	2,250	6,840	4,800
Expanders /Compressors (HP) ea.	1445 / 770	1,794 / 1,024	None
Air Cooler Services	5	5	2
Pumps	None	None	2 services (with spares)
Feed Gas Flow (MMSCFD at 75°F & 180 PSIA)	58.00	15.32	15.27
Refrigeration Flow (Lb/Hr)	70,500	190,300	98,350

PROJECT COSTS

The curve shown in Figure 8 is used for planning purposes and facility studies based on B&V's PRICO® technology. These curves represent the 2008 total installed cost for CO_2 removal, dehydration, liquefaction, refrigerant makeup and boil off gas compression. Table 3 shows a % cost comparison for the three LNG technologies discussed in this technical paper and are estimates based on the comparative cases. The final cost of an LNG facility is site specific and must be developed and adjusted for local conditions and scope of supply.

Process Comparisons – Table 3

	capital cost		energy cost	
	Turbine	**Electric**	**Turbine (% FG)***	**Electric (KW)**
PRICO	**MM $32.5**	**MM $27.0**	**9.5%**	**3,500**
N_2 Expander	**(+) 15% to 20%**		**14.6%**	**5,000**
Open Expander (With Feed Gas Compression)	**(+) 25% to 30%**		**5%**	**1,600**

** Based on the total LNG production, %FG represents the percentage of the gas not liquefied because it is consumed as fuel by the gas turbine driving the liquefaction process.*

CONCLUSION

The LNG peakshaving market is a strong market with many new developments and technical challenges. As the industry grows and pushes into new areas, proven designs and technology will be relied on to continue the industry's proud tradition of innovation and safety. While costs are escalating, it is important to focus on the scope of the project. While liquefaction systems are similar for most facilities, new developments have driven the industry to provide increased flexibility in feed gas and LNG product streams.

All LNG processes including the three discussed in this paper have their pros and cons. There is no such a thing as a perfect process. Therefore, the selection of the most appropriate LNG technology is a very important and difficult process. In the long run the decision will be based on the individual needs of the operator, City and/or State regulations, and/or site specifics. Efficiency, ease of operation, as well as capital and maintenance costs will also play an important role during the selection process.

REFERENCES

1. Energy Economics Research at the Bureau of Economic Geology, Introduction to LNG, http://www.beg.utexas.edu/energyecon/lng/documents/CEE_INTRODUCTION_TO_LNG_FINAL.pdf

2. Brian C. Price, Shawn D. Hoffart: "Small Scale LNG Facility Developments" AIChE Spring National Meeting April 22-26, 2007.

AUTHORS

Nancy C Ballout is a Process Engineer for Black & Veatch Corp. in Houston TX, USA. She has extensive experience in LNG operations and process design and has operated and assisted in numerous liquefaction facilities startups in the US and internationally. Prior to joining BVPI, she worked as an Operations Engineer for the largest peakshaver plant in the US in Philadelphia as well as a Gas Processing Engineer for Intevep PDVSA in Caracas, Venezuela. Mrs. Ballout is a member of AIChE, GPA, and WEN. She holds a BS degree in Chemical and Petroleum Refining Engineering from the Colorado School of Mines.

Brian C. Price is Vice President for LNG Technology for Black & Veatch Corp. in Overland Park, KS, USA. He is in charge of technology development and process design for LNG production facilities, LNG import terminals and related gas processing and NGL recovery facilities. Mr. Price has over 34 years experience in gas processing and related technology areas. Prior to joining BVPI, he worked for ARCO Oil & Gas Co. in various positions including Manager of Process Engineering and Projects Manager. Mr. Price is a member of AIChE and is active in the Gas Processor Suppliers Association. He currently serves as Chairman of Editorial Review Board for the GPSA Engineering Data Book and is past Chairman of the Technical Committee for GPA. Mr. Price has BS and MS degrees in Chemical Engineering from Oklahoma State University.

Figure 1
Open Expander Cycle with no Pre-Boost Compression and Combined Pre-treatment

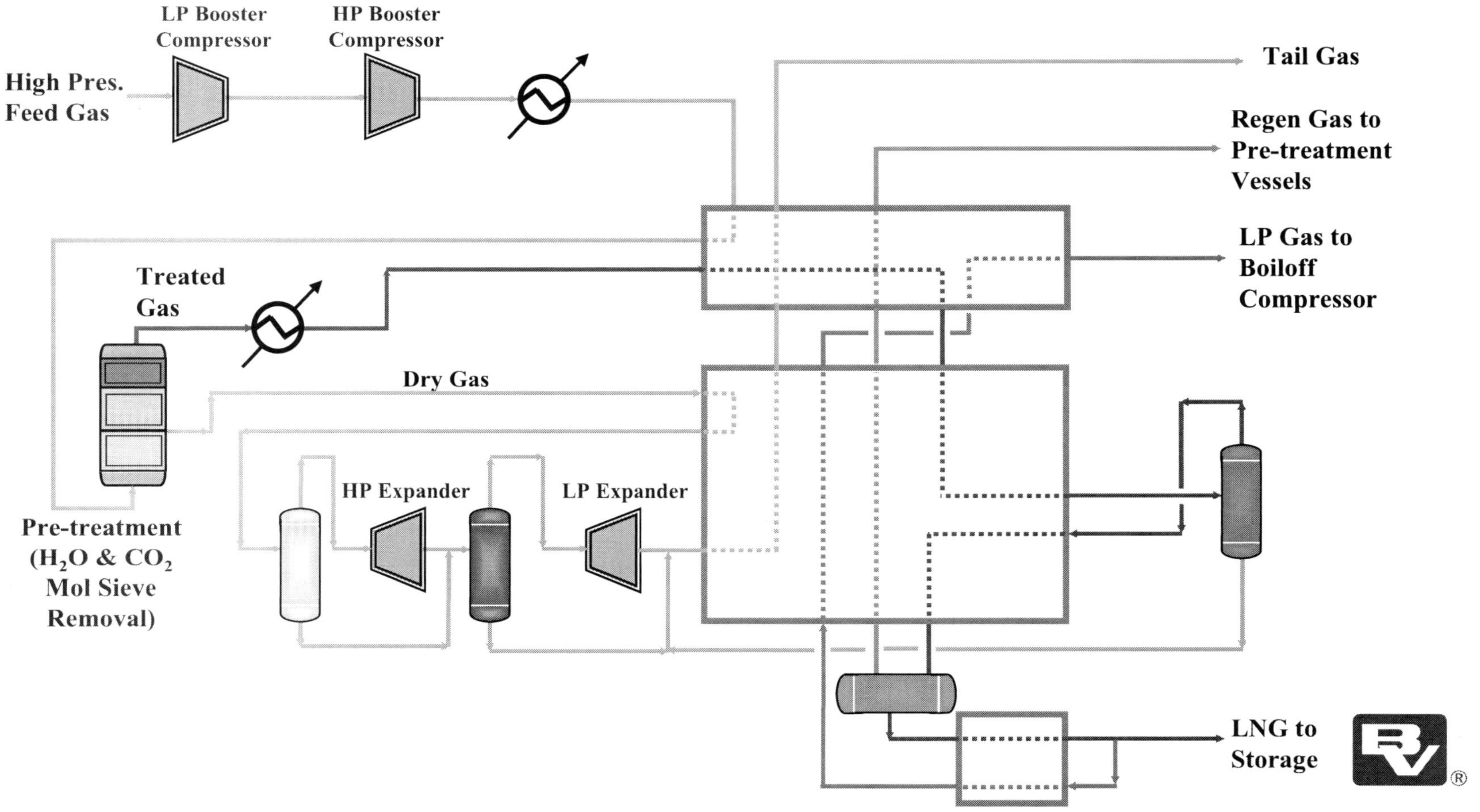

Figure 2
Open Expander Cycle with Individual Pre-treatment Systems

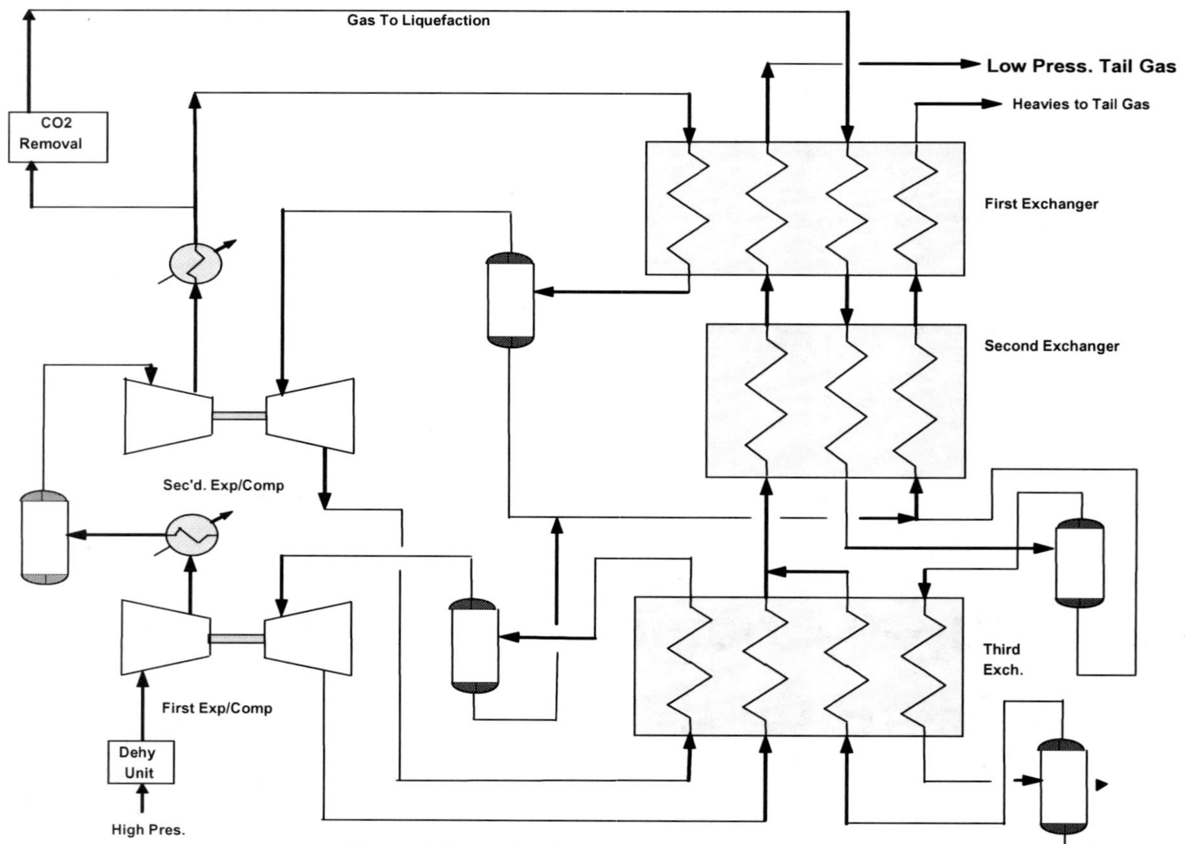

Figure 1 Expander Process

Figure 3
NGL Recovery for BTU Control

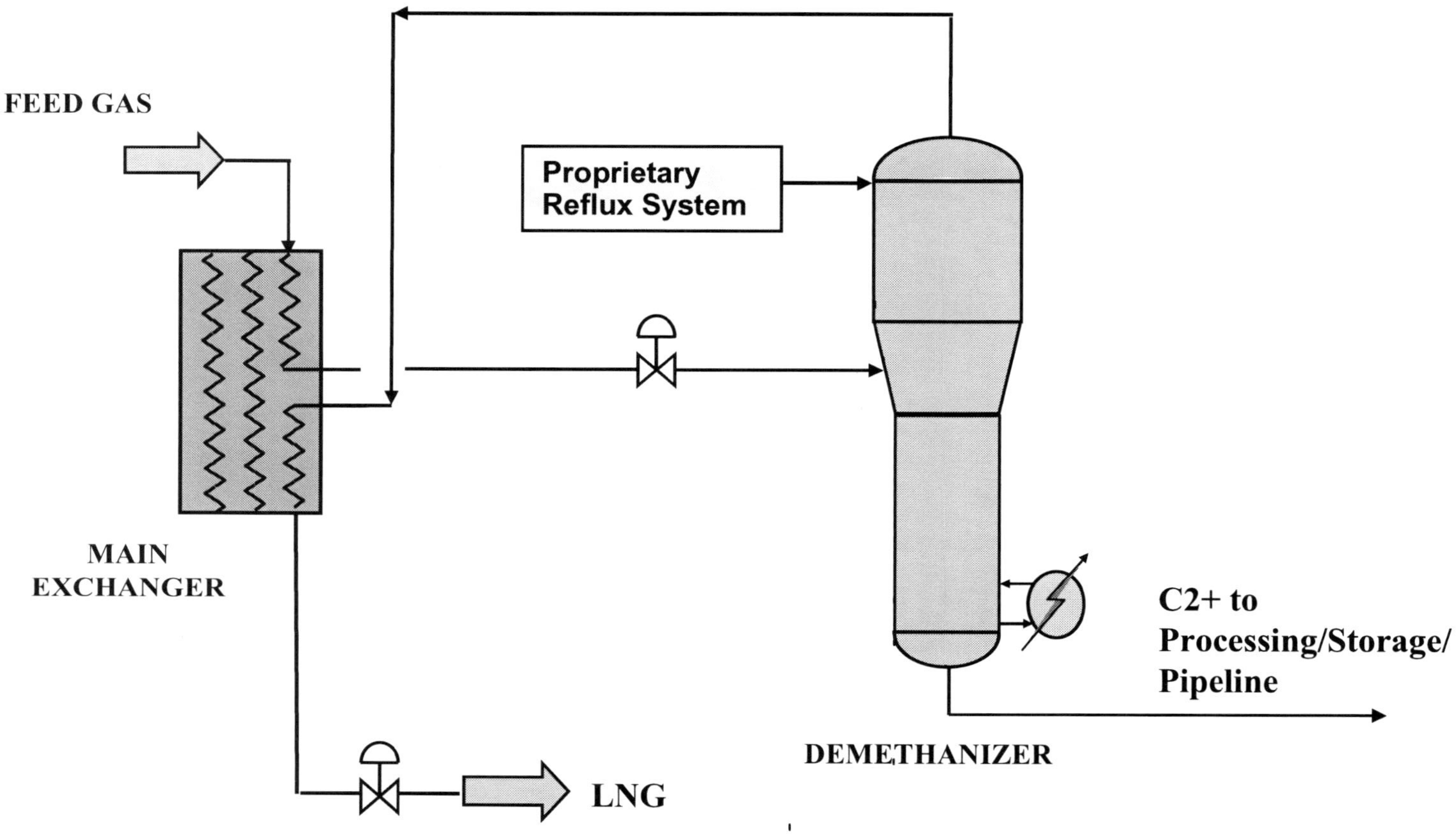

Figure 4
Nitrogen Refrigeration / Expander Process

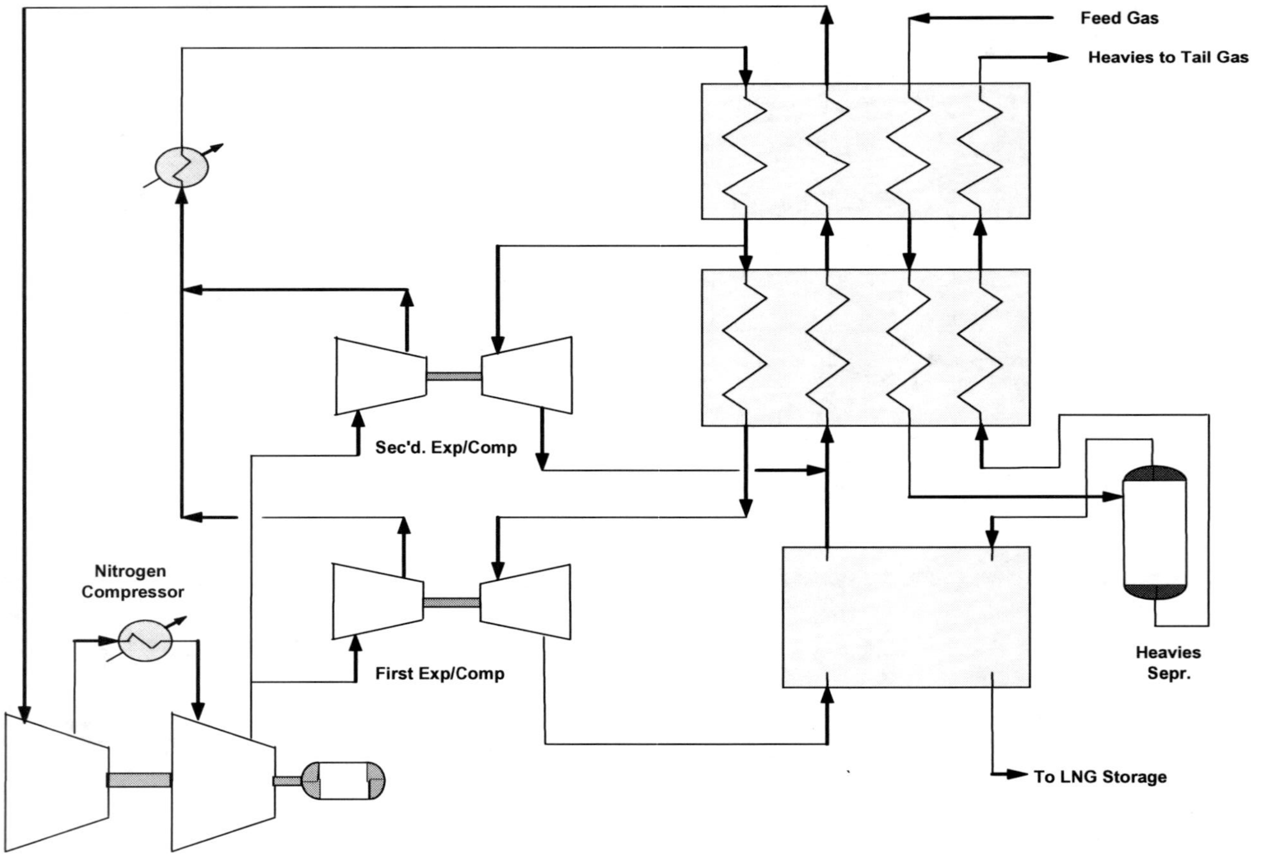

Figure 5
Single Mixed Refrigerant PRICO® Process

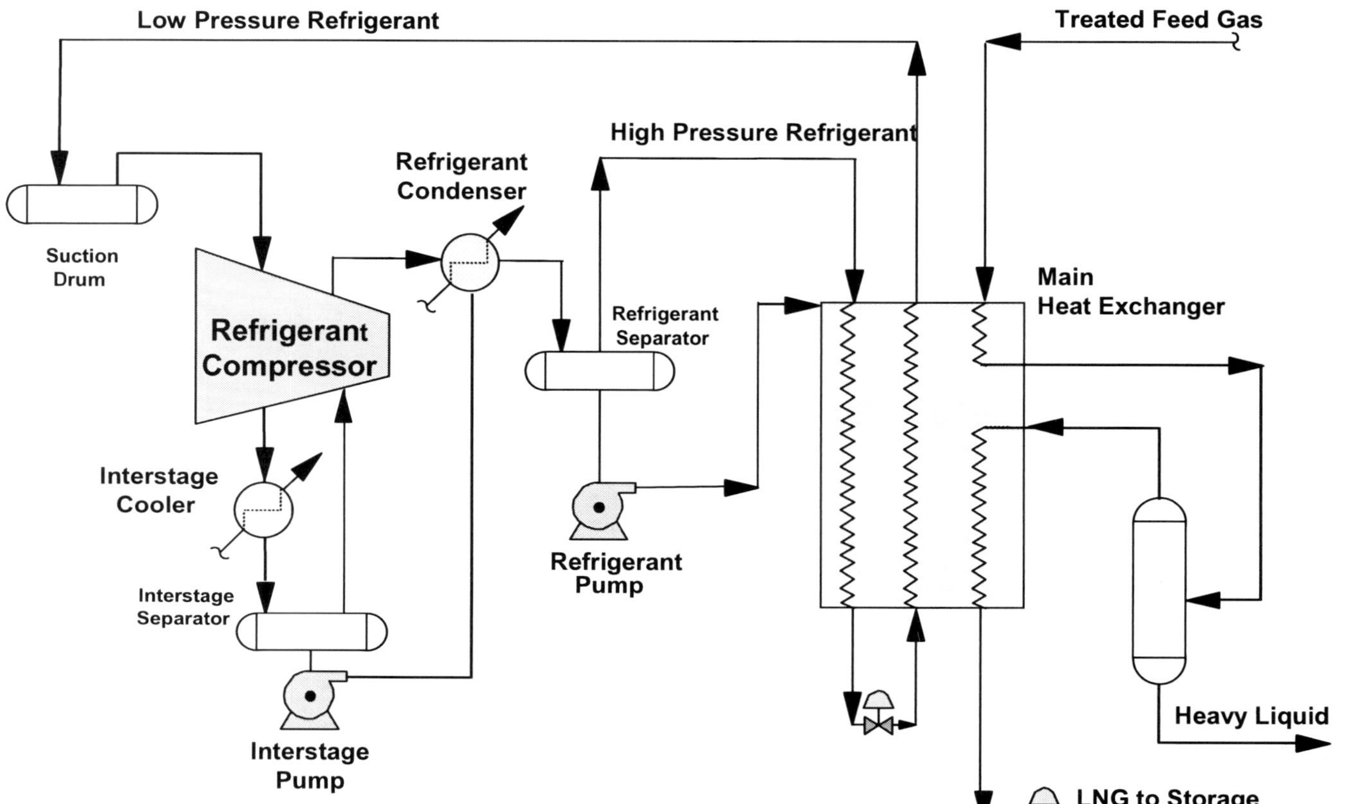

Figure 6 Wrenshall, MN Process Replacement

Modern Plate Fin Box Design

Figure 7
12 MMSCFD Alagasco LNG Plant

Figure 8
White Martins LNG Plant

Figure 9
Indicative LNG Liquefaction Plants Cost

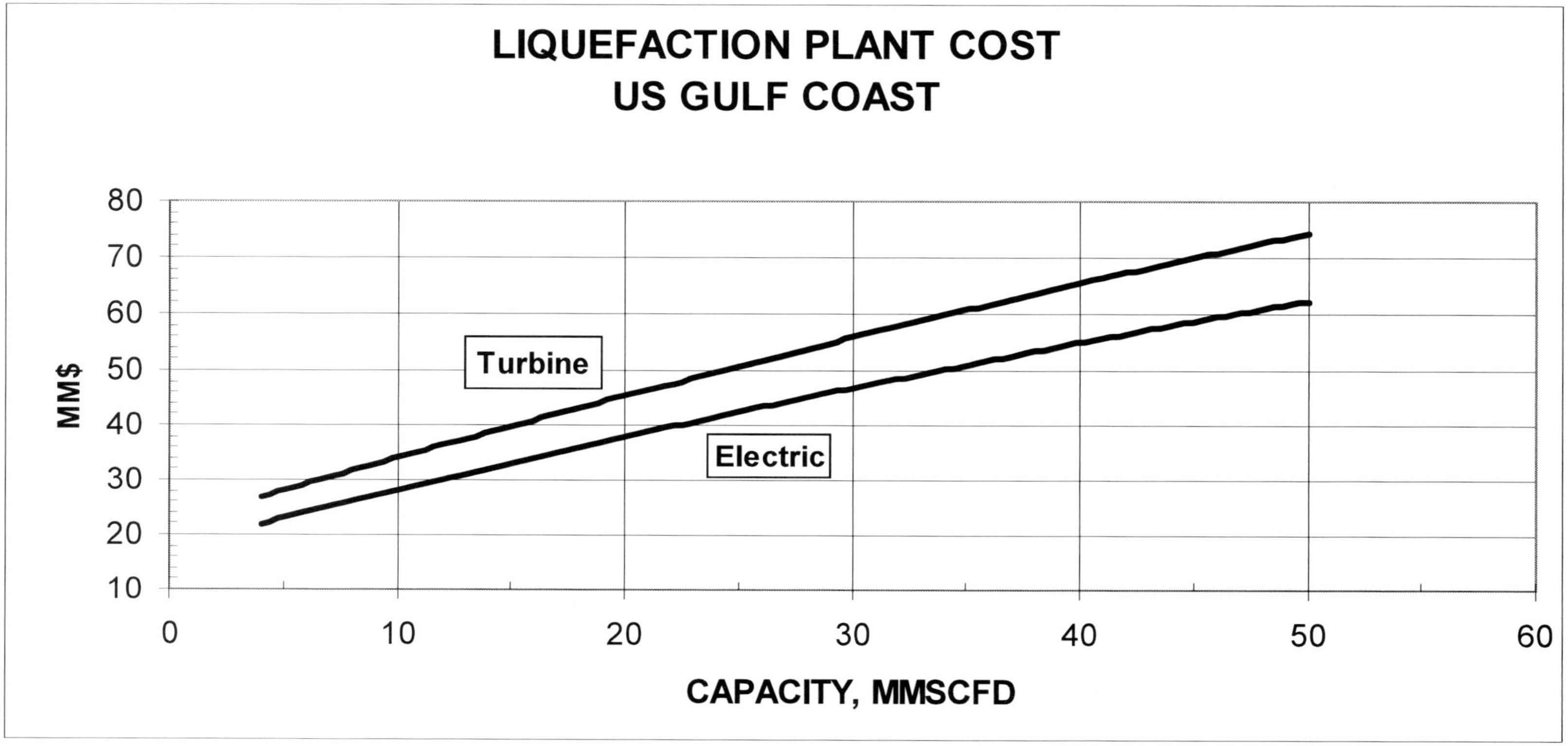

An Innovative Control Scheme of a C3/MR LNG Plant

James C. Bronfenbrenner, Matthew J. Okasinski, and Scott Trautmann
Air Products and Chemicals, 7201 Hamilton Boulevard, Allentown, PA 18195

A new approach to control of a LNG plant utilizing a C3/MR process was developed and tested using a proprietary dynamic simulator. The system employs flow ratios to manipulate the refrigeration supply and monitors system constraints to maintain conditions within a safe operational envelope. The set points for the control system are fundamental to the operation of the process and therefore intuitive to the operator. The system enables the operator to independently specify the desired LNG production and the LNG temperature exit the refrigeration unit, i.e. MCHE, to satisfy the facility fuel balances. This paper describes the key control pairings, constraints and process responses to various potential disturbances and set point changes.

Cost Effective and Practical NGL Stripping Processes

Joseph H. Cho, Ph.D., P.E.
LNG Technology Manager

Justin Capouch
LNG Technology Manager

Chicago Bridge and Iron Company (CB&I)

AIChE Spring Meeting, April 2008
8th Topical Conference on Natural Gas Utilization
New Orleans, LA, April 6-10, 2008

ABSTRACT

Nitrogen dilution has been widely used to reduce imported LNG High Heating Values (HHV). However, pipeline gas specification limits place considerable restriction on nitrogen injection's ability to adjust regasified LNG HHV and component requirements. This process also requires high operating costs in nitrogen generation.

While securing LNG supply through long-term contracts will continue to play an important role, acquiring spot LNG from scores of foreign producers will also be a popular practice. Since imported LNG has wide compositional variations, many LNG receiving terminals need to have the heating value lowered by heavy component (C_2+) stripping. Natural Gas Liquid (NGL) stripping from LNG is more versatile than nitrogen injection and can generate additional revenue if suitable demand for higher value extracted components exists.

Many NGL stripping processes have been developed and are commercially available. This paper presents an overview of several processes and discusses their advantages and limitations. Operability and flexibility with respect to inlet LNG feed variations is evaluated.

INTRODUCTION

Imported LNG accounts for roughly 3% of the total U.S. gas demand in 2006. LNG imports are projected to increase to 15 % by 2020 or about 3.5 TScf annually as domestic gas production declines[1,2]. The issue of "gas interchangeability" becomes important as the imported LNG composition may differ from the local pipeline gas specifications. Typically imported LNG heating values are higher than that of the local pipeline gas.

With few exceptions, most LNG producers favor LNG specifications of the Asian Pacific market where about 65% of LNG has been consumed[2]. Traditionally, it can be said that LNG is a single gas source in the Asian market resulting from the cost effectiveness of LNG's gas transport efficiency.

One of the distinctive trends in LNG trading is the move away from dedicated LNG sources and destinations. More LNG is available from the LNG spot market than in the past. LNG importers want to be able to negotiate LNG purchases without being restricted by compatibility issues with the regasified LNG. This mostly affects the US and UK markets that require a lower heating value gas than the Asian markets. Only a few LNG producers can meet those specifications without additional treatment.

There are several ways to achieve gas interchangeability of the imported LNG at a receiving terminal. This typically includes either inert gas injection (nitrogen, air or other inert gases) or stripping out heavier NGL components from the LNG. Inert gas injection is limited to very few LNG supplies and requires considerable operating cost. The NGL removal process is the most versatile choice but it requires considerable capital investment. Extracted NGL products may also have higher market values that provide additional revenue.

Several NGL stripping schemes have been proposed in recent years and some are protected by patents. All NGL stripping schemes rely on volatility differences between the components. This paper classifies existing schemes into three categories based on their handling of the residue gas after heavy component stripping. Advantages and disadvantages of each category are discussed.

CLASSIFICATION OF NGL STRIPPING SCHEMES

The main purpose of NGL stripping is to reduce ethane and heavier hydrocarbons and produce vaporized LNG that meets pipeline gas specifications. It also produces LPG products for export and produces an ethane rich cut for the terminal's fuel or for export.

In a typical NGL stripping facility, the stored LNG is pumped to a recondenser where it is combined with the tank BOG. The combined LNG and recondensed BOG is then pumped to the NGL stripping facility where it is partially vaporized in a heat exchanger before it is fed to the Demethanizer (DeC1). The DeC1 separates the partially vaporized stream into a C_2+ bottoms liquid product and an overhead residue gas stream consisting mainly of methane.

The DeC1 is the heart of any heavy component stripping scheme. The upper limit in operating pressure for a typical DeC1 is about 45 bar, near the critical pressure of methane. Lower operating pressures not only reduce column cost but also facilitate component separation that, in turn, reduces reboiler duties.

The residue gas must be brought to pipeline pressure specifications that are region-specific and may run as high as 100 bar for long-distance transportation. Existing heavy component stripping schemes differ on how this is accomplished and may be classified into three categories based on their handling of the residue gas:

- Residue Gas Compression
- Residue Gas Condensing without Compression
- Residue Gas Condensing with Compression

Other equipment may be added to enhance recovery and efficiency by the generation of internal reflux and stripping streams. The NGL products may be further fractionated and treated to meet product specifications.

Residue Gas Compression

Schemes in this category directly compress the residue gas to the pipeline delivery pressure. The scheme shown in Figure 1 is straightforward and has good process flexibility as it is insensitive to LNG inlet temperature, composition and terminal send-out limitations.

The column can operate at low pressures enhancing C_2+ recovery and reducing column cost. The choice of heat transfer fluid requires attention at lower operating pressures because of freezing possibilities during normal operation and upsets. The scheme also allows the use of a simple heat exchanger for the LNG preheater service.

Direct residue gas compression to pipeline pressure has high power requirements. The increased capital and operating costs associated with the large compressor make this scheme less cost effective.

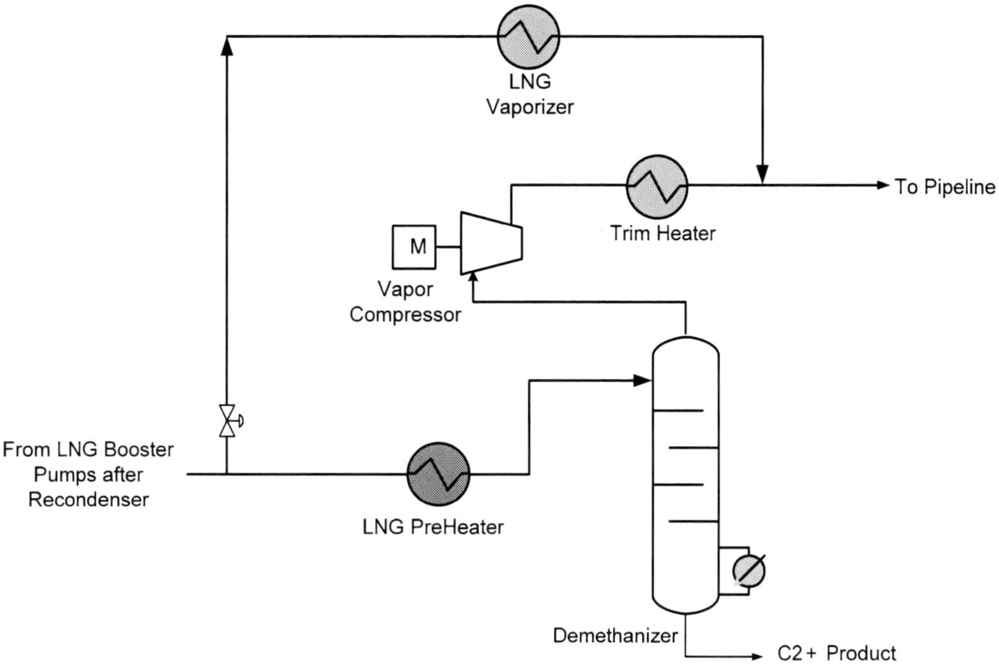

Figure 1 – Residue Gas Compression Schematic (without Condensing)[3]

Residue Gas Condensing without Compression

This category's key element is that the residue gas is recondensed through heat exchange with the feed LNG, pumped to pipeline pressure and vaporized. Figure 2 illustrates an example of a scheme in this category. Since neither a residue gas compressor nor a heater is required, the process is simple and produces a reliable operation as long as these are only small variations in feed composition[4].

While eliminating the residue gas compressor has capital and operating cost advantages, it significantly limits the process' operational flexibility. The DeC1's high operating pressure and close temperature approaches in the LNG preheater/recondenser make it difficult to cope with variations in LNG inlet temperatures, compositions and terminal operation modes. The scheme also has difficulty meeting the LNG sendout pumps' subcooling requirements when supplied with lean or high nitrogen content LNG.

In some process schemes without the residue gas compressors, operating at high pressure may not be practical as it nears the critical pressure of methane. Higher operating pressures also tend to increase the DeC1 capital cost.

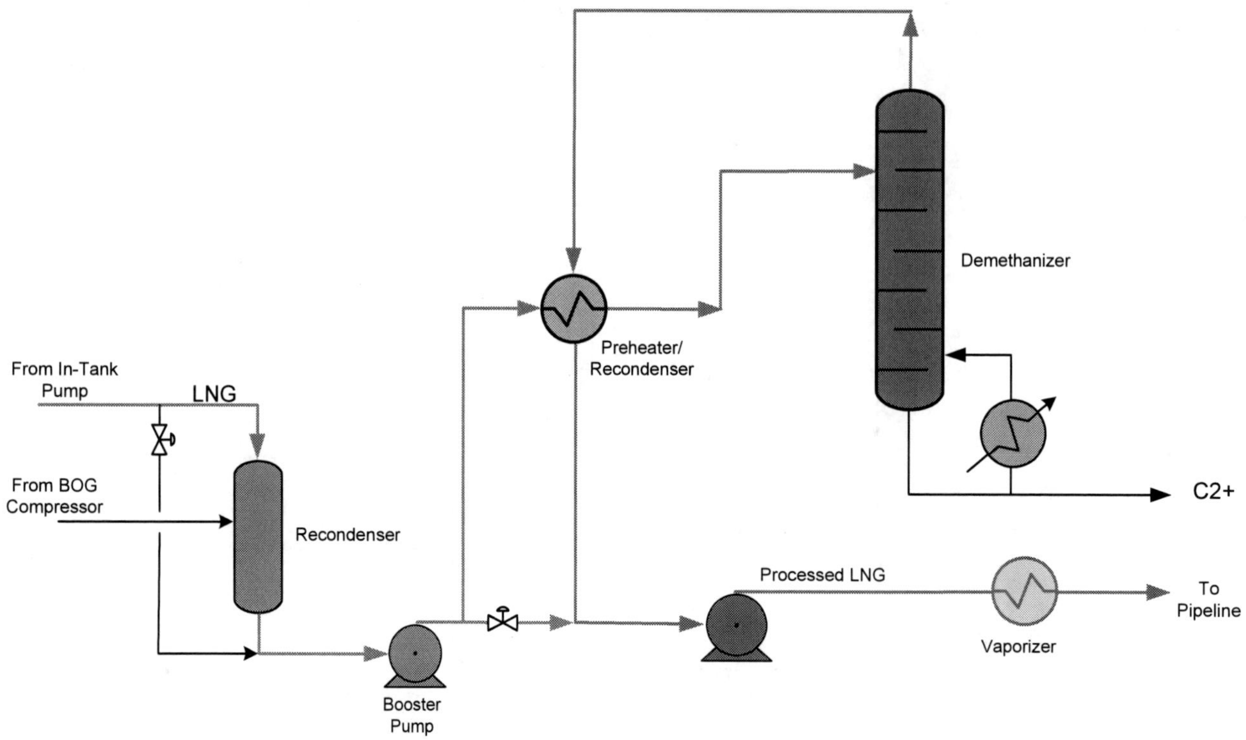

Figure 2 – Residue Gas Condensing without Compression Schematic

Residue Gas Condensing with Compression

Processes in this category combine residue gas compression and condensing to provide good operational flexibility while minimizing compression capital and operating costs. The compression step is a critical feature in any NGL stripping process. It shifts the residue gas dew point and condensation curve. This provides more flexibility with a wide range of feed LNG compositions and operating conditions. The residue gas condensing step allows the sendout to be pumped to pipeline pressure and vaporized; thus reducing compressor size and power requirements.

The discharge pressure is optimized based on the LNG feed variance expected and degree of residue gas subcooling required for the LNG sendout pumps. This allows processes of this type to adequately handle lean and high nitrogen content LNG feeds.

The compressor reliability is also very important as the whole train will be out of service if the compressor breaks down. This issue can be effectively addressed by either:

- using 2 x 50% capacity compressors with a warehouse spare (rotors and key elements), or
- providing an installed spare compressor as a pre-investment towards future expansion.

Figure 3 provides an example of a process in this category and is one of several patented NGL stripping processes that CB&I offers[5]. The feed LNG is partially vaporized in the LNG pre-heater/condenser and sent to the flash column. The flash column overhead vapor is compressed by the residue gas compressor and sent to the LNG pre-heater/condenser where it is condensed and subcooled against feed LNG. A small portion of the condensed residue gas is fed back to the top of the flash column as reflux to increase the NGL recovery rate. The flash column bottoms liquid is pumped, heated by the DeC1 overhead vapor and sent to the DeC1. The DeC1 overhead vapor is partially condensed and fed to the flash column.

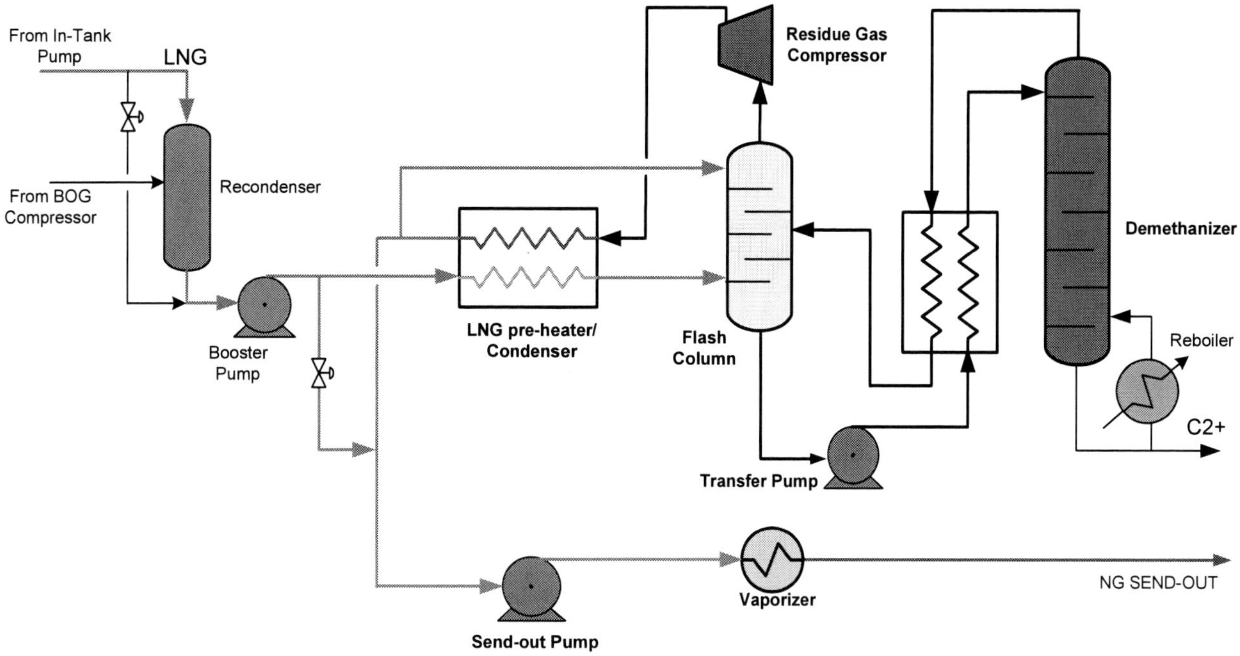

Figure 3 – Residue Gas Condensing with Compression Schematic[5]

DESIGN ISSUES

Heat Exchanger

The LNG pre-heater / condenser operation and design conditions are critical to the overall reliability and performance of the NGL extraction facility control unit.

The process that does not have the compression step will have a very close temperature approach in the LNG pre-heater / condenser and can be very challenging to maintain operation as the feed composition varies. High DeC1 operating pressures are required to recondense the residue gas and it is quite difficult to achieve the proper amount of subcooling with lean and high nitrogen content LNG feeds. A typical cooling curve of a process without residue gas compression is shown in Figure 4.

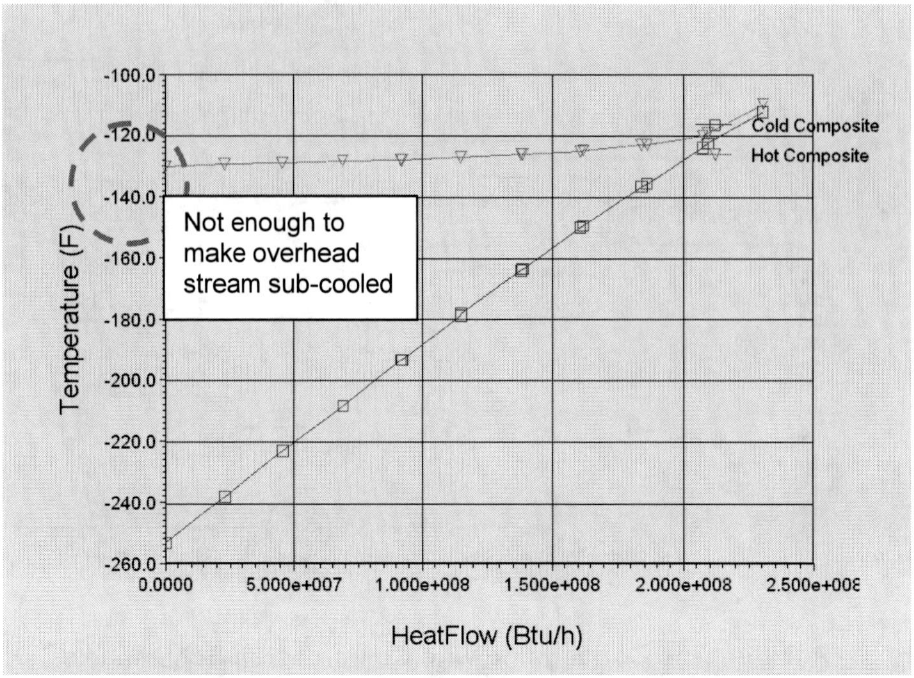

Figure 4 – Typical Cooling Curve of a Process without Residue Gas Compression

The compression step provides the flexibility of shifting the residue gas dew point and widens the temperature approach in the LNG pre-heater / condenser. As a result, the unit's operation is stable over a wide range of feed conditions. The design is able to easily handle variations in both feed composition and temperature that can occur from different LNG sources and terminal operating modes. It also simplifies the exchanger design and allows the use of well proven shell & tube exchangers in some applications.

A typical cooling curve for a process with residue gas compression is presented in Figure 5.

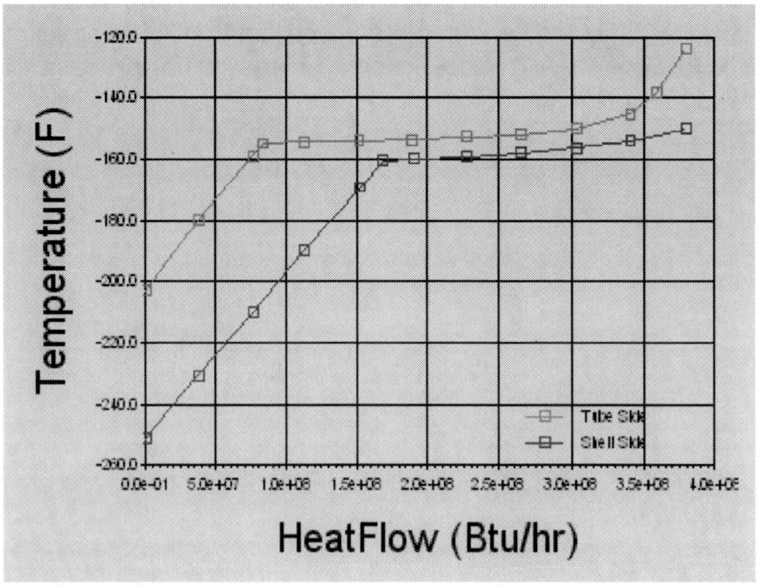

Figure 5 – Typical Cooling Curve of a Process with Residue Gas Compression

Operation Pressure of DeC1 and Reboiler

A high DeC1 operating pressure requires a high temperature level heat source for the reboiling duty. Conversely, low operating pressure requires a low temperature level. The heat supply and control is quite difficult if the required temperature level is too low. Therefore, a certain level of heat source temperature is preferable.

Several parameters need to be considered in designing the reboiler heat supply system regardless of the operating pressure. These include commissioning, stand-by operation under a pressurized condition (for short periods), long periods of shut down and decommissioning.

The heat transfer fluid (HTF) directly impacts the reboiler heat transfer areas and its installation costs. HTF also impacts operation costs. The proper selection of the HTF is important for system optimization and the type of the reboiler. The following points need to be considered in selection of the HTF:

- High flash point
- High auto ignition point
- Viscosity of the temp range
- Low density
- Low coefficient of thermal expansion
- High max film temperature
- High thermal conductivity
- Low thermal oxidation
- Low fouling
- Low vapor pressure
- Environmental consideration

CONCLUSIONS

NGL stripping processes that feature residue gas compression allow importers to secure LNG supplies from a wider range of sources. The most cost effective and practical NGL stripping solutions combine residue gas compression and condensing to offer good operational flexibility while keeping compression capital and operating costs in check.

The compression step provides the flexibility of shifting the residue gas dew point and widens the temperature approach in the LNG pre-heater / condenser. This allows the NGL stripping process to easily handle variations in both feed composition and temperature that can occur from different LNG sources and terminal operating modes.

REFERENCES

1. Energy Information Administration, Annual Energy Outlook 2008, http://www.eia.doe.gov/oiaf/aeo/pdf/earlyrelease.pdf.
2. Energy Information Administration, International Natural Gas and LNG Imports and Exports 2006 Data, http://www.eia.doe.gov/emeu/international/gastrade.html.
3. Rambo, C. I., Wilkinson, J. D., and Hudson, H. M., "Liquefied Natural Gas Processing", U.S. Patent 5,114,451 to Elcor Corporation (1992).
4. Cho, J., Vega, F.D., Kotzot, H., Durr, C.: "Innovative Gas processing with various LNG Sources" LNG Journal Jan/Feb Issue 2005, p. 23-27.
5. Schroeder, S. and Reddick, K., "Cryogenic Liquid Natural Gas Recovery Process", U.S. Patent 6,907,752 to Howe-Baker Engineers, Ltd. (2005).

BIOGRAPHY of SPEAKER

Joseph H. Cho is a LNG Technology Manager at CB&I. He is a registered professional engineer and has been in the LNG industry for Nineteen years. He was involved in process design of LNG liquefaction plants and import terminals for various regions as a process engineering lead. He graduated from the University of Oklahoma with a Ph.D. degree in Petroleum Engineering.

Email : Jcho@cbi.com

Comparison Study on Small-Scale LNG Plants in China

Yang, Xiaodong [1], Gu, Anzhong [2], Xi, Wenku [1], Lu, Xuesheng [2], Gao, Ting [2]

[1] Chemtex Engineering(China) Engineering Co.,Ltd. 1000, ZHangheng Road, Shanghai, China
[2] Shanghai Jiao Tong University. 1954 Huashan Road, Shanghai, China. 200030

AIChE Spring Meeting, April 2008
8th Topical Conference on Natural Gas Utilization
New Orleans, LA, April 6-10, 2008

Introduction

1.1 Shanghai Peak shaving LNG Plant

Shanghai LNG Plant is the first peak-shaving LNG plant in China. It serves as a backup station of the East China Sea Natural Gas Project with a capacity of 100,000Nm^3/day for emergency peak-shaving purpose.

The refrigeration process is integral incorporated cascade cycle (CII). The basic concept of CII process is shown in Figure 1. [1,2] The main equipment of the process are three stage compressor, the fractionor and the integrated cold box that can be divided into two parts, the upper one, in which the NG is percooled, and the lower part, where the NG liquefaction and sub-cooling occurs. The mixed refrigerant comprises N_2 and hydrocarbons. The low pressure refrigerant back from cold box is compressed to a medium pressure and is directed to the fractiontor after cooling. In the fractionator, it is separated into two streams: the heavies from the top flows to the up section of cold box and used as precooling of the feed NG and refrigerant. The vapor exiting from the top of the fractionator is directed into the up section of cold box and is partially condensed after being cooled. The liquid serves as the reflux, while the vapor is further compressed to high pressure and cooled, then is directed to the cold box to be pre-cooled, which results in the partial condensing. Both vapor and the condensate serve as the cooling source for the NG feed and the refrigerant.

1.2 Xinjiang Guanghui LNG Project

The natural gas resource is from Tuha gas field. The output of Xinjiang Guanghui LNG plants is 1.5×10^6 Nm^3/d in natural gas. The simplified diagram of the mixed refrigerant cycle developed by Linde for the Xinjiang Guanghui LNG plant can be found in [1].

The three stage centrifugal compressor with a suction pressure of 0.2MPa and a discharge pressure of 4.0MPa is driven by the gas turbine. The refrigerant from the second stage discharge of compressor partially condensed after the cooler and is combined with the liquid from the condensate of the third discharge stream after cooling to 40°C. Then it is separated into vapor that is compressed further and liquid that provides the cooling capacity for the first section of the main heat exchanger, a spiral-wound type with a temperature of -19°C before the J-T. Vapor exiting the third stage discharge separator will serves as the cooling source for streams of both the second and third section of heat exchanger.

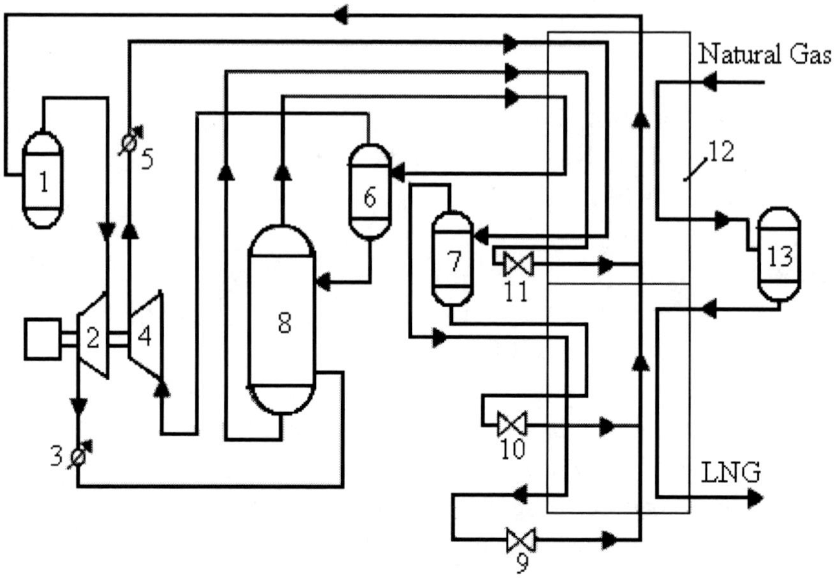

1,6,7,13- Separator 2,4- Compressor 3,5- Cooler
8- Fractionator 9,10,11- J-T Valve 12- Cold Box
Fig 1 CII Process Diagram for Shanghai Peakshaving LNG Facility

1.3 CNOOC Zhuhai LNG Plant

The CNOONGL LNG Unit is designed to liquefy 583,000 Nm^3/d of natural gas. The liquefaction section of the plant uses the PRICO™ process developed by the Gas, Oil, and Chemicals Division of Black & Veatch to liquefy the natural gas feed and produce the LNG product.

The feed NG is cooled when flowing downward through the Refrigerant Exchanger, and is interrupted at the midpoint of the heat exchanger and directed into the Heavies Separator to remove any heavy hydrocarbons that might be present in the feed caused by variations in pipeline gas quality. The overhead gas from the heavies separator is returned to the cold box, and continues its downward flow through the core.

The PRICO™ design uses a simple closed loop refrigeration cycle. The refrigerant is certain mixture of pure components. The low pressure refrigerant back is compressed in the first stage of the refrigerant compressor and then after being partially condensed in the interstage cooler, then the vapor is separated in the interstage KO drum, and directed to the second stage of the compressor. The liquid refrigerant is pumped to the second stage discharge pressure of the refrigerant compressor by the interstage refrigerant pumps. The liquid is combined with the second stage compressor discharge vapor before flowing to the refrigerant condenser, where the high pressure two-phase refrigerant stream is cooled, and then flows to the refrigerant discharge separator and is separated there. The high pressure refrigerant vapor and liquid from the refrigerant discharge separator flow though separate lines to the refrigerant exchanger and are recombined internally in the refrigerant exchanger as they enter the brazed aluminum heat exchanger core and exit the core at the bottom under a totally condensed condition. It then flows through the J-T valve, which reduces the pressure and becomes low temperature stream.

The cold, low pressure refrigerant reenters the main heat exchanger at the cold end and flows upward, serving as the cooling stream and then back to the suction of the compressor.

1.4 XinAo Gas Beihai LNG Plant

Beihai XinAo Beihai Weizhou Island LNG plant is a LNG plant with a design capacity of 150,000Nm^3/d, which use a closed expander refrigeration loop developed by SALOF Refrigeration. The feed flows to a separator, and then is compressed and is directed to a acid removal and dehydration process. After the treatment of amine unit and molecular sieve beds, it flows into the brazed aluminum heat exchanger, and is cooled and exits the heat exchanger at the a mid-point and enters into a demethanizer to get rid of the heavies. The vapor from the top of demethanizer then flows back into heat exchanger to be further cooled to a low temperature and then flows through the J-T valve to get a LNG product with a pressure of 3.4bar. The vapor flashed serves as regeneration gas for the molecular beds, and then to the fuel header and burnt as fuel for the engine that used to drive the reciprocating compressors for both feed and the refrigerant.

The suction pressure for the refrigerant compressor is 2.9bar, and then is compressed after the 3 stage compression. The discharge refrigerant is then further compressed by 2 compressor sides of the expander-compressor units, and after compression, it is pre-cooled by a air cooler and by the main heat exchanger respectively. After pre-cooling, the refrigeration gas flows through 2 expanders and reach a pressure of 3.4bar after 2 stage expansion. Afterwards the refrigerant mixes with BOG from the tank and flows into main heat exchanger to be heated and back to the suction drum of the refrigerant compressor.

Processes Comparisons and Selection of LNG Process

2.1 Features of each process

2.1.1 CII for shanghai LNG Plant

The three stage compression is adopted and a frantionator is introduced after the second stage compression to distill the mixed refrigerant, which theoretically can make the separation of heavy and light hydrocarbons more effective, therefore can help on improvement the energy efficiency. This should be a good way to improve the performance of plant with a big capacity.

Main heat exchanger has designed to have a relatively big margin which makes the process more flexible, i.g. could be more effectively in response to the variations of discharge pressure or flow rate of mixed refrigerant.

2.1.2 Suggestions for CII Process for Shanghai LNG Plant

A refrigerant storage measure should be introduced to store the refrigerant that will be directed to flare under current operation when system shut-down. In view of the characteristic of a peak-shaving facility, the make-up and storage system is even more important due to a relatively frequent shutdown and restart up of the plant.

The pure C_2H_6 adopted by the original process as a component of mixed refrigerant can be replaced by C_2H_4, which has a much better availability in China. Our Aspen simulation results show that the performance could be even better after optimization, which decreases the discharge pressure of refrigerant by this replacement. Also, the simulation indicates that the discharge pressure of refrigerant compressor can be decreased from designing figure 46bar to 35bar while the performance of the plant can be maintained even be improved by the optimum components. This has been testified by the real operation data of the plant.

Comparisons have been investigated between process with fractionator and the revised process with fractionator being eliminated based on Aspen simulation. The results suggest that the contribution of fractionator to the overall energy efficiency of the plant is very limited. Through components optimization and control of temperature after J-T, an equivalent effectiveness can also be achieved, yet with better simplicity.

2.1.3 MRC process for XinJiang Guanghui LNG Plant

The process employs three J-T valves of refrigeration system in temperature control and through regulations of these J-T valves, the temperature approach can be maintained more precisely, which has a positive effect on energy efficiency, but also makes the system more complex and therefore increase the costs for the control system and instrumentation as well as system operation and maintenance.

The three stage compression is adopted and is implemented by a centrifugal compressor with the gas turbine as the driver while the waste heat is recovered by the hot oil system to provide the heat for molecular sieve beds regeneration and amine reboiler. This design can reduce the overall energy consumption of the plant.

2.1.3.4 PRICO process for CNOOC Zhuhai LNG Plant

The PRICO process has an advantage of simplicity and reliability as a result of using a single refrigeration system that may reduce the costs for operation and maintenance of the plant. The control system and instrumentation can be simplified hence the reliability can also be increased.

The PRICO process has a design of containing the refrigerant during the showdown periods, which requires the relatively high design pressure selection for those even under low pressure operation vessels and pipes. As a result, the system can be shut down for extended periods without loss of refrigerant. Thus the cost for refrigerant make-up can be saved and time required for the re-start can be reduced significantly.

The temperature approach control in the main heat exchanger is probably a key know-how of the PRICO process, which is highly confidential.

The overall performance of the PRICO^{TM} process has made it a very attractive solution for Chinese customers and makes it very commercially successful in Chinese small-scale LNG market. Five other plants that use PRICO ™ process are under construction or at the designing stage in China.

2.1.3.5 Expander process for Beihai XinAo LNG Plant

This process has an advantage of compactness of the equipment pieces that are all skid-mounted. Thus the plot plan requirement can be minimized.

The refrigerant loop of the process is merely treated natural gas that can be made up from the plant feed gas after treatment, therefore, the system of refrigerant storage and make-up can be eliminated, and the accessibility of the refrigerant is good.

The irreversibility of the heat transfer is high as a result of gas refrigeration without phase change encountering. Hence, the energy consumption is high. The positive side is the start-up of the refrigeration system is very rapid as the working fluid in the loop is simply natural gas.

2.2 Process selections Principles

Many studies have been focused on the process selection for LNG production facilities. [2, 3-7] It is believed that any thorough optimization of the LNG process would inevitably include a trade-off between cost (capital and operating) and energy efficiency. [8]

Through investigation of these four LNG plants in China, we conclude that the main considerations of Chinese investors are (1) investment capital (2) operation cost (3) gas supply condition (4) capacity (5) flexibility (6) reliability (7) simplicity and automation level (8) recovery and restriction of waster effluents (9) LNG product (10) compressors and drivers.

Generally the final plan will be a compromising one. Therefore, a comprehensive comparison study of investment capital, energy consumption and requirement for operation and maintenance based on the given design basis and battery conditions must be accomplished for the selection of a suitable LNG process. One of our next tasks is to develop a comprehensive mathematic model to evaluate the specified LNG process developed for Chinese clients in terms of both technical and economic aspects.

Exergy Analysis

3.1 Introduction

Although having drawbacks of focusing on the equipment units rather than the flowsheet level and being lack of strong link between exergy and cost [9], exergy analysis has been widely used in the LNG process performance analysis. The advantage of exergy analysis is the inherent capability of including all stream properties (temperature, pressure and composition).

Remeljej and Hoadley [8] investigated four different LNG processes through exergy analysis, based on which the performance of and areas for improvement for each process was given. Since the processes presented in [8] are targeting the offshore facilities instead of more frequently installed onshore ones, the study fails to cover more typical LNG process. Moreover, except for PRICO SMR process, which has a more than 20 years proven operation history and is suitable for both offshore and onshore facilities [10,11], the other three LNG scheme are merely models in concept without real plants in operation.

Aspelund et al [9] proposed a new methodology that combines the traditional pinch analysis and enexgy calculation for process synthesis. This tool was used to minimize energy requirements by optimizing the process compression and expansion work as well as the work necessary for cooling utilities. In addition, this methodology was used to develop a novel LNG process, where it is believed the exergy efficiency can be increased significantly. However, the process mentioned in [9] has a considerably restricted application scenario, and is lack of proven operation. It seems that an analysis on those typical and operation data available LNG processes that have been used widely including applications in China would be necessary to testify the effectiveness of proposed methodology.

Cao et al [12] simulated two skid-mounted types of small-scale NG liquefaction process, and compared the key parameters of the processes by using HYSYS software. However, a more thorough exergy analysis is still expected.

3.2 Exergy Analysis for a LNG Processes in China

3.2.1 Thermodynamic Fundemental of Exergy Analysis

Exergy is a measure of the maximum amount of useful energy that can be extracted from a process stream when it is brought to equilibrium with its surroundings in a hypothetical reversible process. [13] When ignoring the kinetic and potential energy effects under the steady-state units operations, the exergy, or useful available work of a stream can be therefore expressed as

$$e = (H - H_0) - T_0(S - S_0) \tag{1}$$

While the heat exergy can be expressed as

$$E_Q = (1 - T_0 / T)Q \tag{2}$$

Here subscript '0' normally can be considered as the state of environment. Based on the energy conservation equation, the exergy balance equation for an open system can be given by

$$\sum \dot{E}_{Qi} = \left[\sum \dot{W}_i - p_0 \frac{\Delta(mv)_{c.v}}{\Delta\tau}\right] + \Delta(\dot{m}e)_i + \frac{\Delta(me_u)_{c.v}}{\Delta\tau} + \sum \dot{II}_i + \dot{E}_{Q_0} \tag{3}$$

For a steady-system, $\frac{\Delta(mv)_{c.v}}{\Delta\tau} = 0$ and $\frac{\Delta(me_u)_{c.v}}{\Delta\tau} = 0$, therefore the above equation can be transformed to

$$\sum \dot{E}_{Qi} = \sum \mathring{W}_i + \Delta(\dot{m}e)_i + \sum \dot{II}_i + \dot{E}_{Q_0} \tag{4}$$

3.2.2 Exergy Analysis for a LNG Processes in China

The equipment enclosed in the small-scale LNG plant in China, i.g. Shanghai LNG plant, comprise compressor (expander), J-T valve, stream mixer, and multi-stream heat exchanger. When ignoring the kinetic and potential energy effect, we can deduce a simplified exergy balance equation as below

$$E_{in} = W_S + E_{out} + E_Q + \sum II_i \tag{5}$$

Above equation can be used in each major piece of equipment, and we can calculate exergy loss on every piece of equipment. We illustrate this by analyzing the CII process for Shanghai Peak-shaving plant.

The result shows that the exergy loss for this process is mainly on heat exchanger and compressor, then on condenser and J-T valve in below table. The exergy efficiency is 20.3%, a figure that locates between the typical C3+MRC process and expander one.

Table 1 Exergy analysis of CII liquefaction processes

Item Name	The CII Process	
	II_i / KW	Ω_i & η
Exergy Loss for MR Compressor	5542.6	24.0%
Exergy Loss for Heat Exchanger	5963.9	25.8%
Exergy Loss for MR Condensor	3871.3	16.8%
Exergy Loss for J-T and Mixer	3014.6	13.1%
LNG Exergy	4696.4	η =20.3%

3.2.3 Suggestions on Exergy Efficiency Improvement

From above statement, the exergy loss on heat exchanger is fairly high, and therefore, ways should be taken to reduce the loss on heat exchanger.

The exergy loss during heat transfer can be expressed as

$$\delta \mathring{II} = T_0 \mathring{q}_l\, dx[\frac{T_1(x) - T_2(x)}{T_1(x)T_2(x)}] - T_0[\frac{\mathring{m}_1 v_1}{T1(x)} dp_1 + \frac{\mathring{m}_2 v_2}{T_2(x)} dp_2] \tag{6}$$

When ignoring the pressure loss, the above equation can be simplified as

$$\frac{\delta \mathring{II}}{\delta \mathring{Q}} = T_0[\frac{T_1(x) - T_2(x)}{T_1(x)T_2(x)}] \tag{7}$$

Where $\delta \mathring{Q} = kdF(T_1(x) - T_2(x))$ therefore,

$$\delta \mathring{II} = T_0 kdF \frac{[T_1(x) - T_2(x)]^2}{T_1(x)T_2(x)} \tag{8}$$

From equation 8, we conclude that the exergy loss of heat transfer is direct ratio to square of the temperature approach, which means with temperature approach increasing, the exergy loss will increase significantly, while inverse ratio of multiplier of temperatures, which means the exergy loss will be more sensitive to low temperature, and that is why we should lower the temperature approach by means of using more compact heat exchanger like brazed aluminum heat exchanger especially in cryogenic applications.

Conclusions

Four LNG process in China has been introduced, and certain suggestions and analysis have been made.

References

[1] Anzhong Gu, Yumei Shi, Rongshun Wang. A Review of LNG Industry in China. In: Proceedings of the International Congress of Refrigeration 2003, Washington D.C. 2003.

[2] G. Zhu. Study on Transportation properties of Natural Gas and Optimization on Liquefaction Processes of LNG Peak-shaving Plants. Doctoral Dissertation,

[3] Tariq Shukri, LNG Technology Selection, Hydrocarbon Engineering, 2004, 9(2): 71-76

[4] U.Rathmann, Which Liquefaction Process Suits Best for LNG Peak-shaving Plant? , Cryogenic Process and Equipments, 1984

[5] Gilbert B.R. Cost Reduction Ideas for an LNG Project. Chemical Engineering Progress, April, 1985:17

[6] Robert N. D., LNG Peak-shaving Plants Require Careful Cost Estimation. Pipeline and Gas Journal , May, 1978:28

[7] Edwards T.J., Harris C.F., Analysis of Process Efficiency for Base-load LNG Production. Cryogenic Processes and Equipment,1984,

[8] C.W.Remeljej, A.F.A Hoadley, An exergy analysis small-scale liquefied natural gas (LNG) liquefaction process, Energy 31 (2006) 2005-2019

[9] Audun Aspelund, David Olsson Berstad, Truls Gundersen, An Extended Pinch Analysis and Design Procedure Utilizing Pressure based exergy for Subambient Cooling, Applied Thermal Engineering 27 (2007) 2633-2649.

[10] Price BC. PRICO -a simple, flexible proven approach to natural gas liquefaction. In: Proceedings of the 17th international LNG/LPG conference, Gastech' 96, Vienna, 1996.

[11] Price BC. Small - scale LNG facility development. Hydrocarbon Process 2003; 82(1):37-9.

[12] Wen-sheng Cao, Xue-sheng Lu, Wen-sheng Lin, An-zhong Gu, Parameter comparison of two small-scale natural gas liquefaction processes in skid-mounted packages, Applied Thermal Engineering 26 (2006) 898-904.

[13] Seider WD, Seader JD, Lewin DR. Process design principles. New York: Wiley; 1998. p. 207–42.

Biography

No.7 Building,Lane 1000, Zhangheng Road, Shanghai, P.R.China
Phone: 86-21-58306766, Fax: 86-21-50810345, Cell: 86-13911505274.
E-mail: andy.yang@chemtex.com.cn; meierdani@263.net

Yang, Xiaodong is the Leading LNG process engineer in Chemtex (China) Engineering Co., Ltd. He has been working in gas processing industry for over 6 years and is working for 5 LNG projects in China now. He attended Shanghai Jiao Tong University in 1998 and received his Ph.D degree in cryogenic engineering in 2002.

Utilization of Atmospheric Heat Exchangers in Lng Vaporization Processes: A Comparison of Systems and Methods

Tom Dendy
SPX Thermal Equipment & Services, SPX Corporation, 7401 W 129th St, Overland Park, KS 66213 and Rajeev Nanda, Technip USA Inc, 11700 Old Katy Road, Suite 150, Houston, TX 77079.

Recent developments in LNG vaporization systems have relied upon atmospheric heat exchange to capture heat for the vaporization process. While atmospheric heat exchange for cooling purposes is universally understood, so far there are not many published quantitative comparisons of the various available techniques and the equipment used for atmospheric heat exchange. The LNG industry is generally slow in accepting the new technologies due to the large investment cost and the risk involved.

The paper examines many factors associated with the use of atmospheric vaporization systems, including capital and operating cost and risk. The basis for comparison in this paper is an LNG vaporization terminal in the U.S Gulf Coast. Vaporization system options evaluated include; SCV only, two-loop system with direct contact heat exchangers, single loop system with indirect contact heat exchangers (with various working fluids), and hybrid systems combining SCV's, ORV's and direct contact heat exchangers.

The paper also discusses the variations and weather patterns and how these affect plant design and heat exchanger operations, including sensitivity analysis for multiple site variables.

SIMAR Applications in Gas Processing – Challenges in Supercritical Regions

Carlos Yengle
Lead LNG Engineer
Chevron Energy Technology Company
Houston, Texas

Stanley Huang
Staff LNG Process Engineer
Chevron Energy Technology Company
Houston, Texas

AIChE Spring Meeting, April 2008
8th Topical Conference on Natural Gas Utilization
New Orleans, Louisiana, April 6-10, 2008

Introduction

SIMAR, which stands for System Intrinsic Maximum Recovery, is a hypothetical operating scenario in which an expander-based C2+ recovery system can draw refrigeration freely from the environment. SIMAR operation is approximated when the feed stream to the system is Liquefied Natural Gas (LNG), instead of natural gas (NG) streams. Since LNG contains abundant refrigeration that it requires no further help from external refrigeration for all practical purposes.

Figure 1 shows a generalized scheme for C2 recovery based on expander-technology [Huang et al., 2006a]. The process is intended to strip the inlet NG of its heavier components. The residue gas is recompressed and returned to the pipeline. Sweet, dry NG goes through an inlet chilling section, where the gas is chilled to a suitable level before entering the heart of the plant: an expander-based C2 recovery scheme. The refrigeration of the inlet chilling section is mainly provided by the returning residue gas and supplemented by side-draws from the demethanizer (DeC1, side-draws not shown). Depending on actual requirements, external refrigeration may be required (not shown).

The main components of the expander section include a separator, expander, and DeC1. A subcooler is usually provided for improved refrigeration integration in the low temperature regions. The exact design of this section is an art. Since the 1970's, quantum progresses and incremental improvements have pushed the art to an unprecedented level in efficiency and has become the mainstay technology in gas processing. As an example, there are more than one hundred filed patents in the U. S. alone with individually claimed advantages. SIMAR was developed to provide a systematic approach in promoting this art for general users [Huang, 2007]. Developments in process optimization, to further simplify the methodology, have been recently publicized [Yan et al., 2007]. The process configurations described in the open literature or filed patents can be classified into four categories. For those configurations without compressors for internal circulation, characteristic SIMAR curves can be constructed following a few simple steps. The constructed SIMAR curves serve as useful tools for evaluations, comparisons, and in actual process design works. Generally the inlet feed gas stream comes into the system at conditions (below critical point) where it can be chilled into a two-phase region before it is directed to the expander based facilities, and SIMAR curves are simply evaluated. However, when the inlet natural gas stream reaches supercritical conditions, the distinctions between vapor and liquid phases are blurred, and special considerations and procedures need to be addressed while evaluating SIMAR curves. It has been noted that standard HYSYS setup for Figure 1 would fail when the inlet gas pressure is at supercritical conditions. This work presents special precautions that should be followed when dealing with this type of systems.

Technical Background

Since SIMAR is a hypothetical operating scenario, in which an expander-based C2+ recovery system can draw refrigeration freely from the environment, it would help visualizing the system functions if LNG is used as feed during initial discussions. Figure 2 shows such a realization [Huang et al., 2006b]. The LNG feed is pressurized by a pump to a suitable level, heated by a heater, before entering the separator. The residue stream from the DeC1 top goes directly to the re-compressors without refrigeration recovery. The omission of refrigeration recovery greatly simplifies the process flow and significantly reduces the compression duty. For any given configuration, the separator temperature and the DeC1 pressure are the two independent parameters which sensitively affect the C2 recovery levels.
The expander schemes are grouped into three categories: XPDR 1 through 3. Figure 3 shows the schematics of those schemes. Specifically, the differentiation is based on the handling of the vapor stream of the separator, namely, separator top not rectified (XPDR 1), separator top fully rectified (XPDR 2), and separator top self-rectified (XPDR 3). Other type of schemes include the heat pump design, which can be recognized by the use of a compressor, a cooler for rejecting heat to a high-temperature sink, a Joule-Thompson (JT) valve or a second expander and, optionally, a second exchanger to take heat from the low-temperature source. The heat pump design is not included in this work, Figure 4 shows this type of scheme included for information purposes only.

Basis of Discussion

Table 1 lists two feed stream compositions used in model case simulations [McCartney 2003]. The phase envelopes of both Gases 1 and 2 are shown in Figure 5. The richer Gas 2 has an enlarged two-phase region toward higher temperatures and pressures. Unless specified otherwise, Gas 1 is the default feed composition in this work.

Also listed is an industrial cases designated as Case 1. Case 1 is taken from the open domain [Wilkinson and Hudson, 1992]. All simulations of process configurations are performed using HYSYS. Unless specified otherwise, the parameters listed in Table 2 are used in the simulations. In order to hold discussions simple and conceptual, this work adopts thermodynamic heat pumps to account for external refrigeration supplies. Typically, external refrigeration implies closed-loop designs of propane or cascade designs of propane + ethylene circuits. The adopted thermodynamic heat pumps are assumed to be 50% of Carnot efficiency. For example, to move 1 MW of heat from -60 °C (213 K) to 40 °C (313 K), a Carnot engine requires 0.47 (= 313/213-1) MW of compression duty. In this work, twice of the amount, or 0.94 MW, is required due to the reduced efficiency.

Procedural Definition Of SIMAR Curves And Applications

In addition to the phase envelopes, Figure 5 also indicates a procedure which will be used to define SIMAR. The procedure starts from a relatively high set temperature for the separator and the temperature is systematically lowered, as shown by the solid arrow. Physically, this is achieved by controlling the heat input to the LNG pre-heater in Figure 2. The lower temperature limit of the above steps is arbitrarily set at -100 °C. Since the specified compositions can only exist in liquid phase at and below that limit, turbo-expander will be rendered unusable. Whenever it is impossible to operate the turbo-expander, it is replaced by a JT valve and the procedure continues. It is understood that that the expander-driven compressor should be taken out of service and all compression work will be assumed by the motor-driven unit.

Figure 6 plots the C2 recovery level against the separator temperature for the three categories. Both XPDR 1 and XPDR 2 show monotonic trends of improvement as the temperature decreases. The trend continues even after the inlet gas is totally liquefied, when the expander is replaced by a JT valve and no expansion work is recovered. XPDR 3 shows a different trend, though. As the separator temperature decreases, the C2 recovery goes through a maximum. In other words, blindly increasing the refrigeration to the inlet chilling section may not always improve the C2 recovery.

A procedural definition of SIMAR at a specified DeC1 operating pressure is summarized below. For the cases showing maxima, such as XPDR 3, the SIMAR is defined as the maximum of the trend. For the cases showing monotonic trends, such as XPDR 1 and XPDR 2, the SIMAR is defined at the temperature where the separator fluid has 30% of vapor fraction (VF). The choice of 30% is based on a practical consideration that no expanders would be installed if the gas flow is below this level. The same procedure is repeated at another specified DeC1 operating pressure until the entire range of interest is covered. The collection of all SIMAR points forms a SIMAR curve for the process configuration. Figure 7 depicts the SIMAR curves for the three types of process schemes. It is a tradition in gas processing to show compression duties in the abscissa, in lieu of DeC1 operating pressures. Some observations of the three categories can be made. (1) The reflux stream makes XPDR 2 more efficient than XPDR 1 throughout the entire range, which corresponds to the operating pressures of DeC1 from 17 to 42 bar. (2) The efficiencies of XPDR 2 and XPDR 3 are comparable, while each configuration has its advantages over certain ranges. Examples using patented processes were presented elsewhere [Trinter and Huang, 2007].

SIMAR Applied To Natural Gas Processing

The SIMAR curves represent the limiting-case scenario where refrigeration is freely available from the environment. This scenario is approximated by Figure 2. The true merits of the SIMAR procedures will be demonstrated when it is extended to Figure 1, where natural gas is the feed stream. The main difference is the addition of a pre-chiller to chill the inlet gas so that the separator can operate at SIMAR conditions. The pre-chiller draws refrigeration from the returning residue stream. Additionally, it takes refrigeration from the DeC1 bottom section. This is achieved by side-boilers. If there is still deficiency in refrigeration, then external refrigeration should be provided. The detailed pre-chiller configuration does not impact the SIMAR curve, but it will impact the calculation of total compression duty. In the following discussions, the pre-chiller is represented by an integrated design using brazed aluminum heat exchanger, which incorporates all streams participating in the energy exchanges. It has been shown when the integrated design of pre-chiller is replaced by individual exchangers, such as thermal siphons, there is a penalty in thermodynamic efficiency [McMullen and Huang, 2007]. However, sometimes a split may be required due to limitations in equipment sizes or special operational considerations.

Figure 8 presents typical results corresponding to Figure 1. On the left side is the SIMAR curve, which is taken from the XPDR 3 curve in Figure 7. Second from the left is the recompression curve after the refrigeration in the residue stream is recovered by the pre-chiller. The gap between the two curves is the impact of refrigeration in the residue gas. Total compression using side reboilers has been evaluated and results are showing with the two thick curves on the right side of the graph, these curves present the total compression work by using one or two side-re-boilers. The case of two side re-boiler requires a lower total compression work because the provided refrigeration matches better with the chilling requirements. The gap between the total compression curve and the recompression curve is the provision of external refrigeration. Moving upward along any curve in Figure 8 implies decreasing DeC1 operating pressures. As the DeC1 operating pressure decreases, the recompression work increases. Correspondingly, the refrigeration generated by the expander increases. Hence, the demand for external refrigeration can be reduced. Practically speaking, the external refrigeration can be replaced by increasing expander contributions at high C2 recovery levels. This is demonstrated by the convergence of the curves of recompression and total compression. Moving downward along the total compression curves in Figure 8 implies increasing DeC1 operating pressures. The curves also show an increased demand for external refrigeration. This is caused by the lack of refrigeration contributed by the expander. Hence, the external refrigeration is required to make up the shortage.

Although facilities operating at SIMAR conditions make the full utilization of DeC1 potentials, not all gas processing facilities are designed to operate as such. The major consideration is the provision of external refrigeration. To reach SIMAR conditions, the demand for external refrigeration may be excessive. The designer may want to sacrifice the DeC1 efficiency in exchange for savings in the refrigeration systems. Figure 9 plots the SIMAR and sub-SIMAR operations (defined as separator temperatures above SIMAR evaluated temperatures) for Gas 1. Two sub-SIMAR cases are presented with the separator temperature being progressively raised by about 10 °C each. The thin curves on the left side are recompression requirements. As anticipated, the C2 recovery suffers with each raise of the separator temperature. As a compensation, the external refrigeration duty decreases with each raise in the separator temperature. The combined effect is that total compression remains within a narrow range, as shown by the thick curves on the right side. This is an important observation that total compression requirement remains roughly the same regardless of deviations from SIMAR operations. In other words, once the process configuration is fixed, the total compression required to achieve a specified C2 recovery level is insensitive to pre-chilling levels. The choice of the separator operating conditions should be optimized by other criteria, such as economics. For example, if operating a system at SIMAR level requires a supply of external refrigeration at -50 °C. The decision may shift toward sub-SIMAR operations in order to avoid cascaded external refrigeration circuits including ethylene. Ethylene refrigeration systems are relatively expensive to provide. Therefore, given the conditions of an inlet gas, SIMAR can serve as an expedient tool to establish the relations between C2 recovery and total compression work. Then, the exact pre-chilling level will be determined by other factors defined by the projects.

SIMAR Applied To Supercritical Region

When the feed is coming at a very high pressure, the stream resides in a supercritical region, as shown by the dashed arrow in Figure 5. Since phase identification in this region is meaningless, the standard SIMAR procedures may fail. The SIMAR procedure is performed as usual, with the separator in place. Initially the fluid is vapor-like, the entire stream would be channeled to the vapor side. As the testing temperature in the separator continues to decrease, the separator fluid eventually would be identified to be liquid and channeled to the bottom of the separator. Thus, the expander receives no fluid and the HYSYS calculations would stop. At this moment, the separator fluid has to be manually channeled to the top of the separator and normal calculations would resume. In real designs, the expander vendors should be consulted regarding the expander operability in the supercritical regions.

A stream at supercritical conditions is evaluated in HYSYS with vapor fraction value of either zero or one. Theoretically, this value doesn't have any physical meaning for a supercritical fluid, since there is no distinction of liquid or vapor phases in a supercritical region. However, it is important to determine if a supercritical fluid is liquid-like or a vapor-like fluid for calculation purposes. This is because some of the properties reported in HYSYS are calculated using certain sets of specific phase models. In other words, phase identification has to be carried out in order to decide which model to use to calculate these properties. To identify phase type for a stream under supercritical conditions HYSYS applies the following criteria when Peng –Robinson cubic equations of state are used:

1. If the compressibility factor (Z) is greater than 0.3, and the isothermal compressibility factor (beta) is greater than 0.75, a vapor fraction of 1.0 is assigned to the stream.
2. If Z is greater than 0.75 and the sum of composition of light compounds (NBP<230K) is greater than the sum of composition of heavy compounds, a vapor fraction of 1.0 is assigned to the stream.
3. Otherwise vapor fraction of zero and liquid phase correlations will be used.

Figure 10 shows the C2 recovery level when plotted against the separator temperature for the XPDR3 category. Different DeC1 operating pressures affect the level of C2 recovery, but not the shapes of curves. Results show monotonic trends of C2 recovery improvements occurring as the temperature decreases. The traditional way of determine SIMAR temperature, as shown in Figure 6 cannot apply. It is recommended that the true critical temperature of the gas, as calculated by HYSYS, should be used as the SIMAR temperature based on the following reasons:

1. The C2 recovery is relatively flat around this region.
2. This is a convenient reference temperature for preliminary designs. The final determination of process conditions should include the entire refrigeration balances of the plant, as detailed in Section 5.
3. In the practical example shown in Section 7, it becomes clear that no external refrigeration would be required for supercritical applications. Therefore, the exact "separator temperature" is non-consequential as long as the refrigeration can be balanced.

Industrial Case Showing Supercritical Applications

An industrial case is presented below to show the methodology of SIMAR reproduces existing industrial designs [McMullen and Huang, 2007]. The case uses XPDR 3 design and the feed conditions are listed in Table 1. This case demonstrates the supercritical application. The reference contains another industrial case of application in sub-critical regions. Industrial Case 1 has a high-pressure feed stream and is in the supercritical region. As a result, the expander generates sufficient refrigeration for the entire facilities. The energy balances in the pre-chiller can be fulfilled without any external refrigeration. Figure 11 shows both SIMAR curve and the recompression curve. The recompression curve coincides with total compression because no external refrigeration is needed. SIMAR methodology provides entire sets of feasible operating conditions for a given process configuration. The final choice of optimal operating condition should be based on other criteria, such as plant economies and product desirability.

Conclusions

The conclusions of this work are listed as follows:

1. Although SIMAR is defined as a limiting-case scenario, it is a powerful tool for evaluating and comparing performances of expander-based processes for C2+ recovery.
2. For supercritical applications, phase identification becomes problematic which also related to calculations of physical properties. The process configuration has to be modified to channel the fluid through suitable route as well. The selection of SIMAR point requires further investigation. Recommendations were presented for dealing with systems in supercritical regions.
3. An industrial case was presented showing that results acquired following SIMAR procedures are consistent with traditional designs.

References

1. Huang, S., Chen, R.J.J., Cook, D., Elliot, D.G., "Processes for High C2 Recovery from LNG – Part III: SIMAR Applied to Gas Processing", Paper presented in AIChE Spring National Meeting, Orlando, Florida (2006a).
2. Huang, S., Chen, R.J.J., Yao, J., and Elliot, D.; "Processes for high C2 recovery from LNG – Part II: Schemes based on Expander Technology", paper presented in AIChE Spring National Meeting, Orlando, FL (2006b).

3. Huang, S., "Improve Recovery of Intermediate Components Using LNG (Process Methodology)", Hydrocarbon Processing, issue of July (2007).
4. McCartney, D.; "Gas Conditioning for Imported LNG", paper presented in 82nd GPA Annual Convention in San Antonio, TX, 2003.
5. McMullen, C., Huang, S., "SIMAR Applications in Gas Processing: 2. Optimal Design of Expander-based C2+ Recovery Processes", paper presented in AIChE Spring National Meeting, Houston, Texas (2007).
6. Trinter, J., Huang, S., "SIMAR Applications in Gas Processing: 1. Evaluating Expander-based C2+ Recovery Processes", Paper presented in AIChE Spring National Meeting, Houston, Texas (2007).
7. Wilkinson, J. D., Hudson, H. M., "Improved NGL Recovery Designs Maximize Operating Flexibility and Product Recoveries", paper presented at the 71st Annual GPA Convention, Anaheim, CA (1992).
8. Yan, W., Bai, L., Yao, J., Chen, R., Elliot, D., Huang, S., "Optimum Design of Turbo Expander Process", paper presented at GPA 86th National Convention in San Antonio, Texas, March 11-14 (2007).

List Of Tables

Table 1: Gas compositions used in models for this work

	Model Cases		Industrial Case
Equivalent LNG	**Medium**	**Rich**	**Case 1**
Inlet temperature, C	40	40	40
Inlet pressure, bar	50	65	73.40
Critical T, C	-63.45	-48.40	-63.69
Critical P, bar	58.63	75.69	63.89
Components	Mole Fraction	Mole Fraction	Mole Fraction
Nitrogen	0.00127	0.0000	0.0027
Methane	0.90700	0.8695	.9247
Ethane	0.08617	0.0792	.0423
Propane	0.00552	0.0332	.0132
i-Butane	0.00003	0.0074	.0039
n-Butane	0.00001	0.0055	.0032
i-Pentane	0.00000	0.0037	.0015
n-Pentane	0.00000	0.0015	.0009
n-Hexane	0.00000	0.0000	.0026
n-Heptane	0.00000	0.0000	0.00000
n-Octane	0.00000	0.0000	0.00000
Carbon Dioxide	0.0	0.0	.0027
TOTAL	1.00000	1.0000	1.00000
Btu/SCF	1,082	1,165	1087
C2+ content, GPM (Note 1)	2.45	3.63	1.43
C3+ content, GPM	0.15	1.52	0.58

Notes:

1. GPM stands for Gallons Per Thousand SCF.

Table 2 Simulation parameters used in model cases

Natural Gas inlet	Unit	Value
Plant capacity, dry & sweet NG at inlet	MMscfd	100
Send-out residue gas pressure	Bar	100
Send-out residue gas temperature	°C	40
General		
Expander polytropic efficiency		80%
Compressor polytropic efficiency		80%
Pressure drop through plate-fin exchanger	Bar	0.2-0.3
Pressure drop through air coolers	Bar	0.6
Air cooler discharge temperature	°C	40
Minimum delta T in plate-fin exchanger	°C	2.5 ± 0.2
DeC1 operating pressure	Bar	17 to 42
Number of trays in DeC1 column		15
C1 composition in DeC1 bottom product		1%
Minimum delta T in plate-fin exchanger	°C	2.5 ± 0.2
LMTD	°C	5.5 ± 1.0
Max impact on C2 recovery by introducing side boilers		-0.005
High-temperature (HT) sink	°C	40
Medium-temperature (MT) sink	°C	-10
Low-temperature (LT) sink	°C	-35
Low-Low-temperature (LLT) sink	°C	-45

List Of Figures

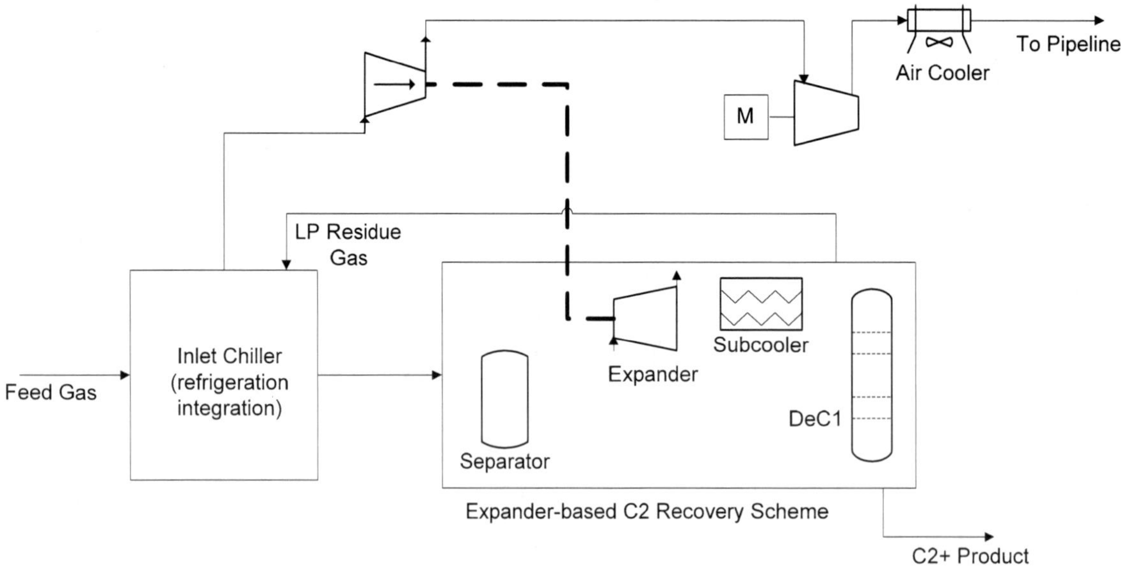

Figure 1 Generalized gas processing scheme for C2 recovery

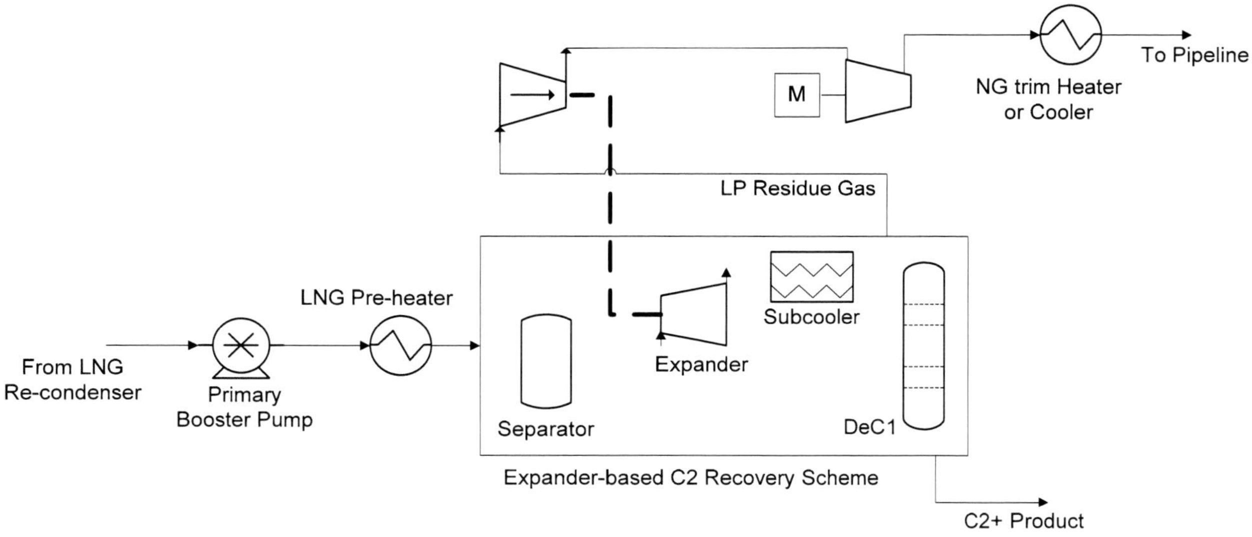

Figure 2 Residue Compression, combined with expander-based C2 recovery schemes

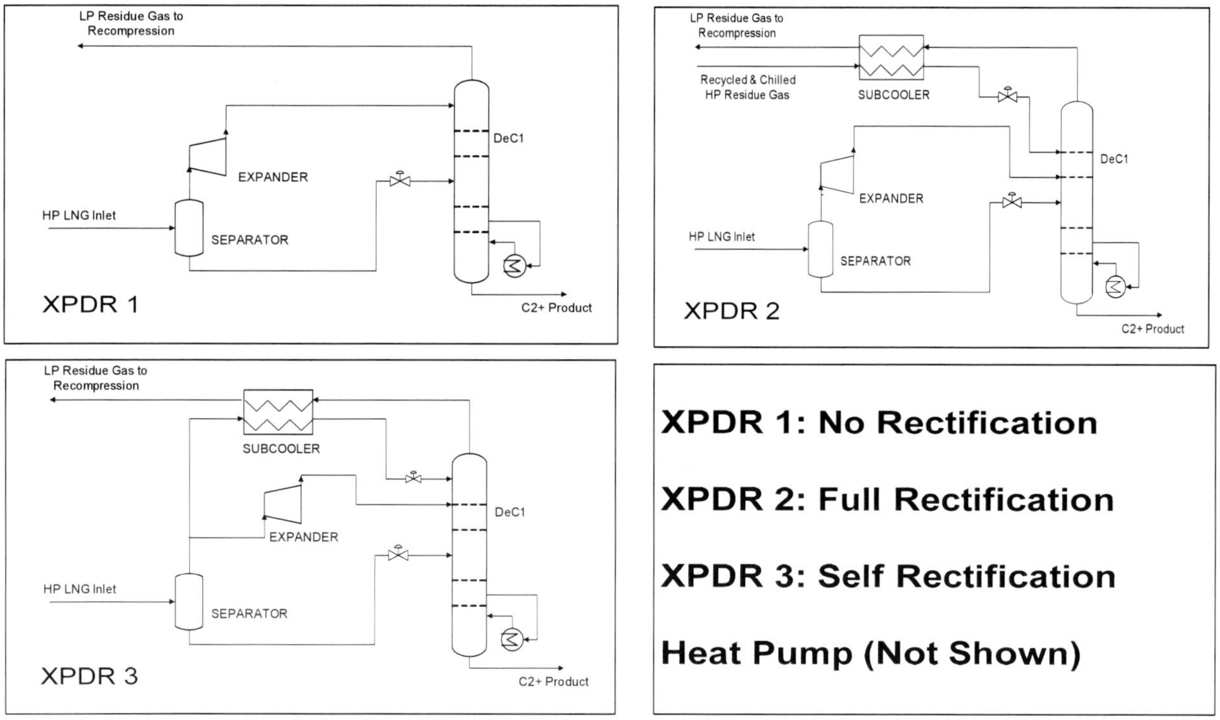

Figure 3 Categorizing expander-based schemes

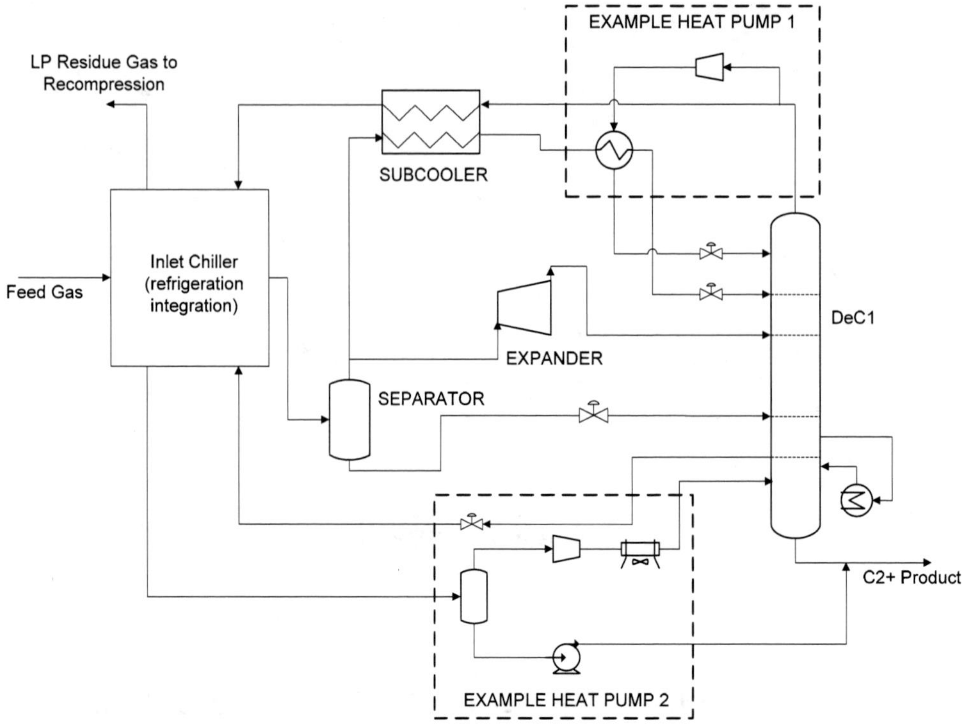

Figure 4 Example heat pumps embedded in XPDR 3

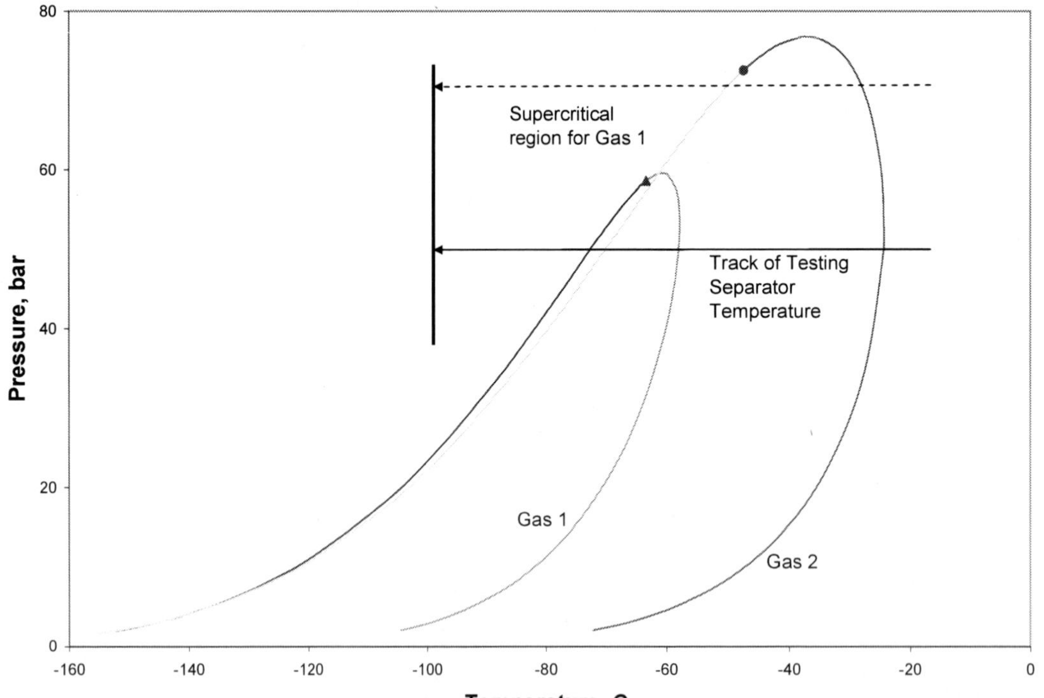

Figure 5 Procedure definition of SIMAR

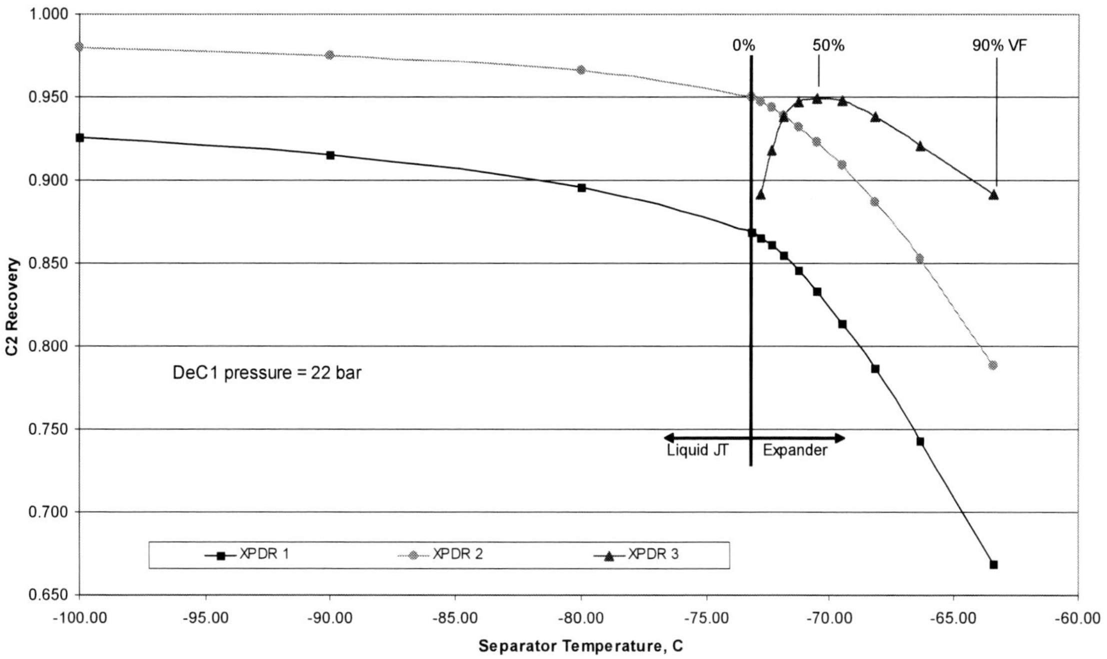

Figure 6 Two types of behavior for different categories. VF stands for vapor fraction.

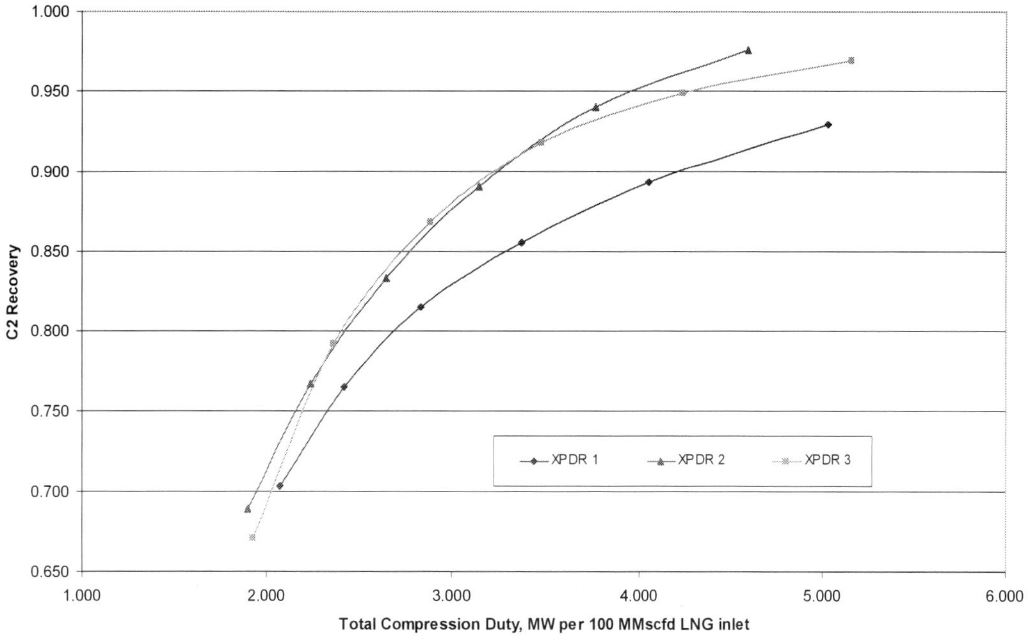

Figure 7 Comparing SIMAR curves for three XPDR categories, LNG feed

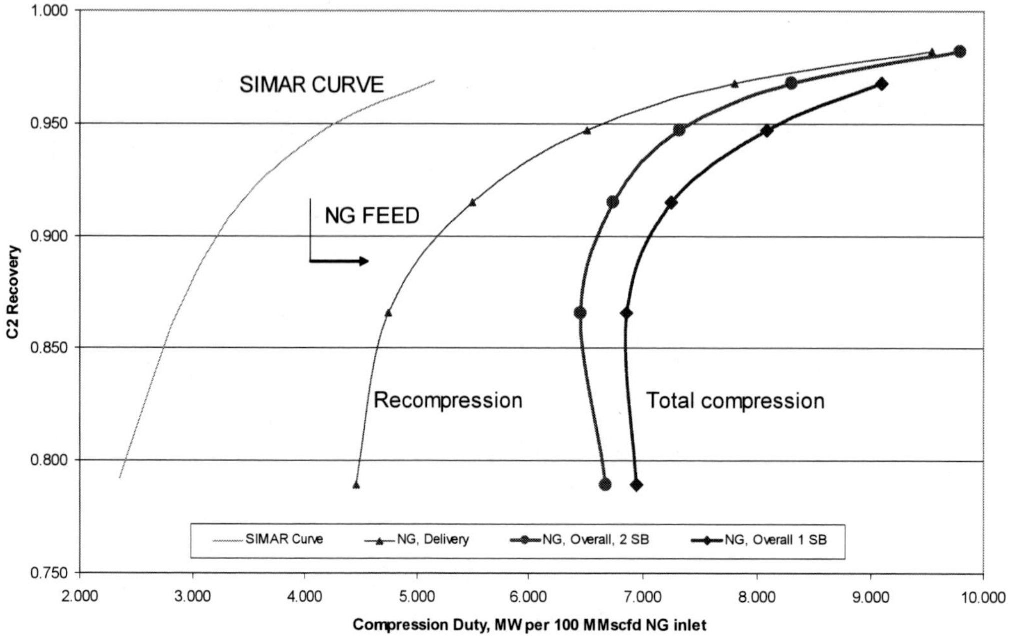

Figure 8 Comparing compression duties and Impact of side-boilers (SB) based on XPDR 3

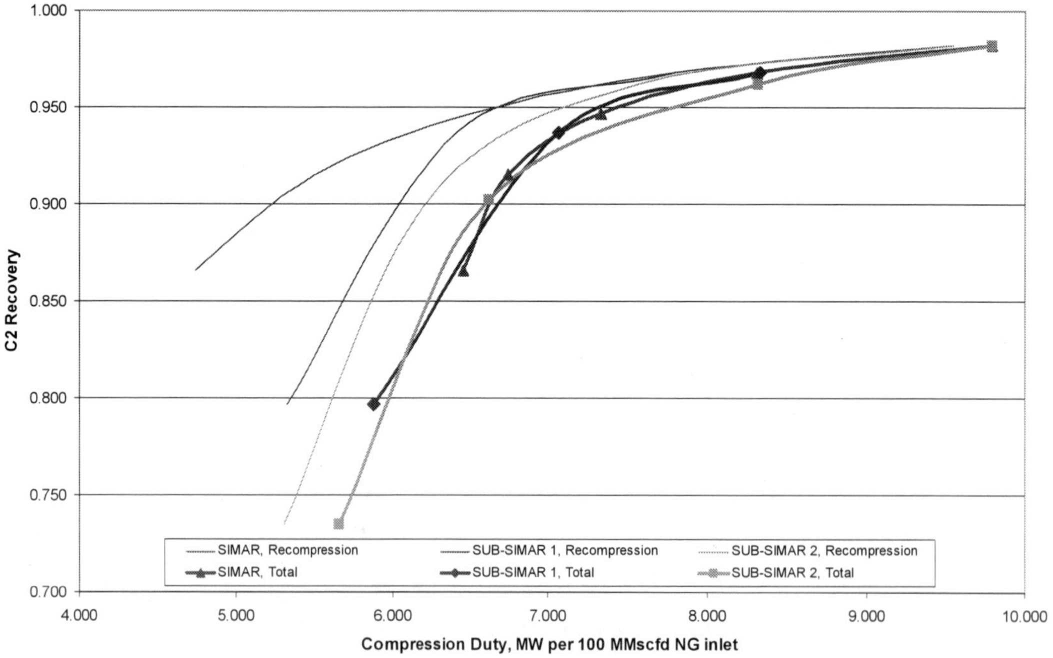

Figure 9 Sub-SIMAR operations for Gas 1

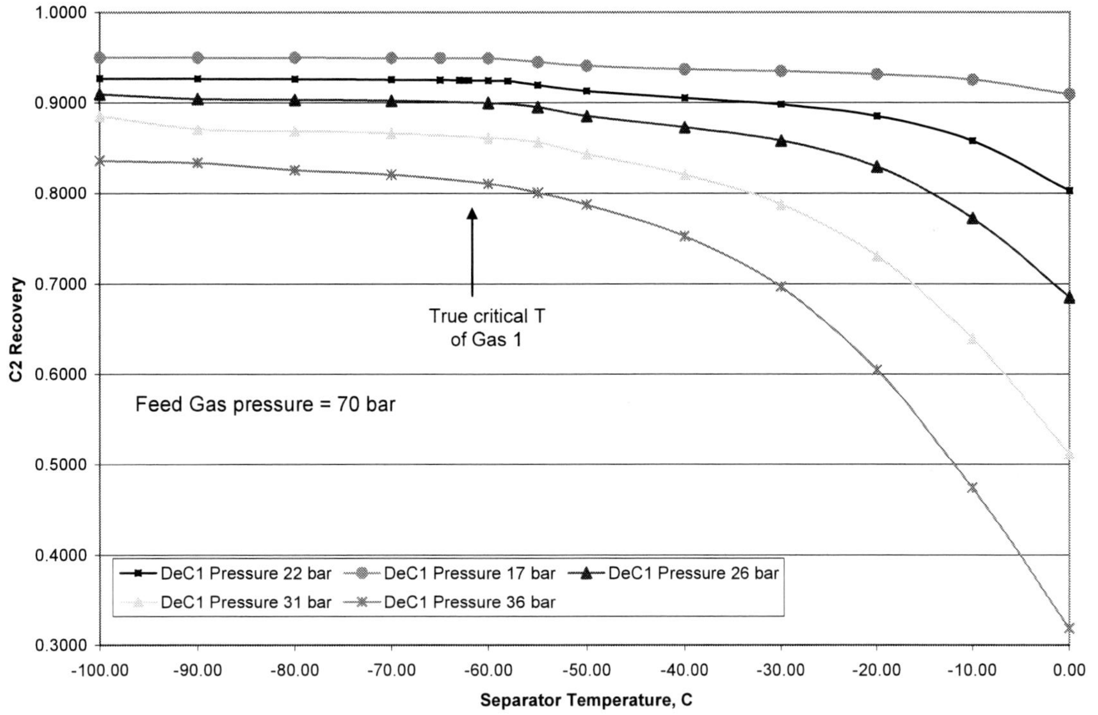

Figure 10 Constructing SIMAR in supercritical regions

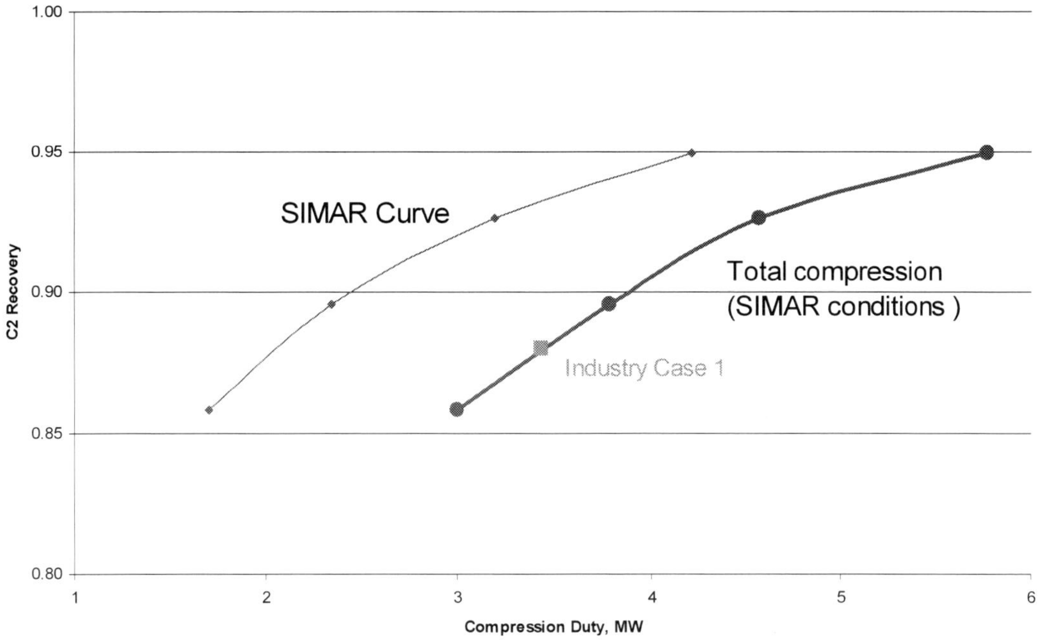

Figure 11 SIMAR operations for Industrial Case 1

Biography

1600 Smith St. Houston, TX 77002, USA
Phone: 713-754-4657, Fax: 713-754-3968, Cell: 713-859-0847
E-mail: cmyengle@chevron.com

Carlos Yengle is a Lead LNG Engineer for Chevron Corporation, He resides in Houston, TX. He has expertise in Process Design, and Steady and Dynamic Modeling and Simulation, including gas processing and liquefaction. Carlos began his professional career in the Process Engineering group at Bechtel Corporation, moved into LNG Technology Development with Chevron Corporation. He holds a BS in Chemical Engineering from "Universidad Nacional de Trujillo, Peru" and an MS in Chemical Engineering from University of South Florida, Tampa

LNG Pump Applications with Variable Speed Motor Controls

Jeff Lovelady

Chief Engineer – Rotating Equipment
Aker Kvaerner
Houston, Texas

AIChE Spring Meeting, April 2008
New Orleans, LA USA April 6 - 10, 2008

Introduction

Due to the increased demand in natural gas, a number of new LNG re-gasification terminals will be constructed in the U.S. These terminals store LNG and re-gasify to supply gas to the gas transmission lines where the demand of natural gas varies from season to season and also during the day resulting in variation of pipeline delivery pressure and export gas rate from the terminal.

Cryogenic pump applications running in LNG vaporization or liquefaction services typically run on fixed speed motors and produce a single flow versus head curve at that speed. Control of these pumps is accomplished by throttling the discharge pressure with a control valve or bypass of a portion of the discharge flow to a secondary flow stream. Pumps that are run at a fixed speed have a single Best Efficiency Point (BEP) at which the pump is designed to run at the rated production point. Pump flows that are throttled by reduction of the discharge pressure run off of the BEP.

Efficiency of pump operation at flow points that are off design can be improved by varying the pump speed to a point on the pump hydraulic curve which is closer to the BEP for the desired flow. Pump hydraulic geometry is designed with to a specific speed that meets the required flow and head at a predetermined speed. The BEP pressure for a fixed geometry is a function of the impeller speed and the required discharge flow. The variable frequency drive (VFD) is used to control the speed of the induction motors applied in LNG pump service. Variation of the pump speed allows the required head and flow to be achieved with better efficiency than that of the same pump run on a fixed speed curve.

Pump Design:

Cryogenic pump designs in LNG service today are the submerged motor type running 2 or 4 pole motor designs at a fixed speed. The motors are directly coupled to the pump section with no intermediate gear for speed changes. Performance characteristics of the pump components are selected by considering the head and flow needed at the rated condition as specified by the process design. The manufacturer selects the impeller / diffuser combination based on the best efficiency at or near the rated flow point. Each hydraulic combination has an optimum speed which by design matches either a 2 or 4 pole motor speed (1800 or 3600 rpm @ 60 Hz). All pumps are required to meet multiple performance points that are off of the rated point to meet changing liquid properties or production requirements. Reaching these off-design conditions requires that suction or discharge conditions be changed to meet the required performance since the speed and hydraulic design are fixed. The performance parameters of each hydraulic combination of impeller and diffuser are characterized by non-dimensional coefficients in order to compare design performance as a function of pump operating conditions such as speed, flow, or head.[(1)] Table 1 lists the most common non-dimensional numbers used in hydraulic performance evaluation.

Table 1 – Pump Hydraulic Non-dimensional Numbers

Flow Coefficient	$\varphi = Q/ND^3$
Head Rise Coefficient	$\psi = g\Delta H/N^2D^2$
Power Coefficient	$\overline{P} = P/\rho N^3 D^5$
Efficiency	$\eta = \varphi\ \psi/\overline{P}$
Specific Speed	$Ns = NQ^{1/2}/(g\Delta H)^{3/4}$

Where: Density – ρ (kg/m^3); Head Rise – ΔH (m); Impeller Diameter – D (m); Flow Rate – Q (m^3/s); Rotational Speed – N (rpm); Shaft Power – P (W)

One important point to note from these coefficients is that while the flow is directly proportional to the speed, the power is proportional to the cube of the speed. There is still some loss at the discharge control valve but the reduction in discharge head, from the pump, at the lower speed results in lower system losses across the valve.

One distinct advantage that the pump designer has in selecting hydraulic combinations for variable speed drives is to use speeds higher than that of the synchronous line speed. Higher operating speed can reduce the number of stages which results in an improvement in efficiency of approximately ½ percent per stage that is removed. The improvement in efficiency due to de-staging the pump is limited to approximately 3-4 percent. This limit is due to the additional losses incurred by the higher re-circulation losses due to the increase in the head produced by each stage. Most LNG plant operations require that the pump be capable of operating between 30 to 120 percent of the rated flow point. The discharge pressure is required to have a continuous rise from the maximum to minimum flow due to the pressure controller on the discharge flow. The pump design can accommodate these flow points much more efficiently by varying the speed thus eliminating the need for the continuously rising head curve. Variation in the running speed allows the pump design to be optimized by operating closer to the best efficiency point (BEP) at each speed – flow point. Improvements in efficiency due to impeller – diffuser speed optimization are approximately 2 percent.

Figure 1- Cut-away of a High Pressure Send-out Pump Mounted in the Suction Vessel

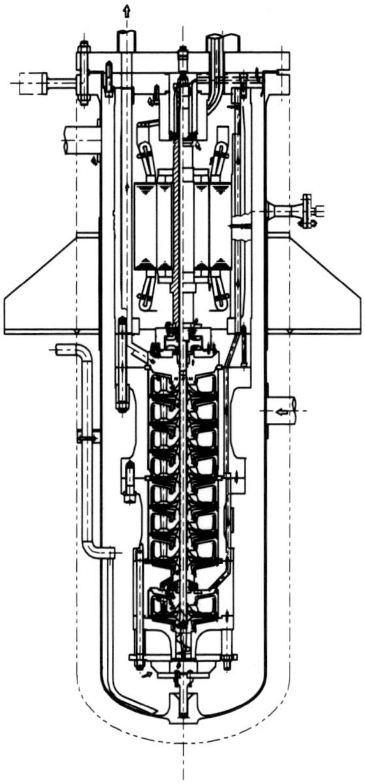

System Description:

There are two different applications for LNG pumps in re-gasification facilities as shown in Figure (2). Low pressure (LP) in-tank pumps, lowered into columns in the storage tank, deliver LNG to the suction header of the high pressure (HP) send-out pumps.

The in-tank pumps are designed to provide the required flow rate to the send-out header at an optimum suction pressure for the HP send-out pumps, typically between 8 to 11 barg. Variable speed control allows the pump to develop a lower net positive suction head (NPSH) at reduced speed. Lower NPSH permits pump down of a storage tank to much lower levels to reduce the heel during a shut-down. Reduced pump down levels lower the operating level of the tank thus reducing the overall require height of the tank. Speed changes also allow the flow to be increased to meet process requirements or to increase the plant capacity. The discharge pressure from the pump has to be controlled to meet process requirements as the speed is adjusted. Design of the trim diameter for the rated condition must consider the range of speed of operation in order to have proper head control at the desired flow rates.

Start-up and pump-down operations of the LP in-tank pumps benefit by running at lower speeds during these operations. Output pressure does not need to meet the send-out pump requirements during these operations; therefore, a reduced speed can be run resulting in a reduction in the pump down (or start-up) liquid level requirements by as much as 50 percent.

The HP send out pumps supply liquid to a vaporizer arrangement at a pressure sufficient to meet the gas pipeline supply header requirement. Raising the pressure, of the natural gas, in the liquid state is much more cost effective than that required by a pipeline compressor downstream of the vaporization units. One of the difficulties in the use of VFD's in LNG service is the fact that there is a reduction in head with a turn down in speed to reduce the flow. In order to maintain the required send-out supply pressure the downstream control valve is needed to reduce the flow which will result in a rise in discharge head. The reduction in speed can then be used to reduce the flow and bring the head back down to that of the rated discharge head; conversely, the any increase in flow results in an increase in head with a corresponding change in speed. The pressure adjustments that are necessary due to the system pressure requirements are less than 2 % of the pump pressure at the adjusted speed, compared to up to 8% with a fixed speed machine. Finding the optimum pressure - flow condition for a VFD driven system requires an iterative approach balancing the flow control valve position with the required reduction in pressure and flow from the pump. Improvements in efficiency range from 3% at low flow to less than one percent as you approach the rated point with equivalent hydraulic designs. Improvements in the hydraulic design based on the use of a VFD arrangement can improve the efficiency by as much as 4% at the rated condition.

Figure 2 – LNG Land Based Re-gasification Block Flow Diagram

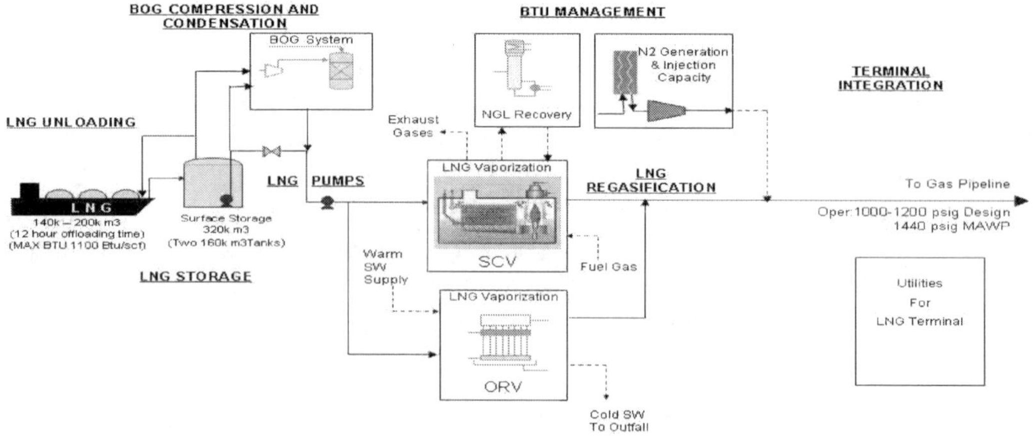

A fixed pump speed control scheme is based on throttling of the discharge liquid in the discharge line from the pump in order to meet the required flow at the resulting pressure on the pump head curve as seen in Figure (3). Since the speed is fixed the only method for controlling the flow (at fixed suction conditions) is to close a control valve in the discharge line until the required flow is met. The black line indicates the rated condition with a fixed speed machine as well as the full rated speed condition of a VFD motor arrangement. The hydraulic characteristics of the fixed speed pump result in a higher pressure, at reduced flows. The reduced speed conditions are indicative of the same hydraulic geometry run at lower speeds for equitable comparison purposes. Reduction of the head by decreasing the speed results in a 3-5% savings in power.

Figure 3 - Performance Characteristics of Fixed and Variable Speed Pumps

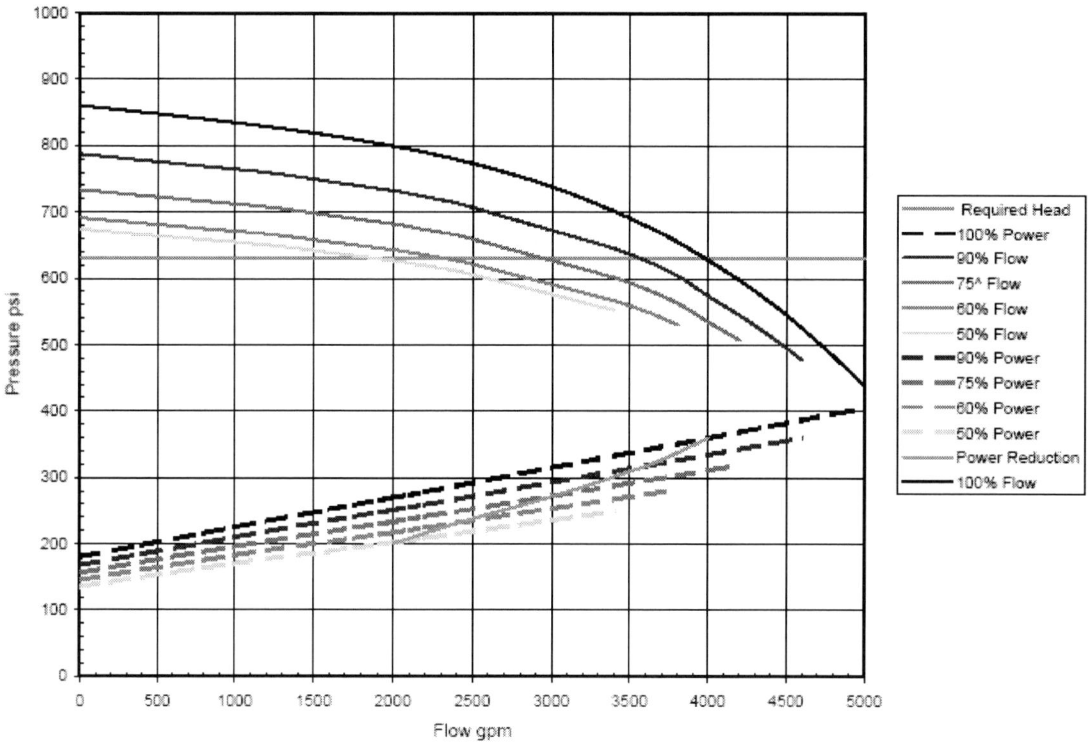

Electrical and Mechanical Benefits:

Fixed speed motors that are started with direct power from the electrical grid undergo abrupt increases in torque and starting current due to the starting inertia of the pump. The pump motor characteristics shown in Figure 4 are indicative of the start of all fixed speed motors. The start-up current for the motor shown is 1180 amps compared to the operating current requirement of 600 amps. The motor curve also indicates the high transient torque that the pump sees during start-up. The high transient torque results in high unit loadings on the bushings and bearings. Damage to bearing components usually occurs during start-up and shut down torque excursions. Initiation of pump operation at a lower frequency allows the motor to be "soft started" thus reducing the mechanical and electrical load on the pump units. A direct cost benefit is the elimination of the motor starter with VFD drives. The cost of the VFD units is approximately 50 percent more than that of the motor starters when the VFD is packaged and run in parallel for all of the pumps in the train.

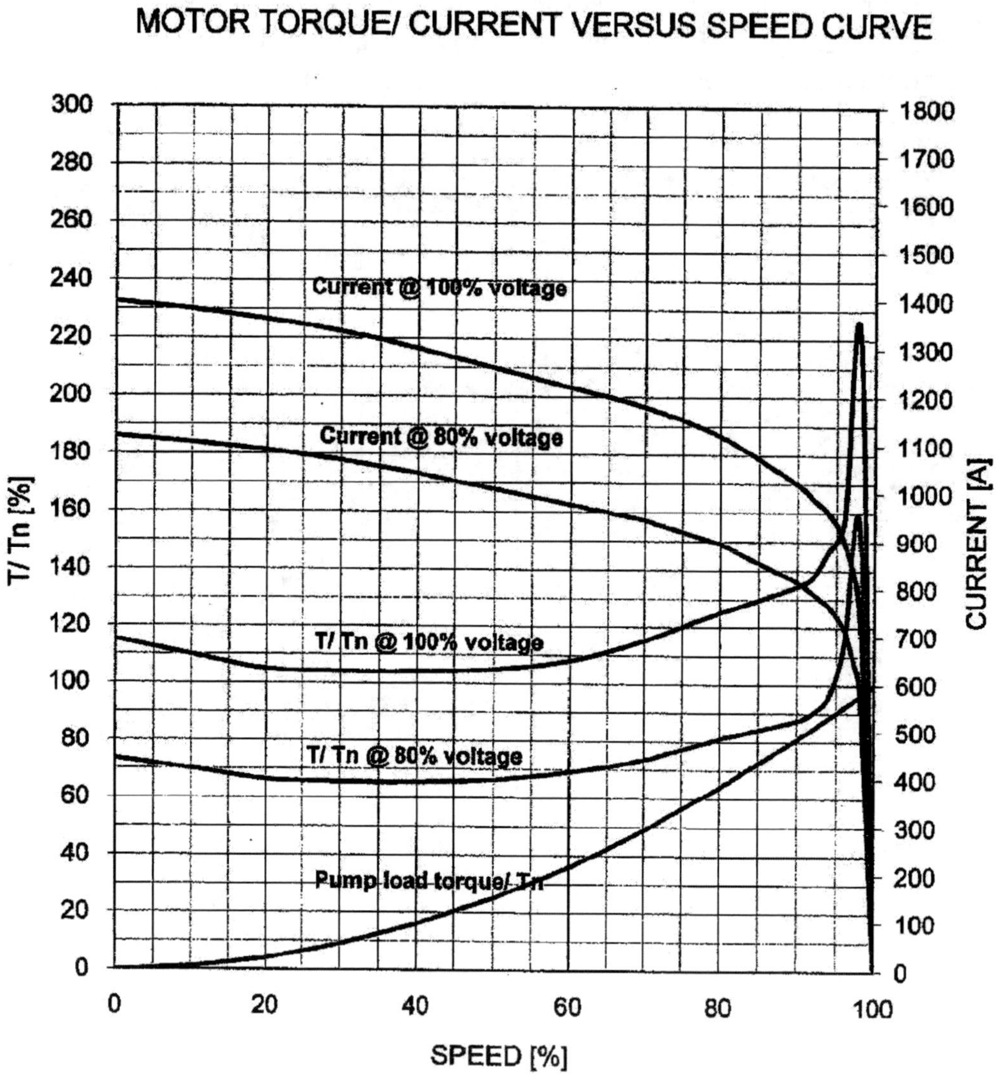

Figure 4 Start-up Motor Current

Starting the in-tank pump at a fixed speed results in rapid fill of LNG into the discharge column and downstream piping. The resulting transient pressure rise can lead to piping or component damage if the piping bypass and start-up procedures are not properly designed. A simple solution to the transient pressure problem is to start the pump at a low speed thus reducing the initial head until the piping is full and recirculation in the discharge bypass has begun. Running the in-tank pumps at reduced speed, during the terminal cool down, also results in a lower overall consumption of LNG due to the lower recirculation rate in the bypass lines.

Conclusion:

Variable speed operation of cryogenic pumps results in improved operating costs with more effective control of the output flow and head. Table 2 below shows a summary of the efficiency improvements that can be made by designing the LNG pump system to run with Variable Frequency drives.

Table 2 – Pump System Efficiency Improvements

Efficiency Improvement Description	Percent
Pump De-staging	3
Impeller – Diffuser Matching	2
Flow –Speed Control	5
Motor and Drive Electrical	1
Total	11

The additional costs of speed controls can be offset by the energy savings as well as reducing the number of units required by running the rated point at a speed greater than the synchronous line speed of most fixed speed motors.

Initial costs of the VFD drive units can be offset by a re-design of the system taking advantage of the elimination of the starter units and a reduction in the number of piping components required for recirculation flow control. The operating cost improvements through better efficiency make the cost justification even more practical.

Efficiency improvements offer a direct economic impact to the operator but a significant side benefit is the flexibility of the flow and discharge pressure that can be accomplished by varying the motor speed. The degree of improvement is dependent on the system design and redundancy requirements but the conservative estimates shown above can be seen in most applications. The use of VFD units offers a cost effective improvement to LNG pump operating costs and improvement in reliability.

Acknowledgments:

I would like to acknowledge the help that Mr. Fabien Wahl of Nikkiso Cryo Inc. provided in verifying the cost and performance details of the VFD pump configurations.

References Cited

1 Stepanoff, A.J., Centrifugal and Axial Flow Pumps: Theory, Design, and Application, Second Edition 1992.

2 Neumann, B., "The Interaction Between Geometry and Performance of a Centrifugal Pump", 2005

Biography of Speaker

Jeff Lovelady is the Chief Engineer for Rotating Equipment, Aker Kvaerner Inc., Part of the Aker Group, Houston, Texas. He has been working with Aker Kvaerner, Inc. for 3 years establishing standards for rotating equipment purchase, managing equipment design and integration for chemical and LNG facilities. Prior to joining Aker-Kvaerner he was the Chief Engineer at Ebara International – Cryodynamics Division. He was also in development and design engineering for dry screw and centrifugal compressor products at GE Oil and Gas, A-C Compressor. He attended Texas A&M University receiving his BS and MS degrees in Mechanical Engineering.

E-mail: jeff.lovelady@akerkvaerner.com

Considerations for Ambient Air Based Technologies for LNG Regasification Terminals

Kamal Shah, P.E, Technical Vice President
Judy Wong, P.E., Process Manager
Bill Minton, P.E., Process Consultant
Aker Kvaerner, Inc.
Houston, Texas

AIChE Spring Meeting, April 2008
8th Topical Conference on Natural Gas Utilization
New Orleans, Louisiana, April 6-10, 2008

Introduction

Increasing energy demand with declining domestic production has resulted in increased energy cost. Most current operations at U.S. LNG terminals utilize submerged combustion vaporizers (SCV) or fired heaters in a closed loop intermediate fluid system with specially designed shell and tube heat exchangers to re-gasify LNG. These gas fired systems utilize between 1.3 to 1.6% of the inlet LNG as fuel, reducing the natural gas available for the market and generating environmental emissions as combustion products. Increased energy costs associated with recent developments in environmental regulations have challenged LNG terminal developers to look to ambient air based technologies instead of traditional technologies based on fuel gas firing. The socio-political directives in the state of California and the agreement reached by the northeast signatory states have also pushed for regulations towards greener technologies.

Various papers[1,2] published previously presented differences between the fuel gas fired SCV, open rack vaporizers (ORV) utilizing sea water and specific ambient air based technologies. It is generally well understood that the recovery of low grade heat from ambient air to re-gasify LNG results in lower fuel consumption and operating cost. In addition to the savings in operating cost, the reduced fuel gas use will result in lower air emissions making the overall terminal more environmentally friendly. It is also generally understood that the capital cost for ambient air based technologies is higher because of the duplication or addition of equipment to supplement heat during lower ambient temperature conditions to provide the needed reliability for terminal operations. The warmer climate in the U.S. Gulf Coast and Pacific West Coast regions results in higher thermal efficiencies and better payback using ambient air based technologies. Various innovative technologies have been used successfully in other cryogenic industries. Other ambient air based technologies have been developed recently to demonstrate the overall savings in fuel gas and reduced environmental emissions. This paper summarizes various ambient air based technologies and provides a comparison of thermal efficiencies for the two main ambient air based technologies, limitations, and other design considerations including operations in low ambient temperature conditions.

Ambient Air Technologies

Various approaches have been proposed recently for use of ambient air as a source of heat to re-gasify LNG. The approaches can be categorized into two basic methods for recovering low level heat from ambient air. They are indirect contact and direct contact. The indirect contact approach utilizes a closed loop intermediate fluid that is heated with ambient air and transfers the heat from this intermediate fluid to re-gasify LNG in specially designed shell and tube heat exchangers. The direct contact approach utilizes ambient air directly to re-gasify LNG in an exchanger designed for cryogenic fluids.

Two methods using indirect contact have received a great deal of attention lately. One of the methods is Air-Fin Exchangers. This method utilizes traditional air-fin exchangers with forced draft fans to push the ambient air through the air fin tubes to heat the intermediate fluid which is used to re-gasify LNG. The second method is generally referenced as Air Tower or reverse acting cooling tower where induced draft ambient air heats water in direct contact with circulated water which can then be directly or indirectly transferred to re-gasify LNG. Both indirect contact processes are described in further detail in the subsequent section.

There are two types of direct contact ambient air vaporizers. One uses a Natural Draft Ambient Air Vaporizer (NDAAV). The design of this exchanger utilizes the natural draft from the difference in density for warm and cold ambient air. Ambient air enters the top of the exchanger, exchanges heat directly to the cold LNG fluid flowing in the exchanger's specialized finned tubes and the air cools producing a down draft. The cooler air then flows out through the bottom of the exchanger. The second method uses a Forced Draft Ambient Air Vaporizer (FDAAV), and utilizes a similar exchanger design except fans are installed on the ambient air vaporizers to force a higher mass of ambient air across the fins of the exchanger for direct heat exchange with cold LNG. Because of the similarity in exchanger design, only the FDAAV process is further described in the subsequent sections.

Indirect Contact Ambient Air Technologies

One of the two methods described above is the Air-Fin Exchanger based system. Figure 1 shows a schematic of this process where ambient air is utilized as the heat source to re-gasify LNG via an intermediate fluid. The intermediate fluid (such as propylene glycol, propane, aqueous potassium formate based low temperature heat transfer fluids[3], etc.) is circulated in a closed loop system which flows inside the finned tubes of the exchanger. The warm air is forced outside the tubes using fans allowing heat transfer from the warm ambient air to the intermediate fluid. The cold air as well as the condensed water outside the tubes flows in a top to bottom down flow direction perpendicular to the tubes. The warm intermediate fluid flows to the LNG vaporizers, which are specialized low temperature differential shell and tube exchangers, to re-gasify LNG. The cold intermediate fluid flows back to the Air-Fin Exchangers for reheating. The process generally is provided with a supplemental heating system to provide heat during low ambient temperature conditions where the available ambient heat supply is not adequate to vaporize the design throughput. The source of supplemental heat and size of the system can vary widely depending on the variation in ambient conditions, availability of other waste heat sources in the terminal or the configuration of the system which may be optimized for the specific LNG terminal.

If the expected ambient temperatures are below 35°F, the supplemental heating system may require 100 percent redundancy to maintain design throughput. The overall intermediate fluid system requires a large circulation rate because of large amount of heat transferred through a low temperature differential. The cold fluid temperature is maintained above 30°F to avoid formation of ice outside the Air-Fin Exchanger tubes, which would restrict the overall temperature differential between the warm and the cold intermediate fluid. Higher humidity and higher air temperature enhances the overall performance of the system. The process produces significant amount of water due to the condensation of moisture from the air providing additional heat input to the system. The water produced is clean, making the process more environmentally friendly. Plot space for the ambient air based system requires special attention. The ambient conditions utilized for design of the system to overcome a certain lower ambient temperature differential may lead to a larger plot space requirement and increased capital expenses. The air-fin exchanger design for lower temperature ambient conditions and its supplemental heating system (many times requiring 100% redundancy) should be studied carefully for overall plot space and capital cost optimization.

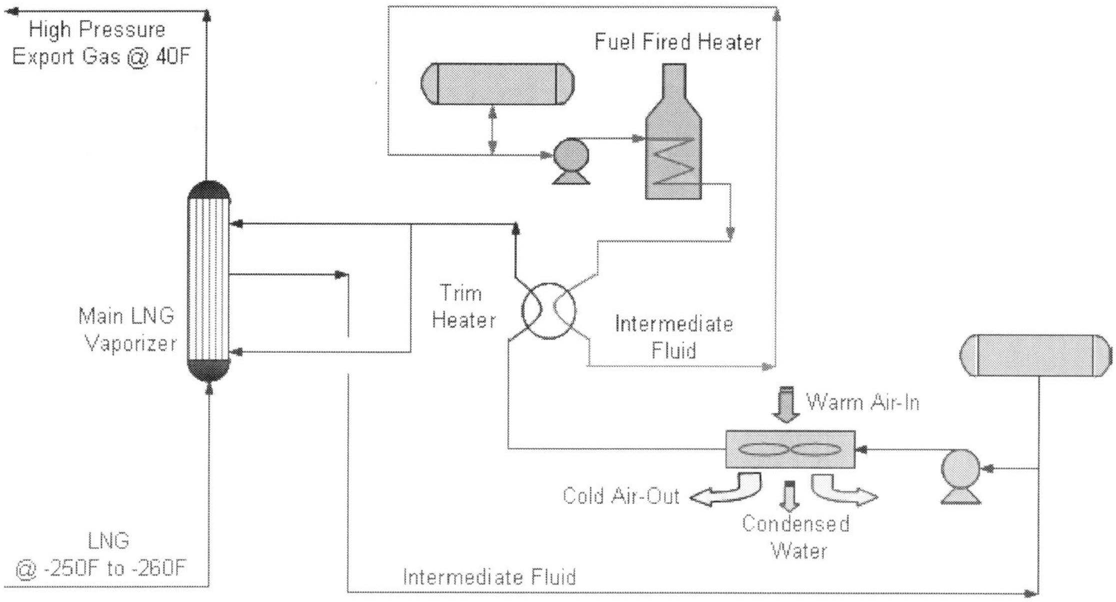

Figure 1 – Air-Fin Exchanger Based LNG Re-Gasification System

The terminal site meteorological data play an important role in optimizing the process. The selection of the intermediate fluid is also important because of freeze up concerns inside the shell and tube LNG vaporizers.

Such a process has been in operation since 2004 at a base load LNG terminal located in Dahej on the west coast of India. The warm ambient air temperature above 80°F in most seasons at this location provides ideal conditions for this system. Figure 2 provides a photograph showing part of this terminal and overall general arrangement of air-fin exchangers.

Figure 2 – Dahej LNG Terminal (Courtesy of IHI)

The other indirect contact ambient air technology is known as Air Tower or reverse acting cooling tower. Figure 3 provides a general schematic for the overall system. In this process, the air tower exchanges heat from the warm ambient air to the cold circulating water in a direct contact air tower. The warmer air from the lower part of the tower enters the side of the air tower and moves up where it comes in direct contact with the cooler water flowing down the tower. The moisture in the air will condense as it becomes colder. Similar to the air-fin exchanger process described above, the condensation of moisture provides a significant contribution to the total heat duty transferred from the air to the water. The cold air exits the tower through an induced draft fan located on top of the tower. The closed loop circulating warm water system will transfer heat to another intermediate fluid system. The intermediate fluid system design is similar to the air-fin exchanger based system described above. The warm intermediate fluid vaporizes the LNG in the specially designed shell and tube vaporizers and returns cold fluid to the water / intermediate fluid cross exchanger. This intermediate system is designed with fuel fired heaters to supplement heating requirements during low ambient temperature conditions.

Similar to the air-fin exchanger system, the terminal site meteorological data is very important in the optimization of plot plan. The circulating water is in direct contact with ambient air and may dissolve some carbon dioxide and salt due to the large water circulation in the air-tower system. Special corrosion and microbial growth related issues must be taken into account during the design. The condensed water produced from moisture in the air is generally clean and can be used for beneficial purposes, making the process more environmentally friendly.

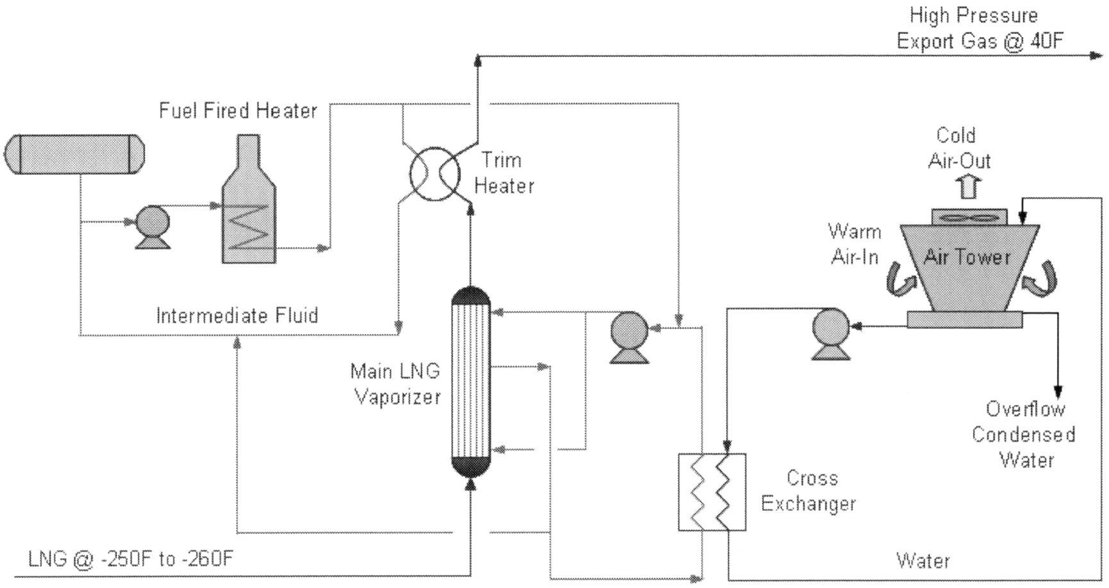

Figure 3 – Air Tower Based LNG Re-Gasification System

One LNG terminal using the Air-Tower design is Freeport LNG Terminal located near Freeport, Texas and is scheduled for start up later this year[4]. This system is designed with a 100% back-up for low temperature ambient conditions.

Direct Contact Ambient Air Technologies

In this type of vaporizer, warm ambient air enters from the top of the exchanger, exchanges heat directly to the cold LNG flowing in the specialized finned tubes to vaporize LNG. The air is cooled and flows downward along the outside of the finned tubes and leaves from the bottom of the exchanger.

Figure 4 below shows the radial heat transfer mechanism in a direct contact ambient air vaporizer, the surrounding ambient air begins to cool while giving up its sensible heat to the cryogenic LNG.

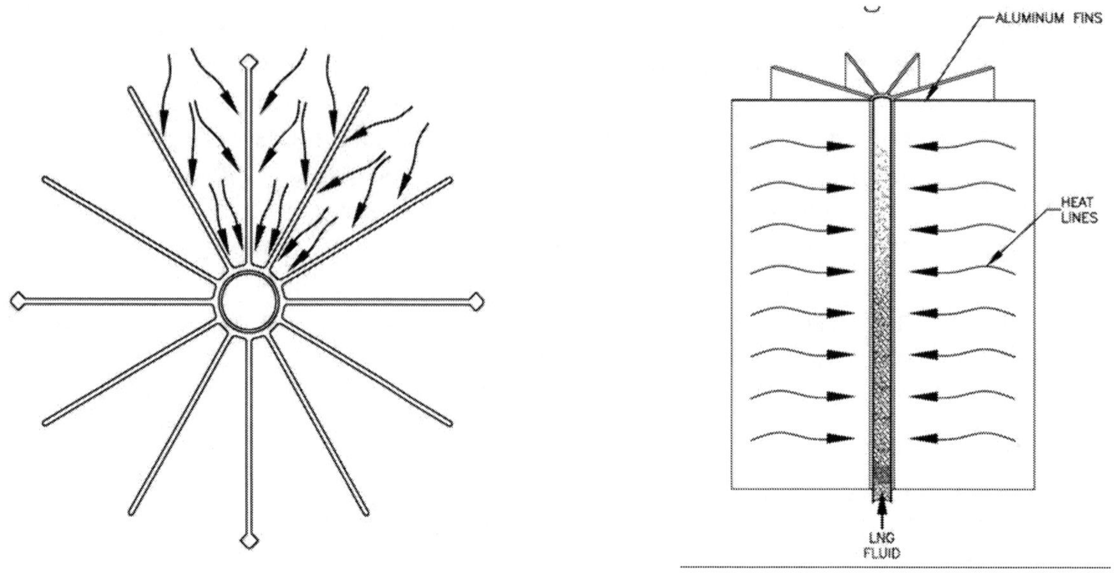

Figure 4 – Finned tubes showing heat transfer mechanism
(Courtesy of Cryoquip)

Forced Draft Ambient Air Vaporizers (FDAAV) are equipped with fans installed on top of the vaporizers. They also have shrouded sides providing more capacity for a given shell to force a higher mass of air through the shell for heat exchange with cold LNG. The forced air units require power to operate the fans, but results in a more compact layout. Figure 5 shows FDAAV units and the finned tubes arrangement.

Figure 5 – FDAAV Units and Finned Tubes Arrangement
(Courtesy of Cryoquip)

As the air flows down the vaporizer and transfer heat through the fins to the LNG flowing in the tubes, it cools and reaches its dew point and starts to condense water from the moisture in the air. As the temperature drops further, the condensed water forms a snow-like frost on the vaporizer cooling fins. This frost or ice formation process supplies additional energy to the system from the heat of fusion. As frost accumulates on the fins, it slowly reduces the performance, requiring the unit to be regenerated when it reaches a predetermined heat transfer level. The unit is regenerated by switching it off to allow for melting of the ice. In order to maintain the overall availability of the LNG vaporizer capacity, additional units are installed to overcome vaporizer regeneration activity. At a given time approximately 30% of the total units operate in regeneration mode.

Figure 6 shows a general process schematic of this process. The export gas outlet temperature in the system will vary based on the system design approach from site ambient conditions. This approach may vary between 20°F to 30°F depending on the design of the FDAAV system. During low temperature ambient conditions, the export gas temperature can fall below the desired pipeline specification. A supplemental heating system is provided for low ambient conditions to maintain the export gas temperature.

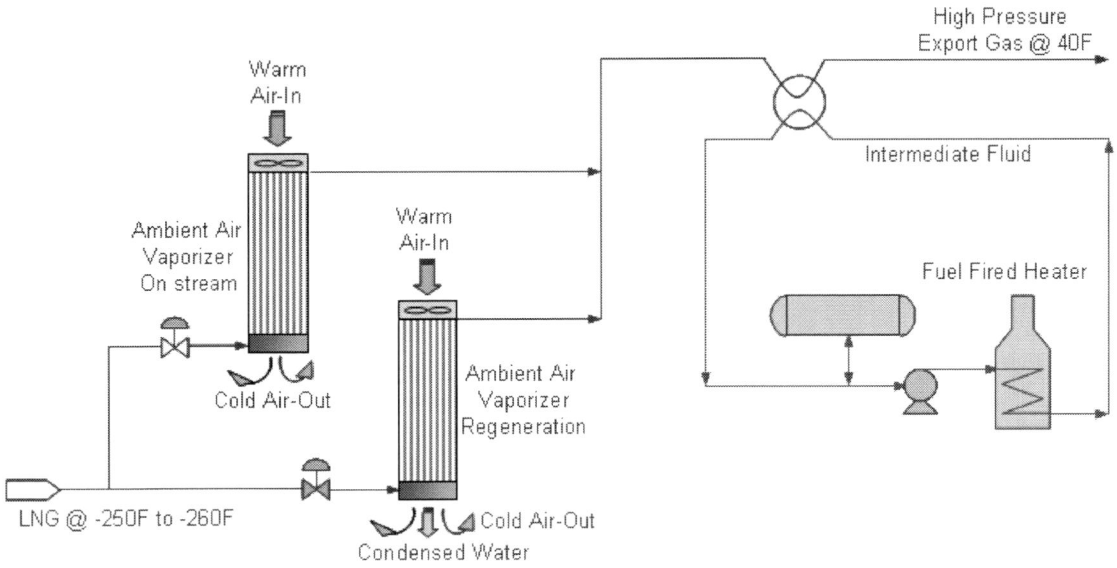

Figure 6 – Forced Draft Ambient Air Vaporizers (FDAAV) Based System

FDAAV units have been used for heating cryogenic liquids in smaller sizes over three decades. Significant progress has been made in increasing capacities over the last 15 years. Each of the FDAAV units shown in Figure 5 are grouped into a bank of 4 cells and have shrouded sides with generally 4 small fans on the top of each cell to provide higher capacity than the NDAAV units. The units are 40 feet tall. Full scale testing for a number of such units is planned at Cheniere's Sabine Pass LNG facility in Louisiana later this year.

One of the advantages of direct contact ambient air vaporizers over the indirect contact ambient air vaporizers (Air Tower or Air-Fin Exchanger technology) is their ability to continue to transfer significant amount of sensible heat from the ambient air to cryogenic fluids even when the ambient conditions fall below 32°F or below freezing temperatures[5]. Obviously, the regeneration of the FDAAV requires some means to heat the air to bring temperatures above 32°F during these low ambient conditions.

Figure 7 shows a schematic of such a process. A heating coil can be installed in each of the FDAAV to heat the ambient air from low ambient conditions to above 32°F to allow regeneration of the units. The moisture content in the air during the low ambient condition is low and it takes much longer to accumulate ice on the vaporizer fins. This allows for longer switching cycles and provides plenty of regeneration time. A closed loop intermediate fluid system provides the needed heat input to the air heating coils for regeneration. The amount of external heat input required to regenerate reduces the thermal efficiency, but overall thermal efficiency is still significantly higher and provides much higher overall fuel savings compared to indirect contact vaporizer technologies.

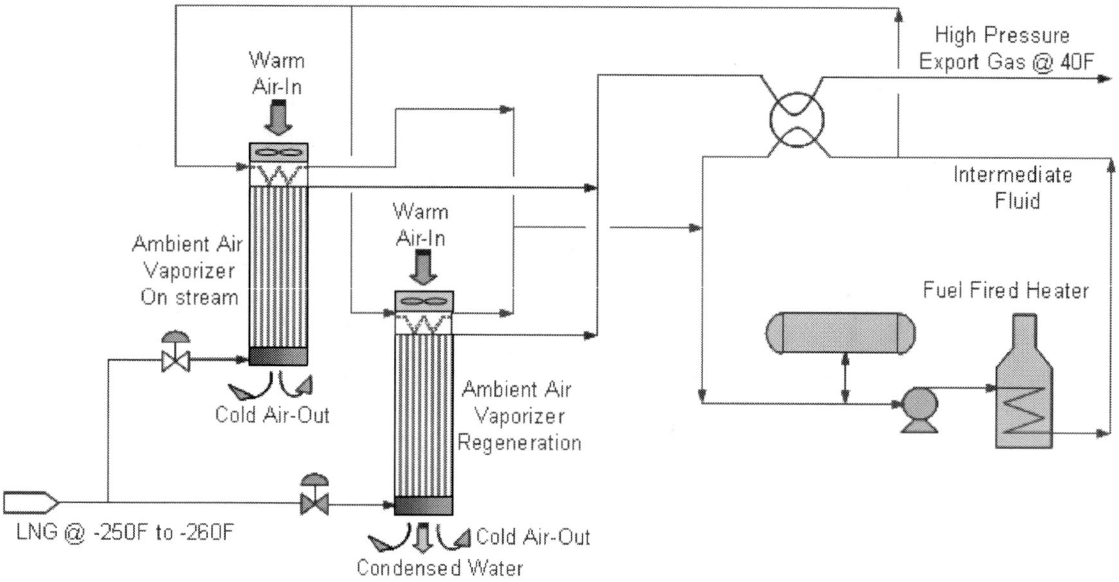

Figure 7 – FDAAV System with Air Heating Coils

The thermal efficiencies of direct contact FDAAV are generally higher than the indirect contact ambient air technologies such as Air Tower or Air-Fin Exchanger vaporizers. An internal study was carried out using a specific FDAAV design point with ambient air at 70°F and 80% relative humidity (RH). Various thermal efficiencies were calculated to determine the effect of ambient temperature and % RH. The design point is considered at 100% thermal efficiency where no supplemental heat is required.

Figures 8A and 8B provide the performance. The performance shows the effect of ambient temperature and relative humidity on overall thermal efficiency for FDAAV. The difference in thermal efficiencies between the direct contact ambient air technologies (i.e. FDAAV) and the indirect ambient air technologies (i.e. Air Tower, Air-Fin Exchanger) is from the fact that the direct contact FDAAV relies on a single approach temperature compared to multiple approach temperatures in the indirect contact ambient temperature technologies. Also, the direct contact with cryogenic LNG provides a higher overall log mean temperature differential (LMTD) providing significantly higher sensible heat in comparison to indirect ambient air technologies, even at lower ambient temperatures and relative humidity.

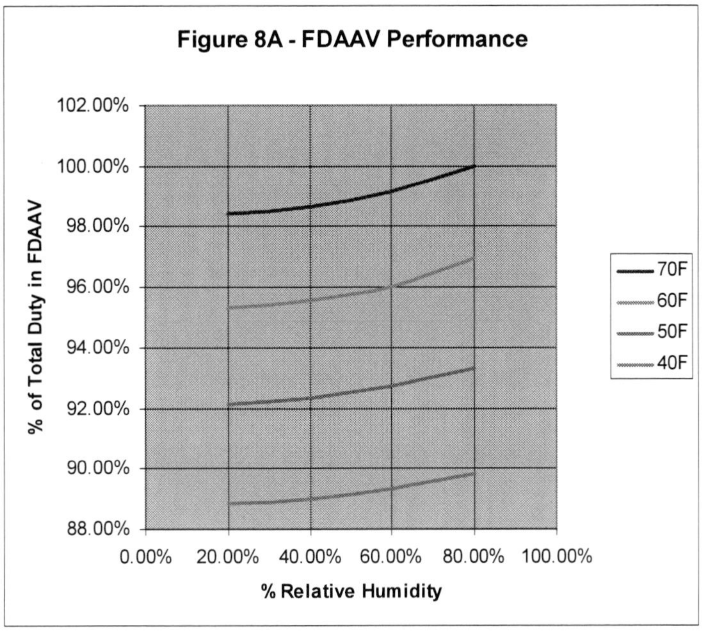

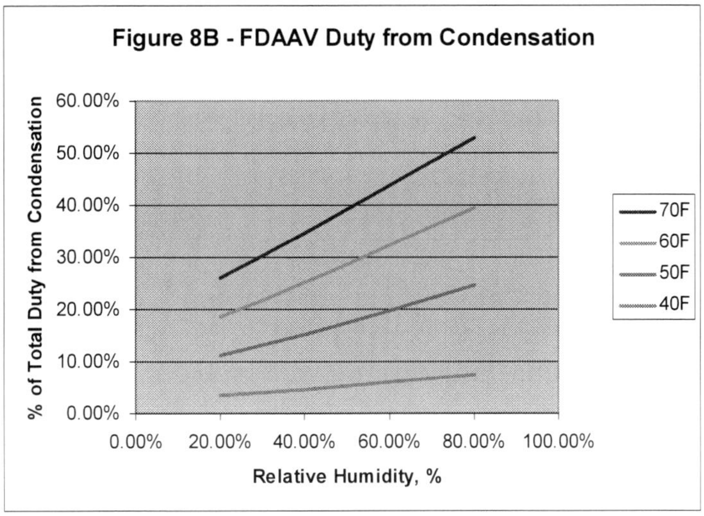

Comparison of indirect and Direct Contact Ambient Air Technologies

Various internal studies, input from equipment suppliers and recent publications[6,7,8] were utilized in the comparison of direct and indirect ambient air vaporizer technologies. For the specific comparison, Air-Fin Exchanger based LNG vaporization for indirect contact ambient air technology and FDAAV based LNG vaporization for the direct ambient air technology was evaluated. The design point for comparison was 1 BSCFD (Billion Standard Cubic Feet per Day), ambient air at 70°F and 80% RH, with export gas delivered at 40°F to the pipeline. The design point is where both processes provide adequate equipment to meet 100% duty with ambient air without any supplemental heat requirement.

Figure 9 shows a chart for comparison of thermal efficiencies between the Air-Fin Exchanger based vaporizers representing the indirect contact and FDAAV representing the direct contact ambient air technologies.

Figure 9 - Thermal Efficiency Comparison

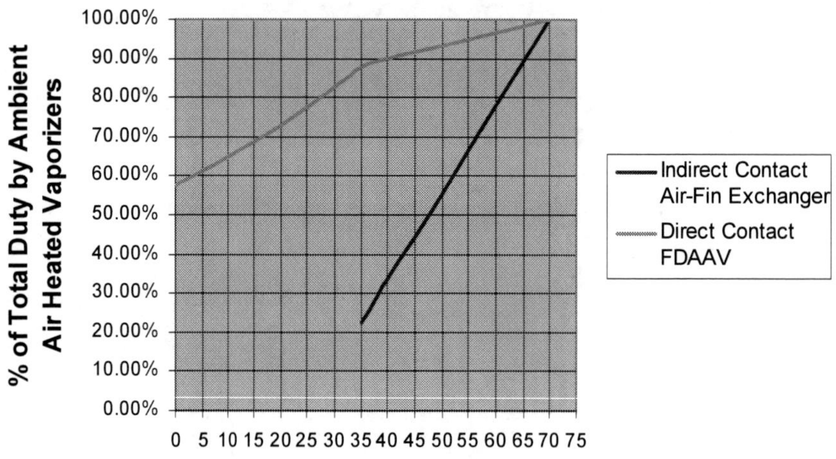

The thermal efficiencies for the indirect contact Air-Fin Exchangers based technology declines significantly as the ambient temperature falls below the design point. The relative humidity is kept constant at 80%, but the amount of moisture available in the air reduces significantly with lower ambient temperatures and so does the condensation heat available. The sensible heat available is also reduced significantly with lower than design ambient temperature. Conversely, the direct contact FDAAV based technology will maintain significantly higher sensible heat available due to the large LMTD available.

The overall thermal efficiencies for the indirect contact Air-Fin Exchanger based system approaches zero once the ambient temperature reaches less than 35°F, requiring 100% duty from the supplemental heaters. The FDAAV process is always forming ice which is melted by ambient air during the regeneration cycle. Below freezing temperatures, ambient air vaporizers cannot be regenerated except by heating the air flowing into the air heating coil located below the fan to approximately 35°F to 40°F. This supplemental heat or additional energy input into the system reduces the overall efficiency of the FDAAV process and is the reason for the break in the performance line shown in Figure 9. The direct contact FDAAV based system efficiency still remains high even at the lower ambient temperatures below 35°F.

The overall plot areas were compared for these two technologies in a previously published paper[9] and our internal studies indicate similar conclusions. The area required for direct contact FDAAV based technologies is much lower compared to the indirect contact Air-Fin Exchanger based technology. Our estimate suggests the area required by FDAAV based direct contact vaporizer system can be 30 to 50% lower. The FDAAV provides much higher capacities and a more compact design with 40 feet tall units, requiring less plot area.

The overall capital cost for all ambient air based LNG vaporizer systems is higher in comparison to SCV or fuel fired based technology. The ambient air based technology cost has been estimated and published in various papers referenced above. Our internal estimates for various clients indicated capital cost for ambient air based LNG vaporization system vary between 1.4 to 2.0 times the cost of SCV based vaporizer system.

The capital cost for indirect or direct ambient air based technology is in direct relation to the design ambient conditions, the availability of vaporizer capacity during the lowest ambient conditions and the level of redundancy provided. The estimated capital cost difference between the direct ambient air based FDAAV system and Air-Fin Exchanger based vaporizer systems for the Gulf Coast location with the design point conditions listed above shows small difference in favor of direct contact FDAAV based technology. Both systems were designed with 100% supplemental heat duty for winter operations. The expected capital cost difference increases, in favor of FDAAV based technology, with lower ambient temperature conditions as the lower available LMTD, results in a larger intermediate fluid circulation rate and size of associated equipment. It should be noted that even at temperatures below freezing, it is not necessary to install 100% supplemental heat duty for the direct contact FDAAV system when installed with air heating coils needed for regeneration. It will still maintain high overall thermal efficiency during winter months when temperatures are approaching 0°F.

The overall operating cost difference between the indirect and direct contact ambient air vaporizer system is dependent on the specific site meteorological conditions. The higher thermal efficiency for the direct contact FDAAV system results in a lower energy consumption. The fan horsepower requirements for the direct contact FDAAV is approximately 10-15% higher than Air-Fin Exchangers, but the intermediate fluid circulation system for indirect contact Air-Fin Exchanger system is much larger requiring higher circulation pump horsepower making up for the difference in fan horsepower. The overall net energy requirement for the direct contact FDAAV based system indicates savings of 35% to 50% over the indirect contact Air-Fin Exchanger based vaporizer system for a specific Gulf Coast location.

As indicated earlier, the indirect contact Air-Fin Exchanger system thermal efficiency approaches zero when the ambient temperatures fall below 35°F, while the direct contact FDAAV based system can provide significant overall thermal efficiency and fuel savings by adding air heating coils further increasing fuel savings.

While the application of FDAAV in colder ambient conditions is a positive aspect and a major difference between the indirect and direct contact vaporizer systems, there are issues that must be taken into account in design, where the site location can expect snow or ice accumulation which restricts the exit air from the unit. The temperature of the cold air exiting the unit can vary between 35°F to below -45°F depending on the ambient conditions. This is important from the design of the surrounding equipment and facility that can be exposed to this cold air. Depending on the site conditions, issues related to fog generation from the colder air exiting from the unit requires CFD analysis for mitigation. In general, these issues can be mitigated with proper safety analysis followed by appropriate design.

The direct contact FDAAV system requires the switching of units for regeneration on a regular basis requiring many actuated on-off switching valves. The operation is similar to industrial molecular sieve or Pressure Swing Adsorption units. Because of the number of these valves in operations constantly, reliability and maintenance operations must be carefully evaluated to maintain desired availability of the system.

Conclusion

Various indirect and direct ambient air based LNG vaporization technologies have been described with their benefits and limitations. All the ambient air based technology described is either operating or is being built and waiting to start-up in a near future. The industry experience in this area is expanding because of increased energy costs and regulatory changes. All ambient air based technologies will have positive environmental impacts by the reduction in fuel consumption and greenhouse gas production. All the ambient air heated vaporizer systems produce clean water by condensation of moisture in the air that requires very little or no treatment and can be used for beneficial purposes.

The capital cost for implementing any ambient air based system is higher compared to the conventional SCV based or fuel gas fired system. The difference in capital cost is slightly lower for direct contact FDAAV system compared to the indirect Air-Fin Exchanger based system for a Gulf Coast area location. The capital cost difference is expected to increase with lower than design point ambient conditions listed above, favoring the direct contact FDAAV based system.

The overall thermal efficiencies and energy savings for direct contact FDAAV based technology is higher compared to the indirect contact Air-Fin Exchanger based technology. The difference in thermal efficiency, between the direct contact FDAAV based system and the indirect contact Air-Fin Exchanger based system, increases when the ambient conditions fall below the design point, favoring the direct contact FDAAV based technology.

References

1. M.J Rosetta, B.C.Price, and L. Himmelberger, "Optimize energy consumption", Hydrocarbon Processing, January 2006.
2. J. H. Cho, R. Femat, H. Kotzot, C. Durr, "Optimization of ambient air vaporization processes", AIChE Spring Meeting, 6th Topical conference on natural gas utilization, Orlando, Florida, April 23-27, 2006
3. "CPTherm™ F-Series bulletin", CPI Engineering Services, Inc. Midland, Michigan
4. R. Nanda, J. Rizopoulos, "Utilizing air based technologies as heat source for LNG vaporization", 86th GPA annual conference, San Antonio, Texas, March 11-14, 2007
5. P. Billman, D. Granger, L.Himmelberger and S. Wiley, "LNG vaporization turns green", 87th GPA annual conference, Grapevine, Texas, March 2-5, 2008
6. R. M. Sharp, N. Baudat, F.P. Ross, "An overview of Mustang Engineering's LNG Smart® vaporization technology" AIChE Spring Meeting, 6th Topical conference on natural gas utilization, Orlando, Florida, April 23-27, 2006
7. M. Soudek, "LNG projects hinging on vaporizer technologies, presentation at 5th annual conference Ecology and & Environment, New York, September 2006
8. M. Soudek, "Mustang in midstream, LNG Smart™ vaporization presentation, California, December 2006
9. J. Davies, "Understanding ambient LNG vaporizers", AIChE Spring Meeting, 6th Topical conference on natural gas utilization, Orlando, Florida, April 23-27, 2006

Biography of Speaker

Kamal Shah is a Technical Vice President, Aker Kvaerner Inc., Part of the Aker Group, Houston, Texas. He has been working with Aker Kvaerner, Inc., Houston office for 20 years and has been involved in the process design, management and technology development in many oil and gas and energy projects. Prior to joining Aker Kvaerner, he was chief process engineer at Key Engineering, Inc. in Houston, Texas. He attended Michigan State University and received his BS and MS degree in chemical engineering. He is a registered professional engineer in Texas.

E-mail: kamal.shah@akerkvaerner.com

Optimized Vaporization Process with Unfavorable Design Conditions

In-Soo Chun
Senior Executive Vice President

Seungchul Lee
General Manager

Hyundai Engineering Co., Ltd.

AIChE Spring Meeting, April 2008
8th Topical Conference on Natural Gas Utilization
New Orleans, LA, April 6-10, 2008

ABSTRACT

There are several types of vaporization schemes which have been commercially developed, such as Open Rack Vaporizer (ORV), Fired heater with Shell and Tube Vaporizer, Submerged Combustion Vaporizer (SCV), Intermediate Fluid Vaporizer, Ambient Air Vaporizer, etc. Selection of the right vaporizer system is the most important process in LNG Terminal design as the regasifying costs contribute major portion of an LNG terminal operation.

High reliability with low operating costs of the regasifying system is a key parameter for a successful operation of an LNG receiving terminal. Incheon LNG terminal, one of the biggest LNG terminals in the World, has successfully optimized vaporization facility for the most unfavorable weather conditions. The terminal's vaporization involves ORV, SCV and seawater heater (SW HTR).

Seawater temperature conditions at Incheon terminal during winter is often below the ORV's design temperature. However, during the same time the gas demand reaches at its peak. The operation flexibilities and production capacity reduces extremely under this unfavorable conditions.

Optimized vaporization has been achieved by combination of SCV and ORV with SW HTR. SW HTR enables seawater to be used as a heating medium, even though its temperature is below the design temperature. Without seawater heating by SW HTR, ORV may not be operated. Even it can be operated with lower seawater temperature, vaporization capacity is quite limited.

This paper reviews ORV performance and optimum heat recovery temperature under harsh seawater temperature. This paper discusses practical optimization of vaporization with the unfavorable seawater conditions in winter. Economic comparison between conventional vaporization and optimized vaporization, which are under operation, is also discussed.

INTRODUCTION

Inchoen terminal, which is the second LNG terminal after Pyeong-taek terminal, was planned to provide base load of natural gas for the Seoul metropolitan area. Gas demand in the metropolitan area is higher than any other areas in the country. The main advantage of this terminal is to shorten the gas transportation distance resulting in lower transportation cost. However, there are several constraints in the terminal design. These include site preparation and temperature conditions of seawater, which is used as vaporization heat source. The terminal was built on a reclaimed site, where the tide level difference is quite high (about 10 meter from high-high tide and low-low tide).

In the winter season, seawater temperature may go down well below the design temperature of ORV. The warmer cooling water at a higher temperature coming out from the steam condenser of a thermal power plant can be used for the vaporization of LNG. Hence, the advantage is using the seawater for re-gasification in winter. However, the terminal is not integrated with any other thermal power plant and thus the advantage of thermal integration can not be used in designing the vaporization system.

The west coast of Korean peninsula, facing Yellow Sea, has low seawater temperature because of cold ocean current in winter. This cold seawater temperature is one of significant design constraints in optimizing the vaporization facility.

Low temperature does not allow to operate the ORV, which have the lowest operating cost. It results in the lowest life cycle cost compared with any other conventional vaporization scheme. The design conditions forces to use other vaporization options, such as Submerged Combustion Vaporizer (SCV), which uses a small portion of vaporized LNG as fuel gas. Since it burns the vaporized LNG, the operation cost is normally high.

This paper discusses the detail design constraints in designing the vaporization system and methodology of optimization of LNG vaporization with the unfavorable seawater conditions. The paper also presents operation results of this vaporization configuration with economic analysis.

DESIGN CODITIONS

Gas Demand Projection

Gas demand projection used for the basic design and actual gas supply are tabulated in Table 1 [1,2]. The gas demand projection established at the beginning of this project is compared with the actual gas supply records [2,3]. The design basis for the phase I of the terminal was 5 Million Tonnes per Annum (MTPA) of LNG. It was considered that the balance will be provided by the first terminal (Pyeong-taek) and third LNG terminal (Tong-Young).

Table 1 – Projected Gas Demand and Actual Gas Supply Data

(Unit: 1000 t/y)

Year	Planned Demand (1)	Acutal Supply (2)	Difference (%)
1992	3,390	3,524	3.8
1993	3,580	4,402	18.7
1994	4,720	5,860	19.5
1995	6,020	7,087	15.1
1996	7,400	9,241	19.9
1997	8,110	11,325	28.4
1998	8,280	10,645	22.2
1999	8,930	12,961	31.1
2000	9,570	14,557	34.3
2001	9,830	15,990	38.5
2002	10,610	17,768	40.3
2003	11,390	18,610	38.8
2004	12,640	21,809	42.0
2005	13,350	23,350	42.8
2006	14,050	24,090	41.7

Notes

1. These are planned data, which have been used for basic design of Incheon terminal, based on gas demand projection in 1992 [1]
2. This is actual gas supply volume data provided by KEEI [2].

Since the terminal was designed for base load terminal, the monthly basis send-out gas load distribution was established during the basic design. This pattern also has been changed during actual operation. Figure 1 illustrates comparison of the monthly gas send-out distribution.

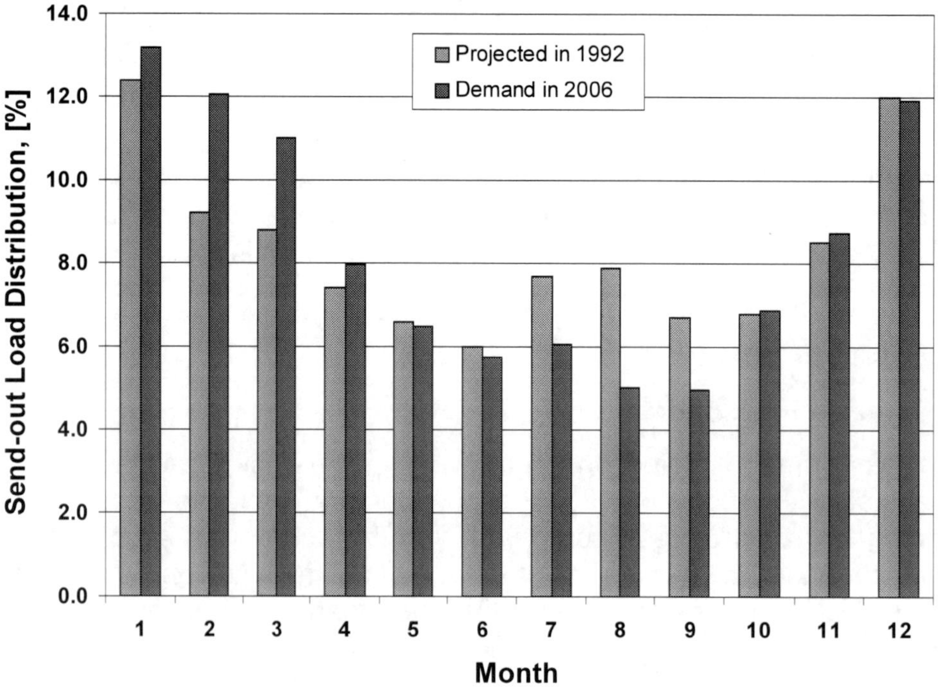

Fig. 1 – Monthly Send-out Gas Load Distribution in 2006 [1,2]

The send-out gas loads in winter season (January to March) are higher than the 2006's load projected in 1992. The projected load distribution estimates that the load distribution in summer season (July and August) would be higher by 1 - 2% than June or September. However, the actual monthly send-out gas load distribution is almost same in June and July. The main differences of send-out gas load are in February and March.

For winter season gas send-out load, it is generally affected by winter ambient temperature. If it is cold winter, the send-out load increases and it will be vice versa.

The vaporizer's economic analysis and optimization was performed based on the planned data at the beginning of the terminal Front End Engineering Design (FEED). The results have been applied to design configuration of vaporization.

Seawater Temperature Conditions

In winter season seawater temperatures from the intake facility are shown in Fig. 2. This water temperature is measured 2m below the seawater surface measured at 6:00 in the morning each day.

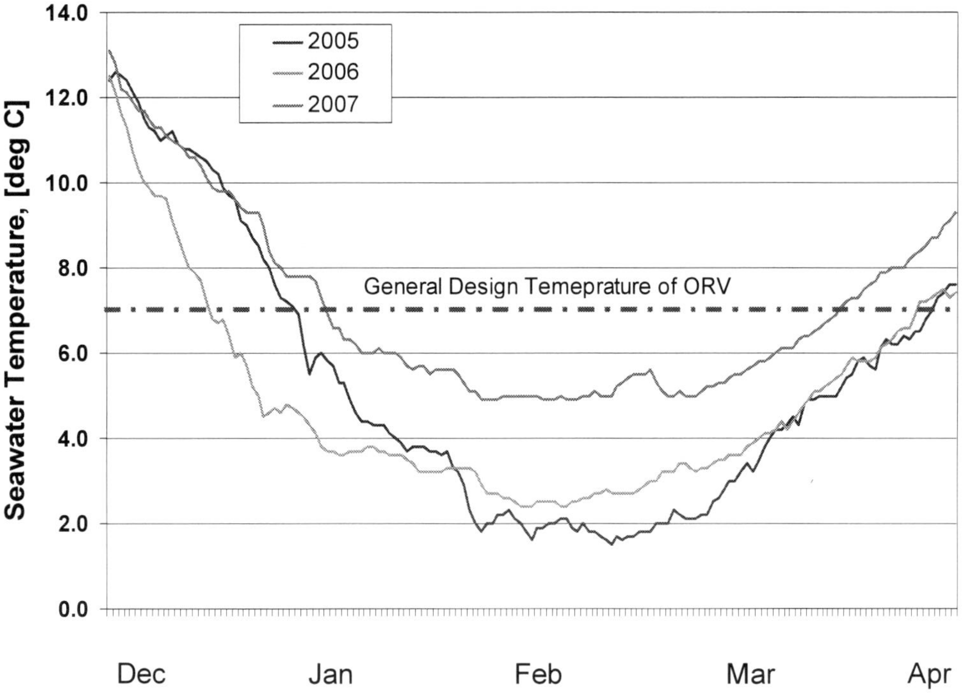

Fig. 2 – Seawater Temperature Profile in Winter Season [4]

Seawater temperature in winter drops below the possible ORV seawater design temperature ranges (5 – 7 deg C). In some cases, seawater temperature drops below the freezing point. When the seawater temperature is low, the gas demand is high. Actual supply records show that this has a quite firm relationship.

Generally, the design seawater temperature ranges between 7 to 9 deg C with 5 to 7 deg C temperature drop. In some cases, the design temperature drop will be higher than 7 deg C if the seawater temperature is above 9 deg C in winter. Since the seawater temperature is unfavorable at the site, the design temperature was evaluated based on operation economics.

Seawater Intake Facility

Seawater is lifted by the seawater pumps before it is supplied to the seawater vaporizer, ORV. Five pumps were installed during Phase I and subsequently four pumps were added during expansion Phases. The capacity of the seawater pumps are 10,000 m^3/h and 12,000 m^3/h for Phase I and expansion, respectively.

Arrangement of the seawater pump station is shown on Fig. 3. The seawater outfall is located opposite side of the terminal in order to avoid any cold seawater recirculation.

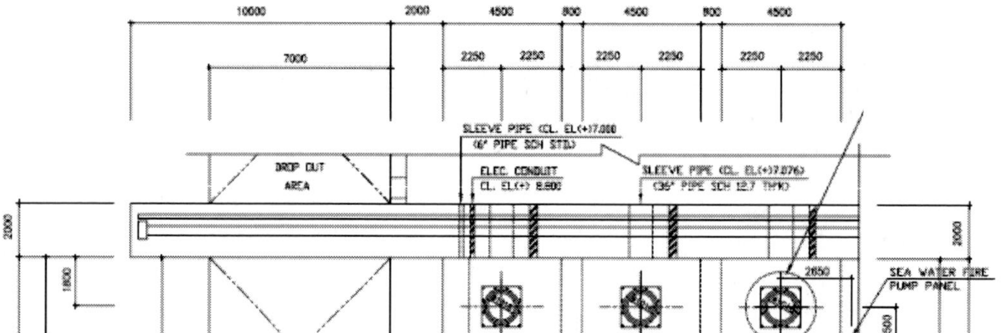

Fig. 3 – Seawater Intake Facility Layout

OUTLINE OF INCHEON TERMINAL PROCESS

Figure 4 illustrates the simplified process scheme of Incheon LNG terminal. LNG is transported by LNG tankers from various import sources. LNG is unloaded to the Terminal's storage tanks through unloading arms and unloading lines. The design unloading rate is 11,000 m^3/hr with three 16-in. full bore liquid unloading arms. At the initial facility design, return gas blowers were relocated from other terminal. Later it was found that the return gas to the LNG ship was sent back by a pressure difference between the tank and ship without using blowers.

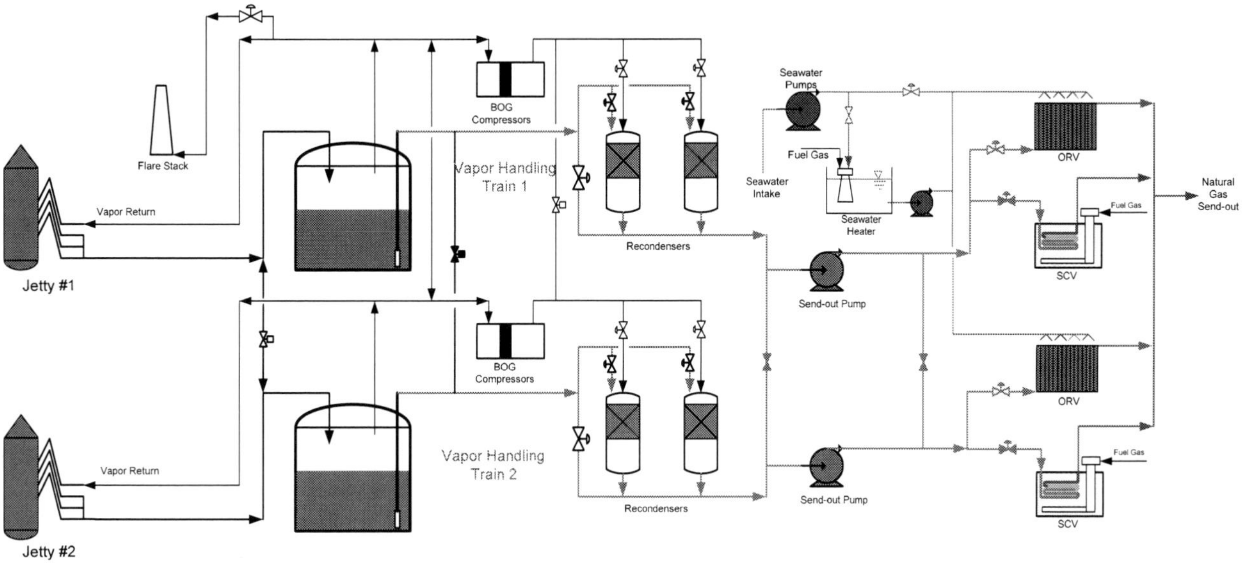

Fig. 4 – Incheon Terminal Process Scheme

The terminal has two berths allowing dual unloading or single unloading from either berth. This configuration also allows one ship unloading while another berth is on stand-by. The stand-by ship can perform vapor breathing operations to prevent over pressuring by extruding high pressure of vapor to onshore BOG header. It minimizes methane emission during unloading operation. Two berth systems can provide storage area zoning operation: LNG unloaded from one jetty can be stored in the above-ground tanks (8 tanks 100,000 m^3 each), and LNG unloaded from the other jetty can also be stored in 2 Above-ground tanks and 8 In-ground tanks (total 1.7 million m^3 including the 4th expansion). Intentionally, all stored LNG can be blended to minimize density difference between tanks and achieve even calorific values of send-out gas.

It is generally considered that one large unloading line will be cheaper than two equal-sized smaller lines. However, there is a very real advantage in dual unloading lines as in case one line to have any operational problems, then the other line is still available for unloading the ships at half capacity or higher. Also, most terminals require the recirculation of LNG in the unloading line during the period a ship is not unloading. This is normally required so that the recirculated LNG picks up the heat out of the terminal via the send-out. Since the terminal has two unloading lines (2 x 32-in.) from each berth, this configuration allows recirculation to be accomplished without the addition of a smaller second line.

At the initial stage, the terminal planned to build above-ground full containment tanks. As the terminal expanded, it needed more storage capacity to cover the demand gap between summer and winter as shown in Fig. 1. As tankage area extends, the required hydraulics increases at the ship manifold. Significant factors in determining hydraulics are tank static head and frictional losses across the unloading arms. Underground storage tank could alleviate the hydraulic limitations. A large in-ground storage tank makes use of high cost reclaimed island more efficiently than conventional smaller above-ground tanks. In-ground tanks allow adherent positioning of tanks compared to above ground tanks. This also saves plot space because of elimination of the dike area.

Boil-off gas (BOG) from storage tanks is gathered in a BOG header and pressurized by BOG compressors. The pressurized BOG is then routed to the recondenser where it is recondensed with sub-cooled LNG discharged from in-tank pumps. The recondensed liquid is further pressurized by the send-out pumps and re-gasified by vaporizers.

The BOG rates are 0.075%/day and 0.1%/day for above-ground tank and in-ground tank, respectively. The large storage volume makes considerably higher BOG rates than other terminals. Ten (10) BOG compressors are installed to handle the BOG during unloading operation.

Open Rack Vaporizers (ORV's) are the base-load vaporizers, and Submerged Combustion Vaporizers (SCV's) are used to back up ORV's during peak shaving operation. However, the available seawater temperature drops below the design points during winter. The ORV performance then decreases considerably as seawater temperature decreases. In order to compensate the seawater temperature drops, twelve (12) Seawater Heaters (SW HTR's) are operated.

OPTIMIZED VAPORIZATION AND TECHNICAL CHALLENGES

In order to achieve low cost vaporization and minimize impact on the environment, ORV's and SCV's are combined as a back up and peak shaving was chosen from various vaporization technologies. A seawater facility requires a long seawater intake pipe (about 1,500m) since the tide difference is high (about 5m – M.S.L.). The seawater discharge temperature drops to 0°C after it is used for LNG vaporization, which is generally not recommended for ORV operation specially during the winter. Figure 5 illustrates the vaporization scheme of the terminal. LNG from the recondensers from each train is fed to the send-out pumps and the pressurized LNG is routed to the vaporizers. Seawater is supplied to ORV's. When the temperature of seawater supplied to ORV's is lower than the design point, it is mixed with the heated seawater from the SW HTR's.

As shown in the seasonal gas demand pattern in Fig. 1, gas demand in the winter is higher than in the summer. However, the seawater temperature as a heat source is lower in the winter. Since LNG regasification is the most important parameter for lowering terminal operating costs, optimization of LNG vaporization has been performed against these unfavorable conditions.

The following parameters were investigated for the vaporization optimization:

- Performance of ORV with the lower seawater temperature;
- Optimum seawater operating temperature;
- Economic combination of ORV and SCV;
- Economic justification of seawater heating;
- Overall economic assessment of vaporization facility.

In the ORV, the relationship between the number of heat exchange panel blocks and seawater requirement has been investigated for optimization and found that ORV manufacturers recommended a seawater temperature difference based on the manufacturer's equipment optimization, which resulted in a less number of heat exchange panels. By analyzing the ORV performance curves with the various operating conditions, the relationship was extracted as shown in Fig. 6. The recommended seawater temperature differences are based on a 42MW heat duty (equivalent to 180t/h – 200t/h of LNG vaporization). Manufacturers confirmed that the minimum outlet temperature from the ORV was 1.5°C, which is obtained when seawater inlet temperature is 3°C.

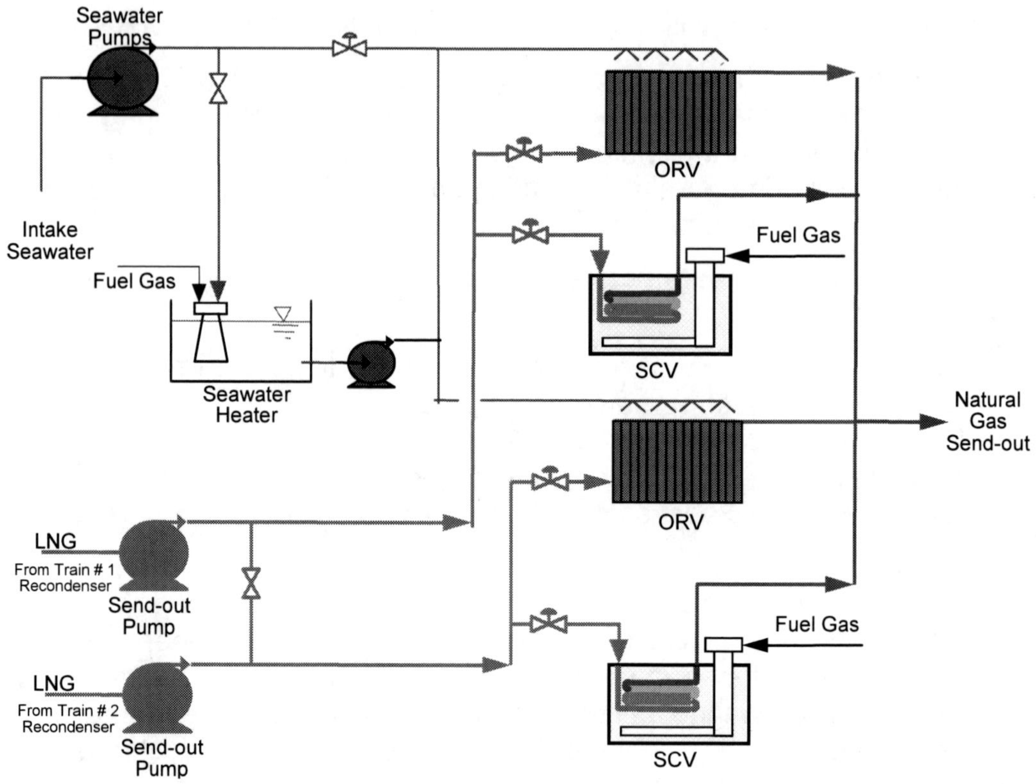

Fig. 5 – Vaporization Scheme

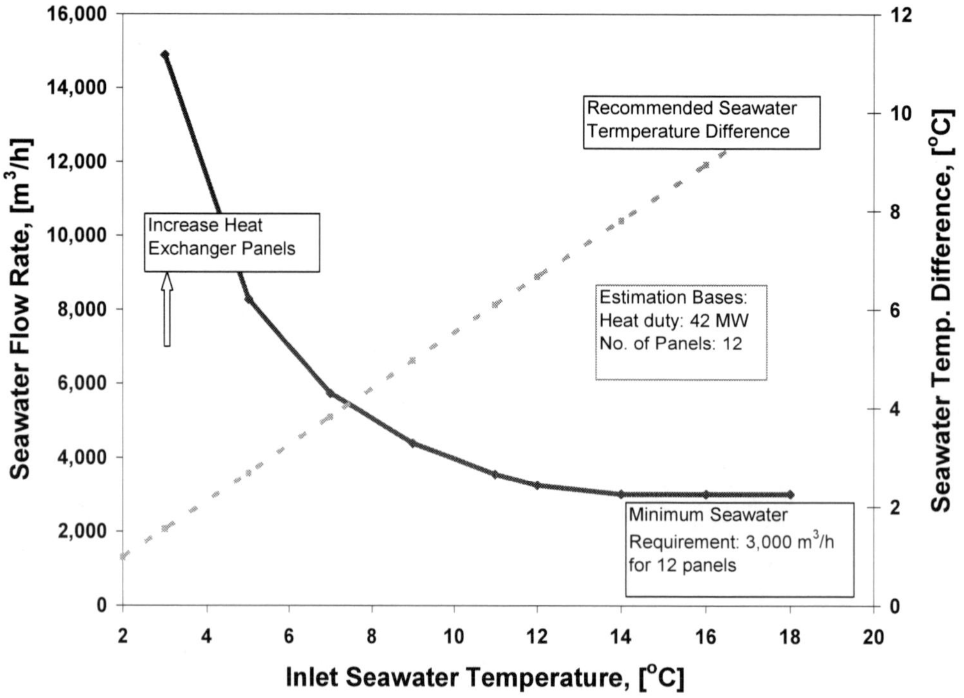

Fig. 6 - ORV Performance with Seawater Temperature

In general, seawater requirement abruptly increases as inlet seawater temperature decreases, while seawater requirement is constant after a certain point even though its temperature increases. This is because seawater should be provided to maintain the minimum water film on the surface of the heat exchange tubes. The seawater requirement varies depending on the number of the heat exchanger panels and tubes.

The higher inlet seawater design temperature requires a smaller ORV and seawater lift structure, resulting in less capital and operation costs. However, these are only valid when the seawater temperature is high enough through out the year. As seawater temperature decreases below design temperature, the SCV operation and seawater heating process need to be operated, resulting in an increase in operating costs. During the winter, the fuel gas requirements for the operation of SCV's and SW HTR's increase linearly as the recovered heat from seawater decreases (seawater temperature drops). The economic analysis results are shown in Fig.7.

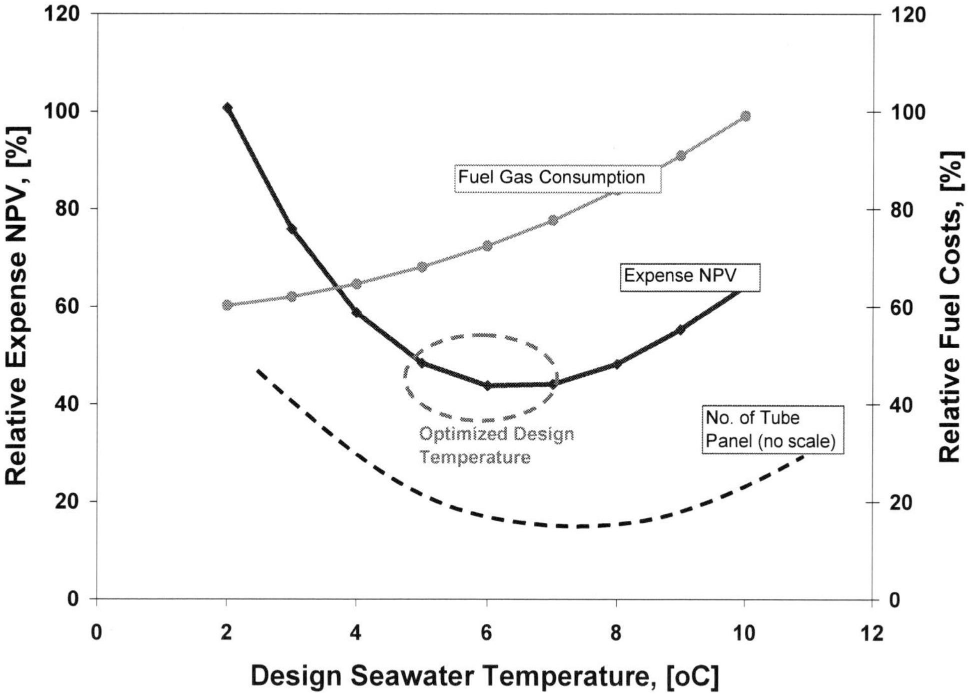

Fig. 7 – Economic Analysis of Different Seawater Temperature

In order to optimize the design seawater temperature, expense-based Net Present Value (NPV) was calculated. From the economic analysis results, design seawater temperature was set at 5°C. The number of tube panels required was also determined based on design seawater temperature and expected delta temperature. The parameters used in economic analysis are summarized in Table 2. It was assumed that fuel cost was $7/MMBtu. Considering current fuel cost, the optimum design temperature may be shifted slightly to the lower point.

Table 2 – Economic Parameters

Description	Parameters
Facility life time	25 years
Discount rate	12%
Electricity cost	7¢/kWh
Fuel cost	7 $/MMBtu
Inflation rate	3%
Tax rate	Not considered

ECONOMIC JUSTIFICAITON OF SEAWATER HEATING

Based on Life cycle costs, economic evaluation on vaporization costs for gas send-out has been performed during the year 2006 on the following Options:

- Base case: ORV+SCV+SW Heater (5 deg C design temp + SW heater)
- Option 1: ORV + SCV (7 deg C Seawater design temp. + no SW heater)
- Option 2: ORV + SCV (5 deg C Seawater design temp. + no SW heater)
- Option 3: All SCV (current the US terminal operation basis)

The terminal operation cost estimates are based on Rich LNG with 1% of maintenance costs for main equipment. Other details specified in Table 2 have been applied to the economic evaluation. Unloading, storage, vapor handling, and other costs are not considered because the same conditions are applied to all options. Gas sale revenues are also not considered. Instead, only incremental costs of vaporization are applied for evaluation purpose.

The economic analysis results are shown in Fig. 8 for relative total installed costs and operation costs to Base case. Option 1 requires slightly lower capital expenditure (Capex) than the Base case, while Option 2 shows almost the same Capex as the Base case. Option 3 with all SCV requires the lowest Capex.

Figure 9 presents the relative Incremental expense-based Net Present Value (NPV) of three Options to the Base case. The vaporization costs of the Base case are the lowest compared to any other Options. Main benefits in the NPV are heat recovery from low temperature seawater, which can not be used for LNG vaporization with the conventional ORV's without the SW HTR's.

For Option 1, the capital expenditure is slightly lower than the Base case because of its high delta temperature in ORV design. Operating cost is higher by about two fold of the Base case, mainly due to lower heat recovery from cold seawater. The NPV difference to the Base case is about 5% due to extra heat requirements for LNG vaporization during winter season.

Capital cost requirement of the Option 2 is almost the same as the Base case as it does not require installation of the SW HTR's. However, operating costs are higher by a 50% than the Base case. NPV is slightly lower than the Base case, mainly due to lower heat recovery from cold seawater.

Since the Option 3 uses the SCV's for LNG vaporization, the required capital expenditure is the lowest than any other Options. However, its operating cost is by 8 times higher than the Base case because main heat for LNG vaporization is provided by fuel combustion. The main economic driver of the Option 3 is the lowest Capex, quick construction and US government approval process compared to seawater options.

Because the Expense-based NPV's of Option 1 and 2 are not significantly different from the Base case, the overall economic evaluation is primarily affected by heat recovery from cold seawater and fuel gas burning for supplementary heat supply. The detail configuration and economic justification should be performed based on the project design conditions.

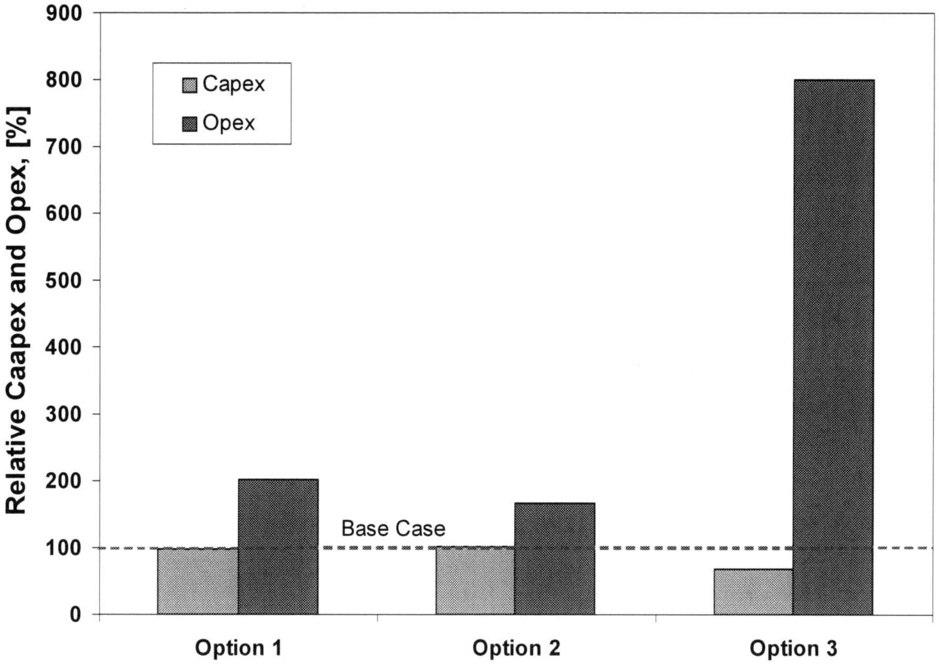

Fig. 8 – Comparison of Relative Capex and Opex

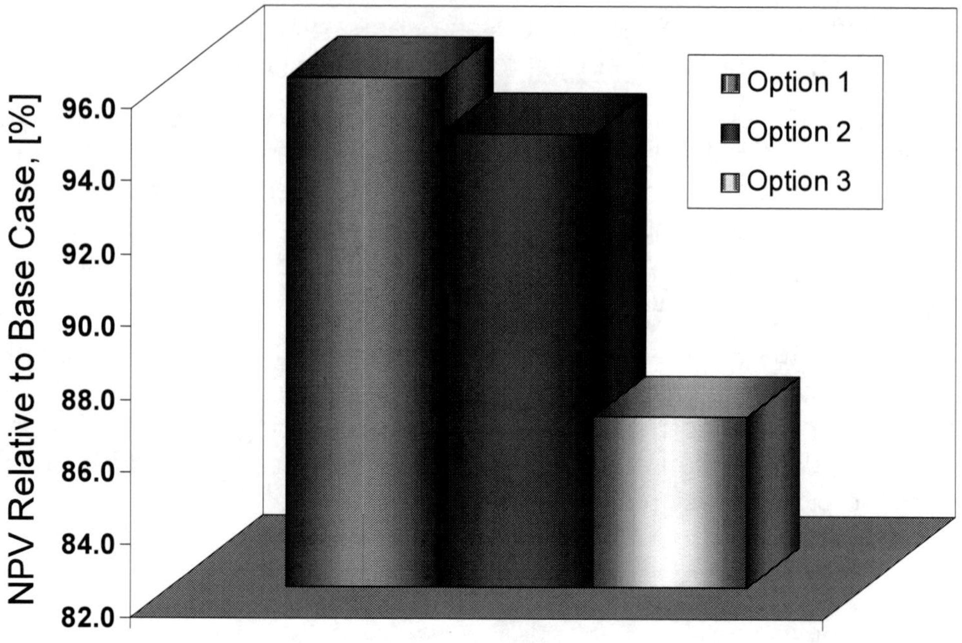

Fig. 9 – Relative NPV of Vaporization Options

CONCLUSIONS

The design temperature of the Open Rack Vaporizer may be different from the manufacturer's recommended design temperature and optimum operation point, which will be justified by the individual project conditions. Optimized design temperature of the seawater vaporizer can be determined by overall economics based on the site operation and design conditions.

In optimum vaporization selection, design constraints with unfavorable seawater temperature have been successfully solved with heating up seawater with the seawater heater before it is provided to the ORV. This results in an increase in heat recovery from cold seawater. Unless otherwise, the seawater can not be used when it is lower than the design temperature.

Seawater heater increases total installation cost of the vaporization system. However, the warming up seawater temperature decreases by 50% of fuel gas consumption compared to the use of SCV vaporization configuration with the same design temperature, but without seawater heating up.

ACKNOWLEDGEMENT

The authors wish to thank Korea Gas Corporation (Kogas) for permission to prepare and present this paper. Special thanks are also extended particularly to Dr. J. H. Cho, CB&I, for his kind assistance and valuable discussions.

REFERENCE CITED

1. Technical and Economical Summary of Basic Design Report for Incheon LNG Terminal, Kogas/Hyundai Engineering Co/MWKL, Unpublished document, February 1993.
2. A Prospect of Gas Demand in Korea (2006-2011), Korea Energy Economics Institute, ISSn 1599-9009, Vol. 8, December 2006.
3. 8th Long Term Natural Gas Supply and Demand Plan, Ministry of Commerce Industry and Energy, Korea, December 2006.
4. Incheon Seawater Temperature, Korean Meteorological Administration, Korea.

BIOGRAPHY of SPEAKER

In-Soo Cheon is a Chemical Engineer with a B.S.ChE degree from Seoul National University, Korea. He is currently senior executive vice president of Hyundai Engineering Co., Ltd.

He has worked as project manager for LNG distribution system in Seoul Metropolitan Area from 1983 to 1989. He has also worked as Project Director for Inchon LNG Receiving Terminal Project in Korea from 1991 to present. His career in LNG and gas processing spans over 39 years and has been involved in various gas monetization projects, ranging from conceptual design to commissioning. Through his vast hand-on experience in LNG projects, he has gained extensive knowledge in efficient and practical design towards safe operation of the LNG terminal.

E-mail: ischun@hec.co.kr

An Old Technology and a New Application in Processing Rich LNG to Meet Pipeline Gas Specification at LNG Receiving Terminals

C.C. Yang and Zupeng Huang
Foster Wheeler USA Corporation
2020 Dairy Ashford Road
Houston, Texas 77077

AIChE Spring Meeting, April 2008
8th Topical Conference on Natural Gas Utilization
New Orleans, Louisiana, April 6-10, 2008

Keyword: LNG, Natural Gas, LNG Cold, Reforming

ABSTRACT

Most existing LNG baseload plants produce LNG with higher heating value, i.e. higher contents of C_2 and heavier components, than is suitable to serve the North American market. As the LNG buyers have multiple LNG specifications, there are many benefits for reducing C_2 and heavier components at the LNG Receiving Terminals.

A process scheme had been developed by FWUSA [1,2,3] to reduce C_2 and heavier components from the send-out LNG by using LNG cold at LNG receiving terminals while minimizing or eliminating gas compression. The extracted C_2 and heavier components, after consumed in the terminal, may have potential markets.

In the North America, however, the extracted C_2 and heavier components are often difficult to sale or handle economically in most regions.

By processing a minimum amount of LNG, extracting the C_2 and heavier components and then by pre-reforming the extracted C_2 and heavier components, a lower calorific Substitute Natural Gas is produced and can be re-injected into the LNG terminal send-out gas to control its heating value.

INTRODUCTION

LNG receiving terminals must comply with strict requirements on send-out gas characteristics, including composition, calorific value, Wobbe Number and quality in order to satisfy the customers' specific needs.

For the North American market, the C_2^+ content and heating value specifications for the pipeline gases are lower than most LNG produced from existing LNG baseload plants. Cost effective management of these components at LNG receiving terminals enhance the LNG value chain, thereby increasing flexibility in sourcing LNG cargoes.

LNG terminal owners are currently focusing on the ways to deal with these specification requirements. To accept cargoes of rich LNG, the receiving terminals must be able to manage the C_2^+ components economically. At the terminals, currently there are two main process schemes are used in adjusting the export gas specifications:

1. Nitrogen and Air dilution
2. Extraction of C_2^+ components

Nitrogen and air dilution are viable options, however, they are expensive in equipment cost with a high operating cost. The installed facility costs are about $100 million for a typical 1 BSCFD send-out gas terminal for injecting 3 mol% nitrogen. And the nitrogen plant will use about 14 MW electric powers, while this can only provide a minor Wobbe Index adjustment.

There are many benefits of extracting C_2^+ at the LNG receiving terminals, including:

- Flexibility to process LNG from multiple suppliers which allows terminal owners to maximize future LNG spot market purchases.
- LNG cold may be effectively utilized.
- Flexibility in meeting multiple customer gas specification requirements.

Foster Wheeler presented a process solution to this concern at the AIChE 2003 Spring National Meeting entitled "A cost-effective design for reducing C_2^+ at LNG receiving terminals".

The design described reduces C_2^+ from the send-out gas by using LNG cold while minimizing or eliminating gas compression. Using low-temperature LNG as a cooling medium for the column overhead can achieve an economic separation. The extracted C_2^+ components can either be exported or used as fuel for either the power generation, if supplied on site, or for the submerged combustion vaporizers, if they are used.

Only minor modifications to the typical LNG receiving terminal equipment are required, including modification of the casing of the send-out pumps to a higher pressure rating, but no additional compression is required. This results in both lower capital and operating costs.

To minimize the capital and operating costs of the facility, the FW design is intended to process a minimum amount of LNG in extracting the C_2 and heavier components from the LNG terminal while targeted to meet the send-out gas specifications. The design also provided an option to bypass the facility when the purchased LNG has met the send-out gas specifications. The highlights of the designs are:

- Using fractionation column(s) for C_2^+ reduction.
- Using a gas direct contact condenser.
- Using LNG as a cooling medium for condensing.
- Pumping condensed LNG to gas send-out pressure for regasification.
- Eliminating the compressors.
- A cost effective design to reduce C_2 and heavier components.

After a part of the extracted C_2 and heavier components is consumed as fuel in the LNG terminal, the remainder may have a potential market.

In North America, however, the extracted C_2 and heavier components are difficult to sell or handle economically in most regions, while the consumption of the fuel for the LNG terminals is often much less than the extracted C_2 and heavier components from the imported rich LNG. In the following discussion, through the handling of the extracted heavier components which utilizes pre-reforming technology for producing low calorific Substitute Natural Gas, and then by re-injecting into the LNG terminal send-out gas, its heating value therefore be controlled.

CONVERION OF C_2^+ COMPONENTS TO LOWER CALORIFIC GAS FOR BLENDING TO SEND-OUT GAS

Each receiving terminal that is importing LNG from a source where the composition does not meet the end market specification will need a strategy to manage the C_2 and heavier components of the gas. The following block flow diagram (Figure 1) shows an overall schematic diagram to process part of LNG and using pre-reforming technology to convert the C_2 and heavier components to a lower calorific gas before recombining with the by-passed send-out gas.

Figure 1 LNG Regasification Block Flow Diagram

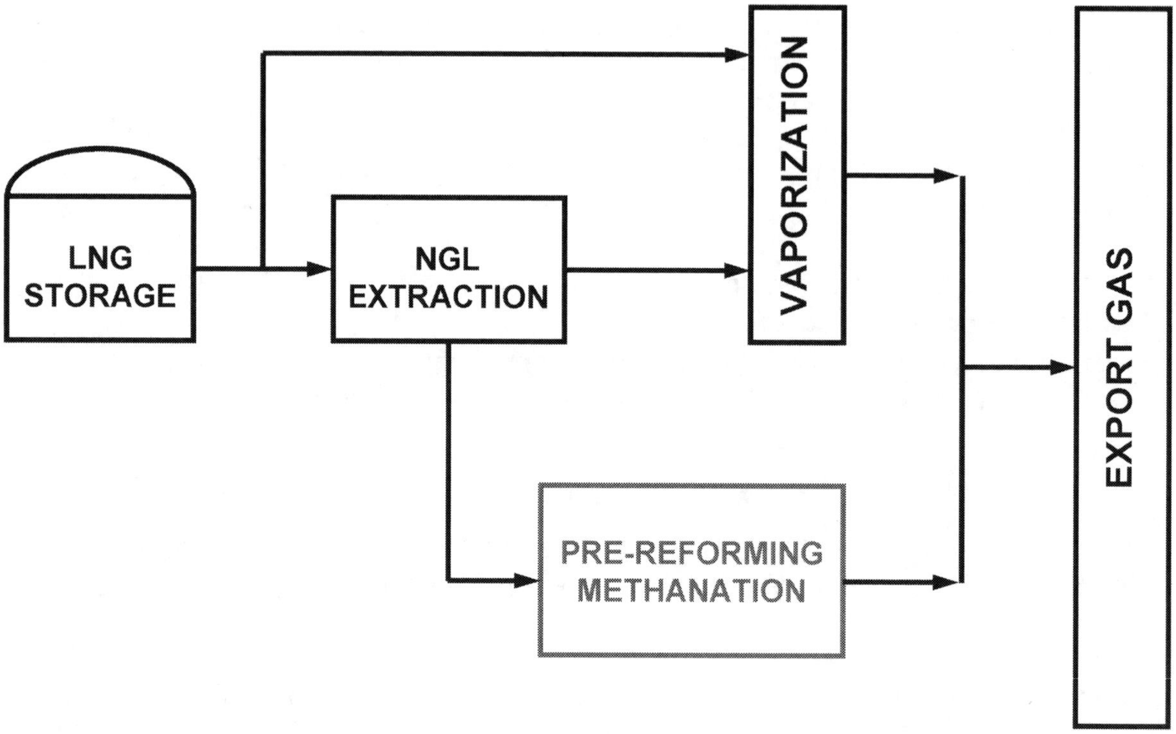

When processing the extracted C_2 and heavier components to produce the lowest calorific gas, cares are required in choosing the pre-reforming catalyst and conversion conditions. Milder conversion to reducing C_3^+ should be considered. This will result in requiring a minimum amount LNG to be processed in the C_2^+ extraction facility. The pipeline gas specification is also a major consideration, particularly the maximum allowable concentrations for hydrogen, carbon dioxide, carbon monoxide.

Catalytic Rich Gas (CRG) technology was developed by British Gas to produce a medium calorific value Town gas. The first plant was commissioned in 1966 based on the catalyst developed by The Gas Council (later to become British Gas). The pre-reforming outlet gas is typical in the range of methane 61 mol%, hydrogen 17 mol%, CO 1 mol% and carbon dioxide 21 mol%. With reformer and CO-shift, it will produce a typical town gas, with a much higher concentration of hydrogen, up to 50-mol%. For our purpose, hydrogen content will be reduced. The CRG methanation SNG plant was developed in the 1970's to reduce the hydrogen to about 4 mol% and carbon oxides content to 23 mol%. A further methanator stage containing CRG catalyst is incorporated after water removal when very high purity SNG is specified. This is a dry methanation, which reduces the hydrogen content to less than 0.5 mol%.

The CRG process is licensed and guaranteed by Davy Process Technology and fully supported by a number of licensed contractors.

The basic scheme consists of feed purification followed by hydrocarbon breakdown and methanation steps. A number of reaction stages may be adopted depending upon the final gas specification required.

In this case a process must be developed to convert the C_2 and heavier stream, extracted from LNG to a methane rich stream (SNG), which will satisfy the export pipeline specification when recombined with send-out gas. The designer may pursue a number of technical areas in order to achieve maximum optimization and integration of the process scheme. These are:

1. The requirement for feed gas pre-treatment
2. CRG pre-reformer feed heating
3. Hydrogen source (for start-up and normal operation)
4. Steam generation
5. Produced CO_2 reduction

PROCESS OVERVIEW (Figure 2)

As stated previously, the pre-reformer is the central part of this process scheme. Key features of the pre-reformer include:

- Conversion of C_2^+ components to methane rich stream by steam reforming reactions
- Nickel based catalytic reactor
- Hydrogen and carbon dioxide are produced
- High inlet temperature (hence heat requirement)

The reforming reactions are shown below:

Endothermic steam reforming reactions

$C_nH_m + nH_2O \leftrightarrows nCO + (n+½m)H_2$ (Higher hydrocarbon-steam reforming) (1)
$CH_4 + H_2O \leftrightarrows CO + 3H_2$ (Steam methane reforming) (2)

Exothermic methanation and shift reactions

$CO + H_2O \leftrightarrows CO_2 + H_2$ (Shift Reaction) (3)
$CO + 3H_2 \leftrightarrows CH_4 + H_2O$ (Methanation) (4)

The following schematic diagram shows process units for converting the C_2 and heavier components to a lower calorific gas including a number of the major processes:

Figure 2 Conversion Process for C_2 and heavier components

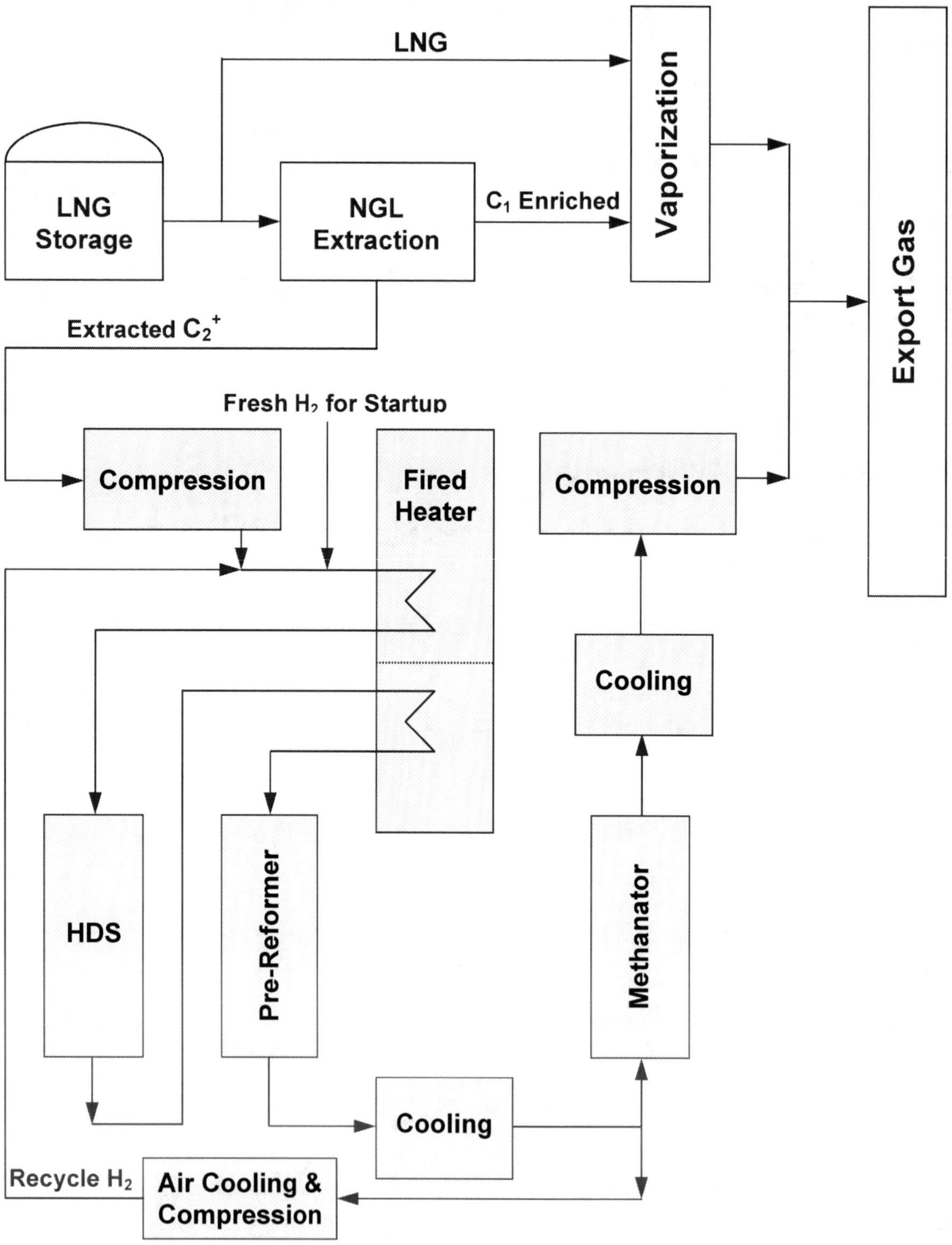

PRE-TREATMENT

The feed stream from the C_2 plus extraction process is pumped to around 40 barg, and then heated. The vaporized feed is then mixed with recycle gas which contains hydrogen from the pre-reforming reactor for hydrodesulphurization (HDS). The feed stream after mixing will be heated to 220 °C in a process-fired heater. Further heating is required to raise the temperature of the vapor to 370 °C before entering the HDS.

The HDS is comprised of three separate catalytic layers designed to treat the feed gas before entering the pre-reformer. In the first layer, any organic sulfur and chloride compounds preset are converted to hydrogen sulfide and hydrogen chloride and any unsaturated hydrocarbons become saturated. Since the feed is extracted hydrocarbons from LNG, it is unlikely that there will be any unsaturated hydrocarbons present in the feed.

The treated gas then passes through the second layer, where the hydrogen sulfide and hydrogen chloride is absorbed by reaction to produce zinc sulfide and zinc chloride.

The third layer is provided to ensure that total sulfur slippage is less than 50 pbb. Catalyst life in the HDS ranges from two to six years. The changeover will take up to 2 days. During changeover, the HDS may be bypassed so as not to disrupt LNG send out. Alternatively, the HDS shutdown may be scheduled by exploiting opportunities when the terminal is receiving a lower calorific LNG.

PRE-REFORMING

The desulfurized feed is then mixed with superheated high-pressure steam and preheated to 400°C in the radiant section of the process-fired heater. The design steam to carbon ratio for the pre-reformer is 1.8 kilogram of steam per kilogram of feed. This ratio is advised by Davy Process Technologies as the supply of excess reactants moves the equilibrium to favor products.

The mixed feed then enters the pre-reformer (CRG Reactor) where it flows over a bed of CRG catalyst. Here the hydrocarbon feed is reformed adiabatically to produce a product gas containing methane with little or no heavier hydrocarbons and equilibrium levels of hydrogen, carbon oxides and steam. The reactions taking place over the reactor may be considered as the hydrocarbon reforming reactions (1) & (2) and the shift reaction (3).

Reactions (1) and (2) are endothermic as shown from left to right and reactions (3) and (4) are exothermic. Reactions (2), (3) and (4) are reversible and once the heavier hydrocarbons are reformed (reaction (1)), the carbon monoxide and hydrogen produced will then react together through the reverse of reaction (2). The methanation reaction (4) is highly exothermic in the forward direction. Reaction (1) is also reversible but under the reaction conditions adopted the equilibrium value lies very heavily to the right and hence virtually all heavier hydrocarbons are removed. Overall the reactions taking place in the CRG reactor are exothermic in nature with the final gas analysis leaving the reactor being determined by an approach to the thermodynamic equilibrium of the above reactions.

METHANATION

In order to meet the export gas specification for hydrogen and carbon oxides' content, the pre-reformed gas must pass through a methanation process. Hydrogen, carbon monoxide and carbon dioxide react to form methane in accordance with the methanation reaction (4). Again, the exothermic reaction takes place adiabatically and due to the reduced inlet temperature of operation the methane level in the gas is increased over the reactor. It should be noted that the removal of water is favourable so that reaction (4) is driven in the forward direction.

EXPORT GAS

Heat contained in the hot gases leaving the methanator is recovered and the gas is cooled to a lower temperature to ensure the water moisture content after combining with the unprocessed send-out gas comply with the pipeline gas specification, typically 7 lbs/MMSCF gas. Process condensate from the knockout drums are collected and sent to a deaerator. The dry gas is compressed to export pipeline pressure and combined with the vaporised LNG and methane from the extraction unit.

STEAM AND CONDENSATE

Condensate that has been produced in the process is reused to produce steam to satisfy the steam to carbon ratio at the inlet to the pre-reformer. The process condensate is of good quality as there is no nitrogen present in the feed, hence no ammonia is produced. There is also no shift reactor; hence no methanol will be formed. The condensate will be sent to the deaerator to produce high quality boiler feed water (BFW).

However, the condensate does not produce enough steam to fulfill the process requirements; hence additional make-up water is needed. The make-up water is combined with the condensate and sent to the deaerator where the CO_2 is stripped. A BFW dosing package, oxygen scavenger, boiler chemicals and pH control chemicals are injected into the BFW system.

BFW water is taken from the sump of the deaerator and pumped via BFW pumps to a steam generation package. Steam is generated from waste heat from the outlet of the pre-reformer. The remaining heat is provided by LNG fuel. High pressure superheated steam will be provided upstream of the pre-reformer at 37 barg and 450°C.

CATALYSTS AND OPERATING CONDITIONS SELECTION STRATEGIES

In applying the CRG technology for converting the C_2 and heavier components to a lower heating value gas, the following design strategies are to be taken into consideration:

1. The pipeline gas specification is a major consideration, in particularly, the maximum allowable concentrations for hydrogen, carbon dioxide, and carbon monoxide. Producing the lowest calorific gas is preferred in choosing the CRG catalyst and conversion conditions. Milder conversion to reducing C_3 and heavier components should be considered. This will result of requiring a minimum amount of LNG to be processed in the C_2 plus extraction facility. Some pipeline gas has a higher maximum carbon dioxide specification amount to 4 mol% and a maximum inert gases specification amount to 7 mol%. In some cases, pipeline gas has a higher hydrogen specification. The pipeline specification has a great impact to the conversion requirements.
2. Minimizing hydrogen yield is the most critical issue in applying the CRG technology. CO is likely not an issue, as it will be in compliance with the piping line gas specifications after the converted gas blended with the send-out gas. Avoid using methanator, if possible, to reduce the cost of the system.
3. Other alternatives to convert/reduce the Hydrogen content from the CRG Reactor outlet, such as to convert hydrogen to CO and CO_2 selectively, using a milder methanator or possibly using Fischer-Tropsch Synthesis with a milder conversion.
4. There is a potential to reduce the requirement for methanation reactors. This centers on the feasibility to extract hydrogen that could be used in the process. Another alternative would be to use one wet and one dry methanator as opposed to two wet methanators.
5. Other technologies in separating hydrogen from the syngas such as using membrane or absorption processes by metal hydride silica separation may also become attractive as these technologies are further developed.
6. There is an opportunity for the C_2^+ case to operate with a lower steam to carbon ratio. The reduced steam demand will reduce the capital cost of the steam related equipment and an overall reduction in the operating costs.

CONCLUSIONS

Cost effective management of the C_2^+ content and heating value specifications components can enhance the LNG value chain, thereby increasing flexibility in sourcing LNG cargoes. Several alternatives for managing C_2^+ components in the LNG value chain exist. It has been seen that extraction can occur at either the receiving terminal or upstream of the liquefaction plant. In the cases where it may not be cost-effective to reduce the C_2 and C_3 at the baseload plant, we have demonstrated that a proposed C_2^+ management process design for send-out gas at the LNG receiving terminal could be economically viable under certain circumstances.

REFERENCES CITED

[1] Yang, CC, Kaplan, Al, and Huang, Zupeng, “Cost Effective Design Reduces C_2 and C_3 at LNG Receiving Terminals,” Oil & Gas Journal, May 26, 2003

[2] Yang, CC, Kaplan, Al, and Huang, Zupeng, “Method and Apparatus for Reducing C_2 and C_3 at LNG Receiving Terminals,” US Patent Number 7,278,281, October 9, 2007 (US Patent Trademark Office).

[3] Yang, CC, Huang, Zupeng, and Barclay, Mike "Processes Help Turn Rich LNG into Lean Gas," LNG Journal, P11-12 June, 2006.

[4] Yang, CC, and Bothamley, Geoffrey, “Maximising the Value of Surplus Ethane and Cost-effective Design to handle Rich LNG,” LNG 15 Poster, April 23-April 27, 2007, Barcelona, Spain.

AUTHOR INDEX

AUTHOR INDEX

AUTHOR INDEX

AUTHOR INDEX

AUTHOR INDEX

American Institute of Chemical Engineers

Sustainable Engineering Forum Conference

Topical Conference at the
2008 AIChE Spring National Meeting

April 6-10, 2008
New Orleans, Louisiana, USA

Volume 2 of 2

Printed from e-media with permission by:

Curran Associates, Inc.
57 Morehouse Lane
Red Hook, NY 12571
www.proceedings.com

ISBN: 978-1-60560-226-4

Some format issues inherent in the e-media version may also appear in this print version.

ISBN: 978-1-60560-226-4

American Institute of Chemical Engineers
Proceedings
Three Park Avenue
New York, NY 10016-5991
Phone: 212-591-8100

www.aiche.org

American Institute of Chemical Engineers

Sustainable Engineering Forum Conference
2008

TABLE OF CONTENTS

Volume 1

Volume 2

Expansion Foam 3-D Temperature Profile during Application on LNG Experimental Results

Jaffee Suardin (Speaker)
Benjamin Cormier
Morshed Rana
Yingchun Zhang
Dr. M Sam Mannan

Mary Kay O'Connor Process Safety Center, Artie McFerrin Department of Chemical Engineering, The Texas A&M University System, 3122 TAMU, College Station, TX 77843-3122

AIChE Spring Meeting, April 2008
8th Topical Conference on Natural Gas Utilization
New Orleans, Louisiana, April 6-10, 2008

Introduction - Expansion Foam application on LNG

Liquefied Natural Gas (LNG) hazards include LNG vapor dispersion and LNG pool fire thermal radiation. Due to its characteristics, a LNG spill creates a LNG liquid pool and LNG vapor that is able to disperse and cover a wide downwind area. Between 5% and 15% in volume concentration in LNG/air mixture, the vapor becomes flammable. Once it ignites, fire occurs and flashes back to the vapor source creating pool fire. A large LNG pool fire emits high thermal radiation thus preventing fire fighters from approaching and extinguishing the fire. This may increase the risk of damage and additional LNG release from nearby equipment or storage. One of the strategies used in the LNG industry and recommended by federal regulation National Fire Protection Association (NFPA) 59A is to use high expansion foam (HEX) to suppress LNG vapors and to control LNG fire by reducing the fire size. This in turn reduces the heat radiation allowing fire fighters to approach the pool fire and be able to extinguish it with dry chemical [1]. In its application, HEX effectiveness heavily depends on HEX application rate, generator location, and LNG containment pit design. While the theoretical study of the effects of HEX application on LNG vapor and fires is important, the complicated phenomena involved increases the needs for medium to large LNG experiments. This also leads to the fact that previous studies completed in the past have not covered and quantified all the necessary and important parameters. NFPA 11 A.8.20.3 suggests HEX application on LNG should be established by specific test under controlled environment [2]. To align with that, this proposed research attempts to investigate HEX application on LNG through novel medium scale LNG field experiments.

An attempt to match the real LNG facilities is prepared by conducting the experiments at Texas A&M University's Brayton Fire School where the containment pit is specifically designed to have similar characteristics to the ones installed at LNG facilities. Additionally, a commercial HEX system is utilized while all measurements are performed with research grade or modified industrial equipment.

In this experiment, the foam generator produced foam with expansion ratio of 250:1 instead of 500:1 (high expansion ratio). Thus, in this experiment, foam will be called expansion foam (EX). The 3-D temperature profile associated with the EX depth, spreading rate, and methane concentration at different downwind distances and heights are studied. 3-D temperature profile correlated with fire occurrence and pool fire thermal radiation reductions are investigated as well. The proposed research is intended to provide knowledge to improve guidelines for LNG vapor dispersion and LNG pool fire thermal radiation suppression.

This paper discusses the experimental setup and some of the data analysis. However, detail data analysis will be presented during the AIChE presentation.

Experiment Objectives

This experiment was aimed to:

- Examine EX 3-D temperature profile – concentration - cloud characteristics during LNG vapor dispersion
- Study the EX 3-D temperature profile - thermal radiation characteristics during LNG pool fire.

Experiment Setup

The experiment was conducted at Texas A&M University Brayton Fire School where the LNG fire training facility was developed under sponsorship from BP. The experiment was performed on the largest pit which has dimension of 33 ft x 22 ft x 4 ft, as shown in Figure 1. The experiment was divided into 4 phases and data were recorded continuously with no interruption in between phases. Equipment utilized includes a hydrocarbon camera to study the methane cloud movement, fire resistant thermocouples, gas detectors, regular camera, weather station, flow meters, pressure gages, and an estimated 10,000 ft of cables. EX system employed was Angus Fire foam proportional system and Angus Fire Turbex Foam Generator.

The first phase was the free LNG vapor dispersion on concrete. Sixty thermocouples were placed inside the pit at 5 different heights (z direction) and 4 different placements in x direction, and 3 different locations in y direction. Wooden walls were placed at the edge of the pit providing a total wall height of 8 ft from the bottom of the pit. Three thermocouples were located on each wall (total 12 thermocouples) and shared the same z direction with the top three thermocouples located inside the pit. In addition, there were 9 thermocouples and 40 gas detectors placed outside the pit in the downwind direction, as shown in Figure 2. Once the equipment was ready, 90 GPM of

LNG was spilled into the pit for 57 minutes to reach a 6 inch LNG pool inside the pit. Flashing and evaporation during the spill resulted in a total of 5058 gallons of LNG used.

Figure 1. Large pit used in the experiment

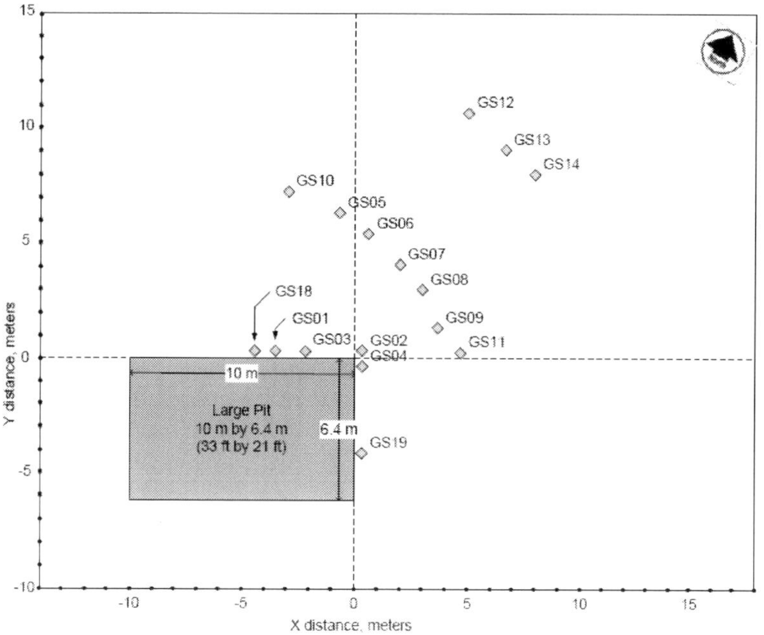

Figure 2. Thermocouples and gas detectors placement - top view

During the spill, LNG vapor concentration in the air and temperature were recorded. This provides data for the condition before EX was applied as mitigation system. The next phase was the EX application. The LNG spill was stopped after 57 minutes and EX was applied until it reached the top of the walls. This phase completed the experiment of 3-D temperature profile of EX during its application on LNG vapor dispersion (no fire occurrence).

The third phase was the preparation for EX application on LNG pool fire experiment. Walls, gas detectors, thermocouples located outside the pit were removed to safe place, leaving all thermocouples inside the pit. Then, the fourth phase started by igniting the LNG vapor to create a LNG pool fire. EX was applied into the pit to control the fire. Once enough data was recorded, fire was extinguished using dry chemical. All of these phases were shown in Figure 3.

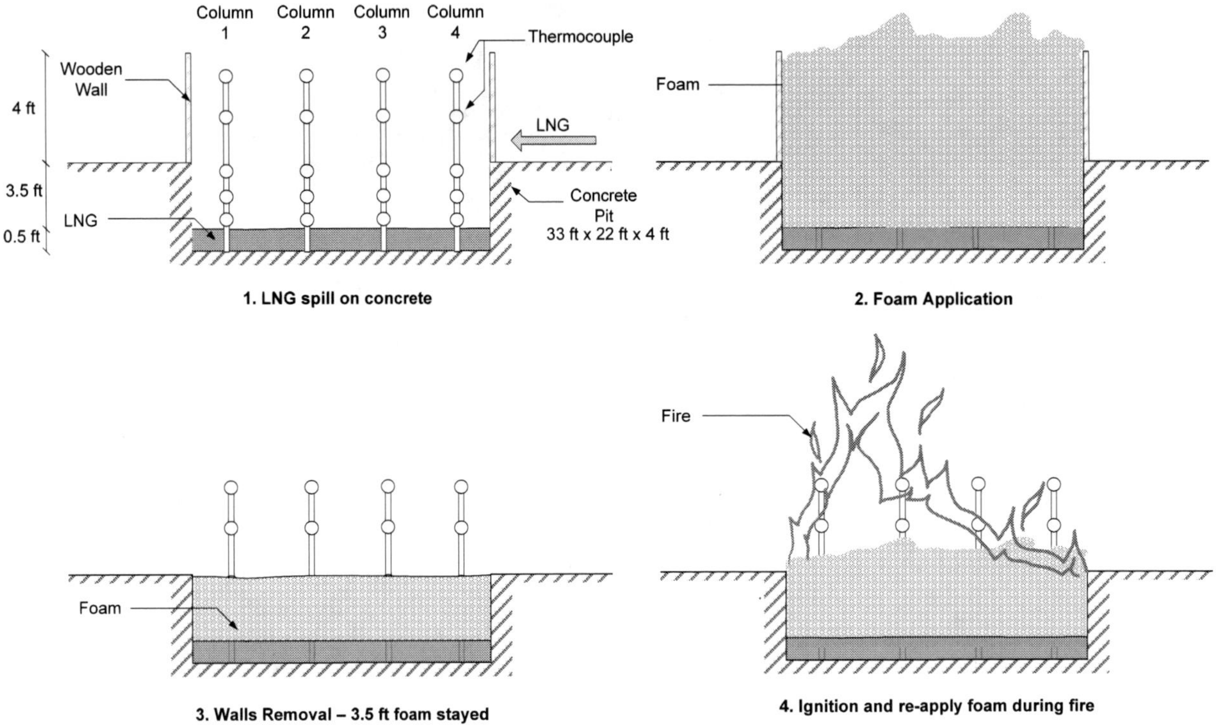

Figure 3. Experimental setup

Summary

The medium scale LNG field experiment has been conducted at Texas A&M University's Brayton Fire School. It was intended to investigate the 3-D temperature profile on the application of expansion foam to mitigate LNG vapor dispersion and LNG pool fire. Experimental setup was provided in this paper however the results will be presented during AIChE conference on April 9th, 2008.

Acknowledgement

The author would like to appreciate BP Global Gas SPU for funding support for this research. In addition, authors would like to acknowledge Texas A&M University's Brayton Fire School for providing experiment setup and location and for assisting conducting experiment safely.

References

1. “Introduction to LNG Safety”, John Alderman, Process Safety Progress, 24(3), p.144-151, 2005.
2. “NFPA 11 Standard for low-, medium-, and high-expansion foam”, National Fire Protection Association, 2002.

Biography of Author

JAFFEE SUARDIN

Research Assistant
Mary Kay O'Connor Process Safety Center, Texas A&M University
Cell: 979-739-2251, Office: 979-862-1310
Email: jaffee.suardin@chemail.tamu.edu
Web: psc.tamu.edu

Jaffee Suardin is a PhD candidate at Artie McFerrin Department of Chemical Engineering at Texas A&M University. He obtained his bachelor degree in chemical engineering in October 2002 at Bandung Institute of Technology in Indonesia. Then he continued his study at Texas A&M University (TAMU) in College Station, Texas and graduated with Master of Science degree in chemical engineering in June 2005.

He joined Mary Kay O'Connor Process Safety Center in January 2004 and began conducting research under the guidance of Dr. Sam Mannan. His master degree research was the integration of Dow Fire and Explosion Index into Process Design and optimization. His current research is the study of LNG vapor dispersion and pool fire mitigation technology using High expansion foam and Foamglas®PFS.

In addition to conducting research in process safety area, He has been involved in several industrial projects including leading HAZOP team, developing P&ID and PFDs, and performing consequence modeling. He interned with Shell International Exploration and Production (SIEP) of Shell Oil Company on summer 2006 and 2007 where he conducted projects on Fire and explosion assessment on FPSO and action tracking system. He developed Dow Fire and Explosion and Chemical Exposure Indices software used in senior project in process design class at TAMU. Currently, He is teaching AspenPlus process design software for undergraduate at Artie McFerrin Department of Chemical Engineering.

A Study on Quantitative Risk Assessment Model for LNG Storage Tank

Meejin Kang
PhD Student, Dept. of Energy Safety Engineering,
Graduate School of Energy & Environment,
Seoul National University of Technology, Korea

S.R Lee
Korea Gas Safety Corporation, Seoul, Korea

Young-soon Lee¶
Professor, Dept. of Safety Engineering,
Seoul National University of Technology, Korea

AIChE Spring Meeting, April 2008
8th Topical Conference on Natural Gas Utilization
New Orleans, Louisiana, April 6-10, 2008

Introduction

As the production and consumption of LNG has recently increased, the installation and construction of LNG terminals has increased, and the capacity of LNG facility is getting bigger and bigger. And with improved technology for LNG storage tank being developed, the relevant industry much prefer two storage types of full-containment and membrane tank that don't need any more a bund wall, also called bunding, and the underground membrane tank as well.

Because the capacity of LNG storage tank is relatively large ranging from 140,000 to 200,000 m3, accidental release of LNG could cause such probable hazards as cryogenic damage, large flammable vapor cloud and potential physical explosion due to the rapid evaporation of flammable liquid, and fire/explosion of flammable liquid and gas, etc. There are lots of existing studies about safety and/or risk of storage tank while the various tank types have been developed. Those studies tend to estimate and compare the reliability of each type of tank using FTA (Fault Tree Analysis) based on the same accident such as release and spill. Besides the studies, a great deal of the qualitative and/or quantitative risk assessments have been carried out to meet legal requirements, standards, codes, or self-requirement.

LNG storage facilities have some distinguished characteristics, because the contained material is only natural gas, methane, and the relevant operation activities are limited to a series of routine works. Accordingly most of the LNG storage facilities show much similarity in hazards or potential risk. This study presents a quantitative risk

¶ Corresponding author
LYSOON@snut.ac.kr

assessment model based on this similarity. This model also addresses important considerations required to perform QRA for each facility so that QRA can be performed more easily using this model.

Recent QRA for LNG Facilities

The hazards of LNG due to its physical characteristics are summarized as follows.

- Potential physical explosion due to rapid evaporation
- Potential pool-fire in case LNG is ignited just after pool formation.
- Potential flash-fire or VCE in case the ignition is delayed after pool formation
- Potential asphyxiation due to decrease of oxygen in case of the evaporation of pool and/or release of natural gas
- Potential jet-fire in case the natural gas is ignited just after jet release through piping or pressure relief valve

The accident of LNG storage facility usually starts from the release of LNG. When LNG is spilled on land, the pool is formed because freezing the moisture in the soil blocks LNG's percolation into the soil. Only if there is a bund wall the spreading and evaporation of pool is limited. (ioMosaic Corp., LNG Code of Practice – Supporting information report, 2005)

LNG release from full-containment tank and membrane tank could happen only when loss of containment occurs simultaneously at both inner container and outer container. Especially the leakages from outer container can be divided into vapor leak and liquid leak. Therefore the major hazards are expressed as the following 3 types. (N Ketchell, R Robinson & P Genoud, Quantification and Comparison of the risks of LNG storage concepts, 1998)

- International leak
- External vapor leak
- External LNG leak

According to the results of FTA and HAZOP for LNG facilities, the credible accident scenario or potential hazards are categorized as shown in table 1. Generally the accident frequency has been analyzed using FTA, the consequence is estimated using LNGFIRE and DEGADIS, which are recommended by NFPA, and other equations or calculation models that are adopted worldwide. NFPA 59A, EN1473 and risk assessment guidelines and codes provide the guidance involving weather condition and evaluation basis, etc. The guidance will help to calculate the consequence and to analyze the effect.

QRA for LNG Facilities

In order to judge the risk of LNG storage facility, the frequency and consequence has to be evaluated after identification of possible accident scenarios. This study suggests the procedure of determining the acceptance of risk as shown in figure 1. This consists of FAM (Frequency Analysis Module), CAM (Consequence Assessment Module) and EAM (Effect Analysis Module).

Figure 1 QRA Procedure for LNG facilities

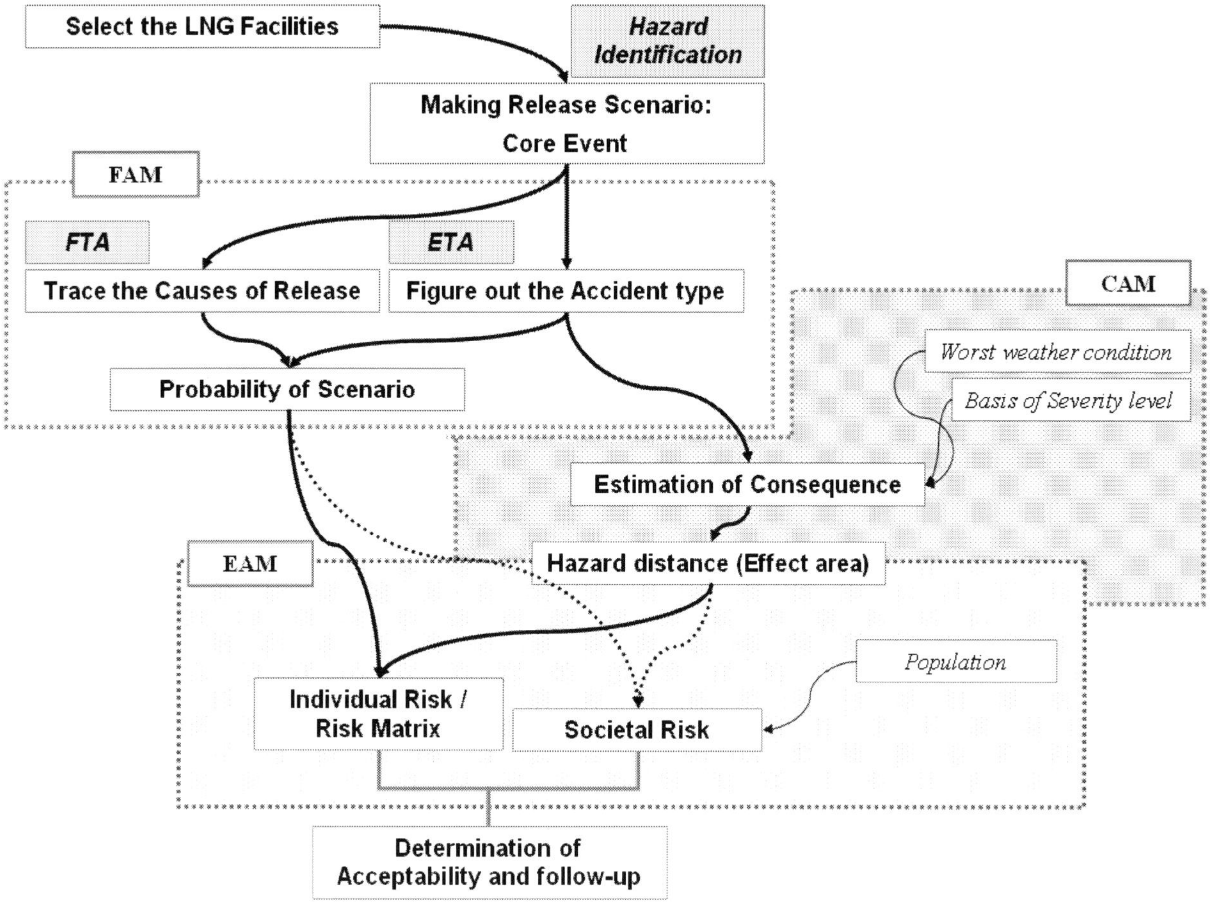

Also this study suggests the possible accident scenarios of LNG storage facility based on the existing studies and hazard evaluation results, as shown in table 1. Each of the scenarios is employed as core event of Hazard identification step in figure 1.

Table 1 Predictable accident scenarios in for LNG storage tanks

Type of LNG storage tank	Predictable accident scenarios	Main causes of scenario
Above tank: Membrane type Above tank: Full containment type Underground tank: Membrane type	1. BOG release through PRV due to the overpressure	- Rollover caused by additional unloading of LNG that has different composition - Wrong operation: Over-filling, Temperature control failure, etc.
	2. BOG release from connection points due to the overpressure	- PRV failure when the pressure is higher than normal.
	3. BOG release due to the failure of roof including various fittings and piping	- Pressure control failure during the loading - Natural disasters: Heavy snow, Strong wind, etc.
	4. External impact	Air craft, Missile, etc.
	5*. External defects with internal impact	- Internal impact: Pump dropping, Cool-down failure - External defects: Wrong maintenance, vehicles crash, etc.
	6. Tank damage due to out of balance	- Heating system failure - Wrong construction of pile, etc.
	7. LNG/BOG release due to maintenance error	
	8. BLEVE due to external fire	
	9. Internal explosion in case of air-intake and when ignition source exists inside the tank	

Note: Number 5* is not appropriate for the underground tank.

FAM is a module that analyzes the frequency, which is referred hereinafter as MTA (Merging tree analysis) in this paper. Besides this module, we suggested a modified LOPA that involves complementary PFD and HD (Hazardous Distance) in our previous study. This modified LOPA needs additional verification of the relevant penalty and credit items and their weight values. MTA applies the bow-tie diagram that is introduced in various studies these days, and adopts the concept of core event.

As shown in figure 2, MTA is divided into two subordinate modules, FTAM (left side) and ETAM (right side), with the core event that is the release of LNG or NG. FTAM consists of various basic events and several protection barriers; ETAM spreads its branches like ETA according to the success or failure of mitigation barriers.

As shown in figure 3, the final events of ETAM correspond to jet-fire, pool-fire, flash-fire, asphyxiation and VCE. However, BLEVE and pressurized vessel rupture show the different structure even with above events, because in those cases there is no core event like LNG/NG release.

Figure 2 Structure of Merging Tree Analysis

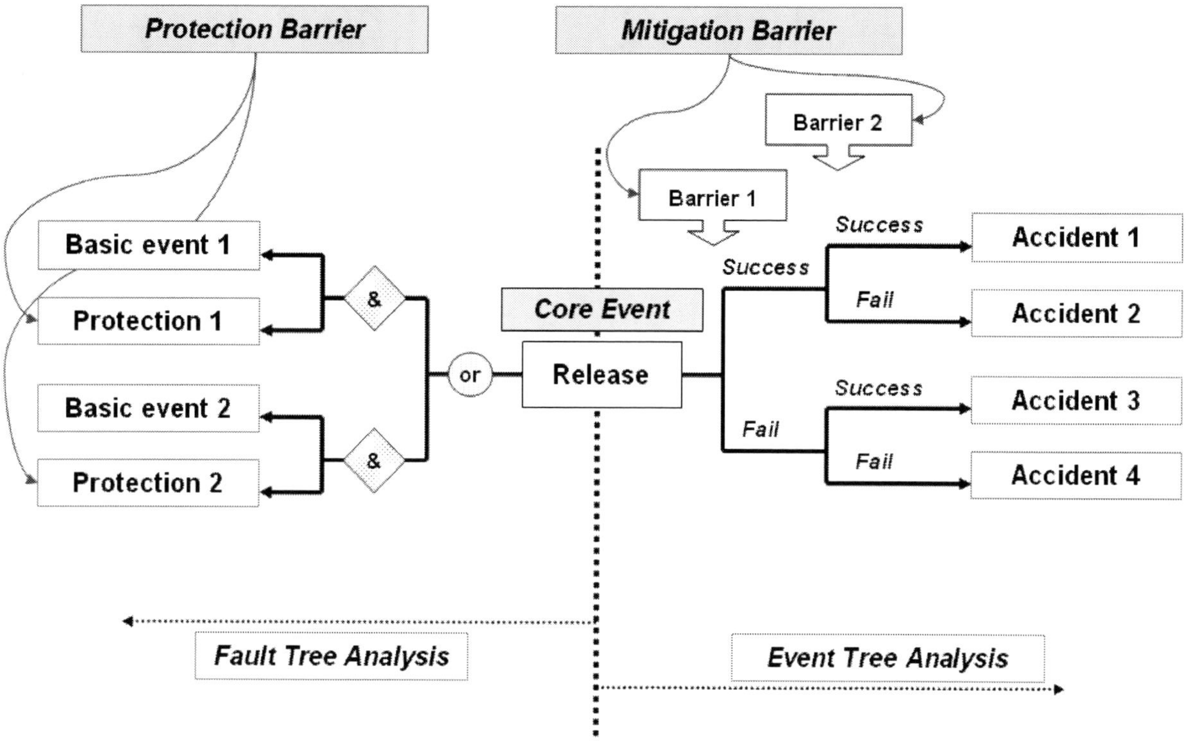

Figure 3 Detailed tree of right wing of Merging Tree Analysis

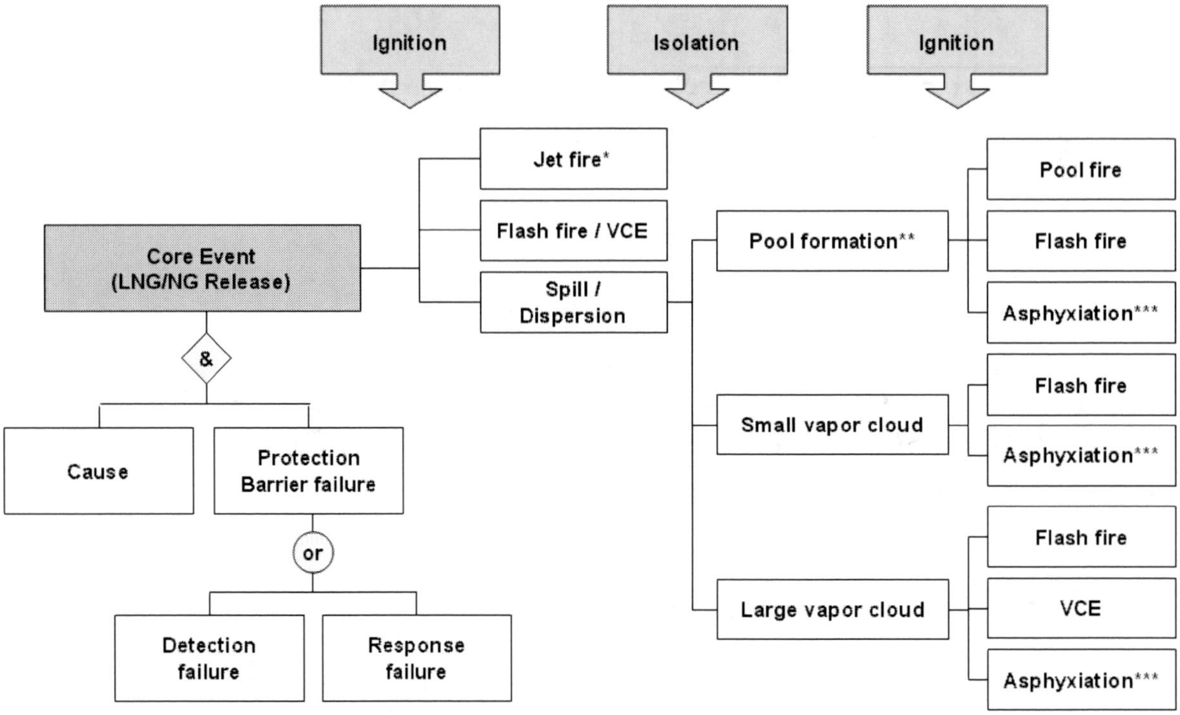

Frequency Assessment Module

When the frequency is evaluated it is necessary to know the number and size of valve, instruments and pipe with the reliability, the probabilities of natural disasters such as seismic, heavy snow, etc. and the following operation specifics.

- Frequency of unloading: fresh unloading and additional unloading
- Frequency of loading and circulation
- Frequency of maintenance: for the empty tank or for the occupied tank

In order to apply the human error, such information as safety management level and appropriate manual, etc. is required.

Consequence Assessment Module

This study presents the worst weather condition that may cause the biggest consequence in case of major possible accidents such as dispersion, jet-fire, pool-fire, flash-fire, VCE and BLEVE. Also the isopleths limit is suggested to conclude the effect of harmful overpressure, thermal flux and concentration.

Table 3 Worst weather conditions for LNG release and LNG fire

Items	Weather condition for dispersion model	Weather condition for fire model
Solar radiation	0 W/m2	600 W/m2
Stability (Pasquill-Gifford stability)	F (Very stable)	F (Very stable)
Ambient temperature	30C	30 C
Wind speed	1.5 m/s	4 m/s
Humidity	50 %	50 %
Reference height	10 m	10 m

Table 4 Contour line limits

Level	Concentration	Thermal impact	Overpressure impact
1	2.5% (50% of LEL of methane)	5 kW/m2	2.78 kPa
2	5% (LEL of Methane)	12.4kW/m2 (for fatality)	6.9 kPa
3	15% (UEL of Methane) 85% (for fatality)	30 kW/m2	15.858 kPa 52.43 kPa-sec (for fatality)

Note: For fatal concentration, under the 15% of oxygen, people can die.

For fatal thermal impact, the probability of death is 90% when people are exposed to12.4 kW/m2 for 1 minute.

For fatal overpressure impact, the probability of death is 90% in case of overpressure impulse at 52.43 kPa-sec.

Note: 5kW/m2 is acceptable thermal impact for public (NFPA 59A) and control room (EN1473).

Effect Analysis Module

The result of frequency analysis and consequence assessment is expressed as an individual risk, societal risk and risk matrix. This result helps to determine whether the accident scenario is fully acceptable, conditionally acceptable or not acceptable. Table 5 shows the acceptable risk criteria for an individual risk and a societal risk, based on reviews of the various standards or codes of Korea, The Netherlands, England and USA.

Table 5 Acceptable risk criteria

Acceptability	Individual risk	Societal risk
Acceptable	F ≤1.0E-06/yr	F-N2 ≤1.0E-02/yr
ALARP region	1.0E-06/yr < F ≤1.0E-05/yr	1.0E-02/yr < F-N2 ≤ 1.0E-01/yr
Not Acceptable	F > 1.0 E-05/yr	F-N2 > 1.0E-01/yr

Note: F means the probability of fatality and N means the number of fatalities

Conclusion

Because the LNG storage facilities are different from other manufacturing facilities in ways of handling material and relevant operation activities, they are in need of a suitable risk assessment model. Therefore this study suggests an appropriate evaluation model specific for the risk assessment of these facilities so that they can perform QRA more easily and concretely using this model.

The final step of QRA is the determination of the acceptance of risk or scenario, so this paper suggests a QRA model consisting of three modules of FAM (Frequency Assessment Module), CAM (Consequence Assessment Module) and EAM (Effect Analysis Module), and furthermore presents the basic input conditions and evaluation basis for each module. The various legislation, standards and codes require the hazard evaluation for LNG facilities to be performed in both cases of their construction and operation. We expect that the result of this study help to perform more easily the hazard evaluations of LNG facilities, especially QRA.

References

1. "A Study on Risk Criteria for semi-quantitative risk assessment of LNG storage tank using modified LOPA and Consequence assessment:", Meejin Kang, et al, WCOGI 2007
2. "Quantitative risk assessment of LNG aboveground tanks based on past operating records of LNG regasification terminals and life cycle assessment", Miyazaki Shinichi, Japan
3. "LNG Code of Practice – Supporting information report", ioMosaic Corp., 2005
4. "Quantification and Comparison of the risks of LNG storage concepts", N Ketchell, R Robinson & P Genoud, 1998
5. "Layer of Protection analysis: Simplified process risk assessment", CCPS, AIChE, USA 2001
6. "Introduction of an easy-to-use risk assessment tool for natural gas transmission pipelines", Jeroen Zanting, et al, Gasunie Research, The Netherlands
7. "Guidelines fro Chemical Process Quantitative Risk Analysis", CCPS, AIChE, USA, 2000
8. "Code of Practice: Nova Scotia", Canada,
9. EN1473, BS, 2007
10. prEN14620, CEN, 2003
11. NFPA 59A: 1996

Biography

Department of Safety Engineering,
Seoul National University of Technology,
172, Gongneung-dond, Nowon-gu, Seoul, 139-743, Korea
Phone: +82.2.970.6935, Fax: +82.2.974.4654,
E-mail: MJKang@snut.ac.kr
Cell : +82.11.781.4093

Ms. Meejin Kang is a PhD student of the department of energy Safety Engineering, Graduate School of Energy & Environment in the Seoul National University of Technology. She had worked for the petrochemical company as a process engineer (1988-1991). She has many experiences in hazard evaluation of the chemical industry and semiconductor industry, etc. (1996-present) and has participated in some projects that intended to develop and/or research about process safety management and risk assessment (2000-present).
She has a certificate of Professional Engineer in Chemical Process Safety.

Experimental Study of Effective Water Spray Curtain Application in Dispersing LNG Vapor Clouds

Morshed A Rana, Benjamin Cormier, Jaffee Suardin, Yingchun Zhang,
M. Sam Mannan*.
Mary Kay O' Connor Process Safety Center, Artie McFerrin Department of Chemical Engineering, Texas A & M University, College Station, TX 77843.
mannan@tamu.edu.

AIChE Spring Meeting, April 2008
8th Topical Conference on Natural Gas Utilization
New Orleans, Louisiana, April 6-10, 2008

Introduction

Water spray curtain is recognized as an effective engineering measure to mitigate various types of hazards in the industries because of its availability, simplicity of use, efficiency and adaptability for different hazards such as gas dispersion, absorption, and fire inhibition [1]. The physical processes involved in using a water curtain to reduce the gas concentration of a vapor cloud include forced dispersion, dilution, and heat transfer. A water spray directed vertically through a gas cloud will have four effects: (i) mechanical effects of creating a barrier to the passage of a gas cloud, imparting momentum to the gas cloud to disperse it, moving gas cloud upwards, (ii) dilution of the gas cloud by air entrainment; (iii) thermal effects between the gas cloud, water, and entrained air; and (iv) physico-chemical effects of absorbing gases with or without chemical reaction [2]. The actions of a water spray in mitigating a vapor cloud may consist of all or any combination of these mechanisms, if properly designed.
Water curtain has also been suggested as one of the most economic and promising LNG vapor mitigation techniques, which can reduce the size of the "exclusion zone" effectively, by forced dispersion. Because of low molecular weight and extremely low temperature, LNG vapor cloud dispersion behaves differently from other dense gases. Water curtains are expected to enhance LNG vapor cloud dispersion mainly through two mechanisms: mechanical effects of forced dispersion, dilution effects and thermal effects. As the solubility of natural gas in water is minimal, the physico-chemical effects can be ignored. LNG Spill experiments by University Engineers for US Coast Guard in the late 70's [3], and wind tunnel and small scale spill tests by Factory Mutual Research Corporation for Gas Research Institute in the early 80's [4] are the major experimental works conducted so far in controlling LNG vapor cloud through water curtain application. The former one simulated spills on LNG ships and used fan-shaped sprays; the later one considered spill contained in a dike and used solid cone and flat sprays, to determine the effectiveness of water spray curtain. Though water curtain represents one promising technique to mitigate the hazard of accidental LNG release, the actual phenomena involved in LNG vapor and water drop interaction was not clearly established from these researches. Other key issues, like (i): the extent of LNG vapor concentration reduction by water curtain; (ii) water curtain effectiveness as a function of spray configuration; (iii) the optimum distance between the water curtain and potential

spill site; (iv) the effective height and width of the curtain; and (v) the highest wind speed for which water curtain remains effective; are yet remained undetermined [4]. That is why a comprehensive design guideline of effective water curtain system for LNG vapor control could not be developed. Indeed, it is also true that there is no simple set of optimum parameters, rather there will be a specific set of design parameters for an effective water curtain system for each particular type of scenario [5]. This paper presents the experimental methodology and some results of the first phase of a research program to study the physical phenomena as well as the effectiveness of a water spray curtain to disperse LNG vapor cloud.

Experimental

This paper is based on the first set of the out-door experiment, conducted on November 16, 2007, at the LNG Emergency Response Training Facility of the Brayton Fire Training Field (Emergency Service Training Institute (ESTI)) in College Station, TX.. LNG spill tests were conducted on the flat concrete ground of the LNG training facility. The experimental set-up and procedure are described including different parts involved in the tests. The experiment consists of the LNG source, the water spray nozzle and water curtain, placements of the measurement points and meteorological measurements. The procedure was to perform natural and forced dispersion of LNG to evaluate the effectiveness of the water spray curtain in terms of concentration and temperature change.

Setup

LNG spill scenario

The scenario for the experiment was continuous LNG flow through a pipe. LNG was supplied from a tanker truck (11,000 gal. capacity) to the spill location with a flexible discharge hose connected to an insulated, fixed Aluminium pipe line. The flow of LNG from the tanker was manipulated by manually controlling the opening of the outlet valve from the tanker to achieve a steady flow rate. The flow rate was continuously measured with a cryogenic flow meter connected between the pipe line and the discharge hose during the tests.

Water Curtain

Water was supplied from the water line to the water spray curtain with a 2 ½" OD fire hose. A flow meter and a pressure gauge were connected at the exit of the fire hose to measure the water flow rate and water pressure at the inlet of the curtain during the tests. Water flow was turned on and off with the ball valve already connected to the water supply line. Two types of portable spray curtains were employed and both were upward directed. One was constructed of seven BETE Fog 1" TF 48 NN nozzles (full cone, spiral, 60°), connected to a 20 ft long 2 inch OD carbon pipe. The other one was Hydro-Shield spray device available through the Brayton Fire School. With this spray head water flows out through a pipe, which has half circular end, and hits a flat plate at the exit. The obstacle created by the flat plate produces 180° flat fan shaped spray pattern. Figure 1 shows both water spray curtains

Figure 1. Photographs of water spray curtains used: (a) full cone, (b) flat fan

Concentration and temperature measurement

34 IR point gas detectors and 32 K-type temperature sensors were used and installed on 14 tripods at different heights to measure the LNG vapor concentration (1.5 ft, 4 ft and 7 ft above ground) and temperature (1.5 ft, 3.5 ft and 6.5 ft above ground) of air-LNG mixture. The tripods were placed at different downwind distances from the LNG spill area to measure the concentration and temperature both upstream and downstream of the water curtain. The instruments were placed according to the predicted wind direction of the experiment day. Figure 2 shows the sensors, detectors and water curtain placements.

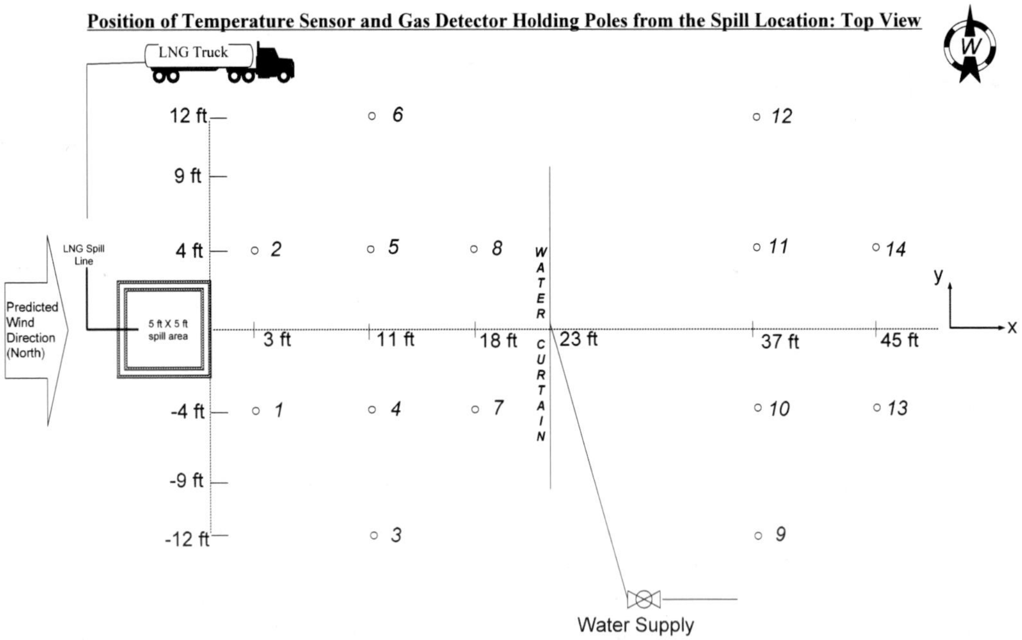

Figure 2. Experimental set-up

The gas detectors were not physically installed on the tripods, but were installed in a wooden vacuum chamber. Long Teflon tubes were connected to one end of each detector to flow methane sample from the field to the detectors in the chamber. The other end of the tube was placed on the tripod pole in the test area. A vacuum pump was used to pull LNG vapors to the detector through the tubes and then released to the environment. Figure 3 shows photograph of tripod poles positions.

Figure 3. Photograph of equipment placement

Other equipment

The meteorological conditions were measured during the experiments with three weather stations and three anemometers. These stations measured the wind velocity, direction, temperature, humidity and heat index. Two hydrocarbon (H/C), one IR and two regular digital cameras were used to capture the whole experiment. The hydrocarbon and IR cameras were used to observe the actual dispersion of invisible flammable hydrocarbons (mainly methane). The H/C cameras were placed away from the visible cloud region and at 90° to each other, so that they can capture all the moments.

Data acquisition

Gas detector, temperature sensor, flow meter and pressure gauge data were acquired at an interval of 1 second with a microprocessor based data acquisition system and recorded simultaneously in two computers. The weather stations and the anemometers have built-in data record system, which could save data for three days. The weather and wind data were transferred to computers after the experiments.

Procedure

Four similar spill tests were conducted to evaluate the effectiveness of two types of water spray curtains in controlling an LNG vapor cloud. Each curtain (full cone and flat fan) was tested twice with separate spill. The first two tests were conducted with the full cone spray curtain and the second two were with flat fan curtain.
Water curtains were directed vertically upward and positioned perpendicularly to the prevailing wind direction, at a fixed downwind distance from the spill location. All of the tests consist of following two measurements: (i) natural dispersion of LNG vapor when no action was taken to control the vapor; and (ii) forced dispersion of the LNG vapor when water spray curtain was applied to control the vapor.

Each test started with continuous spill of liquid LNG onto the concrete ground. As soon as the liquid touched the concrete ground it vaporized and dispersed naturally towards the prevailing wind direction. After achieving dense vapor cloud flow, the water spray curtain was turned on to disperse the cloud forcefully. The flow of LNG was still continued for several minutes during forced dispersion. When the LNG pool reached about 1" to 1½" height, the LNG flow was turned off while continuing the water curtain operation. The curtain was kept on until the visible white cloud disappeared.

Results and Discussion

During the tests, the wind speed averaged 4.5 miles per hour and blew mainly from the south-west to south-east region. The average ambient temperature during the tests was 22.4° C and humidity was 25.5%. A total of 4 tests were completed in a 1½ hour period. Total usage of LNG and water was approximately 1400 and 4700 gallons. Average water pressure for the flat fan spray, at the nozzle head, was approximately 90 psi and for the full cone spray, at the pipe inlet was about 40 psi.

During natural dispersion, LNG disperses upward and gets diluted as it travels downwind. Gas concentrations at 3 ft, 11 ft, 37 ft and 45 ft downwind distances from the spill and 1.5 ft, 4 ft and 7 ft heights were measured in each test. Data shows that concentrations 1.5 ft above the ground and at downwind distances closer to the spill location are higher than concentrations at 4 ft and 7 ft heights. As the vapor travels away from the spill, concentration at 1.5 ft becomes lesser than higher level (4 ft and 7 ft) concentrations. The following figures show examples of natural dispersion state of LNG vapor cloud after 253 sec of continuous spill (flow rate = 56 gpm) at different heights where the unit of concentration is % v/v (methane-air mixture).

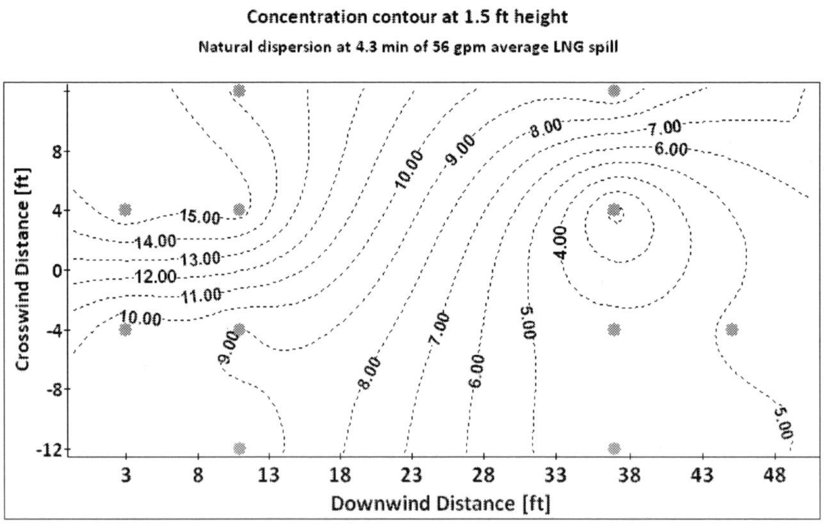

Figure 4. Concentration contour at 1.5 ft above ground during natural dispersion

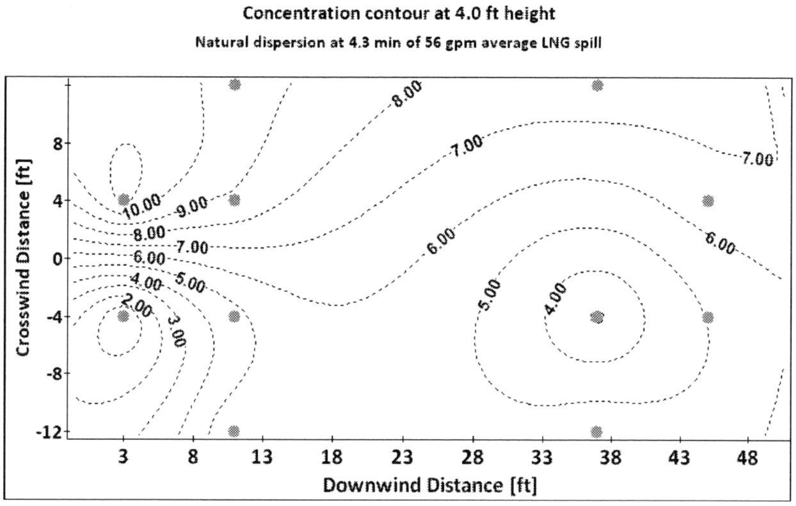

Figure 5. Concentration contour at 4.0 ft above ground during natural dispersion

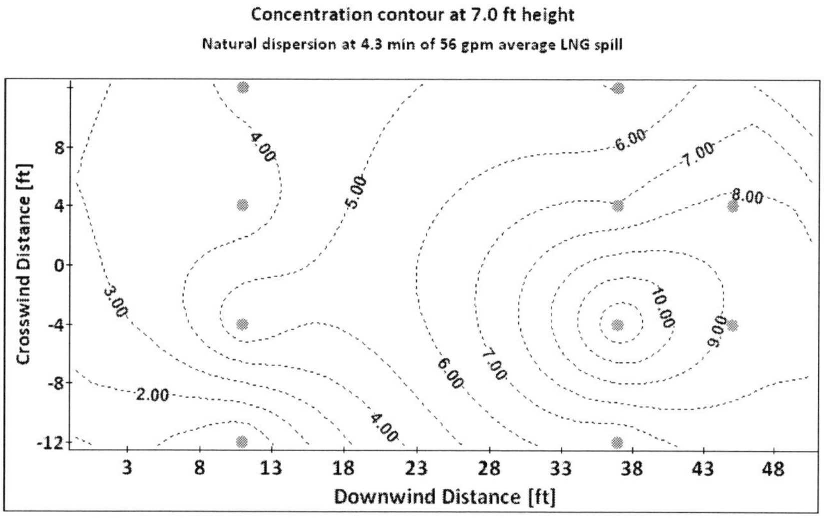

Figure 6. Concentration contour at 7.0 ft above ground during natural dispersion

As there was time lag between the LNG flow from the pipe and detector reading, all the data were edited before studying to minimize the time lag. The gas detector readings show fluctuations. The wind gustiness and crosswind turbulence cause the fluctuation in the methane concentration reading. The fluctuation observed close to the spill location was mainly due to the turbulence created by the momentum of the spill. LNG flow from the pipe was still continued for some time after turning the water spray on. Therefore in preliminary data analysis average readings of the detectors placed at three different heights and each downwind distance are considered. Water curtain effectiveness was determined in terms of percentage average concentration change (%CC) by the spray. To determine %CC, it was first assumed that: (i) vapor cloud travels horizontally at wind velocity, w (ft/sec); (ii) in the spray region, due to entrained air velocity in the spray, v_a (ft/sec), vapor travels at velocity u (ft/sec), where $u = \sqrt{v_a^2 + w^2}$; and (iii) vapor again travels at wind velocity outside the spray region. These assumptions were adapted from Moore's paper [6]. In the calculations, w was known and v_a was determined for each spray from correlation presented by McQuaid [7]. The total travel time, $t(=t_f-t_i=t_1+t_2+t_3)$ for a vapor cloud slice to travel from 11 ft downwind to 37 ft downwind distance through the spray was calculated based on the velocity. Then %CC was determined as

$$\%CC = \frac{C_{\text{avg at 11ft at } t=t_i} - C_{\text{avg at 37ft at } t=t_f}}{C_{\text{avg at 11ft at } t=t_i}} \times 100\%$$

; where C_{avg} is the average concentration in % v/v (air-methane mixture) at different height. As it is a measure of average concentration change, positive value of %CC means concentration reduction and negative value means concentration increase after the spray region. Figure 7 summarizes the %CC calculation procedure.

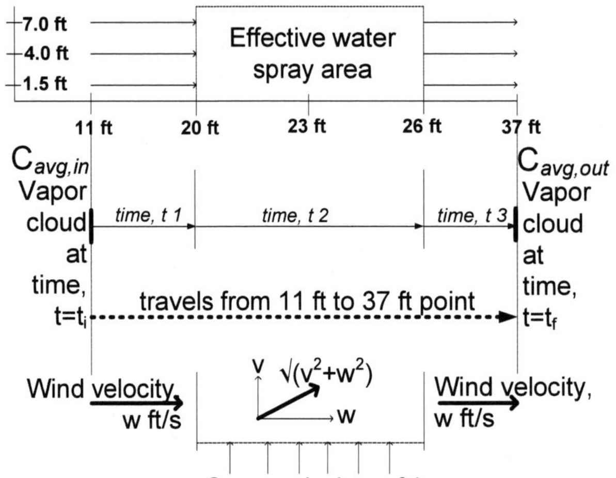

Figure 7. Downwind travel of a LNG vapor cloud slice

For analysis, %CR was determined for each height considering that vapor cloud passed through the spray from 11 ft point (at t_i sec) to 37 ft point (at t_f sec) in t_f-t_i sec Figure 8 and 9 show percent (%) average concentration change by full cone and flat fan spray respectively at different time.

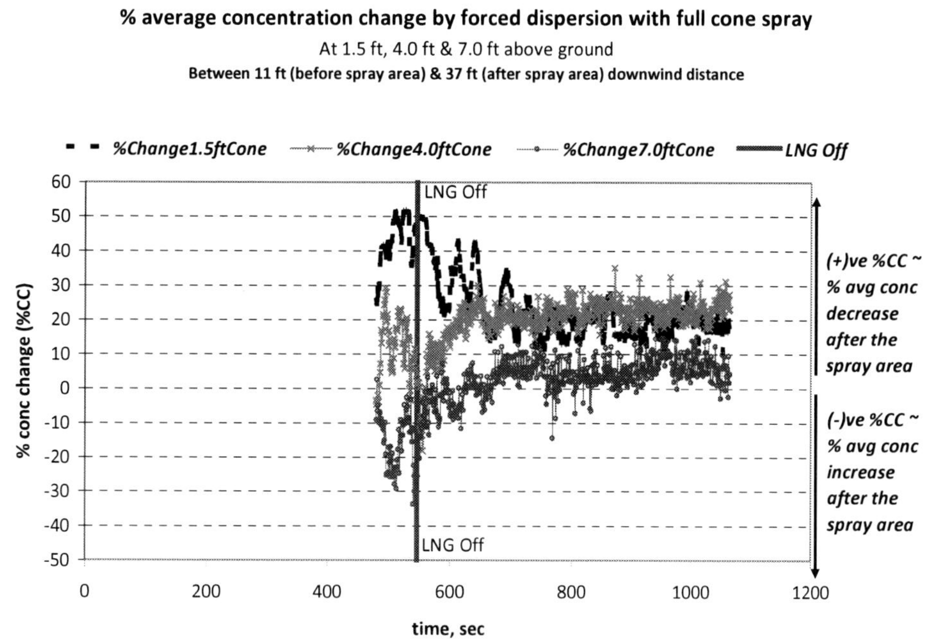

Figure 8. %CC by full cone spray

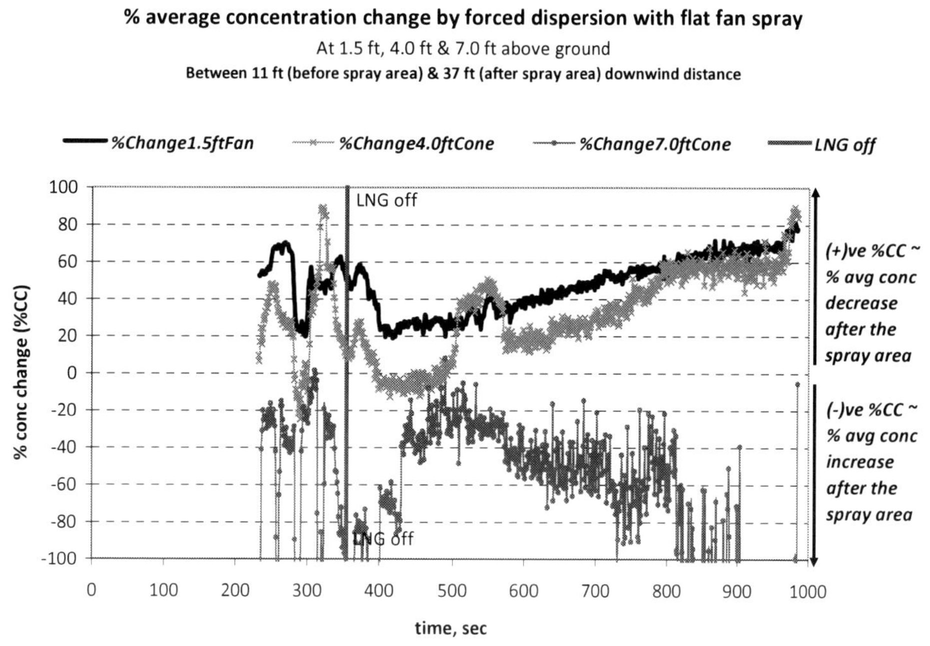

Figure 9. %CC by flat fan spray

All of the test data showed change in concentration of LNG vapor downwind of the spray area. Flat fan curtain reduced the concentration significantly at lower levels when the LNG flow was turned off but it increased concentration at 7 ft height (negative %CC). This implies that it pushed the vapor upward more than diluting it. The average concentration change by the full cone spray decrease but the curves become flat over time when the LNG spill was stopped. Concentration at 7 ft height became lower after the spray region. That means the concentration change could be mainly due to dilution. It could not push the vapor upward efficiently.

At the first phase of forced dispersion, the concentrations at the both distances were still increasing. The water curtain was positioned 23 ft downwind from the spill location. The wind's frequent change of direction allowed the LNG vapor to disperse without passing through the water spray region many times and caused fluctuations in the methane concentration measurements. However, the rate of concentration increase was reduced during the water curtain action, though LNG was still flowing from the pipe. Gas concentration started to decrease at both the distances during the second phase of forced dispersion after stopping the LNG flow. However, the concentration behind the curtain area never reached below one half of the LFL, 2.5% (v/v), by forced dispersion, in any of the tests.

Conclusion

The tests conducted in this research indicate that the action of water sprays reduced the LNG concentration as well as pushed the vapor upward. However, the tests could not show a significant change in concentration at higher level above the ground than the ground level. The full cone spray is more effective than the flat fan in creating turbulence closer to the spray region. The flat fan is effective in reducing the ground level concentration by creating a barrier in the path of the cloud as well as pushing it upward. The tests were not performed for a long enough period of time to reach more decisive quantitative results. Each test had to stop within a short time because the amount of water used was large to cause problems to some of the instruments closer to the test area. This also indicates that to control a large spill the required system has to provide enough water which might not be very cost effective if it runs for a longer period of time. Draining the water can be a problem. However, to establish a definitive engineering guideline on water curtain design for LNG cloud dispersion, experiments for longer time and at different wind conditions are also needed.

Acknowledgement

The authors are grateful to **BP Global Gas SPU** for their financial support of this study. They would also like to thank the **Brayton Fire School and LNG Emergency Response Training** team for their help and contribution in this research.

References

1. Uzanski, D.T. and Buchlin, J.M., "Mitigation of Industrial Hazards by Water Spray Curtains", The 5th Conference of the International Emergency Management Society (TIEMS), Washington, DC, (May 1998), 19-22.
2. Dusserre, G., Dandrieux, A., Thomas, O., "The DVS Model: a New Concept for Heavy Gas Dispersion by Water Curtain", Environmental Modeling and Software, 18 (2003), 253-259.

3. Martinsen, W.E., Muhenkamp, S.P., “Disperse Vapors with Water”, Hydrocarbon Processing, (1977), 260-267.
4. Heskestad, G., Meroney, R.N., Kothari, K.M., “Effectiveness of Water Spray Curtains in Dispersing LNG Vapor Clouds”, Proceedings of the American Gas Association, Transmission Conference Paper, 83-T-69, 2-4 May 1983, Seattle, Washington, T169-T183.
5. Harris, N. C.; “The Design of Effective Water Sprays – What We Need To Know”, North Western Branch Papers, IChemE, (1981), 7.1-7.5.
6. Moore, P.A.C.; Rees, W.D.; “Forced Dispersion of Gases by Water and Steam”, North Western Branch Papers, IChemE, 4.1-4.14, 1981.
7. McQuaid, J.; Fitzpatrick, R.D.; “The Uses and Limitations of Water-Spray Barriers”, North Western Branch Papers, IChemE, 5, 1.1-1.13, 1981.

Biography of the Speaker

Morshed A. Rana
Research Assistant, Mary Kay O'Connor Process Safety Center
Department of Chemical Engineering, Texas A&M University, 3122 TAMU,
College Station, TX 77843-3122
Phone: (979) 862-1310, Fax: (979) 458-1493
E-mail: Morshed.rana@chemail.tamu.edu

Rana, Morshed A. is a PhD student of the Chemical Engineering Department of Texas A&M University advised by Dr. Sam Mannan and a research assistant of the Mary Kay O'Connor Process Safety Center. He attended Bangladesh University of Engineering and Technology and received his BS in chemical engineering. He obtained MS degree in chemical engineering from the University of South Alabama, Mobile, AL in 2004.

Mr. Rana started his career in a plastic products manufacturing company and worked there for one and half year as a junior chemical engineer. During his MS study he got involved working in research areas. His MS research was rapid determination of adsorption isotherms using Gas Chromatography. He also worked as the lead research assistant in Fuel Cell project and worked on waste heat utilization.

Mr. Rana joined Mary Kay O'Connor Process Safety Center in August 2005 and began conducting research under the guidance of Dr. Sam Mannan. His current research topic is to study LNG vapor dispersion behavior using mainly water curtain. In addition to conducting research in process safety area, he is also involved in several industrial projects which include HAZOP study, P&ID and PFDs developments, and consequence modeling. He interned with Whiting Refinery, BP in the Operation Canadian Crude (OCC) project in summer 2007 where he was involved in projects on H_2S dispersion and mitigation.

Managing Escalation Risks in LNG Terminals

Jørgen Bartnes
Principle Consultant, Scandpower
Houston, TX

Baris Arslan
Consultant, Scandpower
Houston, TX

AIChE Spring Meeting, April 2008
8th Topical Conference on Natural Gas Utilization
New Orleans, Louisiana, April 6-10, 2008

Abstract

Over the past decade, the global LNG industry has grown rapidly both in terms of liquefaction and regasification capacity. Seeing a great number of proposed and ongoing LNG projects around the world, one can easily conclude that the LNG industry will most probably keep on growing in the near future. In LNG industry, a great attention is given to safety aspects partly due to strict regulations and negative public opinion. For this purpose, both qualitative and quantitative risk assessments are usually conducted depending on the project phase, applicable standards and prevailing regulations. One of the biggest challenges encountered in most LNG risk analyses, is the evaluation of the escalation risks. The challenge of adequate evaluation of escalation risks is essential with regards to transparency and usability of the risk analyses. This paper aims to demonstrate the evaluation and management of internal and external escalation risks in LNG terminals. Understanding and managing the internal escalation risks is a demanding task for all LNG-terminals, but especially for Floating Storage and Regasification (FSRU) facilities. External escalation risks are, on the other hand, more relevant to discuss in case of onshore LNG facilities. The object of this paper is to illustrate how to manage the escalation risks by help of case-specific examples from two ongoing LNG projects.

Introduction

Preventing major accidents is one of the most important safety concerns in LNG terminals as well as in any other complex facility, handling flammable chemicals. However, this can by no means be achieved without correctly understanding the escalation risks in a facility. Escalation analysis is therefore an integral part of all risk analyses.

Prevention of major accidents involving dangerous substances has been a subject to a great number of international standards and directives. For example, Seveso Directive, prepared upon Seveso Disaster in Italy, is a detailed one with focus on protecting the community in an effective manner. The Seveso accident took place in a chemical plant in Italy and resulted in the highest known exposure to tetrachloridibenzo-p-dioxin (TCDD) in residential populations. Due to Seveso accident, up to a few kilograms of TCDD was released over an area of 18 km^2, affecting thousands of residents and animals. Upon Seveso disaster in Italy, European Community enforced more strict regulations, called Seveso directive.

Other international and national standards and requirements specify evaluation of escalation potentials and escalation prevention measures. The Norwegian regulations for both for onshore and offshore activities have such requirements and various ISO- standards like 10418 and 13702 specify evaluation methods and propose preventive measures.

Today, for the new development projects, an escalation analysis has to be carried out in order to evaluate the risks for domino accidents. As the local authorities need to ensure that decisions made appropriately balance risks to people, project costs and public benefit, it is highly important to estimate the contribution of escalation risks to the overall risk picture.

In this paper, a brief description of methodology for risk evaluation with focus on escalation risks is given. Thereafter, two case studies are described to allow the reader comprehend how the probability of failure and potential consequences of domino accidents are addressed during safety evaluation of an LNG-plant. The case studies are specifically chosen from two ongoing projects. The first example is from a FSRU (Floating Storage and Regasification Unit) where the internal escalation is the major concern. The second example deals with an onshore LNG terminal where the external escalation is the primary concern.

Evaluation of Risk, Special Focus on Escalations

Basic Methodology

The basic methodology applied for evaluation of risks of many types of facilities or installations, including LNG- terminals, are described below. The method includes the following main steps with description of tasks relevant for LNG- terminals or other oil and gas process plants:

HAZID (HAZard IDentification)
A HAZID of the operation of the facility based on the available information is often the starting point of and risk assessment. In this meeting hazardous incidents and the scenarios leading up to them are discussed and identified. A selection of these hazards and scenarios are further analysed in the risk assessment.

Frequency Estimation
The leak frequencies of the selected scenarios from the HAZID are then calculated based on available information as P&IDs, design basis, etc. Leak rates and ignition probabilities are calculated based on relevant statistical information.

Consequence Analysis
Based on the scenarios selected from the HAZID, dispersion, explosion and fire simulations are performed using various computer software. More comprehensive analyses use up to date CFD- software.

Risk Analysis and Risk Assessment
The frequency estimation and the consequence analysis are then combined in a risk model using event trees. The results are then assessed against the acceptance criteria. Risk reducing measures are suggested, based on the ALARP principle.

The risk assessment methodology is shown in the figure below:

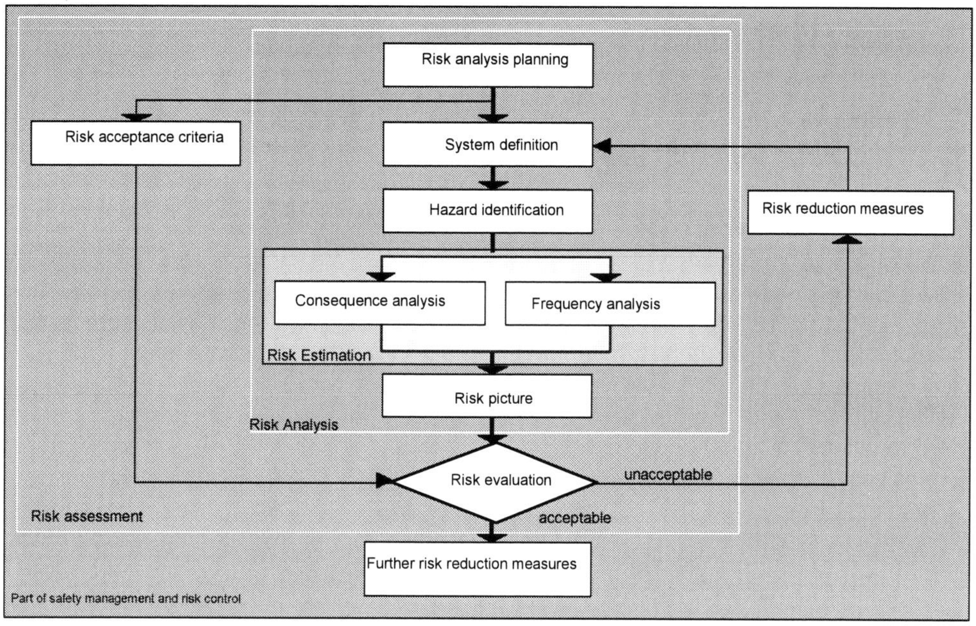

This basic methodology can be used for all kind of risk evaluations from very detailed quantitative analyses to coarse qualitative analyses. The risk picture can be assessed against various acceptance criterions as human risk criterion (1st party fatality risk, 3rd party risk), criterion specifying impairment of safety function, criterion specifying environmental impact or loss of assets and others.

One of the main challenges is to model the real situation or operation in a way that reflects the various barriers or measures established to avoid the undesirable events. The following figure shows the bow-tie methodology that illustrates a meaningful way of looking at the evolution of accidents/undesirable events. This figure shows among else the event tree with barriers and branches fitted in the evolution of an accident. If we stick to this visualization of the evolution of an accident and are able to describe the accident at the right level, there is a good starting point for the analysis.

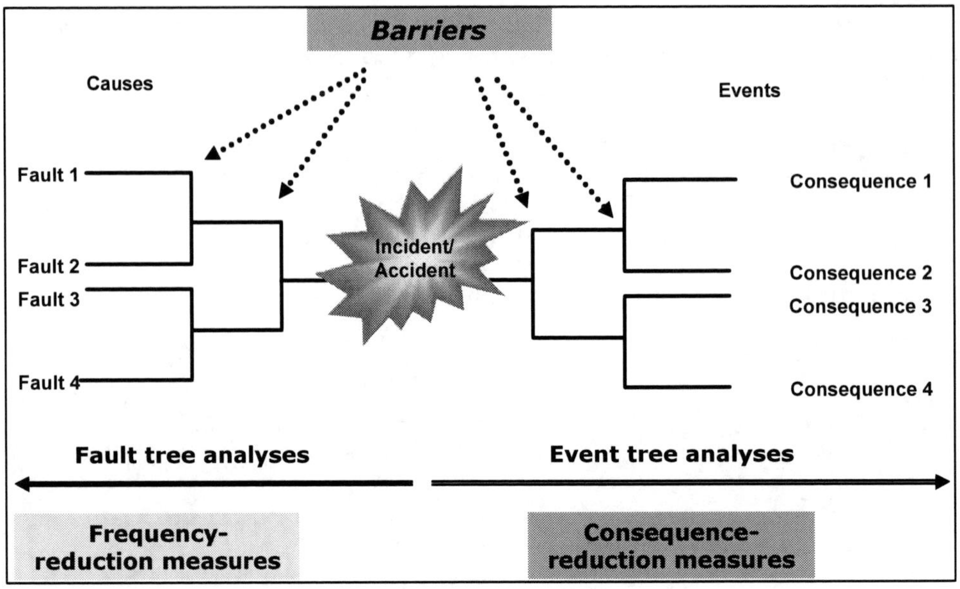

The various consequences of possible hydrocarbon (in this case LNG or gas) leakages are important to evaluate. The main question related to these consequences is if there is an ignition source or not. If a leak ignites, it can result in following three consequences:

- Deflagration of a gas cloud
- Heat radiation from pool or jet fires
- Explosions

Simulations of these effects are often conducted to be able to evaluate the effect on humans or equipment.

Potential outcomes of an unignited leak can be as following:

- Cryogenic effects (only LNG)
- Rapid phase transition (only LNG)
- No effect at all

These effects are often described in more general terms based on available knowledge.

Ignited leaks and hence heat radiation from pool or jet fires and deflagration of gas clouds are usually assessed to have most influence on the risk picture for LNG terminals or FSRUs. Explosions are less likely to occur in most LNG terminals as there are usually few enclosed areas and also due to a fairly good ventilation condition.

Evaluation of Escalation Risks

An escalation analysis is a systematic approach to find out how a primary event can escalate and finally endanger the plant personnel or the residents in the area. In case of an LNG terminal, it is carried out in several steps. First of all, escalation hazards are identified by what the primary event can escalate to damage the vital sections / systems in an LNG terminal. Thereafter, both active and passive safety systems are evaluated in the terminal in order to find out how the domino accidents can be prevented. Additionally, probabilistic models are created for all credible escalation scenarios.

The escalation risks can simply be divided into two categories as internal- and external escalation risks. In case of an internal escalation, a primary event escalates from one section to another section within the borders of the terminal. On the other hand, an external escalation refers to the fact that a primary incident escalates to another facility and leads to a major accident. In fact, a primary incident can first escalate internally and thereafter externally.

Modeling of the escalation risks, reflecting the established barriers in the design can be performed by event trees with several branches related to the various escalation possibilities. The evaluation of the escalation is then done step by step in the event tree. The probability for each branch is found from an evaluation of the combination of consequences/impact (heat radiation levels, duration) and the protection measures (inherent safety design/protection, distances, operational procedures, maintenance/inspection, etc.). The probability for:

- Consequences that go beyond the established barriers, or
- That the established barriers are not properly maintained

will then give the likelihood for the various credible escalation events to happen. The figure below illustrates a model of barriers preventing an accident to happen and reducing the losses. This model shows several barriers at different stages in the evolution of an accident to prevent an accident to occur and to reduce the losses. In this case, barriers can be as following:

- Physical barriers as a part of the construction
- Maintenance/inspection barriers conducted by personnel on regular basis
- Operational procedures (barriers) conducted by personnel due to specific signal or events

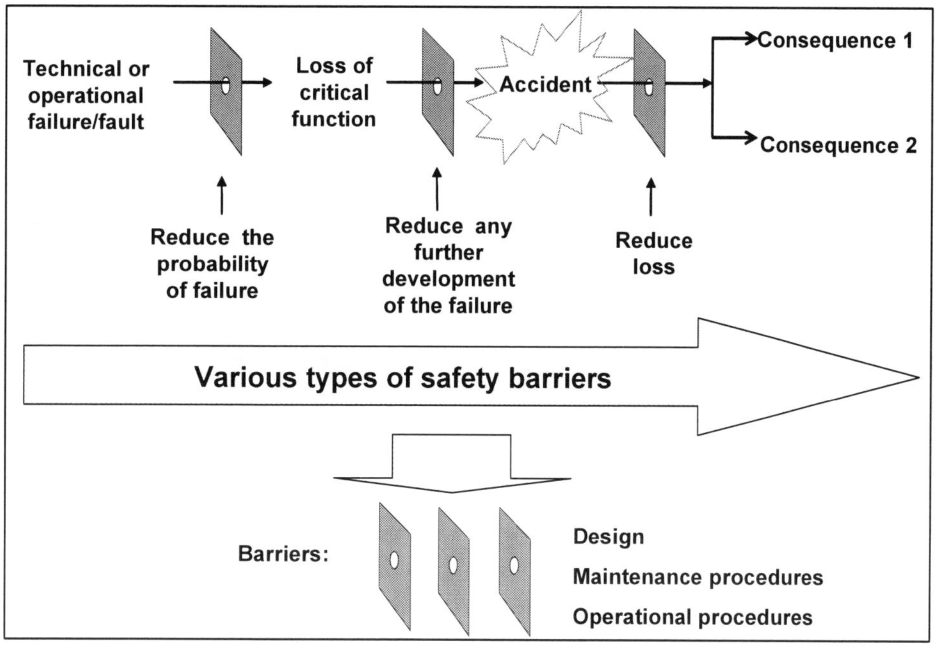

The combination of "bow-tie" evaluation of undesirable events together with an evaluation of the existing barriers provides input to an event tree, created to analyze escalation possibilities. An example of that is shown in page 9.

Risk-based Approach for Managing Escalation Risks in LNG Terminals

The aim of this chapter is to illustrate the performance of escalation analysis for two different LNG-facilities. The case studies are specifically chosen from two ongoing projects. The first example is from a FSRU (Floating Storage and Regasification Unit) where the internal escalation is the major concern. The second example deals with an onshore LNG terminal where the external escalation is the primary concern.

The case study for the Floating Storage and Regasification Units (FSRU) is based on a Quantitative Risk Analyses (QRA) conducted in 2007. The second case study, regarding the onshore LNG import terminal is based on a number of risk assessments for a new development project in Sweden. (See references)

Floating Storage and Regasification Units

The possible escalation scenarios

The FSRU analysed was a conversion of an LNG carrier owned by Golar LNG, and with approximately 125,000 m^3 LNG storage capacity. The FSRU has a total length of 287.5 m and a design draft of 11.7 m.

The conversion of the LNG carrier into an FSRU was focused on the installation of a regasification plant and the associated utility and safety systems. The FSRU will discharge high pressure natural gas from a gas manifold, for export of the gas to shore via export arms, installed at a jetty.

Illustration: Golar LNG

With a location in a busy port, but with good spacing to other activities and ship traffic, the main concern related to escalation was internal escalation from an initial event. Escalation to other facilities was almost impossible due to far distances.

Fire and explosion simulations

The simulation results of large jet fires due to initial leaks in the regasification and export area is shown in the figures below. The first figure shows that a leak directed towards the tanks from the regasification units attains radiation levels of up to 428 KW at time approximately 30 seconds. The second figure shows that a leak directed towards the tanks from the export manifolds attains radiation levels of up to 364 KW at time approximately 30 seconds.

The simulated leak rates are large (66 and 128 kg/s) related to the inventory of the various ESD- segments. That results in fairly short duration (less than 2-3 min.) of these high radiation levels.

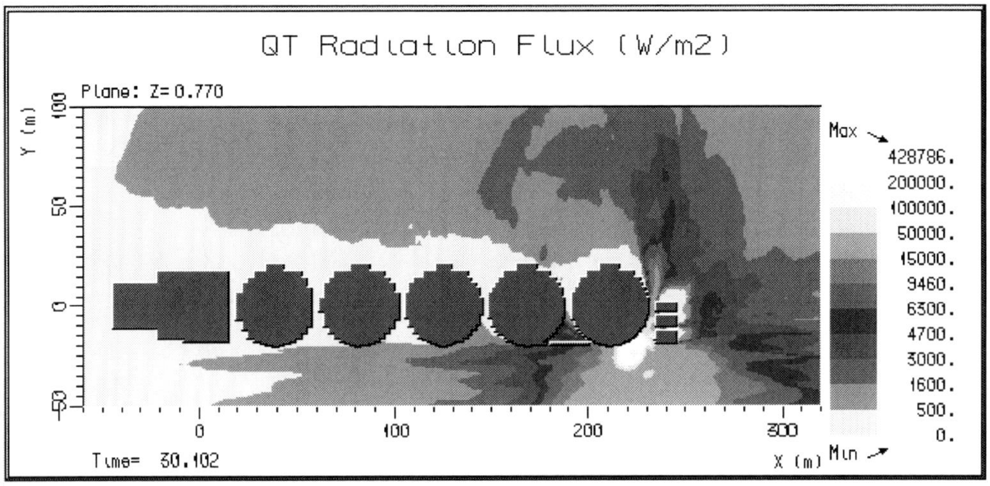

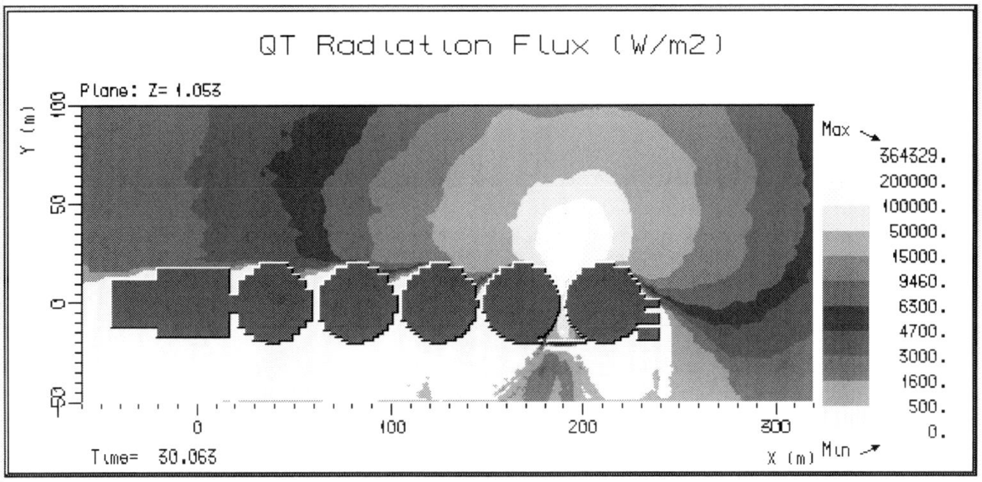

A study of the explosion conditions in the regasification units was performed. This study calculated the side-on pressure in two stages:

- FLACS simulations to determine the source strength see the figure below
- Multi-energy method, which uses the source strength, obtained from FLACS simulations to give a plot of decay of blast loads with distance from the regasification units.

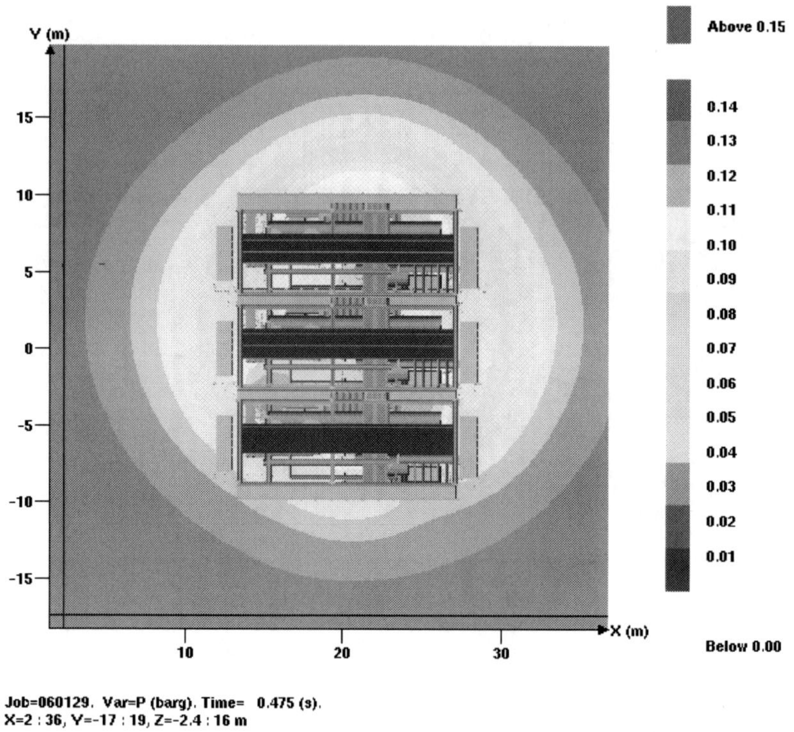

The calculations show that the pressure does not exceed 0.2 barg either inside the Regasification unit or at any distance outside the unit. The Design Accidental Load for the regasification units and other equipment in the vicinity is at least 0.2 barg side-on pressure.

Evaluation of escalation probabilities

The evaluation of escalation probabilities is conducted as a part of the event tree calculation. Escalation due to either explosion or fire or combination thereof is included in the evaluation according to the methodology described in prior sections. An example of an event tree is demonstrated here:

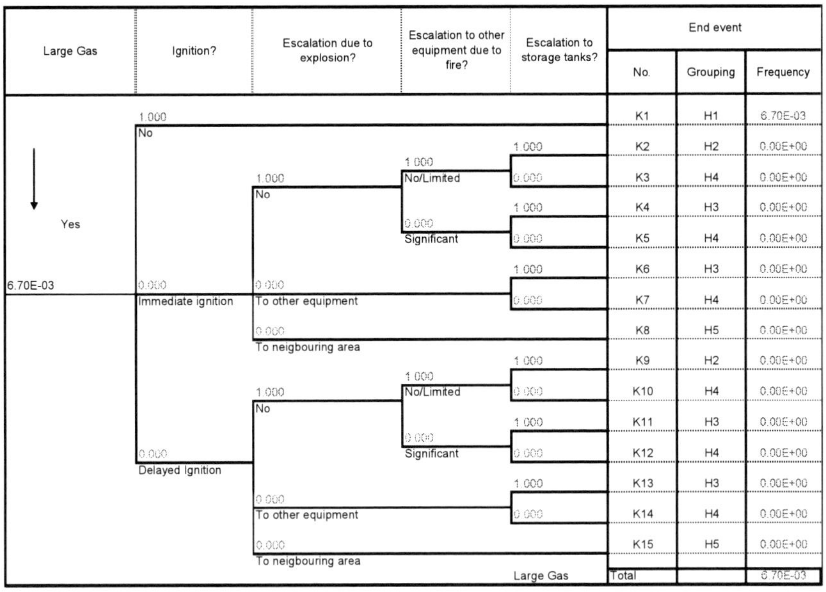

This evaluation shows fairly low probability for escalation. The underlying assumption was that an ignited gas leak has occurred and its likelihood of ignition was also small. The summary of the escalation probabilities are shown in the following tables:

Probabilities for escalation to other equipment assuming an ignited gas leak

Escalation due to:	Area	Probability for escalation (%)							
		Gas				LNG			
		Small	Medium	Major	Large	Small	Medium	Major	Large
Explosion	Regas			0.2 - 0.4	1.0 - 2.0			0.2 - 0.4	1.0 - 2.0
Fire from LNG leaks	Imp/Storage					0.6	2.1	4.2	4.8
	Regas					0.6	2.1	4.2	4.8
Fire from gas leaks	All	0.9	4.6	5.9	5.2				

Probabilities for escalation to the storage tanks assuming an ignited gas leak

Medium	Probability (independent on leak sizes)
Gas	0.02
LNG	0.03

By applying the above described methodology, the frequencies for escalation to other equipment or to the storage tanks were calculated. The resulting frequency for the fire escalated to other equipment is 5.54E-05 whilst the frequency for the fire escalated to storage tank is 3.44E-05

These frequencies for escalation to other equipment and to storage tanks are below annual frequency limits commonly in the range of 1.0 E-04 to 5.0E-04.

Onshore LNG Terminals

The LNG import terminal, discussed in this paper, will be built in Brunsviks holme, an island, neighbouring an existing refinery located approx. 70 km south of Stockholm. The planned LNG terminal will have a storage capacity of 20 000 m^3 LNG.

In this LNG project, analysing escalation risks had an important role since the terminal is located adjacent to an existing refinery. The distance between the refinery area and LNG storage tank is about 500 m. A possible external escalation may lead to a chain-accident, endangering both the personnel in the refinery and residents in the area.

An escalation analysis has been conducted based on two conservative scenarios in order to find out if the "worst credible escalation scenario" may endanger the refinery. Due to customer restrictions, some details regarding escalation analysis cannot be revealed here.

Following cases are studied:

- Large natural gas leakage from LNG storage tank
- Large LNG leakage during unloading

For the simulations, a CFD simulation tool, KFX is used. KFX is a very advanced CFD code for gas dispersion, used for both design and analysis purposes. KFX is developed by ComputIT in Norway.

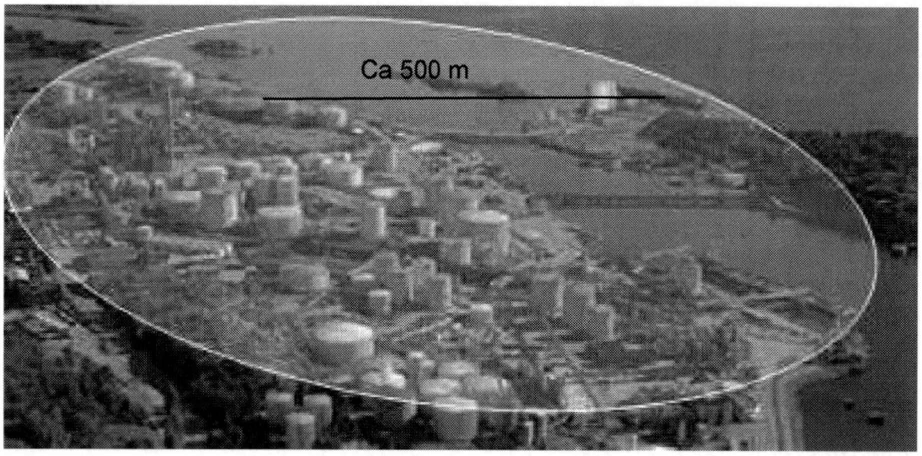

Scenario 1: Large natural gas leakage due to rollover in the LNG-storage tank
This escalation scenario is based on a rollover incident in the LNG storage tank, causing rapid release of large volumes of natural gas vapor to the atmosphere. In this case, it is assumed that the relief arrangement in the storage tank fails to handle the generated boil-off gases. The reader should bear in mind that this is a very conservative scenario, as there are various rollover prediction and protection systems available today.

In this escalation analysis, it was of vital importance to answer the following two questions:

- Can the flammable gas cloud, formed due to rapid natural gas release from the LNG tank, reach to the refinery area?
- If the gas cloud can reach the refinery area, how long time does it take?

The first question is important as it enables us to understand the extension of escalation risks. The second question is also very important to ensure proper organization of emergency preparedness plans and resources in the refinery area and LNG terminal.

The results of CFD simulation show that a flammable gas cloud, formed due to rapid natural gas release from the LNG storage tank never reaches to refinery area under analyzed conditions. The dynamic simulations show that the gas cloud will quickly dilute and go to the concentrations where it is no longer flammable.

The pictures below show a gas cloud which covers a great part of the terminal area and also reaches the neighboring refinery. However, this gas cloud is not flammable due to its low methane concentration. Pictures below are given for illustrative purposes.

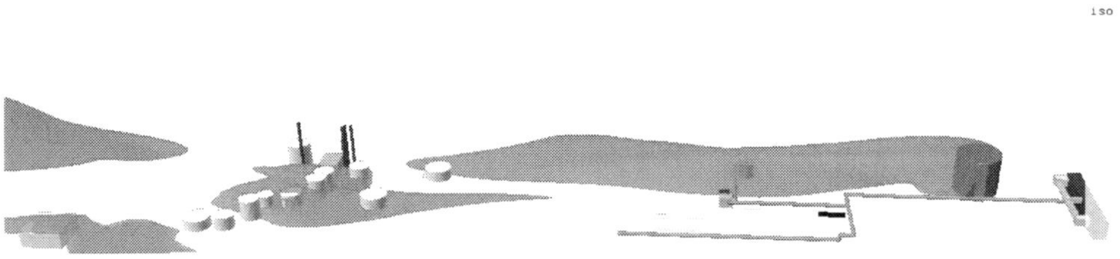

Scenario 2: Large LNG-leakage (jetty area) during unloading of LNG-ship

In this case, a large LNG-leakage during unloading operation is assumed to take place. It should be noted that this event is unlikely to happen, as of the first case in this paper, from a probabilistic point of view. However, our main aim was to investigate a "credible" worst-case scenario. Selection of credible scenarios is done in a HAZID-session, described previously in this paper.

In this case, the CFD simulation showed that no flammable gas cloud will reach the refinery and possess a danger. Therefore the escalation potential for this case was evaluated to be remote. The gas cloud, illustrated in green color in the pictures, is not flammable due to its low methane concentration.

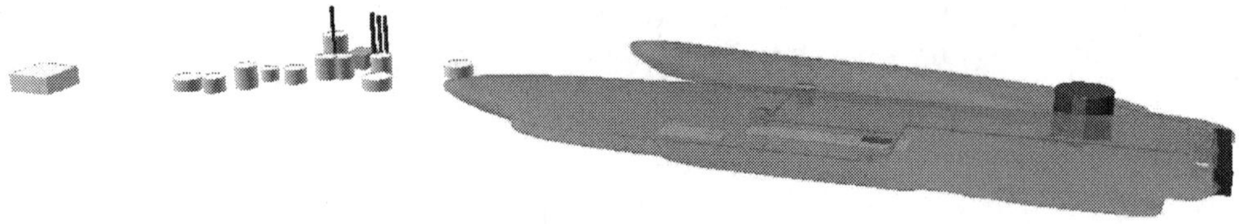

For Case 1, the simulation showed that the gas cloud will enter steady-state after ca 3 minutes. The flammable concentrations were found only in the close proximity of the storage tank within a very limited area. For Case 2, the gas cloud will enter steady-state after ca 6 minutes. The simulations showed also that the gas cloud in Case 2 will expose a greater area than Case 1.

To conclude, it can be stated that the gas clouds will most probably effect the LNG ship and the LNG terminal, and not the refinery. Even if the gas clouds reach the refinery in an extremely unlikely situation, it will take at least 3 minutes. It means that both the refinery and LNG-terminal will have sufficient time to take necessary measures to prevent a potential domino accident.

Discussions

The purpose of the escalation analysis in case of the onshore LNG-terminal is not exactly the same as in a FSRU. In the FSRU-case, the risk analyst does, in fact, know that when a primary accident takes place, domino accidents may be inevitable in such a compact plant if not the escalation hazards are addressed correctly. Therefore the focus in the escalation analysis here is actually to generate recommendations in order to improve the current design and eliminate escalation hazards by that way.

In the onshore LNG-terminal case, the main purpose was to learn if there is an escalation hazard. Knowing that escalation risk is remote or not, emergency preparedness resources can be allocated correctly. In our case, the risk of a domino accident between the LNG terminal and the refinery is not considered to be relevant. Therefore, there is no need of evaluating additional measures, specifically targeting potential domino hazards. If the simulations had indicated such a hazard, the escalation analysis, would recommend a number of barriers, improving the current design, emergency preparedness plan, maintenance and operational procedures etc.

The escalation analysis in the FSRU-case has, however, yielded a number of recommendations. Two of the recommendations are given below:

- Implementation of additional PSD/ESD functions to empty LNG from the booster pump suction drum upstream the regasification units. This addition would prevent the possible rapid boiling of the liquid in the drum and consequently decrease the escalation risk.

- Recommendation to increase the height of the new vent mast for the regasification units. An increased height reduces the likelihood for ignition of a gas release and reduces the possible impact from an ignited release, too.

Conclusions

As an integral part of a QRA, the main purpose of an escalation analysis is to finally highlight potential domino accident hazards and recommend suitable solutions to reduce/eliminate escalation risks. As this paper clearly states, a systematic approach is a must in order to evaluate all credible escalation hazards in an escalation analysis. The systematic approach, demonstrated in this paper, incorporating barrier analysis and advanced computational modeling, is in line with common practice in oil and gas industry.

The results of these case studies emphasize the importance of enhancing the focus on barriers. That can clarify to all what is important, to achieve good safety and no escalation events. By enhanced focus, the authors mean:

- Implementation of right adjusted physical barriers
- Performance of safety and reliability oriented maintenance/inspection
- Performance of safety and reliability oriented operational procedures

References

Brenda Eskenazi; Paolo Mocarelli, Marcella Warner, Larry Needham, Donald G. Patterson, Jr., Steven Samuels, Wayman Turner, Pier Mario Gerthoux, and Paolo Brambilla. "Relationship of Serum TCDD Concentrations and Age at Exposure of Female Residents of Seveso, Italy". Environmental Health Perspectives

G.L.L. Reniers (2005), The assessment of risk caused by domino effect in quantitative area risk analysis

Arslan, B., Wiik, A. (2007), The Use of QRA for LNG Terminals, paper published in Aiche 2007

Valerio Cozzani (2005) Escalation Thresholds in the assessment of domino accidental Events

Arslan, B., Wiik, A. (2006) Quantitative Risk Assessment of LNG Import Terminal in Nynäshamn, Scandpower

G.L.L. Reniers (2006), DomPrevPlanning: User-friendly software for planning domino effects prevention

Arslan, B. (2005) Design and Safety Aspects of LNG Import Terminals, MSc Thesis, Lund Institute of Technology, Sweden

Osbrink, Peter (2006) Technical Description of LNG Terminal in Nynäshamn, SWECO

SIGTTO, LNG Operations in Port Areas - essential best practices for the industry

Valerio Cozzani (2006), Prevention of domino effect: From active and passive strategies to inherently safer design

European Standard (1997) EN1473:1997 Installation and equipment for liquefied natural gas - design of onshore installations

NFPA International (2001), NFPA 59A - Standard for the Production, Storage, and Handling of Liquefied Natural gas

Abrahamsson, M. (2002) Uncertainty in Quantitative Risk Analysis - Characterisation and Methods of Treatment

Biography of Authors

Name: Bartnes, Jørgen
Organization: Scandpower
Position/Title: Principle Consultant
Mailing Address: 1 Houston Center, 1221 McKinney, Suite 3450, Houston, TX, 77010
Phone: +1 832 607 2793
Fax: +1 713 654 1903
E-mail: jba@scandpower.com

Jørgen Bartnes is a Principal Consultant in Scandpower and has extensive experience with risk, safety, emergency preparedness- and environmental projects. He is an experienced project leader for projects and analysis in the market of gas distribution (natural gas, LNG, LPG, etc.).

Name: Arslan, Baris
Organization: Scandpower
Position/Title: Consultant
Mailing Address: 1 Houston Center, 1221 McKinney, Suite 3450, Houston, TX, 77010
Phone: + 1 713 382 3663
Fax: +1 713 654 1903
E-mail: baa@scandpower.com

Baris Arslan holds a MSc. degree in Mechanical Engineering from Lund Institute of Technology in Sweden. Mr. Arslan has a special knowledge within the area of LNG-safety. Arslan works primarily with risk analysis and technical safety in oil & gas industry.

The Design of Optimally Safe Onshore LNG Receiving, Storage, and Regasification Terminals

William W. Cohea, Judy Clark Davis, Vijay K. Nayyer, Pat Watts, and Victor H. Edwards (1)
Aker Kvaerner, Inc.

Michael Livingston
WS Atkins, Inc.

W. Phillip Guy
AMG Engineering, Inc.

Christopher Richardson-Child
Consultant to ExxonMobil

Rebecca Wyman
Project Technical Liaison Associates, Inc. (2)

1 – Correspondence: 3600 Briarpark, Houston, TX 77042-5206,
e-mail: vic.edwards@akerkvaerner.com
2 – Current Address: PME-Mustang Process and Industrial

AIChE Spring Meeting, April 2008
8th Topical Conference on Natural Gas Utilization
New Orleans, Louisiana, April 6-10, 2008

Abstract

Assuring the design of safe onshore LNG receiving, storage, and regasification terminals requires careful consideration of numerous site- and facility-specific factors. This paper will present an overview of design strategies to eliminate, mitigate, and control risks.

Examples of site specific factors include the following:

- Facility footprint; equipment spacing and layout
- Existing infrastructure such as pipelines for natural gas and LPG
- Single site vs. Split site terminal
- Site accessibility to external waste heat sources such as power plants, refineries, etc.
- Natural hazards; seismic activity, severe weather or sea states
- Stability of subsoils; settlement during facility life
- Proximity to the public, shipping lanes, and air traffic
- Global and internal security

This paper will focus on the following considerations:

- Effect on rollover potential of feedstock composition and variability
- Identification of design basis LNG spill scenarios, spill collection and disposal
- Methods of BTU control and potential for LPG spills
- Unloading facilities including the design of unloading arms and headers to minimize spill size
- Design of LNG spill containment systems to minimize thermal radiation and flammable gas exclusion zones
- Choice of insulating concrete properties
- Judicious use of secondary and tertiary containment
- Overboard discharge of LNG during a spill event, including RPT overpressures
- Type of piping (for example, above ground or buried, single or vacuum-jacketed, etc.) including upgraded specifications for FRP piping
- High pressure LNG pipe work blow down and isolation
- Type of LNG storage tanks and ancillary equipment
- Effects of LNG high pressure pumping system on safety
- Selection of vaporizers and heating media including waste heat recovery systems
- Vapor fences
- Use of water curtains to disperse cold LNG vapors
- Integrated fire protection and risk management strategies

Where appropriate, advantages of design solutions will be illustrated with consequence modeling.

Introduction

Demand for natural gas is growing in Europe and North America, where domestic production is declining, and in many developing countries with rapidly growing economies. At the same time, there are many parts of the world where natural gas supply far exceeds local markets. Often, natural gas co-produced with oil is being flared.

Liquefied Natural Gas (LNG) offers a mechanism to meet growing demand, to utilize wasted natural gas, and to reduce global production of greenhouse gases.

The LNG industry has an excellent safety record; there have been no losses of life by the public due to LNG regasification facilities since 1944. Despite this exemplary record, every effort must be made to ensure continued safety in the future.

Figure 1 is a block flow diagram illustrating the principal components of an LNG terminal. LNG is received from an LNG carrier and pumped to an LNG storage tank. LNG is pumped by in-tank pumps to high pressure LNG send out pumps that discharge to LNG vaporizers. Natural gas from the LNG vaporizers is sent to users through a natural gas pipeline. Natural gas vapors from the storage tanks and from the LNG transfer operations at the LNG carrier dock are sent to a boil-off gas compressor for compression, recondensation, and forward to the suction of the high pressure LNG Sendout pumps.

Depending on the higher heating value of the received LNG, Wobbe Index adjustment may be needed. For those receiving terminals that separate C2+ hydrocarbons from the LNG to control the BTU content of the product natural gas, liquefied petroleum gas (LPG) is generally exported to users by pipeline.

This paper focuses on design strategies to eliminate, mitigate, and control risks in onshore LNG receiving, storage, and regasification terminals.

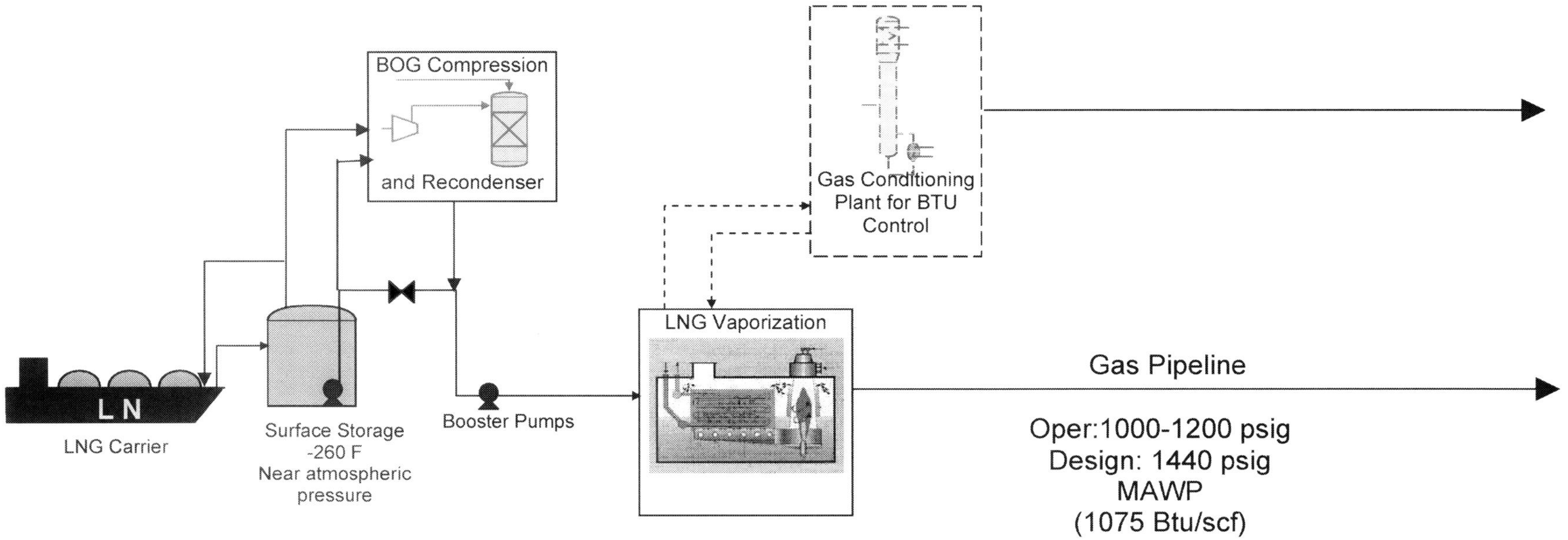

Figure 1 – Typical LNG Receiving, Storage, and Regasification Terminal

Site Specific Factors

Table 1 summarizes some of the factors affecting the siting of LNG receiving, storage, and regasification facilities.

Table 1 – Factors Affecting the Siting of LNG Regasification Facilities

- Regulatory environment. For example, in the USA, the Joint Federal Energy Regulatory Commission (FERC), Department of Homeland Security, Coast Guard (USCG), and Department of Transportation (DOT) exercise oversight through an interagency agreement for LNG facilities.
- Existing infrastructure such as existing pipelines for natural gas and LPG facilitate safe sendout of products.
- Seismic zone, hurricanes, unstable soils, and other natural hazards present challenges that can be overcome by appropriate design and construction measures. For example, full containment LNG storage tanks with seismic isolators are more resistant to hazards. Dikes can be provided to mitigate storm surge from hurricanes.
- Access to marine transportation facilitates safe delivery of LNG. Considerations such as berth, slip, or jetty offloading affect the potential hazards of LNG carrier docking, offloading, and departure, as well as facility cost. Marine traffic will affect (and will be affected by) access by LNG carrier movements.
- Environmental issues such as potential effects on wetlands and atmospheric emissions.
- Limited site size (land availability) may dictate smaller LNG facility size and more rigorous LNG spill prevention, mitigation and control.

Limit the Potential Size of Releases

Full or double containment tanks can be used instead of single containment tanks. The thermal radiation isopleth distances for the tank top fire required for full or double containment tanks is much less than that required for a full dike fire (for a single containment tank). Smaller tank diameters result in smaller isopleths.
The unloading flow from the LNG carrier to the LNG storage tanks can be divided among several parallel pipelines, each from an unloading arm. Because one of the controlling variables for the calculation of vapor dispersion is spill rate, dividing the unloading flow among parallel pipelines requires a smaller vapor dispersion isolation zone. Each parallel pipeline will also have a smaller inventory than a single large pipeline, so spill containment structures can be smaller. Parallel pipelines offer the additional advantage that when not offloading an LNG carrier, LNG can be recirculated more easily to keep lines cold.

Potential Effects of Nearby Assets
Proximity of the site to populated areas and to buildings and other industrial facilities have a direct impact on potential risks. The NFPA59A guidelines and the 49CFR193 federal regulations are conservative for this very reason. The scenarios that must be calculated to site an LNG facility in the USA have as large or greater an impact on the surrounding area than any credible terrorist attack. Facilities provide limited access and constant surveillance to keep people away that have no business near the terminal. The new federal regulations on chemical and hydrocarbon processing plant security (6 CFR Part 27, 2007) apply to LNG regasification terminals and will require each U. S. facility to assess vulnerabilities and to develop adequate Site Security Plans (SSPs).

Proximity to airports increases potential risks, but a large aircraft being hijacked and used as a weapon to crash into an LNG tank or an LNG ship is not considered a credible scenario. This is due, in part, to increased security measures enacted after September 11, 2001, but primarily because most LNG tanks or LNG ships in service generally do not meet the criteria for being an attractive terrorist target. Although a small aircraft would be more easily obtained, it would not have the combination of mass and speed necessary to create the impact energy required to cause either a cargo release from a ship or an LNG release from a tank.

Proximity to high profile terrorist targets increases potential risk. LNG storage tanks and LNG ships, the facility assets critical to public safety, have little potential for a significant release in the event of any credible terrorist attack. There are several levels of security that would have to be overcome in order for an attack to be successful in circumventing the basic design, emergency shutdown mechanisms, facility security and USCG oversight. All of the consequences of a credible terrorist attack produce exclusion zones that fall within the code-required consequence criteria, which are used to permit a facility. These exclusion zones have been established and analyzed as part of the siting and permitting process.

The Department of Homeland Security sets requirements for facility security measures as regulated per the Marine Transport Security Act (MTSA) and the Chemical Facility Anti-Terrorism Standards (CFATS). Sites will differ in methods of providing security which include CCTV cameras, fencing, intrusion detection and communications to protect an LNG terminal.

Susceptibility to Extreme Weather
Onshore LNG terminals provide more reliable offloading during extreme weather conditions. An offshore terminal has the added risks of ship collision, varying degrees of movement during unloading, and less protected areas during extreme weather. Aside from safe docking, an onshore terminal has no ship movement to worry about. Many ship designs are not able to sail with a partial cargo. An unberthed carrier with partially filled cargo tanks might be unstable and might flounder in rough seas. An offshore terminal would be more susceptible to having the unloading operation interrupted than an onshore terminal.

Process Considerations

Feedstock Variability and Rollover Potential

Feedstock composition and variability contribute to the potential for rollover. Rollover can damage the LNG storage tank and its internals. Careful monitoring of feedstock composition and LNG storage tank composition, as well as temperature and density profiles are essential (Uznanski and Versluijs, 2007). LNG storage tank recirculation pumps and both top and bottom entry points for LNG should be used to prevent a pre-rollover stratified condition.

Rollover can occur in a tank that has stratified. This means that the LNG in the tank has separated into layers of different density. When this happens, the circulation patterns caused by heat loss do not cross layers. The two different density layers heat at different rates. The density difference between the two layers crosses and the tank contents "turn over", generating a vapor spike. The resultant vapor spike has the potential to be much larger than the vapor recovery system design capacity. However, the relief systems on the storage tank are designed to handle these potential volumes. Proper placement of new LNG stocks (heavy on top and light on bottom) can inhibit layer formation by promoting mixing in the tank as it is filled. Careful monitoring of temperature and density will also help to let the operator know that a rollover condition may have developed. Circulation between tanks can also help prevent stratification, or mitigate the effects of a rollover situation.

Other Process Safety Concerns with LNG Rich in C2+ Hydrocarbons

Higher hydrocarbons in LNG delivered to a terminal can present potential problems in the sendout gas. C5+ hydrocarbons are of particular concern for condensation of liquids in pipelines. Liquids in pipelines can present safety hazards. High concentrations of C2+ hydrocarbons can increase the higher heating value (HHV) of the product gas above acceptable levels. Control of HHV and Wobbe Index of send out gas will be necessary to meet site-specific pipeline specifications. This can be achieved by distillation to remove LPG and/or by injection of nitrogen or air. However, injection of nitrogen is limited by maximum allowable nitrogen content of sendout gas. And injection of air increases the potential formation of a flammable mixture. Removal of NGL may also have potential disadvantages. NGL markets may not exist at site or may be less profitable than sale as natural gas. Often though the separated LPG products can be used as fuel on site or sold for use as fuel or as a petrochemical feedstock.

Separation, storage and disposition of LPG can present hazards because LPG vapors are much more likely to explode than methane. Confinement and/or congestion increase the likelihood of explosion of hydrocarbon vapors. Methane is very unlikely to explode without confinement and/or congestion. After an LNG spill, methane-rich gas evaporates first, followed by a higher proportion of C2+ hydrocarbons later; thus the risk of an explosion at an LNG spill is greatest later during the spill incident. This is one of the reasons that rapid application of high expansion foam coverage of spilled LNG is advisable.

The local regulations for LPG fuels are often more stringent than the federal LNG regulations. For example, in California an LPG tank must be located such that the explosion of the entire contents does not affect neighboring properties.

If nitrogen is used to adjust the HHV of the product gas from an LNG regasification facility, it may be necessary to build an on-site air separation plant. Nitrogen presents asphyxiation hazards and cryogenic air separation plants have other specific hazards as well.

Proper Choice of Metallurgy for Cryogenic Service

Proper choice of metallurgy for LNG piping and equipment is essential. The tragedy in 1944 in Cleveland was caused by not using a correct steel alloy. Good choices include LNG storage tanks made from 9% nickel steel and stainless steel piping.

FERC requires that any materials that will be in contact with cryogenic temperatures be rated for that service, including supporting structures. There are stringent requirements for the construction of LNG tanks. When a permit is filed for permission to build a terminal, a great deal of attention is given to the tank construction details and materials of construction.

Vaporizer Selection and Design

Table 2 lists some of the safety considerations affecting the choice of LNG vaporizers.

Table 2 – Safety Considerations in the Selection and Design of LNG Vaporizers

- Shell-and-tube vaporizers have the potential for loss-of-containment due to rapid vaporization of LNG following tube rupture.
- Submerged Combustion Vaporizers (SCVs) can serve as an ignition source in the event of an LNG spill.
- The low discharge temperature of combustion gases from SCVs requires reheating prior to selective catalytic reduction. The reheat combustion system on Selective Catalytic Reduction (SCR) units can serve as an ignition source. SCRs may be required on SCVs to control atmospheric emissions of NOx.
- Tube rupture in an Open Rack Vaporizer (ORV) leads to immediate loss-of-containment of LNG and/or natural gas to the atmosphere.
- Loss of heating medium to vaporizers can lead to cryogenic LNG rapidly cooling the downstream carbon steel natural gas pipeline to cryogenic temperatures where the pipeline could embrittle, risking failure and loss-of-containment.
- Use of indirect heat transfer fluid can present an ignition source if the fluid is heated by a fired heater.
- Indirect heat transfer fluids chosen should be environmentally friendly.
- Use of cooling water from adjacent process plant to provide heat for vaporization is excellent approach; consider the use of an indirect heat transfer fluid to minimize the risk of freezing.
- Ambient air vaporization processes are promising and some can produce water as a by-product (Cho, et al., 2006; Hubbard, et al., 2007; Nanda and Rizopoulos, 2007; Shah, et al., 2008).

Technologies that use air as a heat source for LNG vaporization processes use much less fuel than any scheme that requires a fired heater as part of the loop. Potential savings in energy costs can be on the order of $50 million per year, but capital costs are higher (Nanda and Rizopoulos, 2007). A back-up system is required to trim the system when the ambient air temperature drops below operating requirements, which will happen more than half of the year in a cold weather climate. The problem with a water byproduct in the amounts that are produced by an LNG vaporization process is that if there is no nearby use, the water may be too cold to be directly returned to the environment.

There are three main technologies that use air as the heat source to vaporize LNG (Nanda and Rizopoulos, 2007):

1. Air Tower: Using air to heat water in a reverse-acting cooling tower, and in turn, vaporizing LNG directly or indirectly using an intermediate heat transfer fluid.
2. Ambient Air Vaporizers: Using air to directly vaporize LNG (Both natural draft and forced air vaporizers have been used.)
3. Air-Fin Exchangers: Using air to heat an intermediate heat transfer fluid

Figure 2 – Forced Draft Air Heater Vaporizers at Dahej Terminal, India (courtesy of IHI)

Design Release Scenarios and Minimizing Releases

Design release scenarios have an important impact on the design of LNG spill containment and mitigation systems. European regulations call for a risk-based approach (Deetjen and Brown, 2007). US guidelines and regulations specify a leak from the largest single source, which is currently interpreted as a 10 minute spill for a guillotine break of the largest process line (49 CFR 193; NFPA 59A; Qiao, et al., 2007).

A 10 minute spill from a guillotine break of the largest line carrying LNG is much larger than any credible spill from normal operations. Sizing the impoundment for this requires the facility to have a large isolation distance from the public.

Continuous monitoring and selective placement of emergency isolation valves can shorten spill duration substantially. Isolation valves on LNG systems can be closed in less than 30 seconds in almost all cases; transient fluid dynamic analysis should be used to select closing times that prevent excessive pressure surges (Wink and Marcum, 2007).

An Emergency Shutdown System (ESD) is required for an LNG facility to minimize the risk of fire and explosion and to prevent the escalation of any incidents. The total isolation time is the total time required to detect the leak, plus the response time of the control system, plus the time for the isolation valves to close. Although a full guillotine break is an unlikely scenario, most facilities do not take into account the volume of LNG that is located in the unloading lines between the shut-down valves. When taken into account for large diameter piping and long lengths of pipe, this is a significant volume.

PERCs (Powered Emergency Release Coupling) are used on LNG carrier loading lines to limit spill size at jetty. PERCs cut off flow in the loading lines and seal the LNG cargo tanks to keep LNG from spilling into the air. In the event of an emergency PERC separation, a minimal amount of LNG (the volume of LNG trapped between the two isolation valves of the PERC) will be released (This volume is approximately 1 to 2 gallons).

Guillotine breaks are very unlikely because flanges will be minimized and piping will be 100% X-rayed. Flange leaks or failure of a small instrument line are far more likely than guillotine breaks of a large LNG line. The only flanges should be at the top of storage tanks, at unloading platforms, and on the high pressure send out pumps.

LNG spills onto spill pads at LNG carrier unloading should be mitigated as well as at process areas. LNG spills onto water evaporate much more quickly than on land, especially if land spills are contained by insulating material such as insulating concrete. More rapid evaporation increases the size of the resulting flammable cloud or fire. Therefore, dockside facilities should be designed to divert LNG spills to land-based containment systems and to quickly cover with foam. Also, LNG spilled into water from an elevated source can produce small physical explosions called RPTs (Rapid Phase Transition); these are not expected to be strong enough to damage the LNG carrier or the LNG offloading system, but they should still be avoided by diverting LNG spills to land where possible (Nedelka, et al., 2003). In some circumstances, such as offshore facilities, discharging overboard may be a better solution. In the event that it is determined that discharging overboard is the safer solution, then the potential for RPTs and cryogenic embrittlement should be evaluated.

Proper design of curbing on spill pads can limit the surface area of spilled LNG, thereby minimizing evaporation rates. Any LNG spilled should be directed to an impoundment area which has been sized and placed to allow any spill to vaporize without impacting the surrounding public areas.

Protection from Thermal Radiation

Consequence-based Thermal Exclusion Zones

Unintended releases of LNG pose fire hazards, with attendant thermal radiation (Haines, 2007). Both federal regulations (49CFR193) and US industry guidelines (NFPA 59A) specify consequence-based methods for protection from thermal radiation (Qiao, et al., 2007).

For example, 49CFR193.2057, Thermal Radiation Protection, states that each LNG container and each LNG transfer system must have a thermal exclusion zone. The regulation also specifies use of specific computer software to compute the size of the exclusion zones.

Table 3 summarizes one approach to the calculation of the size of the exclusion zones in LNG receiving, storage, and regasification facilities.

Table 3–Federal Regulations Define the Size of Exclusion Zones in LNG Facilities*

Provisions shall be made to prevent thermal radiation flux from a fire from exceeding the following limits when atmospheric conditions are 0 (zero) wind speed, 70°F (21°C) temperature, and 50 percent relative humidity.

(1) 1600 Btu/hr/ft^2 (5000 W/m^2) at a property line that can be built upon for ignition of a design spill

(2) 1600 Btu/hr/ft^2 (5000 W/m^2) at the nearest point located outside the owner's property line that, at the time of plant siting, is used for outdoor assembly by groups of 50 or more persons for a fire over an impounding area containing a volume, *V*, of LNG**

(3) 3000 Btu/hr/ft^2 (9000 W/m^2) at the nearest point of the building or structure outside the owner's property line that is in existence at the time of plant siting and used for occupancies classified by *NFPA 101*®, *Life Safety Code*®, as assembly, educational, health care, detention and correction or residential for a fire over an impounding area containing a volume, *V*, of LNG**

(4) 10,000 Btu/hr/ft^2 (30,000 W/m^2) at a property line that can be built upon for a fire over an impounding area containing a volume, *V*, of LNG**

Notes:

* The federal regulations also define alternate methods for calculation of the required size of exclusion zones for thermal radiation.

** The volume, V, of LNG is calculated as the quantity of LNG spilled in ten minutes at the design basis spill flow rate.

Minimize Thermal Radiation Hazards by Choice of Storage Tanks

LNG storage tanks include three major types:

- single containment
- double containment
- full containment.

The thermal radiation isopleths must be calculated for the tank impoundment. For a single containment tank, this requires thermal radiation calculations for total loss of containment by the tank, leading to fire within the containment dike surrounding the tank. This leads to very large thermal radiation exclusion zones. Because the outer wall of a double or full containment tank provides the code-required impoundment, double and full containment tanks require tank top fire calculations only. Although tank top fire is very unlikely for a double containment tank, it is even less credible for full containment tanks. Thermal radiation exclusion zones for tank top fires in double or full containment tanks are smaller than dike fires for single containment tanks, but may still define part of the plot plan.

Membrane tanks that provide full containment are now also finding application (Lee, et al., 2007).

Figure 3 – Typical Full Containment LNG Storage Tank (Courtesy of IHI)

Thermal Radiation on Equipment and Piping on Top of Storage Tanks
LNG storage tanks frequently have only top penetrations because top penetrations are inherently safer than side penetrations. This means that there is a large amount of piping and instrumentation on the top of a tank that has the potential to be damaged by heat. This piping and instrumentation must be protected from the intense thermal radiation that may be generated by a neighboring tank top fire.

NFPA 59A (NFPA, 2006) specifies that double or full containment containers with concrete secondary containers shall have a separation distance to limit the incident thermal radiation flux from a fire within the primary or secondary container of an adjacent tank as follows:

(1) Steel wall and roofs: 4,700 BTU/ft2 hr
(2) Concrete walls: 9,500 BTUft2 hr

In this case normal firewater provisions, elevated monitors and a monitor at grade are considered adequate fire protection. In the case where the tanks are inside the NFPA 59A isopleths, a water spray or deluge system may be used to limit the thermal radiation flux. Typical full-containment style storage tanks may have an elevation from grade to the tank roof platform of approximately 180 ft. When designing coverage with monitor systems, the actual capability of the ground based and even elevated monitors must be carefully reviewed to confirm that the design flow rate will be capable of reaching the tank platforms when potentially affected by adverse wind conditions.

Thermal radiation from flares and from ignited pressure relief valve vent streams must also be considered.

Fire Exposure/Cryogenic Exposure to Structural Steel
NFPA 59A requires that an evaluation be performed to evaluate the need for the protection of equipment and structures from fire effects. Fire exposures in LNG facilities, as with other hydrocarbon facilities, can generally fall into one of three categories: 1) jet fire, 2) spray fire, 3) pool fire. Structural steel can fail within a matter of minutes when exposed to hydrocarbon fires. The failure of structural steel can result in the release of additional volumes of LNG that potentially could result in the total volume of LNG released to exceed the design basis release scenarios. A comprehensive review of the fire scenarios must be developed to evaluate the potential effects of the different fires. The scenarios should not only focus on the perceived worst case of a full line rupture (which will result in the release of the volume of LNG quickest), but also on smaller size releases that would have an extended duration. The longer duration of the fire for the smaller release sizes may require that the passive fire protection system be designed for a longer duration.

Although not specifically identified in NFPA 59A, cryogenic exposure can be as catastrophic of an exposure as the noted fire exposures. When structural members and equipment are exposed to LNG, they may become embrittled, and subsequently fracture. Brittle fracture failures from cryogenic exposure may be even more rapid than potential failures associated with severe fire exposures. In general, when evaluating the structural members that are considered critical, and would be required to be protected by passive fire protection; those same members may be assumed to be critical with regards to cryogenic protection as well (if they may be potentially exposed to cryogenic liquids).

To evaluate the need to protect structural steel, the potential fire exposed areas from the design scenarios, as well as the cryogenically exposed areas must be identified.

Size and Arrangement of Impoundments:
Single containment tanks require a dike that will contain 110% of the contents of the single largest tank (NFPA 59A). For an impoundment serving more than one LNG tank, the impoundment volume shall contain 100% of the combined volumes of all tanks within the impoundment or 110% of the largest tank volume, which ever is greater. Other requirements also apply to the dike.
Double and full containment tanks shall be designed and constructed such that in the case of a spill and secondary containment fire, the secondary containment wall will contain the LNG for the duration of the fire (NFPA, 2006). Although total loss of containment is even less credible for double and full containment tanks, a berm will be provided that gives tertiary containment of 100% of the contents of the single largest tank. This is not written in the NFPA code, but has been a requirement for the general site safety in several US permit applications.

Thermal Radiation from Flares:
Pressure relief discharges from process equipment such as vaporizers are often routed to flares. Ground flares are not recommended because they could serve as an ignition source for an LNG spill. Elevated flares must be high enough to prevent ignition of venting from LNG storage tanks, as well as high enough to keep thermal radiation at acceptable levels on equipment and personnel.

Flammable Gas Dispersion

U. S. Regulations and Industry Guidelines - Flammable Gas Exclusion Zones:
Federal regulation 49CFR193.2059, Flammable vapor-gas dispersion protection, defines U. S. requirements for protection of the public from flammable gas clouds that could result from an LNG spill. When spilled, LNG vaporizes, releasing cold natural gas. Natural gas mixed with air is flammable in concentrations of approximately 5% by volume natural gas to 15% natural gas. Very close to a spill, the natural gas concentration is too rich to burn, but further away from the spill, additional mixing with air produces a flammable cloud of natural gas. Natural gas formed by vaporization of LNG is initially very cold and thus much denser than the surrounding air. The resulting vapor cloud therefore spreads out like a pancake close to the surface near the spill. As the cloud moves downwind, mixing with air and heat transfer with the ground warm the vapor cloud. At some distance from the spill, the cloud becomes lighter than air and it lifts off. Ultimately, mixing with air dilutes the cloud below the flammable range.

The federal regulation and NFPA 59A both require that flammable gas dispersion from potential LNG spills not extend beyond the property line of the LNG facility. More specifically, the regulation calls for an average flammable gas concentration of no greater than 2.5% (one half of the lower flammable limit [LFL] of methane in air) at a receptor height of 0.5 m. Using one half LFL of methane is a safety factor that accounts for the lower LFL's of C2+ hydrocarbons which may be present, as well as the limitations of modeling of atmospheric dispersion.

Current federal regulations and NFPA 59A suggest use of the dense gas dispersion software DEGADIS for prediction of flammable gas dispersion from LNG spills, and also suggest the use of FEM3A, an early computational fluid dynamics (CFD) program. Use of alternate software is also allowed if it has been validated with data and commercial CFD software can be successfully adapted to impoundments and trenches (Gavelli, et al., 2007). Impoundments and trenches should be located and designed to minimize the risk of fire exposure to any vessels, piping, equipment, and support structures that are not protected from fire exposure.

Reducing the size of the flammable gas exclusion zone:

The two controlling variables for vapor dispersion from an LNG pool are the spill rate and the surface area. Smaller design spill rates result in smaller calculated vapor dispersion zones. Minimizing surface area is also recommended; a deeper but smaller containment footprint is desired.

- Geometry of piping installation affects exclusion zone. For example, multiple loading arms are commonly required to offload the LNG carrier. By headering these together in an underground valve box and providing check valves and isolation valves in each unloading line, credible spill size can be reduced.
- Underground piping, preferably with vacuum-jacketed pipe, can be used to reduce the amount of trenching between the jetty and the LNG storage tanks.
- Use of insulating concrete instead of ordinary concrete for LNG spill containment structures can halve flammable gas dispersion zones. This is important for sites near populated areas. It is important to design an insulating concrete mix that provides a good combination of insulating value and durability.

LNG containment systems - additional considerations:

There are several considerations that must be accounted for when designing LNG containment systems. As noted above, calculations must be made to determine the effective distances for both radiation, and gas dispersion from the LNG spill containment system. This includes an evaluation of the LNG drainage trenches, as well as the impoundment basins.

Due to site constraints, many facilities being designed currently struggle to locate the LNG impoundment basins such that the radiation levels from a fire, and the gas dispersion distances to ½ of the LFL are maintained in accordance with the applicable regulations.

By making the impounding area deeper, with a smaller plan area, the reduced LNG surface area results in a lower boil off rate. However, this same approach also prolongs the overall event time since it will take a longer period of time to weather off, or burn off the released LNG. Therefore, a balance must be reached between the acceptable radiation and gas dispersion levels, and acceptable event duration.

Additionally, constructability issues will have a significant impact on the design of the LNG impoundment basin. While a very small surface area for the impoundment basin has a positive effect on the fire radiation and gas dispersion distances, it may or may not be a constructible solution depending on the site elevation, soils, and ground water elevation.

As noted below, the use of high expansion foam can have a greatly beneficial effect on the gas dispersion and fire radiation from LNG drainage trenches and impoundment areas. For the foam system design to be effective, the trenches and impoundment areas must be designed to allow for a minimum of 1+ meter of free board space on which to apply the high expansion foam.

One of the options to reduce the overall boiloff rate of the LNG in the impoundment system is the use of insulating concrete or other surface materials with a low heat transfer coefficient. The use of these materials can significantly reduce the vaporization rate of the released LNG. One of the concerns with the use of insulating concrete is the balance between the lower heat transfer coefficient, and maintaining a concrete design density that would be rugged enough to withstand the anticipated day to day operations at a facility for an extended period of time. With light weight concretes, the decrease in the heat transfer coefficient has a corresponding reduction in the density of the concrete. Due to durability concerns, it is normally not recommended to use concretes with densities of less than 60- 70 lb/ft^3.

Vapor fences-pros and cons:

A vapor fence can be used to "hold up" the cold dense LNG vapor in one location until it is lighter than air and able to disperse. The amount of gain is comparatively small and this option makes construction more expensive, but can be useful if the site has restrictive geometry.

Mitigation via steam or water curtains:

Properly designed water curtains can effectively disperse flammable vapors to below the lower flammable limit by mixing and warming them. This option is useful for a small space. The disadvantages are that a large amount of water is required and the area must be graded to drain the water curtain away from the LNG spill. This option may not be practical for areas around a large storage tank.
Steam curtains could achieve the same objective with fewer drainage issues, but is potentially more expensive. And the plant's steam system must be designed to minimize loss of steam header pressure when the steam curtain is activated (non-normal operation).
Application of water fogs is another promising approach to warming and dispersing flammable vapors.

Flammable gas dispersion from high pressure leaks:
Because of the jet release, flammable vapors from high pressure LNG leaks, such as a leak between a high pressure sendout pump and a vaporizer have the potential to migrate further downwind than vapor from an LNG pool. Good engineering practice includes modeling of this type of release in addition to flammable vapors from LNG pools. Mitigation measures include minimizing inventories of high pressure liquids and providing rapid leak detection and isolation.

Flammable gas dispersion and thermal radiation control the plot plan:
Figure 4 shows the thermal radiation isopleths and the flammable gas dispersion exclusion zones for a typical plot plan during design development of an LNG regasification terminal. The cross-hatched area corresponds to the additional plot plan needed if ordinary concrete instead of insulating concrete is used for the LNG spill containment systems.

Figure 4 – Flammable Gas Dispersion and Thermal Radiation Control the Plot Plan

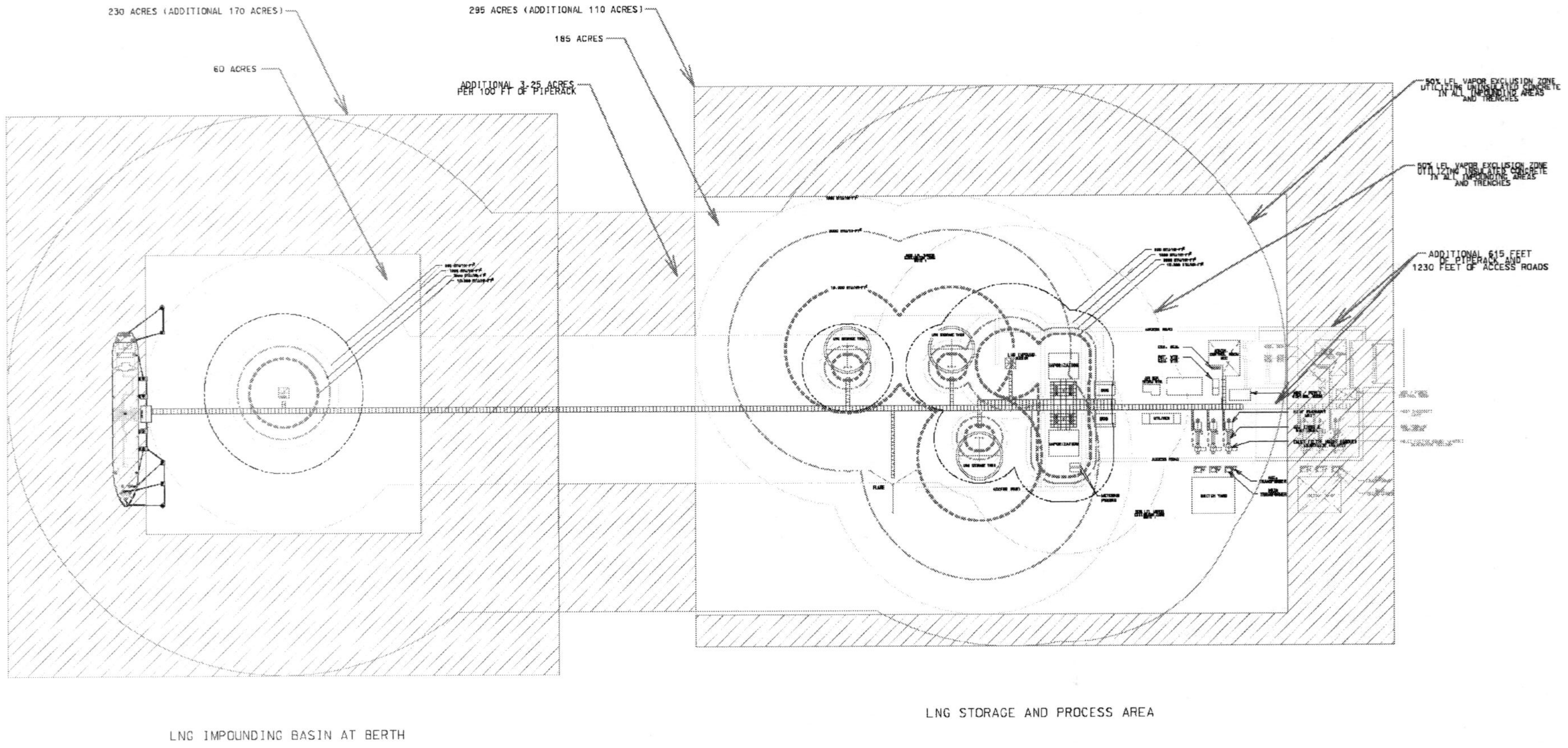

Fire Protection

NFPA 59A provides guidance for fire protection in LNG facilities. Fire and gas detection is required around any potential ignition source. Process equipment containing LNG, refrigerants, flammable liquids, or flammable gases shall be located at least 50 ft from sources of ignition, a property line that can be built upon, control rooms, offices, shops, and other occupied structures. In general grating and sloped pads and trenching should be used where possible to avoid potential pooling of hydrocarbon leaks. ESD shutdown is initiated in terminals upon confirmation of gas or fire detection.

Flammable gas and fire detection, alarm, and response:

The methods of flammable gas and fire detection for LNG facilities are similar to other petrochemical facilities. Generally, point type and open path type combustible gas detectors are utilized to detect vaporized LNG. Fire detection is commonly done through a combination of UV/IR or Triple IR type optical flame detectors, and heat detectors (rate-compensated, fusible-plug type pneumatic heat detectors, etc.).

Detecting potential LNG or gas releases and/or subsequent fires quickly allows for a more timely response to the incident. For facility siting purposes, a 10-minute release volume is generally used. However, for more probable scenarios, with current isolation systems, a release scenario would likely be isolated in less than 5-minutes from detection. To limit the size of the release, it must be detected quickly. To do that, a combination of point-type gas detectors, open path gas detectors, and low temperature detectors are commonly used. Typically, on confirmed detection by facility gas detectors and low temperature detectors, process shutdown is automatically initiated.

Gas dispersion calculations may be used to determine probable locations for dispersed concentrations of gas from anticipated release sources. This allows a more quantitative approach for determining necessary locations of point type and open path gas detectors.

Generally, point type gas detectors must be located in close proximity to the potential leak source to effectively detect gas releases. Gas detectors of the open-path type cover a wider area, by having a transmitter and a receiver. Such detectors have a typical range from 20 to 200 m. In areas where it is necessary to cover a long distance and line of sight is possible between the transmitter and receiver, open-path type detection is a good choice. Open-path detectors work well when utilized to establish a boundary around the protected equipment, providing detection but without identification of location. This allows gas detector coverage to be provided along an extended distance. If point-type detectors were used, it may not be easy or cost-effective to locate and provide detector coverage for such a large area. In most facilities, a combination of point type and open-path type of gas detectors must be used to provide effective coverage of the equipment and storage areas.

Optical flame detectors may be used to detect fires at LNG facilities. The benefit of optical flame detection is that it normally provides faster response times than standard heat detectors. Additionally flame detectors have a defined field of view that may be arranged to cover specific pieces of equipment.

Low temperature detectors (LTDs) are an effective means of detecting LNG releases. LTDs may be used to identify the release of a cryogenic liquid such as LNG. Such detectors would be located at ground level in drainage trenches, impoundment basins, sumps, and/or in the discharge of pressure relief valves (e.g., storage tank roof-mounted emergency vents). Due to the low false alarm rate associated with this type of detection, an alarm typically may be considered a confirmed release for purposes of initiating executive actions such as equipment shutdown, and mitigation system initiation. Additionally, LTDs may be used to inhibit the operation of water sump pumps that would normally be installed in LNG impoundment basins to pump out rain water. Upon initiation of the LTDs in the LNG drainage channels leading to the impoundment basin and/or the LTDs in the basin, the sump pumps could be disabled to prevent them from accidentally transferring spilled LNG out of the impoundment system (assuming that direct contact from the LNG hadn't already resulted in a pump failure).

Emergency shutdown system (ESD):

LNG facilities are required to be provided with an Emergency Shutdown (ESD) System. The purpose of the ESD system is to provide the capability to isolate sources of LNG, flammable and combustible liquids and gases, and other equipment whose continued operation could add to or sustain an emergency. (NFPA 59A, Section 12.3.1).
The emergency shutdown system (ESD) provides protection of personnel and facilities from hazardous situations such as the unplanned release of LNG or other hydrocarbons.

Proper design of an ESD System is essential to the overall facility design, as the response time of the system, and volume that may be isolated between the shutdown valves directly relates to the potential size and duration of the emergency incident.

The ESD System should be designed such that wherever possible it fails to the safe position. NFPA 59A requires that ESD System components that do not failsafe that are located within 50-ft of the equipment that they protect to be either be installed such that they can not be exposed to fire, or to be protected from the effects of a fire for a minimum period of 10-minutes.

An emergency shutdown may be initiated by either of the following:

- Manually by an ESD push button
- Automatically by the safety instrumented system (SIS) upon detection of a hazardous situation such as a hydrocarbon release or a fire

Normally, the ESD actions include the following:

- Activation of ESD alarms and public announcement
- Start-up of emergency back-up power system
- Automatic activation of a process shutdown
- De-energize electrical equipment (except for emergency powered equipment)
- Isolation of the natural gas pipeline
- Isolation of LNG sources such as carrier, LNG pipelines, tanks, etc.
- Shutdown of ignition sources
- Automatic activation of blow down valves (BDVs) to depressure and deinventory system

While the current industry standards address the potential effects of fire on the ESD System, no criteria exists for the potential cryogenic exposure effects from released LNG. ESD Systems should be protected from the effects of potential cryogenic exposure the same as they should be protected from potential fire exposure. Such protection can take the form of spray shields, insulating coatings, etc.

Choice of extinguishing agents:

Dry Chemical

Due to their quick knock down power dry chemical extinguishing agents are the most effective means of actually extinguishing an LNG fire.

Current federal regulations require that each marine unloading area be provided with a minimum of one (1) dry chemical hand hose line, and one (1) dry chemical turret system. During the design of the unloading facilities, it is critical to review the potential fire scenarios, and confirm that in the event of a release that the operators, and or responding fire department/brigade personnel would be able to reach the dry chemical equipment to use it. A review of this type will typically find that additional dry chemical hand line stations or monitors are required to provide effective coverage.

Additionally, dry chemical systems have a limited amount of agent. If the fire is not extinguished prior to the agent supply being exhausted, there is normally not a redundant supply of agent, and expellant gas. A representative dry chemical monitor package may be designed for an application rate of 50 lb/sec, and have a 45 to 60-second supply. There is concern about designing the dry chemical monitor packages as remote controlled monitors. The concern stems from the fact that it takes a few seconds to direct a remote controlled monitor onto the target, the longer the time operating the monitor to get it on the fire, the more agent that is expended on something other than the fire. Therefore, some dry chemical equipment suppliers recommend against equipping the dry chemical monitors with remote control capability.

Water Spray/Water Curtains

Due to the significant temperature difference between released LNG and water, any contact from the fire suppression system water with spilled LNG causes an increase in the vaporization rate. The resultant increase in vaporization rate causes an increase in the fire intensity. Water is an effective agent for radiation exposure protection for equipment, but care must be taken in the drainage system design and method of water application to minimize the additional heat transfer from the water to the LNG.

Water curtains may also be used effectively to reduce the downwind gas concentration (Cormier, et al., 2006). Water curtains function to add heat to the vaporized natural gas, and thereby make it more buoyant; reducing the downwind gas concentrations. Limited published testing is available to quantify the overall performance of water curtains on downstream gas concentrations. Although generally acknowledged as being effective, quantitative means of designing the water curtains and the effects of ambient conditions on the performance are not available. Current industry testing results indicate that a significant reduction in the downstream gas concentrations is possible with proper design (Cormier, et al., 2006). Moderate to high wind conditions that would be anticipated to reduce the effectiveness of water curtains for vapor dispersion control also cause increased heat transfer to LNG vapors, and result in shorter dispersion distances.

High-Expansion Foam

Although high-expansion foam is not an effective extinguishing agent for LNG fires, industry testing has determined that properly designed high expansion foam systems can provide significant benefits for the mitigation of vapor dispersion and radiation mitigation for LNG releases and fires (Suardin, et al., 2007). The use of high expansion foam on LNG pool fires can reduce the fire radiation levels by 90% or more (Worlmad, 1982).

The application rate of the foam is critical to achieve control of the fire. The time to fire control must be integrated into the overall facility fire and loss prevention design basis to assure that the control of the fire may be achieved in adequate time to prevent unacceptable exposure to personnel and equipment.

High-expansion foam is applied over the surface of released LNG to form a foam blanket on top of the liquid LNG. The foam blanket freezes quickly due to the extreme cold of the LNG, and serves to support the foam layer above without submerging into the LNG.

Fire fighting foams are categorized by their expansion ratio as Low (<20:1), Medium (20-200:1), and High-Expansion (>200:1). The manufacturers recommended optimum expansion ratio for use of high-expansion foam for LNG is 500:1. Expansion ratios of less than this amount contain higher volumes of water, and result in increased vaporization from the spilled LNG. Expansion ratios greater than 500:1 are more susceptible to wind and other environmental effects, and tend to breakdown from the fire.

The high-expansion foam blanket functions both for vapor dispersion, and fire radiation mitigation. As the LNG continues to boil beneath the foam blanket, the evolved vapors must rise through the foam. This has a warming effect, which increases the buoyancy of the released vapors, and serves to reduce the dispersion distance. Additionally, the foam blanket creates an insulating barrier against flame re-radiation back to the

LNG pool for ignited releases. The reduction of the incident radiation to the LNG pool surface reduces the boil-off rate during the fire, resulting in a reduced flame height and associated radiation exposure to adjacent areas.

It should be noted that for the high-expansion foam to remain effective in controlling vapor dispersion and fire radiation, it must be reapplied on an ongoing basis to the protected area. Therefore, adequate high-expansion foam concentrate, and site fire water supplies must be available for the entire duration of the anticipated fire scenario.

Emergency response:

One of the key factors in determining the type and extent of fire protection systems that must be provided for LNG facilities is the available municipal fire department, and/or facility fire brigade response.

Responding fire service personnel must be trained in how to deal with the specific properties of LNG, and how it burns. Additionally, depending on the type and arrangement of systems provided at the facility, the responding fire department personnel may require specialized fire fighting equipment.

Extinguish or burn?

When establishing the required types of equipment and procedures for responding to an LNG release, and potential fire, a determination must be made as to if the primary response in the event of a fire will be to extinguish the fire or to allow it to burn.

The initial thought when discussing an LNG release, and subsequent fire would probably be to extinguish the fire as quickly as possible. However, as with any flammable gas fire, extinguishing the fire could lead to a more severe situation since the LNG will continue to vaporize.

Due to the high liquid to vapor expansion ratio for LNG (approximately 600:1), and the fact that the evolved vapors are initially heavier than the ambient air, it is possible to have low level vapor clouds extending several hundred feet for normal design scenarios, and further for catastrophic releases. Based on this, extinguishing a confined release of LNG may not be the best solution unless means of controlling the dispersed vapors is available.

As noted above, potential methods of mitigating the vapor dispersion profile of an LNG release are the provision of high-expansion foam systems to cover the LNG pool, and/or the provision of water curtains designed to warm the LNG vapors, thereby making them more buoyant.

When making the decision whether to extinguish the fire or not, the anticipated burn time of the incident must also be evaluated. Based on an anticipated unloading rate of 12,000 m^3/hr to 18,000 m^3/hr, a 10-minute release of LNG would result in approximately <u>2,000 to 3,000 m3 of LNG being released.</u> Assuming that the hypothetical facility LNG impoundment system contains all of the released LNG, a fire associated with the release would be expected to burn for several hours. Since the vaporization rate of the LNG pool would be less in a steady state non-burning situation, the time to boil off that volume of LNG would be even longer. Any provided vapor mitigation or radiation mitigation systems must take into account the potential time of the incident.

NFPA 59A indicates that the fire water supply for facilities must be designed to fulfill the design flow rate for a minimum of 2-hours. It should be stressed; that this is the minimum permitted water supply duration. However, many designers currently do not take into account the anticipated extended duration of an event when considering the total volume of water and or foam that may be required.

To address this, many new facilities are being designed with the primary fire pumps utilizing a freshwater supply, but having a backup water supply utilizing the available seawater for marine based terminals. With regards to the foam supply, facilities may be designed utilizing atmospheric foam concentrate storage tanks located in protected areas that would allow the foam concentrate supply to be refilled from available stores in the case of a prolonged event.

Threat and Risk

Vulnerability analysis uses the following two criteria to determine if a specific terrorist attack on the proposed facility is credible or "worthy of belief":

- Can a specific attack scenario be successfully carried out? (Possibility?)
- What is the likelihood that such an attack will happen for a particular location? (Probability?)

Location of site and regulatory requirements:

Security for each LNG facility will be regulated by either the Marine Transportation Safety Act (MTSA) or the Chemical Facility Anti-Terrorism Standards (CFATS) or both based on agency review. The U. S. regulations for chemical plant security which may also apply to LNG receiving, storage and regasification facilities are 6 CFR Part 27, 2007 (CFATS) and the 33 CFR Part 105, Marine Transportation Safety Act 2002, Navigation Inspection Circular (NAVIC) 05/05.

These regulations will help to insure security of LNG regasification terminals by requiring the following:

- Completion of Security Vulnerability Assessments (SVAs).
- Development and Implementation of Site Security Plans (SSPs).
- Waterway Suitability Assessments (WSA).
- Access control to facility with TWIC compliant card readers.
- Successful attainment of risk-based performance standards.
- Attaining U. S. Department of Homeland Security (DHS) approval.

The DHS has developed web-based tools for secure completion of SVAs and SSPs by covered facilities and must review and approve these applications. Note that the DHS rules are performance based and are not prescriptive. Each owner of an LNG facility will be responsible for development of effective risk-based security plans that meet DHS goals, but DHS does not say how this must be done.

Traffic in waterway:

All LNG vessels in international service are required to comply with various requirements of the International Maritime Organization (IMO), International Convention for the Safety of Life at Sea (SOLAS) and the International Gas Code (IGC). When trading in United States waters, the United States Coast Guard (USCG) will issue a Certificate of Compliance after a plan review and inspection. LNG ships are also required to notify authorities ninety-six hours before arrival with the last ports of call, crew identity and cargo information. The ship will be boarded and inspected before it is allowed entry to US ports. The post 9/11 Maritime Security Transportation Act (MSTA) gives the USCG additional security powers. All of these requirements provide some degree of comfort that the ship and crew will operate safely in route to the facility.

Security at site:

LNG storage tanks and LNG ships, the facility assets critical to public safety, have little potential for a significant release in the event of any credible terrorist attack. There are several levels of security that would have to be overcome in order for an attack to be successful in circumventing the basic design, emergency shutdown mechanisms, facility security and USCG oversight. All of the consequences of a credible terrorist attack produce exclusion zones that fall within the code-required consequence criteria, which are used to permit a facility. These exclusion zones should be established and analyzed as part of the siting and permitting process.

An analysis indicates the following for onshore LNG facilities:

- The LNG tanks are the only components within the LNG facility which have the potential for a release large enough to create a significant hazard that might extend beyond the facility boundary or exclusion zones.

- The contents of an LNG tank (liquid or vapor) would not react (burn or explode) within the tank because of the lack of oxygen, even with an ignition source present.

- The combination of the outer tank, thick insulation, and 9 % nickel steel inner tank is structurally robust and resistant to the credible blast scenarios that could cause an LNG release. A steel outer tank provides significant penetration resistance against most shaped charge penetration weapons and protects the inner tank from direct blast effects. A concrete tank does an even better job. To inflict significant damage to an LNG inner tank would require a very large, sophisticated weapon, which would be very difficult for a terrorist to obtain and use

- Access to a site for an explosive attack would be difficult. The facility would typically have fences, security gates, security personnel, and closed circuit TV monitoring to stop site-unauthorized access to the tanks. The dikes surrounding the facility are also a significant barrier.

- Large aircraft, if hijacked and weaponized would be capable of doing significant damage but such an attack scenario is not credible and the LNG storage tanks are not a likely target.
- Any blast or penetration attack would be an ignition source. Therefore, the consequence would be an immediate fire rather than a flammable vapor cloud.

References

Joseph H. Cho, Robert Femat, Heinz Kotzot, and Charles Durr, "Optimization of Ambient Air Vaporization Processes", paper presented at 6th Topical Conference on Natural Gas Utilization, Spring National Meeting of the American Institute of Chemical Engineers, Orlando, Florida (April, 2006).

Benjamin Cormier, Yanjun Wang, Michael Moore, Harry H. West, and M. Sam Mannan, "LNG Mitigation Experiments – Preliminary Results", paper presented at the Mary Kay O'Connor Process Safety Center 2006 International Symposium, College Station, Texas (October, 2006).

Caroline J. Deetjen, Tatiana Brown, "Comparison of US Risk-Based Approach with the Rest of the World for LNG", paper presented at 7th Topical Conference on Natural Gas Utilization, Spring National Meeting of the American Institute of Chemical Engineers, Houston (April, 2007).

Filippo Gavelli, Melissa Chernovsky, Edward Bullister, Harri Kytomaa, "Validation of a CFD Model for Vapor Dispersion from LNG Spills into an Impoundment", paper presented at 7th Topical Conference on Natural Gas Utilization, Spring National Meeting of the American Institute of Chemical Engineers, Houston (April, 2007).

Stephen Haines, "Right Sizing Fire Protection", paper presented at the 9th Process Plant Safety Symposium, Spring National Meeting of the American Institute of Chemical Engineers, Houston (April, 2007).

Brad S. Hubbard, Randy Sharp, Ned Baudat, Scott Worthington, "Floating Storage and Regasification Concepts Using the LNG Smart Air Vaporization Technology", paper presented at 7th Topical Conference on Natural Gas Utilization, Spring National Meeting of the American Institute of Chemical Engineers, Houston (April, 2007).

J. C. Kuo, "How Big is the Area Required for LNG Receiving Terminal and Their Design Considerations", paper presented at 7th Topical Conference on Natural Gas Utilization, Spring National Meeting of the American Institute of Chemical Engineers, Houston (April, 2007).

Seung-Rim Lee, Boo-Kil Kwon, Seung-Hyun Lee, Young-Myong Yang, Heung-Seok Seo, Young-Soon Lee, "Safety Assessment and Code Development for Application of Kogas Membrane Type LNG Storage Tank", paper presented at 7th Topical Conference on Natural Gas Utilization, Spring National Meeting of the American Institute of Chemical Engineers, Houston (April, 2007).

Kevin B. McGrattan, Howard R. Baum, and Anthony Hamins, "Thermal Radiation from Large Pool Fires", NISTIR 6546, National Institute of Standards and Technology, Washington, DC (November 2000).

G. A. Melhem, H. Ozog, and A. S. Kalelkar, "Understanding LNG Fire Hazards", paper presented at 7th Topical Conference on Natural Gas Utilization, Spring National Meeting of the American Institute of Chemical Engineers, Houston (April, 2007).

Rajeev Nanda and John Rizopoulos, "Utilizing Air Based Technologies as Heat Source for LNG Vaporization", paper presented at 86th Annual Gas Processors Association Convention, San Antonio, TX (March, 2007).

National Fire Protection Association, NFPA 59A – Standard for the Production, Storage, and Handling of Liquefied Natural Gas (LNG), NFPA, Quincy, MA (2006).

D. Nedelka, V. Sauter, J. Goanvic, and R. Ohba, "Last Developments in Rapid Phase Transition Knowledge and Modeling Techniques", OTC 15228, paper presented at the 2003 Offshore Technology Conference, Houston, TX (May 2003).

Anna Qiao, Caroline Deetjen, David Pertuz, "The Employment of NFPA 59a in LNG Facility Siting", paper presented at 41st Loss Prevention Symposium, Spring National Meeting of the American Institute of Chemical Engineers, Houston (April, 2007).

Kamal Shah, Judy Wong, and Bill Minton, "Considerations for Ambient Air-Based Technologies for LNG Regasification Terminals", paper presented at 8th Topical Conference on Natural Gas Utilization, Spring National Meeting of the American Institute of Chemical Engineers, New Orleans, LA (April, 2008).

Jaffee Suardin, Yanjun Wang, Mike Wilson, M. Sam Mannan, "High Expansion Foam Application for Controlling LNG Pool Fire -- Experiment Results and Analysis", paper presented at 7th Topical Conference on Natural Gas Utilization, Spring National Meeting of the American Institute of Chemical Engineers, Houston (April, 2007).

U. S. Code of Federal Regulations, 6 CFR Part 27 - Chemical Facility Anti-Terrorism Standards, Washington, DC (2007).

U. S. Code of Federal Regulations, 33 CFR Parts 105 – Maritime Facility Security, Washington, DC (2007).

U. S. Code of Federal Regulations, 33 CFR Parts 148, 149, and 150 – Deepwater Ports, Washington, DC (2007).

U. S. Code of Federal Regulations, 49 CFR Part 193 – Liquefied Natural Gas Facilities: Federal Safety Standards, Washington, DC (2007).

Dominik T. Uznanski, Pieter Versluijs, "LNG Expert : the Latest Evolution in LNG Storage Tank Management", paper presented at 7th Topical Conference on Natural Gas Utilization, Spring National Meeting of the American Institute of Chemical Engineers, Houston (April, 2007).

Brett Wink, Stephen Marcum, "Transient Flow Analysis in LNG Regasification Terminal Design Safety", paper presented at 7th Topical Conference on Natural Gas Utilization, Spring National Meeting of the American Institute of Chemical Engineers, Houston (April, 2007)

"The Extinguishment of Natural Gas Fires", Form No. F-75158-1, Wormald U.S., Inc. (1982).

Biography of Author

Vic Edwards

Victor H. Edwards, Ph. D., P. E. is Director of Process Safety for Aker Kvaerner, Inc., 3600 Briarpark Drive, Houston, TX 77042 (713/270-2817; Fax: 713/270-3195; e-mail: vic.edwards@akerkvaerner.com). Responsible for process safety, Vic has received five DuPont awards for safety and environmental engineering excellence. In 1998, Kvaerner Engineers and Constructors named him employee of the year. Vic has contributed more than sixty publications to the engineering literature. He chaired the 1992 Process Plant Safety Symposium and the 1997 International Plant Operations and Design Conference. Vic received the 2003 Service Award from the Process Safety Center at Texas A & M University, where he chairs the Technical Advisory Committee. Edwards earned his B. A. from Rice University and his Ph. D. from the University of California at Berkeley. A registered professional engineer in Texas, he is an AIChE Fellow, and a member of ACS, AAAS, NFPA, NSPE, and the New York Academy of Sciences.

LNG Vapor Cloud Exclusion Zone Requirements Need Review

Jerry Havens, Tom Spicer, and Heather Walker
Chemical Hazards Research Center
Ralph E. Martin Department of Chemical Engineering
University of Arkansas, Fayetteville, Arkansas 72701
Primary contact: jhavens@uark.edu

AIChE Spring Meeting, April 2008
8th Topical Conference on Natural Gas Utilization
New Orleans, Louisiana, April 6-10, 2008

ABSTRACT

LNG spills on land in the United States are required to be contained in purpose-built impoundments (sumps and/or dikes) to limit spreading of LNG in case of loss of containment. The presence of sumps and/or impoundment walls necessitates consideration of the associated effects on vapor holdup and dispersion of the vapor cloud from a spill. The DEGADIS and FEM3A models are approved by 49 CFR 193 for determination of vapor cloud exclusion zones. DEGADIS is limited to prediction of dispersion from area sources over flat, obstacle-free terrain, and hence is not directly applicable to this important dispersion scenario. FEM3A, which was designed to account for such effects, has not been applied to date in any applications to FERC for LNG terminal approval. Ad hoc methods continue to be used to determine input conditions for DEGADIS to estimate dispersion of LNG vapor downwind of an impoundment system. There is confusion about the specification of design spills in the federal regulation 49 CFR 193 requirements and the requirements specified in NFPA 59A which is incorporated in the parent regulation. As a result, the determination of exclusion zones for the LNG import terminals currently being considered have not been consistent, and in some cases have resulted in exclusion zone determinations that do not provide for public safety as intended by 49 CFR 193. This paper provides examples and quantifies errors that the authors believe have been made in determinations of vapor cloud exclusion zones - using specific vapor cloud exclusion zone determinations reported in the final environmental impact statement (FEIS) for the Weavers Cove LNG terminal application and in the more recent draft environmental impact statement (DEIS) for the Bradwood Landing LNG terminal application. As It appears that no provision is being made currently for flashing of LNG upon release from transfer lines, and as the flash vapor component could increase the LNG vapor source term, provision for the effects of flashing upon release, as well as associated questions of whether such releases will be effectively confined to the associated impoundment, should be considered.

INTRODUCTION

The principal hazards attending storage and handling of liquefied natural gas (LNG) are pool fires, which occur when LNG is released from containment and the vapors produced mix with air, catch fire and burn back to the liquid pool; and vapor cloud fires, which occur when the vapors produced travel significant distances while mixing with air before being ignited. Vapor cloud explosion (damaging overpressure) hazards, although theoretically possible are considered highly unlikely unless either the gas/air mixture is contained (constrained in volume), as in a building, or in the rare situation where the LNG vapor contains more than about 15 – 20% of heavy gas components such as propane or higher molecular weight hydrocarbons. This paper deals exclusively with vapor cloud exclusion zones required to limit the hazards of vapor cloud fires.

Dilution of the vapor cloud formed at the source of the release, where the LNG vapor concentration can approach 100%, occurs primarily by turbulent diffusion in the atmosphere; molecular diffusion (in the atmosphere) is negligible in comparison. It is this dilution with air accompanying dispersion of LNG vapor from an LNG spill into the atmospheric boundary layer that is predicted with regulation-specified models to determine requirements for vapor cloud dispersion exclusion zones. Although DEGADIS[1] and FEM3A[2] are approved by 49 CFR 193 for determination of vapor cloud exclusion zones, DEGADIS is limited to prediction of dispersion from area sources over flat, obstacle-free terrain, and hence is not directly applicable to this important dispersion scenario. FEM3A, which is potentially applicable, was approved in 2000, but to these author's knowledge has not been applied in any of the applications to FERC for LNG terminal approval to date. It is critically important to understand that either of the dispersion models currently specified require as input the rate at which LNG vapor is released from the spill – the "vapor source input", which we believe is being specified incorrectly in current applications (using SOURCE5) to FERC.

As a result of current regulatory requirements, LNG spills on land in the United States are invariably into dike/impoundment areas volumes designed to limit spreading of LNG spills. The sump excavation or impoundment into which the design spills are required to be directed necessitate consideration (in the prediction of dispersion from a spill) of the effects on vapor holdup by the impoundment as well as alteration of the atmospheric boundary layer (wind) flow into, over, and around the impoundment. Although the phenomenology of vapor holdup and vapor mixing with air in an impoundment has been studied in wind tunnel and field experiments, and the fallacy of using an overly simplistic estimation of the time and rate of vapor cloud overflow from the impoundment has been demonstrated, ad hoc (and inappropriate) methods continue to be used to determine input conditions for DEGADIS to estimate LNG vapor dispersion downwind of an impoundment system.

There is further confusion about the specification of “design spills” in the federal regulation 49 CFR 193 requirements and the requirements specified in NFPA 59A which is incorporated in the parent regulation. As a result of such inappropriate specification of design spills and the methods used to determine the effects of vapor holdup by the impoundment systems, the methods currently used have resulted in exclusion zone determinations that do not provide for public safety as intended by 49 CFR 193.

This paper quantifies these errors that the authors believe have been made in determinations of vapor cloud exclusion zones reported in the final environmental impact statement (FEIS) for the Weavers Cove LNG terminal application and in the draft environmental impact statement (DEIS) for the Bradwood Landing LNG terminal application.

REGULATORY REQUIREMENTS

The federal regulation 49 CFR 193 [Amdt. 193-17, 65 FR 10959, Mar. 1, 2000] specifies the following requirements for the determination of LNG vapor cloud exclusion zones:

193.2059 Flammable vapor-gas dispersion protection.

Each LNG container and LNG transfer system must have a dispersion exclusion zone in accordance with section 2-2.3.2 of ANSI/NFPA 59A with the following exceptions:

(a) Flammable vapor-gas dispersion distances must be determined in accordance with the model described in the Gas Research Institute report GRI-89/0242, ``LNG Vapor Dispersion Prediction with the DEGADIS Dense Gas Dispersion Model." Alternatively, in order to account for additional cloud dilution which may be caused by the complex flow patterns induced by tank and dike structure, dispersion distances may be calculated in accordance with the model described in the Gas Research Institute report GRI 96/0396.5, ``Evaluation of Mitigation Methods for Accidental LNG Releases. Volume 5: Using FEM3A for LNG Accident Consequence Analyses". The use of alternate models which take into account the same physical factors and have been validated by experimental test data shall be permitted, subject to the Administrator's approval.

(b) The following dispersion parameters must be used in computing dispersion distances:

(1) Average gas concentration in air = 2.5 percent.

(2) Dispersion conditions are a combination of those which result in longer predicted downwind dispersion distances than other weather conditions at the site at least 90 percent of the time, based on figures maintained by the National Weather Service of the U.S. Department of Commerce, or as an alternative where the model used gives longer distances at lower wind speeds, Atmospheric Stability (Pasquill Class) F, wind speed = 4.5 miles per hour (2.01 meters/sec) at reference height of 10 meters, relative humidity = 50.0 percent, and atmospheric temperature = average in the region.

(3) The elevation for contour (receptor) output H = 0.5 meters.
(4) A surface roughness factor of 0.03 meters shall be used
...
(c) The design spill shall be determined in accordance with section 2.2.3.3 of ANSI/NFPA 59A.

NFPA 59A, Standard for the Production, Storage, and Handling of Liquefied Natural Gas (LNG) 2001 Edition, as incorporated in the federal regulation 49 CFR 193 (above), specifies the following requirements for the determination of LNG vapor cloud exclusion zones: (NFPA 59A sections applicable to the determination of exclusion zones for spills from tanks which are not equipped with over-the-top connections are excluded from consideration in this paper, since all LNG import terminals for which applications are pending incorporate over-the-top connections.)

2.2.3.3 *The spacing of an LNG tank impoundment to the property line that can be built upon shall be such that, in the event of an LNG spill specified in 2.2.3.5, an average concentration of methane in air of 50 percent of the lower flammable limit (LFL) does not extend beyond the property line that can be built upon, in accordance with calculations using one of the following:*
(a) The model described in Gas Research Institute report GRI 0242, "LNG Vapor Dispersion Prediction with the DEGADIS Dense Gas Dispersion Model"
(b) The model described in Gas Research Institute report GRI-96/0396.5, "Evaluation of Mitigation Models for Accidental LNG Releases, Volume 5; Using FEM3A for LNG Accidental Consequence Analysis"
(c) A model that incorporates the following:
(1) Takes into account physical factors influencing LNG vapor dispersion, including, but not limited to, gravity spreading, heat transfer, humidity, wind speed and direction, atmospheric stability, buoyancy, and surface roughness
(2) Has been validated by experimental test data appropriate for the size and conditions of the hazard to be evaluated
(3) Is acceptable to the authority having jurisdiction
The computed distances shall include calculations based on one of the following:
(1) The combination of wind speed and atmospheric stability that can occur simultaneously and result in the longest predictable downwind dispersion distance that is exceeded less than 10 percent of the time
(2) The Pasquill-Gifford atmospheric stability, Category F, with a 4.5-mph (2-m/s) wind speed
The computed distances shall be based on the actual liquid characteristics and the maximum vapor outflow from the vapor containment volume (the vapor generation rate plus the displacement due to liquid inflow).
...
Provisions shall be made to minimize the possibility of a flammable mixture of vapors from a design spill specified in 2.2.3.5, as appropriate, reaching a property line that can built upon and that would result in a distinct hazard.

...

2.2.3.5 *The design spill shall be determined in accordance with Table 2.2.3.5.*

...

Table 2.2.3.5 Design Spill

Container Penetration	***Design Spill***	***Design Spill Duration***
...		
Containers with over-the top fill, with no penetrations below the liquid level.	*The largest flow from any single line that could be pumped into the impounding area with the container withdrawal pump(s) considered to be delivering the full rated capacity.*	*The largest flow from any single line that could be pumped into the impounding area with the container withdrawal pump(s) delivering the full rated capacity: (1) for 10 minutes if surveillance and shutdown is demonstrated and approved by the authority having jurisdiction. (2) For the time needed to empty a full container where surveillance and shutdown is not approved.*
...		
Impounding areas serving only vaporization, process, or LNG transfer areas.	*The flow from any single accidental leakage source (emphasis added)*	*For 10 minutes or for a shorter time based on demonstrable surveillance and shutdown provisions acceptable to the authority having jurisdiction.*

GENERAL APPLICATION PRINCIPLES

For all of the terminal applications we have reviewed, including the Weavers Cove and Bradwood Landing applications considered herein, the process of determining vapor cloud exclusion zones has been reduced to the following steps:

1. Determine the impoundment volumes required
2. Determine the design spills
3. Perform DEGADIS calculations

Although there are site specific differences in the layout plan of the various facilities, including variations in the position and volumes of tanks and spill impoundments, the requirements of the regulation do not address such specificity. The following idealized configuration of a single tank and impoundment system therefore represents all of the variations considered, as the exact locations of the system components does not affect the determinations of the vapor cloud exclusion zones so long as the DEGADIS model, which assumes dispersion from a ground level surface over flat, unobstructed terrain, is used.

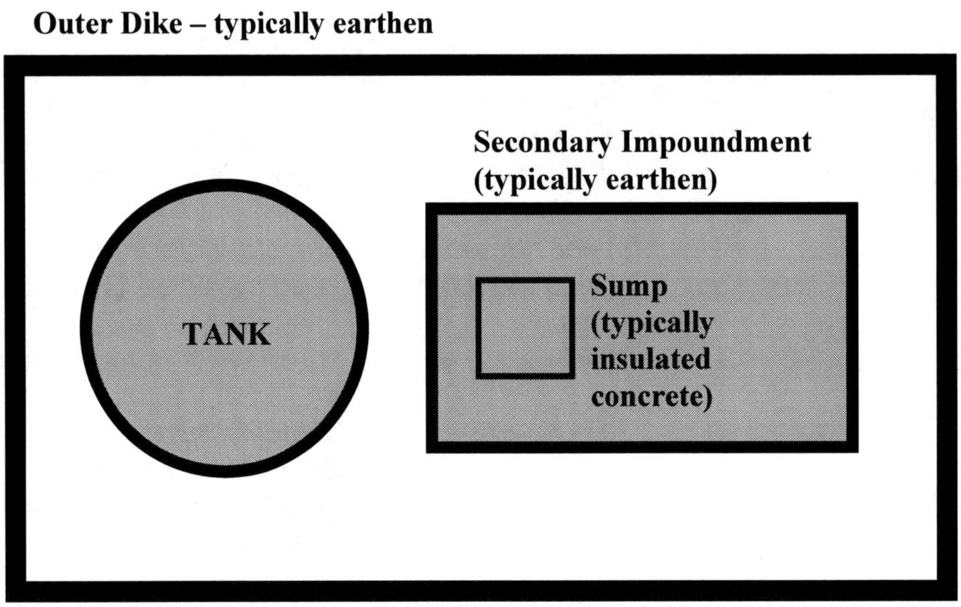

Figure 1. Idealized LNG Import Terminal Site Plan

It is noted that with these simplifying assumptions, which have been made in all of the applications we have reviewed, any effects which could be due to the presence of the tank(s) and/or impoundment walls have been excluded from consideration; as a consequence, all dispersion calculations (using DEGADIS) account only for factors of dispersion assuming flat, obstruction-free terrain.

Significantly, all determinations to date of input LNG vapor flow rates required by DEGADIS appear to have been made using the SOURCE5 model[3] to determine the rate of gas overflow that would result from a design spill directed into the impoundment, assuming that the LNG vapor overflows as pure LNG vapor, i.e. without any mixing of the vapor with air in the impoundment. Our research group has repeatedly asserted[4,5,6,7] that this practice can lead to significant errors because the assumption that the LNG vapor from the spill does not mix with air delays its overflow from the impoundment, during which time the evaporation rate decreases significantly due to the cooling of the impoundment floor and walls in contact with the liquid LNG. As will be shown below, the importance of this assumption is site-specific - because the different impoundment system designs result in different amounts of vapor holdup (accumulation of pure vapor before overflowing the impoundment system), resulting in different values of the LNG vapor (source) term to DEGADIS.

Perhaps most importantly, inconsistency in specification of design spills has led to a wide variation in the vapor cloud exclusion zones determined. Zinn[8] has described the evolution of the inconsistency in the regulations. The result of incorporation of NFPA 59A in 49 CFR 193 has been that while retaining the requirement for providing impoundments sufficient to contain the spill from a specified pipe assumed to suffer a guillotine rupture with full flow continuing for 10 minutes, the design spill for determining the vapor cloud exclusion zone for the same impoundment and for the same line has been changed to require the flow from "a single accidental leakage source" rather than the previously required full line rupture.

WEAVERS COVE

The Weavers Cove application, arguably the first terminal application for siting in an urban location (Fall River, MA) since the Boston (Everett) terminal was sited in the Seventies, continues to be hotly contested.

The Weavers Cove terminal layout can be idealized as shown in Figure 1.

Weavers Cove DEIS

Table 1 gives the pertinent data for the impoundments specified by Weavers Cove in the DEIS, including a concrete sump sized for a tank withdrawal system design spill, a secondary earthen impoundment sized for a ship-unloading-line spill, and a tertiary earthen impoundment around the outer boundary of the facility. Table 2 gives the design spills specified by Weavers Cove in the DEIS.

Table 1. Impoundment specifications* – Weavers Cove DEIS

Source	Spill Size, gallons	Impoundment System	Impoundment Size, gallons
LNG storage tank	52,834,410	Outer container concrete wall Outer earthen berm	59,884,318 1,894,761,819
In-tank pump withdrawal header (4 pumps)	98,900	Impoundment sump, concrete	104,158
Ship unloading line	608,000	Impoundment, Earthen	837,248

*Only the controlling (largest) spills are considered

Table 2. Design spills – Weavers Cove DEIS

Source	Design Spill Size, gallons	Impoundment Size, gallons
LNG storage tank	NA	NA
In-tank pump withdrawal header (4 pumps)	98,900	104,158
Ship unloading line	4,350 gallons*	837,248

*accidental leakage source – a l inch diameter instrument connection on the discharge header of the unloading arms

In the draft environmental impact statement[9] FERC approved Weavers Cove's determination that no vapor cloud would extend beyond the outer earthen berm specified in Table 1. This determination was made on the basis that the total amount of vapor formed from the 10-minute pump withdrawal header, assuming a liquid to vapor expansion of 1:235, would, as a pure vapor, occupy only 841,465,111 gallons (~45%) of the total capacity of the outer earthen berm, and would therefore remain within the confines of the facility boundary.

Public comments challenged the FERC approved procedure[10], which assumes that there would be no expansion of the cold LNG vapor (due to warming) and, most importantly, no mixing with air within the outer earthen impoundment. The same public comments questioned the propriety of using the much smaller "accidental leakage source" assumed from the ship unloading line rather than the same spill that was used to size the ship unloading line impoundment.

Weavers Cove FEIS

Table 3 gives the modified specifications for the impoundments specified by Weavers Cove in the final environmental impact statement[11] including a concrete sump, a secondary earthen impoundment sized for the ship-unloading-line spill, and a tertiary earthen impoundment around the outer boundary of the facility. Table 4 gives the design spills specified by Weavers Cove in the FEIS.

Table 3. Impoundment specifications* – Weavers Cove FEIS

Source	Spill Size, gallons	Impoundment System	Impoundment Size, gallons
LNG storage tank	52,834,410	Outer container concrete wall Outer earthen berm	59,884,318 1,894,761,819
In-tank pump withdrawal header (5 pumps)	124,391	Impoundment sump, concrete	124,391
Ship unloading line	627,127	Impoundment, Earthen	837,248

*Only the controlling (largest) spills are considered

Table 4. Design spills – Weavers Cove FEIS

Source	Design Spill Size, gallons	Impoundment Size, gallons
LNG storage tank	NA	NA
In-tank pump withdrawal header (5 pumps)	124,391	124,391*
Ship unloading line	52,257 gallons**	124,391

* FERC recommended increasing the depth of the concrete sump by approximately 9 inches, thus increasing the volume of the sump from 104,158 gallons to 124,391 gallons

**accidental leakage source – a 4" diameter instrument connection on the discharge header of the unloading arms

FERC responded to public comments regarding the impropriety of assuming no mixing of the LNG vapor in the outer berm by "performing a supplementary vapor dispersion analysis for the design spill by conservatively assuming no earthen structure on the plant perimeter", coupled with a revised requirement that the concrete sump volume be resized to accommodate the 5-pump withdrawal header spill of 124,391 gallons. FERC also increased the ship unloading line spill to 52,257 gallons, reflecting a revised assumption of a 4 inch diameter instrument line failure instead of the 1 inch line specified in the DEIS.

With the tank withdrawal header spill then being the controlling (largest) design spill, FERC recalculated the vapor dispersion exclusion zone. However, although they neglected any vapor holdup (as pure LNG vapor) in the outer earthen berm, they now assumed that the secondary earthen berm, with 837,248 gallons volume, filled with pure LNG vapor before overflowing, thus still determining the input source rate to DEGADIS erroneously. This approach is unjustified.

FERC's Approved SOURCE5 and DEGADIS Calculations – Weavers Cove FEIS

Using SOURCE5, FERC estimated that "… the vapor would overtop the earthen impoundment within 6 minutes with a vapor production rate of 10.9 kg/s". A transient DEGADIS simulation, with the atmospheric conditions specified in Table 5, then predicted that 2.5 percent average gas concentration would extend 640 feet from the center of the concrete sump.

Table 5. Input Environmental Conditions* for DEGADIS – Weavers Cove FEIS

Average gas concentration in air denoting lower concentration hazard limit.	2.5 % (by volume)
atmospheric temperature	Not specified, 49 CFR 193 requires regional average
wind speed @ 10 m elevation	2 m/s
relative humidity	50 %
elevation for contour (receptor) output	Not specified, 49 CFR 193 requires 0.5 m
Surface roughness	0.03 m

** 49 CFR 193 requires that dispersion conditions are a combination of those which result in longer predicted downwind dispersion distances than other weather conditions at the site at least 90 percent of the time, based on figures maintained by National Weather Service of the U.S. Department of Commerce, or as an alternative where the model used gives longer distances at lower wind speeds, Atmospheric Stability (Pasquill Class) F, = 4.5 miles per hour (2.01 meters/sec) at reference height of 10 meters, = 50.0 percent, and = average in the region. FERC approved Weavers Cove's application which opted for the latter.*

FERC determined that this flammable vapor exclusion zone (640 feet), based on the revised impoundment sump, "would not extend offsite". This is where the matter stands as this paper is written.

CHRC Calculations

The Chemical Hazards Research Center attempted to repeat FERC's calculations reported in the FEIS, requesting of FERC the exact input conditions for SOURCE5 and DEGADIS that had been used in the DEIS and FEIS. Responding to an initial request by CHRC for the properties of the concrete sump used, FERC stated that the properties used were as follows: thermal conductivity = 1.44 W/mK, density = 2,323 kg/m^3, and heat capacity = 1,260 J/kgK. However, using the original SOURCE5 model developed by the Gas Research Institute as well as the Breeze LFG^{TM} model which incorporates SOURCE5 and DEGADIS, CHRC was still unable to repeat FERC's predictions exactly. Requests for verification of additional DEGADIS and SOURCE5 input data to assist CHRC in verifying FERC's calculations were denied on grounds that the information was protected as Confidential Energy Infrastructure Information (CEII).

CHRC performed independent calculations, using SOURCE5, of the design spill/vapor cloud dispersion scenario described by FERC (above) and estimated that the vapor would overtop the earthen impoundment in just over 6 minutes with a vapor production rate of 10.5 kg/s. Then, assuming the values required by 49 CFR 193 for the environmental input data (temperature = 7.4 C, determined from specifications on the internet of average annual temperature in Fall River, and atmospheric pressure = 1 atm) and assuming SOURCE5 default (earth) dike floor properties for the ship unloading line impoundment, CHRC predicted that the 2.5 percent average gas concentration would extend 653 feet from the center of the concrete sump.

CHRC then calculated two additional scenarios, which we believe are the more correct ones applicable to the intended requirements of 49 CFR 193:

- WC Scenario 2 - CHRC recalculated the scenario just considered, neglecting the presence of the (secondary) earthen dike surrounding the concrete sump. We believe that this is the proper way to make the calculation, using DEGADIS, as it neglects holdup of pure vapor in the earthen impoundment.

- WC Scenario 3 - CHRC calculated the vapor cloud exclusion zone for the scenario of the full ship unloading line spill into the ship unloading line impoundment, and neglected the presence (as FERC has done) of the outer earthen berm – We neglected the volume of the concrete sump (in comparison with the volume of the earthen impoundment) and assumed that the spill was directed into the earthen impoundment.

Table 6 summarizes the results, including CHRC's estimate of the scenario considered by FERC.

Table 6. CHRC dispersion calculations – Weavers Cove

Scenario	Vapor overflow time, s	Vapor overflow rate, kg/s	Distance to 2.5% vapor concentration, ft
Scenario WC1	363	10.5	653
Scenario WC2	12	37.6	955
Scenario WC3	19	135.2	1745

Bradwood Landing

Consideration here of the Bradwood Landing application is limited to analysis of the draft environmental impact statement[12], as the FEIS has not been released at time of writing. As for the Weavers Cove terminal, the Bradwood Landing layout can be idealized as shown in Figure 1.

Table 7 gives the pertinent data for the impoundment specified for Bradwood Landing in the DEIS - a concrete sump sized for a tank withdrawal system design spill and ship unloading line spill (not simultaneously). Table 8 gives the design spills specified by Bradwood Landing in the DEIS.

Table 7. Impoundment specifications* – Bradwood Landing DEIS

Source	Spill Size, gallons	Impoundment System	Impoundment Size, gallons
LNG storage tank	45,805,758	Outer container concrete wall	52,118,481
Ship unloading line	529,091	Impoundment, Concrete	538,632

*Only the controlling (largest) spills are considered

Table 8. Design spills – Bradwood Landing

Source	Design Spill Size, gallons	Impoundment Size, gallons
LNG storage tank	NA	NA
Tank pump discharge header	70,590	538,632
Ship unloading line	140,320 gallons**	538,632

**accidental leakage source – a 6 inch diameter purge line or a 6 inch diameter recycle line attached to the unloading line

<u>FERC's Approved SOURCE5 and DEGADIS Calculations – Bradwood Landing</u>

FERC predicted, using SOURCE5 and DEGADIS, with the atmospheric conditions specified in Table 5, that 2.5 percent average gas concentration would extend 243 feet from the center of the concrete sump.

Table 9. Input Environmental Conditions for DEGADIS* – Bradwood Landing

Average gas concentration in air denoting lower concentration hazard limit.	2.5 % (by volume)
Atmospheric temperature	52 F
wind speed @ 10 m elevation	2 m/s
Relative humidity	50 %
elevation for contour (receptor) output	Not specified, 49 CFR 193 requires 0.5 m
Surface roughness	0.03 m

** 49 CFR 193 requires that dispersion conditions are a combination of those which result in longer predicted downwind dispersion distances than other weather conditions at the site at least 90 percent of the time, based on figures maintained by National Weather Service of the U.S. Department of Commerce, or as an alternative where the model used gives longer distances at lower wind speeds, Atmospheric Stability (Pasquill Class) F, = 4.5 miles per hour (2.01 meters/sec) at reference height of 10 meters, = 50.0 percent, and = average in the region. FERC approved Bradwood Landing's application which opted for the latter.*

FERC determined that this flammable vapor exclusion zone, although extending off-site into the navigation channel, where there were no prohibited land uses (within the 243 feet distance), would satisfy the vapor exclusion zone requirements of 49 CRF 193.

CHRC Calculations

The Chemical Hazards Research Center attempted to repeat FERC's calculations reported in the Bradwood Landing DEIS, requesting of FERC the input conditions for the concrete sump used. FERC stated that the properties of the concrete sump would be revised (from "similar to Epoxy Concrete" specified in the DEIS) for the FEIS as follows: thermal conductivity = 0.32 W/mK, density = 801 kg/m^3, and heat capacity = 2,323 J/kgK. CHRC performed independent calculations of the vapor cloud dispersion scenario described in the Bradwood Landing DEIS (above) using the sump properties that were specified for the (as yet unreleased) FEIS. CHRC estimated that the vapor would overtop the earthen impoundment in just under 15 minutes with a vapor production rate of 3.6 kg/s. Then, assuming the values shown in Table 9 for the environmental conditions required for DEGADIS, CHRC predicted that the 2.5 percent average gas concentration would extend 325 feet from the center of the concrete sump.

CHRC then calculated one additional scenario, which we believe is the correct one applicable to the intended requirements of 49 CFR 193. In this scenario (Scenario 2), CHRC calculated the vapor cloud exclusion zone for the scenario of the full ship unloading line spill into the ship unloading line sump.

Table 10 summarizes the results, including CHRC's estimate of the scenario considered by FERC in the DEIS.

Table 10. CHRC dispersion calculations – Bradwood Landing

Scenario	Vapor overflow time, s	Vapor overflow rate, kg/s	Distance to 2.5% vapor concentration, ft
BL Scenario 1	888	3.6	325
BL Scenario 2	348	9.9	646

CONCLUSIONS

Vapor cloud exclusion zones determined independently by our research group indicate that the values reported in the Weavers Cove FEIS and the Bradwood Landing DEIS are deficient:

- The assumption has been made (for both determinations) that the vapor evolved from the design spill overflows the impoundment as pure vapor at the temperature of the LNG. This assumption results, for Fall River, where a secondary impoundment is responsible for additional (to the sump) vapor holdup, in a 32 % underprediction of the vapor cloud exclusion zone.
- Determination of the exclusion zone that would result if the design spill were specified as the 10 minute duration full line rupture of the ship unloading line into its specified impoundment increases the extent of the vapor cloud exclusion zones for Fall River (from the value specified in the FEIS) by a factor of 2.7 and for Bradwood Landing (from the value specified in the DEIS) by a factor of 2.

These calculations do not take into account the following factors which we believe should be considered in the determination of vapor cloud exclusion zones.

- In all cases we have reviewed to date, applicants have opted to determine the vapor cloud exclusion zones for the low wind speed, stable (2 m/s, F stability) condition. As 49 CFR 193 specifies that this alternative is appropriate only where the model used gives longer distances at lower wind speeds, this assumption does not appear to be defensible, because the transient release rates being considered the maximum distance determined could be at a higher wind speed. This is particularly true if the effect of heat transfer to the cloud, as currently practiced by all applicants, is considered.

- All of the vapor cloud determinations we have reviewed are for releases from transfer lines. It appears that no provision is being made currently for flashing of the LNG upon release from transfer lines. In order to estimate the potential importance of flashing in such releases, we considered the release conditions treated above for Bradwood Landing Scenario 1. For a line pressure of 1 barg, the flash fraction would be approximately 0.3 %. Therefore, the total release rate of 374 kg/s would generate approximately 260 m^3/s of vapor, and this (flash) vapor, if collected in the sump, would overflow in roughly 8 seconds. In contrast, the overflow time predicted by SOURCE5 (with no flashing) is 888 seconds. The earlier overflow time (reflecting flashing) would cause the overflow rate to be correspondingly higher. It is our understanding that ship unloading transfer lines can operate at approximately 20 psig or higher, and that tank withdrawal pump transfer lines can operate at approximately 1000 psig or higher. As the flash vapor component will add to the magnitude (and possibly the density due to aerosol formation) of the LNG vapor source term, provision for consideration of the effects of flashing upon release, as well as associated questions of whether such a flashing jet will be effectively confined to the associated impoundment should be considered.

Literature Cited

1. Havens, J.A., and T.O. Spicer, "LNG Vapor Dispersion Prediction with the DEGADIS Dense Gas Dispersion Model", Gas Research Institute Report GRI- 89/0242, 1990.
2. Spicer, T.O., J.A. Havens, and H.L. Walker, "Evaluation of Mitigation Methods for Accidental LNG Releases: Volume 5/5—Using FEM3A for LNG Accident Consequence Analysis", Topical Report for Gas Research Institute, April 1997.
3. Atallah, S., M. Sirdesai, and D.M. Jennings, "Source 1: A Model for Creating a DEGADIS Input File", Gas Research Institute Report GRI-92/0534, January 1993.
4. FERC Docket CP04-36 (Weaver's Cove LNG).
5. Havens, J. and T. Spicer, "LNG Vapor Dispersion in Impoundments", Plant Safety Progress, Vol. 24, No. 3, September 2005.
6. Havens, J. and T. Spicer, "Error in FERC Environmental Impact Statement Determinations of LNG Vapor Cloud Exclusion Zones: Failure to Account for Air Mixing in Vapor Impoundments", Mary Kay O'Connor Process Safety Center Symposium, College Station, Texas, October 24, 2006.
7. FERC Docket CP06-365 (Bradwood Landing LNG).
8. Zinn, C.D., "LNG Codes and Process Safety", AICHE Spring National Meeting, Paper 109e, Atlanta Georgia, April 2005.
9. Draft Environmental Impact Statement – Weaver's Cove LNG Project, Federal Energy Regulatory Commission, August 2004.
10. FERC Docket CP06-36 (Weaver's Cove LNG).
11. Final Environmental Impact Statement – Weaver's Cove LNG Project, Federal Energy Regulatory Commission, May 2005.
12. Draft Environmental Impact Statement – Bradwood Landing Project, Federal Energy Regulatory Commission, August 2007.

BIOGRAPHY

Dr. Jerry Havens

Maurice E. Barker Distinguished Professor
Chemical Hazards Research Center
University of Arkansas
700 Research Center Boulevard
Fayetteville, AR 72701

Phone: 479-575-2055
Fax: 479-575-8718
Email: jhavens@uark.edu

Professor Havens holds the Maurice E. Barker Chair in Chemical Process Safety and the Environmental Fate of Pollutants at the University of Arkansas. He is recognized internationally for his expertise in atmospheric dispersion of hazardous chemicals and fire/explosion phenomena. The DEGADIS and FEM3A dispersion models developed and or validated by Dr. Havens and his colleague Dr. Tom Spicer against wind tunnel experiments at the University's Chemical Hazards Research Center are specified for use in LNG plant siting applications in the United States.

106c Predicting Ozone Formation in the Troposphere Using Mechanistic Modeling

Linda J. Broadbelt, Department of Chemical and Biological Engineering, Northwestern University, 2145 Sheridan Road E136, Evanston, IL 60208 and Shumaila Khan, Department of Chemical Engineering, Northwestern University, 2145 Sheridan Road, Evanston, IL 60208.

Ozone is a major component of photochemical smog, an important air quality problem. Emission of volatile organic compounds (VOCs) into the atmosphere leads to increases in the ambient ozone concentration. It would be extremely valuable to have the ability to determine how significantly a particular VOC contributes to ozone formation for both environmental and industrial purposes. A promising strategy is to assemble knowledge of the kinetics and photochemistry into detailed mechanistic models from which predictions of ozone concentrations may be obtained. Automated mechanism generation was applied. A group additivity approach was developed to estimate absorption cross sections over the wavelength region of tropospheric interest. Mechanisms were then generated automatically for various systems using different criteria for halting generation to control the explosive nature of the chemistry. A range of VOC mixtures was investigated. The models were compared to experiments, and the models were able to extrapolate well to different conditions.

106a Prescreening for Environmentally-Important Properties through Computational Chemistry

David A. Gallagher, CAChe Research, 13690 SW Otter Lane, Beaverton, OR 97008

The ability to pre-screen for environmentally-important properties of candidate compounds and products, particularly before they are synthesized, can save significant laboratory time and expense. Although, R&D cost savings are important, the associated benefits of shortening time to market and minimizing environmental risk, offer the greatest potential for long-term profitability.

Many environmentally-important properties are closely related to physical and chemical properties that can be predicted from just molecular structure information using quantum chemistry calculations. The latest generation of computational tools now makes it relatively simple to pre-screen large libraries of hypothetical candidate compounds for environmentally-important properties, using a regular Windows-based computer.

The author will review various approaches to the prediction of environmentally-important properties with examples covering aquatic fate, persistence, atmospheric fate, atmospheric lifetimes of halocarbons, soil remediation rates, and the design of ionic liquids for greener processes, using readily available tools based on quantum chemistry, statistical thermodynamics and related methods.

106f New Mathematical Method for Prediction of Pollutant in River Simulated by Gas Chromatography Measurment

Esmail Jamshidi Sr. and Alireza Bahramian III. Department of chemical engineering, Amir kabir university of technology, Hafez, somaye, Tehran, Iran

Abstract

Prediction of pollutant concentration in river has been studied for a long time. Quall II solves time-dependent mass diffusion and convection simultaneously. Quall II is using numerical solution where the errors of prediction are 30-70%. These errors arise from large step in space (Δx). In this study a new mathematical method has been used to solve the second order partial differential equation where the pollutants along with diffusion and convection respect to time can be adsorped and desorped in down stream. An exact measurement can be used to verify accuracy of any prediction or calculations. The Gas Chromatography is exactly ideal system for finding the accuracy of these calculations. Since governing equation and physical characters are same in both cases, theses similarities can be tested by chemical measurement. The new mathematical method applied to solve the governing equation, and results shows only 7% error in predicted concentration respect to measured concentration in Gas Chromatography.

106e Phosphorus Fractions and Phosphate Sorption Characteristics of Sediments from the Three-Gorges Reservoir (China) in Relation to the Environmental Conditions I

Ying Wang, Zhenyao Shen, and Hongyuan Wang. Beijing Normal University, Beijing, 100875, China

The phosphorus contents and fractions of four sediment samples collected at the influx of four main branches (CunTan, XiaoJiang, DaNing and XiangXi River) of the Three-Gorges Reservoir on the Yangtze River in China were analyzed by sequential chemical extraction. Adsorption characteristics of sediments for phosphate were studied systematically in batch experiments. The effects of environmental conditions such as pH, temperature, concentration of sediment and salinity on P adsorption were also investigated. The results showed that inorganic P was the major fraction of total P. P adsorption kinetics and the adsorption isotherm could be well described by Simple Elovich model and modified Langmuir model, respectively. P adsorption on the sediments varied with pH. During the range from 6 to 8, P adsorption exhibited a maximum. Higher temperature (278K•A293K•A303K) accelerated the adsorption of phosphorus, however, higher salinity restrained the adsorption of phosphorus because of the competitive effect.

Mathematical Modeling of Leaching of Heavy Metals from Bio-Film Coated Cement Based Waste Forms

Norman W. Loney, Mojdeh Tabatabaie and Rachna Jaiswal

Otto H. York Department of Chemical Engineering
New Jersey Institute of Technology
University Heights, Newark NJ

Abstract

Solidification/stabilization processes have been used to immobilize heavy toxic metals in contaminated wastes for decades. These processes are designed to protect the environment from the harmful impacts of disposed hazardous wastes. Cement based solidification/stabilization of wastes containing heavy metals has been widely practiced. The stability of the captured heavy metals in cement based forms has been investigated under different physical, chemical and biological conditions. Experimenters have observed that under certain acidic conditions, a significant number of heavy metals in substantial amounts can be leached from refined biofilm-coated cement based waste forms. In order to explain the controlling mechanism in this leaching process, a mathematical model is proposed. The proposed model is developed for cylindrical geometry with the Michaelis-Menten kinetics as the primary mechanism controlling the leaching process. The primary species concentration profile will be derived analytically by first using a regular boundary perturbation followed by separation of variables or Laplace transforms as appropriate.

Introduction

Heavy metals are usually classified as those elements included between copper and bismuth on the Periodic Table that have specific gravities greater than 4.0 (Duffus, 2002; wikipedia, Sept 15-2007). While trace amounts of many of these metals such as iron, cobalt, copper, manganese, molybdenum, vanadium, strontium, zinc, and chromium are necessary for a wide range of metabolic activities in human body and other living organisms (Mertz, 1969; 1993; 1998), excessive amounts of these metals can be detrimental to their health (USEPA, 1998). Solidification/stabilization (S/S) processes have been used to immobilize heavy toxic metals in contaminated wastes for decades. Cement-based S/S has been a dominant practice to contain heavy metals in contaminated wastes (USEPA, 1999a & 1999b). While immobilization of wastes containing heavy metals have reduced spread of their presence in ground and surface water, studies have shown that under specific pH (Olmo et al, 2003) and/or presence of certain acid forming microorganisms (Idachaba et al, 2001) such as thiobacillus thiooxidans (T.t.) and thiobacillus ferroxidans (T.f.) entrapped species can become soluble and leach out. T.t and T.f. have been used for waste treatment processes in mining industry (Vachon et al, 1994).

Much research have attempted to address factors affecting leaching of heavy metals from cement or cement like materials and some proposed mathematical models (Bishop, 1986; Halim et al, 2005; Malviya and Chaudhary, 2006;). There is, however, very limited study on modeling of biological leaching of heavy metals from cementitious waste forms. Such models can be useful in identifying predictive measures when developing S/S strategies. It is the intent of this paper to develop a working model for bioleaching of heavy metals from cementitious matrices using well known Michaelis-Menten kinetics (Fogler, 2005; Shuler and Kargi, 2002).

Theory

Figure 1 shows a model sample that is examined in this study; a long circular solid cylinder with heavy metal contaminant distributed evenly throughout along the axis. The outer surface is coated with an acid forming bio-film capable of dissolving the stabilized heavy metal on the surface, hence, creating a large enough concentration gradient that causes a uniform, radial diffusion of entrapped metals in the matrix toward the outer surface.

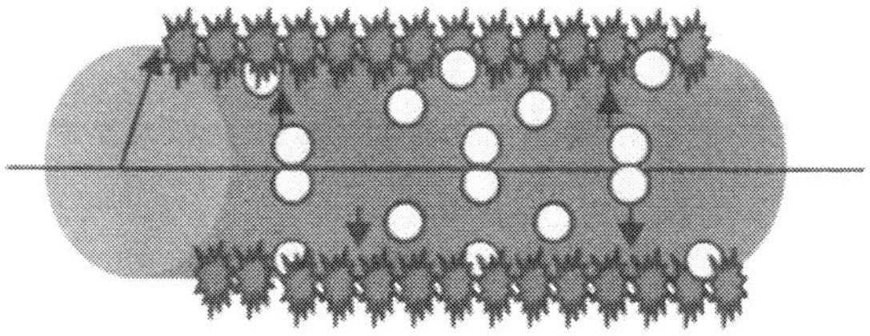

Figure 1. A cylindrical solid sample with evenly distributed heavy metal contaminant along the axis coated with bio-film .

Considering only radial transport of heavy metal species along the solid cylinder, concentration change with time can be described by Fick's second law of diffusion given by the PDE (1) (Crank, 1975; Bird, Stewart and Lightfoot, 2002):

$$\frac{\partial C_A}{\partial t} = D_{AB}\left(\frac{\partial^2 C_A}{\partial r^2} + \frac{1}{r}\frac{\partial C_A}{\partial r}\right) \qquad (1)$$

where D_{AB} is the diffusion coefficient of A (a heavy metal) in the solid matrix B.

Assuming the concentration of A, C_A, throughout the solid cylinder is uniform at time t=0 and stays finite at the center of the cylinder (r=0) at all times, PDE (1) would be subject to the following initial and boundary conditions:

$$C_A(r,0) = C_{A0} \tag{2}$$

$$C_A(0,t) \ \ is \ \ finite \tag{3}$$

Furthermore, assuming the rate at which species diffuse to the surface be the same as the rate of species dissolution by bio-film via Michaelis-Menten kinetics on the surface, the following boundary condition can be applied:

$$-D_{AB}\frac{\partial C_A}{\partial r} \cong \frac{k_1 C_A}{1+k_2 C_A} \quad (\text{mol/cm}^2\text{s}) \tag{4}$$

Method

Let

$$\Psi(\xi,\tau) = C_A/C_{A0}; \quad \xi = r/R; \quad \tau = \frac{D_{AB}t}{R^2} \qquad (5), (6), (7)$$

Then (1) becomes:

$$\frac{\partial \Psi}{\partial \tau} = \left(\frac{\partial^2 \Psi}{\partial \xi^2} + \frac{1}{\xi}\frac{\partial \Psi}{\partial \xi}\right) \tag{1A}$$

Eq. (2) becomes:

$$\Psi(\xi,0) = 1 \tag{2A}$$

Eq. (3) becomes:

$$\Psi(0,\tau) \ \ is \ \ finite \tag{3A}$$

Eq. (4) becomes:

$$-\frac{\partial \Psi}{\partial \xi} = w\left[1 - \varepsilon\Psi + \varepsilon^2\Psi^2 - \cdots\right], \quad at \ \ \xi = 1 \tag{4A}$$

with

$$w = \frac{Rk_0}{D_{AB}} \tag{8}$$

Then if

$$\Psi = \Psi_0 + \varepsilon\Psi_1 + \varepsilon^2\Psi_2 + \cdots \qquad (9).$$

Equations (1A) and (2A) can be restated as:

$$\frac{\partial \Psi_0}{\partial \tau} + \varepsilon \frac{\partial \Psi_1}{\partial \tau} + \varepsilon^2 \frac{\partial \Psi_2}{\partial \tau} + \cdots = \frac{\partial^2 \Psi_0}{\partial \xi^2} + \varepsilon \frac{\partial^2 \Psi_1}{\partial \xi^2} + \varepsilon^2 \frac{\partial^2 \Psi_2}{\partial \xi^2} + \varepsilon^3 \frac{\partial^2 \Psi_3}{\partial \xi^2} + \cdots + \frac{1}{\xi}\left[\frac{\partial \Psi_0}{\partial \xi} + \varepsilon \frac{\partial \Psi_1}{\partial \xi} + \varepsilon^2 \frac{\partial \Psi_2}{\partial \xi} + \cdots\right] \tag{10}$$

and:

$$\Psi_0 + \varepsilon \Psi_1 + \varepsilon^2 \Psi_2 + \varepsilon^3 \Psi_3 + \cdots = 1 \qquad at \;\; \tau = 0 \tag{11}$$

While Equations (3A) and (4A) become:

$$\Psi \;\; is \;\; finite \Rightarrow \Psi_i \;\; is \;\; finite; \;\; i = 0,1,2,\cdots \;\; at \;\; \xi = 0 \tag{12}$$

$$-\frac{\partial}{\partial \xi}\left[\Psi_0 + \varepsilon \Psi_1 + \varepsilon^2 \Psi_2 + \varepsilon^3 \Psi_3 + \cdots\right] = w\left[1 - \varepsilon(\Psi_0 + \varepsilon \Psi_1 + \cdots) + \varepsilon^2 (\Psi_0 + \varepsilon \Psi_1 + \cdots)^2 - \varepsilon^3 (\Psi_0 + \varepsilon \Psi_1 + \cdots) + \cdots\right], \;\; at \;\; \xi = 1 \tag{13}$$

Equating powers of ε results in a series of linear problems such as:

ε^0 :

$$\left.\begin{array}{l} \dfrac{\partial \Psi_0}{\partial \tau} = \dfrac{\partial^2 \Psi_0}{\partial \xi^2} + \dfrac{1}{\xi}\dfrac{\partial \Psi_0}{\partial \xi} \\[2ex] \Psi_0 = 1 \quad at \;\; \tau = 0 \\[2ex] \Psi_0 \;\; is \;\; finite \;\; at \;\; \xi = 0 \\[2ex] -\dfrac{\partial \Psi_0}{\partial \xi} = w \;\; at \;\; \xi = 1 \end{array}\right\} (I)$$

ε^1 :

$$\left.\begin{aligned}
&\frac{\partial \Psi_1}{\partial \tau}=\frac{\partial^2 \Psi_1}{\partial \xi^2}+\frac{1}{\xi}\frac{\partial \Psi_1}{\partial \xi}\\
&\Psi_1=0 \quad at \;\; \tau=0\\
&\Psi_1 \;\; is \;\; finite \;\; at \;\; \xi=0\\
&\frac{\partial \Psi_1}{\partial \xi}=w\Psi_0 \;\; at \;\; \xi=1
\end{aligned}\right\} (II)$$

The solution of (I) by Laplace transform method - Let

$$L\{\Psi_0(\xi,\tau)\}=u(\xi,s) \tag{14}$$

Then

$$s\,u(\xi,s)-1=\frac{d^2u}{d\xi^2}+\frac{1}{\xi}\frac{du}{d\xi} \tag{15}$$

subject to:

$$u(\xi,s) \; is \; finite \; at \;\; \xi=0$$

(16), (17)

$$-\frac{du}{d\xi}=\frac{w}{s} \;\; at \;\; \xi=1.$$

This produces a general solution:

$$u(\xi,s)=a_1J_0\left(i\sqrt{s}\xi\right)+a_2Y_0\left(i\sqrt{s}\xi\right)+\frac{1}{s} \tag{18}$$

which reduces to:

$$u(\xi,s)=a_1J_0\left(i\sqrt{s}\xi\right)+\frac{1}{s} \tag{19}$$

because $Y_0\left(i\sqrt{s}\xi\right)$ is unbounded as $\xi \to \infty$, a_2 must be chosen as zero.

$$-\frac{du}{d\xi}=\frac{w}{s} \Rightarrow a_1=\frac{w}{s\,i\sqrt{s}J_1\left(i\sqrt{s}\right)} \tag{20}$$

Such that:

$$u(\xi,s)=\frac{wJ_0\left(i\sqrt{s}\xi\right)}{si\sqrt{s}J_1\left(i\sqrt{s}\right)}+\frac{1}{s} \tag{21}$$

Inverting:

$$\Psi_0(\xi,\tau)=1+L^{-1}\left\{\frac{wJ_0\left(i\sqrt{s}\xi\right)}{si\sqrt{s}J_1\left(i\sqrt{s}\right)}\right\} \tag{22}$$

Consider the fraction:

$$\begin{aligned}\frac{wJ_0\left(i\sqrt{s}\xi\right)}{si\sqrt{s}J_1\left(i\sqrt{s}\right)}&=\frac{w\left[1+\frac{s\xi^2}{2^2}+\frac{s^2\xi^4}{2^2\bullet 4^2}+\frac{s^3\xi^6}{2^2\bullet 4^2\bullet 6^2}+\cdots\right]}{-s\left[\frac{s}{2}+\frac{s^2}{2^2\bullet 4}+\frac{s^3}{2^2\bullet 4^2\bullet 6}+\cdots\right]}\\&=\frac{w\left[1+\frac{s\xi^2}{2^2}+\frac{s^2\xi^4}{2^2\bullet 4^2}+\frac{s^3\xi^6}{2^2\bullet 4^2\bullet 6^2}+\cdots\right]}{-s^2\left(\frac{1}{2}+\frac{s}{2^2\bullet 4}+\frac{s^2}{2^2\bullet 4^2\bullet 6}+\cdots\right)}\end{aligned} \tag{23}$$

But

$$\frac{\left[1+\frac{s\xi^2}{2^2}+\frac{s^2\xi^4}{2^2\bullet 4^2}+\frac{s^3\xi^6}{2^2\bullet 4^2\bullet 6^2}+\cdots\right]}{-\left(\frac{1}{2}+\frac{s}{2^2\bullet 4}+\frac{s^2}{2^2\bullet 4^2\bullet 6}+\cdots\right)}=A_0+A_1s+A_2s^2+\cdots \tag{24}$$

or

$$\left[1+\frac{s\xi^2}{2^2}+\frac{s^2\xi^4}{2^2\bullet 4^2}+\frac{s^3\xi^6}{2^2\bullet 4^2\bullet 6^2}+\cdots\right]=(A_0+A_1s+A_2s^2+\cdots)\left(-\frac{1}{2}-\frac{s}{2^2\bullet 4}-\frac{s^2}{2^2\bullet 4^2\bullet 6}-\cdots\right)$$

$$=A_0\left(-\frac{1}{2}-\frac{s}{2^2\bullet 4}-\frac{s^2}{2^2\bullet 4^2\bullet 6}-\cdots\right)+A_1s\left(-\frac{1}{2}-\frac{s}{2^2\bullet 4}-\frac{s^2}{2^2\bullet 4^2\bullet 6}-\cdots\right)+A_2s^2\left(-\frac{1}{2}-\frac{s}{2^2\bullet 4}-\frac{s^2}{2^2\bullet 4^2\bullet 6}-\cdots\right)\cdots \tag{25}$$

Then equating coefficients of both sides result in:

$$1 = \frac{-A_0}{2} \Rightarrow A_0 = -2; \; \frac{\xi^2}{2^2} = -\frac{A_0}{2^2 \bullet 4} - \frac{A_1}{2} \Rightarrow A_1 = \frac{1}{2}\left(\frac{1}{2} - \xi^2\right); \; \frac{\xi^4}{2^2 \bullet 4^2} = -\frac{A_0}{2^2 \bullet 4^2 \bullet 6} - \frac{A_1}{2^2 \bullet 4} - \frac{A_2}{2}$$

$$A_2 = \frac{1}{4^2 \bullet 6} - \frac{\left(\frac{1}{2} - \xi^2\right)}{2^2 \bullet 4} - \frac{\xi^4}{2 \bullet 4^2}$$

(26), (27), (28)

Therefore:

$$\frac{w J_0\left(i\sqrt{s}\xi\right)}{s i\sqrt{s} J_1\left(i\sqrt{s}\right)} = \frac{w\left(-2 + \frac{s}{2}\left[\frac{1}{2} - \xi^2\right] + s^2\left[\frac{1}{4^2 \bullet 6} - \frac{1}{2^2 \bullet 4}\left(\frac{1}{2} - \xi^2\right) - \frac{\xi^4}{2^2 \bullet 4^2}\right] + \cdots\right)}{-s^2} \tag{29}$$

Then the residue at $s = 0$ is given by:

$$\rho_0(\tau) = e^{0\tau}\left[B_1 + \tau B_2\right]; \; B_1 = \lim_{s \to 0} \frac{d}{ds}\left[\frac{s^2 w J_0\left(i\sqrt{s}\xi\right)}{s i\sqrt{s} J_1\left(i\sqrt{s}\right)}\right]$$

$$= \lim_{s \to 0} \frac{d}{ds}\left[s^2 \frac{w\left(-2 + \frac{s}{2}\left[\frac{1}{2} - \xi^2\right] + s^2\left[\frac{1}{4^2 \bullet 6} - \frac{1}{2^2 \bullet 4}\left(\frac{1}{2} - \xi^2\right) - \frac{\xi^4}{2^2 \bullet 4^2}\right] + \cdots\right)}{s^2}\right] \tag{30}$$

$$= \frac{w}{2}\left(\frac{1}{2} - \xi^2\right)$$

$$B_2 = \lim_{s \to 0}\left[\frac{s^2 w J_0\left(i\sqrt{s}\xi\right)}{s i\sqrt{s} J_1\left(i\sqrt{s}\right)}\right]$$

$$= \lim_{s \to 0}\left[s^2 \frac{w\left(-2 + \frac{s}{2}\left[\frac{1}{2} - \xi^2\right] + s^2\left[\frac{1}{4^2 \bullet 6} - \frac{1}{2^2 \bullet 4}\left(\frac{1}{2} - \xi^2\right) - \frac{\xi^4}{2^2 \bullet 4^2}\right] + \cdots\right)}{s^2}\right] = -2w \tag{31}$$

Therefore the residue at $s = 0$ is:

$$\rho_0(\tau) = e^{0\tau}[B_1 + \tau B_2] = \frac{w}{2}\left(\frac{1}{2} - \xi^2\right) - 2w\tau \tag{32}$$

For the case $s \neq 0$:

$$\frac{wJ_0\left(i\sqrt{s}\xi\right)}{si\sqrt{s}J_1\left(i\sqrt{s}\right)} = \frac{P(\xi,s)}{Q(s)};\; Q(s) = si\sqrt{s}J_1\left(i\sqrt{s}\right) = 0 \Rightarrow J_1\left(i\sqrt{s}\right) = 0 \tag{33}$$

Then making use of:

$$i\sqrt{s} = \lambda \Rightarrow s = -\lambda^2 \Rightarrow J_1(\lambda_n) = 0,\; n = 1,2,3,\cdots \tag{34}$$

Also,

$$\begin{aligned} Q(s) &= si\sqrt{s}J_1\left(i\sqrt{s}\right) \Rightarrow Q'(s) = \frac{d}{ds}\left[si\sqrt{s}J_1\left(i\sqrt{s}\right)\right] \\ &= \frac{3}{2}i\sqrt{s}J_1\left(i\sqrt{s}\right) + is^{3/2}\frac{1}{2}s^{-1/2}\frac{d}{ds}\left[J_1\left(i\sqrt{s}\right)\right] \end{aligned} \tag{35}$$

But,

$$J_1'(x) = 1/2[J_0(x) - J_2(x)];\; J_2(x) = \frac{2}{x}J_1(x) - J_0(x) \Rightarrow J_1'(x) = J_0(x) - \frac{1}{x}J_1(x) \tag{36}$$

such that,

$$\begin{aligned} Q'(s) &= \frac{d}{ds}\left[si\sqrt{s}J_1\left(i\sqrt{s}\right)\right] = \frac{3}{2}i\sqrt{s}J_1\left(i\sqrt{s}\right) - \frac{s}{2}\left[J_0\left(i\sqrt{s}\right) - \frac{1}{i\sqrt{s}}J_1\left(i\sqrt{s}\right)\right] \\ &= \frac{s}{2}J_0\left(i\sqrt{s}\right) \equiv \frac{\lambda^2}{2}J_0(\lambda) \end{aligned} \tag{37}$$

Therefore,

$$L^{-1}\left\{\frac{wJ_0\left(i\sqrt{s}\xi\right)}{si\sqrt{s}J_1\left(i\sqrt{s}\right)}\right\} = \frac{w}{2}\left(\frac{1}{2} - \xi^2\right) - 2w\tau + \sum_{n=1}^{\infty}\frac{wJ_0(\lambda_n\xi)}{\frac{\lambda_n^2}{2}J_0(\lambda_n)}e^{-\lambda_n^2\tau} \tag{38}$$

Finally,

$$\Psi_0(\xi,\tau) = 1 + \frac{w}{2}\left(\frac{1}{2} - \xi^2\right) - 2w\tau + \sum_{n=1}^{\infty}\frac{wJ_0(\lambda_n\xi)}{\frac{\lambda_n^2}{2}J_0(\lambda_n)}e^{-\lambda_n^2\tau} \tag{39}$$

is the first estimate. Following similar procedure, subsequent estimates can be derived. Upon complete solution of the model, it will be tested against available experimental data in literature, Tiruna-Barna et al. (2005) and Idachaba et al. (2004). We anticipate this model to simulate these data reasonably well.

Conclusion

An analytical method to model bioleaching from cemntitios waste forms containing heavy metals is presented. Using Michaelis-Menten kinetics to describe biological activities, the model provides a more realistic approach for developing predictive strategies in containment of heavy metals.

Nomenclature

A_n, a	constants
B_n	constant
C_A	element concentration, mol/cm^3
D_{AB}	diffusion coefficient, cm^2/s
J	Bessel function notations
$k_0, k_1, and\ k_2$	rate constant
r	cylindrical coordinate, cm
R	cylindrical column mean radius, cm
s	Laplace variable
t	time variable, s
u	variable defined in Eq.(14)
w	dimensionless group defined in Eq.(8)

Greek symbols

ψ	dimensionless concentration defined in Eq.(5)
ξ	dimensionless radius defined in Eq.(6)
τ	dimensionless time defined in Eq.(7)
ε	a small parameter
λ	eigen value

References:

1. Bird R. B., Stewart W. E. and Lightfoot E. W., Transport phenomena, 2nd edition, John Wiley & Sons, Inc. 2002.
2. Bishop P. L. (1986). Prediction of heavy metal leaching rates from stabilized/solidified hazardous wastes. Toxic and Hazardous Waste, Proceedings of the Mid-Atlantic Industrial Waste Conference, 18, 236-252.
3. Crank J., The Mathematics of diffusion, 2nd edition, Oxford Press, 1975
4. Duffus, J. H. (2002). Heavy metals – a meaningless term? Pure Appl. Chem., 74 (5), 793-807.
5. Fogler H.S. (2005). Elements of Chemical Reaction Engineering, 4th edition, Prentice Hall.
6. Halim, C. E., Short, S. A., Scott, J. A., Amal, R., and Low, G. (2005). Modeling the leaching of Pb, Cd, As, and Cr from cementitious waste using PHREEQC. J. of Hazardous Materials-A, 125, 45–61.
7. Idachaba M.A., Nyavor K., Egiebor N.O., Rogers R.D. (2001). Stability evaluation of a cement based waste form to microbially induced degradation. Waste Manage Res, 19, 284-291.
8. Idachaba, M. a., Nyavor, K., Egiebor, N. O. (2004). The leaching of chromium from cement-based waste from via a predominately biological mechanism. *Advances in Environmental Reseach.* 8, 483-491.
9. Malviya, R., Chaudhary, R.(2006). Leaching behavior and immobilization of heavy metals in solidified/stabilized products. Journal of Hazardous Materials, 137 (1), 207-217.
10. Mertz, W. (1969). Chromium occurrence and function in biological systems. Physiol Rev, 49,163-239.
11. Mertz, W. (1993). Chromium in human nutrition: a review. J Nutr, 123, 626-33.
12. Mertz W. (1998). Interaction of chromium with insulin: a progress report. Nutr Rev, 56,174-7.
13. Olmo, I. F., Chacon, E., and Irabien, A. (2003). Leaching behavior of lead, chromium (III), and zinc in cement/metal oxides systems. J. of Environmental Engineering, ASCE / JUNE 2003, 532-538
14. Retrieved September 15, 2007 from Wikipedia, Web site: http://en.wikipedia.org/wiki/Heavy_metals
15. Shuler M. L. and Kargi F. (2002). Bioprocess engineering: basic concepts, 2nd edition, Prentice Hall.
16. Tiruta-Barna, L., Rethy, Z., & Barna, R. (2005). Release dynamic process identification for a cement based material in various leaching conditions. Part II. Modeling the release dynamics for different leaching conditions. Journal of Environmental Management, 74, 127-137.
17. US Environmental Protection Agency (1998). Toxicological review of hexavalent chromium. Cas No. 18540-29-9, Washington, Dc.
18. United States Environmental Protection Agency (1999-a). Presumptive remedy for metals-in-soil sites. Office of Solid Waste and Emergency Response, EPA 540-F-98-054, OSWER-9355.0-72FS, PB99-963301.

19. United States Environmental Protection Agency (1999-b). Solidification/ stabilization resource guide. Office of Solid Waste and Emergency Response (5102G), EPA/542-B-99-002.
20. Vachon, P., Tyagl, R. D., Auclalr, J. C., and Wliklnsont, K. J. (1994) Chemical and biological leaching of aluminum from red mud. J. Envlron. Sci. Technol., 28, 26-30.

Prediction of Global Warming Potentials through Computational Chemistry – Testing Robustness of Methodology through Experimental Comparisons

Paul Blowers and Kyle Hollingshead
Department of Chemical and Environmental Engineering
The University of Arizona

Abstract

Global warming is a scientifically based concern regarding addition of naturally occurring and anthropogenic chemicals to the troposphere where the species can trap infrared energy. Predicting global warming potentials requires highly accurate rate constant measurements for the reactions of the chemicals with hydroxyl radicals, which is the first and rate limiting step in tropospheric degradation. Radiative forcing, the amount of energy that can be captured by the chemicals per square meter of exposed area for a given concentration, requires spectroscopic information about peak locations and intensities, which are then aggregated into absorption cross sections. These values are then used in atmospheric modeling simulations to determine the radiative forcing. Both kinetic and spectroscopic measurements have many potential experimental difficulties, which makes predicting global warming potentials (GWPs) from theory attractive.

We build on our previous work by examining an emerging class of compounds, fluorinated ethers, using theoretical chemistry to predict GWPs. Previous work investigated CH_2F_2 and found excellent comparison to experiment for predicting all intermediate steps for GWPs, including kinetic degradation rates with hydroxyl radical under low temperature tropospheric conditions, atmospheric lifetime estimates, radiative forcing in the atmospheric window, and overall GWPs at 20 year, 100 year, and 500 year time horizons.

We find good agreement for all parameters for the hydrofluoroethers compared to experimental values. Radiative forcing estimates are also in good agreement with available experimental results. Finally, we now have a larger database of chemicals where we have verified our methodology of accurately predicting global warming potentials using theory.

Introduction

Hydrofluoroethers (HFEs) are an emerging class of compounds poised to be used widely in industry as replacement materials for chlorofluorocarbons (CFCs), fluorocarbons (FCs), and hydrofluorocarbons (HFCs). Thirty-three patents on HFE use were filed between December 2004 and June 2006[1, 2], and another 52 have been filed in the same number of months since then. These substances are attractive to industry because they are not ozone depleting and may have shorter atmospheric lifetimes and thus have lower global warming potentials (GWPs). It is also expected that their toxicities will be lower, although data is not yet publically available on this topic[3].

For details of the complete methodologies, please see prior publications regarding this work[1, 2, 4, 5]. We use the B3LYP/6-31g* combination of method and basis set to predict optimized geometry parameters for different conformers of HFEs that have been previously experimentally or industrially synthesized. We then predict molecular spectroscopy using frequencies scaled by 0.9613[6]. The scaled frequencies are used in statistical thermodynamic calculations to estimate the weighting factors for different conformers before being combined with predictions of radiative forcing using Pinnock's binning method[7] as extended by Papasavva[8, 9] and then further extended in our prior work. Where kinetic degradation rate information is available for HFE reactions with hydroxyl radical in the gas phase, we use that to predict atmospheric lifetimes and then global warming potentials for a large range of HFEs. Where kinetic data were not available, we instead use a correlation attributed to Urata, et al.[10], to estimate the rate constant and atmospheric lifetime for those species.

One of the greater concerns about the larger HFEs is that their environmental degradation products are unknown and that there may be many stable end-points during degradation. After investigating the HFE parent species, we predict radiative forcing values for some of the daughter species hypothesized by Wallington, et al.[11-19] to be formed. We find that radiative forcing values in some cases are larger for the daughter species than for the parent species.

This is problematic because the atmospheric lifetimes of these daughter species are often unknown and the removal mechanisms from the troposphere are also unknown. Further work on the degradation routes and final species formed is needed in order to estimate the atmospheric implications of increasing HFE usage.

Results and Discussion

Extensive work in our laboratory on a variety of gas phase systems have repeatedly found that the B3LYP/6-31g* method and basis set combination accurately predicts both molecular structures and vibrational frequencies, particularly when vibrational peak locations are scaled by 0.9613[6]. In general, vibrational frequencies are within 34 cm^{-1} of experimental results when this scaling factor is used.

In one of our prior publications we predicted radiative forcing values for 27 HFEs and plotted the predicted values (in W m^{-2} $ppbv^{-1}$) versus values previously determined using a cruder computational approach and versus previously modeled values using experimentally determined adsorption cross sections. Figure 1 here shows the agreement between our predictions and previous values.

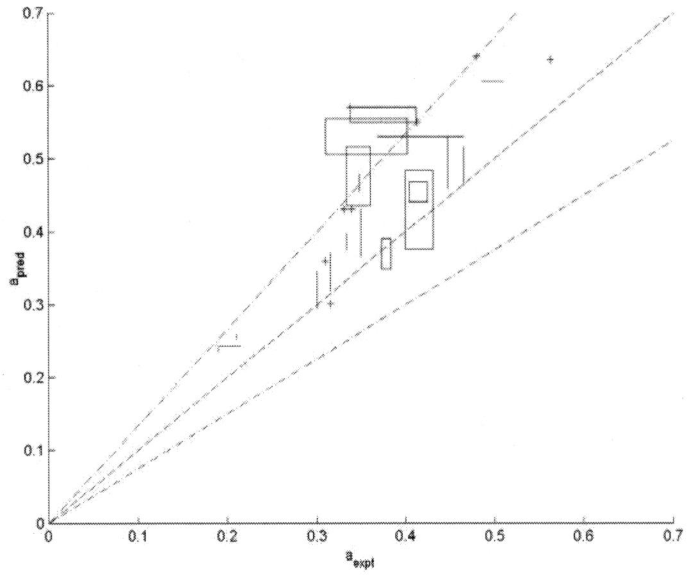

Figure 1 - Radiative forcing (in W m^{-2} $ppbv^{-1}$) predicted from quantum chemistry at the B3LYP/6-31g* level using Pinnock's method combined with Papasavva's methodology and Elrod's spreadsheet tool compared to experimental cloudy sky values for 27 hydrofluoroethers under consideration as replacement chemicals in industry. The center 45° dashed line represents perfect agreement, while the other two dashed lines represent errors of plus and minus 25% as discussed in the text.

In Figure 1, the dashed 45° line represents perfect agreement between prediction and previous results. The upper and lower dashed lines represent the expected best agreement between two high quality modeled results using experimental spectroscopy, which is about a 25 percent difference between values for a particular compound[20]. Each cross, line, or box represents a different HFE. Crosses indicate that only one conformer calculation was done and that only one modeled result was available in the literature. Horizontal lines indicate that there have been multiple results modeled from experimental data. Vertical lines indicate that there is more than one stable conformer possible for a given structure with different radiative forcing values. For example, CHF_2OCHF_2 has 4 different stable conformeric forms, as shown in Figure 2 on the following page. Boxes indicate that multiple radiative forcing values have been determined from experimental data and that more than one conformeric form can exist.

The results in Figure 1 indicate that, in general, the predicted values for radiative forcing using our efficient and inexpensive computational chemistry approach are within 25 percent of previously modeled results. While the predicted values are on average higher than the modeled results, only two of the species lie outside the error estimates that we expect to see between two high quality modeled results. Since our original work was published, we communicated with the authors of the work where we had the largest discrepancy and they responded that our higher prediction was actually closer to their value when they performed their analyses again (unpublished). This leaves only one species where we are outside the 25 percent error estimate.

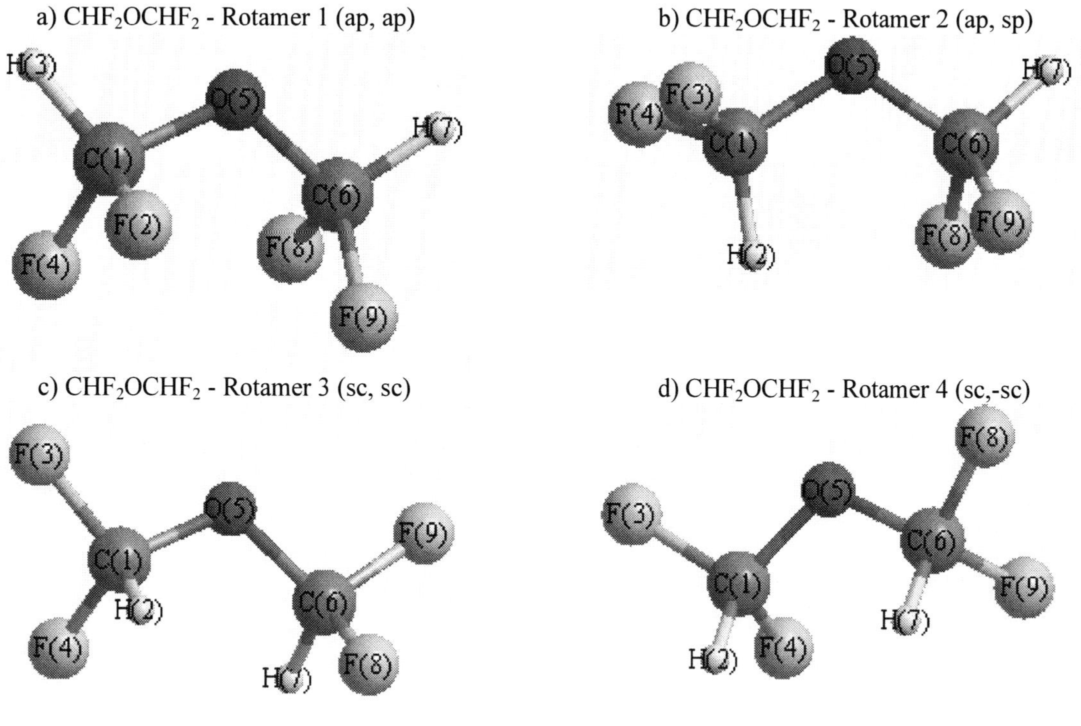

Figure 2 - Example of a HFE that has four different conformers that are stable and their optimized structures at the B3LYP/6-31g* level of theory. Nomenclature was taken from Ulic and Oberhammer.

As we discussed briefly regarding Figure 2, there are several rotameric or conformeric forms of some species possible. This is particularly true when there are H- and F-atoms collocated on the same carbon-center. Different conformers are also possible when one has larger HFEs. In this work we refer to the HFEs by the number of carbon atoms on each side of the ether linkage. For example, the 1,2-HFEs have one carbon on one side of the oxygen atom and 2 carbon atoms on the other. Each extra carbon atom on a side leads to the possibility for more conformers as one can now start to have rotations about each C-O or each C-C bond, leading to both gauche and planar forms along the C-O-C backbone.

We performed early work[21] on the 1,1-HFEs that found that different conformers could have predicted radiative forcing values that were different from each other by as much as 22 percent. For the 1,2-HFEs, the differences in radiative forcing between conformers were as much as 25.1 percent. Trends were analyzed in the predicted radiative forcing values for the 1,2-HFEs, and it was noted that the radiative forcing values were highest for the conformers where the C-O-C-C atoms all lay in one plane as opposed to being bent into a gauche formation[1]. Therefore, to minimize computational effort necessary for larger species with many possible conformers, we analyzed only the structures and vibrational frequencies for the linear forms. This provided an upper value for radiative forcing and lead to a higher GWP prediction. We felt it was better to err on the side of safety in evaluating global warming with these higher values.

Table 1 compares our radiative forcing values and global warming potentials predicted with computational chemistry to experimentally modeled values for all HFEs that have been previously synthesized. For the 1,1-HFEs where no kinetic data were available, we used an empirical correlation and extensive calculations to estimate the rate constants for degradation by hydroxyl radical[21]. One sees from the table that our predicted GWPs are generally close to the values modeled from experiment.

Because of the anticipated difference of up to 25% in radiative forcing values, and because kinetic information often leads to atmospheric lifetimes that differ by a factor of 2, it is expected that values for GWP will vary from research group to research group by about a factor of about 0.4 to 2.5 times the expected value. We examine variations between different modeled results from experimental data first to highlight these trends. For convenience and consistency, we have always divided the highest result by the lowest result at a given time horizon.

For the 1,1-HFEs, Orkin, et al.[22], and Good, et al.[23], report values for CF_3OCH_3 that differ by factors of 1.7, 1.8, and 1.5 at the 20, 100, and 500 year time horizons, respectively. For CHF_2OCHF_2, the factor of difference was 1.1 at all time horizons between Good, et al's.[23], results and those of Myhre, et al.[24] Similar differences were found between Good's[25] results and Christidis, et al's[26] for CF_3OCHF_2, where difference factors of 1.18, 1.16, and 1.20 were found for the 20, 100, and 500 year time horizons.

For the 1,2 HFEs, three sets of comparisons can also be done with results being reported for $CHF_2OCH_2CF_3$ by Christidis, et al.[26], Sekiya and Misaki[27], and Oyaro, et al.[28]. Here, the factor of difference between the highest and lowest results is 1.77 at the 20 year time horizon and 1.87 at 100 years. Oyaro[28] and Christidis[26] both modeled GWPs for $CHF_2OCHFCF_3$ at the 100 year time horizon with factor of difference between their results of 2.58. Factors of 1.85 and 3.83 at the 20 and 100 year time horizons were found for the differences between the results of Oyaro, et al.[28], and Jain, et al.[29], for $CF_3OCHFCF_3$.

Only one larger HFE,$CHF_2OCF_2CHFCF_3$, has been studied by more than one research group. For this 1,3 HFE, Wallington, et al's results differ from Oyaro, et al's. by factors of 3.93 and 7.87 at the 20 and 100 year time horizons, respectively. We now turn to comparing existing data to our predicted values of GWP when we use computed radiative forcing values from our work and either atmospheric lifetimes or kinetic data reported in the experimental literature.

Looking at the data in Table 1, we find that our values for GWP are within a factor of 1.75 of the previously modeled data for the 1,1-HFEs, CF_3OCH_3 and CHF_2OCHF_2. However, the factors for CF_3OCHF_2 are larger, with values of 6.22, 2.64, and 1.72 because we used an updated kinetic value that leads to an atmospheric lifetime that is 77.9 years versus the 165 years previously used[2] and because our predicted radiative forcing value is higher.

For the 1,2-HFEs, the worst factors of difference between our results and a previous result for each species at any time horizon are: 2.33 for $CH_3OCH_2CF_3$ at the 100 year time horizon; 1.88 for $CF_3OCH_2CH_3$ at the 20 year time horizon; 5.47 for $CH_3OCF_2CHF_2$ at 100 years; 2.28 for $CHF_2OCH_2CF_3$ at 100 years; 1.08 for $CH_3OCF_2CF_3$ at 100 years; 4.66 for $CHF_2OCHFCF_3$ at 100 years; 2.17 for $CF_3OCHFCHF_2$ at 100 years; and 4.29 for $CF_3OCHFCF_3$ at 100 years. For species where multiple experimentally modeled results are available we have reported the worst value for comparison instead of averaging the previous experimental results or discarding an outlier when the other values were in agreement, as is the case for $CF_3OCHFCF_3$. If we had discarded the last values on that species, our new worst-case scenario factor of difference would have been 1.27. For all but two HFEs in this class, we are within the expected factor of difference of 2.5.

The GWP data available for the larger HFEs are much more limited. For $CH_3OCF_2CF_2CF_3$, our factor of difference is 1.29; for $CHF_2OCF_2CHFCF_3$, it is as high as 3.34 compared to Oyaro, et al's.[28], data, but only 2.36 compared to Wallington, et al's[13]., 2004 results. For $CF_3OCF_2CHFCF_3$, results are only available from Wallington, et al.[13], where we differ by a factor of 3.48 at the 20 year time horizon and 2.43 at 100 years. Our factor of difference is 1.19 for $CH_3OCF(CF_3)_2$ and 3.04 for $CH_3CH_2OCF_2CF_3$ at 100 years compared to Sekiya and Misaki's results[27]. A difference factor of 3.16 was found for $CH_3OCF_2CF_2CF_2CF_3$ when comparing to Wallington, et al's[11], earlier work. Finally, comparing to the results of Oyaro, et al.[28], we find the largest factor of difference for $CF_3CHFOCF_3CF_2CF_3$ is 3.62 at 20 years, and for $CF_3CH_2OCF_2CHFCF_3$, 2.34 at 100 years.

Table 1- Atmospheric lifetime estimates, instantaneous radiative forcing, and global warming potential results for HFEs at time horizons of 20, 100, and 500 years compared to previously published results

Species	τ (years)	Radiative forcing (W m^{-2} $ppbv^{-1}$)	GWPs			τ (years)	GWPs			Reference
			20 year TH	100 year TH	500 year TH		20 year TH	100 year TH	500 year TH	
1,1 ethers										
CH_3OCH_2F	0.2[c]	0.154[d]	94	29	9					
CH_3OCHF_2	1.1[c]	0.210	517	158	49					
CH_2FOCH_2F	0.9[c]	0.242	513	157	49					
CF_3OCH_3	3.3[a]	0.243	1456	446	139		1300	360	130	[22]
							2200	656	202	[25]
CH_2FOCHF_2	0.9[c]	0.310	541	166	52					
CHF_2OCHF_2	15.5[a]	0.484	8645	3648	1141		9760	5720	1830	[25]
						11.3	10700	6300	2000	[24]
CH_2FOCF_3	4.9[c]	0.326	2499	778	243					
CF_3OCHF_2	77.9	0.470	63124	31739	13053	165.0	11800	1400	9120	[25]
							10137	12015	7596	[26]
1,2 ethers										
$CH_3OCH_2CF_3$	0.1	0.240	70	21	7	0.44	36	9		[28]
$CF_3OCH_2CH_3$	0.3	0.259	284	87	27		151	47		[28]
$CH_3OCF_2CHF_2$	0.3[a]	0.345	842	257	80	0.49		47		[27]
$CHF_2OCH_2CF_3$	3.2	0.389	4254	1304	407		1871	570	171	[26]
							3319	1064		[28]
							1870	570		[28]
						4.9		712		[27]
$CH_3OCF_2CF_3$	3.3	0.448	1844	566	177	6.5		522		[27]
$CHF_2OCF_2CHF_2$	16.2	0.554	7076	3042	952	14.4[c]				
$CHF_2OCF_2CH_2F$	7.4	0.455	3766	1224	382	6.7[c]				
$CHF_2OCHFCF_3$	5.9	0.529	9091	2875	897		4544	1591		[28]
						4.2		617		[27]
$CF_3OCHFCHF_2$	5.6	0.479	7909	2849	777		3801	1311		[28]
$CF_3OCF_2CF_2H$	16.9	0.568	19305	7996	2504	18.8[c]				
$CF_3OCH_2CHF_2$	1.3[c]	0.366[e]	592	181	57					
$CF_3OCH_2CF_3$	2.9[c]	0.439[e]	1417	434	136					
$CF_3OCHFCF_3$	33.3[c]	0.528[e]	7992	5147	1692		>7235	>6516		[28]
							6275	4606		[28]
							3400	1200	370	[29]

Species	τ (years)	Radiative forcing (W m^{-2} $ppbv^{-1}$)	GWPs			τ (years)	GWPs			Reference
			20 year TH	100 year TH	500 year TH		20 year TH	100 year TH	500 year TH	
1,3 ethers										
$CH_3OCH_2CF_2CF_3$	0.06	0.320	60	18	6	0.1[c]				
$CH_3OCH_2CF_2CHF_2$	0.04	0.221	34	10	3	0.1[c]				
$CHF_2OCH_2CF_2CHF_2$	2.4[a]	0.447	1080	330	103					
$CH_3OCF_2CF_2CHF_2$	2.8[c]	0.401[e]	1154	353	110					
$CH_3OCF_2CF_2CF_3$	3.3[a]	0.434	1316	404	126	6.4		522		[27]
$CHF_2OCF_2CHFCF_3$	4.1	0.633	5997	1842	576		1828	551		[28]
							7180	4340		[13]
$CF_3OCH_2CF_2CF_3$	1.8[c]	0.525[f]	812	248	78					
$CF_3OCF_2CHFCF_3$	26.7	0.641	22977	11012	3521		6600	4530		[13]
$CHF_2OCH_2CF_2CF_3$	3.7	0.479	1641	504	158					
$CH_3OCH(CF_3)_2$	0.22	0.360	81	25	8					
$CH_3OCF(CF_3)_2$	2.6	0.439	1046	320	100	4.9		380		[27]
$CH_2FOCH(CF_3)_2$	0.51	0.424	202	62	19					
$CF_3OCH(CF_3)_2$	117.0	0.563	61134	28403	13383					
2,2 ethers										
$CH_3CH_2OCF_2CF_2H$	0.12	0.370	150	46	14					
$CH_3CH_2OCF_2CF_3$	0.31[b]	0.488	470	143	45	0.31		47		[27]
$CF_3CH_2OCH_2CF_3$	0.24[a]	0.398	99	30	9					
$CHF_2CF_2OCH_2CF_3$	4.0[a]	0.516	1935	596	186					
$CF_3CH_2OCF_2CF_3$	4.2[c]	0.531[g]	1899	586	183					
1,4 ethers										
$CH_3OCF_2CF_2CF_2CF_3$	4.0	0.552	4434	1366	426		1388	432		[11]
$CHF_2OCH_2CF_2CHFCF_3$	0.6[c]	0.486[h]	235	72	22					
$CH_3OCF_2CH(CF_3)_2$	2.8[c]	0.489	1103	338	105					
2,3 ethers										
$CHF_2CF_2OCH_2CH{=}CH_2$	0.002	0.431	1	0	0					
$CHF_2CF_2OCH_2CF_2CHF_2$	2.0[c]	0.529[h]	854	261	82					
$CHF_2CF_2OCH_2CF_2CF_3$	2.3[c]	0.582[h]	1001	306	96					
$CH_3CH_2OCF_2CHFCF_3$	0.26	0.431	290	89	28		127	38		[28]
$CF_3CH_2OCF_2CHFCF_3$	4.3	0.593	1885	582	182	2.8[c]				
$CF_3CHFOCF_2CF_2CF_3$	38.0	0.636	23871	12276	4122	>67	>6589	>5936		[28]
3,3 ethers										
$CF_3CHFCF_2OCH_2CF_2CHF_2$	3.0	0.621	1232	377	118	1.8[c]				
$CF_3CHFCF_2OCH_2CF_2CF_3$	8.8	0.670	3291	1121	350					

Species	τ (years)	Radiative forcing (W m^{-2} $ppbv^{-1}$)	GWPs			τ (years)	GWPs		Reference
			20 year TH	100 year TH	500 year TH		20 year TH	100 year TH	
1,5 ethers									
$CH_3OCH_2CF_2CF_2CF_2CHF_2$	0.1[c]	0.406[h]	31	9	3				

[a]averaged rate constant value from numerous measurements
[b]used given atmospheric lifetime from reference due to lack of kinetic data reported
[c]from [10]
[d]averaged value for different rotamers not reported in previous work.
[e]averaged value for different rotatmers reported in our previous work[30].
[f]only the linear rotameric form calculated in this work for the first time.
[g]only one rotamer found in this work (see supplementary information).
[h]average of more than one rotamer shown in supplementary information in this work.
.

In general, our GWP predictions using radiative forcing values obtained through computational chemistry methods compare favorably to those from modeled experimental results. Our earlier work summarized in Figure 1 showed that radiative forcing values are within 25% error compared to values based on experiment, so the larger errors in GWPs come from our use of updated kinetic rates from newer experiments when justified. With our comparison between our results and earlier modeled results complete, we note that there are many HFEs where there were previously no GWPs reported at any time horizon. We have populated a database of GWPs for this class of compounds in Table 1.

We turn now to discuss the atmospheric implications of the introduction of HFEs into widespread use. Wallington, et al., have done extensive work on HFEs and their potential degradation products[11-13, 17-19, 31]. A summary of the degradation mechanisms suggested by this large body of work is laid out here using some specific reactions and other general characteristic steps.

The first step of the degradation pathway is hydrogen atom abstraction by a hydroxyl radical present in the troposphere, for example[13]:

$$CF_3CFHCF_2OCF_3 + {\bullet}OH \rightarrow CF_3CF{\bullet}CF_2OCF_3 + H_2O \qquad \text{(reaction 1)}$$

where we represent the position of the radical by a dot.

The second step, paralleling the extensive compendia of atmospheric chemistry summaries by Atkinson[32], is the addition of molecular oxygen to the radical:

$$CF_3CF{\bullet}CF_2OCF_3 + O_2 + M \rightarrow CF_3CFOO{\bullet}CF_2OCF_3 + M \qquad \text{(reaction 2)}$$

where we have included the implied collision partner, M, which would need to be present to carry away excess energy from the product species. This reaction is generally assumed to be very rapid so the radical, in this case $CF_3CF{\bullet}CF_2OCF_3$, would not have a large environmental impact itself.

While Wallington, et al., found that the peroxy radical species would self react to form oxyradicals, this is not likely to occur in the atmosphere and we instead parallel their other work[11] where the peroxyradical instead encounters NO, NO_2, or another generic peroxy radical $RO_2{\bullet}$, where R is any organic group:

$$CF_3CFOO{\bullet}CF_2OCF_3 + NO{\bullet} \rightarrow CF_3CFO{\bullet}CF_2OCF_3 + NO_2 \qquad \text{(reaction 3a)}$$

$$CF_3CFOO{\bullet}CF_2OCF_3 + NO{\bullet} + M \rightarrow CF_3CFONO_2CF_2OCF_3 \qquad \text{(reaction 3b)}$$

$$CF_3CFOO{\bullet}CF_2OCF_3 + NO_2 + M \rightarrow CF_3CFOONO_2{\bullet}CF_2OCF_3 + M \qquad \text{(reaction 3c)}$$

$$CF_3CFOO{\bullet}CF_2OCF_3 + HO_2 \rightarrow \text{products} \qquad \text{(reaction 3d)}$$

$$CF_3CFOO{\bullet}CF_2OCF_3 + RO_2 \rightarrow \text{products} \qquad \text{(reaction 3e)}$$

Unfortunately, it has not been possible to measure the kinetic data for these reactions so their relative importance in the tropospheric degradation pathways remains unknown.

For the specific case used as an example so far, carbon-carbon bond scission is suspected once the oxyradical is formed, leading to:

$$CF_3CFO{\bullet}CF_2OCF_3 + M \rightarrow CF_3CFO + CF_3OCF_2{\bullet} \qquad \text{(reaction 4)}$$

where the resulting radical continues to degrade through oxygen addition, oxyradical formation, and carbon carbon scission:

$$CF_3OCF_2{\bullet} + O_2 + M \rightarrow CF_3OCF_2O_2{\bullet} + M \qquad \text{(reaction5)}$$

$$CF_3OCF_2O_2{\bullet} + NO{\bullet} \rightarrow CF_3OCF_2O{\bullet} + NO_2 \qquad \text{(reaction 6)}$$

$$CF_3OCF_2O{\bullet} + M \rightarrow CF_3O{\bullet} + COF_2 + M \qquad \text{(reaction 7)}$$

and finally,

$$CF_3O{\bullet} + CF_3O{\bullet} + M \rightarrow CF_3OOCF_3 + M \qquad \text{(reaction 8)}$$

Other fully fluorinated species like $CF_3OCF_2{\bullet}$ [17] were also suspected of undergoing carbon-carbon scission to form COF_2 until $CF_3O{\bullet}$ degraded to form HF and one final COF_2 molecule.

$$CF_3O{\bullet} + \text{H-containing species} \rightarrow HF + COF_2 + \text{radical form of H-containing species} \qquad \text{(reaction 9)}$$

Other possible degradation pathways exist when there are two hydrogen atoms remaining on the carbon atom nearest the oxygen radical after reaction 3a. One possibility is the formation of a formate species through hydrogen abstraction hypothesized by Wallington, et al[11], which accounted for 100% of their product formation in that work. To move to a new example species, they followed the pathway of HFE-7100 up to the oxyradical formation as in reactions 1-3 above, before the following reaction led to formation of a formate species:

$$C_4H_9OCH_2O{\bullet} + O_2 \rightarrow HO_2{\bullet} + C_4H_9OCHO \qquad \text{(reaction 4')}$$

where the formate was then stable. Another possibility was the same carbon-carbon bond scission found in reactions 5-7, which did not appear to experimentally happen in this case. Similar experimental results were found for degradation of n-$C_3F_2OCH_3$[18] and CF_3OCH_3[19].

After reaction 4', Wallington, et al., hypothesized that the formate may then oxidize to COF_2 but would be unreactive to OH. Atmospheric removal of all stable intermediates was often assumed to happen through wet or dry deposition, but the speed of those phenomena are unknown so the atmospheric lifetimes for these stable species cannot be determined.

In another work, Wallington, et al.,[12] followed $CF_3CH_2OCHF_2$ through pathways similar to reactions 1-3 until one of two new possibilities occurred:

$CF_3CO{\cdot}HOCHF_2 + M \rightarrow CF_3{\cdot} + HC(O)OCHF_2 + M$ (reaction 4a")

$CF_3CO{\cdot}HOCHF_2 + O_2 \rightarrow HO_2{\cdot} + CF_3COOCHF_2$ (reaction 4b")

where this mechanism was used to explain the formation of trifluoroacetic acid through reaction 4a". This substance is currently accumulating in the environment and has no non-anthropogenic sources.

If a chain forms where an -OCHFO▪ radical is created, like when reactions 1-3 for CH_2FOCH_2F occur, then other new pathways open up, one of which is carbon-oxygen scission:

$CH_2FOCHFO{\cdot} + M \rightarrow CH_2FO{\cdot} + CHFO$ (reaction 4a"')

Another possibility from this scenario is that a hydrogen atom is ejected to give:

$CH_2FOCHFO{\cdot} + M \rightarrow CH_2FOCOF + H{\cdot} + M$ (reaction 4b"')

Or, the final possibility is that molecular oxygen can remove a hydrogen atom to give a fluorinated analog to formate:

$CH_2FOCHFO{\cdot} + O_2 \rightarrow CH_2FOCOF + HO_2{\cdot}$ (reaction 4c"')

In Goto, et al's[15]., work, the relative rates of these three pathways were 48%, 31% and 21%, respectively for 4a"', 4b"', and 4c"'.

Now that the mechanisms for degradation have been discussed, we report radiative forcing values for some of the potential end-point daughter species of HFE and hydrofluoroalcohol degradation. The results are summarized in Table 2.

Table 2 - A preliminary summary of predicted radiative forcing values for some potential daughter species from HFE degradation.

Daughter Species	Parent Species		a ($W\ m^{-2}\ ppbv^{-1}$)		
COF_2	any HFE that forms an F-saturated oxyradical during degradation	0.121			
HCOF	CH_2FOCH_2F [15]	0.083			
CF_3OH	some fluoroalcohols [33]	0.273			
$CF_3C(O)H$[a]	$CF_3CH_2OCH_2CF_3$ [34]	0.170	0.170		
$CHF_2C(O)H$	some fluoroalcohols [33, 35]	0.115	0.112		
$CF_3C(O)F$	CH_2FOCH_2F [15]	0.237			
$CH_2FC(O)H$[a]	some fluoroalcohols [35, 36]	0.040	0.037		
$CH_2FOC(O)F$[a]	CH_2FOCH_2F [15]	0.265	0.250	0.194	0.226
CF_3CH_2CHO[a]	some fluoroalcohols [33]	0.173	0.187		
CF_3CH_2OCHO[a]	$CF_3CH_2OCH_2CF_3$ [34]	0.413	0.464		
$CH_3OC(O)CF_3$[a]	$CH_3OCH_2CF_3$	0.298	0.293		
$CF_3C(O)OC(O)H$	$CH_3OCH_2CF_3$	0.483			
$CF_3C(O)OCH_2CF_3$[a]	$CF_3CH_2OCH_2CF_3$ [34]	0.464	0.477		
$CF_3C(OH)_2CF_3$[a]	fluoroalcohol	0.353			
$C(O)HOCH(CF_3)_2$	$CH_2FOCH(CF_3)_2$	0.438			

[a]These species have more than one rotameric form possible and the value of each stable conformer is reported without thermodynamically based weighting factors.

We next summarize some of the implications to this work in light of the new results for daughter species formed during degradation. First, there is much that is unknown about the degradation pathways and the intermediates and final products formed after potential HFE release into the atmosphere. Almost no kinetic information is known about many of the intermediate steps, making it impossible to estimate the daughter species' contributions to global warming. While some species may be short lived due to dry or wet precipitation and have correspondingly low GWPs, the claim that atmospheric removal is quick is not yet supported by experimental evidence. The concentration of COF_2 in the stratosphere was already on the rise in 1994, indicating that COF_2 may also be accumulating in the troposphere[37].

The formation of the daughter species is of concern because some of these species have higher radiative forcing values than their parent species. For example, CF_3CH_2OCHO, with a radiative forcing around 0.43, may be generated through the decomposition of $CF_3CH_2OCH_2CF_3$, which has a radiative forcing value of around 0.40. This indicates that the current definition of GWP as being the time integral of the radiative forcing combined with kinetic destruction profiles for the parent species is not a true representation of a species' GWP. Instead, we advocate the introduction of a summation over all concentration profiles for not just the parent species, but all daughter species that have atmospheric lifetimes long enough to cause a potential impact on global warming:

$$GWP_i = \sum_i \frac{\int_0^{TH} a_i \left[x_i(t)\right] dt}{\int_0^{TH} a_{ref} \left[x_{ref}(t)\right] dt}$$

where TH is the time horizon for GWP the species will be considered over, a_i is the radiative forcing due to a unit increase in atmospheric concentration of species i, $[x_i(t)]$ is the time dependent concentration of a pulse of species i, while the corresponding quantities for a reference gas are in the denominator and can be represented as the absolute global warming potential (AGWP) of the reference species.

Larger HFEs that have long carbon backbone chains saturated by fluorine atoms will generate one COF_2 molecule for every carbon atom as the species degrades. This means that a species' global warming impact may increase significantly as the species degrades because the radiative forcing of COF_2 is 0.121. If the atmospheric lifetime of COF_2 is long enough, as suggested by the accumulation of this species in the stratosphere, then the GWP of the parent species is a poor predictor of the overall impact. For example, a HFE like $CF_3CHFOCF_2CF_2CF_3$ has a radiative forcing of 0.636 and will likely fragment to form four or more COF_2 molecules or other similar species, with each stable degradation product contributing to the overall global warming impact. Clearly, there is much work needed in order to understand the complex degradation pathways, atmospheric lifetimes, removal mechanisms and overall GWPs of this emerging class of HFE compounds. Knowledge of this information will allow for better informed decisions about whether HFEs are truly more environmentally benign than the HFCs they are already replacing in the manufacturing sector.

References

1. Blowers, P., Moline, D. M.; Tetrault, K. F.; Wheeler, R. R., Tuchawena, S. L., *Prediction of Radiative Forcing Values for Hydrofluoroethers Using Density Functional Theory Methods.* J. Geophys. Res. Atmos., 2007. **112**: p. D15108.
2. Blowers, P., Moline, D. M.; Tetrault, K. F.; Wheeler, R. R., Tuchawena, S. L., *Global Warming Potentials of Hydrofluoroethers.* ES&T, accepted 2007.
3. Newsted, J.L., Nakanishi, J.; Cousins, I.; Werner, K.; Giesy, J. P., *Predicted distribution and econological risk assessment of a "segregated" hydrofluoroether in the Japanese environment.* Env. Sci. Technol., 2002. **36**: p. 4761-4769.
4. Blowers, P., Tetrault, K. F., Trujillo-Morehead, Y. K., *Estimation of gas-phase heat capacities for hydrofluoroethers with two carbon atoms.* Ind. Eng. Chem. Res., (accepted).
5. Blowers, P., Tetrault, K. F., Trujillo-Morehead, Y. K., *Estimation of Heat Capacities for Hydrofluoroethers With Two Carbon Atoms.* Ind. Eng. Chem. Res., submitted 2007.
6. Scott, A.P., Radom, L., *Harmonic vibrational frequencies: an evaluation of Hartree-Fock, Moller-Plesset, Quadratic Configuration Interaction, Density Functional Theory, and semiempirical scale factors.* J. Phys. Chem., 1996. **100**: p. 16502-16513.

7. Pinnock, S., Hurley, M. D.; Shine, K. P.; Wallington, T. J.; Smyth, T. J., *Radiative forcing of climate by hydrochloroflurocarbons and hydrofluorocarbons.* J. Geophys. Res. Atmos., 1995. **100**(D11): p. 23227-23238.
8. Papasavva, S., Tai, S.; Esslinger, A.; Illinger, K. H.; Kenny, J. E., *Ab initio calculations of vibrational frequencies and infrared intensities for global warming potential of CFC substitutes: CF_3CH_2F (HFC-134a).* J. Phys. Chem., 1995. **99**: p. 3438-3443.
9. Papasavva, S., Tai, S.; Illinger, K. H.; Kenny, J. E., *Infrared radiative forcing of CFC substitutes and their atmospheric reaction products.* J. Geophys. Res., 1997. **102**: p. 13643-13650.
10. Urata, S., Takada, A.; Uchimaru, T.; Chandra, A. K., *Rate constants estimation for the reaction of hydrofluorocarbons and hydrofluoroethers with OH radicals.* Chem. Phys. Lett., 2003. **368**: p. 215-223.
11. Wallington, T.J., Scheneider, W. F.; Sehested, J.; Bilde, M.; Platz, J.; Nielsen, O. J.; Christensen, L. K.; Molina, M. J.; Molina, L. T.; Wooldridge, P. W., *Atmospheric chemistry of HFE-7100 ($C_4F_9OCH_3$): Reaction with OH radicals, UV spectra and kinetic data for $C_4F_9OCH_2$(dot) and $C_4F_9OCH_2O_2$(dot) radicals, and the atmospheric fate of $C_4F_9OCH_2O$(dot) radicals.* J. Phys. Chem. A, 1997. **101**(44): p. 8264-8274.
12. Wallington, T.J., Hurley, M. D.; Fedotov, V.; Morrell, C.; Hancock, G., *Atmospheric chemistry of $CF_3CH_2OCHF_2$ and $CF_3CHClOCHF_2$: Kinetics and mechanisms of reaction with Cl atoms and OH radicals and atmospheric fate of CF_3C(O(dot))$HOCHF_2$ and CF_3C(O(dot))$ClOCHF_2$ radicals.* J. Phys. Chem. A, 2002. **106**: p. 8391-8398.
13. Wallington, T.J., Hurley, M. D.; Nielsen, O. J.; Andersen, M. P. S., *Atmospheric chemistry of $CF_3CFHCF_2OCF_3$ and $CF_3CFHCF_2OCF_2H$: Reaction with Cl atoms and OH radicals, degradation mechanism, and global warming potentials.* J. Phys. Chem. A, 2004. **108**(51): p. 11333-11338.
14. Christensen, L.K., Sehested, J.; Nielsen, O. J.; Bilde, M.; Wallington, T. J.; Guschin, A.; Molina, L. T.; Molina, M. J., *Atmospheric chemistry of HFE-7200 ($C_4F_9OC_2H_5$): Reaction with OH radicals and fate of $C_4F_9OCH_2CH_2O$(dot) and C_4F_9OCHO(dot)CH_3 radicals.* J. Phys. Chem. A, 1998. **102**: p. 4839-4845.
15. Goto, M., Kawasaki, M.; Wallington, T. J.; Hurley, M. D.; Sharratt, A. P., *Atmospheric chemistry of CH_2FOCH_2F: Reaction with Cl atoms and atmospheric fate of CH2FOCHFO* radicals.* Int. J. Chem. Kinet., 2002. **34**(3): p. 139-147.
16. Andersen, M.P.S., Nielsen, O. J.; Wallington, T. J.; Hurley, M. D.; DeMore, W. B., *Atmospheric chemistry of $CF_3OCF_2CF_2H$ and $CF_3OC(CF_3)_2H$: Reaction with Cl atoms and OH radicals, degradation mechanisms, global warming potentials, and empirical relationship between k(OH) and k(Cl) for organic compounds.* J. Phys. Chem. A, 2005. **109**: p. 3926-3934.
17. Inoue, Y., Kawasaki, M.; Wallington, T. J.; Hurley, M. D., *Atmospheric degradation mechanism of CF_3OCF_2H.* Chem. Phys. Lett., 2001. **343**: p. 296-302.
18. Ninomiya, Y., Kawasaki; Guschin, A.; Molina, L. T.; Molina, M. J.; Wallington, T. J., *Atmospheric chemistry of n-$C_3F_7OCH_3$: Reaction with OH radicals and Cl atoms and atmospheric fate of n-$C_3F_7OCH_2O$(*) radicals.* Env. Sci. Technol., 2000. **34**: p. 2973-2978.
19. Christensen, L.K., Wallington, T. J.; Guschin, A. ; Hurley, M. D., *Atmospheric degradation mechanism of CF_3OCH_3.* J. Phys. Chem. A, 1999. **103**: p. 4202-4208.
20. Fuglestvedt, J.S., Berntsen, T. K.; Godal, O.; Sausen, R.; Shine, K. P.; Skodvin, T., *Metrics of climate change: Assessing radiative forcing and emission indicies.* Climatic Change, 2003. **58**: p. 267-331.
21. Blowers, P., Tetrault, K. F., Trujillo-Morehead, Y. K., *Global warming potential prediction for hydrofluoroethers with two carbon atoms.* Theor. Chem. Acc., accepted.
22. Orkin, V.L., Khamaganov, V. G.; Guschin, A. G.; Kasimovskaya, E. E.; Larin, I. K., *Development of atmospheric characteristics of chlorine-free alternative fluorocarbons: Report on R-134a & E-143a.* 1993, Oak Ridge National Laboratory: Oak Ridge, TN.
23. Good, D.A., Francisco, J. S., *Structure and vibrational spectra of chlorofluorocarbon substitutes: An experimental and ab initio study of fluorinated ethers CHF_2OCF_3 (E125), CHF_2OCHF_2 (E134), and CH_3OCF_3 (E143a).* J. Phys. Chem. A, 1998. **102**: p. 1854-1864.
24. Myhre, G., Nielsen, C. J.; Powell, D. L.; Stordal, F., *Infrared absorption cross section, radiative forcing, and GWP of four hydrofluoro(poly)ethers.* Atmos. Environ., 1999. **33**: p. 4447-4458.
25. Good, D.A., Francisco, J. S.; Jain, A. K.; Wuebbles, D. J., *Lifetimes and global warming potentials for dimethyl ether and for fluorinated ethers: CH_3OCF_3 (E143a), CHF_2OCHF_2 (E134), CHF_2OCF_3 (E125).* J. Geophys. Res. Atmos., 1998. **103**(D21): p. 28181-28186.
26. Christidis, N., Hurley, M. D.; Pinnock, S.; Shine, K. P.; Wallington, T. J., *Radiative forcing of climate change by CFC-11 and possible CFC replacements.* J. Geophys. Res., 1997. **102**(D16): p. 19597-19609.

27. Sekiya, A., Misaki, S., *A continuing search for new refrigerants.* Chemtech, 1996. **12**: p. 42-48.
28. Oyaro, N., Sellevag, S. R.; Nielsen, C. J., *Atmospheric chemistry of hydrofluoroethers: Reaction of a series of hydrofluoroethers with OH radicals and Cl atoms, atmospheric lifetimes, and global warming potentials.* J. Phys. Chem. A, 2005. **109**(2): p. 337-346.
29. Jain, A.K., Li, Z.; Naik,V.; Wuebbles, D. J.; Good, D. A.; Hansen, J. C.; Francisco, J. S., *Evaluation of the atmospheric lifetime and radiative forcing on climate for 1,2,2,2-tetrafluoroethyl trifluoromethyl ether* ($CF_3OCHFCF_3$). J. Geophys. Res., 2001. **106**(D12): p. 12615-12618.
30. Blowers, P., Moline, D. M.; Tetrault, K. F.; Wheeler, R. R., Tuchawena, S. L., *Prediction of Radiative Forcing Values for Hydrofluoroethers Using Density Functional Theory Methods.* J. Geophys. Res. Atmos., (submitted).
31. Andersen, M.P.S., Nielsen, O. J.; Wallington, T. J.; Hurley, M. D.; DeMore, W. B., *Atmospheric chemistry of* $CF_3OCF_2CF_2H$ *and* $CF_3OC(CF_3)_2H$*: Reaction with Cl atoms and OH radicals, degradation mechanism, global warming potentials, and empirical relationship between k(OH) and k(Cl) for organic compounds.* J. Phys. Chem. A, 2005. **109**: p. 3926-3934.
32. Atkinson, R., *Gas-phase degradation of organic compounds in the troposphere.* Pure Appl. Chem., 1998. **70**(7): p. 1327-1334.
33. Sellevag, S.R., Kelly, T.; Sidebottom, H.; Nielsen, C. J., *A study of the IR and UV-Vis adsorption cross-sections, photolysis and OH-initiaited oxidation of CF3CHO and CF3CH2CHO.* Phys. Chem. Chem. Phys., 2004. **6**: p. 1243-1252.
34. Wallington, T.J., Guschin, A.; Stein, T. N. N.; Platz, J.; Sehested, J.; Christensen, L. K.; Nielsen, O. J., *Atmospheric chemistry of* $CF_3CH_2OCH_2CF_3$*: UV spectra and kinetic data for* $CF_3CH(dot)OCH_2CF_3$ *and* $CF_3CH(OO(dot))OCH_2CF_3$ *radicals and atmospheric fate of* $CF_3CH(O(dot))OCH_2CF_3$ *radicals.* J. Phys. Chem. A, 1998. **102**: p. 1152-1161.
35. Papadimitriou, V.C., Prosmitis, A. V.; Lazarou, Y. G.; Papagiannakopoulos, P., *Absolute reaction rates of chlorine atoms with* CF_3CH_2OH, CHF_2CH_2OH, *and* CH_2FCH_2OH. J. Phys. Chem. A, 2003. **107**: p. 3733-3740.
36. Sellevag, S.R., Nielsen, C. J.; Sovde, O. A.; Myhre, G.; Sundet, J. K.; Stordal, F.; Isaksen, I. S. A., *Atmospheric gas-phase degradation and global warming potentials of 2-fluoroethanol, 2,2-difluoroethanol, and 2,2,2-trifluoroethanol.* Atmos. Environ., 2004. **38**: p. 6725-6735.
37. Zander, R., Rinsland, C. P.; Mahieu, E.; Gunson, M. R.; Farmer, C. B.; Abrams, M. C.; Ko, M. K. W., *Increase of carbonyl fluoride* (COF_2) *in the stratosphere and its contribution to the 1992 budget of inorganic fluorine in the upper stratosphere.* J. Geo. Phys. Res., 1994. **99**(D8): p. 16737-16743.

SCALE UP OF THE EXPOXIDATION OF H_2-O_2-C_3H_8-C_3H_6 MIXTURES IN JET-LOOP REACTOR SYSTEM

Feng Shubo[1,2] *Li Shuyuan*[1] *Zhao Dishun*[2] *Wu Chunmin*[3]
[1]College of Chemical Science & Engineering, China University of Petroleum, Beijing, 102200, PRC; [2]College of Chemical and Pharmaceutical Engineering, Hebei University of Science & Technology, Shijiazhuang, 050018, PRC; [3]Shijiazhuang Refineries and Chemical Co., Ltd.

1 Introduction

Propylene oxide (PO) is an important organic feedstock. PO finds its application in the production of polyethers useful for manufacture of polyurethanes, for synthesis of mono- and dipropylene glycol and some other heavy organic synthesis products. But, the existing production process results in a large amount of enviromental polluents. The development of sustainable technologies and environmentally friendly process will give an important contribution to human society. There are several green processes being developed, such as TS-1 and H_2O_2 process and molecular oxygen epoxidation process. As the increasing cost of propylene, a cheap feedstock has being researched for. Propane is an available material from refineries. Propylene oxide can be produced from propane. Propylene can be obtained by the dehydrogenation of propane in a high temperature, but the separation of the resulting gas mixture needs a considerable amount of heat energy. The present paper is to study the green epoxidation of C_3H_6 in H_2-O_2-C_3H_8-C_3H_6 mixtures without separating propylene from the gas mixture, Pd/TS-1 being used as a catalyst and methanol being used as a solvent. Based on experimental data and analysis of the complex epoxidation reactions competing with each others, it may be suggested that the epoxidation process is composed of five molecular reactions in H_2-O_2-C_3H_8-C_3H_6 mixtures:

$$C_3H_6 + H_2O_2 \rightarrow C_3H_6O \quad (1)$$

$$C_3H_6 + H_2 \rightarrow C_3H_8 \quad (2)$$

$$H_2 + O_2 \dashrightarrow H_2O_2 \quad (3)$$

$$H_2O_2 \dashrightarrow H_2O + O_2 \quad (4)$$

$$H_2 + O_2 \dashrightarrow H_2O \quad (5)$$

The first part of the present study is the epoxidation of H_2-O_2-C_3H_8-C_3H_6 mixtures in a batch stirred reactor of 300mL. The epoxidations were investigated under the following conditions: solvent is methanol, temperature of 50-70 ℃, pressure of 0.3-0.5MPa, ratios of H_2-O_2-C_3H_8-C_3H_6 of 9:10:73:8--5:7:80:8(mol/mol).Experimental results indicate that PO yield is up to 8% per pass, PO selectivity 53% and H_2 utilization 50% when propylene conversion is ca. 30%. The process is as follows. The propane is first dehydrogenated in a dehydrogenation reactor into C_3H_6 and H_2. And then O_2 is added to the resulting gas mixture. The mixture is then entered into an epoxidation reactor. Epoxidation reaction was studied in a continuous slurry cyclinder reactor of 300mL. In laboratory stage, experiments were performed in a magnetic stirred reactor of 300mL. The used bifunctional catalyst Pd/TS-1 was prepared by wet impregnation of Pd $(NH_3)_4Cl_2$ solution. It is found that oxygen concentration has a considerable effect on PO yield per pass. The higher oxygen concentration in the gas mixture, the more PO formed in the epoxidation is. However, too high oxygen concentration in the system will lead to the risk of explosiveness. The improvement of PO yield is limited to some extent. Because oxygen occurs in feed gas mixtures, there will be a safty problem in an epoxidation of H_2-O_2-C_3H_8-C_3H_6. Therefore, Decrease in oxygen concentration has an important effect in the commcercialization of the process. It is reported that the solubility of H_2-O_2 in methanol solvent is very small. Our idea for scale up of the process is that an efficient reactor type is selected and studied. In order to make the process to a stage of commercialization, a scale up experiment was made in jet loop reactor system designed by ourselves.

The reactor has a volume of 8000mL and made of stainless steel. Experimental results indicate that epoxidation effect in jet-loop reactor is considerably better than that in the case of stirred reactor of 300ml. Turbulent effect caused by jet loop flow make the liquid film in the catalyst surface thin, reducing mass transport obstacles and enhancing the solubility of hydrogen and oxygen in methanol. In this way, required oxygen is decreased greatly and process safty improved.

It is reported that the direct formation of H_2O_2 from H_2 and O_2 is a controlling factor in synthesizing PO in H_2-O_2-C_3H_8-C_3H_6 mixtures. And to enhance the solubility H_2 and O_2 in the liquid is a crucial step of controlling the formation of H_2O_2 and PO formation. Formation rate of H_2O_2 decides the formation rate of PO in H_2-O_2-C_3H_8-C_3H_6 mixtures.

1.1 The consideration of a suitable reactor type

For many processes in chemical industry or biotechnology the homogeneous distribution of solids and gases in a continuous liquid phase is necessary. For the gas-liquid-solid three phase reactor system, jet-loop reactors are widely used to ensure a homogeneous mixing with low power inputs.

Multiphase flows with gases and solids suspended in a liquid phase become more and more important for industrial applications. In addition to the suspension of solids it is often necessary to dissolve gas bubbles because most processes require the presence of soluble gases (e.g. oxygenation, hydrogenation). The homogeneous distribution of gases and solids in liquid flows demands high power inputs and straight unimpeded flows to overcome the differences in density. For this requirement loop reactors have found a widespread acceptance in industry.

In recent years, hydrodynamics and mass transfer characterization of a jet-loop reactor with a downflow liquid jet reactor has been studied extensively. It is characterized with high efficiency of mass transfer.

Another potential limitation in this three-phase reaction system is that of mass transport, which may either involve the rate of solubility of the gases into the liquid phase or the rate of diffusion across a boundary layer at the surface of the catalyst particle.

For an oxygenation with O_2, it is reported that the results of experiment on mass transfer performance of the jet loop reactor showed, oxygen transfer rate was 8 to 10 times of that of the traditional aeration.

1.2 The configuration of reactor system and ejector

The set up consisted of a cyclindrical 8 liter stainless steel reactor of diameter 150mm and height 450mm, a liquid recirculation pump, a heat exchanger, flow meter and an ejector. The liquid was recirculated through the external liquid loop with the help of a pump transporting. The high-velocity jet, discharging from the nozzle, creates an underpressure in the gas suction chamber which entrains the gas phase into the ejector. Both the liquid jet and the entrained gas phase enter the reactor and somewhere in the mixing tube the gas phase is dispersed into the liquid phase. The main holding vessel can be considered as a conventional bubble column, with the ejector used as a special gas distributor. Reaction mixture is heated by electrical heating system. Liquid velocity was adjusted by the needle value in outlet of the circulation pump. Certainly, variable speed motors is the best one.

The objective of the present study is to explore the possibility of performing epoxidation of H_2-O_2-C_3H_8-C_3H_6 in jet-loop reactor system. An emphasis was put on the epoxidation reaction efficiency, not on mass transfer characterization. The following indexes, propylene conversion, propylene oxide selectivity, PO yield, H_2 utilization and O_2 concentration, are investigated in our experiments. Considering that reaction mixture might be in the explosive limit, it is important how to decrease O_2 concentration in feedstock gas mixture under circumstances of obtaining the same PO yield as bubble column reactor. Intense mixing in jet-loop reactor results in the decrease of O_2 concentration requirement in feedstock gas mixture.

The purpose of the present study was to explore a number of different factors that may influence the net formation rate of H_2O_2 and PO, the selectivity for the conversion of H_2 to H_2O_2, and the stability of the system. Such factors include the presence of an acid, the effect of temperature and pressure and the influence of $Cl^-$and Br^-.

The schematic chart of the experimental set-up is shown in Fig.1. All the experiments were carried out in the semi-batch flow arrangement. The quality of gas-liquid dispersion formed in the jet-loop reactor was characterized by the average values of PO yield produced in the gas mixture.

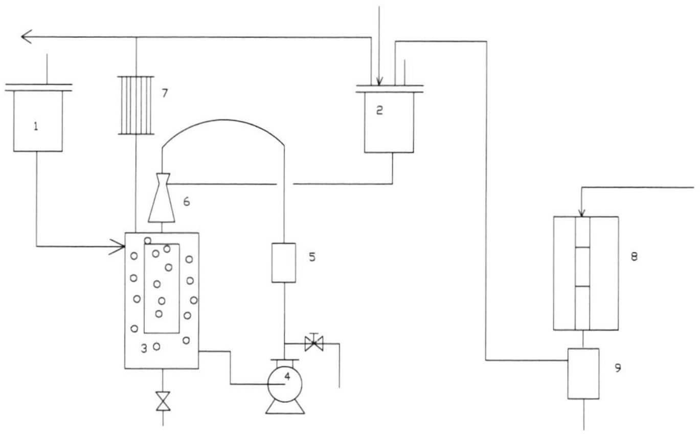

Fig.1 Schematical diagram for epoxidation of H_2-O_2-C_3H_8-C_3H_6 mixture
1-liquid feeding tank; 2-gas mixer; 3-loop reactor; 4-circulation pump; 5-liquid flow meter; 6-ejector; 7-cooling heater; 8-tubular fixed reactor; 9-condenser;

Table 1 Reactor geometry parameters

Description	Value
reactor diameter (m)	0.15
reactor height (m)	0.45
Working volume (m3)	0.008
Draft-tube length (m)	0.35
Inner diameter of draft-tube (m)	0.09
Circulation pump pressure	0.5MPa
Ejector PS-1-25	DN 25, PN 1.0Mpa

2 Experimental

2.1 Reagents and apparatuses.

Methanol, AR grade; Pd $(NH_3)_4Cl_2$, AR grade; Hydrogen, oxygen and nitrogen are of 99.9% purity; TS-1 was purchased from Chinese hunan catalyst Co., Ltd. Propane of 95% purity is provided by beijing gas Co.; H_2PtCl_6, AR grade; $SnCl_2$, AR grade; γ-Al_2O_3, AR grade.

2.2 Preparation of the catalyst Pd/TS-1

The difunctional catalyst system is Pd/TS-1-CH_3OH. Pd $(NH_3)_4Cl_2$ is first loaded into TS-1 using wet impregnation method, undergoes vacuum drying, and then reduced in a flow tubular reactor of electrical heating in H_2/N_2 atmosphere without going a high temperature calcination.

2.3 Preparation of propane dehydrogenation Pt-Sn/Al_2O_3 catalyst

A commercial γ-Al_2O_3, previously 40–60 mesh sieved and calcined in air at 550 ℃ for 4 h, was used as support. Pt/γ-Al_2O_3 catalyst was prepared by impregnating γ-Al_2O_3 with excess aqueous solution of the Pt precursor (H_2PtCl_6). After impregnation, the catalyst was dried at 120 ℃ overnight and then calcined at 520 ℃ in air for 6 h. For Pt-Sn/γ-Al_2O_3, Sn was first deposited by impregnating $SnCl_2$ ethanol solution, then dried and calcined, and finally the Pt component was added from aqueous Pt solution in a similar way.

2.4 The operating procesure of experiments in the jet-loop reactor system and product analysis

Add 0.55L methanol, 100 mL water, and 50g the prepared catalyst, 30mL 0.05MHBr solution, 25mL 0.05MH_2SO_4 solution, into the stainless steel reactor of 0.8L. Pass the gas mixture of H_2-O_2-C_3H_8-C_3H_6 into the slurry reactor. The gas effluents are analyzed using gas chromatography. Thermal conductivity detector and flame ionization detector are connected in series. Thermal conductivity detector is used for measuring the components of H_2, O_2, C_3H_6 and C_3H_8. The flame ionization detector is used for measuring the components of methanol, water, propylene oxide, propylene, propane. The content of various components is calculated using a specific method.

In order to increase the selectivity of H_2O_2 and PO, a ceratin amount of HBr and H_2SO_4 was added to reaction system.

2.5 Evaluation of the jet-loop reactor system performance

Three specific data, PO yield, PO selectivity and H_2 utilization were investigated, respectively. Their definitions are given as follows.

Y = PO molar amount in the effluent / C_3H_6 molar amount in the feedstock;

S = PO molar amount in the effluent / C_3H_6 molar amount into PO and propane;

$H_{2,u}$ = (1- H_2 molar amount into water)/consumed H_2 molar amount in feedstock gas mixture.

3 Results and discussion

A significant difference was observed PO yield for two different reactor operation modes, with and without jet-loop. Experimental data obtained under conditions of the jet-loop reactor clearly demonstrated an increase of PO yield with increasing gas feed pressure, shown in Fig.3. Variation of gas feed flow rate was studied. The too high gas flow will reduce PO yield per pass. Within the whole range of our experimental conditions, efficiency of gas phase utilization is enhanced considerably.

3.1 Propane dehydrogenation process in fixed catalytic bed reactor

The adjustment of the proportions in the mixture gas is rather difficult because the unstability of the conversion of propane dehydrogenation and the uncertainty of coking amount. Therefore, In order to make experiments reproducibile, concise control of catalyst preparation and reaction conditions is a necessity. Hydrogen amount in the mixture gas is easy to go a high value, propane conversion is adjusted to a low value,10%, in each run of experiment. Reaction conditions are as follwos: reaction temperature, 580℃; residence time, 0.2 sec.; catalyst volume, 15g. In addition, run time is short because quick coking in catalyst surface. In order to maintain the reproducibility of experimental data, a constant shift of the catalyst is a necessity.

3.2 Investigation of reaction temperature

Effect of reaction temperature to PO yield was studied in the range of 45-65□. Experimental results in the Fig.2 indicate that 60□ is the best point corresponding to highest PO yields.

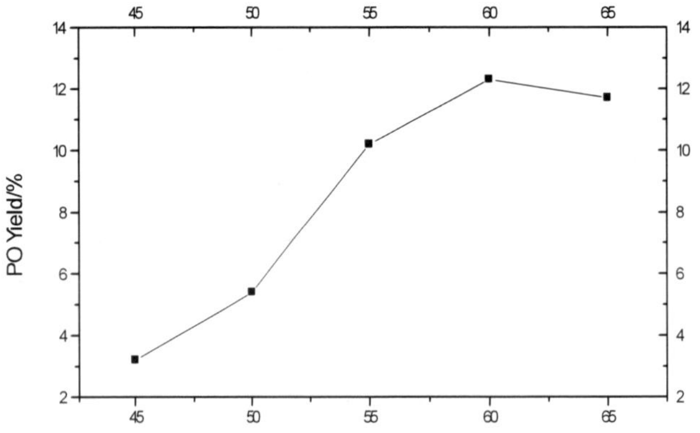

Fig.2 Effect of reaction temperature on propylene oxide yields

3.3 Investigation of reaction pressure

Effect of reaction pressure to PO yield was investigated and results are shown in Fig.3. It can be seen that there is an optimal reaction pressure which corresponds to highest PO yield. Effect of reaction pressure can be explained in two ways. First, pressure increase is favorable to the solubility of H_2-O_2-

C_3H_8-C_3H_6 in methanol solvent. In another way, pressure increase is also favorable to the formation of water from H_2 and O_2. Therefore, an optimal pressure, 0.3MPa, is rationalized. This is similar to the results obtained in laboratory stirred reactor of 300mL.

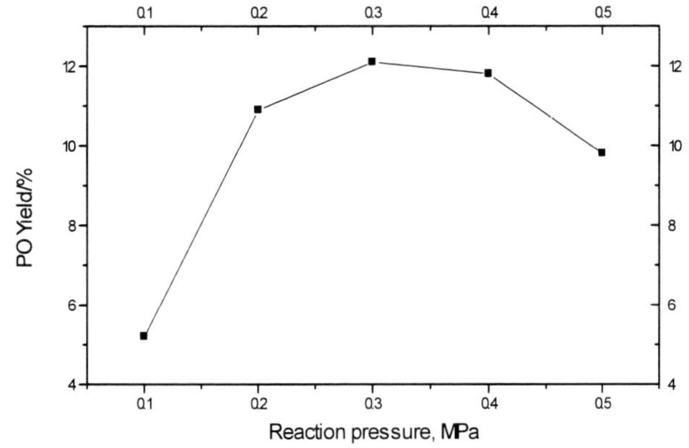

Fig.3 Effect of reaction pressure on propylene oxide yields

3.4 Investigation of gas mxture flow rate

Effect of flow rate to PO yield is shown in Fig.4. When flow rate exceeds a certain value of 1600mL/min, PO yield starts to decrease with the increase of flow rate.

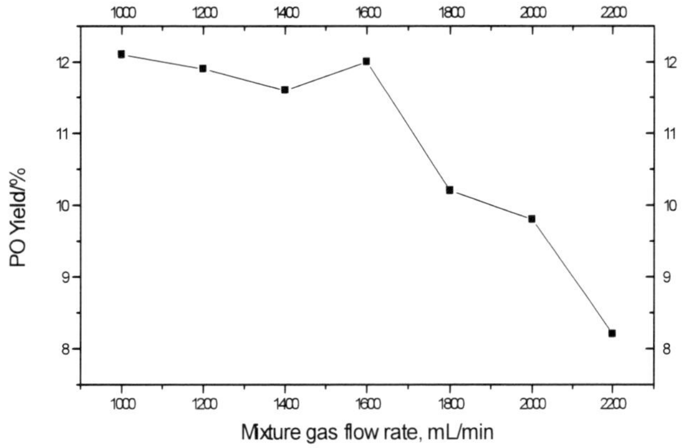

Fig.4 Effect of feedstock gas mixture flow rate to PO yield

3.6 Investigation of gas proportions

Composition of feedstock gas mixture has a considerable effect to epoxidation reaction. Experimental results are shown in Table 2 and Fig.6. It can be seen that a better PO yield was observed in the ratio of O_2/H_2 large than 1. A typical result is listed in Table 4. It can be seen that a trend similar to that in the case of H_2-O_2-C_3H_8-C_3H_6 mixtures. Due to difficult adjustment of proportions of each component in mixture gas resulting from the change of conversion of propane and selectivity of propylene, the regularity of PO yield may show some error.

Table 2 Effect of gas proportions on propylene oxide yields

No.	gas proportions H_2	O_2	C_3H_6	C_3H_8	H_2 utilization	selectivity	PO yield，%

1	7	8	8	77	45%	40	11.2
2	5	12	8	75	51%	48	10.6
3	5	8	8	79	58%	51	9.3
4	7	12	8	73	47%	45	12.1

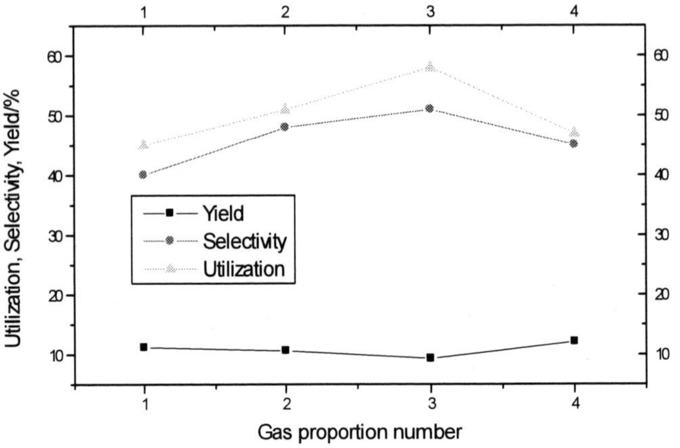

Fig.5 Effect of gas proportions on propylene oxide yields

3.7 Investigation of stability of the catalyst formulation

In a typical condition, the stability of the catalyst formulation was observed and shown in Fig.6. It can be seen that the catalyst behaves a good stability.

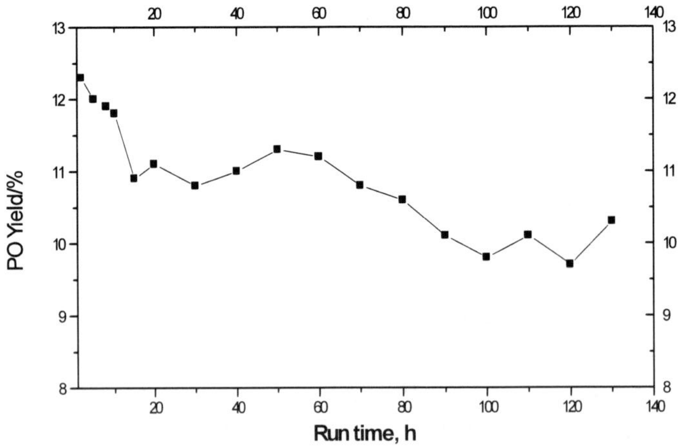

Fig.6 Stability of the catalyst system Pd-TS-1

The transfer rate of components in the gas mixtures through gas-liquid interface controls PO formation rate. In the same composition of H_2-O_2-C_3H_8-C_3H_6, it is observed that the more thin liquid film, the less the content of O_2 in the gas mixture. This indicates that forced circulation of liquid methanol solvent containg solid molecular sieve TS-1 promotes O_2 mass transfer through liquid film. The effect of liquid solvent circulation flow rate and liquid pressure on the mixing result was investigated.

The jet loop reactor technology combines the advantages of both jet and loop flow, has the characteristics of high mass transfer efficiency, short hydraulic retention time, high volume loading and capacity of anti impact load etc.It showed that the liquid average velocity, the turbulent violence and the solid hold-up increased with the increase of jet liquid flow rates. Inside the draft tube, the liquid average velocity indicated approximately the distribution of tube flow. Downflow liquid velocity in downflow jet loop reactor varies with the increasing amount of jet liquid flow rates. Tubulent violence increases with the increasing amount of jet liquid flow rates, leading to the improvement of PO yield.

Scale up of the epoxidation is a necessary way to commercial stage of the process. By the investigation of the epoxidation in a reactor of having larger volume, a certain amount of valueable data for further scale up was obtained. Especially for the epoxidation process in H_2-O_2-C_3H_8-C_3H_6 mixture, explosive risk occurs. Since exact explosive limit data for the gas mixture are difficult to obtain, oxygen concentration as low as possible is favorable to the commercialization of the process. The jet-loop reactor system is noted for its high efficiency of mass transfer of gas-liquid and low power input. So this reactor type was selected as experimental reactor.

Potential diffusion limitation across a boundary layer at the surface of the palladium is more difficult to ascertain and distinguish from transport across the gas–liquid interface. For example, a larger stirring rate could also decrease the thickness of the boundary layer and thereby increase the rate of reaction. This is the reason why a jet-loop reactor is selected as candidate reactor type for the epoxidation.

Conversion of propylene to PO is increased when propane formation is inhibited. This is an advantage of high propane concentration in the mixture. If operated in low propane concentration, propylene concentration go to a high value, hydrogen concentration in the mixture gas will be too high, resulting in difficult realization of suitable gas proportions.

4 Conclusions

1) The present data provides a reference of further scale up of the process.Based on experimental results of the present paper, following conclusions are concluded.

2) Preparation of propylene oxide in H_2-O_2-C_3H_8-C_3H_6 mixtures is feasible. The highest yield of PO product in the present study is up to 12.5 wt% in jet loop reactor. The optimal reaction conditions are as follows: temperature, 60□; gas velocity, 1600mL/min; liquid circulation velocity, 3000mL/min; reaction pressure, 0.3Mpa; catalyst system Pd/TS-1-CH_3OH.

3) The hydrogen utilization seems to be ths same tendency in both reactor system.

4) Both in jet-loop reactor and in stirred tank reactor, oxygen solubility in methanol sovent is a controlling step of the epoxidation. Compared with stirred tank reactor, low oxygen concentration can meet the needs of epoxidation reaction in jet loop reactor, and same epoxidation effects are obtained.

References

[1] Meiers R, Doingerdissen U，Hölderich W F, *Journal of Catalysis*, **1998**,*176(2)*, 376-386.
[2] Laufer W, Hoelderich W F, *Applied Catalysis A: General,* **2001,** *213(2),* 163-171.
[3] Uphade B S, Yamada Y, Akita T, *Applied Catalysis A: General,* **2001,** *215(1-2)*, 137-148.
[4] Jenzer G, Mallat T, Maciejewski M, *Applied Catalysis A: General,* **2001**， *208(1-2),* 125-133.
[5] Hayashi T, Tanaka K, Haruta M, *Journal of Catalysis,* **1998,** *178,* 566-575.
[6] Beckman E J, *Green Chemistry,* **2003**, *5,* 332-336.
[7] Liu Y Y，Murata K，Inaba M, *Green Chemistry,* **2004**, *6,* 10-515.
[8] Min J S, Ishige H, Misono M，et al. *Journal of Catalysis*, **2001,***198,* 116-121.
[9] Park S E, Huang L, Lee C W，Chang J S, *Catalysis Today*, **2000,** *61,* 117-122.
[10] Mul G, Zwijnenburg A, et al. *Journal of Catalysis*, **2001**, *201,* 128-137.
[11] EP 943,611（1999）.
[12] WO 99 52,884.
[13] WO 99 52,885.
[14] US 5,973,171（1999）.
[15] Michael Schlqter, Sfren Scheid, Saskia John, Norbert R7biger, *Power Technology,* **2005,** *151*, 68–76
[16] Ali Fadavi, Yusuf Chisti, *Chemical Engineering Journal,* **2005**, *112*, 73–80.
[17] P. Havelka!, V. Linek, J. Sinkule, J. ZahradnmH, M. FialovaH, *Chemical Engineering Science,* **2000,** *55,* 535-549.

142b Pilot Plant Scaling: Engineering and Chemistry Considerations

Jesse Black, Shell Global Solutions (US) Inc, P. O. Box 1380, Houston, TX 77251

Rules of thumb and best practices in scale up and scale down of unit operations in pilot plants, directed toward the practicing chemical engineer and process chemist.

Chemistry and Catalysis in Process Development

Joseph B. Powell, Ph.D.
Chief Scientist, Chemical Engineering
Shell Global Solutions US Inc.
Westhollow Technology Center
Houston, Texas 77082

Overview

A surprising percentage of new process developments in the fuels and petrochemicals arena are centered around catalyst performance, whether for synthesis of a new commercial product, or yield and rate improvements for an existing process route. An ability to understand and optimally study the "art of catalysis" is thus a key skill for the process development engineer. Process safety is also a key consideration for all development programs, and a broad knowledge of potential reactivities, reactive hazards, and/or materials compatibility is a valuable aid to safe implementation. Success in process development and support can often depend upon ability to get needed analytical information on process stream compositions in a timely manner for use in controlling units or diagnosing new problems in exploratory studies or in troubleshooting existing operations. Finally, trace chemistry, or the chemistry or reactions which are unexpected, should not be present, but nonetheless are, provide some of the greatest challenges in process development and commercial support. The practicing process development engineer is well served via a detailed grounding in the fundamentals of chemistry and catalysis, and via being paired in R&D teams with certified chemist support. Ditto for a process development chemist's grounding in chemical engineering principles. The following presents some overviews and examples of engineering, chemistry and catalysis interactions of importance in process development, which may influence skill-set selection for process development teams, or process development career path training and skill development for engineers and chemists.

Catalysis

"Catalysis" involved nanotechnology before nanotech became popular. It is an exciting field for generation of intellectual property (patents). However, the mindset for establishing initial reaction feasibility may not always be entirely compatible with process development goals. Heterogeneous catalysts almost invariably deactivate, and initial activity is often anomalous relative to average performance in rate and selectivity. Catalyst testing for process development is thus relegated to the drudgery of life testing. Multi-throughput or enhanced experimentation techniques provide opportunities for substantial cost savings and relief, but represent challenges in scale down. Many of these challenges requires rigorous application of chemical engineering principles, and good opportunities for chemist-engineer interaction.

A second feature of this interaction is to bring in a process cost mindset. Many exotic catalysts, ligands, promoters, reducing agents, and other stoichiometric reactants, whether for heterogeneous or homogeneous reactions, can be found to work quite well for a given batch synthesis. If this approach is superimposed on a process design where separations and catalyst recoveries are difficult, then performance advantage may disappear relative to a less-exotic system.

A third feature, especially for homogeneous catalysis, is to consider the implications of a continuous process, if / when required to achieve target economics. Feed impurities, operational upsets, undesired side reactions can move the catalyst system away from target compositions. An ability to rebalance, reformulate, and recover from anticipated upsets can be a key (often required) feature for commercial viability. If there are upset scenarios that have catastrophic and irreversible implications on catalyst and the latter cannot be readily reformulated in situ, then these must be identified during process development, in order to engineer away the threat, or move to a more robust alternative.

Fitting reaction kinetics to arbitrary, power law models results in limited gains in understanding needed to improve selectivity and yields, or ability to extrapolate to new conditions without additional experiments. Development of kinetic models supported via chemically viable reaction pathways and intermediates can be a time saver in process optimization.

Overall, the overriding importance of catalyst performance for fuels and petrochemical process development makes this a highly important skill for the engineer's or chemists arsenal.

Process safety

Process development can proceed through a bewildering array of chemicals, catalysts, ligands, promoters, solvents, reactants before finding an optimal configuration. A process or reactive hazards evaluation exercise can generate a wide range of scenarios, including off-normal high temperature operating conditions not typically studied during process screening. While chemical handbooks and MSDS sheets can provide some idea of incompatibilities, and testing every possible chemistry via reactivie calorimetry prior to implementation at any scale is of course advised as a best practice, a rudimentary knowledge of chemical reactivity is of great benefit to a process development engineer or chemist, in assuring process safety. It is generally advised to have at least a second colleague review the safety aspects of any proposed implementation, and to make sure that engineering and chemistry skills are present for any review team or process, in evaluating these risks. Constructing a detailed model of reaction pathways and kinetics include side reactions and trace chemistries, can be valuable in modeling and assessment of the hazards of upset scenarios.

Analytical chemistry

When exploring new process options, it often is not possible to get large amounts of support from external groups in developing new analytical methods, in concert with process screening. Ditto for a new problem arising in commercial support requiring immediate resolution. Short turnaround times between conceptualization of an analytical need, and analysis, is often needed. Pilot plants typically need nearly real time data for use in process control. Whether for process screening, pilot control, or commercial process troubleshooting, the optimal solution may often be a rapidly implemented, lower accuracy (approximate) method that may not be the preferred or certified method of analysis.

Some examples of this chemistry / engineering interaction in analyses are:

- development of a low cost, rapid turnaround colorimetric method for homogeneous catalyst concentration measurement (with atomic spectroscopy via ICP being the recognized standard method);
- UV/vis color analysis to examine color body rejection in distillation for use in on-line control (with GS-MS characterization of all components the standard method);
- Use of stopwatch and rule to time separation of phases in extraction (vs. laser light scattering) to diagnose cause of emulsion problems;
- Use of published GC response factors for off gas analysis (vs. rigorous calibration via standards which may not be present)
- Examination and interpretation of *in situ* spectra to assess actual catalyst species present in bulk solution, vs. condensation / precipitation of enriched phases on probes.
- Distillation and weighing of heavy ends components which are too oligomeric to allow quantification by chromatographic means.
- Troubleshooting a process GC which stops giving peaks in the middle of the night.

A rudimentary grounding in analytical chemistry is thus quite valuable for the process development engineer, while an analytical or process chemist with a mindset for engineering and broad range of chemistry experience can be of extreme benefit in rapid response to process development and support challenges.

Trace chemistry

If catalysis is often at the heart of initiating a new process development, trace chemistry is often the reason for failure, or "commercial surprises". Testing of authentic feeds, closing recycle loops especially for heavy recycles where analyses may be incomplete or difficult, are two ways to address this threat. Trace chemistry can present unusual challenges for analyses, and may require engineering assessments to assure sample validity. Brainstorming of reaction mechanisms and plausibility is a valuable example of chemistry and engineering interactions. Most trace chemistry resolutions reveal reactions and pathways that are quite obvious, after the fact. Deliberate brainstorming to anticipate unexpected trace chemistry reaction pathways and the possible process implications, is a great use of engineering and chemistry skill early in process development, preferable before deployment.

142e Production of Laccase from *Flammulina Velutipes* and Its Application in Biobleaching of Kraft Bamboo Pulp

Mengru Liu, Shiyu Fu Sr., Huaiyu Zhan Sr., **Hailong Li**, and Hao Liu. State Key Laboratory of Pulp and Paper Engineering, South China University of Technology, No.381 Wushan Road, Guangzhou, China

A laccase has been produced from Flammulina velutipes in liquid substrate fermentation using potato juice and glucose as the prime substrates. The nutrient conditions on the production of laccase, such as nitrogen sources and its amount, trace elements of Cu^{2+} and Mn^{2+}, were optimized in this study. The maximum laccase yield is over 26 IU/mL. Combined with HBT, the laccase was applied to Totally Chlorine Free (TCF) and Elemental Chlorine Free (ECF) bleaching of Kraft bamboo pulps. The results indicated that Kraft bamboo pulps can be bleached to a brightness of 83.1 % ISO using TCF bleaching sequence and 85.8%ISO using ECF bleaching sequence in combination of LMS treatment. Also the physical properties of paper were improved after LMS bleaching. The ECF bleaching pulps had slightly lower burst index and tensile index, but higher brightness than those of the TCF bleached pulps. Also the enzyme-assisted bleaching before QP stage is proved to be more efficient than its use as a pre-bleaching agent.

142a Hydroformylation and Carbonylation Processes: New Trends in Synthesis of Fine Chemicals and Pharmaceuticals I

Raghunath V. Chaudhari, Department of Chemical & Petroleum Engineering, University of Kansas, 1501 Wakarusa Dr, Room 112B, Lawrence, KS, KS 66047-1803

Hydroformylation of olefins is one of the well known examples of homogeneous catalysis operating on industrial scale for the last few decades. Its application for the manufacture of C3 to C20 oxo alcohols from linear olefins is well established and the major improvements in these processes were primarily due to breakthrough in catalysis. The wider applicability of this technology has been limited due to difficulties in catalyst-product separation; however, recent developments in biphasic catalysis and immobilized metal complex catalysis have shown a way to expand the applications of this technology. Similarly carbonylation of olefins, alcohols and oxidative carbonylation of amines and phenols by homogeneous catalysis have a potential application in developing novel catalytic routes including non-phosgene alternatives to many industrial products. Particularly, the newly emerging applications in fine chemicals and pharmaceuticals may prove to be very attractive to develop competitive and environmentally benign processes for complex organic products with aldehydes, carboxylic acid and organic carbonate functionality. The aim of this lecture will be to present the current state of development of industrial hydroformylation and carbonylation technologies, including new concepts in catalysis and catalyst-product separation. Several examples of applications in fine chemicals, pharmaceuticals and specialty products will be discussed in comparison to their conventional routes. Some of the examples are: hydroformylation of vinyl aromatic compounds as an alternative route for aryl propionic acids (e.g. Ibuprofen and Naproxen), hydroformylation of 1,4-diacetoxy-2-butene to Vitamin-A intermediate, terpenes to and dicyclopentadiene to TCDial (perfumery products).

105c Green Chemistry: Designing Less Hazardous Chemical Products and Processes

Richard E. Engler, Green Chemistry Program, US Environmental Protection Agency, 1200 Pennsylvania Ave. NW (7406M), Washington, DC 20460

Green chemistry is the design of chemical products and processes to reduce or eliminate the use or generation of hazardous substances. US EPA promotes green chemistry throughout the chemical enterprise with a variety of activities. This presentation will provide an introduction to the concept and principles of green chemistry as well as an overview of US EPA's Green Chemistry Program activities, including the Presidential Green Chemistry Challenge.

105a Bridging the Communication Gap between Engineers and Chemists

Osman Aboul-Nasr, Technical Director/Fluor Fellow, Fluor, 100 Fluor Daniel Drive, Greenville, SC 29607-2762

An old definition of a chemical engineer is one who can speak of engineering in front of a chemist, and of chemistry in front of and engineer. This implies that the three disciplines do not speak a common language. This paper will propose methods to bridge the perceived communication gap between chemists, chemical engineers and engineers of other disciplines. The chemical engineer will be the medium, and the focal point, that helps to convert the chemist's bright idea into workable designs that other engineering disciplines, can turn into viable economic enterprise with a social conscious. It will also present success stories that show how communication can cross not just discipline boundaries, but language, country and time barriers.

105b Perspectives on Chemistry and Chemical Engineering in Process Development and Support

Paul R. Weider, Westhollow Technology Center, Shell Chemical Company, P.O. Box 1380, Houston, TX 77251

Anecdotes and perspectives on chemistry and chemical engineering from process innovation, through development and commercialization.

Preparing Product Cost Estimates with Limited Data

John Anderson
Engineering and Process Sciences / Core R+D
The Dow Chemical Company
Midland, Michigan

AIChE Spring National Meeting
April 8, 2008
Session #105d

Topics to be covered:

-Elements of cost: fixed and variable
-Process parameters that affect cost
-Capital depreciation
-The ‘cost’ of capital
-Handling uncertainty
-Sources of data

What are Fixed Costs? Variable Costs?

A Fixed Cost is one that isn't affected very much by the amount of product made.

A Variable Cost is one that IS affected by the amount of product made.

(Note that no cost is *completely dependent* on volume…and no cost is *completely independent* of volume either!)

Process parameters which affect cost...

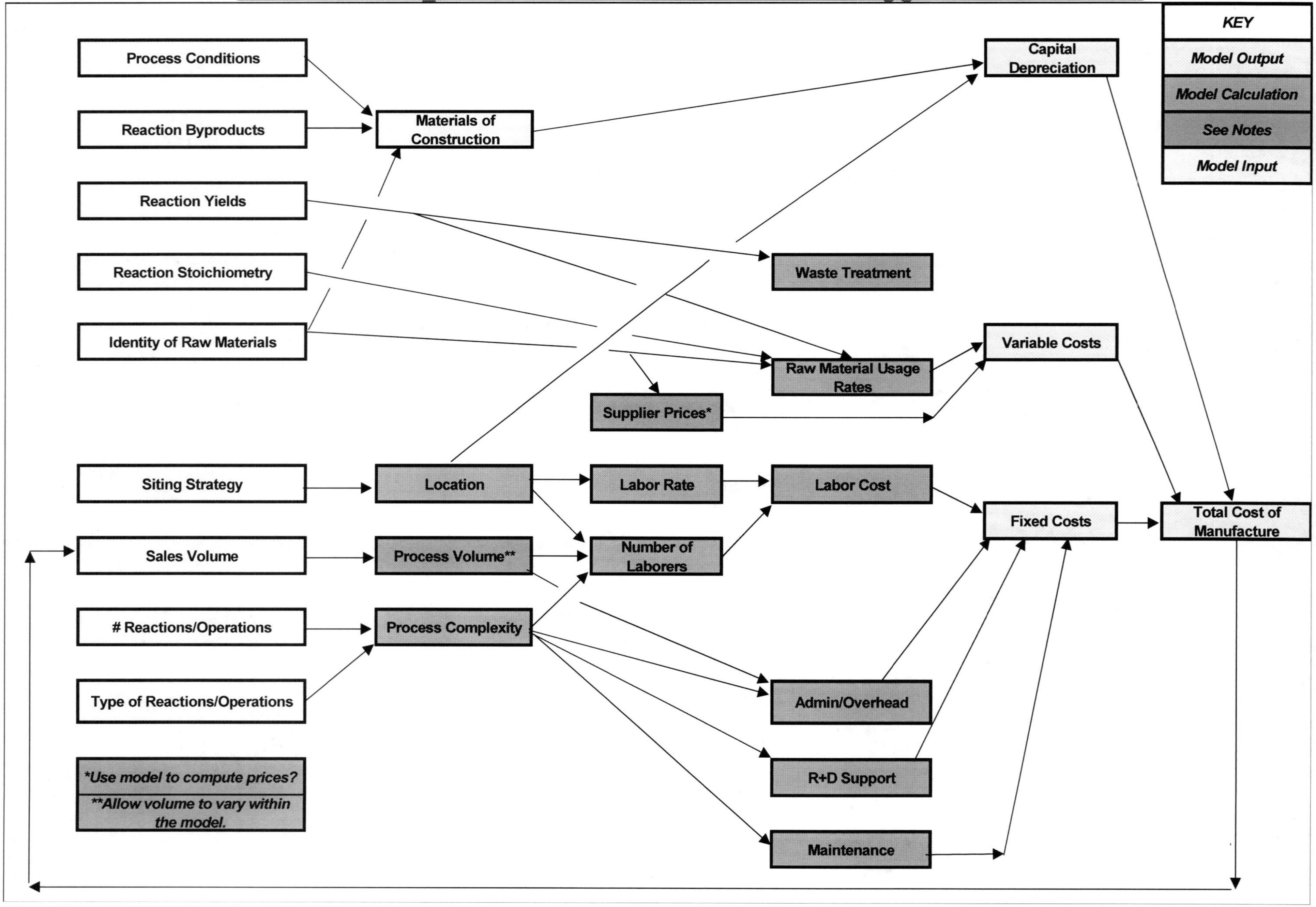

There is one ***special case*** of a Fixed Cost:

Capital

This cost is frequently the most contentious of all costs.

There are several reasons for this…here's an important one:

100% OF THIS MONEY IS SPENT BEFORE ANY PRODUCT IS SOLD, SO THIS COST INFERS *GREATER RISK*.

By necessity, 'quick' methods used for 'early phase' capital estimates:

1. ***Use historical data for processes that bear SOME resemblance to the process being estimated***
2. ***Ignore many of the pertinent variables (e.g. plant location)***
3. ***Are not of sufficient quality for the capital project 'process'***
4. ***Should be combined with other costs to 'dilute' the effect of capital on TOTAL COST***

The most important variable in making a capital estimate is the sheer* number of equipment items required *to operate the process.

Operating conditions are not as important.
Size of equipment is not as important.
Materials of construction are not quite as important*

***Though the cost of high alloys is growing very rapidly.**

ONCE A DECISION IS MADE TO INSTALL AN EQUIPMENT ITEM…ANY ITEM…
THE IMPACT OF THE DETAILS IS SMALL.

OR, IN OTHER WORDS, ONCE YOU DECIDE TO INSTALL IT,
MOST OF YOUR MONEY IS SPENT (OR, AT LEAST COMMITTED).

DEPRECIATION EXAMPLE:
COMMODITY CHEMICAL

ASSUME:

CAPITAL = $163 MM

DEPRECIATE AT A CONSTANT RATE OVER 10 YEARS

VOLUME = 441 MMLBS Product/Year

Cost of Capital Depreciation =
($163 MM / 10 years / 441 MMlbs / year)

= $0.037 / lb Product

Handling Uncertainty

Q: What can we 'know' before knowing everything (or even knowing very much)???

A: We may know what costs are likely to be unimportant / negligible (see below)

Cost Distribution: Fine Chemical

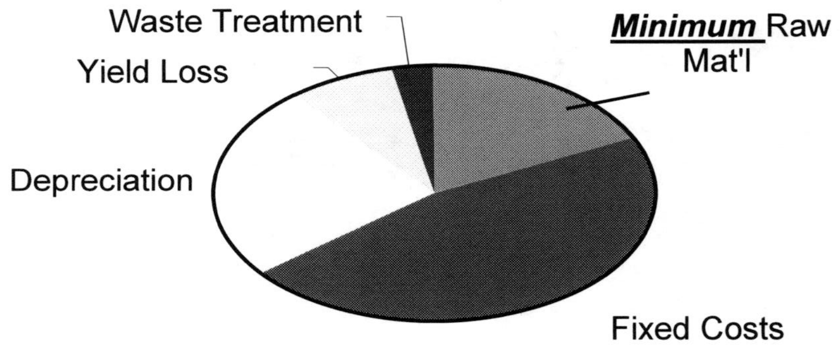

Cost Distribution: Commodity

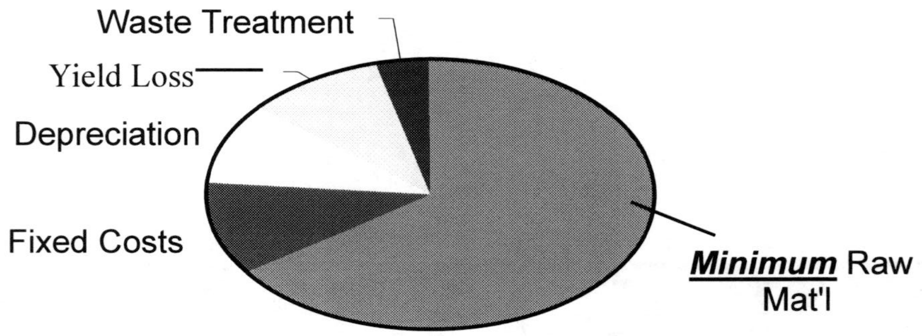

What about the "cost of capital"?

-The cost of capital is different from depreciation.
-Depreciation accounts for the spending of money which would occur even if there was no "time value".
-Depreciation is a REAL cost that MUST be recovered. **It will appear on a plant's cost books.**

-The cost of capital accounts for the time value of money (which, in turn, reflects the value of that money in alternative investments). It is not a real cost, but it can be viewed as a cost. **It will NOT appear on a plant's cost books.**

Sources of data:

Obvious: Textbooks, Patents, Journal articles

Less obvious: Directory of World Chemical Producers, web sites

Non-obvious (?): Company repositories (if you don't have one, build one!!!)

Best of all: Knowledgeable R+D colleagues!!!!

What I hoped to 'prove' is that:

1. **Much useful work on cost can be done with limited data**

2. **Often, more 'data' is available than what one might think**

3. **Project direction can be gainfully altered with even minimal insights about cost**

105e Trace Chemistry and Manufacturing

Jonathan H. Worstell, Higher Olefins and Derivatives, Shell Chemical Company, 3333 Highway 6 South, Houston, TX 77082

Trace chemistry, or the reaction that cannot happen but does, is often the bane of commercial operations. This session presents philosophies and approaches to anticipating and troubleshooting trace chemistry issues in process development and commercialization.

105f Process Modeling and Design in Pharmaceutical R&D

David J. am Ende, Jason Mustakis, and Timothy J. Watson. Chemical R&D, Mail Stop 4016 Bldg 118E room 303L, Pfizer, Inc, Eastern Point Road, Groton, CT 06340

A process model can be defined as a mathematical description of a process, often based on chemical, physical, or engineering principles, used to understand and predict performance within a range of parameters. Typically derived from experimental data, the process model requires a strategic experimental plan for efficient use of resources as well as the mathematical analysis to compute a statistically adequate model. The presentation will provide a perspective of process modeling approaches being employed in development of API. Specifically a Quality by Design (QbD) application using an example from a recent API manufacturing process will be described. The work includes a risk assessment and multivariate design supporting the example of design space. Details of reaction kinetics and Design of Experiments (DoE) will be presented in the context of leveraging the synergy of Chemists and Chemical Engineers in Process Development.

76a Runniness and Randomness of Confined Fluids

Thomas M. Truskett, Department of Chemical Engineering and Institute for Theoretical Chemistry, The University of Texas at Austin, 1 University Station, C0400, Austin, TX 78712

Fluids trapped in small spaces feature prominently in science and technology, and understanding their properties is critical to progress in fields that range from cell biology to the engineering of nanoscale devices. Interestingly, such "confined fluids" behave differently than bulk samples. Properties strongly affected by confinement include (i) how the particles pack together, (ii) how their thermodynamic state changes under heating or squeezing, and (iii) their tendency to rapidly mix or flow. The first two have been reasonably well understood for some time, but a general rule for even qualitatively predicting the third has proven elusive. In this talk, we highlight how recent theory and molecular simulation have discovered links between the entropy ("randomness") and the dynamics ("runniness") of confined fluids. These links provide new physical insights and allow one to forecast how nanoscale confinement will modify the behavior of bulk materials.

76d Disentangling the Stability and Function of Natively Unfolded Proteins

Hank Ashbaugh, Chemical and Biomolecular Engineering, Tulane University, 327 Lindy Boggs Building Tulane University, New Orleans, LA 70118

Natively unfolded proteins are emerging as an important class of biomacromolecules that carry out necessary biological functions despite their lack of well-defined three-dimensional structures, contrary to traditional ideas linking protein shape and function. In this talk, we present a series of computer simulation studies aimed at probing the stability and function of these disordered proteins. Using a coarse-grain model for unstructured polypeptides, we explore the effects of sequence composition (characterized by sequence hydrophobic and charge) and complexity (characterized by sequence non-randomness) on global conformational stability as a coil-globule transition. In addition to being able to probe the physics underlying the experimental charge/hydropathy and complexity correlations, our model provides insights into the joint correlation between orthogonal predictors for disordered sequences. In the final portion of this talk, we discuss ongoing simulations to explore the effects of sequence on the stability of helical peptides, as well as the water-mediated interactions between disordered phenylalaine-glycine repeat sequences that moderate gating of the nuclear pore complex.

76b Investigating the phase behavior of fluids in the vicinity of a surface

Jeffrey R. Errington, Chemical and Biological Engineering, University at Buffalo, 303 Furnas Hall, Buffalo, NY 14260

Fluids in the presence of one or more surfaces exhibit a rich variety of phase transitions that are absent in bulk fluids. Examples include wetting, prewetting, layering, and capillary condensation transitions. Even the simplest of systems display a broad range of phase behavior, with surface phase diagrams depending qualitatively upon the relative strengths and ranges of the fluid-fluid and fluid-substrate interactions and the structural characteristics of the substrate. In this presentation, we describe our recent efforts aimed towards obtaining a better understanding of surface phase behavior through the use of molecular simulation.

76e Adsorption of Microorganisms on Single-Walled Carbon Nanotubes and Other Porous Media

Martha Mitchell, V. K. K. Upadhyayula, and Shuguang Deng. Department of Chemical Engineering, New Mexico State University, Las Cruces, NM 88003

Waterborne outbreaks resulting from contamination of drinking water systems with biothreat pathogens are a cause of major public health concern in many countries today. State of the art technologies, which are primarily designed to treat sewage based contaminations, are not very effective against biothreat agents. Most of the biothreat agents are colorless, tasteless and odorless and are likely to be undetected by existing systems. Carbon nanotube technology has a great potential to make important advancement in water security and protection from biothreat agents. Carbon nanotubes can be used as adsorbent media for effective removal of multiple pathogens from contaminated water, they can also be used as potential biosensors for simultaneous concentration and detection of multiple pathogens in water.

In this research the results of batch adsorption studies of three different microorganisms on single-walled carbon nanotubes, activated carbon and NanoCeram™ are presented and the feasibility of using carbon nanotubes for removal of microorganisms are reported. Batch adsorption experiments are conducted to determine adsorption kinetics and adsorption equilibrium of pure cultures of Bacillus subtilis (B.subtilis), Escherichia coli (E.coli) and Staphylococcus aureus (S.aureus) on SWCNT aggregates. Adsorption of mixed cultures of E.coli and S.aureus on single-walled carbon nanotube aggregates is also tested.

76c Hydrophobic and hydrophilic hydration and interactions relevant to biological systems

Shekhar Garde, Department of Chemical and Biological Engineering, Rensselaer Polytechnic Institute, Ricketts Building, 110 8th St, Troy, NY 12180

Water plays a critical role in mediating many complex self-assembly phenomena in aqueous solutions, including protein folding, micelle and membrane formation, and molecular recognition. Specifically through its structuring around various solution species, water can induce attractive or repulsive interactions between them, depending on their chemistry (hydrophobic vs hydrophilic), shape, and size. Molecular-level understanding of such solvation phenomena requires approaches that treat water explicitly and not simply as a dielectric continuum. I will present theoretical and simulation results on the lengthscale dependence of hydrophobic interactions, and highlight their relevance to condensed matter physics and to biological self-assembly. Knowledge of water-mediated interactions obtained over a broad thermodynamic space of temperature, pressure, and additive concentration can be integrated to understand a variety of biophysical phenomena. I will illustrate this point using an application to pressure denaturation of proteins.

111c Using Thermodynamic Properties from First-Principles Calculations to Accelerate the Search for Metal Hydride-Based Hydrogen Storage Materials

David S. Sholl[1], Sudhakar Alapati[2], Ki-Chul Kim[2], Bing Dai[3], and J. Karl Johnson[3]. (1) School of Chemical and Biomolecular Engineering, Georgia Institute of Technology, Atlanta, GA 30332, (2) Department of Chemical Engineering, Carnegie Mellon University, 5000 Forbes Ave., Pittsburgh, PA 15213, (3) Department of Chemical and Petroleum Engineering, University of Pittsburgh, University of Pittsburgh, 1249 Benedum Hall, Pittsburgh, PA 15261

Metal hydrides provide one of the most promising routes for reversible storage of hydrogen with high capacities, but the number of possible materials that can be considered for this application is very large. We will discuss how a thermodynamics database developed from first-principles calculations has been used to rigorously screen a large collection of metal hydrides and related materials for hydrogen storage. These calculations correctly identify the promising materials that were known from previous experiments, but more importantly, they also identify a number of other promising reactions for which no prior experimental data was available. These results suggest multiple targets for experimental studies. At the same time, they point to several important challenges, including the reliable prediction of crystal structures from limited experimental information and the important role of reaction kinetics in practical applications of metal hydrides for hydrogen storage.

111a Development and Application of Advanced Molecular Simulation Methods for Computing the Properties of Ionic Liquids

Edward J. Maginn, Department of Chemical and Biomolecular Engineering, University of Notre Dame, 182 Fitzpatrick Hall, Notre Dame, IN 46556

Ionic liquids are a relatively new class of materials that have generated strong interest from academic and industrial researchers over the past few years. They are defined as pure salts that are liquid near ambient conditions. They can have many interesting properties such as negligible vapor pressure, excellent solvation power for polar, aromatic and polymeric materials, high thermal stability and an extremely wide liquidus range.

In this talk, I will describe several new simulation techniques that we have developed which enable us to compute properties of ionic liquids that, up to now, were beyond the capabilities of standard simulation methods. These include melting points, crystal polymorph stability, thermal conductivity, viscosity, and gas solubility. I will then show how we are using this information in collaboration with experimental groups to design new ionic liquids with properties tuned for a variety of energy and environmental applications.

111d Using Molecular Modeling to Understand Enzymatic Cellulose Hydrolysis

Clare McCabe, Department of Chemical Engineering, Vanderbilt University, Nashville, TN 37235-1604, Xiongce Zhao, Center for Nanophase Materials Sciences, Oak Ridge National Laboratory, Oak Ridge, TN 37831, and Michael E. Himmel, National Renewable Energy Laboratory, 1617 Cole Blvd, Golden, CO 80401.

Cellobiohydrolase I (CBH 1) from Trichoderma reesei is one of the most active enzymes in the hydrolyzation of cellulose. CBH 1 is a multi-domain enzyme, consisting of a large catalytic module containing an active site tunnel and a small cellulose binding module joined to one another by a 27 residue linker peptide. As an exoglucanase CBH 1 hydrolyzes cellulose in a "processive" manner; the exact mechanism of is not fully understood. As part of an ongoing effort to better understand the hydrolysis process, molecular dynamics simulations that focus on determining the function of the linker peptide have been performed. Although the sequence of the linker peptide is known and it has been shown that the linker is important for enzymatic activity, the role of the linker is yet to be determined. Results from molecular dynamics simulations in which the motion and conformation of the linker adsorbed on a cellulose surface in an aqueous environment will be presented and our simulation data compared with the hypothesis that the linker behaves like a spring, which helps the enzyme to move along a cellulose chain. Since this spring-like motion is a key element in the processive depolymerization of cellulose by Cel7A, our efforts to understand this mechanism, and thus in the future to engineer more capable cellobiohydrolases, is a key element in making the conversion of biomass to biofuels more efficient.

111b Perspectives and Challenges in Adsorption Science and Nanoporous Materials

Krista S. Walton, Department of Chemical Engineering, Kansas State University, 1005 Durland Hall, Manhattan, KS 66506

Adsorption science is pervasive, and advances in this field have contributed to a wide variety of applications that have a direct impact on society. This is especially true regarding the development of novel porous materials, which are crucial for applications such as catalysis, gas storage, chemical and biological sensing, and separations. In recent years, advances in nanotechnology have led to a more targeted molecular design approach for materials development. The emergence of metal-organic frameworks in the past decade is an excellent example of this. Notably, the high crystallinity of these materials is conducive to computational studies to complement experimental investigations of their adsorption properties. Such well-characterized systems can aid the design of better materials for specific adsorption applications. This talk will describe recent advances in adsorption science and nanoporous materials. Theoretical and experimental perspectives on adsorption will be given, and future challenges and opportunities in this field will be discussed.

LNG pool fire models: similarities and differences

Filippo Gavelli, Melissa K. Chernovsky, and Harri K. Kytomaa
Exponent Inc., Bowie, MD USA

Abstract

The analytical models currently used to predict thermal heat flux hazards from an LNG pool fire are based on the solid flame theory: the fire is approximated as a tilted cylinder, whose dimensions depend on the pool fire size, the wind speed and other parameters. This paper compares some of the models currently used for the hazard analysis of LNG pool fires, to describe the effect of model parameters on the hazard distances to selected heat flux levels, particularly for large pool fires (over 100 m diameter). The results indicate that the flame height correlation, the emissive power and the smoke shielding have significant impact on the predicted hazard distances for large pool fires. The impact of the atmospheric transmissivity model, instead, is found to be minor.

Keywords: pool fire; LNG; hazard distance; smoke shielding; flame height

Introduction and Motivation

One of the risks associated with the transport, storage or regasification of LNG is that of a large spill, on land or water, leading to a pool fire that may impact nearby public and property. Heat flux hazards from an LNG pool fire are currently calculated using models based on the "solid flame" theory [1]: the pool fire is modeled as a cylinder, whose surface emits thermal radiation at a specified, constant rate. This model determines the dimensions of the solid flame cylinder (base and height) according to the diameter of the burning liquid pool and other parameters. The cylinder is also assumed to be tilted in the direction of the wind, and the angle of tilt is dependent on the wind speed.

The scope of this paper is to compare some of the solid flame-based models currently used for the thermal heat flux hazard analysis of LNG pool fires, to describe the effect of model parameters on the results (i.e., hazard distances associated with specified heat flux levels), particularly for large pool fires (over 100 m diameter). The model comparison also provides some guidance to those planning to perform large-scale LNG pool fire tests, by identifying the parameters that have the most impact on the results. Experimental efforts should aim to quantify those parameters reduce the uncertainty in heat flux predictions.

LNG Pool Fire Models

The analytical models currently used to predict thermal heat flux hazards from an LNG pool fire are based on the solid flame theory: the fire is approximated as a tilted cylinder, whose dimensions depend on the pool diameter, the wind speed and other parameters. The solid flame theory then assumes heat flux to radiate uniformly from the surface of the cylinder. The thermal heat flux from the solid flame cylinder to a target is determined using the following factors:

- The emissive power of the fire surface;
- The transmissivity of the atmosphere to thermal radiation; and
- The position (distance, direction and orientation) of the target relative to the fire (also known as the "view factor").

The thermal heat flux from a solid flame cylinder to a target at distance *x* is given by:

$$\dot{Q}''(x) = E_f \cdot \tau(x) \cdot VF(x,a) \tag{1}$$

where $\dot{Q}''(x)$ is the thermal heat flux per unit area to a target at distance *x* from the center of the pool fire; E_f is the surface emissivity of the fire; $\tau(x)$ is the atmospheric transmissivity to thermal radiation over a path length *x*; VF is the view factor; and a is the pool fire flame height. The thermal heat flux calculated according to equation (1) decreases with the distance of the target as shown in Figure 1. For the purpose of hazard analysis, the target orientation is taken into account by combining the view factors to a vertical and to a horizontal target.

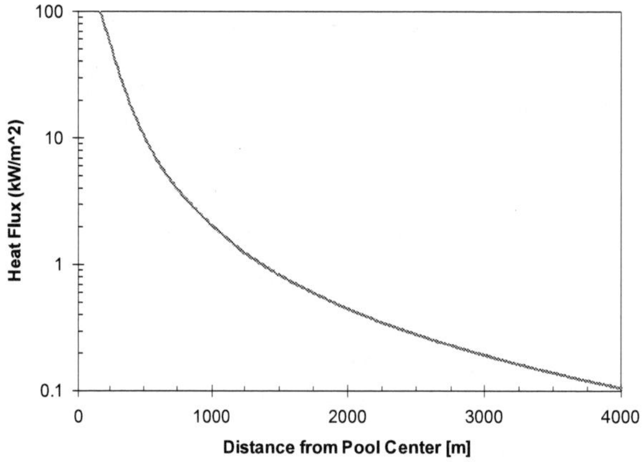

Figure 1. Thermal heat flux from an LNG pool fire, as a function of distance from the center of the pool. Values calculated for 100 m diameter pool fire.

Emissive Power

The heat flux radiated by the solid flame cylinder is defined by the flame surface emissive power. Typical values for the emissive power from LNG pool fire range between approximately 175 and 325 kW/m^2, depending on the model used. The original solid flame theory assumes the emissive power to be uniform over the surface of the cylinder. In some cases, however, the solid flame cylinder is divided into two or more zones to account for the shielding effect of smoke from the upper portions of the fire column. The following types of solid flame models can be found in the literature and are compared graphically in Figure 2:

1. Single-zone models [1], in which the entire flame height is assumed to be unobscured by smoke (i.e., "clear flame");
2. Two-zone models [2], in which the lower part of the cylinder is assumed to be a clear flame, while the upper part is assumed to be partially obscured by smoke;
3. Three-zone models [3], in which the base of the flame cylinder is assumed to be unobscured, the middle portion of the flame is anchored to the base and intermittently

obscured by smoke, and the upper region of the flame is separated from the base into stretching flame puffs which are substantially or totally obscured by smoke.

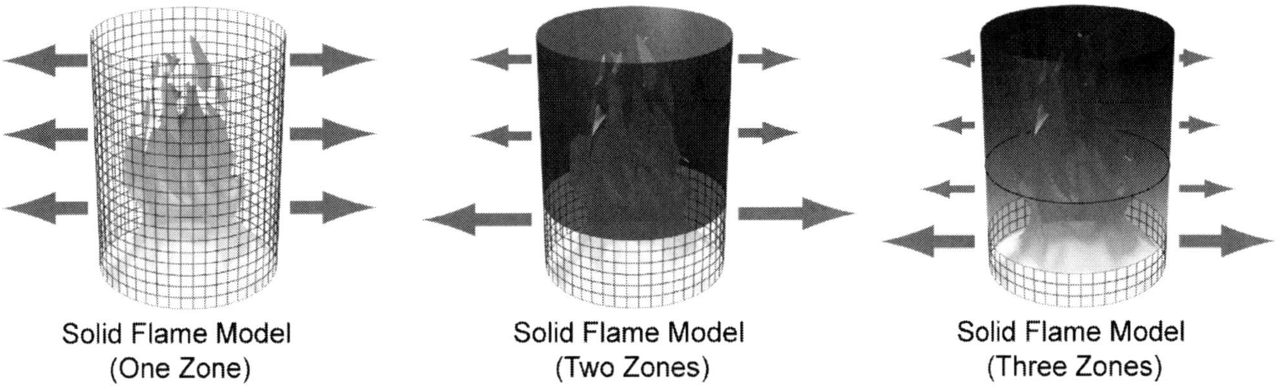

Figure 2. Graphic representation of three types of solid flame models. The red arrows represent the effective surface emissive power.

Atmospheric Transmissivity

The thermal radiation emitted by a pool fire is partially scattered and/or absorbed by the atmosphere. Scattering is due mostly to the presence of particulate in the atmosphere. Absorption is due mostly to the presence of water vapor in the air [3] and therefore, is dependent on the atmospheric relative humidity as well as on the distance between the pool fire and the target. The solid flame theory accounts for thermal radiation scattering and absorption through the atmospheric transmission coefficient. A frequently used correlation for atmospheric transmissivity in pool fire models is the one used by the Federal Energy Regulatory Commission (FERC) [2]:

$$\tau(RH,x) = C_1\left(x \cdot P_v\left(RH,T_{amb}\right)\right)^{-C_2} \qquad (2)$$

where RH is the relative humidity; T_{amb} is the ambient temperature; x is the path length; P_v is the water vapor pressure in air; and C_1 and C_2 are constants.

Other models (e.g., the one used by Sandia [4]) assume a constant transmissivity, independent of relative humidity or distance. The typical transmissivity value used in these models is 0.8, although sensitivity studies are performed with values as low as 0.5. As Figure 3 shows, the constant transmissivity model tends to underestimate the heat flux to a target close to the pool fire (less than approximately 250 m) and to overestimate the heat flux to a target farther away from the fire.

View Factor

The view factor is a complex expression, which takes into account the dimensions of the solid flame cylinder (base, height and tilt), the position of the target (distance, direction and orientation), and other factors such as smoke shielding.

The size of a pool fire is strongly dependent on the size of the burning liquid pool. With a spill into an impoundment, the liquid pool is bound by the walls of the impoundment. In the case of an LNG spill over water, instead, the liquid pool is unconfined and therefore, the diameter of the pool can grow over time, with the growth being dependent on several factors such as the LNG spill rate and the pool burn rate. Once the size of the liquid pool is determined, the view factor depends largely on the flame height. Different authors have developed correlations to predict the flame height based on the liquid pool size and other physical parameters [5-9]. The typical expression for a flame height correlation is:

$$\frac{a}{D} = C_1 \left(\frac{\dot{m}_b''}{\rho_a \sqrt{g \cdot D}} \right)^{C_2} (u^*)^{C_3} \tag{3}$$

where a is the pool fire flame height; D is the liquid pool diameter; $\dot{m}_b''$ is the mass burning rate per unit pool area; ρ_a is the ambient air density; g is gravity; and u* is the non-dimensional wind velocity.

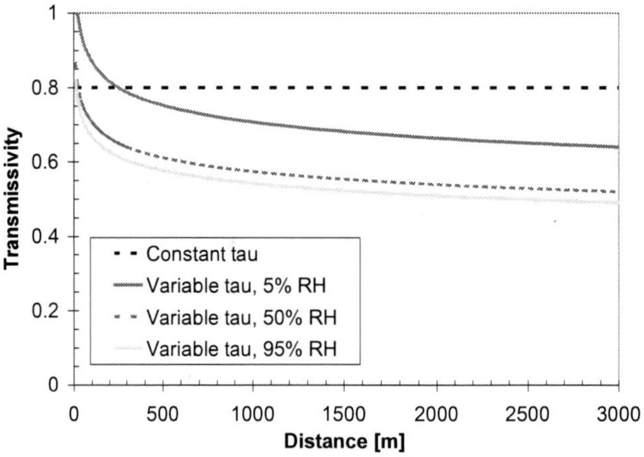

Figure 3. Comparison of atmospheric transmissivity vs. distance as used in different LNG pool fire models. Standard atmospheric conditions are assumed for the comparison.

This paper compares the following flame height correlations for LNG pool fires:

- Heskestad [5], which is based on many fire scenarios including burners using various fuels 5-50 cm in diameter, gasoline pool fires from 0.3 to 23 m in diameter, square pools of JP-4 fuel from 1 to 10 m, and under-ventilated wood cribs;
- Moorhouse [6], which is based on rectangular pool fires of LNG, as large as 13.7 m [7];
- Pritchard and Binding [8], which is based on hydrocarbon (primarily LNG) pool fires, 6-22 m in diameter;
- Thomas [9], which is based on wood crib fires 10 – 200 cm on a side; and
- Zukoski [7], which is based on natural gas burners 10-50 cm in diameter.

The flame height correlations are compared in Figure 4 (for pool diameters up to 70 m) and Figure 5 (for pool diameters up to 700 m). A relatively large spread in flame heights (approximately a factor of 2.2) is obtained for both small and large pool fires. However, the Thomas and Moorhouse correlations – arguably the more commonly used correlations for LNG pool fires – predict very similar flame heights for pool sizes up to approximately 20 m diameter. The same two correlations, however, diverge in their predictions for larger pool sizes: for a 100 m diameter pool fire, the Moorhouse correlation predicts a flame height approximately 35%

taller than the Thomas correlation; the ratio grows to approximately 50% for a 200 m pool diameter, and to approximately 70% for a 500 m pool diameter.

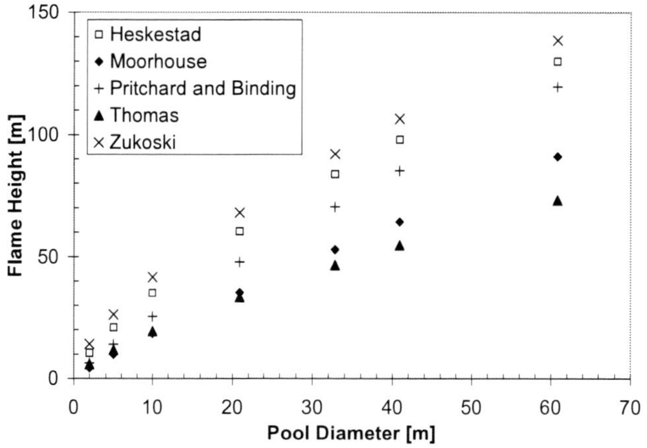

Figure 4. Comparison of flame height correlations by: Heskestad, Moorhouse, Pritchard and Binding, Thomas, and Zukoski. Pool sizes up to 70 m diameter.

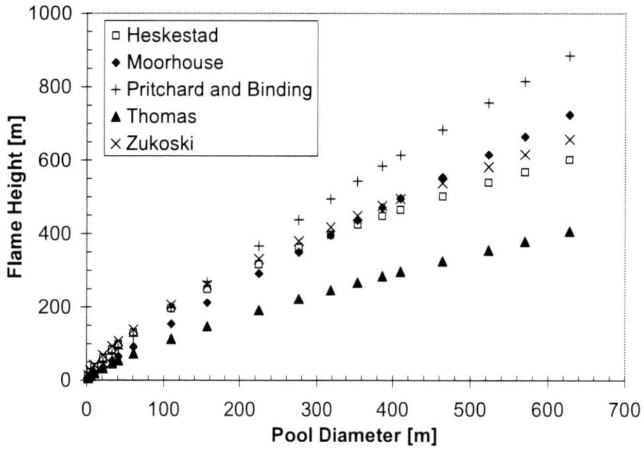

Figure 5. Comparison of flame height correlations by: Heskestad, Moorhouse, Pritchard and Binding, Thomas, and Zukoski. Pool sizes up to 700 m diameter.

The geometry of the pool fire cylinder is also affected by the wind speed, which causes the flame to tilt in the direction of the wind. Also, if the liquid pool is unconstrained, drag from the wind may affect the shape of the liquid pool, elongating it in the direction of the wind and turning a circular pool base into an ellipse. The most widely used flame tilt model is the one developed by the American Gas Association [10], while Moorhouse [6] developed a correlation for flame drag based on LNG pool fire tests. Flame drag is not considered in this paper.

Once the shape and size of the solid flame cylinder are known, the view factor can be calculated, as a function of the position and orientation of a target. Due to the flame tilt caused by the wind speed, the worst-case scenario in terms of heat flux to a target occurs when the target is directly downwind of the fire. The view factor from a tilted cylinder to a target located directly downwind, as currently used in pool fire models, was first calculated by Mudan [11]. As discussed earlier, two- and three-zone solid flame models account for the shielding effect of smoke by calculating the view factor separately for each zone:

$$VF(x,a) = \sum_{i=1}^{N} s_i \cdot VF(x,a_i) \tag{4}$$

where N is the number of zones in the solid flame model; s_i is the fraction of emissive power emerging from the smoke for zone *i* (s_1 = 1); and a_i is the height of flame zone *i*.

LNG pool fire hazard analysis requires the distance to selected heat flux levels to be determined. The specific requirements vary with jurisdiction, both in terms of heat flux levels and in terms of approach to the analysis. For the purpose of this paper, the 5 kW/m^2 heat flux level is used, which is defined as the permissible energy level for personnel performing emergency operations lasting several minutes with appropriate clothing [12], to calculate the hazard distances predicted by the various models.

Other Parameters

Additional physical parameters associated with the composition of the LNG and with ambient conditions (e.g., temperature, humidity, wind speed) play a role in the solid flame model. For the scope of this study, these parameters are maintained constant across all models, as summarized in Table 1.

Table 1. Parameters for solid flame model comparison.

Physical Parameter	Value
LNG Composition	100% Methane
Pool Burn Rate	$3*10^{-4}$ m/s
Ambient Temperature	20 °C
Relative Humidity†	50%
Wind Speed	8 m/s at 10 m elevation
Emissive Power†	220 kW/m^2
Transmissivity†	0.8

† When not the parameter being varied.

Pool Fire Model Comparison

The comparison between solid flame-based pool fire models is performed by varying individual components, to determine their individual impact on the hazard distances. The predicted hazard distances are plotted versus the burning liquid pool diameter, to visualize how results vary with the pool size, and they are measured with respect to the center of the pool fire.

The first parameter addressed is the flame height. Figure 5 shows that, for any given pool size, the predicted flame height can vary by a factor of approximately 2.2, depending on the correlation being used. Figure 6 shows the effect of flame height on the predicted hazard distance to 5 kW/m^2, using a single zone solid flame model with constant transmissivity (equal to 0.8) and a surface emissive power of 220 kW/m^2. The hazard distance trends are consistent with the differences in the flame height correlations, as expected: a taller flame corresponds to a longer hazard distance. The ratio between the longest hazard distance (Pritchard and Binding) and the shortest (Thomas) is approximately 1.4 – that is, the hazard distance does not increase proportionally with the flame height. However, in absolute terms, the predicted hazard distance can vary by as many as 700 m for a 500 m diameter pool radius. Therefore, the choice of flame height correlation can have a significant impact on the consequence analysis for large-scale LNG spills.

Another variable of interest in pool fire models is the atmospheric transmissivity. As discussed earlier, atmospheric transmissivity is a factor (less than or equal to one) that accounts for the reduction in thermal radiation to a target, due to scattering and absorption by the atmosphere. Figure 7 shows the effect of transmissivity on predicted hazard distances to the reference heat flux levels, using from a single zone solid flame model with the Thomas' flame height correlation and a surface emissive power of 220 kW/m^2. The constant transmissivity model is compared with the variable transmissivity model, with the latter applied to different relative humidity values (5%, 50% and 95%). The results show a small difference between all models for pool diameters up to approximately 150 m. The difference increases as the distance traveled by the thermal radiation increases, allowing for scattering and absorption to become more significant. The constant transmissivity model becomes more and more conservative as the hazard distance grows, consistent with the trend shown in Figure 3; the constant transmissivity predicts hazard distance approximately 4% to 15% greater than the variable transmissivity model, for RH = 50% and 95% respectively. Based on these results, the atmospheric transmissivity model is not a critical factor in the hazard distance calculations.

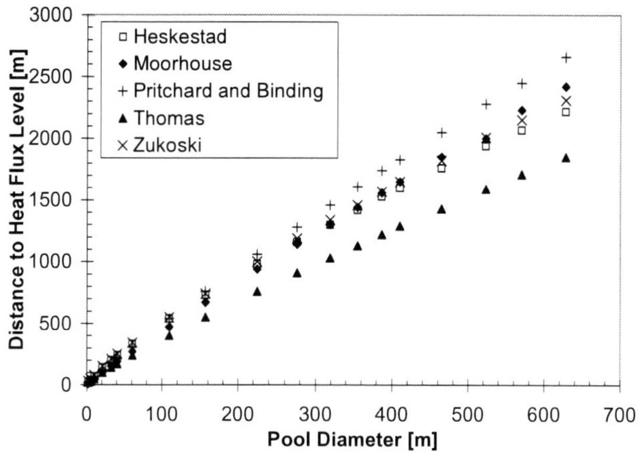

Figure 6. Single zone solid flame model predictions of hazard distance to 5 kW/m^2 heat flux level vs. pool diameter for five different flame height correlations: Heskestad, Moorhouse, Pritchard and Binding, Thomas, and Zukoski.

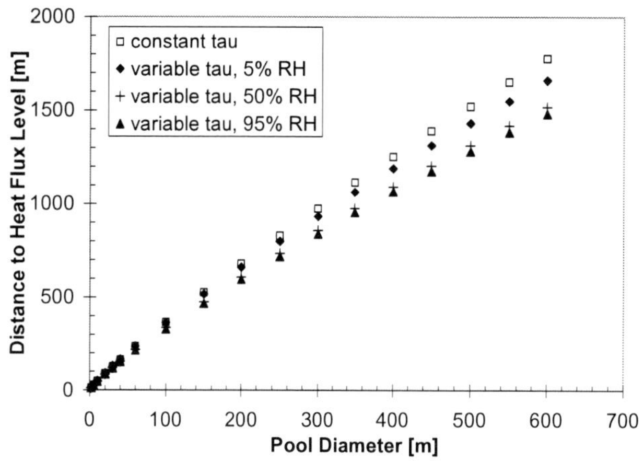

Figure 7. Single zone solid flame model predictions of hazard distance to 5 kW/m^2 heat flux level vs. pool diameter for constant transmissivity and variable transmissivity (with 5%, 50%, and 95% relative humidity).

The amount of thermal radiation emitted by a pool fire is, like the flame height, a quantity not yet well known. This is particularly true for larger LNG pool fires, where the impact of oxygen starvation on the overall fire behavior is still unclear. Consequently, different pool fire models have used emissive power values ranging between approximately 175 and 325 kW/m^2. The effect of emissive power on the predicted hazard distances is shown in Figure 8: the hazard distances to 5 kW/m^2 are plotted, as a function of pool diameter, for a single zone solid flame with the Thomas flame height correlation and constant transmissivity (0.8).

The ratio between the hazard distance predicted using the highest emissive power (325 kW/m^2) and the lowest (175 kW/m^2) is approximately 1.25 – that is, the hazard distances do not grow proportionally to the emissive power. However, in absolute terms, the predicted hazard distance can vary by more than 300 m for a 500 m diameter pool radius. Therefore, the choice of emissive power from the pool fire can have a significant impact on the consequence analysis for large-scale LNG spills.

The final parameter considered in this study is the smoke shielding of the pool fire. Solid flame models with one, two and three zones are compared to determine the effect of the smoke shielding assumption on the hazard distances, as a function of pool fire diameter (see Figure 9. For consistency, all three models use the Thomas correlation for flame height and constant transmissivity. The surface emissive power of the flame was set to 220 kW/m^2 for the one-zone model, 265 kW/m^2 for the two-zone model, and 325 kW/m^2 for the three-zone model.

The three models are in agreement for small pool fires, as expected since LNG pool fires are essentially smoke-free for pool diameters up to approximately 15 m. The effect of smoke shielding becomes progressively more significant as the pool size grows and the fires are predicted to become more sooty. For a 500 m diameter pool, for example, the difference between the longest hazard distance (one-zone model) and the shortest (three-zone model) is approximately 500 m. Therefore, the effect of smoke shielding of the pool fire thermal radiation can have a significant impact on the consequence analysis for large-scale LNG spills.

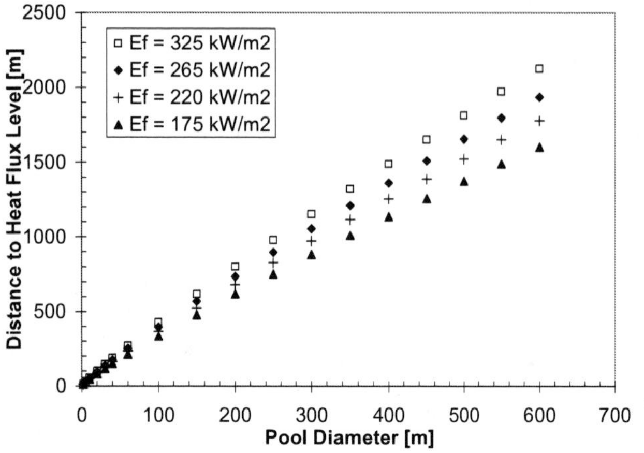

Figure 8. Single zone solid flame model predictions of hazard distance to 5 kW/m^2 heat flux level vs. pool diameter for emissive power values of 175, 220, 265 and 325 kW/m^2.

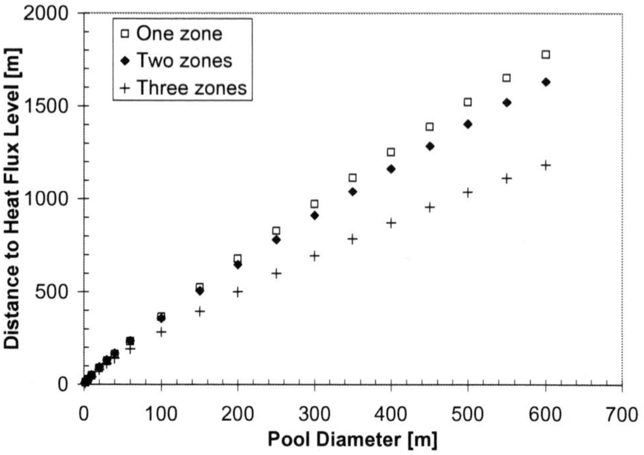

Figure 9. Solid flame model predictions of distance to 5 kW/m^2 heat flux level, as a function of pool diameter for four models: single zone, two zones, three zones, and LNGFIRE3.

Conclusions

The hazard distances from LNG pool fires are currently calculated using solid flame based models. Several models have been developed and used for LNG pool fires, particularly for the hazard analysis of large-scale LNG spills on water. These models, although based on the same general theory, differ in terms of the physical values of certain parameters (e.g., the flame surface emissive power) or the correlations used to define the geometry of the solid flame cylinder (e.g., the flame height).

This paper performed a sensitivity study on the pool fire model parameters, by evaluating the effect on the predicted hazard distances of varying each parameter individually. The parameters considered were: flame height correlation; atmospheric transmissivity; flame surface emissive power; and smoke shielding.

The results indicated that the flame height correlation, the emissive power and the smoke shielding have significant impact on the predicted hazard distances, particularly as the pool size grows beyond approximately 100 m diameter. The impact of the atmospheric transmissivity model, instead, was found to be minor.

The findings of this analysis confirm the need for experimental data on large-scale LNG pool fires. The focus of these experiments should be on characterizing the pool fire flame height and the average surface emissive power from different portions of the fire column, so that smoke shielding effects may also be quantified.

References

1. Beyler, C.L., “Chapter 3-11: Fire Hazard Calculations for Large Open Hydrocarbon Fires,” in SFPE Handbook of Fire Protection Engineering, 3rd Edition, edited by P. DiNenno, National Fire Protection Association, Quincy, MA (2002).
2. FERC, “Notice of availability of detailed computations for the consequence assessment methods for incidents involving releases from liquefied natural gas carriers”, Docket No. A04-6-000, June 29, 2004.
3. Raj, P.K., “Large hydrocarbon fuel pool fires: Physical characteristics and thermal emission variations with height,” Journal of Hazardous Materials, 140 (2007) pp. 280-292.
4. Sandia National Laboratory, “Guidance on risk analysis and safety implications of a large liquefied natural gas (LNG) spill over water”, SAND2004-6258 (2005).
5. Heskestad, G., “Luminous Heights of Turbulent Diffusion Flames,” Fire Safety J. 5 (1983) pp. 103-108.
6. Moorhouse, J., “Scaling criteria for pool fires derived from large-scale experiments,” J. I. Chem. Sym. 71 (1982) pp. 165-179.
7. McCaffrey, B., “Chapter 2-1: Flame Height,” in SFPE Handbook of Fire Protection Engineering, 2nd Edition, edited by P. DiNenno, National Fire Protection Association, Quincy, MA (1995).
8. Pritchard, M.J., and Binding, T.M., “FIRE2: A New Approach for Predicting Thermal Radiation Levels from Hydrocarbon Pool Fires.” I. Chem. E. Symposium Series, 130 (1992) pp. 491-505.
9. Thomas, P.H., “The size of flames from natural fires,” Proc. Combust. Inst. 9 (1963) pp. 844-849.
10. American Gas Association, “LNG Safety Research Program,” Report IS 3-1 (1974).
11. Mudan, K.S., “Geometric View Factors for Thermal Radiation Hazard Assessment,” Fire Safety J. 12 (1987) pp. 89-96.
12. ABS Consulting, “Consequence Assessment Methods for Incidents Involving Releases from Liquefied Natural Gas Carriers,” for the Federal Energy Regulatory Commission, GEMS 1288209 (2004) pp. 29-35.

Comparison of LNG Computational Fluid Dynamics Consequence Modeling with Brayton Field Fire School Test Data – Preliminary Assessment

Benjamin Cormier, Jaffee Suardin, Morshed Rana, Yingchun Zhang, Sam Mannan*
Mary Kay O'Connor Process Safety Center
Artie McFerrin Department of Chemical Engineering Department
Texas A&M University System
College Station, Texas 77843-3122, USA
*(979) 862-3985; mannan@tamu.edu

AIChE Spring Meeting, April 2007
8th Topical Conference on Natural Gas Utilization
New Orleans, Louisiana, April 6-10, 2008

Abstract

Complex scenarios increase the use of Computational Fluids Dynamics (CFD) modeling. Either explosion or gas dispersion, CFD consequence modeling has become an inevitable tool to predict effect of a release. Passive barriers such as dike, containment pit, and vapor barrier have been recommended to be used to control LNG release on land and its associated LNG spreading, vapor dispersion, and pool fire. Because of limited LNG field experimental data, the effectiveness of these barriers are not well understood. The combination of CFD modeling and field test could help the industry to better understand the effectiveness of passive barriers or other mean of hazard mitigation measures as well as to properly calibrate the CFD model for more realistic consequence prediction resulting from an LNG release.

BP Global Gas SPU and Mary O'Connor Process Safety Center (MKOPSC) have joined together in a program to study LNG release hazards and identify potential mitigation systems. This paper reports some of the preliminary results from the LNG field experiment at the Texas A&M University's Brayton fire school (TEEX) along with the vapor dispersion modeling using a CFD model (CFX-11 from Ansys).

Introduction

The paper presents the result of LNG field tests and its associated CFD modeling. LNG field experiment results on 22 ft by 33 ft and 4ft deep containment pit is described in Figure 1.

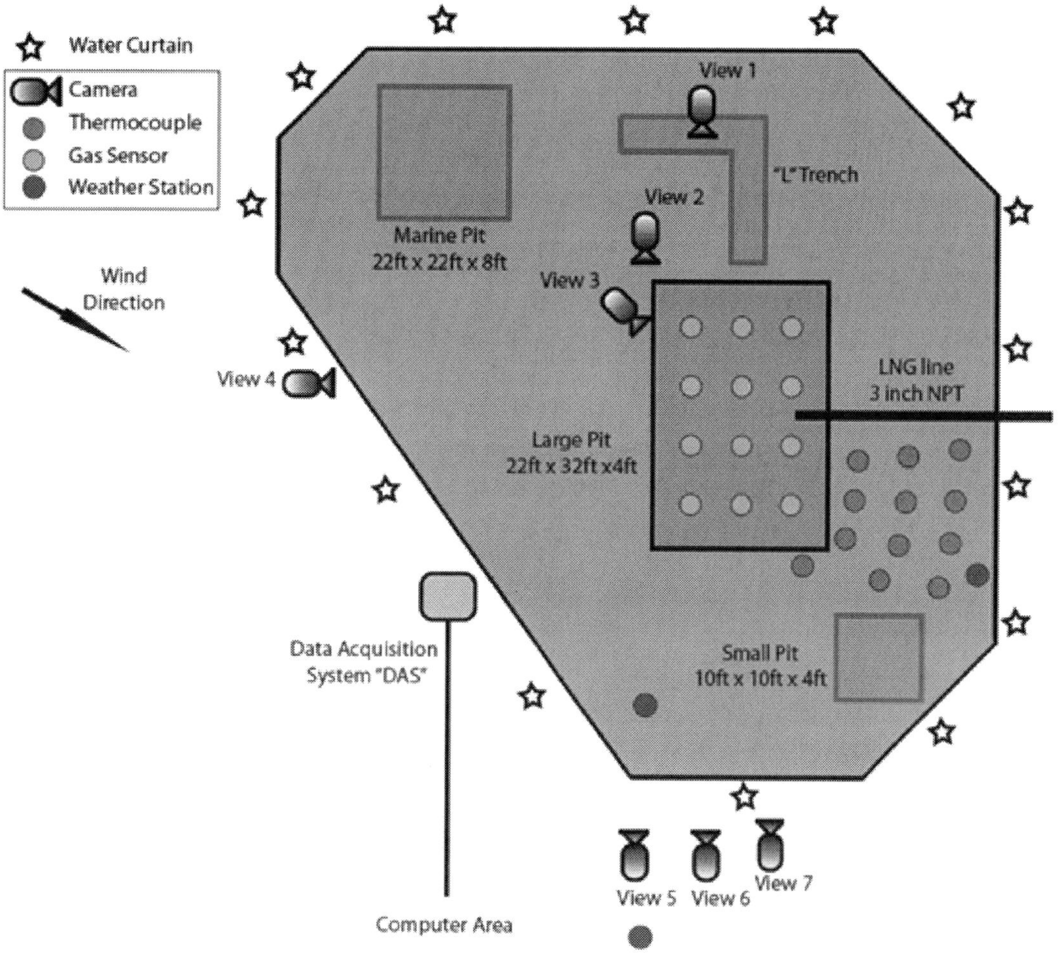

Figure 1. Facility siting at Brayton field fire school LNG project

An extra four feet wall was added as an obstacle to the vapor cloud around the large pit. Two scenarios were developed in this pit including LNG spill on water and spill on concrete inside the pit. Gas concentration and temperature were recorded and compared with CFD simulation. CFD simulation can incorporate the feature of the site, the extra wall and the depth of the pit. Both scenarios have important information that are incorporated into the modeling of vapor dispersion. The details of the two scenarios are presented in Table 1.

Table 1. Spill Scenario summary

Test Name	Type	Spill rate, gpm	Spill Duration, sec
TEEX1	On water	220	600
TEEX7	On Concrete	90	4,000

TEEX1

In this scenario, LNG is spilled on water as described in Figure 2. Prior to the spill, the pit was filled with four feet of water covering three out of the five thermocouples on each column and leaving two thermocouples for measuring the vapor cloud temperature. A wooden wall was added to create a barrier for the vapor of LNG. Additional thermocouples were added (not shown here) horizontally near the landing area in order to record the size of the LNG pool. A ramp to reduce the momentum created by the spill was added to the site. This scenario is repeated twice, TEEX1 and TEEX2. Only TEEX1 is presented in this paper. An array of gas detectors was placed at the last minute in the downwind direction. Forty gas detectors were used.

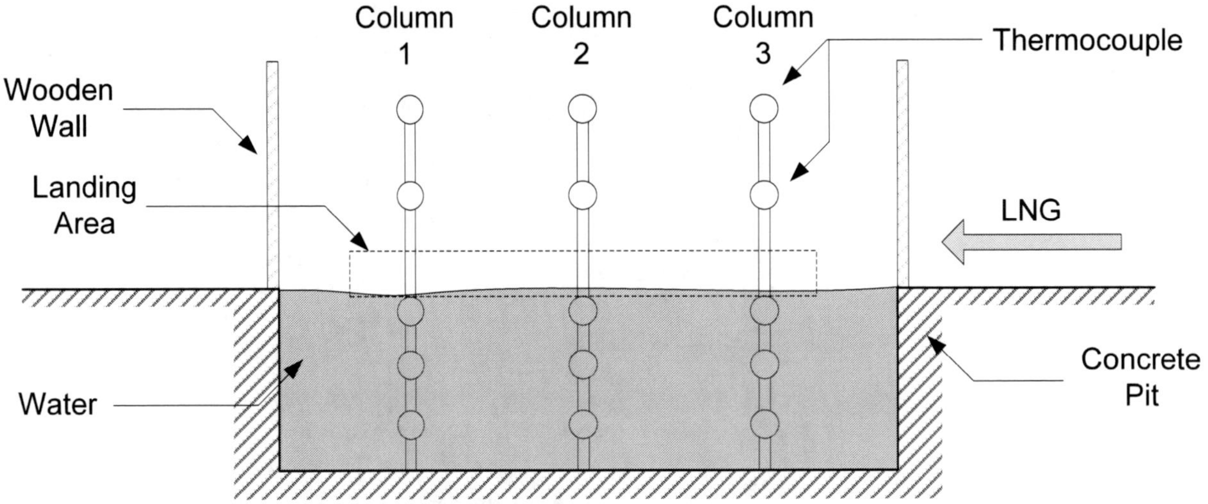

Figure 2. Scenario scheme for spill on water in the large pit, TEEX1

The purpose of this test is to capture the evaporation rate, the turbulence created by the release. After reviewing the data collected, the vaporization from LNG spill onto water seems to have more turbulence than vaporization from LNG spill on concrete. This valuable information will advance the understanding of LNG vapor dispersion modeling.

TEEX7

This test was designed to study the foam application onto LNG pool. The large pit is used to collect data on the foam temperature. This test consists of four steps which Figure 3 shows only the first step. An extra four feet wall was also added to the large pit. The empty pit was filled with six inches of LNG. During this period (usually about one hour with the current LNG delivery system), data were recorded to study the vapor dispersion with a barrier. The depth of the pit and the extra wall represent a barrier to the vapor. Once the LNG reached the desired level, foam was applied onto the boiling LNG. through two foam generators. During the period when a vapor cloud was present, an array of gas detectors and thermocouples was placed downwind.

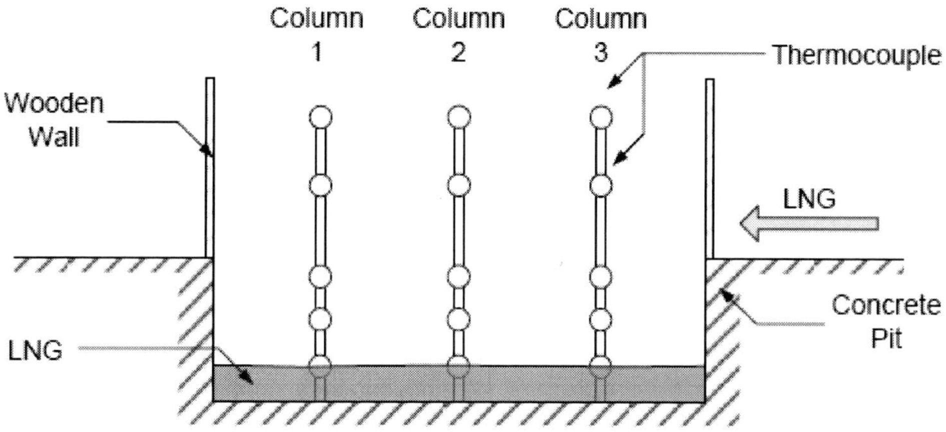

Figure 3. Scenario scheme for spill on concrete in the large pit, TEEX7

Simulation setup

CFX-11 is composed of preprocessor, solver and post-processor for problem definition, problem solution and results processing[1]:

1. CFX-Build, in which the computational domain geometric and mesh are constructed.
2. CFX-Physical processor, where boundary conditions, turbulence models, domain fluids and solution scheme are defined.
3. CFX-Solver, where the code performs calculations towards the final solution.
4. CFX-Post, in which the elaboration of results takes place, with graphical capabilities.

The first two modules are used to define the problem (preprocessor), the third module is used in getting a solution (solver) while the last module analyzes the results with graphical presentation (post-processor). For the vapor model dispersion, CFX-11 uses the Reynolds averaged Navier-Stokes equations and it is based on the finite volume method for the conversion of partial differential equations and auxiliary conditions into a discrete system of equations.

Domain and mesh

The volume of interest is usually called the domain. For tests performed at TEEX, the domain is determined by the volume where the gas sensors and thermocouples were placed. The size of the grid is 25 meters by 20 meters and 7 meters above the ground. A larger grid could be used later on to estimate the methane concentration far downwind, if any. Most of the data collected is situated within this volume. The edge length spacing for mesh is set from 0.05 to 1.04 m (0.17 to 3.4 feet). Minimum edge length value is used for the details such as wall, foam generator, and LNG source. Maximum edge length value is used for far downwind. These two edge length values were used to create the mesh in the domain. The mesh could be refined by changing the minimum or the maximum edge length but this can exponentially increase in calculation time. A sensitivity analysis should be used to estimate the effect of the change in the mesh configuration. The domain size and boundary conditions are illustrated in Figure 4.

Boundary conditions

The Boundary conditions define the input and output of the domain. Figure 4 shows the different boundary conditions used and the geometry of the domain for TEEX1. Each boundary is described in the following sections.

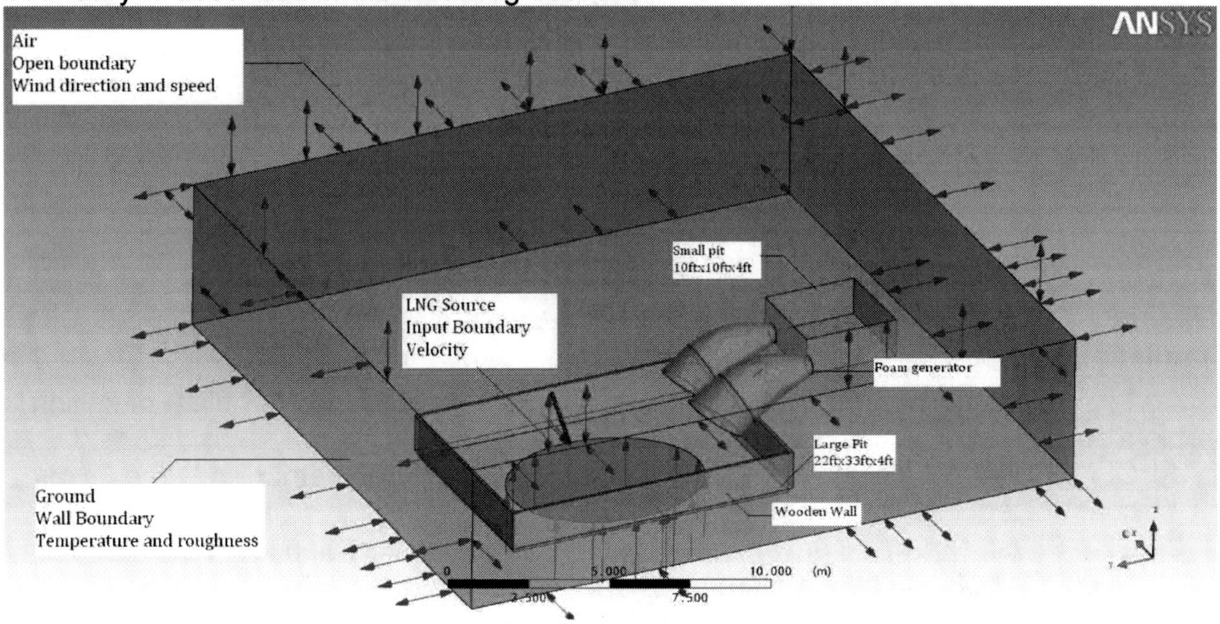

Figure 4. Illustration of the domain with boundary conditions

Air: TEEX1 wind profile

Air boundary condition is defined by the wind speed profile, wind direction, and temperature profile. The wind speed and temperature change as they gain altitude. Each test shows an initial period with constant wind direction, shown in Table 2. The definition of these variables in the simulation is important to represent the turbulence and the heat transfer mechanism involved with vapor dispersion modeling.

Table 2. Wind direction and speed summary

Test Name	Wind Direction	Wind Speed, $m.s^{-1}$	Period, s
TEEX1	South, 180	Between 1 and 3	600
TEEX7	South, 180	From 1.5 to 0.6	3,500

LNG source: LNG line and pool formation

LNG vapors are generated from two sources, LNG pool and LNG flashed out of discharge hose. The thermocouples inside the pit at the water interface give the temperature profile for the LNG pool area. Figure 5 shows the temperature profile during TEEX1 when the pool reached its steady state.

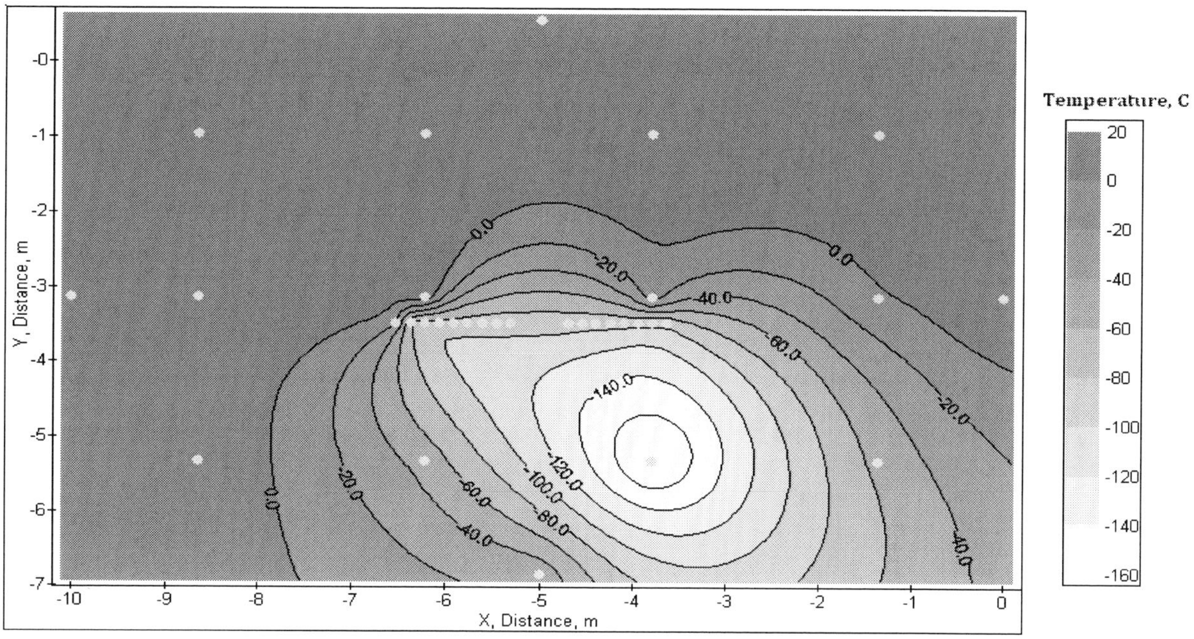

Figure 5. Temperature profile for the interface between LNG and water

The surface of the LNG pool is estimated from the temperature profile and it is 27.6 ± 0.4 m^2 with the flow rate of 6.4 $kg.s^{-1}$ (220 gpm). The location of the pool was against the upwind wall and underneath of the foam generators. The LNG pool did not cover the total available area of 67.5 m^2.

The velocity of the fluid at the discharge hose was estimated at 4.5 $m.s^{-1}$ for LNG and at 12 $m.s^{-1}$ for flashed LNG. Flashed LNG is assumed to be 10% of the total liquid travelling inside the pipeline. This assumption is following the previous work on the LNG delivery system[2]. The different velocities created from the release are the cause of the turbulence inside the wall. Even though a ramp is installed to ease the vertical momentum, the horizontal momentum is still present and creates turbulences above the LNG pool. The location of the pool is explained by the fluid velocity.

This location and the shape are then used in the setup of the simulation. The formation of a LNG pool onto water is added to the domain by using an input boundary that simulated the evaporating pool. From the data analysis of the water, it is possible to establish an evaporation rate distribution over the surface area. Since the data analysis is still in process, an average value for evaporation rate of 0.2 $kg.s^{-1}$ or velocity of 0.0135 $m.s^{-1}$ is used for this paper. The fluid velocity is also added to the simulation through the discharge hose. The details of the LNG source are shown in Figure 6.

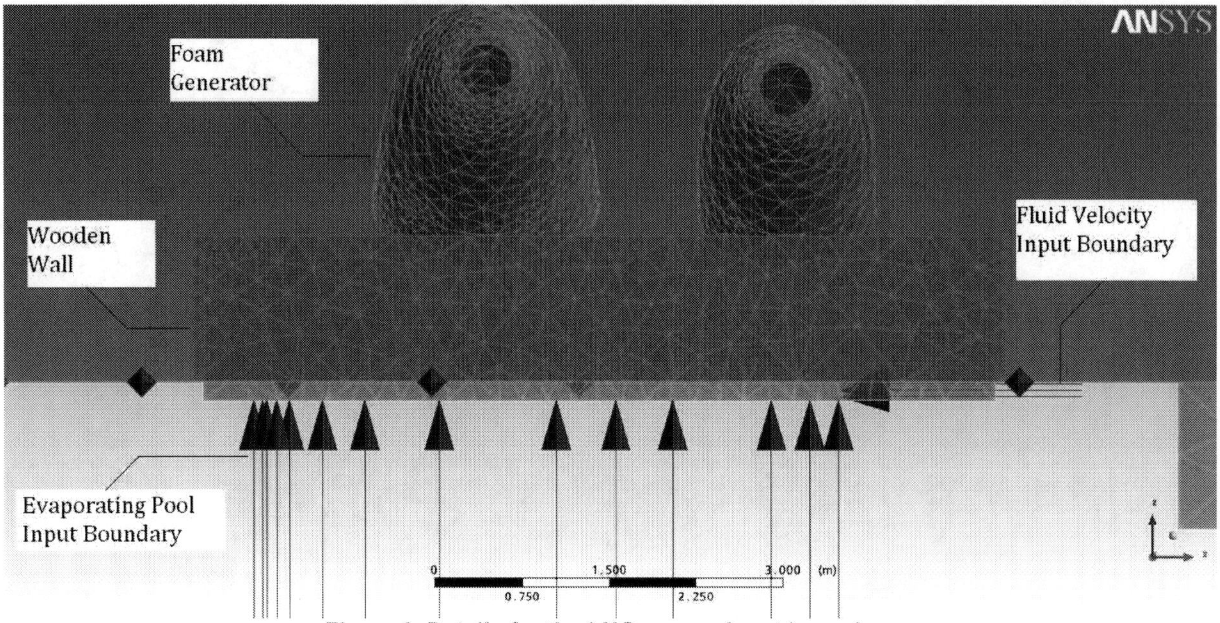

Figure 6. Details for the LNG source input boundary

Ground

The ground around the spill area, the small pit and the inside wall of the large pit are made of concrete. In the setup, these features are set to a wall boundary. A wall boundary does not permit exchange with the domain but features such as roughness and temperature can be added. Therefore a roughness and a temperature are added to account for their effects on the cloud to simulate the heat transfer from the ground. In Figure 4, the ground is illustrated by the color green. The foam generator and the wooden wall have the same boundary conditions.

Obstacles

The interest of using CFD modeling is to incorporate the obstacles in the domain so their effects can be studied. These tests provide a new set of LNG vapor dispersion data with obstacles. The level of details is important in the development of the domain. For instance in Figure 4, the foam generators represent obstacles for the traveling vapor. The wooden wall creates a barrier between wind and evaporating LNG. The purpose of the design for the obstacles is to capture their features with respect to the wind and vapor cloud.

Transient simulation

Complex phenomena have a tendency to include several important variables. Usually they are not constant. In order to correctly account for these variables, a transient simulation should be used. In this case, wind speed, direction, evaporation rate and flash from release are some of the initial variables that need to be incorporated in the modeling. The same methodology is applied for TEEX7.

Results

The results are separated into two sections, inside the pit profile and outside the pit. Only preliminary results are presented here, the data analysis of the tests and the CFD modeling are still in progress. Concentration profile inside the pit for TEEX1 based on temperature measurement and concentration profile in the downwind field based on gas sensor measurement for TEEX7 are shown below.

During the tests, thermocouple sensors are placed inside of the pit, a total of 92 thermocouples are distributed in between the bottom of the pit and the top of the wooden wall. For TEEX1, the pit is filled with almost four feet of water. The thermocouple above the water level recorded the temperature of the LNG cloud. Based on the temperature sensor, the concentration can be interpreted with an energy balance. Figure 7 illustrates the concentration profile obtained during TEEX1 and TEEX7. The wind is blowing from upwind to downwind according to Figure 7.

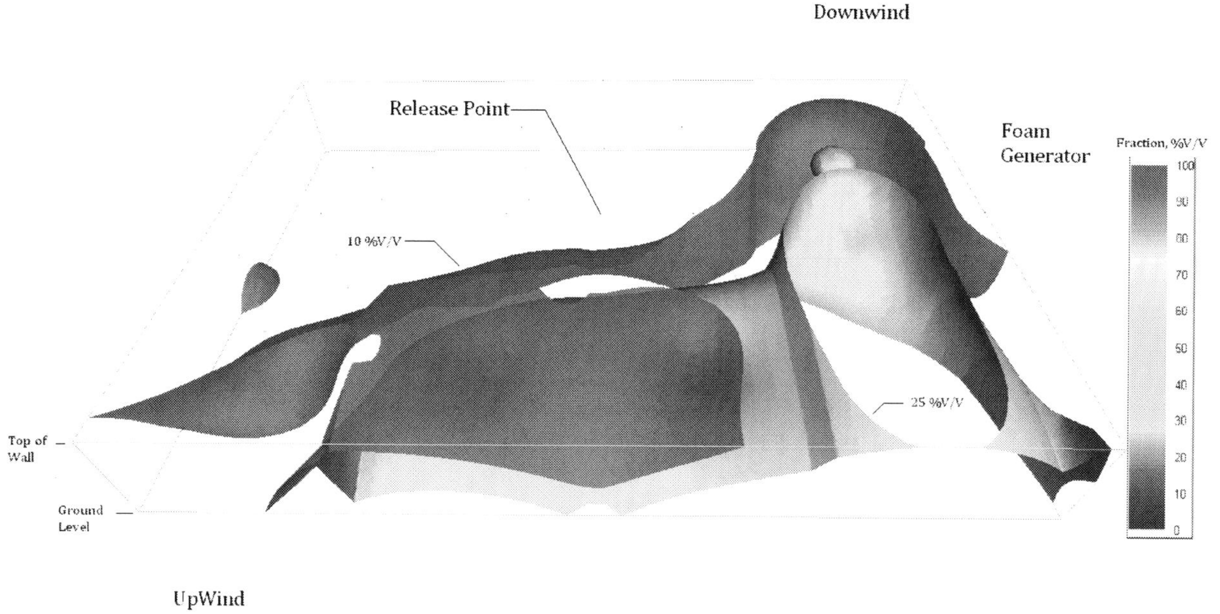

Figure 7. Concentration isosurfaces based on temperature measurement, TEEX1

The shape of the cloud is driven by the turbulences created from the boiling LNG on water and the fluid velocity. This information is useful to compare with CFD simulation results shown in Figure 8. The same concentration isosurfaces are selected for comparison.

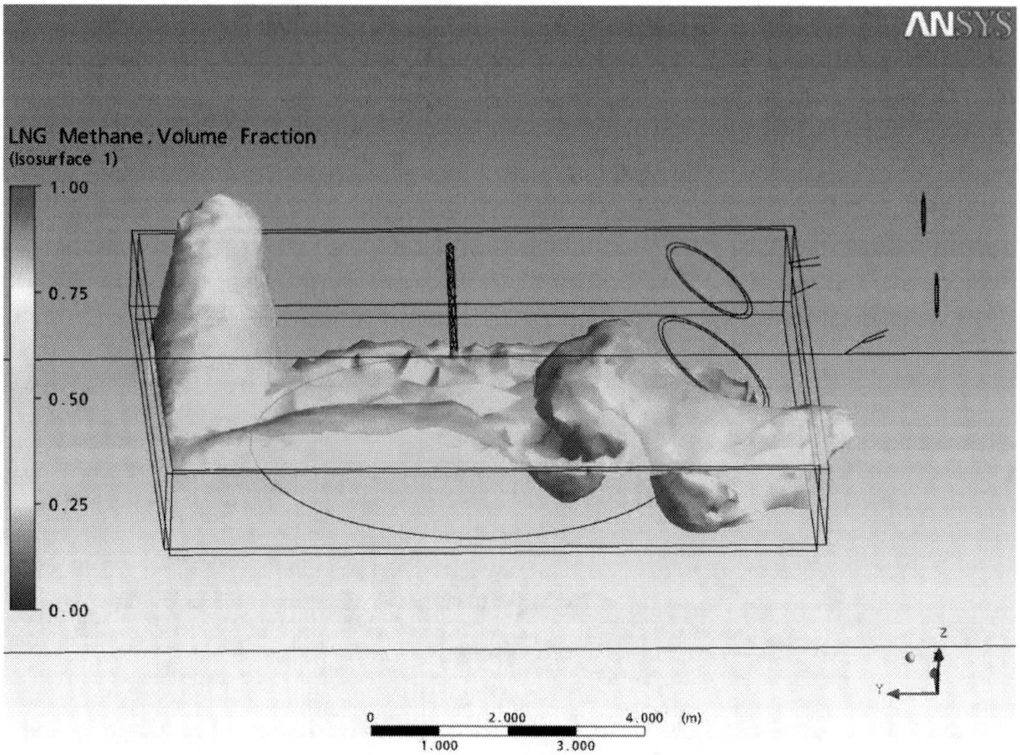

Figure 8. Concentration isosurfaces results from CFD simulation of TEEX1

The results from the simulation show the cloud accumulating on the upwind wall, then it entrains by the wind to the downwind side above the pit and the wooden wall. The shape of the cloud is generally in agreement with the experimental data. Concentration profiles downwind also compare reasonably well.

TEEX7 has a different configuration due to long spill time. It required about one hour filling the pit at 90 gpm. Only six inches of LNG were in the pit at the end of the spill due LNG lost from evaporation and flashed LNG. The evaporation is a function of the heat transfer from the concrete and surrounding to LNG. Therefore, the evaporation profile experienced a maximum then decreased as LNG accumulated in the pit. This test started relatively in the late afternoon with slow wind speed. As illustrated in Figure 9, the wind conditions were from 1 $m.s^{-1}$ to 0.6 $m.s^{-1}$ at constant wind direction. Three sample lines were located at 1.3 meters, 2.3 meters, and 3.3 meters above the ground at the edge of the wooden wall in the centerline for the vapor cloud. This profile reflects the evaporation rate in the pit. Despite a peak at before 1000 seconds, the maximum concentration at this location reaches 40 %V/V. This profile during TEEX7 will be compared with simulation results.

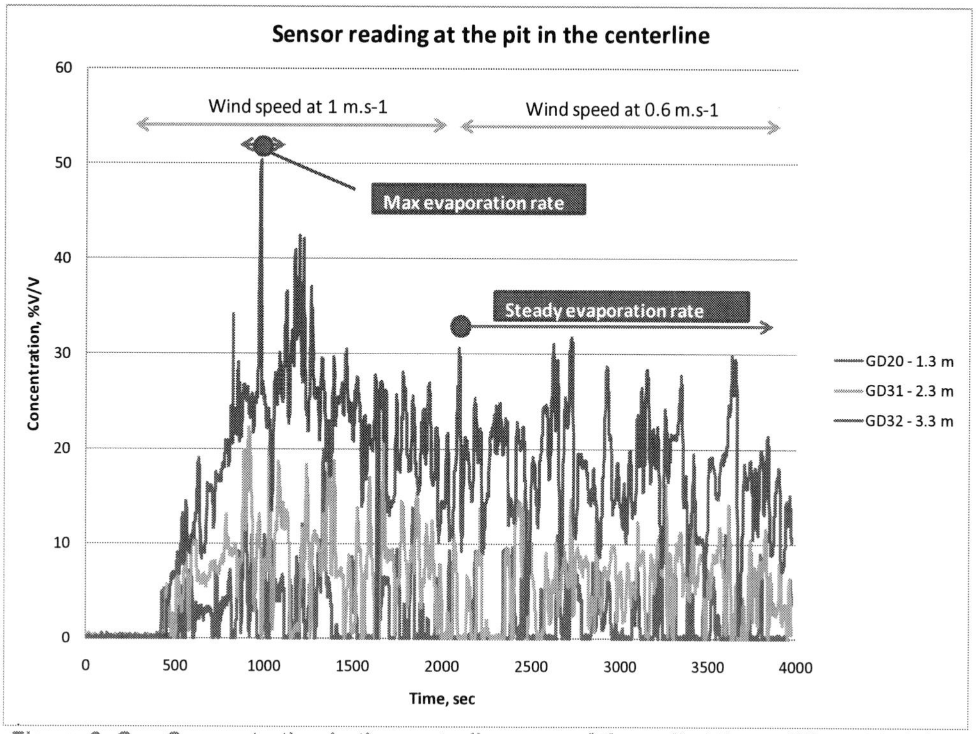

Figure 9. Gas Concentration in the centerline near of the wall at three different heights

Conclusions

The current data analysis is already bringing useful information to advance the current status on CFD modeling. The purpose of this paper is to bring a new insight on using CFD modeling for LNG vapor dispersion modeling. The collected data shows some features of the spill that are not currently modeled in consequence modeling. The information is useful to close some of the gaps associated with LNG vapor dispersion modeling.

The spill on water shows the position of the pool and its vapor cloud against upwind wall. This is mainly due to the fluid velocity created by the release. The spill on concrete has different features compared to the spill on water. In the case of the spill on concrete, the evaporation rate is constant over the surface area but it varies over time following a profile due to the concrete cooling down. In the other hand for spill on water, the evaporation rate is not constant over the surface area due to turbulence from surrounding but it stays on average constant over time because of the infinite heat sink. The concentration profile above the evaporating pool shows interesting features due to the turbulence from the surrounding. For spill on water, the fluid velocity influences the concentration profile above the water. Less turbulence is noticed during the long spill in concrete compared to spill on water.

The current CFD modeling used in consequence modeling has tendency to over simplify the source term. These preliminary results already show the need to improve the source term. Once the features of the source term are captured in the modeling, it should improve the accuracy of the concentration downwind. The current research at MKOPSC in CFD modeling will continue to investigate this by incorporating the fluid velocity into the source term modeling to simulate the turbulence from the release. Most of this work is still in process of analysis.

Acknowledgement

The authors would like to acknowledge BP Global Gas SPU and Brayton Field fire school at Texas A&M - BP Global Gas SPU for sponsoring this research project and Brayton Field Fire School for providing the site and help with the testing.

Reference

1. *ANSYS CFX-11 Manual* 2006.
2. Cormier, B., *Effect of Tank Pressure and Insulation on LNG Liquid Flow Rate - Experience at Brayton Fire School.* Mary Kay O'Connor Process Safety Center Centerlline, 2007. **11**(2).

Biography of the Speaker

Benjamin Cormier

Research Assistant
Mary Kay O'Connor Process Safety Center, Texas A&M University
Office: 979-862-1310
Email: Benjamin.Cormier@chemail.tamu.edu
Web: psc.tamu.edu

Benjamin Cormier is a PhD candidate at the Artie McFerrin Department of Chemical Engineering, Texas A&M University. He attended Virginia Tech where he obtained his Bachelor of Science degree in Chemical Engineering in May 2004. He came to Texas A&M University for a Master of Science degree in Chemical Engineering and completed his Masters in December 2005. Since summer 2005, he has been involved with Mary Kay O'Connor Process Safety Center under the direction of Dr. Sam Mannan.
Benjamin Cormier has been involved with LNG since the beginning of January 2006 and he is sponsored by BP technology. His research involves LNG tests which are usually carried out at the Brayton fire school at Texas A&M University. He has been focusing on the computational fluid dynamics on source modeling for LNG evaporation.
In addition to conducting research in process safety area, He has been involved in several industrial projects including leading HAZOP team, developing P&ID and PFDs, and performing consequence modeling. He interned with DNV inc and Shell International Exploration and Production (SIEP) of Shell Oil Company on summer 2006 and 2007, respectively.

The Risk Assessment of Bayesian-LOPA Methodology for an LNG Importation Terminal

Geun-Woong YUN
Graduate Student/Research Assistant
The Department of Chemical Engineering of Texas A&M University
Mary Kay O'Connor Process Safety Center
College Station, Texas

AIChE Spring Meeting, April 2008
8th Topical Conference on Natural Gas Utilization
New Orleans, Louisiana, April 6-10, 2008

Introduction

The Liquefied Natural Gas (LNG) is one of the fastest growing energy sources in U.S. to diversify the energy portfolio and fulfill the increasing energy demands. In order to meet the increasing LNG demand, many LNG facilities including LNG importation terminals are operating currently. Moreover, many projects for LNG terminals have been proposed to fill the gap between supply and demand of LNG in North America. However, LNG is a highly flammable gas which may result in fire or explosion when it is discharged into undesired conditions. Especially in LNG importation terminals, unexpectedly large consequences may occur due to the huge amount of LNG storage.

Therefore, as the number of LNG facilities is increasing, it is becoming more important to control and estimate the potential risks in LNG terminals to ensure their safety and reliability. For LNG terminals, some risk assessment methods such as LOPA (Layer of Protection Analysis) or QRA (Quantitative Risk Analysis) may be used to evaluate the potential risks quantitatively. Through these methods, the latent risk of new proposed or existing terminals may be reduced in terms of applying some suggested recommendations.

Why need Bayesian-LOPA Methods

One of the most cost effective ways among risk assessment methods to estimate the risk is LOPA because it can produce quantified risk results with less efforts and time than other methods, especially QRA. Thus, LOPA was used for the risk assessment of LNG terminals. For LOPA application, equipment (or facility) failure data are required to compute the occurrence frequency of potential risks. However, the failure data from LNG industry are very scarce and have statistically weak grounds due to relatively short-term operational history and insufficient population of sample data. Bayesian estimation, named after Thomas Bayes who found the theorem, is identified as a good method to be used to compensate for the weaknesses found in the LNG industry's failure data. It can update the generic data gathered and accumulated for a long time from historical experiences in similar industries with plant specific data from the LNG industry or the specific plant. It does mean that the data updated by Bayesian logic can reflect both long-term based historical experiences and plant specific conditions. In other words, the weakness of failure data from LNG industry can be overcome by updating the data with Bayesian estimation. Therefore, for an LNG terminal, the Bayesian-LOPA methodology, which was newly developed in this paper, was used to conduct risk assessments.

Methodology Development

In order to apply the LOPA method, it is necessary to get the frequencies of initiating events and probability of failure on demand (PFD) of each IPL (Independent Protection Layers). As shown in Figure 1, the Bayesian engine to update failure data will be needed in getting both initiating event frequencies and PFDs of IPLs.

Figure 1 The Flow Diagram of the Bayesian-LOPA Method

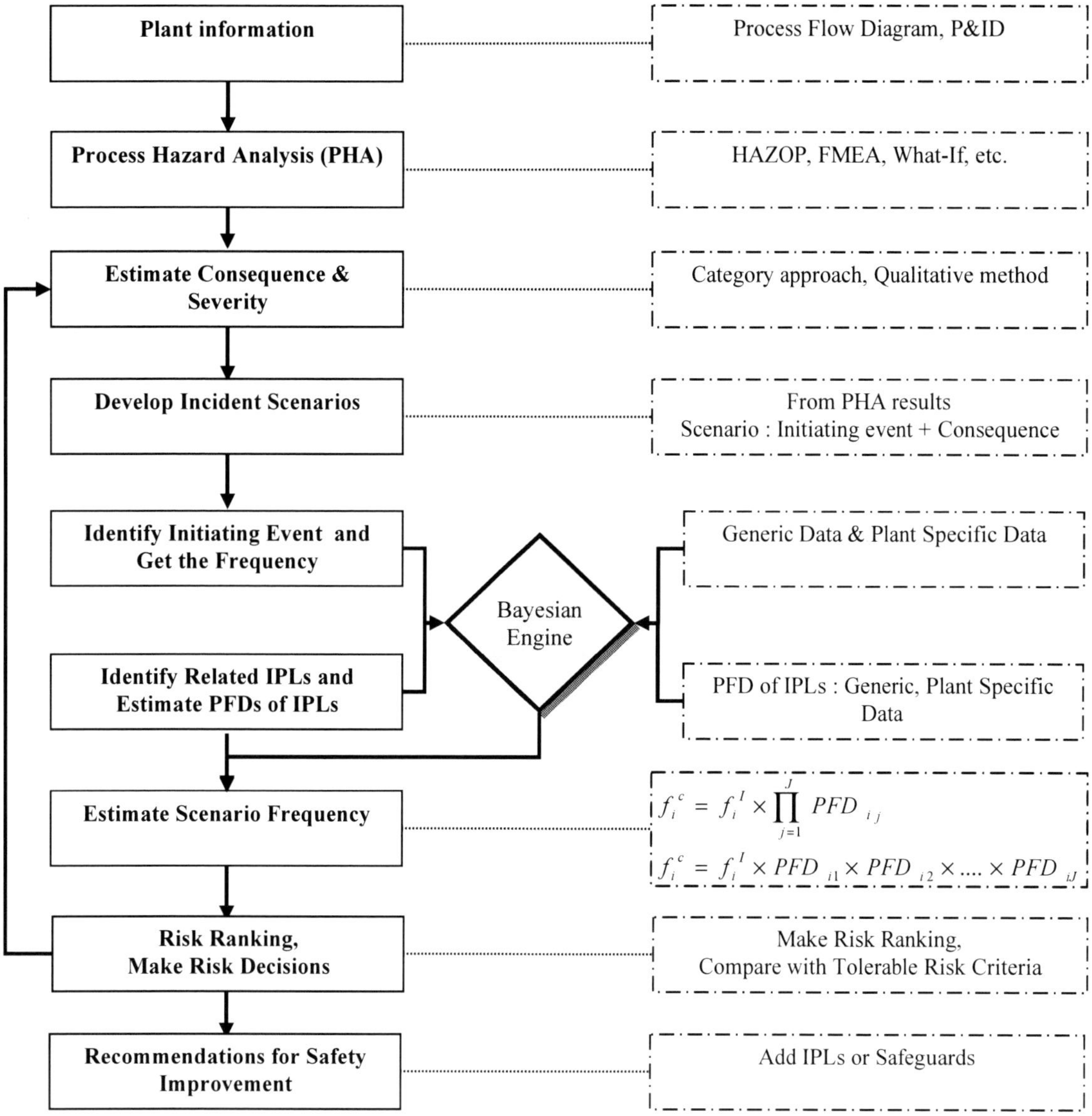

Based on Bayesian estimation, the frequency of an initiating event can be obtained using a conjugate gamma distribution as the prior information and Poisson distribution as the likelihood function as shown in Figure 2. OREDA (Offshore Reliability Data) were used as a prior distribution since they were produced from a gamma distribution. If there is no prior information, Jeffreys noninformative prior may be used. The LNG plant failure database made by J. R. Welker was used as plant specific likelihood information. The PFDs of IPLs can be estimated with the conjugate beta prior distribution and binomial likelihood distribution as shown in Figure 3. EIReDA (European Industry Reliability Data Bank) was used as prior information since it was made from a beta distribution. When EIReDA does not provide failure data, the newly developed Frequency-PFD conversion method may be used instead. The Bayesian-LOPA methodology was developed by the combination of Bayesian estimation and LOPA procedures as shown in Figure 1. The newly developed method was applied to an LNG importation terminal. After applying the method, the found risk values were compared to the tolerable risk criteria given by CCPS (Center for the Chemical Process Safety) to make risk decisions and compared to each other to get a risk ranking in view of probabilistic risk analysis.

Figure 2 The Diagram of Bayesian Estimation for Initiating Event Frequency

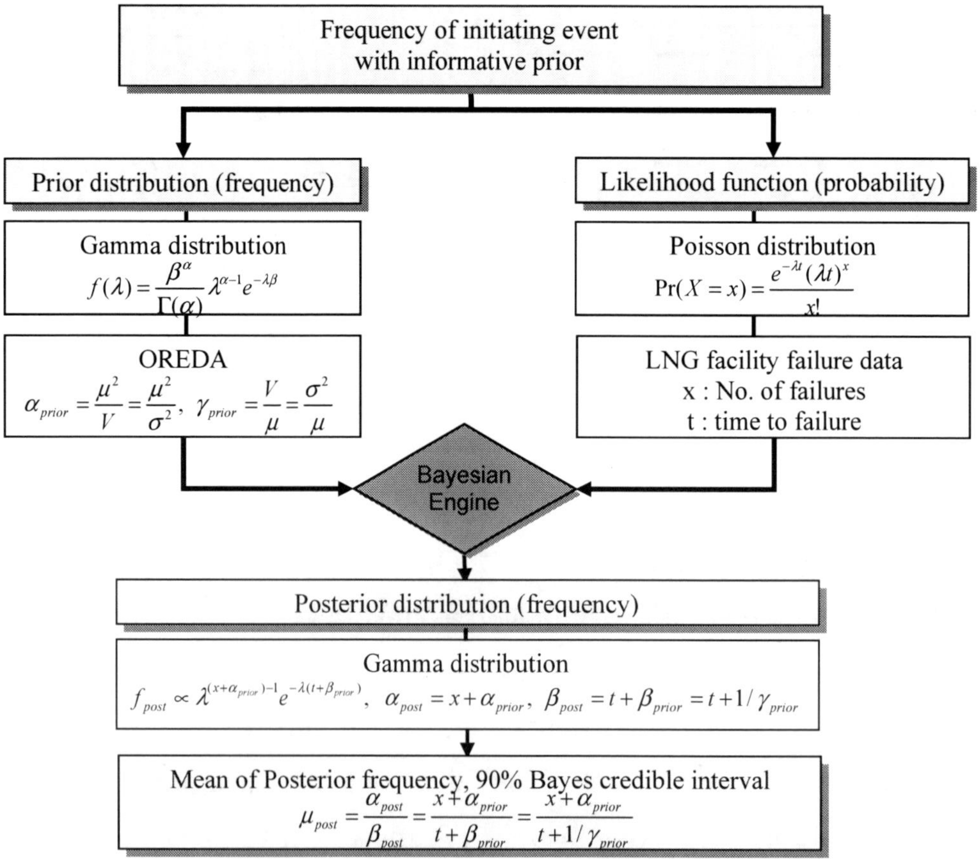

Figure 3 The Diagram of Bayesian Estimation for PFDs of IPLs

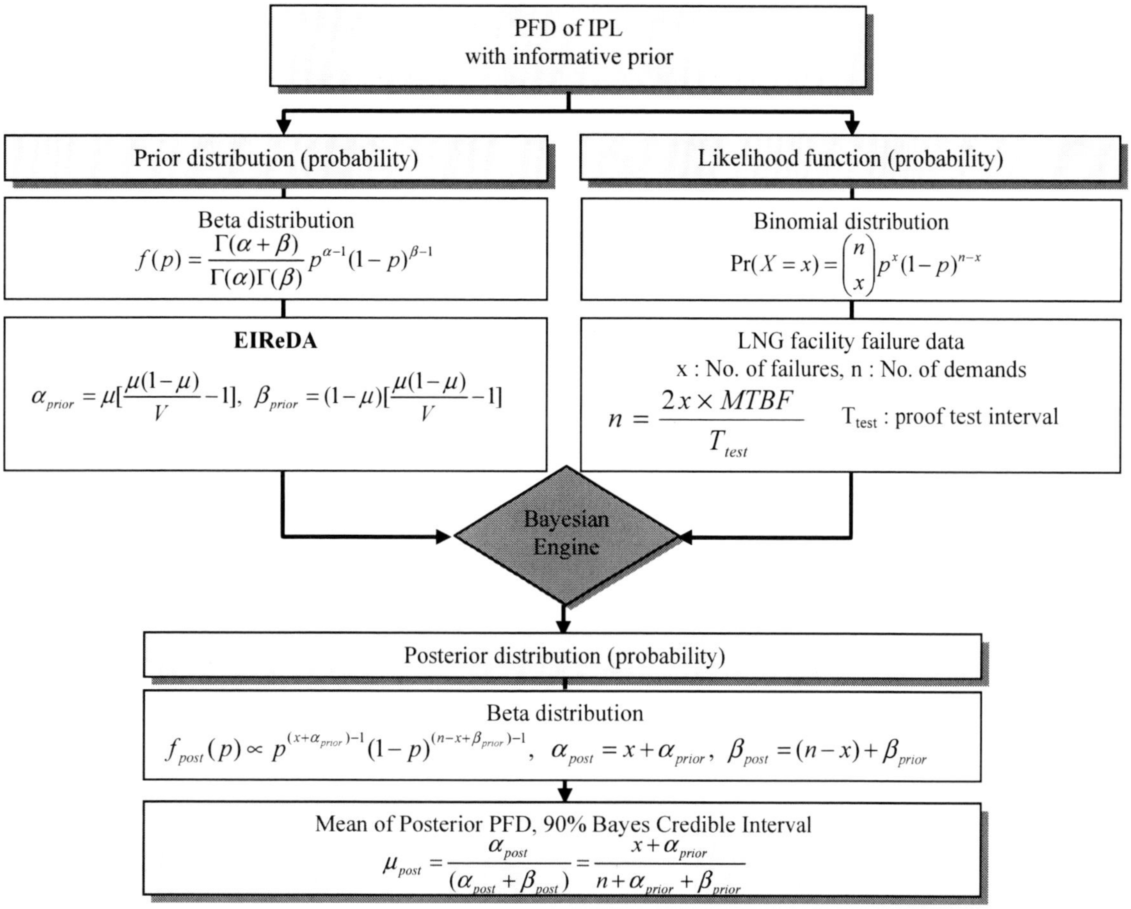

Summary and Conclusion

Through HAZOP study, seven incident scenarios were developed. As an example, scenario 5 is "possible damage of tank from overpressure in tank due to rollover from stratification". For seven incident scenarios including scenario 5, the Bayesian-LOPA method produced valid and well-updated risk values because the posterior values of every initiating event or IPLs were located between prior and likelihood values as shown in Figure 4 and 5. As illustrated in Figure 5, the posterior risk value of incident scenario 5 was below than 10^{-5} given by CCPS, so risk decision could be made that the scenario meet the tolerable criteria.

Consequently, the newly developed Bayesian-LOPA methodology as a risk assessment methodology really works well in LNG importation terminals which have sparse failure data and it can be applied in other industries including refineries, petrochemicals, nuclear plants, and aerospace industries to get more reliable risk values. Furthermore, it can be used with other frequency assessment methods such as Fault Tree Analysis (FTA) and Event Tree Analysis (ETA) to strengthen their results.

Figure 4 The PFD Values of Pressure Alarm in Scenarios 5

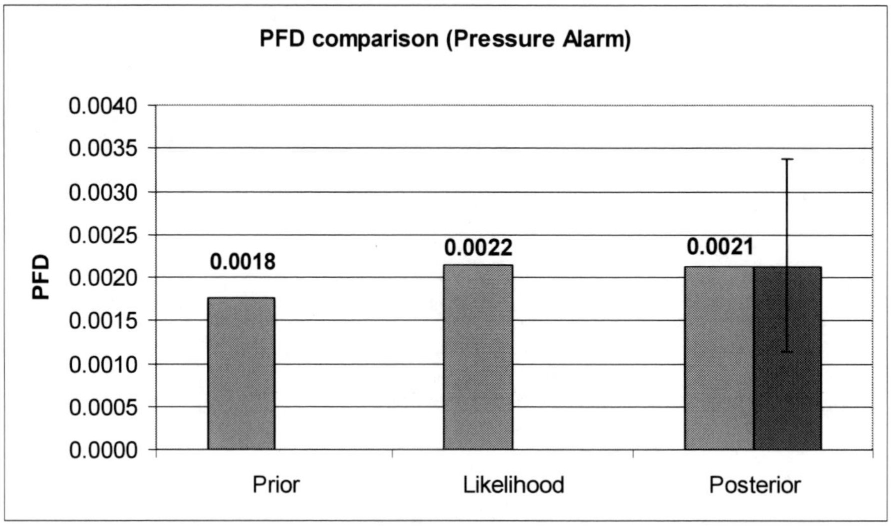

Figure 5 The Risk Value of Incident Scenario 5

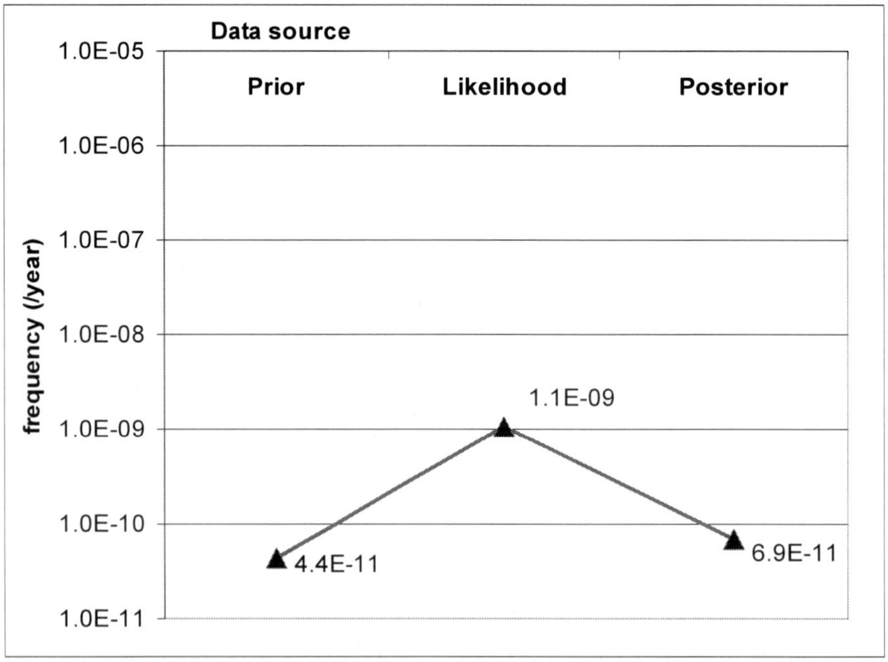

References

1. CEN (Committee of European Nations), Installation and Equipment for Liquefied Natural Gas - Design of Onshore Installations, EN-1473, Paramus, NJ, 1997.
2. Mannan, S.M. and H. West, "LNG Safety Practice & Regulations: From the 1944 East Ohio Tragedy to Today's Safety Record", 2001 AIChE Meeting, Houston, TX, 2001.
3. National Fire Protection Association (NFPA), Standard for the Production, Storage, and Handling of Liquefied Natural Gas (LNG), NFPA 59A, Quincy, MA, 2001.
4. Center for Chemical Process Safety (CCPS), Layer of Protection Analysis - Simplified Process Risk Assessment, pp 1-258, New York, NY, 2001.
5. Dowell, A.M., "Layer of Protection Analysis and Inherently Safer Processes", Process Safety Progress, 18(4), pp 214-220, New York, NY, 1999. etc.
6. Johnson, D.W. and J.R. Welker, Development of an Improved LNG Plant Failure Rate Data Base, Applied Technology Corp., Norman, OK, 1981.
7. Sandia National Laboratories, Handbook of Parameter Estimation for Probabilistic Risk Assessment, Albuquerque, NM, 2003.
8. Korea Gas Safety Corporation (KGS), Comparative Risk Assessment of LNG Tank Designs Training, pp 237-361, Shiheung Si, Korea, 2004.
9. Yoon, E.S., M.S. Mannan, K. Park, and K. Kim, "Safety Analysis for LNG Terminal Focused on the Consequence Calculation of Accidental and Intentional Spills", the Mary Kay O'Connor Process Safety Center Symposium, pp 200-218, College Station, TX, 2005.
10. Markowski, A.S. and M.S. Mannan, Fuzzy Logic Application for LOPA, the Mary Kay O'Connor Process Safety Center, College Station, TX, 2006.
11. Modarres, M., Risk Analysis in Engineering - Techniques, Tools, and Trends, pp 113-183, Taylor&Francis, Boca Raton, FL, 2006.
12. Crowl, D.A. and J.F. Louvar, Chemical Process Safety - Fundamentals with Applications, 2nd edition, pp 448-454, Prentice Hall PTR, Upper Saddle River, NJ, 2002.

Biography

The Department of Chemical Engineering / Mary Kay O'Connor Process Safety Center
Texas A&M University, 3122 TAMU, College Station, TX 77843-3122
Phone: (979) 862-1310, Fax: (979) 458-1493
E-mail: g0y0859@chemail.tamu.edu

Yun, Geun Woong is a PhD student of the Chemical Engineering Department of Texas A&M University advised by Dr. Sam Mannan and a research assistant of the Mary Kay O'Connor Process Safety Center. He attended SungKyunKwan University and received his BS in chemical engineering, and also received a MS degree in mechanical engineering from the YonSei University in Rep. of Korea. Moreover, He obtained a MS degree in chemical engineering from the Texas A&M University in U.S.

Mr. Yun has worked for the Korea Gas Safety Corporation for 9 years as a safety engineer and research scientist for gas related facilities (including LNG terminals) and equipment. He was a national secretary for 5 years for ISO standardization in the field of gas cylinders and bursting discs.

Mr. Yun has the Certified Welding Inspector Certification issued by AWS (American Welding Society) and Design Corrosion Control Certification by NACE (The Corrosion Society of America). He received the Certification of LOPA Training as well as SIL (Safety Integrity Level) Verification by SIS-TECH. Furthermore, he has Gas Safety Expert Certificate and Occupational Safety Expert Certificate by the Human Resources Development Service of Korea.

Modeling Underwater Penetration of LNG Carrier

John L. Woodward, Ph.D.
Baker Engineering and Risk Consultants, Inc.
3330 Oakwell Court, Suite 100
San Antonio, TX 78218

AIChE Spring Meeting, April 2007
8th Topical Conference on Natural Gas Utilization
New Orleans, Louisiana, April 6-10, 2008

Abstract

An underwater breach of a tank in a double-hulled LNG carrier poses several challenging questions to be addressed by an adequate model. The pressure balance between external hydraulic pressures and head pressure in the hull space will influence the inflow of both sea water and of LNG. This pressure balance can be evaluated by stepwise idealizations. First consider the case where only the outer hull is breached. Head pressure in the hull space increases by compression by reducing vapor volume. If the inner hull is also breached, then the interaction of LNG with sea water can lead to stronger pressure build-up, water freezing, and possibly a rapid phase transition (RPT), especially if the inflow rates are high.

This paper describes the modification of a model that accounts for the following phenomena: 1) LNG tank head pressure decreasing by the piston effect of outflow; 2) LNG and sea water filling the hull space to the top of holes in the outer hull; 3) LNG evaporation in the hull space and vapor outflow through the outer breach; 4) LNG outflow and gas or water inflow to the LNG tank at the inner breach; 5) ice formation by LNG evaporative cooling; and 6) the effect of water inflow evaporating LNG and the resulting pressure increase inside the tank. The model does not allow for localized ice build-up at the inner breach constricting flow. Predictions show tank pressure, LNG outflow rates, and water inflow rates over time.

1.0 Introduction

An underwater breech of the double hull of an LNG carrier could possibly occur by grounding, vessel collision, or sabotage. The consequences would be at least sea water intrusion into the hull space from the outer hull breech. If the inner hull is also breeched, LNG outflow is likely, at least into the hull space, and possibly water inflow into the LNG tank. In-rushing water tends to pressurize the hull space and this head pressure, unless relieved rapidly, acts as a back-pressure to limit inflow. LNG outflow from the inner tank adds vapor to the hull space pressure, and increases hull head pressure. It could begin to freeze the water in the hull space, and at high water levels in the hull space, ice could inhibit LNG drainage from the inner tank. The interaction of LNG with water, both in and outside the hull adds further possible consequences, including a rapid phase transition (RPT).

This paper explores the pressure balances that influence inflow of water and LNG to the hull space, vaporization of LNG, outflow of vapor and air, and possible water inflow into the LNG tank. The hull space pressure dynamics are shown to have an important influence on LNG leakage. The appendix summarizes the modifications made to the unreleased $SafeSite_{3G}$® model for blow down to the double hull space.

2.0 Initial Pressure Profiles on LNG Membrane Carrier

A single tank in a membrane LNG carrier of 110,000 tonnes displacement is considered here. A cross-sectional diagram, not entirely to scale, is shown in Figure 1. Assumed dimensions are provided in Table 1 for a 35,000 m^3 tank. Of this, 25,000 m^3 is above the water level, with the balance below water level.

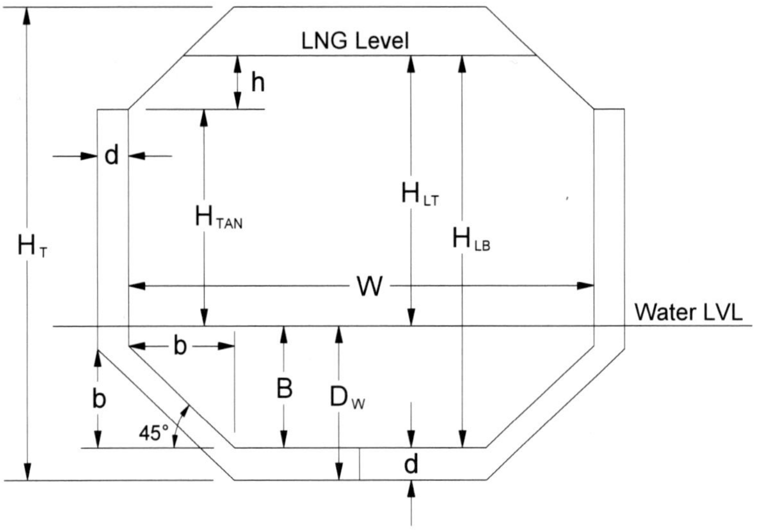

Figure 1. Cross-Section of Typical LNG Membrane Carrier Tank

Table 1. Assumed Dimensions for LNG Membrane Carrier Tank

Dimension	Symbol	Value, m	Dimension	Symbol	Initial Value, m
Ship length	L_{SHIP}	270	Number of Tanks	N_{TANKS}	5
Tank length	L	50.0	Ship draft	D_W	10.12
Tank height	H_T	30.58	LNG level above water level	H_{LT}	17.05
Interior tank width	W	30.0	LNG level from inner tank bottom	H_{LB}	25.084
Interior bevel height	B	6.60	Water level from bottom of inner tank	B	8.12
Double hull separation	d	2.2	Web frame spacing	L_{WEB}	2.8
Outer hull hole diameter	D_{OH}	various	LNG level in upper bevel	h	3.34
Height of top of outer hull hole	H_{OH}	various	Leaked water level in hull	h_W	varies

The composition of LNG varies primarily with the source location of the LNG, and ranges typically from 80% to about 97% methane. A hypothetical middle-range composition is used here and listed in Table 2.

Table 2. LNG Composition Modeled

Component	Mole %	Mass %
Methane	87.0	73.504
Ethane	7.0	11.085
Propane	4.0	9.289
n- Butane	2.0	6.122
Total	100.0	100.0

This mixture has a molecular weight, M_{LNG}, of 18.99 kg/kmole, and a normal boiling point of 113.29 °K. At the normal boiling point, the density of this LNG mixture, ρ_{LNG}, is 469.1 kg/m^3. Taking the sea water temperature as 10 °C, the density of sea water, ρ_W, is 999.7 kg/m^3. This density difference is the reason LNG tanks sit high above the water line. It is also why the LNG hydraulic pressure exceeds that of water until a considerable amount of LNG drains. To see this, we plot hydraulic pressures of LNG and water against the elevation above the bottom of the LNG tanks in Figure 2. Hydraulic pressure, P_{hydf}, is given as a function of the depth of fluid, z, and fluid density, ρ_f, using SI units in kPag (gauge) by:

$$P_{hydf} = g\ z\ \rho_f/1000 \qquad (1)$$

The total pressure at a depth z, P_{Zf}, depends also on the head pressure above the liquid, P_h:

$$P_{Zf} = P_h + P_{hydf} \qquad (2)$$

The coordinates in water depth, z, and height above the LNG tank bottom, H_w, are related by:

$$H_w = D_W - z \qquad (3)$$

We assume that the initial head pressure is atmospheric in both the hull space and in the LNG tank. Temporarily ignoring the effects of the double hull space on pressure balances, consider a puncture of both the outer and inner hull for which the top of the hole is at the same elevation (water depth). The hydraulic pressure difference between LNG and water at depths of 1, 3, 5, and 7 m is plotted in Figure 3. Figure 3 indicates that the LNG would have to drain from 25.17 m to 14.8 m with a hole 7 m deep, or to 10.6 m with a 5 m deep hole before reaching hydraulic equilibrium. These levels are also connected by drawing a horizontal line in Figure 2, as shown by the example with a 5 m deep puncture (water 3.12 m, LNG 10.6 m above the top of the hole).

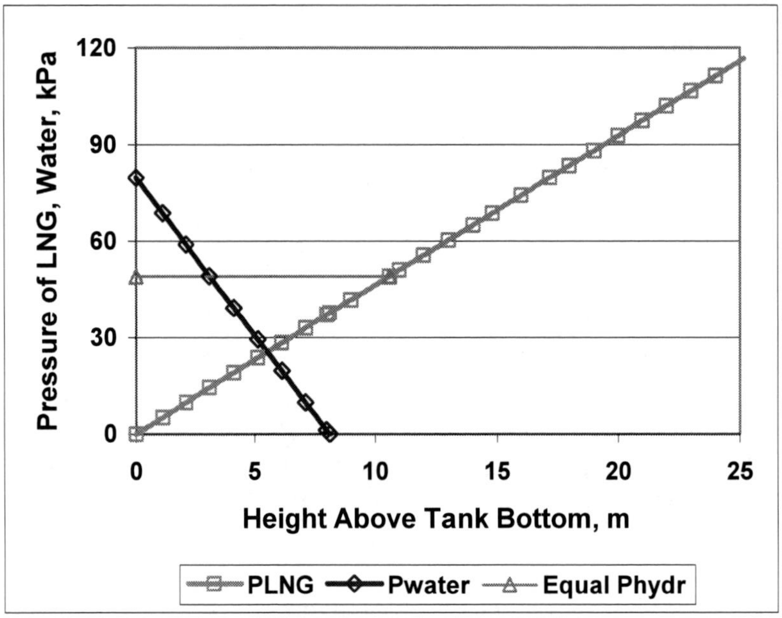

Figure 2. Pertinent Hydraulic Pressure Profiles

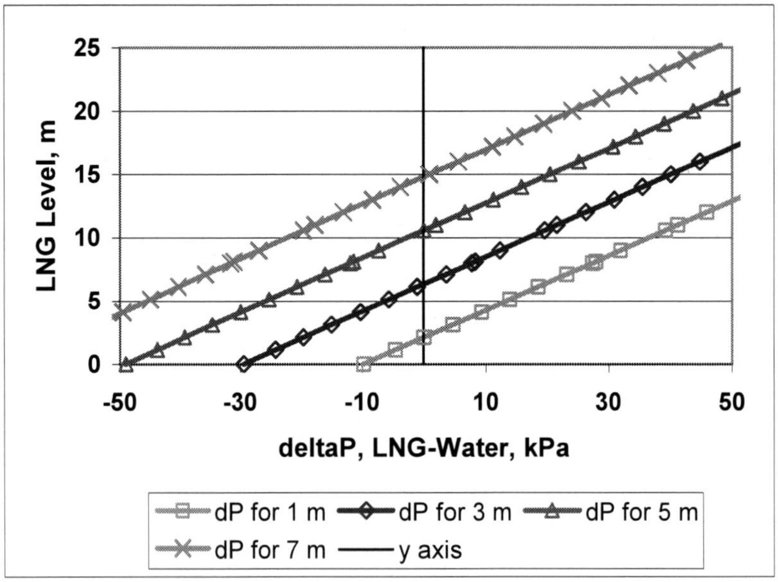

Figure 3. Hydraulic Pressure Difference Between LNG and Water at Water Depths of 1, 3, 5, 7 m

Of course, the double hull effects cannot be ignored. An overview of the phenomena treated is developed next, followed by modeling quantification.

3.0 Overview of Scenarios and Phenomena

An underwater breach of both the outer and the inner hull results in complex interactions between water inflow and LNG outflow. In order to reduce this complexity, the approach taken here is to develop incremental steps with idealized simplifications. One phenomenon at a time is analyzed over a time period step. The step ends when limiting conditions are met, allowing for idealizations such as perfect mixing of LNG and water.

Scenario 1. Outer hull is penetrated below water line at water depths Z_{hTOP}, and $Z_{hBOT.}$ Inner hull is not penetrated.
Consequence: Sea water enters and some air leaks out.

1.1 Early Response Period: External water covers outer hull hole.
Consequence: Water flows into hull space, raising hull head pressure by compressing air.
Limiting Condition: Hull head pressure, $P_{h,}$ increases to equal hydraulic pressure of water at the top of the outer hole, $P_W(Z_{hTOP})$. Call this pressure P_{h1}, the initial "equilibrium pressure".

1.2 Mid Response Period: The next incremental increase in hull head pressure pushes air out the top of the hole in the outer hull.
Consequence: Air flows out and head pressure increases above P_{h1}. Water flows in at an equal volume rate to air outflow.
Limiting Condition: Water reaches the top of the hole in the outer hull. Water inflow stops.

1.3 Late Response Period: Minor fluctuations in water hydraulic pressure caused by wave action or carrier roll cause variations in the average water level at the top of the hole. Small amounts of water enter, and air leaves.

Scenario 2. Both outer and inner hull are penetrated.
Consequence: Liquid LNG and sea water flow into hull space and mix. LNG vaporizes completely and freezes some water.

The height (relative to the bottom of the ship) of the top of the holes in the outer (H_{hTOP}) and inner hull (H_{LTOP}) is important for establishing pressure balances. There are three possibilities:

1) Hole top elevations equal ($H_{hTOP}= H_{LTOP}$)
2) Outer hull top of hole is higher ($H_{hTOP}> H_{LTOP}$)
3) Inner hull top of hole is higher ($H_{hTOP}< H_{LTOP}$)

This paper treats primarily Possibility 1, and briefly treats Possibilities 2 and 3.

2.1 Early Response Period Water covers outer hole. LNG covers inner hole. **Consequences:** Water flows into hull space, raising hull head pressure by compressing air. LNG flows into the hull space and vaporizes. LNG vapors increase hull head pressure.

By the assumption of perfect mixing of liquids, we temporarily ignore the growth of the ice layer as limited by transport properties. We use thermodynamics alone to find the fraction of water frozen. With this assumption, the vaporization of LNG cools all the water in the hull to the freezing point and then part of the water freezes.

As the LNG level in the inner tank drops, the inner tank head pressure drops by the piston effect minus LNG evaporation. The vacuum relief valve admits air to bring the tank head pressure to atmospheric.

Limiting Condition: Hull head pressure, P_h, increases to equal the hydraulic pressure of water at the top of the outer hole, $P_W(Z_{hTOP})$ and the hydraulic plus head pressure of LNG at the top of the inner hole (pressure P_{hI}).

Alternate Limiting Condition: If the vacuum relief capacity is too limited, then the LNG head pressure briefly remains below atmospheric. This decreases the pressure on the LNG side of the inner hole. A gulping flow reversal would admit vapor from the hull head space since only air/vapor exists above the hull water layer in this period.

2.2 Mid Response Period: The next incremental increase in hull head pressure pushes air/vapor out the top of the hole in the outer hull and air/vapor into the LNG tank.

Consequence: Air flows out and head pressure increases since vapor generation exceeds vapor outflow. Water flows in at an equal volume rate to air outflow.

Limiting Condition: Water reaches top of the hole in the outer hull. LNG and water inflow to hull space ceases. If the water level is above the top of some inner hull holes, a small amount of water will flow into the LNG tank. Inflowing water freezes and ice sinks to the bottom of the LNG tank. The ice temperature drops to the LNG temperature, generating 1.08 kg of LNG vapor per kg of water. This added LNG vapor increases the LNG tank head pressure and stops the water inflow until the pressure relief valve can discharge the added vapor.

2.3. Late Response Period A meta-stable balance of three pressures exists ($P_W(Z_{hTOP})$, $P_h(Z_{hTOP})$, and $P_{LNG}(Z_{hTOP})$) with the level of both sea water and LNG above the top of the outer and inner holes.

Consequences: Similarly to Step 1.3, wave action temporary causes small increases in the hydraulic sea water pressure and forces small amounts of sea water and/or LNG inflow and some vapor out flow. If the top of some inner holes are above the top of the outer hole, then hull head pressure fluctuations would allow air/vapor inflow into the LNG tank. Altogether, a slow draw-down of LNG is expected with a long duration for this period.

Alternate Limiting Condition: The depth of ice in the LNG tank or at the inner hole builds up until it interferes with the cycle of water in/LNG out in the LNG tank. These steps are quantitatively elaborated below.

4.0 Double Hull Pressure Considerations- Outer Wall Penetration

Consider Scenario 1 with the three periods defined in Section 3. If only the outer hull is penetrated, water inflow will displace the air in the double hull space and, if the double hull space is tightly sealed, the pressure in this space will rise by compression. In reality, the double hull space is not perfectly tight, but leakage can be treated subsequently. Assuming a tightly sealed hull, we first examine the early response behavior.

The initial mass of air in the hull space, m_{air0}, is given by the hull space volume, V_{hull}, times the ambient air density, ρ_{air}, which at an ambient temperature, T_{amb}, of 10 °C and 99% relative humidity, is 1.210 kg/m^3. The hull space volume is made up of the cross-sectional area of the hull, A_{hull}, times the tank length, subtracting the web spacer volume, V_{web}. The thickness of web frame spacers is assumed to be one inch. The web spacers have large holes to enable personnel passage, and allow the hull space to serve as a ballast tank. The cross-sectional area of the hull is divided into the tank bottom, A_{hB}, the 45° bevel corner, A_{hC}, and the vertical side, A_{hS}.

Formulas are given below for the wetted cross-sectional area at a given level of water, h_{wh}, in the double hull space. These formulas reduce the total cross-sectional area of each space when the water level reaches the top of each subsection. The formulas assume the bottom hull space is divided at the middle by a solid wall:

<u>Water Level in Bottom Hull (Half of Total)</u>

$$A_{hBw} = \frac{h_{wh}}{2}(W - 2b + h_{wh}) \quad \text{if } h_{wh} \le d \tag{4a}$$

and when h_{wh}= d (top of bottom section):

$$A_{hB} = \frac{d}{2}(W - 2b + d) \;\; = 20.90 \text{ m}^2 \tag{4b}$$

<u>Water Level in Hull Corner</u>

$$A_{hCw} = d\left(h_{wh} - d\right) \quad \text{if } d \le h_{wh} \le b + d \tag{5a}$$

and when h_{wh} = b+d (top of corner):

$$A_{hC} = db \qquad\qquad = 14.52 \text{ m}^2 \tag{5b}$$

<u>Water Level in Vertical Side</u>

$$A_{hSw} = d(h_{wh} - b - d) \quad \text{if } h_{wh} \ge b + d \tag{6a}$$

and if h_{wh} = H_T –b (top of vertical side)

$$A_{hS} = d(H_T - 2b - d) \;\; = 33.396 \text{ m}^2 \tag{6b}$$

<u>Total Hull Cross-Sectional Area and Volume</u>

$$A_h = A_{hS} + A_{hC} + A_{hB} = 68.816 \text{ m}^2 \tag{7}$$

$$V_{hull} = A_h L - V_{web} \quad = 3440.8 - 31.46 = 3409.3 \text{ m}^3 \tag{8}$$

or adding the pertinent sections, $A_{hj} = A_{hB}, A_{hC}, A_{hS}$

$$V_{hull} = \sum_j A_{hj} L_h \frac{3409}{3440.8} \tag{9}$$

Where the numerical ratio is a correction coefficient, C, that accounts for webbing.

The volume of air in the hull space, V_{hair}, is decreased by the volume of water intrusion, V_W. The density of gas or vapor in the compressed volume is:

$$\rho_{V2} = \frac{m_{air0}}{V_{hair}} = \frac{m_{air0}}{V_h - V_W} \tag{10}$$

The head pressure in the hull, P_h, assuming no air leakage, can be found either by adiabatic compression formulas or by isothermal compression. Isothermal compression is probably more appropriate, considering the large area of the uninsulated outer hull. This is given by the ideal gas equation of state:

$$P_h = \frac{\rho_{V2} R T_2}{M_{V2}} \tag{11}$$

where M_{V2} is the mole weight of the air and LNG vapor, R, is the gas constant, and T_2 is the temperature after compression (ambient temperature). Thus, the head pressure in the hull space, assuming no air losses, can be plotted as a function of the volume of water intrusion as shown in Figure 4.

Figure 4 also plots the depth of water in the hull space, h_{hW}, as a function of the volume of water intrusion. The water depth is readily found from the volume of water intrusion and the volume in each hull segment, V_i, using Eqs 5(a), 6(a), and 7(a). A discontinuity in the slope of the hull depth curve occurs when the water fills the bottom hull section.

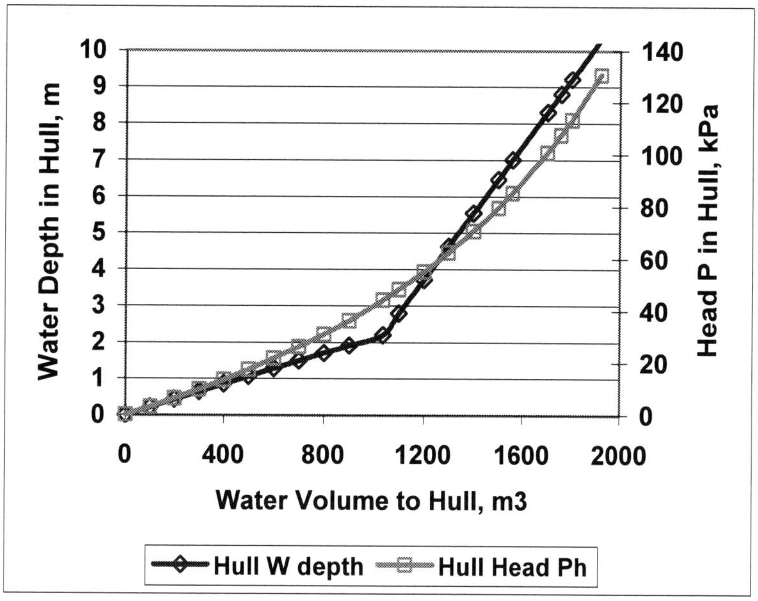

Figure 4. Effect of Water Intrusion in Double Hull Area on Head Pressure in Hull

Pressure profiles are plotted in Figure 5, both to define the nomenclature and to indicate that the hydraulic pressure gradient is the same for water outside and inside the hull space. This fact simplifies finding the equilibrium pressure since hydraulic pressure inside the hull space only affects the balance after the hull space water level, h_{Wh}, rises above the hole bottom, H_{hBOT}. That is, the pressures inside the hull space are given by:

$$P_h = P_{headh} \qquad h_{Wh} \le H_{hBOT}$$

$$P_h = P_{headh} + g\rho_W \left(h_{Wh} - H_{hBOT}\right) \qquad h_{Wh} > H_{hBOT} \qquad (12)$$

The hydraulic water pressure outside the hull is given by:

$$P_W = g\rho_W H_{hBOT} = P_{Wmax} \qquad h_{Wh} < H_{hBOT}$$

$$P_W = g\rho_W \left(h_{Wh} - H_{hBOT}\right) \qquad H_{hBOT} \le h_{Wh} < H_{hTOP} \qquad (13)$$

$$P_W = g\rho_W H_{hTOP} = P_{Wmin} \qquad h_{Wh} > H_{hTOP}$$

Since an interface of air on one side and water on the other is unstable, the stable state has a water depth in the hull at or above the top of the hole. Eq. 13 indicates that the water pressure that influences water inflow when the hole is completely immersed occurs at the top of the hole, P_{Wmin}.

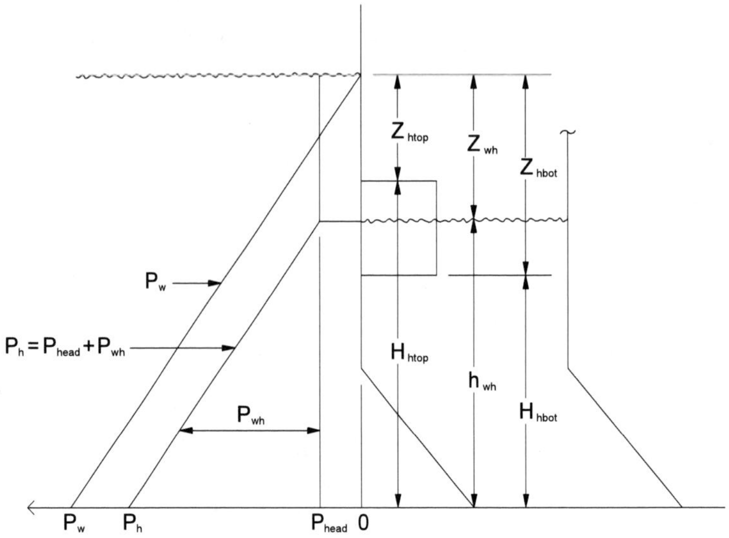

Figure 5. Pressure Profiles Outside and Inside Hull Space

Figure 6, cross plotting from Figure 4, plots the head pressure against water depth in the hull, h_{Wh}. Eqs 12 and 13 are used to find the water depth in the hull at pressure equilibrium (inside and outside the hull), h_{hWeq}. The upper curve in Figure 6 represents the total back-pressure for a 7 m deep hole.

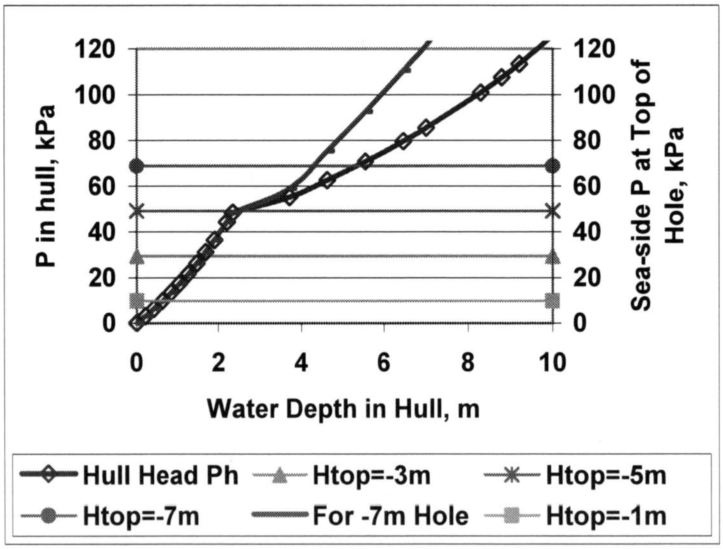

Figure 6. Head Pressure in Hull by Water Intrusion Compression

The horizontal lines in Figure 6 are the "equilibrium pressures" at the end of the early response period. This is the point when the water inflow mechanism changes to constant volume exchange. The hull pressure equals the hydraulic pressure of water and of LNG, so by pressure considerations alone, these inflows should stop. Yet the hull water level is below the top of the outer hole, so air in the inner hull will escape by buoyancy. The mid response period begins in which equal-volume inflow of water and outflow of air occurs until the water depth in the hull space rises to just above the top of the hole.

The intersection of the horizontal lines in Figure 6 with the upper curve gives the hull water depth at the beginning of the mid response period. The conditions at this point are summarized in Table 3.

Table 3. Conditions at the End of Early Response Period

Top of Hole Below Water Level, m	Hydraulic Pressure of Sea at Hole kPa	Water Level in Hull at Initial Pres. Equilibrium m	Final Water Level in Hull m	Initial Water Intrusion Volume m^3	Final Water Intrusion Volume m^3
-1	9.8	0.40	9.32	187	1812
-3	29.4	1.12	7.32	525	1594
-5	49.1	1.74	5.32	818	1376
-7	68.7	2.27	4.26	1070	1260

Figure 7 plots the initial and final water level from Table 3 along with an illustrative hole location. The 7-meter deep hole is the only case where the final water level is well above the top of the hole.

With wave action, the hydraulic sea pressure at the hull penetration varies cyclically. This can cause more air displacement at the low point of the wave amplitude. Because of this, the Table 3 values for final water intrusion are minimal estimates.

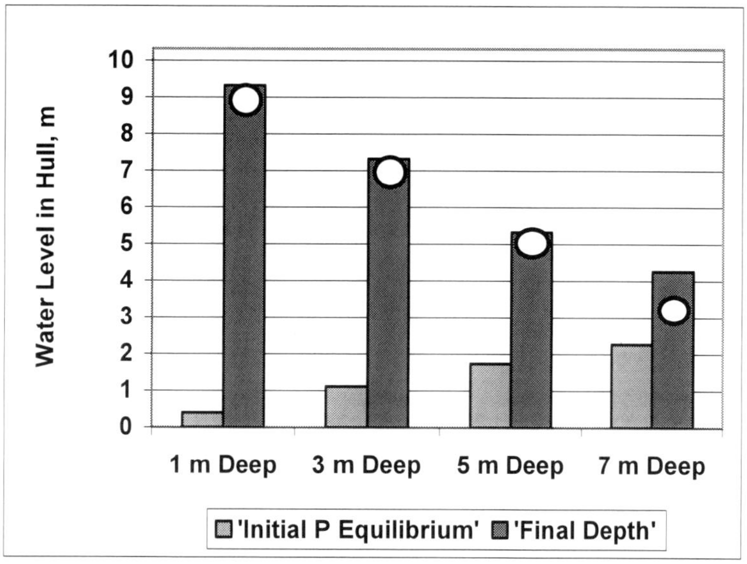

Figure 7. Water Depth in Hull With Outer Hull Penetration

5.0 Dynamic Considerations, Outer Hull Penetration

5.1 Early Response Period

The discharge rate through a penetration below the liquid level is obtained from the Bernoulli equation for subcooled liquid flow. The water inflow rate, F_{Win}, (or LNG inflow rate) is given in terms of the hole area, A_{hole}, discharge coefficient, C_D, water density, ρ_w, and the pressure difference between the hydraulic pressure of water, $P_w(z_{hole})$, and the back-pressure, P_{back}:

$$F_{Win} = C_D A_{hole} \left[2\rho_w \left(P_w - P_{back} \right) \right]^{1/2} \tag{14}$$

For example, the initial rate of inflowing water is shown in Figure 8 for outer hull penetrations with equivalent hole diameters of 0.15, 0.30, 0.45, and 0.60 m at hole top depths of 1, 3, 5, and 7 m.

In the early response period, the initial mass of air in the hull space, m_{air0}, remains constant under the assumption of an air-tight hull. The volume of water inflow increases the back pressure by Eqs 12 and 13. This decreases the inflow rate over time as shown in Figure 9 for a 0.15 m hole diameter. The curve for a 7 m deep hole differs from the others. Recall that the water level for a 7 m deep penetration rises above the hole, adding a hydraulic pressure contribution, and this contribution drives the discharge rate lower more quickly.

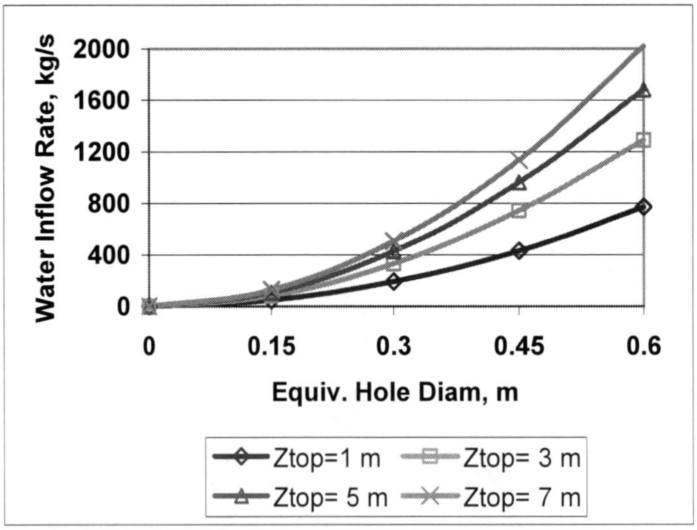

Figure 8. Initial Inflow Rate of Water for Holes with Top at 1, 3, 5, 7 m Depth

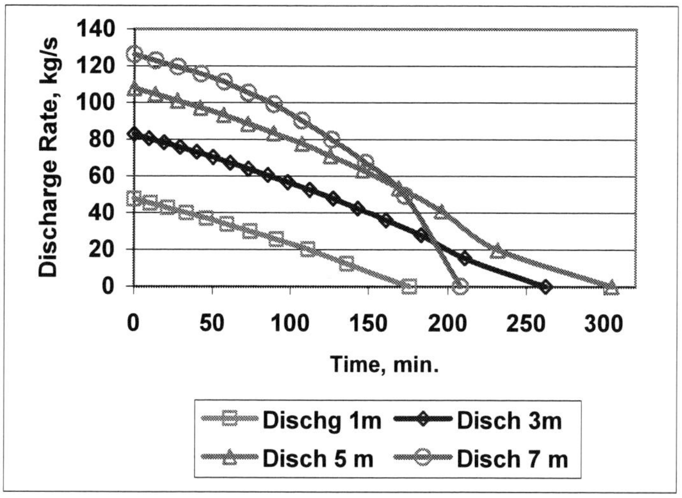

Figure 9. Blowdown Discharge Rates for Outer Hull Penetration, 0.15 m Hole at Depths of 1, 3, 5, 7 m

Figure 10 shows how the back-pressure builds up until it equals the external hydraulic pressure of sea water. The important result of the dynamic analysis is a better estimate of the time to reach equilibrium. This is summarized in Table 4, along with the initial water flux, G_w, (F_{disW} divided by the hole area, A_w).

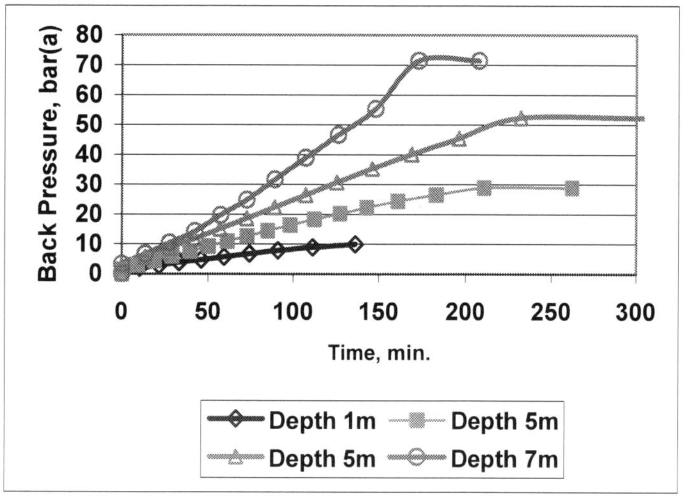

Figure 10. Back Pressure Build up for Outer Hull Penetration, 0.15 m Hole at Depth of 1, 3, 5, 7 m

Table 4. Time to Reach V for 0.15 m Hole in Outer Hull

Depth to Bottom on Penetration, m	Initial Water Inflow Rate, kg/s	Initial Water Flux, kg/(m^2 s)	Time to Init. Equil. at Initial Inflow Rate, min	Time to Init. Equil. by Dynamic Model, min
-1	47.7	2701.0	65	176
-3	82.7	4678.2	106	262
-5	107.7	6094.6	127	305
-7	126.3	7147.1	141	208

5.2 Mid Response Period, Equal Volume Exchange Model

At the end of the early period the hull head pressure balances the hydraulic pressure across the outer hull hole, so only buoyancy affects the vapor outflow. The buoyancy driving force, ΔP_{buoy} is given by:

$$\Delta P_{buoy} = g z_{top} \left(\rho_L - \rho_V \right) \tag{15}$$

Vapor begins to bubble out through the outer hull penetration when the water level drops below the top of the hole. The discharge rate of vapor, F_{disV}, is found for a hole with water on one side and vapor on the other with an overall pressure difference, ΔP_{hole} of:

$$\Delta P_{hole} = P_{head} + \Delta P_{buoy} - P_{back} \tag{16}$$

The vapor outflow flux, G_V, is a function of hull head pressure and vapor density. This can be found by assuming values for head pressure and vapor composition and calculating discharge rate, F_{disV}, using a compressible flow model. The outflow rate divided by the area of the hole, A_v, gives the vapor flux, and similarly for the water inflow flux, G_W. Submerged liquid or vapors with density lower than sea water are known to leak at constant pressure by equal volume exchange. Equal volume discharge of vapor and inflow of water is found using f_W, the fraction of the hole area covered by water, A_w/A_{OH}:

$$\frac{G_w}{\rho_w} f_w A_{OH} = \frac{G_V}{\rho_V} \left(1 - f_w\right) A_{OH} \tag{17}$$

This balance provides the fraction f_w as:

$$f_w = \frac{G_V \rho_w}{G_V \rho_w + G_w \rho_V} \tag{18}$$

After finding f_w, the discharge rate of vapor and inflow rate of water is readily found:

$$F_{Win} = f_w A_{OH} G_w$$
$$F_{disV} = (1 - f_w) A_{OH} G_V \qquad (19)$$

In the mid response period, the water inflow rate drops to a nearly constant rate as shown in Figure 11. Figure 11 is for a hole diameter of 0.15 m at 1 m depth. The mid response period lasts until the water volume in the hull builds to 1812 m^3 and the water level covers the outer breach at 470 minutes. The final response period is shown with zero water inflow, since inflow continues only by wave action and carrier roll disturbances.

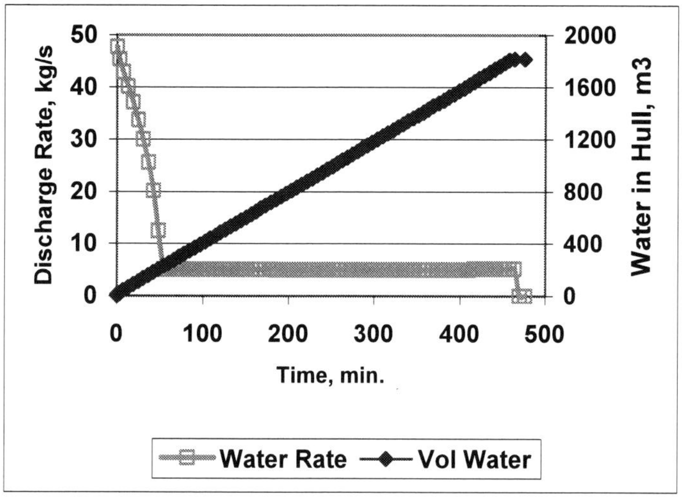

Figure 11. Water Inflow Rate and Volume in Hull Through Mid Period

6.0 LNG/Water Interaction Considerations

Consider Scenario 2. If the inner hull is also breached at the same time as the outer hull, then LNG will leak and interact with the water in the hull space. Since the vapor pressure of LNG increases rapidly with temperature, any LNG in contact with water will boil as limited by the turbulent mixing and/or film boiling heat transfer. Water in the hull space will be chilled and some fraction could freeze. The contact area between LNG and water is key to whether some localized ice build-up could develop and possibly interfere with subsequent LNG/water mixing.

The response can also be divided into three periods defined in Section 3 for Scenario 2.

6.1 Early Response Period Develops Ice/Water Slush

As a first step in analyzing water/LNG interaction, assume unhindered mixing. An enthalpy balance will reveal the effect of LNG boiling on the incoming water. Let m_W and m_C represent the mass fractions of water and LNG in contact over a small time sufficient to achieve thermodynamic equilibrium. Provided adequate mixing, m_w and m_C could also be taken as rates of inflow into the hull space. The enthalpy balance is:

$$\frac{m_w}{m_C} = \left[H_{wL}(T_{amb}) - H_{wL}(T_{eq}) + f_w H_{wFus} \right] = \Delta H_{Cvap}(T_{NBP}) + H_{CV}(T_{eq}) - H_{CV}(T_{NBP}) \tag{20}$$

where:

H_{wL} = enthalpy of liquid water, kJ/kg

H_{wFus} = heat of fusion of water, 333.68 kJ/kg

f_w = mass fraction of the mass m_w that freezes

ΔH_{CV} = heat of vaporization of LNG, 516.2 kJ/kg at T_{NBP} of LNG

H_{CV} = enthalpy of LNG vapor, kJ/kg

T_{amb} = ambient temperature, initial temperature of liquid water, 283.15 ^{o}K

T_{NBP} = normal boiling point of LNG, 113.3 ^{o}K

T_{eq} = equilibrium temperature of LNG:water mixture, taken as 273.2 ^{o}K

Enthalpy values are computed using the DIPPR (Design Institute of Physical Properties Research, Daubert and Danner, 1989) correlations and ideal mixing rules. Water properties were taken as fresh water, recognizing that the freezing point of sea water is several degrees below zero Celsius.

The results of solving Eq. 20 are shown in Figures 12 and 13, plotted side-by-side to emphasize that the two can be considered to be a single graph with a change of the abscissa scale. Figure 12 plots the ratio m_{LNG}/m_w, or the mass of LNG vaporized against the mass fraction of water that freezes. Figure 13 plots the same LNG/water ratio against the degrees of ice sub-cooling. Figure 12 starts with 0.045 kg of LNG vaporized by cooling liquid water 10 ^{o}C to the freezing point. The maximum value in Figure 13 of 1.085 is achieved by sub-cooling ice to the normal boiling point of LNG.

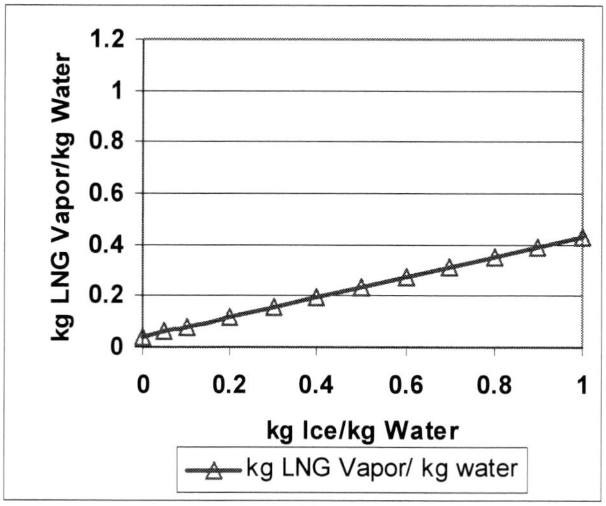

Figure 12. Mass Fraction of LNG Vaporized by Freezing Water

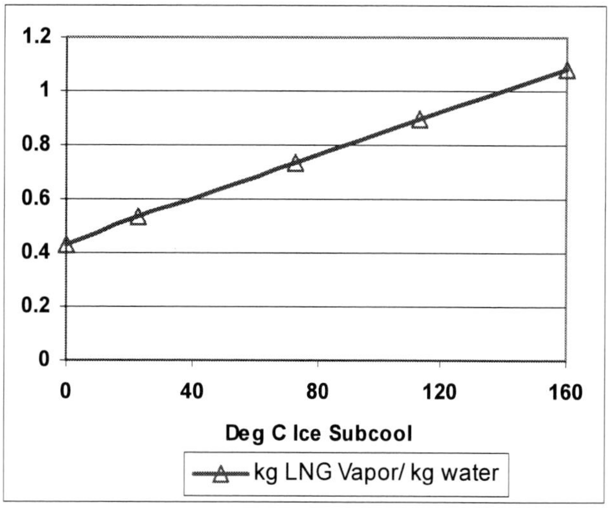

Figure 13. Mass Fraction of LNG Vaporized by Sub-cooling Ice

With initially low LNG/water mass ratios, all of the LNG would vaporize. This vapor adds to the initial mass of air, assuming again that the hull space is tightly sealed. So, the head pressure builds by LNG vaporization, shortening the time to reach the mid response period.

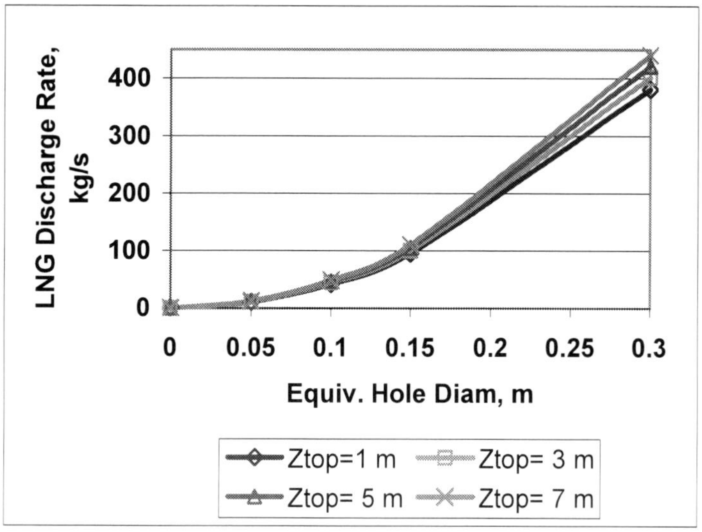

Figure 14. Initial Discharge of LNG into the Hull Space for Holes with Top at 1, 3, 5, 7 m Depth

The initial discharge rates of LNG are shown in Figure 14 as a function of hole size and penetration depth. Compared with Figure 8 for water inflow, the discharge rates for LNG are about the same as those for water; the effects of a larger LNG liquid head being balanced by a lower liquid density compared to water.

Table 5 gives the mass ratio of the initial inflow rates into the double hull of LNG to water at two ratios of the hole diameters (outer hull to inner hull). Inner hull penetrations are expected to be less than outer diameter penetrations; specifically assume a ratio of 3:1 and 2:1. With these ratios, the mass ratio of LNG to water is in the low range in Figure 12. Consequently, a slush (a low ratio of ice to water) is expected in the early response period. This early response period continues until the water level in the hull compresses the back-pressure to the hydraulic pressure of water at the top of the hole and vapor backflow begins.

Table 5. LNG to Water Inflow Ratios and Mass Fraction of Ice Formed In Initial Period Before Vapor Backflow in the Hull

Hole Diameter Ratio, D_{OH}/D_{IH}	Mass Ratio Inflow to Hull Space, m_{LNG}/m_W			
	1 m Hole Depth	3 m Hole Depth	5 m Hole Depth	7 m Hole Depth
$3:1 = \begin{cases} (0.15/0.05 \text{ m/m}) \\ (0.30/0.10 \text{ m/m}) \\ (0.45/0.15 \text{ m/m}) \end{cases}$	0.222	0.135	0.110	0.097
2:1 = (0.60/0.30 m/m)	0.495	0.312	0.250	0.218
	Mass Ratio Ice/Water			
$3:1 = \begin{cases} (0.15/0.05 \text{ m/m}) \\ (0.30/0.10 \text{ m/m}) \\ (0.45/0.15 \text{ m/m}) \end{cases}$	0.128	0.0944	0.0847	0.0797
2:1 = (0.60/0.30 m/m)	0.234	0.0787	0.0749	0.0729

6.2 Liquid Inflows Decrease as Hull Head Pressure Builds

An LNG leak contributes additional vapor and cooling and strongly changes the dynamic. It is reasonable to assume that the outer hull hole area exceeds that of the inner hull. Thus, there is likely to be sufficient water intrusion to vaporize all LNG in the hull space in this early period. The vapor density in the hull space increases by the added mass of LNG vapor, m_{CV}, which modifies Eq. 10:

$$\rho_V = \frac{m_{air} + m_{CV}}{V_h - V_W} \tag{21}$$

The head pressure, given by Eq. 11, requires knowing the vapor temperature, T_{V2}, and the molecular weight of vapor, M_{V2}. M_{V2} is given by the mass fraction of LNG, w_C, the molecular weight of air, M_{air}, and of LNG, M_C:

$$\frac{1}{M_V} = \frac{w_C}{M_C} + \frac{(1 - w_C)}{M_{air}} \tag{22}$$

The vapor temperature is found using the adiabatic mixing curve shown in Figure 15. This is readily found by solving the mass and enthalpy balance.

Example dynamics are shown in Figures 16 and 17 for outer and inner hole diameters of 0.15 and 0.05 m, respectively, (3:1 diameter ratio) at 1 m and 3 m hole depth. They show an initial decrease in back-pressure caused by the cooling effect of LNG. As a result, the water inflow rate temporarily increases. The rapid build-up of LNG vapors greatly shortens the time to reach the initial equilibrium back pressure compared with the times in Table 3 for an external hull puncture. These times are summarized in Table 6, along with the volume of water inflow, the vapor temperature, and vapor mass fraction of LNG.

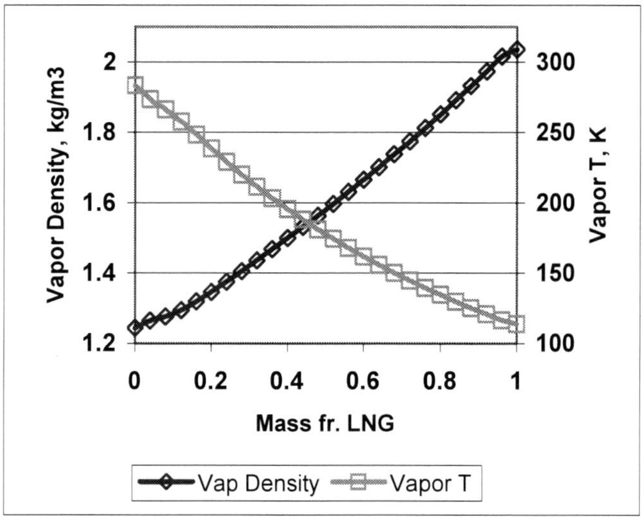

Figure 15. Adiabatic Mixing Curve for LNG Vapor and Air

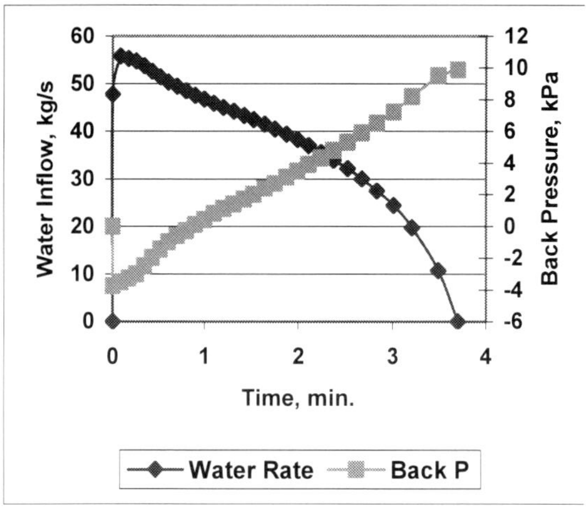

Figure 16. Early Response, 1 m Deep, D_{OH} 0.15 m, D_{IH} 0.05 m, Water Inflow, Back Pressure Increase

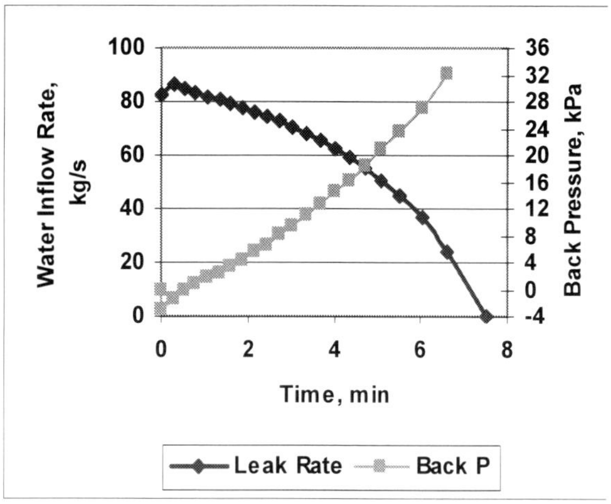

Figure 17. Early Response, 3 m Deep, D_{OH} 0.15 m, D_{IH} 0.05 m, Water Inflow, Back Pressure Increase

The vapor composition and density is very nearly the same for all hole sizes at the end of the early response period, (except for numerical analysis error) depending only on the depth of the outer hull penetration. The mass of LNG vapor introduced by the end of the initial period increases and the volume of water in the hull space decreases with larger LNG tank hole size.

Figure 18 shows that the mass ratio of LNG to water increases slightly in the early response period since the water inflow decrease is initially more sensitive than LNG inflow to back-pressure buildup. The small amount of water inflow plotted in Figure 18 emphasizes that the hull head pressure build-up is largely because of the LNG vapor accumulation rather than volume compression.

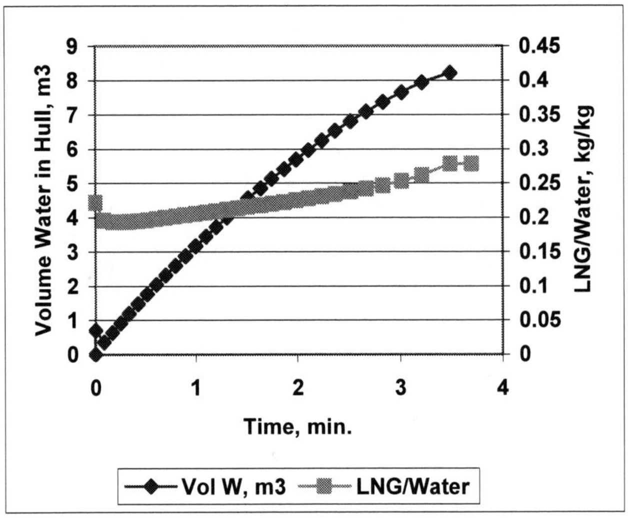

Figure 18. Water In Hull and LNG/Water Mass Ratio for Early Response With 1 m Deep, D_{OH} 0.15 m, D_{IH} 0.05 m Penetrations

6.3 Mid Response Period

Table 6 lists the flux and flow rates for water inflow and vapor outflow found by the SafeSite$_{3G}$® model (unreleased version) at the start of the mid-range period.

By a similar mechanism, after the hull head pressure equals the LNG tank internal pressure at the inner hole, air and vapor enter the LNG tank, and LNG continues to leak. Since the inflow and outflow to the LNG tank are equal volumetric rates, there is little effect on the LNG tank head pressure during this period.

Figure 19 illustrates water and LNG inflow rates predicted with the equal volume exchange model applied at first for vapor/water only. The water inflow rate is shown going to zero at the start of the vapor outflow period. This is a modeling artifice, since the change in outflow would not likely be discontinuous. The LNG rate also remains at zero since the model does not yet provide for equal-volume LNG outflow and vapor inflow to the LNG tank.

Figure 20 provides the change in vapor composition and the mass ratio of LNG to water over time. When the LNG out flow is choked off, these values remain nearly constant.

Table 6. Flux at Beginning of Mid Response Period

Hole D. Ratio, D_{OH}/D_{IH}	Variable	1 m Hole Depth	3 m Hole Depth	5 m Hole Depth	7 m Hole Depth
All	G_W, kg W/(m^{2}s)	2701.	4678	6095	7147
	G_V, kg W/(m^{2}s)	126.5	248.1	351.7	447.9
	f_w, A_w/A_{OH}	0.9665	0.9620	0.9580	0.9554
3:1 D_{OH}=0.15m	F_{disW}, kg water/s	184.5	318.1	412.7	482.7
	$Fdis_V$, kg vapor/s	0.30	0.666	1.04	1.41
	Initial LNG F_{dis}, kg/s	10.6	11.2	11.7	12.2
2:1 D_{OH}=0.60m	F_{disW}, kg water/s	184.2	316.5	408.9	476.1
	$Fdis_V$, kg vapor/s	0.30	0.663	1.03	1.39
	Initial LNG F_{dis}, kg/s	381	401.6	421.1	439.8

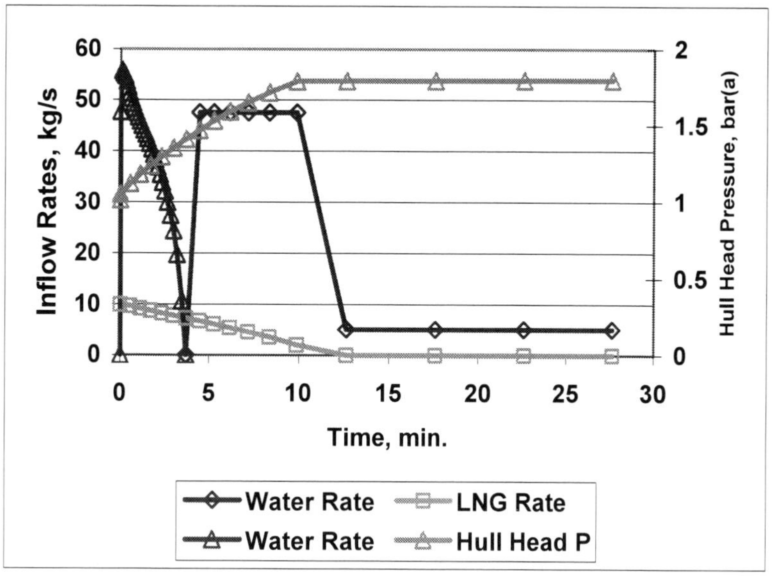

Figure 19. Early Response of Penetration of Outer and Inner Hull, 1 m Deep, 0.15 m Outer Hole, 0.05 m Inner Hole; Leak Rates and Back Pressure

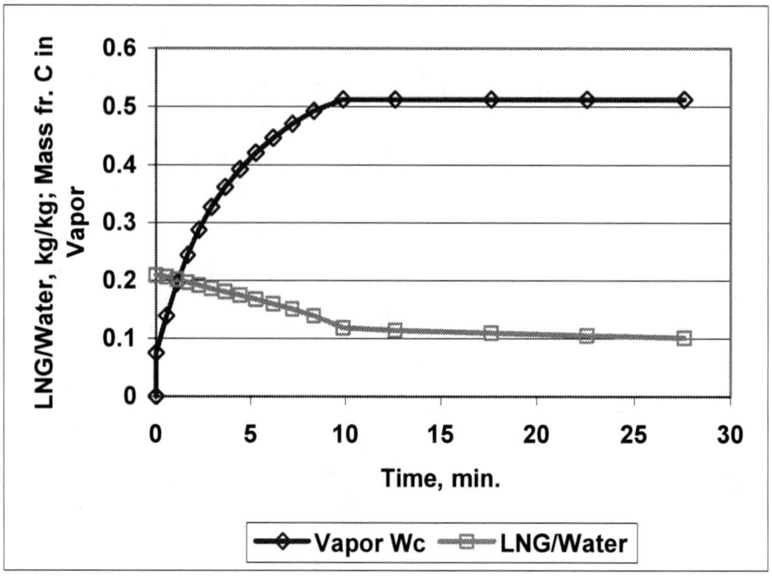

Figure 20. Vapor Composition, LNG/Water Mass Ratio, Volume of Water in Hull, Example Case Mid Response Period

The equal-volume exchange period ends when the volume of water reaches the top of the outer and inner hull holes, as illustrated in Figure 21. Response values in the mid-response period are summarized in Table 7. The maximum pressures listed in Table 7 are likely an upper limit since they do not account for vapor bleed into the LNG tank. The pressures and water levels in the hull at the end of the mid response period listed in Table 7 show numerical overshoot. Correct pressure and final water levels are also listed based on the theoretical considerations discussed earlier.

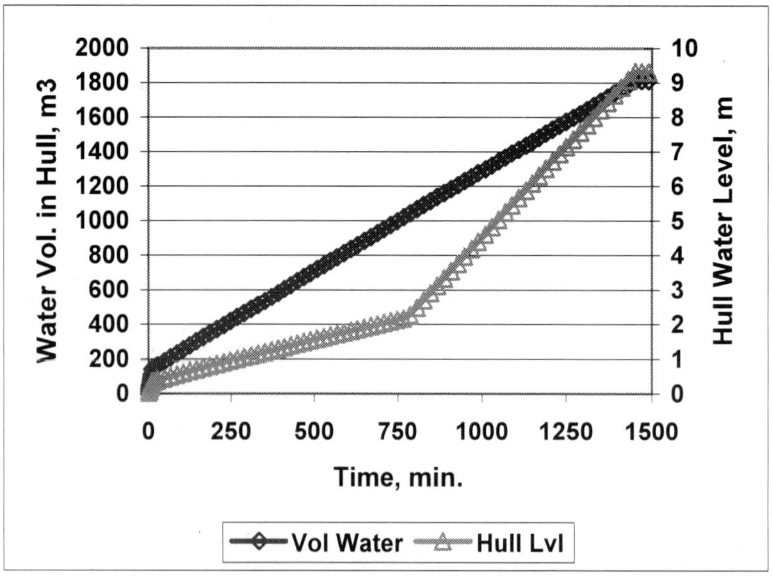

Figure 21. Predicted Water Volume and Level in Hull for Equal Volume Exchange Period, Example Case

Table 7. Conditions at Start and End of Mid Response Period

Hole D. Ratio, D_{OH}/D_{IH}	Conditions at Start of, Vapor Backflow to Water (Mid Response Period)	1 m Hole Depth	3 m Hole Depth	5 m Hole Depth	7 m Hole Depth
3:1 D_{OH}= 0.15m	Time, min.	3.7	7.5	10.5	12.4
	Vol. Water, m^3	8.3	27.5	48.9	68.4
	Mass fr LNG , w_C	0.386	0.554	0.645	0.691
	Vapor Temp, °K	198	169	155	149
	Vapor Mole Wt	24.0	22.40	21.6	21.2
	Vapor Density, ρ_V kg/m^3	1.6215	2.0886	2.5219	2.9215
2:1 D_{OH}= 0.60m	Time, min.	0.12	0.23	0.27	0.35
	Vol. Water, m^3	4.2	11.7	21.0	31.6
	Mass fr LNG, w_C	0.400	0.575	0.628	0.693
	Vapor Temp, °K	196	165	157	148
	Vapor Mole Wt	23.9	22.2	21.8	21.2
	Vapor Density, ρ_V kg/m^3	1.6332	2.1123	2.5000	2.9251
D_{OH}/D_{IH}	**Conditions at Start of Vapor Backflow to LNG**	**1 m Hole Depth**	**3 m Hole Depth**	**5 m Hole Depth**	**7 m Hole Depth**
3:1	Time, min.	10.5	10.5	10.5	12.4
	Head Pressure, bar(g)	0.84	0.94	1.13	1.17
2:1	Time, min.	0.36	0.39	0.45	0.48
	Head Pressure, bar(g)	0.96	0.96	1.29	1.13
Any	cf LNG Pressure, bar(g) at top of hole	0.837	0.930	1.023	1.116
D_{OH}/D_{IH}	**Conditions at End of Mid Response Period**	**1 m Hole Depth**	**3 m Hole Depth**	**5 m Hole Depth**	**7 m Hole Depth**
3:1	Time, min	1450	720	438	284
	Max. Head Pressure, bar(g)	2.01	1.96	1.89	1.69
	Water Inflow Rate, kg/s	5.1	8.4	10.7	12.3
	Vol. Water, m^3	1810	1600	1380	1165
	Water Level in Hull, m	9.3	7.4	5.4	3.4
2:1	Time, min	40.4	25.4	15.5	10.4
	Max. Head Pressure, bar(g)	2.53	2.20	2.19	1.79
	Water Inflow Rate, kg/s	180	304	380	440
	Vol. Water, m^3	1950	1870	1539	1190
	Water Level in Hull, m (Values should be)	10.6 (9.32)	9.9 (7.32)	6.5 (5.32)	3.6 (3.32)

6.4 Effects of Water Inflow to LNG Tank

To evaluate the effect of water flowing into the LNG tank, assume that the elevation of some inner hull holes is below the water level in the hull. This is done in Table 8, assuming a water level above the lower holes of 0.32 m. Table 8 lists the theoretical hull head pressures at the end of the mid period, the water inflow rate, and the LNG vapor generation rate. The latter is 1.08 times the water inflow rate, according to Figure 12.

Taking a pressure relief valve with a one-foot equivalent relieving diameter, we find the head overpressure in the LNG tank to discharge the generated vapor. These overpressures (around 0.35 barg) are far higher than the pressure driving force for water inflow (0.015 barg), indicating that the pressure increase caused by water intrusion into the LNG tank would quickly choke off further water inflow.

Table 8. Overpressure from Water Inflow into LNG Tank

Hole D. Ratio, D_{OH}/D_{IH}	Variable	1 m Hole Depth	3 m Hole Depth	5 m Hole Depth	7 m Hole Depth
3:1 D_{IH}= 0.05m	Top Elevation of Outer Hole	9.32	7.32	5.32	3.32
	Top Elevation of Inner Hole	9.00	7.00	5.00	3.00
	Hull Head Pressure, bar(g)	0.852	0.945	1.038	1.130
	Pressure at Low Hole, bar(g)	0.867	0.960	1.052	1.145
	Delta P Across Hole, bar(g)	0.015	0.015	0.015	0.015
	F_{dis} Water to LNG Tank,kg/s	23.31	23.78	24.42	24.96
	F_{Vap} Vapor Generation, kg/s	25.28	25.79	26.48	27.07
	Overpressure, bar(g)	0.344	0.355	0.37	0.383

7.0 Conclusions

An unreleased version of SafeSite$_{3G}$® was modified to model how the back-pressure build up affects blow-down (time-dependent flows of water and LNG). This modeling provides estimates of the response time and quantification of pressures, flow rates, and volumes of water and LNG leaked into the double hull.

An underwater breach of the double hull of an LNG carrier would increase the pressure in the double hull space by water inflow compressing the vapor volume. Water would cover the outer hull breach until the head pressure in the hull space approximately balances the hydraulic pressure of sea water. This pressure balance point occurs much sooner if the LNG tank is also breached. LNG vaporizes completely by mixing with water, and develops a low mass fraction of ice/water by chilling the water. The build-up of LNG vapor is a strong factor in the rise of hull pressure. A maximum pressure in the hull space is predicted to be about two bar gauge. Checking the strength of the double hull is beyond the scope of this work, but is expected to be sufficient. The effects of a possible rapid phase transition in the hull space is also outside the scope of this study.

After the initial period of pressure rise, air and LNG vapor begins to flow out of the outer hull breach, driven by the buoyancy difference with sea water. Water continues to flow into the hull space on an approximately equal volumetric flow rate basis. LNG also continues to flow into the inner hull until the hull pressure equals the LNG hydraulic pressure and momentarily (theoretically) stops the LNG leak. After that, air and vapor backflow into the LNG tank at a volumetric flow rate equal to LNG outflow. Since the inflowing air/vapor volume equals the out-flowing liquid volume, there is minimal affect on the head pressure for the LNG tank in this response period.

The effect of hole elevation and hole size is shown throughout. Deeper penetrations have higher external hydraulic pressure and higher water inflow rates for the same hole size. Preliminary pressure equilibrium (the end of the early response period) occurs later for deeper penetrations. The final water level in the hull ultimately reaches the top of the outer hull penetration.

With a distribution of inner hull holes, some holes may fall below the final water level. Since the hull head pressure tends to equilibrate with the external hydraulic pressures, there is an initial pressure driving force for water infiltration into the LNG tank. This is assuming the head pressure in the LNG tank is approximately atmospheric, as held by pressure and vacuum relief valves. The intrusion of water, though, generates about 1.08 kg of LNG vapor per kg of water, and this vapor increases the LNG tank head pressure. Preliminary calculations indicate this pressure increase will quickly choke the inflow of water for even small (0.05 m diameter) inner hull penetrations.

Ice could build near the inner hull penetrations and further choke off LNG outflow and potential water inflow. The tendency for ice formation drops off over time since the LNG leak rate is choked off by increases in hull pressure.

Appendix: Blow Down Model of Hull Space

In the unreleased version of the $\text{SafeSite}_{3G}^{®}$ program developed by BakerRisk, blow-down calculations for LNG and water discharge from a tank are coupled with corresponding inflow to the double hull space. Blow down of water inflow, $F_{Win}(t)$ and LNG outflow, $F_{disLC}(t)$ are treated separately in the SafeSite_{3G} program. Only this modification is documented here.

Liquid discharge is treated essentially by Eq. 14 for incompressible flow. Vapor outflow, F_{disV}, is found using compressible flow formulas and the pressures are found with Eqs 1, 2 and 16. The flow rate predictions provide the needed discharge mass fluxes, G_W and G_V, used in Eqs 18 and 19. The mass flux of water, G_w, is a function only of the depth to the top of the outer hull puncture. The mass flux of vapor varies with the head pressure, P_{hullh}. The head pressure is given by Eq. 11. The vapor temperature is found by an enthalpy balance on the vapor space and interpolated from Figure 15. The vapor molecular weight is found from Eq. 22 using the mass fraction w_{CV} from Eq. 26, below. The vapor density is given by Eq. 21.

The component mass balances on the hull space must account for the inflow of LNG and water and the outflow of air and LNG vapor assuming all liquid LNG vaporizes. The component mass balances on the hull space are determined as follows:

$$m_w = \int F_{Win} dt \quad (23)$$

$$m_{air} = m_{air0} - \int (1 - w_{CV}) F_{disV} dt \quad \text{and } m_{air} \geq 0 \quad (24)$$

$$m_{CV} = \int (F_{disLC} - w_{CV} F_{disV}) dt \quad (25)$$

$$w_{CV} = \frac{m_{CV}}{m_{air} + m_{CV}} \quad (26)$$

$$V_W = \frac{m_W}{\rho_W} \quad (27)$$

References

Baker Engineering and Risk Consultants, Inc., "SafeSite$_{3G}$® Theory Manual" Version 1.04, Apr. 28, 2005.

Daubert, T. E. and Danner, R. P, "Physical and Thermodynamic Properties of Pure Chemicals, Data Compilation", AIChE, New York, NY, 1989.

Biography of Speaker

JOHN L. WOODWARD, Ph.D.

Ph.D. in Chemical Engineering, Purdue University
B.E.S. in Chemical Engineering, Brigham Young University
B.S. in Chemistry, Brigham Young University

Dr. Woodward has over 38 years experience in the petrochemical industry, involved in nearly all process units in petroleum refining, and in a number of chemical plants. His career began in the field of process control, and in recent years has concentrated on process safety, risk analysis, and modeling of fires, explosions, and toxic gas dispersion.

Dr. Woodward began his career as a process control engineer for Shell Oil Company. Subsequent positions include Exxon Research and Engineering Company (Group Head in the Environmental Control and Safety Division and Senior Engineer in the Process Control Division), AIChE (Assistant to the Director, CCPS), Ecology and Environment, DNV Technica (Staff Engineer), and Baker Engineering and Risk Consultants, Inc. (Principal Engineer).

Dr. Woodward has been responsible for significant safety and risk analysis projects, and more recently for the development of a major software project utilizing his experience in developing the Unified Dispersion Model (UDM) for consequence analysis as the modern "engine" of the PHAST and SAFETI programs. As a result of developing the theory for discharge rates, Dr. Woodward wrote a section for the 7th and 8th editions of Perry's Chemical Engineer's Handbook and a chapter in a book on "Prevention and Control of Accidental Releases of Hazardous Gases". He has also written two books for AIChE's Center for Chemical Process Safety (CCPS), one on aerosol droplet evaporation and rainout modeling, and one on predicting the flammable mass in a vapor cloud.

An indoor dispersion model developed by Dr. Woodward has proved useful to analyze indoor spills of volatile solvents for chemical and pharmaceutical companies. The approach helps to design the number and type of floor drains and impaction plates to reduce aerosol evaporation.

For cases involving litigation, he has modeled number of accidental releases. Dr. Woodward has modeled chlorine releases and the release of naptha to the roof of a large storage tank when part of the roof sunk from a pontoon failure. Model predictions were compared with measurements of a post-event monitoring program to establish the evaporation rate and mass evaporated.

Dr. Woodward has investigated accidents and developed specialized models to analyze and describe the events. This includes the complete destruction of a propane absorption tower and an explosion in a 26 foot diameter fractionator. He investigated explosions that destroyed two processing buildings at a catalyst company and analyzed causes and damage from an explosion of ditertiary butylperoxide.

Blast Waves Produced by Rapid Phase Transition of LNG on Water

Ernesto Salzano
Istituto di Ricerche sulla Combustione, Consiglio Nazionale delle Ricerche,
Piazzale Tecchio 80, 80125 Napoli (I). Email: salzano@irc.cnr.it

Roberto Bubbico
Dipartimento di Ingegneria Chimica, Università di Roma "La Sapienza",
Via Eudossiana 18 - 00184, Roma (I). Email: bubbico@imp**

AIChE Spring Meeting, April 2008
8th Topical Conference on Natural Gas Utilization
New Orleans, Louisiana, April 6-10, 2008

Introduction

Massive offshore and onshore storage of Liquefied Natural Gas (LNG) have addressed the public concerns on all possible hazards on people and other installations either onshore or on ship. Indeed, after accidental or voluntary (after attack or sabotage) releases, the vapour forms a cold gas cloud which is denser than air and can travel downwind until an ignition source is encountered. As a consequence, several accidental scenarios should be of concern: flash fire, fireball, pool fire, and vapour cloud explosion, the latest provided that confinement and/or congested geometry surround the release point. To these regards, several reviews on risks related to the release of large amount of LNG are reported in the literature (Schubach & Bagster; ABSG, 2004; Sandia, 2004; Shaw et al. 2005; Luketa-Hanlin, 2006; Melhem et al., 2006; Koopman & Ermak, 2007).

Among the possible accidental scenarios, when cryogenic gas as LNG is spilled on water at a very fast rate, the phenomenon of Rapid Phase Transition (RPT), has also to be taken into consideration. RPT is the very fast (explosive) evaporation of cryogenic liquid after liquid-liquid (LNG-water) contact, which does not necessary involve fire or other chemical reactions, but represents a pure physical explosive phenomenon. Its occurrence has been confirmed by the results of different large scale experimental tests reported in the open literature (see for instance Atallah, 1997, for a complete description of experiments).

The analyses of the mechanism of RPT explosion of LNG have been proposed by several authors even recently (Atallah, 1997; Conrado & Vesovic, 2000; ABSG, 2004; Havens & Spicer, 2006). However, the conditions that allow an actual explosive evaporation to occur are still not clear. Moreover, the intensity of blast wave and the propagation in the surrounding atmosphere, the knowledge of which is essential for risk assessment of large scale installations, is not easily predictable, as the total energy involved in the explosion and the peak overpressure at the source term depend on many factors such as the release rate, source localization (above or below the sea level), external environmental conditions, LNG composition, water temperature, and impact velocity of LNG on water surface. In this paper, an acoustic model has been adopted for the definition of the explosion pressure and total energy produced by the RPT of LNG. Results are compared with the conclusions of a recent report by Sandia National Laboratory (Sandia, 2004) which has revisited LNG hazards focused on spills from ships onto water, founding that i) the achievable overpressures from RPT explosions for very large spills (~ 100 m^3/min) and the possible upper bounds of damage to structures have not been evaluated; ii) the most significant impacts to public safety and property from an accidental spill exist within approximately 250 m of the spill and for an intentional release, within about 500 m of the spill. Results of comparison may be used for the definition of exclusion zones, for protection of people and property along the transit route even for on-shore releases, and for public impact.

RPT theory and experiments

A recent report by ABSG Consulting Inc. (2004) states clearly that RPT is a poorly understood phenomenon. The main conclusion of the report is that RPT has to be attributed to superheating of the LNG and to the lack of nucleation sites for the formation of gas bubbles and for the promotion of boiling. Following the same report, RPT may result in two types of effects: the overpressure resulting from the rapid phase change and dispersion of the puff of LNG expelled into the atmosphere. The second effect may increase the downwind distance where the LFL is reached or it may affect the size of a pool fire. In another report, Luketa-Hanlin (2006) recognizes that research on LNG/water expansions has been principally at laboratory scale and that few field tests have been performed, with the exception of the Burro and Coyote test series performed by LLNL (see the report of Luketa-Hanlin and the following paragraphs for more details), which have been the only large-scale test explicitly undertaken for the study of RPTs. Most of the explosions were "early RPTs", i.e., they occurred immediately after the spill even if, in some cases, RPTs continued for the duration of the spill (over a minute). Observed RPT were generally located near the spill point and appeared to be primarily underwater. Delayed RPTs occurred in three tests at the end of the spill and were located away from the spill point. The maximum equivalent free-air, point source TNT explosion that occurred was 6.3 kg for a test performed with LNG spill rate of 18 m^3/min. To this regard, the spill rate has been found to be a fundamental parameter in determining the strength of the vapor expansion. Indeed, increasing the spill rate increases the effective explosive yield of an RPT event: An abrupt increase in the RPT explosive yield was found at around 15 m^3/min. RPTs were also found to increase the distance to LFL by as much as 65% due to the expansion induced by the strong vaporization (see also ABSG (2004). Due to the preferential initial vaporization of methane, delayed RPTs may be associated with ethane enrichment (up to 40% by volume) in the generated 'puffs', due to the higher concentration of ethane in the LNG on the pond. Since ethane has a lower flammability limit (LFL = 3%) than methane (5%), the distance to LFL will be increased (up to about 60%). In the large scale tests, the occurrence of RPTs appears to be correlated with the water temperature and with the depth of penetration of the LNG into the water. Indeed, RPTs has occurred only for water temperature above 290 K, though when an adjustable plate, used to prevent LNG from impinging directly upon the pond, was removed allowing for deeper penetration into the water, RPTs resulted even for water temperature of 285 K. The correlation between water temperature and the superheat temperature (TSH) of LNG required for RPT occurrence is usually expressed as $TSH < T_{water} < 1.1\ TSH$, where TSH for methane, ethane, propane, and butane are respectively 168 K, 269 K, 326 K, and 376 K (Khalil et al., 1988) and the superheat temperature of hydrocarbon mixtures is approximately the mole fraction average of the superheat temperatures of the components. To this regard, Schubach & Bagster (1996), analyzing the open literature for steam explosions (IChemE, 1994), affirm that the maximum energy yield of RPT is limited by the superheat energy of the cold fluid involved and that only a small fraction of the available superheat energy is transferred into the shock wave. Hence, in terms of energy production, RPTs are less efficient than higher explosives at generating blast. Also, Melhem et al. (2006) have surveyed the open literature about LNG rapid phase transitions observing that the hazard potential of rapid phase transitions can be severe, but is highly localized within the spill area, and that until a more detailed model is developed to better represent RPT process, the use of the superheat limit explosion yield of 20.7 kJ/dm^3 is recommended.

The survey of Atallah (1997) on LNG-water releases states that, at the moment, there is no theoretical method to predict the severity of an RPT event involving the release of LNG from an LNG carrier onto water. The review appears the most complete on RPT, for the aims of this work. Citing Enger and Hartman (1972a-d), it states that RPT for gases released on water follows the superheat temperature model, where TSH is about 0.8-0.9 of critical temperature of the gas. Citing Corbin and Reid (1981) and Jazayeri (1975) a peak overpressure related to RPT is obtained experimentally. No explosion was found for LNG on water, unless C3 addition to the gas mixture is consistent. McRae (1982; 1983) and McRae et al. (1983; 1984a; 1984b) have analyzed the effects of impact of LNG for RPT: Underwater spill greatly affects the ability of release to produce an RPT. Indeed, Atallah reports the effect of water depth on equivalent explosive mass calculation. For 30 cm underwater, the explosive mass is between 0.01 and 0.001 times the actual explosive mass. Shaw et al. (2005) have also analyzed the effect of underwater releases, which result in great diminution of blast wave pressure on water surface.

Nedelka et al. (2003) believe that RPT is unlikely to damage large structural elements of a ship. This argument is only speculative and no experiments are presented. More recently, Koopman & Ermak (2007) have defined the main hazards from episodic LNG releases. However they relegated to secondary importance either the Vapour Cloud Explosion, based on the determination that LNG vapors are unlikely to explode unless strong confinement or abnormally large amounts of gases more reactive than methane are present, or RPT, due to scaling considerations, which appear to limit the amount of danger (overpressure) that could occur. However, Havens & Spicer (2006), commenting that work, claim that RPTs may nonetheless cause domino effects on containment failures. Nevertheless, Koopman and Ermak (2007), by reviewing the well known large-scale experimental tests of Burro and Coyote, define two types of RPTs: those that occur at the spill point at any time during the spill, which are correlated to the spill rate; and those that generally occurred out on the edge of the LNG pool, late in the spill, which are correlated with ethane/propane concentration in the main gas. For both cases, the strength of RPTs as reported by the authors was observed to vary from the equivalent of 3.5 kg of TNT down to a few grams of TNT. RPTs were observed to increase the burn area to about two times that estimated without RPTs and to extend the downwind burn distance by 65%. For spills on water, the consequences of RPTs on dispersion distances and burn areas should be considered in risk analyses, as the puffs of LNG vapor created by the RPTs may increase the hazard area significantly. The Burro and Coyote tests established a correlation between spill rate and RPT intensity but they were limited by the China Lake facility to about 19 m^3/min. In Burro tests, RPT explosions with a maximum overpressure (static) of 0.05 bar were measured at 30 m from the RPT itself. The explosions were strong enough to cause damage to the facility.

Explosions and acoustic waves propagation

The present analysis is based on the acoustic approximation as defined by Lighthill (1978), which reproduces the peak overpressure ΔP (in bar) with respect to distance r from the acoustic source point. If ρ is the density, t the time, u the velocity, and the wave potential φ is defined as:

$$\vec{u} = \text{grad}\ \varphi = \nabla\varphi \qquad (1)$$

then conservation equations for mass and momentum can be written as:

$$\frac{d\rho}{dt} + \nabla\rho\nabla\varphi = 0 \tag{2}$$

$$\frac{d}{dt}(\rho\nabla\varphi) + \nabla P = 0 \tag{3}$$

Under iso-entropic and acoustic assumptions, the conservation equations can be written as:

$$\frac{d\rho}{dt} + \nabla\rho\vec{u} = \frac{1}{c_o^2}\frac{dP}{dt} + \nabla\rho\vec{u} = 0 \tag{4}$$

$$\nabla P = -\frac{d}{dt}\left(\rho\vec{u}\right) = -\frac{d}{dt}(\rho\nabla\varphi) \tag{5}$$

Assuming spherical symmetry, these equation can be written as:

$$\frac{d^2P}{dr^2} = \frac{1}{c_o^2}\frac{d^2}{dt^2}rP \qquad \text{(Acoustic wave equation)} \tag{6}$$

$$\frac{d^2\varphi}{dr^2} = \frac{1}{c_o^2}\frac{d^2}{dt^2}r\varphi \qquad \text{(Potential wave equation)} \tag{7}$$

where r is the distance from the acoustic source. In the case of spherical symmetry the solution of the scalar potential wave equation, is:

$$\varphi = \frac{f(t)}{r} \tag{8}$$

where f is an arbitrary function for the wave, the factor (1/r) determines the amplitude and t = (t* - r/co) where t* is the actual time and r/co is the time lag for propagation. Considering now an expanding sphere centered at the origin so that the equation of its surface is:

$$r = a(t) = a_o + a_1(t) \tag{9}$$

and the fluid velocity at the sphere surface is:

$$\vec{u} = \frac{dr}{dt} = \dot{a}(t) = \frac{d\varphi}{dr} \tag{10}$$

If the volume flow rate through the surface of the expanding sphere of radius r is:

$$V'(t) = 4\pi r^2\frac{dr}{dt} \tag{11}$$

it is then:

$$\frac{d\varphi}{dr} = \frac{dr}{dt} = \frac{V'(t)}{4\pi r^2} \tag{12}$$

and by integrating on r:

$$\varphi = -\frac{V'(t)}{4\pi r} \tag{13}$$

At any point (receiver) located at a distance (radius) r from the source, the pressure observed as the wave with form f is emitted at the time t is then:

$$\int_r \nabla P = \Delta P = -\int_r \frac{d}{dt}\rho\nabla\varphi = -\rho\frac{d\varphi}{dt} \tag{14}$$

and finally:

$$\Delta P = \rho\frac{d}{dt}\left[\frac{V'(t)}{4\pi r}\right] = \rho\frac{V''(t)}{4\pi r} \tag{15}$$

or, if the mass flow rate is q'(t), it is:

$$q'(t) = \rho\,V'(t) \tag{16}$$

$$\Delta P = \frac{q''(t)}{4\pi r} \tag{17}$$

This equation says that at a distance r from acoustic source of an expanding sphere, as a consequence of volume flow rate in the volume V, the pressure is P. If the volume or mass flow rate is constant or zero, the pressure gauge is zero. The term q is often defined as the strength of the acoustic source. Finally, from the equation of sound:

$$P_o = \frac{\rho_o c_o^2}{\gamma} \tag{18}$$

it follows:

$$\frac{\Delta P}{P_o} = \frac{\gamma V''(t)}{c_o^2 4\pi r} = \frac{\gamma q''(t)}{\rho_o c_o^2 4\pi r} \tag{19}$$

Given the frequency ω of spherical pressure wave, Lighthill has demonstrated that the equation is affected by source characteristic for values of $\omega r/c_o$ lower than 6. If co is 350 m/s at ambient temperature, and the total duration time of flash is conservatively - in terms of distances for which near field effects are consistent - equal to 1 ms, it can be calculated a wave frequency of 2000 s^{-1} (= 1/0.0005) and the maximum value for r is about 1 m. Quite clearly, the results are affected if long duration releases are of concern. The effect of non-ideality of acoustic solutions will be analysed in next works.

The application of the acoustic equation to LNG release on water

Most of the experimental investigations available in the literature have been performed mainly for the analysis of different phenomena such as vapor cloud explosions and fires. However, during the Burro tests, performed by the LLNL and the Naval Weapons Center at China Lake, California, the occurrence of RPTs was also investigated. A brief summary of some experimental results is reported in Table 1 (for original data see http://www.ac-research.org).

Table 1 – Summary of Burro Series.

	Test 2	Test 3	Test 4	Test 5	Test 6	Test 7	Test 8	Test 9
Month/year	6/80	7/80	7/9/80	7/80	8/80	8/80	9/80	9/80
Time of day for spill	16:00	15:00	14:00	16:20	16:00	18:10	19:10	18:40
Duration (s)	173	166	175	190	129	174	107	79
Spill Rate (m3/min)	11.9	12.2	12.1	11.3	12.8	13.6	16.0	18.4
Spill Volume (m3)	34.3	34.0	35.3	35.8	27.5	39.4	28.4	24.2
Avg. Wind Speed (m/sec)	5.4	5.4	9.0	7.4	9.1	8.4	1.8	5.7

For some of these tests, RPT occurrence was actually observed. The full pressure histories and total duration time of the positive phase for these tests are reported in Atallah (1997) which in turn refers to McRae (1982; 1983; 1984a; 1984b), for the Burro tests 4, 5 and 9. Limiting the analysis to these three tests, the corresponding positive phase duration and maximum pressure are reported in the following Table 2.

Table 2. Positive phase duration and peak pressure as measured along the blast wave pathway on Burro tests as reported by Atallah (1997).

Burro tests	Spill time (s)	Spill rate (m3 min-1)	Volume (m3)	Gauge distance (m) 11.4	24.3	30.0	35.5
				Peak Pressure (bar)			
Test 4	175	12.1	35.3	0.06	0.05		0.02
Test 5	190	11.3	35.8	0.17	0.085		0.034
Test 9	79	18.4	24.2			0.05	
				Positive phase time (ms)			
Test 4	175	12.1	35.3	11.0	12.5		14.0
Test 5	190	11.3	35.8	7.0	11.0		12.0
Test 9	79	18.4	24.2			40.0	

The data reported in the previous tables may be useful either for the evaluation of the initial energy of the explosion or for the definition of the initial explosion pressure. In the present case, the acoustic model will be applied to the calculation of the peak pressure profile generated by the fast evaporation of LNG from water surface, and the above data will be used for validation.

As suggested by van den Berg et al. (2004), in the case of fast evaporation associated to the BLEVE of liquid propane, the volume source strength can be expressed as a function of the volume of the liquefied gas, V, the flash fraction after release, F, and the expansion ratio between liquid and vapor, σ. Based on this assumption, the blast overpressure can be rearranged as:

$$\frac{\Delta P}{P^{\circ}} = \frac{\gamma}{2\pi r} \frac{1}{c_0^2} 2 \frac{VF\sigma}{\Delta t^2} \tag{23}$$

where Δt is the flash time.

Following the conservative approach suggested by van den Berg et al. (2004), in the present case it will be assumed that the flash fraction for liquefied natural gas is F = 1. Furthermore, considering that the liquid density at ambient temperature is 423 kg/m^3 and that of the vapor is 1.819 kg/m^3, at the boiling point, and between 0.6 and 0.8 kg/m^3 at ambient temperature (depending on LNG composition), we obtain an average value for the expansion ratio of about 600. However, as in the case of the BLEVE, no information is available about the time required for the evaporation process to occur. This last parameter will depend on different factors such as the water temperature, the degree of mixing between liquefied gas and water, the heat transfer rate and so on. As a consequence, the value of Δt in Eq. (23), has been varied to find the best line matching the experimental data.

In Figure 1 the experimental overpressures measured at specific distances from the source during the Burro tests 4 and 5 (both carried out at a release rate of about 12 m^3 min^{-1}) are compared with the peak pressure profiles calculated by using the acoustic model.

Figure 1. Calculated (continuous lines) and experimental values (single points), for RPT overpressure vs. distance for release rate of 12 m3 min-1 (Burro tests 4 and 5). The two fit lines refer to flash time of 0.2 ms (top) and 0.7 ms (bottom), respectively.

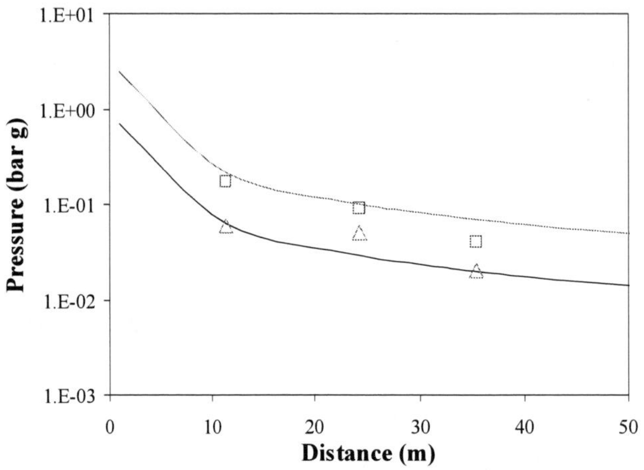

The two fitting lines have been calculated using a flash time of 0.2 ms (upper line) and 0.7 ms (lower line), respectively. This result clearly shows that the calculated values of the pressure wave agree with the experimental ones only when an almost instantaneous (less than 1 ms) and complete flash (F=1) of the released LNG is assumed.

Based on this result, in order to better investigate the influence of the release rate and of the flash time, the calculation of the peak pressure profile as a function of the distance has been repeated for three different values of each of these two parameters, i.e. 10, 25, 50 m^3/min and 0.1, 1 and 10 ms, respectively (Figure 2).

Figure 2 clearly shows the influence of the release rate and flash time and, as expected, the largest values of overpressure will be obtained at large release rates and very short flash times. Furthermore, it can be noticed that values of about 0.01 bar (considered as the minimum pressure for an RPT to be detected) are only observed for flash time lower than 1 ms, in the near field, whatever the release rate.

The maximum flow rate adopted during the Burro test, was about 19 $m^3\,min^{-1}$, however, as also suggested by Sandia (2004), in some cases (for example for intentional breaches, such as after a terrorist attack), quite larger releases are possible.

As a consequence, in order to have even a rough estimation of a "safe distance" corresponding either to an accidental or to an intentional release (for which no experiments are available) of LNG on water, a number of simulations have been performed at different release rates.

In Figure 3, the overpressure profiles at a release rate of 100, 200, 300 $m^3\,min^{-1}$, are reported. Based on the previous results about the flash time after release, an average value of 0.5 ms has been adopted for the calculations. In the Figure, a horizontal line at a reference value of 7 kPa, corresponding to an medium damage level from overpressure (steel/aluminum panels failure, partial demolition of houses), has been reported. It can be seen that only in the case of highly conservative conditions (300 $m^3\,min^{-1}$) the pressure profile is higher than the reference value, up to about 500 m. For the lowest release flow rate investigated, relatively high values of the overpressure are obtained only in the very near field, where the applicability of the model is uncertain. In particular, in order to have an overpressure of 8 kPa at a distance of 10 m from the release point, a minimum release rate of about 10 $m^3\,min^{-1}$ is calculated. However, it must be stressed that, as shown in Fig. 2, these results are heavily dependent on the assumed flash time. The actual value of this parameter will likely depend on a number of different factors, such as:

- characteristics of the liquefied gas (e.g. initial temperature, superheat limit temperature)
- characteristics of water (surface and bulk temperatures)
- extension of the interfacial area and heat transfer rate through the contact surface
- degree of mixing (which will also depend on the release velocity/flow rate, on the impact angle on water, and hence on the penetration of LNG into water, and on water turbulence)
- total water heat capacity (e.g. possibility of freezing).

Figure 2. Calculated acoustic RPT overpressure vs. distance, with flash time of 0.1, 1, and 10 ms and for release rates of 10 (———), 25 (····), and 50 m3 min -1 (----).

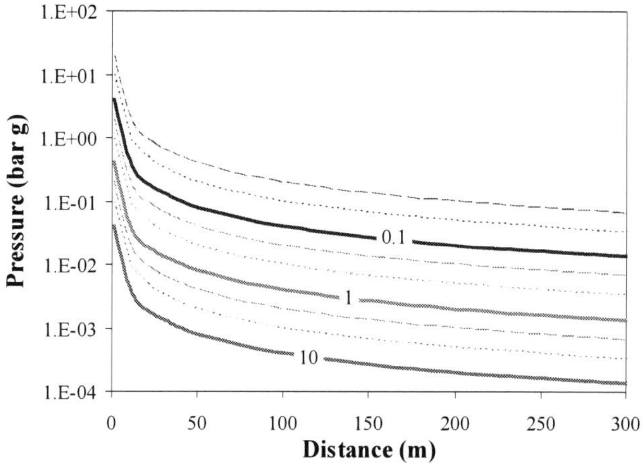

Unfortunately, no information about this parameter is presently available in the literature and thus, as a matter of fact, the velocity of this step of the phenomenon is unknown. Under this respect much work is necessary.

The obtained threshold data agree with Sandia (the most significant impacts to public safety and property from an accidental spill exist within approximately 250 m of the spill and for an intentional release, within about 500 m of the spill) only for ultra-conservative catastrophic spills of 300 m3 min-1, for which however no experiments exist.

Total energy release per second produced by RPT can be calculated starting from maximum pressure calculated at very short distance from spill by means of classical Brode equation:

$$E = \frac{V\Delta P}{\gamma - 1} \qquad (26)$$

These values could be compared to TNT energy equivalence reported in the literature if the total release time is provided, although there are many difficulties in reproducing the experiments on water as ice is formed.

Conclusions

Different types of consequences are likely following the release of liquefied natural gas (LNG), including physical explosion, vapor cloud explosion or flash-fire (in case of delayed ignition), pool fire and some other minor hazards (such as a jet fire), which can nonetheless degenerate into more severe accidents due to domino effects. In the particular case of release on a liquid surface (water), the possibility of an explosive expansion of the vapor generated by the very fast evaporation of the liquefied gas, must be taken into consideration. Results are reported in Table 3.

Figure 3 Calculated RPT overpressure vs. distance for release rates respectively (from bottom to top) of 100, 200 and 300 m3 min-1. Flash time of 0.5 ms. The horizontal line is the typical threshold value for overpressure of 0.08 bar.

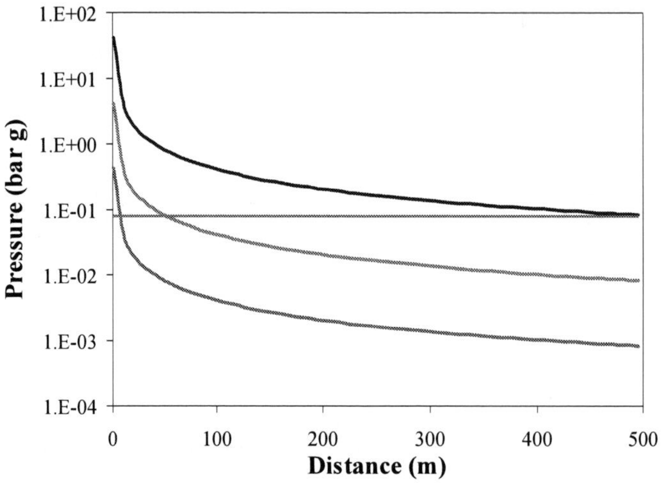

Table 3. Total energy per second of release. Flash time of 0.5 ms.

Release rate (m3 min-1)	Pmax at 5 m from release point (bar)	Energy (MJ) x second of release
5	0.08	2.1E-06
10	0.16	8.5E-06
100	1.62	8.5E-04
300	4.87	7.7E-03

The acoustic evaluation of pressure wave produced by the rapid phase transition of LNG released accidentally on water has demonstrated that the safety distances reported in the literature are confirmed. In the near field, although the recognized limits of near-field assumptions for acoustic fields, the pressure peak and the energy obtained are able to produce damages to very close atmospheric equipment and steel panel even for release rate of 10 m3/min, provided that 0.07 bar is considered a threshold value for structural integrity, but rapid decay of pressure are observed for larger distances.

References

1. ABSG Consulting Inc., 2004. Consequence Assessment Methods for Incidents Involving Releases from Liquefied Natural Gas Carriers, Federal Energy Regulatory Commission Contract n. FERC04C40196.
2. Atallah S., 1997. Rapid Phase Transitions, Topical Report GRI-92/0533, Gas Research Institute.
3. Conrado C., Vesovic V., 2000. The influence of chemical composition on vaporisation of LNG and LPG on unconfined water surfaces. Chemical Engineering Science, 55, 4549-4562.
4. Corbin G.A., Reid R.C., 1981. Liquid Natural Gas Rapid Phase Transitions, Report. GRI-80/0031, Gas Research Institute, Chicago, IL.
5. Enger T., Hartman D.E., 1972a. Explosive Boiling of Liquified Gases on Water. Material Research Council Conference on LNG Importation and Terminal Safety, Boston.

6. Enger T., Hartman D.E., 1972b. LNG Spillage on Water. Technical Progress Report, 1-72, Shell Pipeline Corporation.
7. Enger T., Hartman D.E., 1972c. Mechanics of the LNG-Water Interaction. AGA Distribution Conference, Atlanta.
8. Enger T., Hartman D.E., 1972d. Rapid Phase Transformation during LNG Spillage on Water. LNG-3, Washington.
9. Havens J., Spicer T., 2006. United States Regulations for Siting LNG Terminals: Problems and Potential, In: Salesi I.A., Havens J., Spider T., LNG Safety Research: FEM3A Model Development, Ist Quart. Report 2006, Award No. DE-FG26-04NT42030.
10. IChemE 1994. Explosions in the process industries, 2nd Ed., Rugby, UK.
11. Jazayeri B., 1975. Impact Cryogenic Vapor Explosion, MS Thesis, Chemical Engineering Department, Massachussets Institute of Technology, Cambridge, MA.
12. Khalil A., Fouad M.A., Kamel M.M., 1988. Modeling of confined cryogenic fuel spills over water. Proceedings of the 5th Miami International Symposium on Multi-Phase Transport and Particulate, , pp. 187–199.
13. Koopman R.P., Ermak D.L., 2007. Lessons learned from LNG safety research, Journal of Hazardous Materials, 140, 412-28.
14. Lighthill J., 1978. Waves in Fluids, Cambridge University Press, Cambridge.
15. Luketa-Hanlin A., 2006. A review of large-scale LNG spills: Experiments and modeling. Journal of Hazardous Materials,132, 119–140.
16. McRae T.G., 1982. Preliminary Analysis of RPT explosions observed in the LLNL/NWCLNG Spill Tests, Report n. UCRL-87564, Lawrence Livermore National Laboratory, Livermore, CA.
17. McRae T.G., 1983. Large-Scale Rapid-Phase-Transition Explosions, 3rd Multiphase Flow and Heat Transfer Symposium, Miami FL.
18. McRae T.G., Goldwire H.C., Hogan W.J., Morgan D.L., 1984a. The effects of large-scale LNG/Water RPT Explosion, Environmental Impact of LNG Spills, Winter Meeting of the ASME, Heat Transfer Division, New Orleans.
19. McRae T.G., Goldwire H.C., Koopman R.P., 1984b. Analysis of Large-Scale LNG/Water RPT Explosions, Report n. UCRL-91832, , Lawrence Livermore National Laboratory, Livermore, CA.
20. Melhem G. A., Saraf S, Ozog H., 2006. LNG Properties and Hazards - Understand LNG Rapid Phase Transitions (RPT), IoMosaic Corporation.
21. Napier D.H., Roochland D.R., 1984. Ignition characteristics of rapid phase transition explosions, Proc. Comb. Inst. of Canadian Section, Spring Technical Meeting.
22. Nedelka D., Sauter V., Goanvic J., Ohba R., Last developments in rapid phase transition knowledge and modeling techniques, OTC 15228, 2003.
23. Porteous W.M., Reid R.C., 1976. Light Hydrocarbon Vapor Explosion. Chemical Engineering Progress, 72, 83-89.
24. Sandia 2004. Guidance on Risk Analysis and Safety Implications of a Large Liquefied Natural Gas (LNG) Spill Over Water, SAND2004-6258.
25. Schubach S.A., Bagster D.F., 1996. Some aspects of modelling steam explosions. Journal of Loss Prevention in the Process Industry 9, 193-197.
26. Shaw S., Baik J., Pitblado R., 2005. Consequences of Underwater Releases of LNG. Process Safety Progress 24, 3.
27. van den Berg A.C., van der Voort M.M., Weerheijm J., Versloot N.H.A., 2004. Expansion-controlled evaporation: a safe approach to BLEVE blast. Journal of Loss Prevention in the Process Industries 17, 397–405.

Biography

Dott. Ernesto Salzano
via Diocleziano 328, Napoli, 80124, Italy.
Phone: +39 0817621922, Fax: +39 0817622915, Cell: +39 3475879041
E-mail: salzano@irc.cnr.it

Ernesto Salzano is a researcher of Istituto di Ricerche sulla Combustione since 1993. Main research activities are in the field of industrial safety: domino effects, seismic effects on industrial equipment, gas and dust explosion (experiments and modeling, CFD). He is authors of 30 papers on international specialist journals and books on safety and almost 100 papers in international conferences.

He is the project leader of several research contracts with Italian Department of Civil Protection, University of Bologna, ENI Tecnomare, DSM, and leads an experimental lab for safety parameters of substances at high pressure. He collaborates with Baker Engineering and Risk Consultants in US.

He is the scientific coordinator of EU research projects as: INTERREG; INTEG-RIS, K on multidisciplinary approaches on emergency risk (LNG, Na-Tech, Security, Nanotech).

Energy Based Risk Methodology for Collision Protection of LNG Carriers

Robin Pitblado[1], Erik Skramstad[2], Hans Kristian Danielsen[2], Thor Hysing[2]
1. DNV Energy, 16340 Park Ten Place, Houston TX 77084
2. DNV, Veritasvein, Høvik, Oslo, Norway

AIChE Spring Meeting, April 2008
8th Topical Conference on Natural Gas Utilization
New Orleans, LA, April 6-10, 2008

Abstract

LNG terminal developments are still the subject of major debate associated with potential hazards of marine accidental events. Prior risk studies have used estimates for potential hole size from accidental events that were based to a large degree on judgment, with hole size estimates primarily in the range from 0.75m to 5m.

A major sponsored project has undertaken a review of hole size from a detailed structural calculation methodology. This has used real collision data from the EU HARDER project to map known collision outcomes onto the actual structure of membrane and spherical LNG vessels. This showed that LNG vessels have great strength as the structures between the outer and inner hulls take and redistribute collision loads. However, once the inner hull starts to fail, the resistance to small increments in size of failure are relatively less than before inner hull failure. Results from the analysis concluded that it may be more accurate to establish a specific energy threshold causing the onset of tank failure rather than to nominate a "characteristic" hole size when such characterization is difficult.

This paper presents the results of the Finite Element Method (FEM) analysis for the two LNG vessel types and for several potential collision scenarios (size of vessel, speed, angle of collision). A consequence approach does not work well and only a risk-based approach leads to meaningful decisios on safe operations. An example of the risk based approach is presented.

Introduction

The USA approach to LNG facility siting is strongly consequence based. Onshore reception terminals must comply with DoT 49 CFR 193 and parts of NFPA59A (2006) which are adopted and which identify a number of scenarios to be modeled (Taylor, 2004). The marine situation is less well specified, but a number of recent studies (Sandia, ABS, DNV) have focused on consequences as the primary means to establish safe separation distances. However, regardless of the assumptions or models employed in the various studies, due to the size of vessel and potential hole sizes, none has been able to achieve the equivalent of the on-shore terminal requirement of ½ LFL (lower flammable limit) not leaving the site – interpreted as ½ LFL not reaching occupied shoreline for defined reasonable cases.

Marine shipments of LNG have an excellent operational safety record, with no significant loss of LNG cargo over 40+ years of operations. However, normal risk-based approaches acknowledge the potential for release, even if the operational record shows no such events (CCPS, 1999). The selection of hole size in the absence of real collision event data is the reason that the Sandia, ABS and DNV studies all used different hole sizes. Terrorism events add even greater uncertainty. An EU project, INTEGRISK is now being formed to develop more scientific methods for emerging risks where there are no historical precedents to draw from.

The original DNV work was conducted as a Joint Sponsor Project (Pitblado et al, 2004). The DNV hole sizes are smaller than the other studies – however the ultimate dispersion consequence distance and thermal hazard zone is a combination of several assumptions and models – hole size, discharge coefficient, weather, spread and evaporation models, dispersion models, burning rate, thermal radiation and the consequence results were not as different as the hole size alone might imply (Pitblado et al, 2006).

A second Joint Sponsor project was initiated to address some of the issues identified from the first study. In particular, one aspect was to advance from hole sizes based on judgment to ones derived from realistic collision models using detailed structural dynamics simulations. This technology has become well established in recent years for LNG vessels (Paik et al 2001) and is used for assessments such as prediction of collision outcomes for vessels at moored at FPSO (Floating Production Supply and Off-take) facilities.

Collision Scenarios

The study aimed to cover the full range of vessel collision scenarios so that a probabilistic hole-size result could be obtained. This involved 3 typical serious potential striking vessel sizes with two bow types, and two types of LNG carrier (membrane and spherical tank). Key statistics for the vessels are shown in .

Striking vessel sizes:	90m, 140m, 230m
Striking vessel displacement:	4000, 15,000 and 75,000 tonnes
Striking angles:	90°and 45°
Striking bows:	Raked and bulbous
Striking speed:	up to 19.5 kts (10 m/s)
LNG carriers:	Membrane and Spherical 138,000 m^3 LNG capacity
LNG vessel displacement:	110,000 tonnes
LNG vessel dimensions:	270m length, 45m breadth, 27m high, draft 11.5m
LNG Double hull separations:	Membrane 2.2m, Spherical 2.5m

Real Collision Data – EU HARDER Project

Following a number of serious marine accidents, the EU sponsored a project (HARDER) to establish design criteria for ship stability for vessels involved in collisions (Laubenstein et al, 2001). Collision statistics were used for this purpose, but these had not been updated for 40 years. DNV acted as project secretary for the multi-party project.

The data aspect of the project reviewed existing collision data sets (e.g. from IMO) and added new data (mainly from Classification Societies) and this provided a dataset of almost 3000 collision cases. The aim was to establish a probabilistic external hole size for ship stability, not specifically to address issues of importance for LNG vessel collisions such as depth of penetration or secondary structural damage (such as inner hull damage in a double hull vessel).

HARDER statistics include open water collisions (i.e. unlike restricted speed and port controls as apply to LNG vessels making terminal approaches) and cover all types of vessels (some double hull vessels, but no LNG carrier incidents). Figure 1 shows HARDER data is a good correlation to the total world fleet and the distribution of vessel striking speeds.

Figure 1. Comparison of HARDER data to actual world fleet and distribution of striking speeds

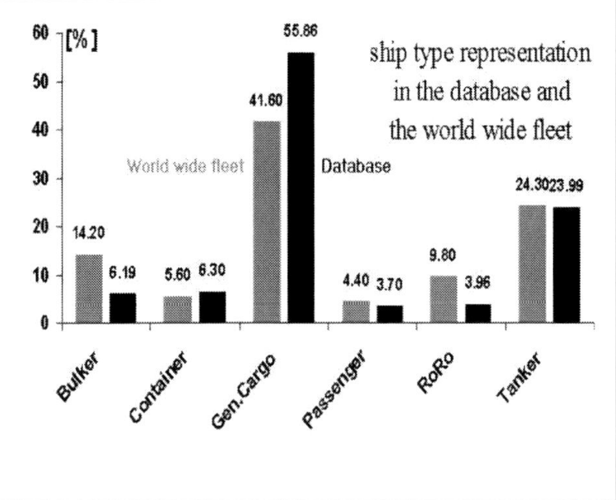

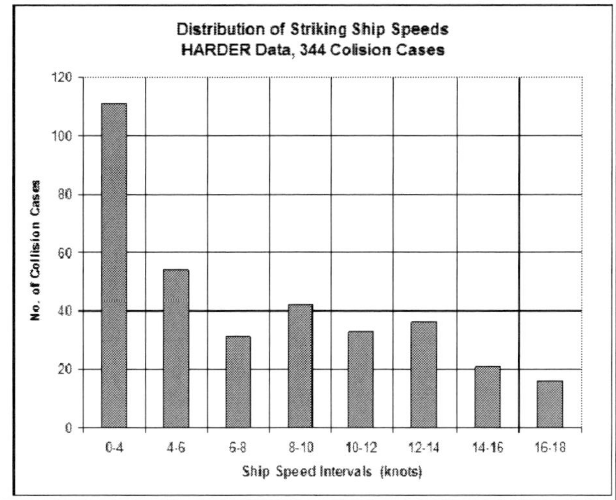

Note: Vessel count is less than total HARDER data set as many events did not record speeds.

A finding of HARDER, not necessarily intuitive but clearly demonstrated, is that the hole size of the struck vessel is not strongly correlated to vessel dimension, and that a normalized hole dimension (hole size divided by ships length) is a reasonable basis for analysis. Many real holes are long tears along the vessel length, not circular holes as demonstrated by normalized hole sizes close to 1.0.

Figure 2. Correlation between normalized hole size and struck vessel length

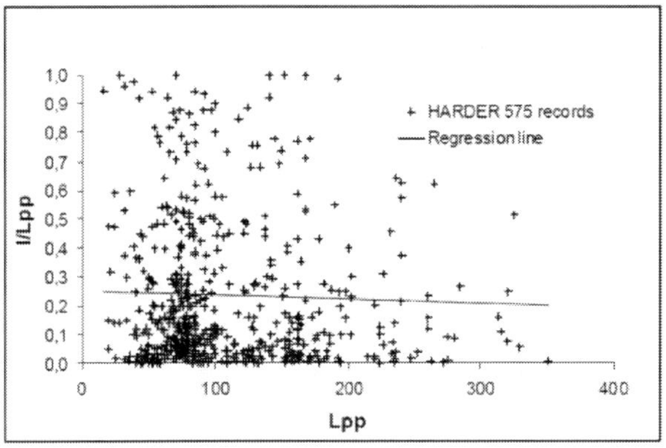

Collisions, non-dimensional damage length versus ship length (575 records)

As HARDER is the best data set, DNV determined a means to use this data but link it to the different aspects of LNG vessels. To do this two corrections were applied:

- The large LNG vessel size will increase the amount of the impact energy absorbed by the colliding ships vs average HARDER
- The deformation resistance of the LNG vessel will be larger than the resistance of most of the struck ships of the HARDER statistics

These two aspects cancel out to a degree, but when applied to the data set a slightly different cumulative distribution of hole sizes results. The effect is that a 138,000m3 LNG vessel will suffer fractionally less damage than the generic average vessel. For example, in HARDER for a hole of normalized dimension 0.05 height / length ratio the cumulative probability of being this dimension or less is 40%, whereas for an LNG vessel the cumulative probability is 72%.

Figure 3. Cumulative probability of hole size less than a normalized hole size

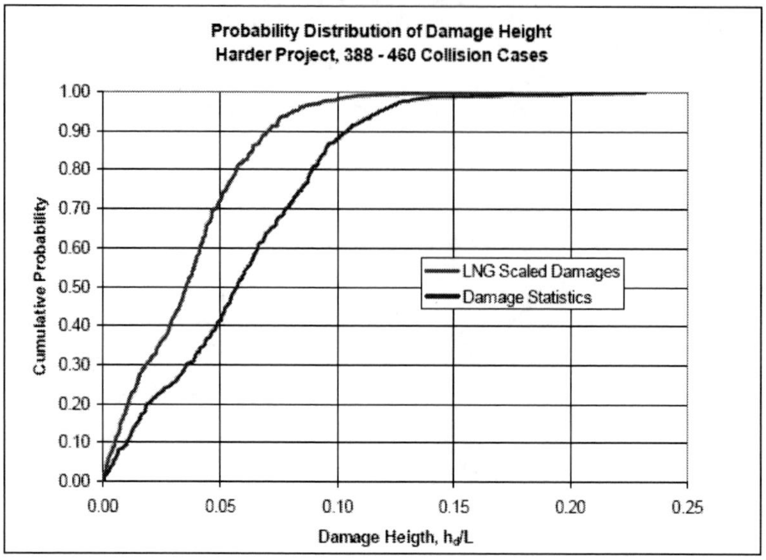

In conclusion, DNV believe that HARDER is the best set of available collision consequence data currently available. With some corrections to allow for LNG vessels that have not suffered any collision damage in the dataset and for the vessel size which puts these at the upper end of vessels, a revised cumulative damage distribution has been obtained which may be used for risk assessments.

Structural Hydrodynamic Simulation

The method of analysis uses HARDER data to define the dimension of hole damage in the outer hull as this is the best available data. However, HARDER does not tell anything about inner hull damage. For this application, DNV has used detailed dynamic finite element methods (FEM) to determine inner hull damage given external damage dimensions specified from HARDER. This requires modeling all the structural elements – inner and outer hull plates, stringers and bulkheads, etc. The analysis follows the collision impacts growing as the striking vessel enters the hull and tears / deforms internal elements. For purposes here, a tear to the inner hull is assumed to equal the tear to the LNG tank. No structural analysis of the tank deformation and additional resistance has been carried out. They may deform somewhat before a tear occurs, thus this analysis should be conservative.

Material properties for the steel used in the analysis are for standard marine steel in Classification Rules. This has a specific gravity of 7.85 and specified yield stress of 235 MPa, potentially 15% less than actual expected values. Average hull plate thicknesses were around 17mm for the outer hull and 15mm for the inner hull. The critical failure strain was calculated to lie between 17-19%, and hull elements will be expected to tear when this critical strain value is exceeded.

The ship models have been created in the DNV SESAM PatranPre and then transferred to the non-linear finite element ABAQUS Explicit commercial code. The bows, boundary conditions, loads and contact areas have been modeled in detail. Key assumptions and parameters for the FEM model are listed below:

The key assumptions all add a degree of conservatism, but not excessive conservatism:

- the bow of the striking ship is assumed to be infinitely stiff (i.e. it does not absorb any energy)
- the struck LNG vessel is rigid in space (i.e. does not move in the water, equivalent to tied to jetty)
- damage is calculated to the inner hull only, not to the LNG containment tank

FEM model mesh details are summarized below and a sample detail is shown in Figure 4.

- Both the membrane and spherical vessel base models cover a length of 3 LNG tanks.
- The fine mesh extends over one tank length, over the total height of the ship and 4 meters inwards

- The fine mesh starts at the section 0.25*tank length aft of the aft bulkhead and runs to the section 0.25*tank length aft of the forward bulkhead.
- The aft bulkhead includes a fine mesh part extending inwards 4 metres from the hull side.
- The fine mesh part include the side structure, frames, stringers, decks and transverse bulkhead,
- The size of the elements in the coarse mesh zone is equal to stiffener spacing (around 2-3m). Stiffeners in this area are modeled with beam elements.
- The size of the elements in the fine mesh zone is 75 mm.
- Two bow shapes, one bulbous and one raked, were modeled using rigid (infinitely stiff) shell elements. The size of the elements in the fine mesh zone (contact zone) is approximately 75 mm.

The scenarios modeled show the magnitude of collision considered. While larger striking vessels exist (e.g. VLCC oil tankers), it is believed that port controls in the USA would prevent interaction of such vessels with an LNG carrier at speed.

Figure 4. FEM Grid mesh used for membrane LNG carrier

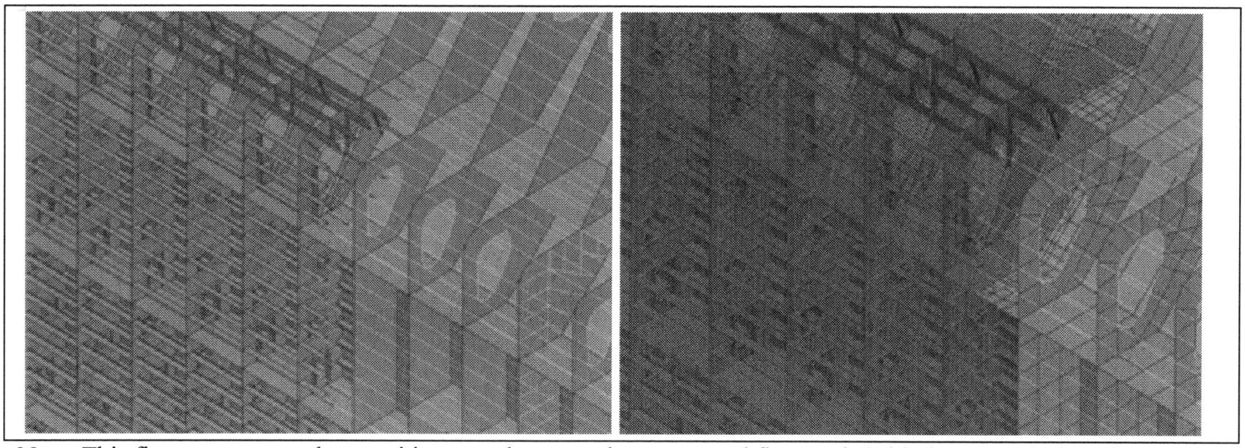

Note: This figure zooms on the transition zone between the coarse and fine mesh (with the upper deck and outer shell removed for visual clarity).

Figure 5. Collision scenarios showing striking (90m, 140m, 230m) and struck vessels (LNG)

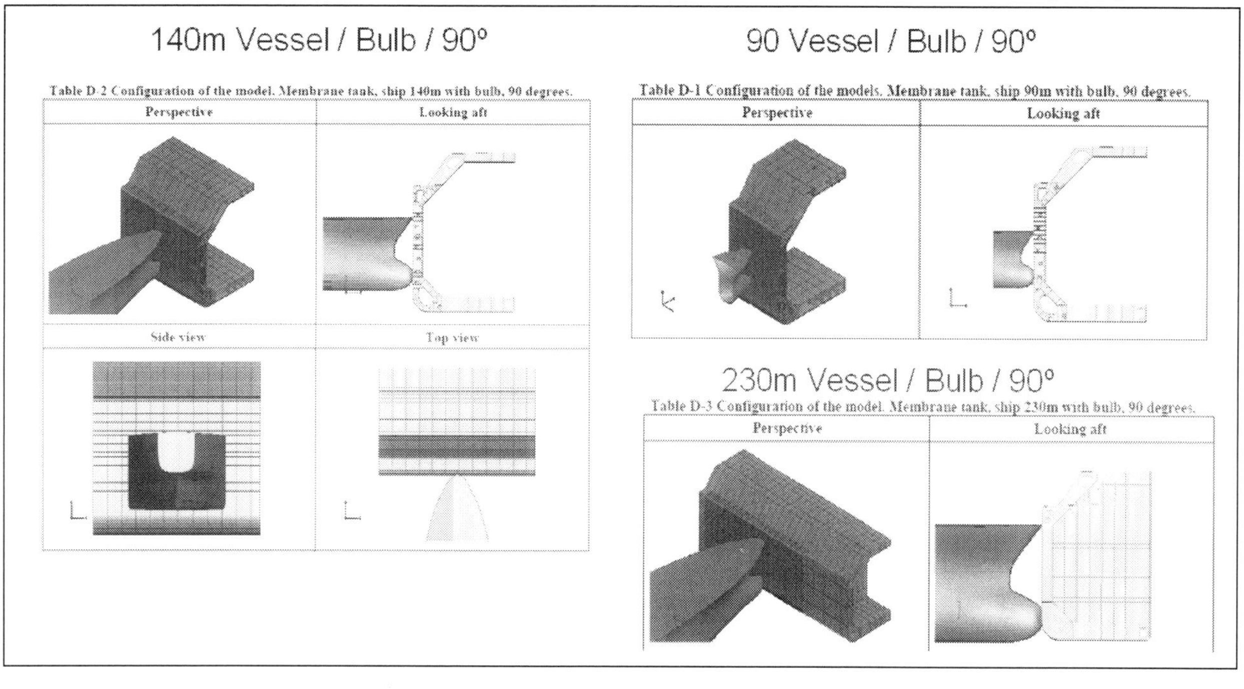

The FEM code allows a time history of the collision to be predicted, showing the progressive damage extent of the LNG carrier as the striking bow overcomes the resistance of the carrier structure. An example time slice is shown for a raked bow intruding into a membrane carrier (Figure 6).

The results obtained allowed a probabilities hole size distribution to be obtained (Figure 7). This shows a rather flat distribution for hole size. This means that once the energy exceeds that necessary to cause an indentation in excess of about 3m in the LNG carrier, then a wide range of hole sizes are similarly likely. There is no peak at specific hole sizes that might define a characteristic hole size.

In simple terms, an LNG carrier is a very strong vessel, and most collisions will be insufficient to overcome all the structures in order breach the inner hull and thus the LNG tanks beyond. However once the outer hull has been breached, and all the stringers and hull structures deformed, and the inner hull has also been breached, then there is relatively little incremental resistance to further intrusion. This leads to an almost equal probability of a small or large hole once the threshold has been exceeded. In DNV's earlier work (Pitblado et al, 2004) it was proposed, based on IMO data, that a hole size anywhere in the range 0 to 1.5m was equally likely. The structural analysis now shows this covers a broader range.

Figure 6. FEM prediction of raked bow intrusion into membrane carrier

Figure 7. Probability distribution for hole size in inner hull (and hence LNG tank)

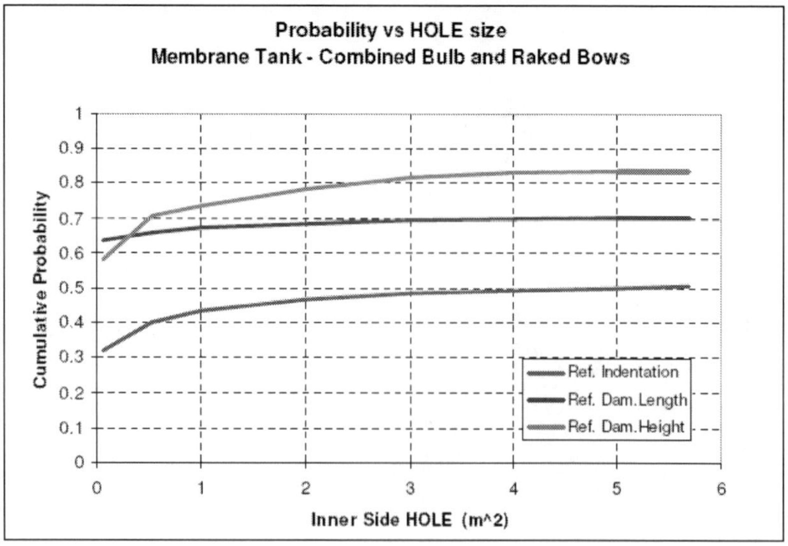

Corresponding to the hole size formation, the FEM analysis gives the energy transmitted in the collision and this is plotted in Figure 8. This figure shows a threshold energy for collision, in this case for both bulbous and raked striking bows into a membrane LNG carrier. The threshold corresponds to the indentation reaching approximately 3m, and after this tearing of the inner hull will occur and thus equivalent LNG tank tearing is assumed.

Figure 8. Damage extent related to energy of collision

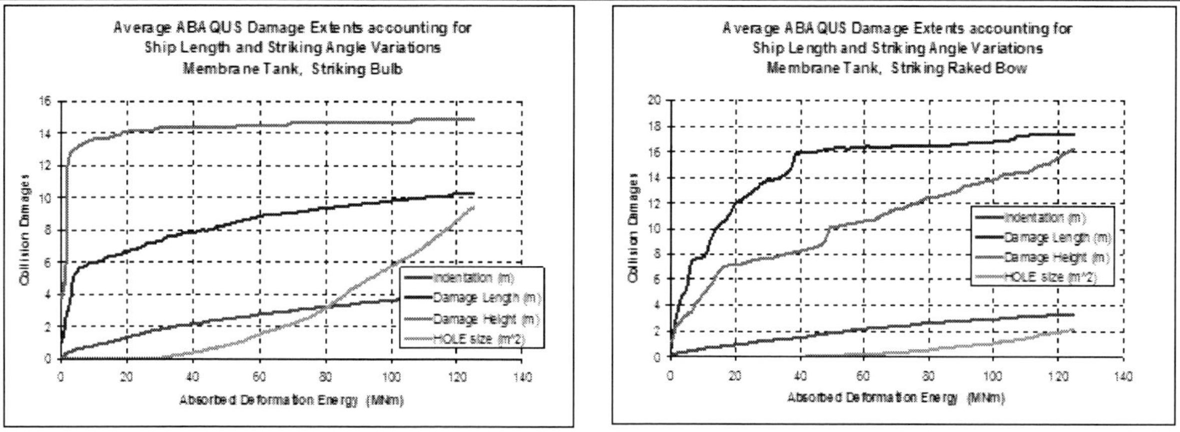

Average Damage Extents to account for Ship Size and Striking Angle Variations, Membrane Tanks – Striking Bulb Bow

Average Damage Extents to account for Ship Size and Striking Angle

Probabilistic (Energy Based) Approach

In order to understand and control risk effectively, it is necessary to fully assess the collision risk in a specific port situation and quantify the effect of risk reducing measures. An alternative approach has been developed for this purpose. The objective of the revised approach is to demonstrate that the probability of collisions which may impair the containment is sufficiently low, rather than demonstrating the safety distance for a generic hole size. This requires that the main focus will be to model and understand the full spectrum of collisions and to differentiate between the majority (low energy with no potential to penetrate the tanks), and the fewer higher energy with the potential to penetrate the containment. This approach requires that

- site specific traffic data are available and used,
- the damage/energy correlation is known for the appropriate pair of striking and struck ships.
- traffic controls and other risk reducing measures (speed controls, route segregation etc)are incorporated in the model and used to select appropriate controls.

The approach is outlined in the Figure 9 and the basic steps of the approach are as follows:

1. The traffic density in the port is collected and categorized according to ship-types and nominal speeds.
2. Specific traffic controls, physical obstacles inhibiting particular collision angles, and distances between shipping lanes and berthing arrangements are identified and included in the model.
3. The collision probabilities for the different ship-types are determined based on an approach assuming that any vessel can have a probabilistic deviation from the planned course reflecting, traffic controls such as requirements for piloting, traffic monitoring etc. This analysis will generate corresponding pairs of probability and maximum impact energy for each pair of ship types.

4. Adapt to local conditions. The results are distributed in a matrix to reflect course and impact angles (breakwaters, shore lines etc) and also experience data for speed reductions before impact (EU-HARDER). The total collision spectrum will thus be represented by a matrix with different speed and directional scenarios.
5. The resulting impact energies are determined for all these scenarios and based on results obtained from our correlation of the HARDER data with FEM structural analyses giving the corresponding indentation depth and inner hull damage.
6. The results, i.e. the pairs of indentations and corresponding probabilities are integrated and a probability distribution for indentation depth is established.

Figure 9. Energy threshold risk-based methodology (numbers refer to method sequence)

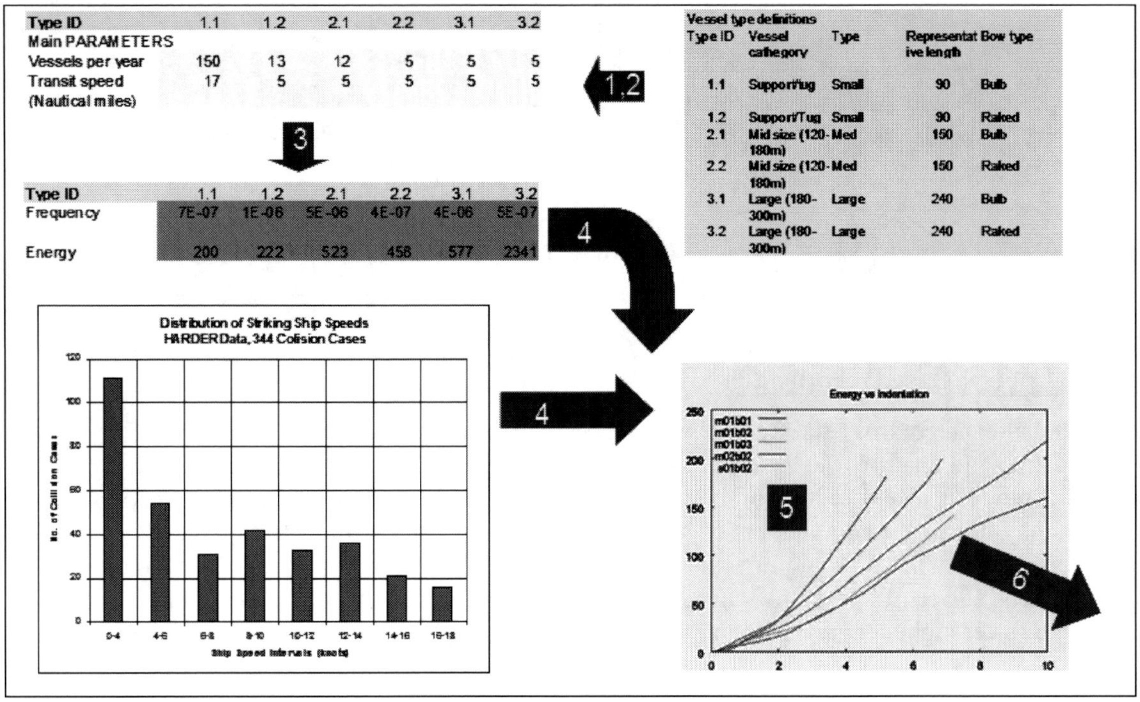

The approach have been tested for two cases

- A terminal located in a busy port with strict traffic controls and a breakwater shielding 270 degrees of possible approach angles
- A terminal located at river waterway with normal ship traffic 1.0 km away from the jetty

Figure 10. Probability of LNG carrier indentation – Sea Port and River Port

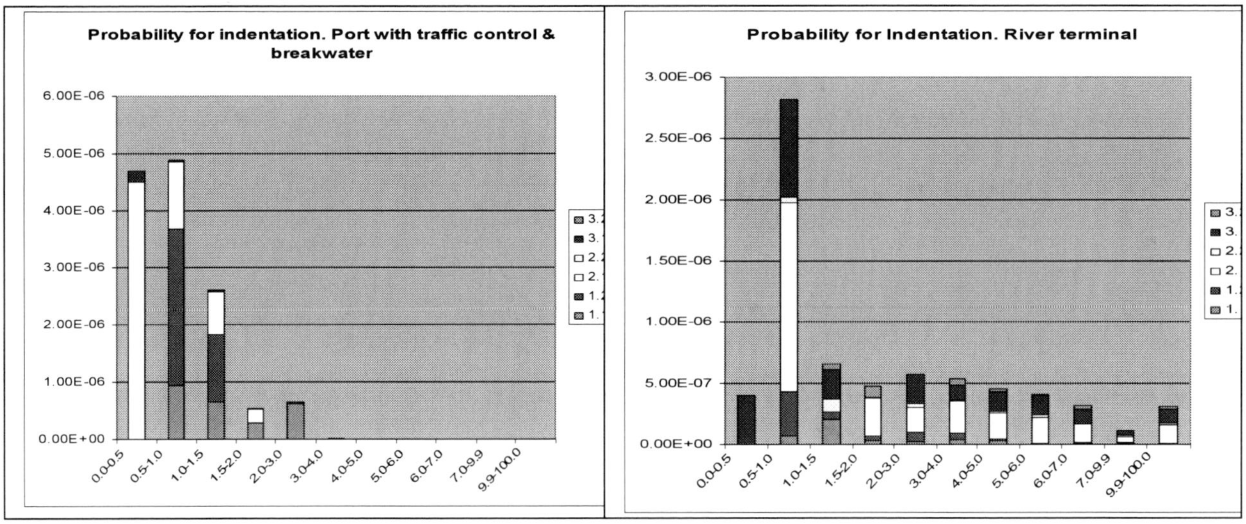

This information is used as discussed above to determine the risk of indentations exceeding the gap between hulls. Where this exceeds about 3m, there is the potential from our FEM analysis for inner hull tear and causing an LNG tank hole. In some cases of traffic density and typical operations there is a small but finite risk of indentation exceeding the critical threshold. A risk based approach could demonstrate from these examples shown that holes are of very low probability due to port operational controls, but may nominate additional port controls to reduce this further. At some point a level of tolerable risk is reached, similar in concept to the (unstated) threshold in NFPA59a that does not require consideration of LNG tank rupture events, although this is also possible.

Conclusions

Based on a detailed FEM analysis of a wide range of collision scenarios (angle of attack, striking vessel size, and membrane and spherical LNG carrier) it is now concluded that a single or even a small collection of holes does not give an accurate representation of potential damage outcomes. An LNG carrier is very strong as evidenced by over 40 years of safe operations without any significant loss of cargo event, even when associated with major accidental events. However, in a collision, once the full set of structural barriers (outer hull, stringers and inner hull) are defeated, then further intrusion meets only incremental resistance and a wide range of hole sizes are possible, almost with equal likelihood. This result is not intuitive and other LNG studies have not proposed this outcome.

This poses a challenge to consequence based approaches for LNG port operations planning. It is difficult to justify a specific hole size for such consequence calculations. Instead, an approach is proposed based on the energy threshold in a collision up to the threshold of no tearing of the inner hull and by inference to the LNG tank. This will identify necessary port controls (speeds, escorts, tug supports, pilotage, etc) that can assure to a predetermined level of confidence (i.e. the risk) that the energy of any collision will be less than the threshold required for leakage of the tank. This approach meets the community desire for protection, not just limitation of outcome. It is also port specific, taking account of port traffic conditions, currents, maneuvering room, etc).

The results of the energy based analysis demonstrate that

1. the majority of collisions represent low energy without potential for penetrating the LNG tanks
2. There is a measurable potential for higher energy collisions with possible penetration of the LNG tanks, but the probability is very low
3. The effects of risk reducing measures such as reduced speeds in the port area, tug support, etc. can be evaluated and the model is thus a powerful tool to assess and select appropriate measures.

Acknowledgements
A Joint Sponsor project work supported the collision analysis aspect of this work. Membership included: Maersk Line Limited, ConocoPhillips, Chevron Global Gas, Statoil Natural Gas LLC, Hydro Oil & Energy (now part of StatoilHydro), Petrobras, DNV Norway, TransCanada PipeLines Ltd, Rabaska, and BP America Inc. The risk-based collision energy extension is DNV's own proposal and was not a part of the JSP. The results and opinions presented here are those of DNV only.

References

ABS (2004) Consequence Assessment Methods for Incidents Involving Releases from Liquefied Natural Gas Carriers, Study for Federal Energy Regulatory Commission GEMS 1288209.

Center for Chemical Process Safety (1999), Guidelines for Chemical Process Quantitative Risk Assessment, 2nd Ed, Wiley-AIChE, New York.

Dept of Transport, Transportation Title 49, Part 193 – Liquefied Natural Gas Facilities: Federal Safety Standards

Hightower M, et al (2004) Guidance on Risk Analysis and Safety Implications of a Large Liquefied Natural Gas (LNG) Spill Over Water, Sandia Report SAND2004-6258

Laubenstein L, Mains C, Jost A, Tagg R, Bjørneboe NK (2001) "Updated Probabilistic Extents of damage based on actual Collision Data", 2nd International Conference on Collision and Grounding of Ships, Copenhagen.

National Fire Protection Association (2006) NFPA 59A Standard for the Production, Storage, and Handling of Liquefied Natural Gas (LNG)

Paik J, I.H. Choe, A.K. Thayamballi (2001) On Accidental Limit State Design of Spherical Type LNG Carrier Structures Against Ship Collisions. 2nd International Collision and Grounding of Ships Conference. Copenhagen.

Pitblado RM, Baik J, Hughes GJ, Ferro C, and Shaw SJ (2004) Consequences of Liquefied Natural Gas Marine Incidents, Plant Operations Progress.

Pitblado RM, Baik J, Ragunathan V (2006) LNG Decision Making Approaches Compared, J Haz Matls 130, 148-156.

Taylor DW (2004) The Role of Consequence Modeling in LNG Facility Siting, Mary Kay O'Connor Process Safety Center Conference, October 26-27.

Modeling of an Industrial NGL-Recovery Unit Considering Environmental and Economic Impacts

Paul N. Sharratt[1], Aurora Hernández-Enríquez[1,*], and Antonio Flores-Tlacuahuac[2]

[1]*School of Chemical Engineering and Analytical Science, The University of Manchester, Manchester M60 1QD, United Kingdom, and* [2]*Departamento de Ciencias e Ingenieria Quimica, Universidad Iberoamericana, Mexico DF 01219, Mexico*

* To whom correspondence should be addressed. E-mail: Aurora.Hernandez@postgrad.manchester.ac.uk
(1) Gas Processing Centre: Cactus Chiapas, Mexico. PEMEX Gas y Petroquimica Basica.

AIChE Spring Meeting, April 2008
8th Topical Conference on Natural Gas Utilization
New Orleans, LA, April 6-10, 2008

In this work, an integrated model is presented that identifies key areas in the operation of a cryogenic NGL-recovery unit. This methodology sets out to provide deep understanding of various interrelationships across multiple plant operating factors including reliability, which could be essential for substantial improvement of process performance. The integrated steady-state model has been developed to predict the economic and environmental impacts of a real cryogenic unit (600 MMCUF/D) during normal operation, and has been built in Aspen Plus TM. Further models are required to test the complete control and reliability systems. The model agrees satisfactorily with the measured performance of the unit. It is used to evaluate the benefits and losses from improvements made to the unit. Model-based simulation reveals considerable opportunities to improve efficiency and control of bypassing, which can be used to improve the existing maintenance programs.

Introduction

This work aims to use mathematical modelling to find opportunity areas in an operating gas processing plant. The approach sets out guidelines to be applied to the industry in general, based on an integrated model which better predicts the behaviour of the processes, and is designed to analyse the set of concerns identified by the management team. By its nature, an operating petroleum company not only operates processing plants for the production of the main products, but also all the supporting services related with their operation such as utilities generation plants, maintenance, safety, environmental wastes disposal, management, strategy and planning support, and technical support; these must be considered in the development of models.

The European Integrated Pollution Prevention and Control Bureau in a Reference Document (BREF) on Best Available Techniques for Mineral Oil and Gas Refineries ((EIPPCB) 2003) identifies that the gas industry has industrial sites that manage huge amounts of raw materials and products and are also an intensive consumers of energy and water used to carry out various processes. They also generate emissions to the atmosphere, water and the soil, so environmental management has become a major factor for refineries. The case study addressed here, is no exception, it is a Mexican cryogenic plant whose purpose is to separate sweetened natural gas into useful saleable products such as Liquid Natural Gas (CH_4) and Cryogenic Liquids (C_2+). It has a maximum capacity to process 600 MMCUFD, and is one of the major energy consumers (30%) of processes in the Gas Processing Centre(1).

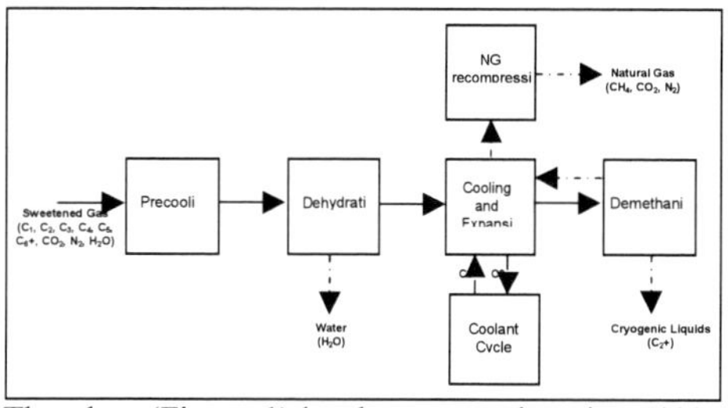

Figure 1. Cryogenic unit plant process flowsheet

The plant (Figure 1) has been operating since 1997 and is divided into 6 principal sections: precooling, dehydration, cooling and expansion, demethanizer, natural gas recompression, and coolant cycle. The cryogenic process uses high levels of cooling through the cooling systems and successive auxiliary expansions to liquify natural and to separate components or groups of components by distillation (demethanizer). In order to do that, it uses: Propane as a coolant, Natural Gas as a fuel in the dehydrating cycle, utility services such as high pressure steam (100 kg/cm^2), cooling water and electrical energy; Coolant oil is used to cool all the compressors and pumps. A considerable proportion of heat removed in the process is done by heat exchange with the outlet Natural Gas from the dome of the demethanizer column and from the dome of the last separator drum, thus reducing the total Propane coolant used.

The required goal of the plant is to get the best liquid recoveries, i.e. higher values of barrels of liquid recovered on the unit per million cubic feet of gas (bbl / mmcuf) during all the expansion and cooling operations. This directly translates into economic profit and products within the specification target.

One of the principal problems faced in the normal operation of the plant is the hydrates formation through the unit. Hydrates are formed due to poor water elimination (i.e. water content of >2 ppm) in the inlet stream. They block the equipment, reducing the process capacity of the plant while the hydrate is being eliminated. Methanol is the antifreeze agent used to eliminate the hydrates formed in the system.

Such energy systems involve a large and diverse number of interactions with the world that are seen when we look beyond their physical boundaries. To solve environmental concerns, the effects of undesired emissions must be evaluated. Traditional design procedures start with an evaluation of the thermodynamic performance of the system subject to physical and technical constraints, resulting in models which include exergetic and monetary costs of mass and energy streams in the system. In addition to that, the environmental economics (environomics) considers the costs related to flows of other resources consumed and the environmental and social costs related to flows of pollutants. Since all these terms are costs, the objective function is still economic. However, specific links may not exist between the environmental impact and economic objectives, e.g. taxes on pollutant emissions are not imposed in many countries, or, if they are imposed, they are often based on the installed power and the relationship with the emission rate is not direct. This is the case of Mexico, where federal environmental regulations only set environmental standards for few emissions in most important industrial areas.

Table 1 presents the related BREF specifications ((EIPPCB) 2003) and Mexican legal limits (MexicanGovernment 2007) for various pollutants that are generated by the Gas Processing Centre.

Parameter	BREF	MX Law[1]
Air emissions		
SO_2 (Ton SO_2 / MM Ton HC)	30-6,000	50
NOx (Ton NOx / MMTon HC processed)	60-500	190
CO_2 (Ton CO_2 / Ton HC processed)	0.02-0.82	
VOC (Ton VOCs / MM Ton HC processed)	50-6,000	
Water discharges		
T range (*C)	10-35	40
Oil (mg/l)	0.05-9.8	15
BOD5 (mg/l)	2-50	30
Suspended Solids (mg/l)	2-80	40
Total N (mg/l)	1.5-100	15
Lead (mg/l)	0.2-0.5	0.2
Waste / energy		
Ton Waste Generated / MTon HC processed	133 - 4,200	
Specific Energy Consumption, GJ / Ton HC processed	1-4	

1 NOM-001-SEMARNAT-1996

Table 1. Gas Processing Centre main pollutants.

The main pollutants generated in the unit studied typically include: A) Air emissions: carbon dioxide (CO_2), carbon monoxide (CO), nitrogen oxides (NO_x), particulates, sulphur oxides (SO_x), and volatile organic compounds (VOC). B) Water emissions: amines, methanol, oil, and metals. C) Land emissions: oil, zeolites, and metals. Unfortunately these data were not available specifically for the unit.

Previous work on unit modelling

A variety of work has been published on modelling of industrial processes. A discussion of dynamic simulation of complex industrial systems and a summary of over a decade of work in the modelling simulation and control of cryogenic separation and liquefaction processes was presented by Mandler (Mandler 2000). This work dealt with the development of tools to facilitate the construction of simulation flowsheets and their efective use in control system analysis and design; The use of Simulink TM and Aspen Dynamics TM commercial packages was reviewed and the question of the required level of modelling detail in dynamic simulation applications was addressed. The LNG process has been studied from optimization perspective in several publications (Nogal, F.L et al. 2005; Jensen and Skogestad 2006; Singh and Hovd 2006); These works developed steady state optimization and there is little literature available on dynamic modelling and control structure design. Singh and Hovd (Singh and Hovd 2006) studied the effect of simplifications of a heat exchanger model on the resulting control structure design for the LNG process under gPROMS and Multi-Flash software; It was found that significant model simplifications such as neglecting heat transfer through conduction along longitudinal direction of metal wall and assuming enthalpy as a conserved property rather than internal energy may be introduced without compromising the control structure design. They later performed (Singh and Hovd 2006) a controllability analysis in the same process and it was compared with established industrial practice. They concluded that compressor speed should be used to control LNG temperature and throttle valve opening should be used to control the degree of superheat at the compressor suction.
A study to evaluate the effects of high nitrogen levels on gas processing operations was developed by Instituto Mexicano del Petroleo (Padron, Rodríguez et al. 2003), the results showed that the sweetening process does not appear to be affected because the diethanolamine solvent does not absorb nitrogen. However for the NGL-recovery process, compression and expansion equipment showed the greatest effects on the increased nitrogen; liquid recovery was affected, and at high nitrogen concentrations, separators can drag liquid into the gas phase. Simulations were performed in TSWEET simulator for the sweetening gas process and PRO/II 5.0 for the NGL-recovery process and Soave Redlich Kwong equation of state. The incorporation of analysis of the dynamic performance of the process into the design and engineering stage of projects was proposed (Gonzalez-Bustamante, Sala et al. 2007) and applied to a natural gas installation in a power plant. Results of the model built under

Simulink TM platform were validated with the actual data obtained from the distributed control system, and permitted to carry out a dynamic analysis of process systems and to design regulation and control systems at a reasonable cost for industrial projects. Mehrpooya1 et al. (Mehrpooya, Gharagheizi et al. 2006) performed simulations in HYSYS and the PRSV equation of state of a natural gas liquids (NGL) recovery unit currently in operation and compared with a real unit. Optimized operating conditions have been determined, in which the objective function is based on cost analysis. Different turboexpansion processes have been analyzed and the best flow sheet selected, based on capital analysis and operating limitations. The best revamping alternative is a turboexpander-exchanger process, in which case the profit is increased by 28 %. In the work of Konukman and Akman (Konukman and Akman 2005) flexibility and operability issues of a HEN were investigated with rigorous simulations using the process flowsheet simulator HYSYS for a HEN-integrated natural gas turbo-expander plant (TEP) operating under ethane-recovery mode. For energy-integrated complex plants like the TEP, the difficulties of automated HEN synthesis and flexibility analysis with process flowsheet simulators were demonstrated.

Most of these models predict performance for a single unit plant considering only either the economic or the energy system but none of them consider environmental issues. Sharratt (Sharratt 1999) stated that particular areas that cause problems are accidental releases, fugitive emissions and off-site emissions, and the modeler should consider whether these are relevant. There has been much work done with algorithms more advanced than those traditionally used to solve multi-objective optimization problems. However, it is difficult to find the right one or the right way to optimize a more realistic (and hence more complex) operational model of the plant. Difficulties always arise when an attempt is made to implement the studies in more complex systems.

Needs identification

Models are developed and used to answer questions; that is their principal purpose. To build a mathematical model for a process is necessary to have a thorough understanding of the underlying concepts. In industry and engineering, it is common practice for a team of multidisciplinary people to work together in building a model. Because of that, a series of surveys were performed among the heads of each department at the gas processing plant. These are the final model users, so it was important to obtain their comments about their needs. The surveys identified two principal causes of not having a model to be used in daily operations. The first one is related with the high level of uncertainty that the forecasting given by the provider (Pemex Exploracion y Produccion) has itself. The second one is that all the optimization models developed to date at the plant are linear and they can not accurately represent reality away from the fixed point where they were built. Another important issue found by the strategic, technology and environmental departments is that the dated models just include economic factors and cannot be used to predict improvements in the environmental perspective, or the damage to be caused as a result of changes in operations.

The main needs of each one of the heads of function are summarized as follows:

- *Principal Manager:* Interested in a more accurate estimation of future investments benefits.
- *Production Manager:* Interested in finding more realistic energy savings with improvements done.
- *Strategy Manager:* Worried about the dynamic changes coming with the feedings and translating its results into tangible yearly targets.
- *Maintenance Manager:* Taking into consideration reliability critical factors of the equipment in unit.
- *Optimization Manager:* Developing a nonlinear model, which parameters (costs and yields) are updated in a short time to give more realistic results.
- *Environmental Manager:* Reducing the water pollution agents because they will soon be the principal problem there. To have a model with an open structure to let considering new regulatory changes in a near future.
- *Financial Manager:* Reducing production costs, that from his perspective, it is the only manner in which an improvement done can be reflected into the financial results, and also in taking into account the worth of the assets of all the equipment in unit versus the maintenance cost of each one.

From all of the comments referred to here, a general view was drawn that all of the issues need to be taken into consideration for the development of the model so that it can be used as a reliable useful tool for the users.

Model development

When a plant model is constructed it would be normal to include all equipment in the model. This leads to very large models, in fact the bigger the size of the plant, the larger the model and indeed the more complex its solution. On the other hand, it can be simplified by making some assumptions in the model, which leads to reduce the size of it and also to facilitate its exportation to other software. One of the assumptions that can be done is taking into account the critical equipment and leaving out those items that do not add value to the model. In that way a selection was made from among all the equipment involved in the operation of the cryogenic 1 plant of those which were relevant to the purposes of this study. Nevertheless, we recognize that by doing that we have to make some valid assumptions in the model and to be sure of their validity in the real context of the plant operation.

Our aim is to get the best economic, environmental and plant reliability performances in a realistic way, using a model that can be solved in a reasonable and practicable time. Based on those three criteria we have to answer the following questions:

1.- Is reliability an important factor?
Here it is important to consider the reliability record and the criteria corresponding to each one of the plant items. Every item has an historical record of the work done to it due to failures. Based on that record, and on the criteria that the company has itself, each item can be evaluated for its criticality.
In our specific case, for PEMEX criticality level A is the highest level of critical reliability. This level A can be categorized based on two criteria:

- Criterion 1. Means a low fail probability, but a high consequence of the same due to the operation of the equipment by itself (process itself);
- Criterion 2. Considers a high fail probability, but a low consequence of the same due to the operation of the equipment by itself (reliability itself).

Because of this we gave priority to all that equipment qualified with level A.

2.- Is money highly important?
It is a question about the importance of the equipment in the economic performance of the plant, it is principally due to the process itself and also includes the policy of the company. In our case, the selected plant cryogenic 1 was seen to be an important energy consumer and there is interest in analyzing the heat exchange equipment. Because of that the model contains all of the heat exchangers included in the plant. In addition to that, the quality and quantity of the products obtained are highly important because those are directly sold to the final customers. So that, we must consider all of those equipment directly related with product recovering and quality, for instance the chillers, distillation tower, compressors and expanders.

3.- Is the equipment highly important environmentally?
This is a question about the importance of the equipment in the environmental performance of the plant and includes operation and maintenance issues. It also includes the policy of the company. The case here analyzed is an important environmental emissions source, to soil, to air and to water. Because of their potential for leaks, heat exchangers and dynamic equipment such as pumps and compressors are considered to be a potential source of pollution to air and water. This is another reason for including in the model all of the heat exchangers in the plant and taking into special consideration compressors, expanders and pumps. For the same reason, all the pipe connections must be taken into account for this purpose but they will be grouped to make estimates of their overall environmental impact.

After answering those three questions it was possible to categorize and choose among all the equipment in the cryogenic unit. The model was built following the design flowsheet, the operation manual, and the equipment data sheets under ASPEN PLUS platform 2004.1. The standard modules available in the ASPEN PLUS library were enough to represent most of the equipment used in the model except from dehydration module, which was not available in the library; It was limited to a variation in water content in the unit inlet stream. Peng-Robinson equation was used to calculate the thermodynamic properties. Some other assumptions done are listed as below.

Modeling Assumptions.

- The Dehydration process completely removes Water from inlet stream (water content of < 2 ppm).
- The Demethanizer column (DA101) contains a pumparound stream which was energetically integrated with heat removed from heat exchangers EA113A and B.
- The Turboexpansors GC-101 and GC-102 were energetically integrated with the compressors GB-102 and GB-101.
- The Propane coolant cycle is fully closed with open inlets every 3 months (on average) to refill 20% of the tank FA-201.
- Compressors GB-201A and GB-201B are operating, the third one GB-201R is in stand by position.
- Compressors GB-103A and GB-103B, are operating, the third one GB-103R is in stand by position.
- Coolers EA-116A and EA-116B are operating, the third one EA-116R is in stand by position.
- Coolers EA-118A and EA-118B are operating, the third one and EA-118R is in stand by position.
- Pumps BA-110A and BA-110B are operating, the third one and BA-110R is in stand by position.
- The amount of fuel gas burned in the furnace BA-101 is held constant.

Results and discussion

Results of the simulation model presented in this work are compared to the average operational data of first 6 months of 2007 in terms of principal streams compositions, liquid recovering, duty required by each compressor, duty required by demethanizer column reboiler, and duty removed in water condensers, all of them are respectively shown in Tables 2 and 3. The dominant operating cost of this system, after the sweetened feeding gas, is the high pressure steam (100 kg/cm^2) that is consumed from the Centre power plant to run the high pressure compressors GB-103A, B, R, this high pressure steam is then transformed inside the unit into medium pressure steam (43 kg/cm^2), which is also used to run compressors GB-201A, B, R, and the last step produces low pressure steam (4.5 kg/cm^2) to be used to run the column reboiler EA-114. The more efficient this energy chain transformation, the less amount of steam consumed by the power plant turbines.

	Inlet gas to unit	C_{2+} product		Residual Gas High Pressure	
% mole		Real	Simulated	Real	Simulated
N_2	2.91	0.0	0	3.33	4.10
CO_2	0.001	0.0	0.02	0.0	0.01
CH_4	77.87	0.22	0.05	93.55	91.86
C_2H_6	10.85	51.45	47.46	3.04	3.93
C_3H_8	4.82	26.51	29.93	0.08	0.10
nC_4H_{10}	1.61	9.01	9.98	0	0
iC_4H_{10}	0.73	4.03	4.59	0	0
nC_5H_{12}	0.46	2.78	2.54	0	0
iC_5H_{12}	0.43	2.43	2.22	0	0
nC_6H_{14}	0.32	3.57	3.21	0	0
T, F	114	104	105	104	110
P, PSI	922	867	867	1,010	1,015
VF	1	0	0	1	1
Volume Flowrate, mmcufd	520			418	397
Volume Flowrate, bpd		58,156	60,400		

Table 2. Principal stream table for the norrnal operating of the Cryogenic unit 2007.

Equipment	Net duty, btu/Hr	Power, HP	Source
Reboiler EA-114	15,305,224	6, 098	Low Pressure Steam (4.5 kg/cm^2)
Compressors GB-201 A, B, R		47,203	Medium Pressure Steam (43 kg/cm^2)
Compressors GB-103 A, B, R		38,953	High Pressure Steam (100 kg/cm^2)
Cooler EA-115	-4,287,735	(1,708)	Cooling Water
Coolers EA-116A, B, R	-42,351,242	(16,873)	Cooling Water
Coolers EA-117 A, B	-17,308,720	(6,896)	Cooling Water
Coolers EA-118 A,B, R	-27,468,683	(10,944)	Cooling Water
Cooler EA-120 A, B	-14,663, 591	(5,482)	Cooling Water
Condensers EA-202 A,B,C,D,E,F	-196,398,058	(78,246)	Cooling Water
Pump GA-101 / R		14	Electric Energy
Pump GA-102 / R		8	Electric Energy
Pump BA-110 A, B, R		1,147	Electric Energy
Total Equivalent Power		**213,932**	

Table 3. Principal energy consumers for the normal operating simulation.

The total equivalent power is the sum of compression power and the equivalent power of the reboiler heat and condensers heat estimated.

This model helps us understand the cryogenic unit operation, reduce energy requirements for liquefying, and understand the effect of inlet stream variations into the distillation column final products.

As seen in Table 2, all the principal streams resulting from the simulation are in good accordance with the operational data, making the model an acceptable representation of reality. The overall liquid recovery calculated from simulation is 116 bbl/mmcuf, which value is not too far from the real one of 111 bbl/mmcuf. In Table 3 it can be observed that the highest power equivalent consumers in the unit are the propane condensers EA-202 A to F, which are cooling water users, and the lowest power consumers are the electrical pumps GA-101s and GA-102s. The steam chain referred in compressors GB103s, GB-201s and reboiler EA114 represent also an important quantity of power consumed in the unit plant.

The model was seen to be suitable as it could give answers to some of the specific questions set by the manager team. These questions were set as cases in the simulation and their results are summarized as follows.

Case 1. Estimated benefit of including an additional turbo-expander in the unit performance (future investments). The inclusion of an additional turbo-expander following GC-101 with another of half its size gives an small increase in the liquid recovery from 116 to 119 bbl/mmcuf and also gives a reduction of the total power consumption of the plant from 213,932 to 205,364 HP. This power reduction is principally derived from an 86% of reduction in reboiler EA-114 and a 7% of reduction in compressors GB-103A, B, R which reduces the low and high pressure steam consumptions. This can be seen in Tables 4 and 5. On the other hand, the investment of buying the new turbo-expander has, an incremental impact on the capital costs which needs to be analyzed using a NPV approach. This issue lies outside the scope of this paper. The important fact to point out here is that the model could be successfully used to predict accurately the benefit for future investments done.

Case 2. Focusing on the bypassing that the plant has. This case simulated the (undesirable) results that some of the valves [FCV200, 201, 202, 203, 204, and 205] could cause in the coolant propane

% mol	Inlet gas to unit		Residual Gas High Pressure				C_{2+} product			
	Case 3a	Case 3b	Normal	Case 1	Case 2	Case 3b	Normal	Case 1	Case 2	Case 3b
N_2	2.91	6.03	4.10	4.12	4.10	6.87	0	0	0	0
CO_2	0.001	0.03	0.01	0.0	0.01	0.03	0.02	0.02	0.02	0.03
CH_4	77.87	76.87	91.86	92.03	91.83	89.00	0.05	0.33	0.05	0.05
C_2H_6	10.85	10.04	3.93	3.74	3.94	4.00	47.46	47.60	47.45	47.88
C_3H_8	4.82	4.24	0.10	0.11	0.10	0.10	29.93	29.67	29.93	31.13
nC_4H_{10}	1.61	1. 22	0	0	0	0	9.98	9.90	9.98	9.49
iC_4H_{10}	0.73	0.65	0	0	0	0	4.59	4.55	4.59	4.64
nC_5H_{12}	0.46	0.33	0	0	0	0	2.54	2.53	2.55	2.67
iC_5H_{12}	0.43	0.30	0	0	0	0	2.22	2.20	2.20	2.33
nC_6H_{14}	0.32	0.29	0	0	0	0	3.21	3.20	3.21	1.78
T, F	114	114	110	110	110	110	105	103	104	104
P, PSI	922	922	1,015	1,015	1,015	1,015	867	867	867	867
VF	1	1	1	1	1	1	0	0	0	0
Volume Flowrate mmcufd	470	520	397	397	397	403				
Volume Flowrate bpd							60,400	62,000	60,400	55,400

Table 4. Principal stream table for the simulation cases: normal, 1, 2 and 3.

Equipment	Power, HP CaseN	Power, HP Case1	Power, HP Case2a	Power, HP Case2b	Power, HP Case3a	Power, HP Case 3b
Reboiler EA-114	6, 098	866	6,098	6,098	4,163	2,791
Compressors GB-201A,B,R	47,203	47,203	52,179	49,686	47,203	47,203
Compressors GB-103A,B,R	38,953	36,271	38,953	38,953	35,319	39,626
Cooler EA-115	(1,708)	(1,698)	(1,708)	(1,708)	(1,544)	(1,719)
Coolers EA-116A, B, R	(16,873)	(16,303)	(16,873)	(16,873)	(16,014)	(17,331)
Coolers EA-117 A, B	(6,896)	(6,883)	(6,896)	(6,896)	(6,963)	(6,812)
Coolers EA-118 A,B, R	(10,944)	(10, 882)	(10,944)	(10,944)	(9,890)	(11,046)
Cooler EA-120 A, B	(5,482)	(5,842)	(5,842)	(5,842)	(5,280)	(5,684)
Condensers EA-202A/F	(78,246)	(78,247)	(83,281)	(78,243)	(78,246)	(78,246)
Pump GA-101 / R	14	12	14	14	13	14
Pump GA-102 / R	8	8	8	8	7	7
Pump BA-110 A, B, R	1,147	1,149	1,147	1,147	1,098	1, 094
Total Equivalent Power	**213,932**	**205,364**	**223,943**	**216,411**	**205,740**	**211,574**

Table 5. Principal energy consumers for the simulation cases: normal, 1, 2, and 3.

cycle when they had a damage to their body which produced internal leaks. These valves were selected because of the elevated incidence on having interventions in their bodies and tending to bypass most of the time in their maintenance records. Results are shown in Tables 4 and 5.

Case 2a) In this case there is an internal leak in the first stage loop, (either in valve FCV-200 or in valve FCV-202) of 1% of the total outlet stream from GB-201A or GB-201B respectively (7.5 MMCUFD). The consequence given by the model is that this internal leak would go towards the separation tank (FA203) increasing the overall temperature of the inlet stream to the first stage of the compressors (GB-201A or B), which would bring a rise of its compression power requiring more steam and also an incremental cooling water required to condense the coolant propane (EA-202A/F) so that to keep the coolant power required in the cryogenic process. All of those effects are translated into a general increase in total equivalent power from 213,932 to 223,943 HP (5% more) that not only represents an economic loss but also an environmental issue regarding both increased greenhouse effect gases emitted to the air (CO_2, H_2O and NO_x) and increased water pollution. The principal unit streams were not affected at all because of the same Propane coolant power was kept.

Case 2b) In this case the internal leak is assumed in the second stage loop (either in the valve FCV-201 or in the valve FCV-203) of 1% of the total outlet stream from compressor GB-201A or GB-201B respectively (7.5 MMCUFD) The consequence given by the model is that this leak would go towards the FA202 tank increasing the overall flowrate of the inlet stream to the second stage of GB-201A or B, which brings the same compression effect of case 2a but in a lower increase in total equivalent power from 213,932 to 216,411 HP (1% more). The principal unit streams again were not affected at all because of the same Propane coolant power is kept.

The understanding of energy loss and environmental effects found here can be used to support the improvement of the existing maintenance programs to reduce the internal leaks in those valves. These cases can be complemented with a calculation of fugitive emissions in the unit with introducing a calculator based on an EPA algorithm that relates inlet gas stream vs. most of the connexions, seals and in general, probable leaking points so that to estimate the energy lost and environmental damage given by them.

Case 3. Quantifying the effects of the low feed flowrates and the variable composition that the centre has recently received. In this case, two separate scenarios were studied (cases 3a and 3b) and then these were joined to see the interaction between them.

Case (3a) was the low feed flowrate, which was reduced from 520 to 470 MMCUFD. The results showed an important effect in the liquid recovery of the unit decreasing 21% from 116 to 92 bbl/mmcuf as well as a reduction of 4% from the total energy consumed in the plant from 213,932 to 205,740 HP. The result also implies difficulties in reaching the required specification of the products. The net effect of these results needs to be estimated in terms of net profit and then it has to be compared with the normal operating case profit, which again falls out of the context of this work. It could readily be done by using either a link to calculate it within other software or an implicit calculator to calculate it inside.

Case 3b) studied the effect of variable composition of the feed to the unit. This problem has to deal basically with the injection of Nitrogen to the gas fields to increase their pressure and thus to increase production. That Nitrogen is carried through the feed gas to the unit, decreasing liquid recovery and making the model more difficult to converge. The simulation results are given in Tables 4 and 5, from where it can be observed that there is a reduction of 9% in the liquid recovery from 116 to 106 bbl/mmcuf, and also there is an minimal decrement of 1% from the total energy consumed in the plant from 213,932 to 211,574 HP, these findings are in agreement with those found by the Instituto Mexicano del Petroleo in 2003 (Padron, Rodríguez et al. 2003). The former change reduces profit through losing liquids in the gas, and besides that it may create instability in the control system too.

From the cases mentioned here, there have been answers given to some of the questions from the head of the plant. However, there still remain a set of questions that can only be partially answered with this model and other which need other kinds of models to solve. These are as follows:

1. An accurate calculation of reducing the production cost of the all the unit processes (energy ones). This can be partially answered with this model, but to be completely answered it needs to be complemented with a calculator that makes the link with real costs and also an optimization analysis in function of profit and in energy.

2. Finding opportunity areas with the heat exchange system to find energy savings in the unit, applying Pinch analysis in the coolers and heaters of it. This model can be used as a platform to do Pinch analysis, it just has to be exported to Aspen Pinch TM and do the analysis there so that this question could be completely answered.

3. Taking into consideration the dynamic changes that come with the most probable failures (ie electrical failures) and translating its results into more accurate estimated lost. The specific answer given to this point can use this model as a platform to be then exported to Aspen Dynamics TM in where it can be dynamically modelled and else, the most probable dynamic changes can be simulated and calculations estimated including their response time.

4. Taking into consideration the critical reliability factors that have been calculated by the maintenance department, and given to most equipment in the unit. In order to do that, it would be necessary to build a stochastic model which can represent the likelihood of failure and then can be linked with the one here exposed so that to develop the calculations required.

5. The plant's current linear planning model need to be linked to real variables so that the parameters (costs and recoveries) can be recalculated and adjusted in a short time (nonlinear model needed). This can be done with this model, the issue here regards with optimization of the plant, which is included inside Aspen Plus TM Model Analysis Tools, and it has the advantage that the model already done is nonlinear and the real variables represented.

6. Reducing the water pollution agents. This can be done with introducing calculators in the model which are related with maintenance activities and leaks resulting in material that affect water pollution.

7. To be a model with an open structure to let considering new regulations in a near future (optimization objectives and calculators). To reach this point, it would be necessary to have more than two objective functions, as a result the convergence of the model would be very difficult, it would then be necessary to link this with a solver which helps the model to converge in a better and faster way.

8. Taking into account the worth of the assets versus the maintenance cost of each one. It can be answered using this model as a base, and then it needs to be complemented with a calculator that makes the link with real maintenance costs and asset of each one of the most critical equipment.

As a consequence of what has been done to date the model can give answers to about 1/3 of all the questions of the end users. Because of that future work must be done so that to give responses in a reliable way to the remaining questions listed above.

Conclusions and future work

A cryogenic unit plant model was developed for predicting the performance of a real one, and based on the actual needs of the final users. The model was built using ASPEN PLUS TM process simulator and consists of several standard modules available in its library. The model prediction was compared to real operating data in terms of the principal streams composition, liquid recovery, and total power consumption, and close agreement was found.

Specifically, from the results obtained in this paper we can conclude that the model built can answer about 30% of the total answers of the users. Future work must be done to solve the questions remaining un-answered. This work should include complementary calculators and external links, dynamic and stochastic models in general.

Methodologically, using the approach proposed in this study, the prediction of the behaviour of large-scale processes at different operating conditions becomes possible in a simulator tool such as ASPEN PLUS. In doing this it is important to pay attention to users' needs. The results of this study could be used to represent the starting point of further models developed to give the answers required.

Literature Cited

(EIPPCB), E. I. P. P. a. C. B. (2003). "Best Available Techniques for Mineral Oil and Gas Refineries (BREF)."

Gonzalez-Bustamante, J. A., J. M. Sala, et al. (2007). "Modelling and dynamic simulation of processes with 'MATLAB'. An application of a natural gas installation in a power plant." Energy **32**( ): 1271–1282.

Jensen, J. B. and S. Skogestad (2006). "Optimal operation of a simple LNG process." Proceedings Adchem 2006 (IFAC symp. on Advanced control of chemical processes), Gramado, Brazil **2-5 April 2006**: 241-246.

Konukman, A. E. S. and U. Akman (2005). "Flexibility and operability analysis of a HEN-integrated natural gas expander plant." Chemical Engineering Science **60**: 7057 – 7074.

Mandler, J. A. (2000). "Modelling for control analysis and design in complex industrial separation and liquefaction processes." Journal of Process Control **10**: 167-175.

Mehrpooya, M., F. Gharagheizi, et al. (2006). "An Optimization of Capital and Operating Alternatives in a NGL Recovery Unit." Chem. Eng. Technol. **29**(No. 12): 1469–1480.

MexicanGovernment (2007). "Diario Oficial de la Federacion (DOF)." Official Journal of the Federation **NOM-001-SEMARNAT-1996**.

Nogal, D., F.L, et al. (2005). "Improved design of mixed refrigerant cycles using mathematical programming." GPA Europe Meeting: GTL and LNG in Europe, Amsterdam.

Padron, A. S., J. E. R. Rodríguez, et al. (2003). "Study evaluates the impact of nitrogen on gas processing efficiency." Oil & Gas Journal **101**(46): 50.

Sharratt, P. (1999). "Environmental criteria in design." Computers and Chemical Engineering **23**: 1469-1475.

Singh, A. and M. Hovd (2006). "Dynamic Modeling and Control of the PRICO© LNG process." Proceedings AIChE Annual Meeting.

Singh, A. and M. Hovd (2006). "Model Requirement for Control Design of an LNG Process." 17th European Symposium on Computer Aided Process Engineering: ESCAPE17.

Reliable Thermodynamic Data for Improving LNG Scrub Column Design

Lindsay Laskowski
LNG Process Engineer, Chevron Energy Technology Company, Houston, Texas

Mohamed E. Kandil
Research Fellow, The University of Western Australia, Crawley, Australia

Eric F. May
Lecturer, The University of Western Australia, Crawley, Australia

Mark A. Trebble
Professor, Curtin University of Technology, Bentley, Australia

Robert D. Trengove
Associate Professor, Murdoch University, Murdoch, Australia

Justin Trinter
LNG Process Engineer, Chevron Energy Technology Company, Houston, Texas

Stanley Huang
Staff LNG Process Engineer, Chevron Energy Technology Company, Houston, Texas

Kenneth N. Marsh
Emeritus Professor, University of Canterbury, Christchurch, New Zealand

AIChE Spring Meeting, April 2008
8th Topical Conference on Natural Gas Utilization
New Orleans, Louisiana, April 6-10, 2008

Introduction

The construction and operation of a liquefaction facility requires a significant expenditure of capital. Simulations of these facilities have the potential to reduce the total cost of ownership if they are sufficiently optimized. Process simulators are used to estimate the operating parameters of the liquefaction plant and to size the equipment. Central to the simulation is the calculation of the thermodynamic properties of the hydrocarbon fluid mixtures at each process condition, which are usually calculated with an equation of state (EOS). The thermodynamic properties of natural gas have been studied extensively and are relatively well known at most non-cryogenic conditions. However, the availability of reliable thermodynamic data for multi-component mixtures at the high-pressure, cryogenic conditions relevant to LNG production is still poor [1]. Accurate vapor-liquid equilibria (VLE), volumetric and calorific data for multi-component mixtures at LNG process conditions would improve the reliability of liquefaction facility simulations. This increased reliability would then allow for an optimization of the facility's design and operating parameters.

The LNG scrub column (sometimes known as a de-methaniser) is an area of particular focus within this study because it operates at both low temperatures and elevated pressures, and with the widest range of components. The scrub column is a specific type of distillation column that is used to separate the lighter components (C_1 and C_2) from the heavier components (C_{3+}) present in a treated natural gas feed stream. Figure 1 shows a high-level schematic of a typical liquefaction facility and the location of the scrub column within the process train. There are three main reasons why the scrub column is used in an LNG train: (1) to prevent freezing (of the heavier components) in the downstream Main Cryogenic Heat Exchanger (MCHE), (2) to control the heating value of the LNG, and (3) to control the hydrocarbon dew point of the LNG upon re-gasification to meet pipeline transport specifications. Reasons (2) and (3) are particularly sensitive to the LPG (C_3, C_4) content of the vapor product from the scrub column. In an optimized scrub column, most of these and the heavier components are recovered in the liquid product stream from the bottom of the scrub column. This liquid stream is usually processed further to extract additional sales products (LPG, condensates) and/or make-up refrigerant (C_2, C_{3+}) for use in the MCHE [2, 3].

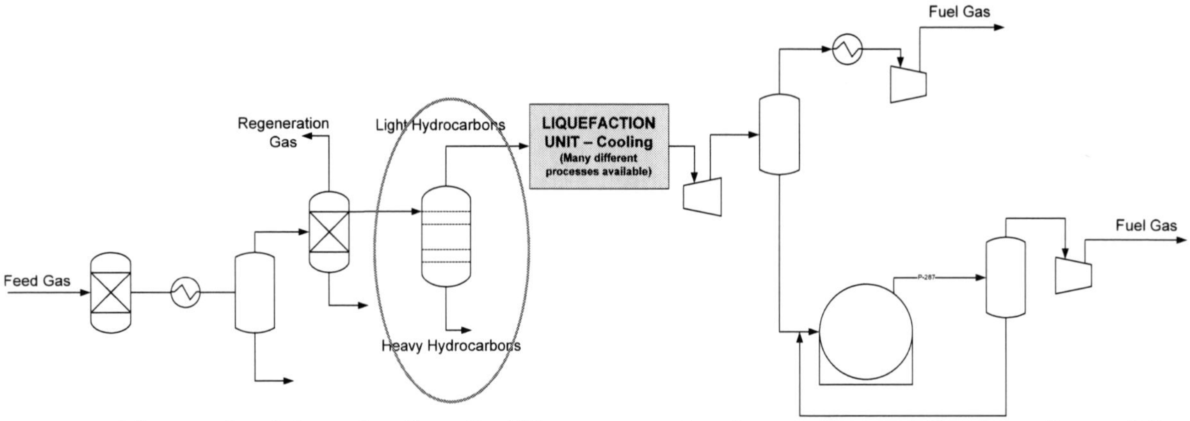

Figure 1 – Liquefaction facility process schematic and the location of the scrub column

We and others [4] have observed significant discrepancies between simulator predictions of scrub column product streams and those observed on-site. We believe that the primary reason for this discrepancy is that the (typically cubic) EOS embedded in the process simulator is not anchored to thermodynamic data representative of the mixtures and conditions present in industrial scrub columns. The uncertainties in the VLE, volumetric and calorific predictions of these EOS are, therefore, unknown which increases the uncertainty of the process simulation. To account for and overcome this uncertainty the engineering design margins for the construction and operation of the plant (e.g. equipment sizing and reflux rates) are increased to ensure safe operation and the delivery of a product that meets delivery specifications.

Chevron has initiated a joint project with the Western Australia Energy Research Alliance (WAERA) to study more extensively and determine experimentally the relevant thermodynamic properties of LNG-like multi-component mixtures at cryogenic conditions. Particular emphasis will be placed on measurements with mixtures at conditions representative of the scrub column. The EOS used by process simulators will in turn be tuned to the new data as well as any other relevant data available in the literature. This should lead to improved EOS predictions of thermodynamic properties and, thus, improved simulation of the scrub column and liquefaction plant. More reliable simulations will lead to optimized plant design and operation, a reduction in the engineering design margins currently used and, consequently, a reduction in the required capital and operational expenditure for the liquefaction facility.

The processed gas which is feedstock to a scrub column, and the LNG product itself, is often considered to be a relatively 'simple' (near ideal) multi-component mixture of similar non-polar components. Engineers, therefore, often find it surprising that significant uncertainties exist in the simulation of LNG liquefaction processes. However, given the context of the expenditure required for an LNG plant, 'significant' means uncertainties of only a few percent. The objective of this paper is to illustrate some examples of simulation uncertainty for LNG scrub columns arising, in particular, from the underlying thermodynamic model, and to estimate the impact of these uncertainties. The paper will also discuss briefly Chevron's alliance with WAERA and the work being conducted to help address this problem.

Technical Background

The scrub column operates essentially at a constant pressure, the value of which depends upon the feed pressure and composition. Typically, the scrub column conditions are around 4800 kPa (700 psia) with its feed gas pre-cooled and partially condensed at approximately 0 °C (32 °F). Separation of the heavy components is achieved by maintaining a temperature gradient along its vertical length and by ensuring intimate contact between the liquid and vapor phases on about 10-20 trays distributed along the columns length. Typically, a reboiler maintains the temperature at the bottom of the column at about 50 °C (122 °F) while the top of the column is maintained at about -30 °C (-22 °F), with an overhead condenser and a reflux system. These conditions are dependent upon the process; in our example below the reboiler temperature is 138 °C (280 °F). In an ideal system, the vapor and liquid phases would reach equilibrium on each tray with the vapor then moving up the column to a colder tray and the liquid moving down to a warmer tray. The vapor is therefore flashed sequentially at decreasing temperatures allowing the heavier hydrocarbons to be stripped out and condensed. The condensed liquid flows down the column and is flashed sequentially at increasing temperatures, liberating any light hydrocarbons that may be dissolved in it.

Simulation of the scrub column requires the simultaneous solution of three sets of equations representing the energy balance, material balance and phase equilibrium on each tray. The solution to these equations for the entire column is achieved iteratively using, for example, the inside/out convergence algorithm by Joe Boston as discussed in Russell [5]. Central to any solution algorithm is the reliable and efficient calculation of the thermodynamic properties of each mixture on each tray: this includes the distribution at equilibrium of the components between the two phases (the $K_i = y_i / x_i$ values), the phase enthalpies, the phase volumes and the phase entropies. For a scrub column simulation, the most important of these are the K_i values and the phase enthalpies. The uncertainty associated with the thermodynamic model used by the simulator to calculate these quantities has a direct impact on the reliability of the simulation and therefore, the required operating margin.

Several authors have examined the impact of thermodynamic uncertainties on the reliability of process simulations [6, 7, 8]. Kister [9] has reviewed several examples of problems associated with the simulation of distillation columns and identified the inability to predict K_i values as the leading cause of problematic simulations. Thus, the selection of the most appropriate thermodynamic model for use in the process simulation is critical.

Equations of state are the most widely used thermodynamic models for engineering applications. For hydrocarbon processes such as LNG production, the cubic EOS of Soave, Redlich and Kwong (SRK) [10] and Peng and Robinson (PR) [11] are used most frequently. These equations are both more than 30 years old and many different EOS have since been developed, (for a recent review see, for example, Valderrama [12]) and many of which are available for selection in modern process simulators such as Aspen HYSYS® from AspenTech V2004.2 [13]. The deficiencies of the SRK and PR EOS are well known and an array of correction techniques (e.g. volume translation [14]) and sophisticated mixing rules (e.g. excess free energy [15]) to extend the applicability of these equations are available. Nevertheless, for predicting the VLE behavior of multi-component mixtures of light hydrocarbons, the (almost) unmodified SRK and PR EOS are still recommended for use in most process simulators [12,13].

Multi-parameter EOS have emerged over the last two decades as thermodynamic models able to represent all of the "good" thermodynamic data available for a substance within their experimental uncertainty: see Span et al. for a review [16]. Such EOS have largely been confined to the description of a select number of pure substances or binary mixtures; however Kunz et al. [1] on behalf of the Groupe European de Recherche Gaziere (GERG) have recently developed a multi-parameter EOS for natural gas mixtures (up to 18 components) valid at temperatures between -183 °C and 177 °C (-298 °F and 350 °F) and pressures to 35000 kPa (5080 psia). The uncertainties of the GERG-2004 EOS are claimed to be equal to that of available experimental data for binary and multi-component mixtures of natural gas components: typically 0.1 % for gas densities, 0.1 – 0.5 % for liquid densities, 0.5 – 2 % for enthalpy differences and 1 – 5 % for VLE properties. These claims are based on the fact that the GERG-2004 EOS was regressed to a large set of binary mixture data and compared to several data sets for multi-component mixtures. We are currently testing how well the GERG-2004 EOS matches these claims for other sets of reliable multi-component natural gas mixture data.

The GERG-2004 EOS is a significant step forward in the description of the thermodynamic properties of natural gas mixtures. However its computational complexity prevents it from being incorporated into current process simulators because they are incompatible with the iterative algorithms (such as that discussed in Russell [5]) needed to solve complex unit operations such as a scrub column. The computational simplicity of cubic EOS is the primary reason for their longevity and ubiquitous use; despite their flaws it is likely that they will be used in LNG process simulations for some time to come.

Kunz et al [1] have identified that the VLE data situation for mixtures of natural gas components is still poor, particularly for multi-component mixtures at cryogenic conditions. More high-quality thermodynamic data for such systems will improve the predictions of both complex multi-component EOS, such as GERG-2004, as well as the cubic EOS used in process simulators. One primary goal of the Chevron – WAERA alliance is to measure such data and anchor the cubic EOS used by simulators to it. In the following section we demonstrate the motivation for this by illustrating the uncertainty caused by the use of conventional cubic EOS in the simulation of an LNG scrub column.

Sensitivity of a Simple Scrub Column Simulation on the EOS choice

Figure 2 shows a simple simulation of an LNG scrub column, generated using Aspen HYSYS® from AspenTech V2004.2 [13] with the SRK EOS. The details in this simulation are representative of, but different to, those in operational plants. This simulation of a scrub column uses the default "reboiler" unit operation but models the condensing system as three separate unit operations. The compositions of the feed and product streams are listed in Table 1. Two compositions are given for each product stream corresponding to the result of the simulation obtained with the SRK or PR EOS.

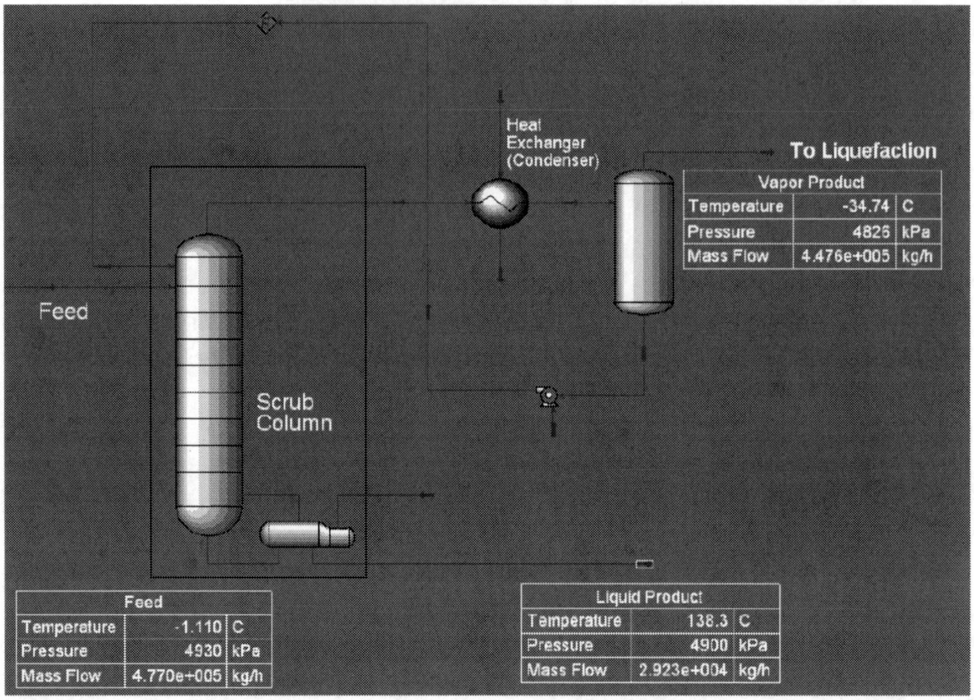

Figure 2 – Process Flow Diagram for a Simple Simulation of LNG Scrub Column using Aspen HYSYS® from AspenTech V2004.2 [13] and the SRK EOS

	Feed *mole fractions*	Vapor Product *mole fractions*		Liquid Product *mole fractions*	
		SRK	PR	SRK	PR
N2	0.00053	0.00054	0.00054	0.00000	0.00000
C1	0.88295	0.90168	0.90066	0.03825	0.04032
C2	0.06524	0.06424	0.06428	0.10914	0.11327
C3	0.03415	0.02877	0.02934	0.27276	0.26923
iC4	0.00560	0.00280	0.00297	0.13271	0.12979
nC4	0.00677	0.00191	0.00212	0.22865	0.22440
iC5	0.00254	0.00005	0.00007	0.11603	0.11730
nC5	0.00169	0.00001	0.00002	0.07802	0.08037
nC6	0.00026	0.00000	0.00000	0.01222	0.01266
nC7	0.00026	0.00000	0.00000	0.01222	0.01267
Mass Flow [kg/hr]	4.770E+05	4.476E+05	4.491E+05	2.923E+04	2.824E+04
	Percent Change		0.3%		-3.4%

Table 1 – Feed Parameters for the Simulation in Fig 2 and the Dependence of the Calculated Product Properties on the EOS Choice

In Figure 3, the calculated dew point curves for the scrub column feed stream are shown for the SRK, PR and GERG EOS. Also shown is the line corresponding to the scrub columns operation; this line passes through a region of the feed's phase envelope where the difference between the SRK and PR EOS dew point curves is near its maximum of more than 2 °C (4 °F). The basic operational principle of a scrub column is to use the phase behavior of the feed to separate the mixture components, yet simulations of that behavior are most discrepant near its operating line. Furthermore, the more accurate GERG EOS indicates that, in this case, the SRK equation is more accurate than the PR equation. Current simulation practice, however, favors the use of the PR EOS [13].

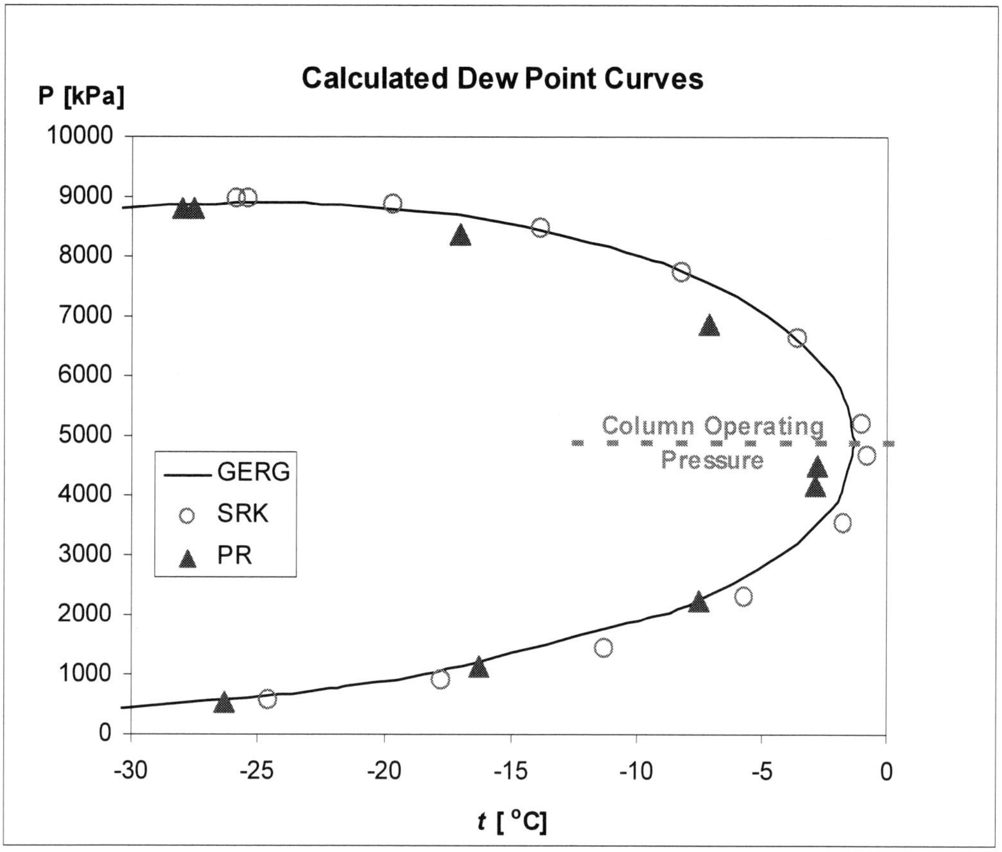

Figure 3 – Dew Point Curves for the Feed Stream in Table 1: Comparison of Predictions from the SRK, PR and GERG [1] EOS

Further discrepancies in the simulation arising from the choice of EOS are listed in Table 1.The mass flow rate of the liquid product, which can be used to derive further sales (LPG, condensate) or as a make-up MCHE refrigerant stream, decreases by 3.4 %. The composition of the vapor and liquid products also varies and this variation is shown graphically in Figure 4. The compositions in Table 1 are quoted to more significant figures than would be measurable by on-site chromatographs. In Figure 4 we show only those relative changes in composition that would be measurable by such a chromatograph. This primarily affects the vapor stream, for which we consider only the hydrocarbons to n-C_4.

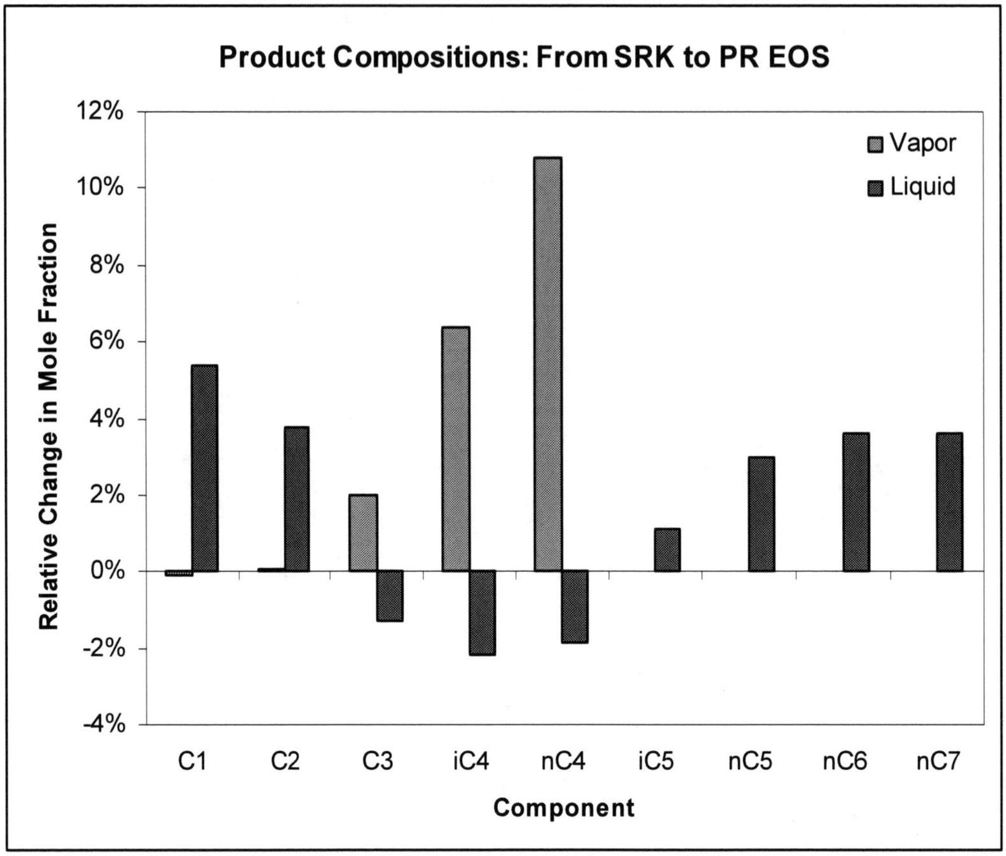

Figure 4 – Percentage Change in the Scrub Column Product Stream Mole Fractions (See Table 1) Caused by Changing from the SRK to the PR EOS (Only changes large enough to be reliably measured are plotted)

It is clear that the greatest discrepancy occurs for the butanes; with the PR EOS about 8 % more of the butanes are carried through to the vapor product. This could have implications for freeze-out problems in the downstream MCHE, or in the final LNG product specifications.

As another test of the simulation's sensitivity to the choice of EOS, the composition of the feed stream in Figure 2 was modified to include a trace amount (10 parts per million) of an aromatic compound (benzene, toluene or p-xylene). The presence of such levels of these compounds in a treated gas feed to an LNG plant is not unheard of, and it is essential that the scrub column remove them prior to the MCHE: even small trace amounts of these compounds can, given sufficient time and flow rates, accumulate in cryogenic equipment causing decreased operational efficiencies, blockages or failures. The ratio of the total molar flow of the aromatic compound in the vapor product to that in the feed stream is called the "component recovery" and, as shown in Figure 5, this is a sensitive function of the column's "reflux ratio", which is the ratio of the molar flow rate returned to the column from the condenser drum to the molar flow rate to the condenser. The reflux ratio plays a key role in determining a column's running cost and its efficacy at delivering a vapor product that meets the required specifications. It is, therefore, also a parameter that is often used to apply design margins to account for process uncertainties. Increasing the reflux ratio by a factor of two, for example, is an approach sometimes taken to ensure that the product specifications are met, even though this will increase the column's operating cost.

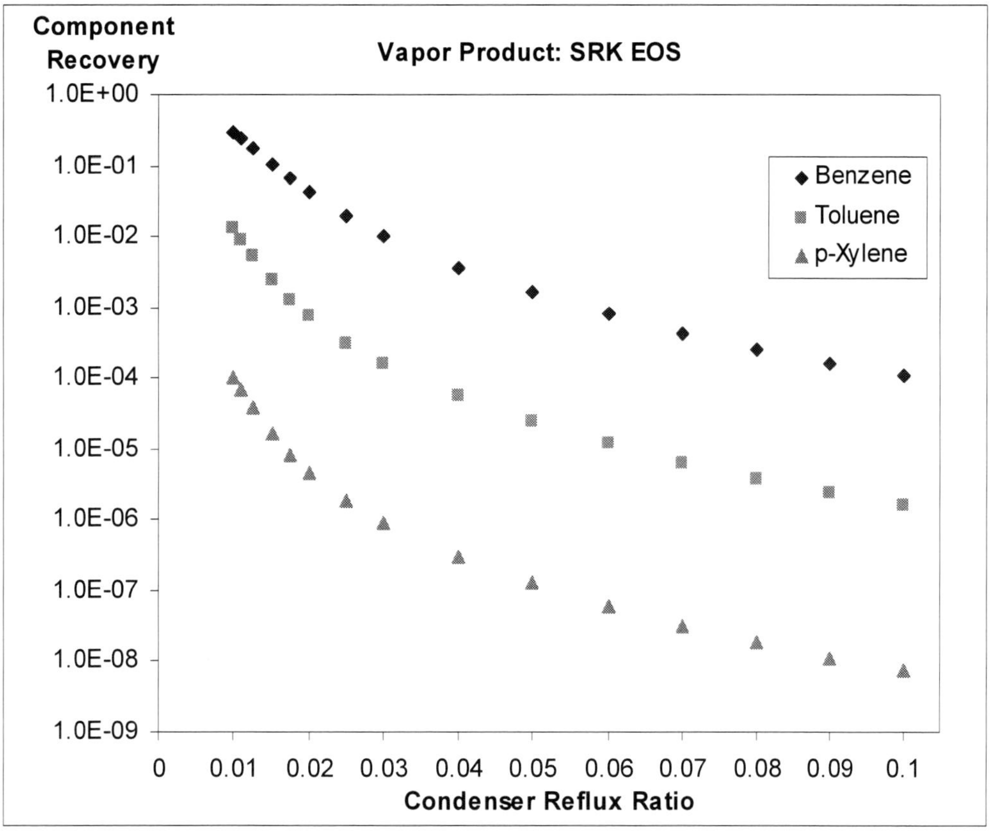

Figure 5 – Simulated Aromatic Component Recovery in the Scrub Column's Vapor Product as a Function of Condenser Reflux Ratio Using the SRK EOS

Very little reliable VLE data exists for low temperature, high-pressure solutions of aromatic compounds in hydrocarbon fluids, which are exactly the conditions of the scrub column condenser. This is reflected in the sensitivity of the component recovery calculations on the choice of EOS used in the simulation. In Figure 6 the ratio of the aromatic component recovery calculated with the PR EOS to that calculated with the SRK EOS, as a function of reflux ratio is shown. For benzene, toluene and p-xylene, the averages of the predicted recovery ratios are 1.5, 2.0 and 4.1, respectively. In other words, the predicted carry-over of benzene from the feed to the vapor product is 50 % larger if the PR EOS is used instead of the SRK EOS, and 410 % larger for p-xylene.

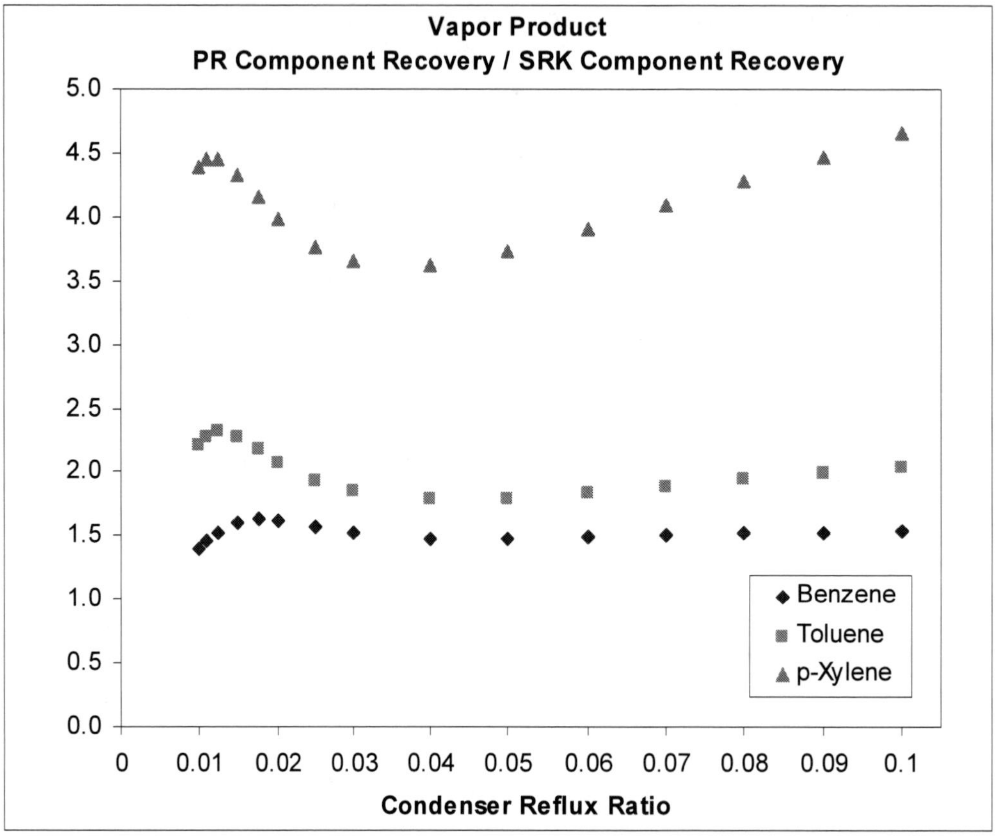

Figure 6 – Ratio of the PR EOS to the SRK EOS Predictions of the Component Recovery in the Scrub Column's Vapor Product

As illustrated in Figure 5, a scrub column operating at a reflux ratio of 0.06 will achieve an aromatic component recovery in the vapor product about 200 % smaller than for a reflux ratio of 0.05. An engineer operating the scrub column might, therefore, decide to absorb the extra cost of a 20 % increase in reflux rate to ensure that the carry-over of the aromatic is a factor of two smaller than it would be if the reflux ratio were left at 0.05. However, as shown in Figure 6, this design margin of 200 % is comparable to or smaller than the uncertainty of the EOS used in the simulation.

Different Results from Different Simulators

The discrepancies discussed in the previous section pertained to the use of different EOS within the same simulator: Aspen HYSYS® from AspenTech V2004.2 [13]. The identical scrub column and parameters was then modeled in PRO/II® Version 8.1 [17], and further discrepancies were highlighted when utilizing the same EOS. Figure 7 shows the same simple simulation in Pro/II that was simulated in Figure 2 in Aspen HYSYS®.

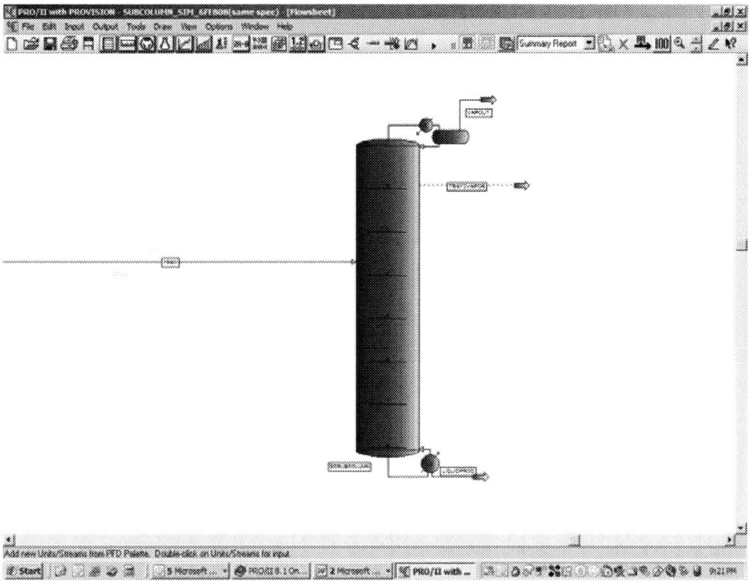

Figure 7 – Process Flow Diagram for a Simple Simulation of LNG Scrub Column using PRO/II® Version 8.1 [17] and the SRK EOS

The scrub column was modeled in Pro/II utilizing the same two design specifications for convergence; the recovery of methane at the column overhead product and the overhead condenser outlet temperature. The column could not converge utilizing the same Aspen HYSYS® specification of a 0.999 methane recovery fraction, but instead converged at a recovery fraction of 0.99. Other discrepancies are shown in Table 2.

	Vapor Product *mole fractions*		**Liquid Product** *mole fractions*	
	HYSYS	**Pro/II**	**HYSYS**	**Pro/II**
N2	0.00054	0.00054	0.00000	0.00000
C1	0.90066	0.90040	0.04032	0.04570
C2	0.06428	0.06440	0.11327	0.10530
C3	0.02934	0.02940	0.26923	0.26180
iC4	0.00297	0.00288	0.12979	0.13590
nC4	0.00212	0.00224	0.22440	0.22360
iC5	0.00007	0.00008	0.11730	0.12060
nC5	0.00002	0.00000	0.08037	0.08130
nC6	0.00000	0.00000	0.01266	0.01290
nC7	0.00000	0.00000	0.01267	0.01290
Mass Flow [kg/hr]	4.49E+05	4.48E+05	2.82E+04	2.78E+04
Percent Change		-0.3%		-1.5%

Table 2 – Comparison of the PRO/II® Simulation Shown in Fig 7 with the Aspen HYSYS® Simulation of the Same Scrub Column. The PR EOS was Used in Both Cases

Impact of the Uncertainties: How much do they matter?

There are many uncertainties associated with the design of an LNG plant which are much larger than those associated with the thermodynamic model or software package used to simulate them. Designers of LNG plants must account, to some degree, for the feed to the plant to vary in composition and mass flow rate. Equipment selection and sizing will be greatly affected by the possibilities for future expansion and the required turn-down. Furthermore, all of these uncertainties are small in comparison to those of the model used to describe the reservoir providing the feedstock for the LNG plant.

The capital expense of an LNG plant is so large that even small uncertainties represent large amounts of money. However, it is arguable that the accuracy of plant simulations matters most in terms of reducing the operational costs of a plant via optimization. Once the plant is constructed and running, optimization of the operational parameters can help to reduce costs and/or increase revenue. Such an optimization is only possible with the use of process simulators and, therefore, uncertainties inherent in the simulation limit the optimization that is possible.

Planned Experiments

While some limited experimental data exists for simple binary mixtures at low temperatures, insufficient reliable thermodynamic data exists for multi-component mixtures at high pressures and low to cryogenic temperatures. This is the principal reason for the uncertainties in LNG simulations resulting from the choice of thermodynamic model. The cubic EOS used in simulators are tuned (by necessity) to data measured at conditions far from those realized in the LNG plant. Furthermore, most EOS are tuned only to binary mixture data.

We have therefore set about constructing experimental systems to measure VLE, volumetric and calorific properties of representative multi-component, "LNG-like" mixtures at temperatures down to -160 °C (-260 °F) and pressures to 5500 kPa (800 psia). The multi-component mixtures will be prepared from the following components: {N_2, CO_2, C_1, C_2, C_3, C_4, C_5, n-C_6, cyclohexanes, benzene, toluene and xylenes}. This will provide relevant thermodynamic data characteristic of the many different mixtures and conditions found in the LNG liquefaction train. A particular focus will be given to mixtures containing trace levels of aromatic compounds and naphthenic compounds at temperatures near -40 °C (-40 °F) and pressures near 5000 kPa (720 psia) because of the importance of, and difficulty with, scrub column simulations.

The VLE experiments will be conducted using a high-pressure cell with active internal mixing, housed in a cryogenic thermostat with two automatic sampling valves. These valves will sample simultaneously the equilibrium vapor and liquid phases in the cell and the samples will be analyzed by a single on-line gas chromatograph (GC). The

GC has been specially designed to be able to monitor all potential components in each phase. It consists of two PoropakQ columns, two flame ionization detectors and one thermal conductivity detector (TCD). By carefully adjusting the flows of the two columns through the TCD, we will be able to use this single detector to monitor the inert compounds in both phases. Between VLE measurements the GC will be calibrated automatically using gravimetrically prepared standards. The GC data will allow the determination of the pressure, temperature and composition dependence of the equilibrium ratios K_i for the mixture. Knowledge of the K_i is essential to the simulation and prediction of VLE.

A second cell will be constructed, similar in design to the first but with two major differences. The vapor and liquid phases in the second cell will be mixed using external circulation loops which pass through vibrating tube densimeters. The equilibrium ratios for the mixtures in the second cell will be measured by the sampling valves and GC in the same way as was done for the first cell. This will provide a check of the K_i data quality and confirm that the two different mixing mechanisms are equivalent. The vibrating tube densimeters in the second cell will provide the mass densities, and therefore the molar volumes, of the vapor and liquid phases. Volumetric data is essential for converting mass flow rates into volume flow rates, which is particularly important for vessel sizing. Furthermore, liquid products are usually sold by volume. The SRK and PR are notorious for their poor prediction of liquid volumes, and the volumetric data we generate will lead to correction schemes like that of Peneloux et al. [14], which are applicable over a wider range of conditions.

Heat capacity data for the mixtures is vital to the simulation of unit operations such as the MCHE and other heat exchangers. We will use a specialized cryogenic calorimeter from Setaram [18], modified for use at high pressures to precisely measure the heat capacity of prepared vapor and liquid mixtures with compositions similar to those of the phases measured in the VLE cells. Enthalpies of vaporization for these mixtures will be studied in detail as these parameters are particularly important in modeling the expansion of a high-pressure liquid into a two-phase mixture using a throttling valve.

Conclusion

The thermodynamic properties of LNG-like mixtures at the temperature and pressure conditions found in LNG liquefaction trains have not been extensively studied. Consequently, the EOS used to calculate the VLE, volumetric and calorific properties of the fluids in simulations of the process introduce uncertainties into the results of the simulations. This in turn affects the ability of operators to optimize the process and possibly reduce operational costs. To address this problem, Chevron and WAERA are undertaking a three year project to experimentally determine the required thermodynamic properties of cryogenic multi-component mixtures. The EOS used in process simulators will be anchored to the new data, reducing the uncertainties of current LNG simulations. This will reduce total cost of ownership and allow for more informed business and technical decisions to be made.

References Cited

1. O Kunz, R Klimeck, W Wagner, M Jaeschke, *The GERG-2004 Wide-Range Equation of State for Natural Gases and Other Mixtures*, GERG Technical Monograph 15 (2007) Fortschr.-Ber. VDI, Reihe 6, Nr. 557, VDI Verlag, Düsseldorf.
2. J M Campbell, *Gas Conditioning and Processing, Vol 2, 7th Ed.* Campbell Petroleum Series, Oklahoma, 1994, 275-280.
3. D Coyle, F. Felix de la Vega, C Durr, "Natural Gas Specification Challenges in the LNG Industry," Paper PS4-7, *KBR Publications*, 2007; also in the *Proceedings of LNG15*, April 2007.
4. J Page, *An Investigation into LNG Scrub Column Thermodynamics at the NWS Onshore Gas Plant*, Honours Thesis, The Centre for Oil and Gas Engineering, University of Western Australia, 2001.
5. R A Russell, "A Flexible and Reliable Method Solves Single-Tower and Crude-Distillation-Column Problems" *Chem. Eng.* (1983) 92, 53-59.
6. E C Carlson, "Don't Gamble with Physical Properties for Simulations" *Chem. Eng. Prog.* (2002) 98, 52-58.
7. C L Rhodes, "The Process Simulation Revolution: Thermophysical Property Needs and Concerns" *J. Chem. Eng. Data* (1996) 41, 947-950.
8. A H Harvey and A Laesecke, "Fluid Properties and New Technologies: Connecting Design with Reality", *Chem. Eng. Prog.* (2002) 98, 34-41.
9. H Z Kister, "Can We Believe the Simulation Results" *Chem. Eng. Prog.* (2002) 98, 52-58.
10. G Soave, "Equilibrium Constants from a Modified Redlich-Kwong Equation of State." *Chem. Eng. Sci.* (1972) 72, 1197.
11. D Y Peng and D B Robinson, "A New Two Constant Equation of State" *Ind. Eng. Chem. Fundam.* (1976) 15, 59-64.
12. J O Valderrama, "The State of Cubic Equations of State" *Ind. Eng. Chem. Res.* (2003) 42, 1603-1618.
13. Aspen HYSYS® from AspenTech V2004.2, (2005) Aspen Technology, Cambridge MA 02141, USA
14. A Peneloux, E Rauzy, R Freze, "A Consistent Correction for Redlich-Kwong-Soave Volumes." *Fluid Phase Equilib.* (1982), 8, 7-23.
15. D S Wong, S I Sandler, "A Theoretically Correct Mixing Rule for Cubic Equations of State." *AIChE J.* (1992) 38, 671-680.
16. R Span, W Wagner, E W Lemmon, R T Jacobsen, "Multiparameter Equations of State. Recent Trends and Future Challenges." *Fluid Phase Equilib.* (2001) 183, 1-20.
17. PRO/II® Version 8.1, (2007) Invensys Systems Inc, Lakeforest CA 92630, USA.
18. Setaram Low Temperature Calorimeter BT2.15.
See http://www.setaram.com/BT-2.15.htm

Speaker Biography

Lindsay Laskowski

Lindsay Laskowski graduated from Texas A&M University with a Bachelors of Science degree in Chemical Engineering in 2002. Lindsay started her oil and gas career as a process engineer with J. Ray McDermott Engineering and obtained her Engineering in Training Certification. She focused on the topside production equipment and process for offshore platform production. In January of 2006, she joined Chevron Energy Technology Company as a LNG Process Engineer focusing on new technology development in the field of liquefied natural gas, LNG. Lindsay assists in major capitol projects but spends the majority of her time developing new processes and products to reduce the total installed cost of the LNG value chain. She currently has a patent pending in liquefaction technology.

Lindsay serves as the Society of Women Engineers – Houston Area Section President as well as volunteering for Special Olympics Texas. Lindsay received the 2008 Young Engineer of the Year Award for her technical and volunteering accomplishments.

Cost and Drivers for Site Preparation and Foundation for LNG Facilities

Pramod Rao, PhD, P.E.
Senior Geotechnical Engineer

J. C. Kuo, P.E.
Senior Advisor

Chevron, Energy Technology Company
Houston, Texas

AIChE Spring Meeting, April 2008
8th Topical Conference on Natural Gas Utilization
New Orleans, Louisiana, April 6-10, 2008

Site Selection and Development

Selection of a plant site has a significant influence on the design, construction and operating costs of LNG Liquefaction Plants or Receiving Terminals. The principal factors that influence site selection and development include the following.

Proximity of the end User–For LNG Receiving Terminals, the proximity to the end user is a critical factor in evaluating a site. The closer the terminal is to the end user, the lesser the requirement for infrastructure and equipment necessary to support the LNG Import Terminal.

Proximity to Population Centers and Local Perceptions - There are potential benefits related to a LNG Facility site being in close proximity to the urban or population centers as it maximizes employment opportunities for local and businesses from project. Yet, the site should be sufficiently far away to minimize disturbances to neighborhoods. Generally, a safety or exclusion zone is prescribed by regulatory authorities around the LNG Facility. This requirement may vary from country to country. Local perceptions about the project also play a big role in the land acquisition process. Community engagement is necessary to identify a site of low environmental and social sensitivity.

Ecological Provisions may include protection of wild life, endangered species, flora and fauna, mitigation of wetlands, disposal of construction generated spoils, quarantine requirements.

Environmental Considerations – All design must be in accordance with the rules and regulations which have been established by Federal, state and local governmental bodies.

Land Availability – Selection of the site will be based on availability of usable land for the LNG plant. In certain situations, land may have to be created by reclamation.

Potential for Future Expansion - In many cases, a phased approach is employed to ramp up the LNG production in the LNG Liquefaction Plant which may require construction of additional LNG trains in the future. In the case of LNG import terminal, the need for increasing the send-out rate for LNG terminal may require additional infrastructure. Additional land should be apportioned for this purpose.

Meteorological Conditions, Bathymetry and Oceanographic Conditions, Snow/Ice Conditions - Meteorological conditions such as wind speed, air temperature, and wind chill index influence the structural design of the proposed structures and also impact the constructability of LNG facilities. Wave data including wave height, period, direction and current data, extreme sea states and bathymetry data dictate the location of the harbor, layout of the turning basin and access channel, design of the near-shore structures such as jetties for product delivery, and sedimentation and dredging requirements. In cold regions, sea ice can influence the maneuvering of ships. The level of ice cover should be studied.

Snow is the key winterization element which has the largest impact on the design, construction and operation of infrastructure and structures in cold regions. Historical snow precipitation data should be used to determine the snow loads on structures. Based on experience and knowledge of these facts, a focus on winterization issues may give valuable input to the general site layout and orientation and shape of buildings to minimize the problems with blowing and drifting snow.

Icing may result from atmosphere icing (precipitation icing) or sea-spray icing due to wind-borne sea-water spray/mist occurring at temperatures below freezing. The formation of layers of ice on the structural components of the plant as well as the process equipment and personnel access facilities is of concern both from a structural design standpoint as well as for safety.

Geotechnical/Soil Data - The in-situ soil conditions shall be analyzed to determine the proper type of foundation and ground improvement method desired.

Seismicity Data - In seismic areas, geophysical surveys and seismic hazard studies shall be performed to identify the presence of any geologic fault traces and to develop site-specific response spectra for Operating Basis Event and Safe Shutdown Events.

Land Topography - Land topography shall be used to determine the physical features which will dictate the drainage design layout and final grading. Determination of final grading and drainage scheme for the LNG facility will be based on a number of considerations including flood plain elevation, size of property in relationship to the size of plant facilities, desirable location on site, and plant access based on topography.

Available Infrastructure and Utilities - Site analysis should extend beyond the project boundaries. All existing and proposed transportation systems to and around the site shall be evaluated for their accessibility. The primary and secondary roadways shall be examined to determine potential access points, traffic loads and safety conditions including potential hazards. All utility systems including water supply, sewer, storm drainage, electrical, gas and telephone systems shall be located at site and capacities evaluated. If an existing system is running at or near its capacity, additional growth in the area may require improvements to the utility systems.

Construction Materials and Labor - Availability of resources including construction materials such as aggregate, rip-rap, steel, cement, borrow fill and local labor in the proximity of the area are critical to reducing project costs and subsequent delays in project schedules.

Factors Influencing Site Preparation Activity and Foundation Alternatives

Site preparation for any project is influenced by several direct factors that relate to the site conditions and functionality of the proposed construction. These factors are very site-specific and may not be all applicable for every site. The purpose of this section is to highlight the role these factors play in developing site preparation strategies.

Functional Layout and Structural Requirements of LNG Facilities – A typical LNG liquefaction plant includes the following areas: (a) All process facilities to treat and liquefy the gas into LNG, (b) All process facilities to treat the condensate for shipment, (c) All facilities to store and load the LNG onto ships, (d) All facilities to store and transport the condensate, (e) All facilities to store and transport the condensate, (f) All facilities to allow ships to approach and dock at the loading berths, (g) All utilities and other facilities to support the process plants, (h) Administration and residential facilities to support plant operation.

The layout for LNG Facilities is dictated primarily by safety requirements, process requirements, and other design considerations.

In some cases, it may be possible to adjust and optimize the layout to minimize its impact on the foundation cost or site preparation activity due to variation in soil conditions at the site, provided the limits established by the above requirements are not violated. For instance, at a site where a portion of the site has competent soils and other areas where the soils are not as competent due to possibility of fill soils, it may be possible to consider ways to realign the layout of the plant to ensure that settlement sensitive equipment such as rotating machines, heater support columns, directly connected reboilers to vessels, and LNG tanks are located on competent soils and lightly loaded equipment and structures are located on less competent soils. In some cases, a re-orientation of plot plan may be sufficient to eliminate costly foundations (deep foundations) or ground improvement costs. In other cases, it may help reduce the extent of ground improvement or use of pile foundations.

In areas with high seismic activity and soft subsurface soils, amplification of seismic waves results in significant lateral loads on heavy and tall structures such as towers and vertical vessels. Foundation costs in highly seismic region could be significantly more than that under normal circumstances. The use of seismic isolators may be required, particularly if the sites are close to active faults. Soils susceptible to liquefaction will have to be treated to reduce the liquefaction potential. In areas with high seismic activity with a rock stratum, heavy and tall structures will require anchoring the foundation to the rock to prevent overturning as using weight of the overburden and foundation to counteract the uplift loads may require deeper excavations that may be cost prohibitive compared to installing rock anchors.

Land Acreage Requirement (LNG Facility Size) - The LNG Liquefaction Plant output (MTPA) impacts the size of the LNG plant required which dictates the plot plan size required to construct the plant.

A good rule of thumb for the initial estimation of the total area could be 9.0 hectares per MTPA, for up to 20 MPTA grass root LNG Plant. Larger plant will require smaller area. It is important to note that other factors such as differences in liquefaction process; client or country requirements can impact the plot size.

Similarly, a correlation between the gas output of billion cubic feet per day (BCFD) and the approximate required terminal plot space can be used. For 1.0 BCFD and up to 5.2 BCFD grass root receiving terminal, 42 hectares per BCFD is a "rule of thumb" for the initial total area estimation for full containment tanks. For single containment tanks, a multiple factor from 4.0 to 2.0 is recommended. Larger terminals will need smaller multiple factor.

Availability of Land Near-Shore - Land availability near-shore can be scarce due to existing local communities, land acquisition issues and regulatory requirements. Reclaiming land offshore may be an option under these circumstances. Reclamation consists of building land by hydraulic placement of imported fill soils. The following are critical factors that influence the site preparation activity at a reclaimed site:

(i) Sources of Fill Materials – The recommended fill soils for any land reclamation activity generally consist of controlled structural fill materials. The availability of the desired gradation and characteristics of the material and the distance of the source from the site are critical for site preparation.

(ii) Availability of Equipment –With increasing LNG plant construction activity around the world, the dredging market is tight. The demand for dredging equipment is high with very few reputable dredging companies. This can be cause for escalation in costs if there are delays in schedule.

(iii) Installation of Temporary Site Drainage – Reclamation activity generates a significant quantity of water that should be discharged and drained properly to avoid flooding the neighboring communities. As a minimum, temporary site drainage would be necessary.

(iv) In-situ Soil Conditions – In-situ soil conditions may influence the performance of the hydraulically placed fill. If the estimated long-term total settlements due to the settlement of the fill, which are in turn due to consolidation of the underlying soils and the settlement of the foundation itself exceeds the design criteria, the site preparation may involve some sort of ground improvement or deep foundation alternatives or a combination of both.

Reclamation activity can increase the time required for site preparation significantly. Time required for site preparation can be on the order of 12 to 18 months for approximately 5 to 7 million cubic meters of fill. Unlike conventional site preparation activity, reclamation drives the cost of site preparation up and requires extensive quality assurance and quality control along with instrumentation for monitoring performance of the reclaimed ground. The potential for impact to schedule is high due to longer time for site preparation, equipment down time and tie-up on other projects, permits not available, site issues such as mud-waves and slope instability during construction.

Finished Site Elevation - The finished grade elevation is governed by functional requirements, flood-plain elevation at site, and civil grading requirements for site drainage. The finished grade elevation dictates the amount of fill required to be placed at a site. In some cases where the fill thickness is extensive with excessive self-weight settlement of fill soils, the fill type will also be important.

Site Topography - The extent of cut and fill required will depend on the desired finished grade elevation and the site topography. As far as possible, the cut and fill volumes at the site should be balanced to ensure minimum import of the fill or disposal of fill. Excavations in soft soils may require adequate lateral support to the walls. In cold regions, consideration should be given to the seasonal variation in temperature and its impact on the frozen soils when designing excavation support systems. Mass excavations in rock may require controlled blasting of the rock to develop a terrace for construction. Particular attention should be paid to noise pollution concerns and damage to adjacent structures due to vibration.

Existing Soil/Rock and Groundwater Conditions - The choice of site preparation method and foundation alternatives is greatly influenced by the subsurface conditions. The subsurface conditions, encountered at a site can be broadly classified as Rock, Firm/Hard Soils, or Soft/Loose Soils. Rock is the most desired geo-material and provides adequate bearing capacity. Most structures can be supported on shallow foundations such footings, block, combined footings and mat type foundation. Minimal settlements are anticipated for structures founded on rock. In some limestone formations, sinkholes, and karsts can occur in which case comprehensive mapping of the entire area will be necessary to identify these locations. The caverns may have to be filled with grout so that the structures are adequately supported or these locations should be avoided.

Firm to Hard soils provide better bearing capacity as compared to the soft soils and therefore, require minimal site preparation. The settlements depend upon the loads and the size of the foundation. But often, settlements are manageable and no special ground improvement is necessary. Both shallow foundations and deep foundations system are suitable foundation types. The choice depends upon the loads and the tolerable settlement criteria.

Soft or Loose soils are the most problematic soils and do not provide adequate bearing capacity and may cause excessive settlements of proposed structures. Extensive site preparation should be anticipated. Ground improvement methods with or without the use of deep foundations may be required. The commonly used ground improvement techniques are listed below: (i) Preloading with/without Prefabricated Vertical Drains, (ii) Vibro-Flotation and Vibro-Replacement, (iii) Dynamic Soil Compaction, (iv) Jet Grouting, (v) Deep Soil Mixing.

Though these methods are widely used, it may be challenging to employ some of these methods in some parts of the world such as West Africa. Some of the challenges that one faces include: (a) Import of equipment and materials can result in high potential for schedule delay due to equipment stranded at Customs, (b) No Local Capability for Performing the Work – Specialist Contractors may be required, (c) Local Labor may not have adequate experience and therefore are required to be trained, (d) Local Content Requirement may have to be satisfied (can be up to 30%), (e) Environmental Concerns such as contaminated ground may prevent the disposal of spoils, (f) Pre-production Testing Required – Laboratory Testing and Field Trials.

Logistics - Logistics such as sources of borrow material, transportation requirements (Land or Sea), Hauling distances, On-site Stock Pile Locations and Available Infrastructure also affect site preparation.

Strength, Weakness, Opportunities and Threats for Typical Ground Improvement Techniques

Plate 1.0 lists the strength, weakness, opportunities and threats for ground improvement techniques routinely used. This list provides general guidance on factors that need to be weighed before the selection of a particular type of ground improvement technique. By no means is this list of factors complete and it should be appropriately modified to address any site-specific issues that come into play, as they are not addressed here-in.

Foundation Types

Foundations primarily consist of shallow foundations and deep foundations. Shallow foundations can be used where there is a suitable bearing stratum near the surface, no highly compressible layers below, and calculated settlements are acceptable. Where the bearing stratum at the ground surface is underlain by weaker and more compressible stratum, consider the use of deep foundations or piles. Deep foundations can be classified further depending upon the material type and type of installation employed.

Driven Piles – Piles installed by impact or vibration

a) Timber Piles
b) Steel – H sections or Pipe Piles
c) Precast Concrete (including pre-stressed)
d) Composite – Steel encased grouted piles
e) Drilled Foundations – Formed by drilling or excavating a hole in the ground

a) Bored Piles or Drilled Straight Shafts
b) Auger-Cast In-Place Piles

Typical Site Preparation &Foundation Installation Schedules and Costs

The earlier section provide factors that directly influence site preparation and foundation alternatives at a site, which collectively influence the baseline costs for site preparation and foundations. However, there can also be factors other than those listed above that may indirectly affect schedule and lead to incremental costs.

(a) Uncertainty in Soil Data - If the soil data obtained during the Front-End Engineering Design is limited and shows considerable spatial variation across the site, there could be considerable uncertainty in the estimate of the engineering parameters at the site which may lead to non-conservative estimates in site and foundation performance. Significant changes in design may become unavoidable during construction leading to schedule delays and ultimately causing escalation in project costs. To mitigate this risk, extensive geotechnical investigation should be carried out during the early phase of the project. If soil variability is high, supplemental geotechnical exploration with tighter spacing between the boring locations should be performed.

(b) Disturbed Samples in Soft Clay Soils - Disturbed samples obtained in the soft clay soils may not provide good estimates of soil parameters leading to conservative designs and increased site preparation costs. To minimize the disturbance of samples obtained in soft clay soils, piston sampling should be employed in order to obtain relatively undisturbed soil samples in the soft clay soils. Obtaining undisturbed soil samples using piston sampler requires experienced and skilled drillers which may not be practical in certain conditions. Alternatively, in-situ testing such as cone penetrometer tests, and T-ball penetrometer tests can be performed.

(c) Delay in geotechnical information can have a negative impact on the project schedule. Timely completion of geotechnical information is critical for design to proceed.

(d) Underestimates in the amount of imported fill in soft soil sites and underestimate of the amount of material that needs to be disposed.

(e) Miscellaneous: Technologies that have time dependencies such as preloading with surcharge can cause schedule creeps and in the extreme case can cause cost overruns, Absence of SME review of geotechnical reports may prove to be costly, External expertise may be limited in some areas (e.g. soft soils) and in some cases the expert brought in by the FEED contractor remains in Phase 4, which may not be in the benefit of CVX as adequate technical oversight may not be possible on a project, Difficulty obtaining permits to do some of the site preparation (e.g., blasting permit in sensitive areas), Environmental groups, NGOs, community issues that impact site preparation, nearby areas (fishing) etc, Seasonal or lack of seasonal issues (dry, rainy or frozen soils), Natural shore erosion issues (impacts on LNG facilities)

An estimate of typical unit costs for various ground improvement technology options is presented in Table 1.0 for comparison purposes. These costs are US based and do not include mobilization or demobilization costs for equipment and labor.

Table 1.0: Estimated Units Costs for Various Ground Improvement Technology Options (US Based / Mob-Demob Not Included) (2005)

Approach	Unit Cost
Excavate and Replace	\$30 - \$45 /m^3
Deep Soil Mixing	\$75 - \$90/m^3
Jet Grouting	\$130 - \$ 150/m^3
Prefabricated Vertical Drain with Preload	\$1.5 - \$ 2/ lineal meter
Sand Drains with Preload	\$5 - \$10 / lineal meter
Vibro-Compaction	\$7-\$10 / m^3
Deep Dynamic Compaction	\$10 - \$22/m^2
Explosive Compaction	\$3 - \$4 / m^3

Table 2.0 provides site preparation cost estimates for some typical projects. Specific details regarding the projects cannot be divulged due to confidentiality reasons and hence, a rough estimate is provided to provide the reader with an idea of the site preparation costs relative to the total project costs.

Table 2.0: Site Preparation and Foundation Cost Estimates* for Typical LNG Liquefaction Plant Sites

Site Description	Schedule	Approximate Site Preparation Costs	Approximate Total Project Costs	Percentage of Total Project Costs
Rocky land terrain in cold regions with significant blasting and terracing to build an LNG plant with 3 x 5.0 MMTPA trains	30 months	\$400 million	\$15 billion	3 %
Firm Soils in Highly Seismic Region – 3 x 5.0 MMTPA train	18 months	\$ 250 million	\$12 billion	1% or less
Soft Soils with Reclamation – 3 x 5.0 MMTPA train	24 months	\$ 600 million	\$ 15 billion	4 %

* Itemized costs could for site preparation and foundation costs could not be obtained

Conclusions

Selection of a plant site has a significant influence on the design, construction and operating costs of LNG Liquefaction Plants or Receiving Terminals. The drivers for any site preparation activity or foundation strategy consist of both direct and indirect factors. Direct factors relate to the site conditions and functionality of the proposed construction and include the functional layout of the LNG plant and structural requirements, land acreage requirement (plot plan size), availability of land near-shore, finished site elevation, site topography, existing soil/rock and groundwater conditions and logistics. These factors are very site-specific and may not be controlling at every site. Selection of ground improvement at a site should consider the relative strengths, weakness, opportunities and threats of the various options. Selection should be based on the most optimal solution with the least risk. Indirect factors may cause schedule delays and lead to incremental project costs.

The cost of site preparation and foundations for a typical 3 x 5 MMTPA train LNG plant could range from about 1 % at sites with competent soils to about 5% at sites with incompetent soils. If the site preparation accounts for future expansion during the initial phase, then the cost of site preparation and foundations could be higher percentage in the initial phase with one train, approximately 8 % to 10%.

References

1. "How Big Is the Area required for LNG Base Load Plant And Their Design Considerations", J.C. Kuo, Presented to the AIChE 2006 National Meeting, LNG section, Apr 23-27, 2006.
2. "How Big Is the Area required for LNG Receiving Terminal And Their Design Considerations", J.C. Kuo, Presented to the AIChE 2007 National Meeting, LNG section, Apr 22-26, 2007.
3. Xanthakos, P.P., Abramson, L. W., and Bruce, D.A., "Ground Control and Improvement."
4. NAVFAC 7.03, "Soil and Foundations Design Manual"

Biography

Dr. Rao has over 12 years of geotechnical consulting experience on a wide range of petrochemical projects both domestic as well as international. He has provided geotechnical consulting services and performed technical reviews on several LNG projects in Texas, Louisiana, and internationally in Mexico, Venezuela, and Angola. He has a Ph.D. in civil engineering with specialization in geotechnical engineering from the University of Houston and is a licensed professional engineer in the State of Texas. He has published several papers in peer reviewed conferences on the subject of deep foundations: auger cast piles, lateral loads on piles/piers, performance of model piles under seismic events and site characterization of settlement analyses of LNG tanks.

J.C. Kuo is a veteran of the Gas Processing/LNG industry, with 34 years experience in the field. He is senior advisor for Chevron.LNG Systems. He has served as the Process Manager for many projects, including the Gorgon LNG, Delta Caribe LNG, Casotte Landing and Sabine Pass LNG terminal projects. He also served as the Technical Reviewer for Angola, Olokola, Algeria and Shtokman LNG projects. Before working with Chevron, he was the Technology Manager for IPSI/Bechtel and served as the Process Manager for the Pemex Catarell offshore project, the Egyptian LNG (Idku) train 1/2 and China Shell Nan Hai Projects. He has contributed to LNG and gas processing technology improvements through a patent, a book and many papers on the subject. He has a BS from Chung Yuan Christian University, Taiwan and a MS from the University of Houston, both for Chemical Engineering. He also holds a MS in Environmental Engineering from Southern Illinois University. He is a registered Professional Engineer in Texas and a member of AIChE. He is the president of the 99 Power Qi Qong Texas divisions.

Plate 1.0: Analysis of Strengths, Weaknesses, Opportunities and Threats of Various Ground Improvement Techniques

Approach	Strengths	Weaknesses	Opportunities	Threats
Excavate and Replace the Unsuitable Soils	• Established, well known procedure with successful track record.	• Replacement under water with granular soils may require densification/improvement to preclude liquefaction	• Utilize shallow foundations	• In offshore reclaimed areas, there could be potential side slope failures and mud flows into the excavated area • Potential liquefaction risk if adequate placement density is not achieved • Silt and mud in water may settle out to bottom of the excavation while open or may be trapped in sand placed
Deep Soil Mixing (granular and cohesive soils)	• Accepted method worldwide • Quantifiable design • Quality control essential during installation	• Has not been used extensively under LNG tanks to improve clay soils. Acceptance by LNG tank vendors would have to be assured	• Utilize shallow foundations • For depths less than 20 m does not need the world's largest DSM equipment	• May end up with unacceptable product if not installed by experienced specialty contractor • High organic content may require special cement/mixing
Jet Grouting	• Quantifiable Design • Eliminates liquefaction potential • Quality control essential during installation	• Size of column variable with soil profile • Potential to generate significant spoils • Requires post construction verification program	• Utilize shallow foundations • Can do selected vertical (of discrete depth intervals) and area improvement	• May end up with unacceptable product if not done by experienced specialty contractor
Prefabricated Vertical Drain (PVD) with Preload	• Proven track record world-wide • Use at most sites would follow conventional application	• Some uncertainty in time rate of settlement • PVDs need to be stored away from sunlight • Requires reliable instrumentation and monitoring • Potential impact on foundation construction start date	• Utilize shallow foundations if preload height and duration are sufficient as verified by careful instrumentation and monitoring	• Rate of primary consolidation may be slower than expected , requiring more fill • Risk of disturbance and slower drainage with improper mandrel
Vibro-Compaction (Granular soils) or Vibro-Replacement (cohesive soils)	• Appropriate method to remediate Liquefaction Potential • Improves overall foundation stiffness • Reduces Liquefaction Potential	• Requires post construction verification • Results sometimes erratic	• Utilize shallow foundations • Reduce risk of liquefaction • Restriction on fines content for below water placement in reclaimed areas could be relaxed using this option which may reduce the reclamation costs (as sand sourcing cost may be lower)	• Ineffective if fines/silt content in fill is higher than ~ 15%
Deep Dynamic Compaction (granular soils)	• Can eliminate liquefaction Potential • Relatively easy/unsophisticated method	• May not improve the entire targeted depth • Requires experimentation to minimize variability in results • Requires post construction verification program	• Same as above	• May be slow if old equipment used • Risk of uneven improvement • Ineffective if fines/silt content in fill is higher than ~ 25 %
Explosive Compaction	• Used for clean granular soils • Reduces Liquefaction Potential	• High degree of control required in upper few meters of soil • Results sometimes erratic	• Same as Above	• Uneven compaction –Need to be careful • Ineffective if fines content in fill > ~ 15% • Difficulty in Permitting in some parts of the world

Impact of LNG Specification on Liquefaction and Import Plants

David Coyle, Felix F. De La Vega, Charles Durr and Frank L. Del Nogal

KBR
601 Jefferson St.
Houston, Texas 77002

AIChE Spring National Meeting 2008
8th Topical Conference on Natural Gas Utilization.
New Orleans, LA, April 06-10, 2008.

ABSTRACT

LNG is primarily methane, but in most cases also contains ethane, propane and heavier components. Until recently, potential buyers of LNG had some flexibility in the ranges of natural gas properties they were able to accept. In many cases the LNG buyer exported electric power and not natural gas, and therefore did not have to meet the needs of natural gas consumers.

But now, new markets are opening up where pipelines to consumers already exist, and the consumers have specifications which lie outside the range of the LNG produced at many locations. The challenge is to understand the requirements of the potential new markets and plan ways to meet the requirements of both the LNG exporters and importers.

To that end, it is useful to review the natural gas specifications of current consumers and those expected in the near future, such as North America and others. These specifications are compared to those of LNG being produced, and the options and problems for conditioning the natural gas on either end of the LNG trade are explored in detail.

INTRODUCTION

Natural gas contains methane, heavier hydrocarbons and inert components which affect burner performance. For this reason, pipeline companies and LNG buyers specify allowable ranges of components and heating values. These requirements vary widely depending on the market location. But despite this, LNG product specifications have not been a major plant design issue. Historically, plant designs were based on long-term contracts to few buyers, and there was little need for flexibility in the plant designs, either on the liquefaction or receiving ends of the trade.

However, the situation is changing as LNG trade becomes more global. The owners of liquefaction plants are targeting more than one market, and new markets have requirements that are not always compatible with existing trades. Furthermore, the growing spot market for LNG provides opportunities for buyers and sellers who have the ability to be flexible on product specifications. As a result, there is now considerable movement towards technical solutions for conditioning LNG on liquefaction and receiving ends.

WORLDWIDE NATURAL GAS SPECIFICATIONS

Natural gas specifications have several purposes, including corrosion prevention, avoiding liquid drop out in pipelines, and burner performance. The specifications for corrosion prevention limit concentrations of CO_2, $H2_S$, mercaptans or total sulfur. LNG facilities treat gas down to 50 ppmv CO_2 to avoid freezing in the cryogenic processing unit, and therefore meet receiving end pipeline specifications. The sulfur specifications are typically consistent with the Japanese market which limits H_2S to 5 mg/Nm^3 maximum and total sulfur content to 30 mg/Nm^3 maximum. Meeting these specifications for Japan will also meet the requirements for Europe and the U.S. (except California which limits total sulfur to 18 mg/Nm^3).

Specifications to prevent corrosion and liquids dropout are fairly consistent, and requirements of the LNG plant itself (i.e. avoiding freezing during cryogenic processing) tend to make those specifications almost universal. The requirements for heating value and gas interchangeability remain. Here the specifications vary significantly, as do LNG products from worldwide sources, and it is interchangeability that is presenting the greatest challenges.

Early LNG trade was primarily to Japan from Pacific Rim and Middle East export plants and to Europe from Northern Africa. The Japanese specifications vary depending on the importing utility company, but typically have a high heating value between 39.7 and 43.3 MJ/Sm^3 (megajoules per standard meter cubed at 1 atm and 15 °C, which converts to 1065 to 1160 Btu/scf for a standard cubic foot at 14.73 psia and 60 °F). This relatively high range permits maximum use of infrastructure by moving greater combustion heat capacity for a given volume. European countries typically allow wider ranges. Spain, for example, allows a range between 35.0 and 44.9 MJ/Sm^3 (940 and 1205 Btu/scf).

Figure 1 shows representative heating value specifications for different countries. In one example, a gas with HHV = 42.6 MJ/Sm^3 is suitable for the Japanese and Korean markets, but is too high for the U.S. or U.K. In the second example, a gas with HHV = 37.2 MJ/Sm^3 meets U.S./U.K. specs but has an HHV too low for Korea or Japan. Both examples, however, are within the ranges allowed for France and Spain.

In Europe, a gas interchangeability index (the Wobbe Index) is also specified. The Wobbe Index is defined as HHV/[Gas specific gravity (relative to air)]$^{1/2}$, and is often more stringent than the heating value specification. British Standard EN ISO 6976 provides all the details on standard conditions, component heating values, ideal versus non-ideal gas calculations and more. The

Wobbe Index is not a completely accurate way to predict the behavior of burners using alternate gases, but it is convenient and accurate enough for most purposes.

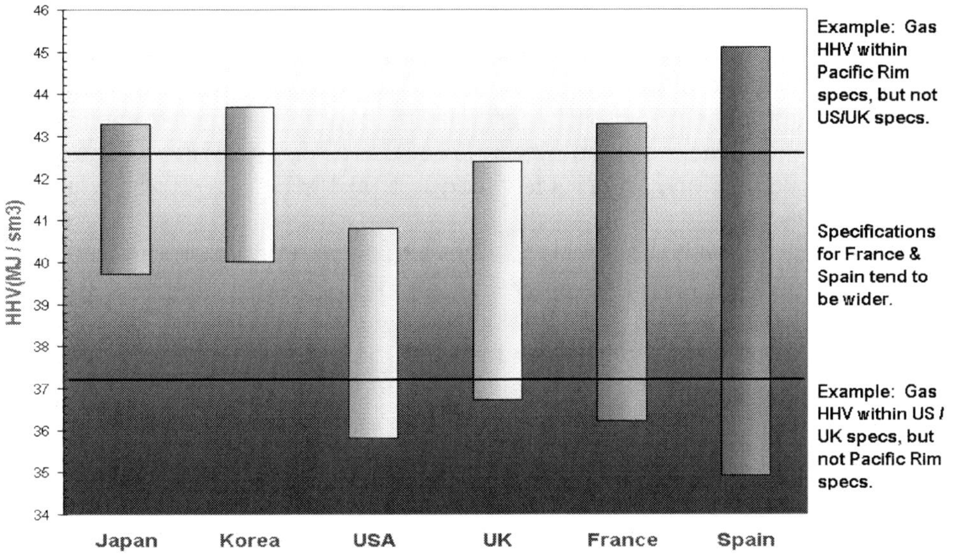

Figure 1. Worldwide heating value specifications.

Generally speaking, a gas in the preferred ranges of HHV and Wobbe Index will burn without combustion problems such as flame lifting, excessive NOx and CO, or yellow tipping. Figure 2 shows the preferred Wobbe Index region and the consequences of operating in objectionable ranges.

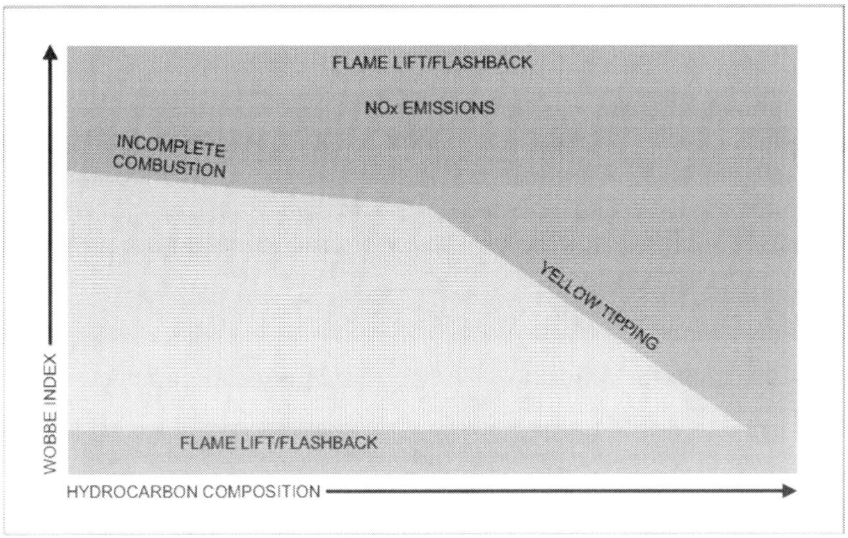

Source: National Petroleum Council

Figure 2. Combustion properties as a function of HHV and Wobbe Index.

The LNG market is changing because of two players in the marketplace: the U.S. and U.K. Ironically, both new players were in the LNG business in the early to mid-1970s but a regulatory crisis in the U.S. and North Sea discoveries in the U.K. resulted in both countries receiving only minimal quantities of LNG. Now, the U.S. is projecting a widening gap between gas production and demand that can only be satisfied with LNG imports projected to grow to 90 million metric tons per year (MTPY) by the year 2025 (US EIA, 2004).

The U.K. interconnecting pipelines once exported gas from the U.K. to the continent, but recently the flow has reversed direction. U.K. gas imports via pipeline and LNG are now expected for the long term, and three LNG import facilities are now under construction with a combined capacity of about 16 MTPY (Haynes and Martin, 2006). LNG capacity increases via expansions of these projects or new projects should provide an additional 8 MTPY by 2013.

Clearly, the U.S. and U.K. demands will have quite an impact on the Atlantic Basin market which is currently at about 49 MTPY. The size of the U.S. and U.K. markets are more than large enough to target, and dealing with gas quality specs is a requirement for participation. This is especially true in the Middle East, where cargoes can go to either Pacific or Atlantic consumers. There are many options for modifying the quality of imported or exported LNG on both ends of the trade, and all parties involved have a high level of interest finding the best solutions for their own situations.

The specifications in the U.S. remain somewhat fluid as there is no national standard for heating value or Wobbe Index. In March 2005, the Natural Gas Council (NGC) issued interchangeability recommendations to the FERC that included a Wobbe Index maximum of 1,400 Btu/SCF (52.1 MJ/Sm^3). FERC made an announcement on 15 June 2006 where they recommended following the NGC criteria on a voluntary basis. While this avoids some disruption of trade, there is still the effect of market uncertainty. Who takes the risk on quality if there is a chance of mandated requirements in the future?

LIQUEFACTION END SOLUTIONS

Modifying heating value at the liquefaction end usually means adding or extracting ethane, propane and butane (LPG), though nitrogen may also play a part. For natural gas supplies rich in LPG components in the Atlantic Basin, a lower high heating value (HHV) is preferred if the U.S. and U.K. markets are to be the consumers. On the other hand, Pacific Rim consumers prefer a gas with increased HHV, and Pacific Rim sources that are lean in LPG components may require upward HHV adjustment. Natural gas sourced in the Middle East can go either direction to the Pacific or Atlantic markets, which raises the possibility of producing two product qualities.

Reducing HHV

Two methods available for reducing HHV on the liquefaction end are LPG extraction and adding nitrogen, as discussed below.

Plant Capacity and LPG Extraction. One of the first considerations a contractor faces when LPG extraction becomes a requirement is the effect the extraction has on the plant capacity. Because of the extracted LPG, the LNG production drops unless more power and capacity is added to the refrigeration systems, and this is one of the reasons published $/ton of LNG can be misleading. The cost of extracting LPG can be significant and yet the product is often not considered in published $/ton/yr numbers.

Table 1 shows a typical natural gas before and after extraction, based on extracting propane and butane products with the recoveries shown and the results are essentially independent of the selected liquefaction process technology. The comparison holds the LNG product rate constant on a tonnage basis. With LPG extraction the total product rate (LNG, propane, and butane combined) increases by 7.7% while total refrigeration power increases by almost 10%. Power thus increases at a greater rate than total product, indicating a greater energy requirement for extracting and fractionating LPG.

Table 1 - Effects of LPG Extraction*

		Without LPG Extraction	With LPG Extraction**	Δ (%)
LNG Rate	(MTPA)	4.6	4.6	0.0
Propane Prod. Rate	(MTPA)	0	0.21	
Butane Prod. Rate	(MTPA)	0	0.17	
Total Prod. Rate	(MTPA)	4.6	5.0	7.7
LNG HHV	(MJ/Sm^3)	42.4	40.6	-4.6
LNG Wobbe Index	(MJ/Sm^3)	52.9	51.9	-2.0
Nominal Compression Power	(MW)	100%	109.8%	9.8%
Specific Power	($MW/MTPA_{LNG}$)	100%	109.8%	9.8%
	($MW/MTPA_{TOT}$)	100%	102.3%	102.3%

* Basis: Molar Feed composition is 6% N_2, 83% C1, 7% C2, 2% C3, 1% C4, 1% C5+, LNG Nitrogen specification of 1 mol%, HHV and Wobbe Index calculated on ideal gas basis.
** At 66% C3 recovery and 83% C4 recovery.

Just to make a point, if the cost of liquefaction is assumed to be linear with power then the $/ton of LNG goes up by the same percentage as the specific power. The $/ton of LNG then appears 10% higher for the plant that extracts LPG even though the benefit of LPG product sales is hidden.

For this reason it is more appropriate to use $/ton of total product including LPG for comparisons to overcome some of the inherent differences between plants with and without extraction. However most published figures on plant costs and capacities only provide the LNG product for reference. A similar thought process should also be applied to specific energy which shows only a 2% difference when based on total product compared to the 10% difference obtained using only LNG production.

Table 1 also shows the magnitude that product properties change as LPG products are extracted. The HHV drops by 1.8 MJ/Sm3 while Wobbe Index drops by only 1.0 MJ/Sm3. The desired Wobbe Index often controls the extent of extraction required and the index is not as sensitive to extraction as heating value (however, as discussed later Wobbe Index is more sensitive than heating value to nitrogen injection).

Another factor a design engineer considers is the optimum amount of LPG extraction. Natural gas specifications in the design basis are usually given as a range with maximum and minimum HHV. It is possible to target a lower HHV limit with deeper extraction, or a higher HHV by allowing more LPG to remain. Usually the LPG netback to the exporter has a higher value as separate LPG products as opposed to components in the LNG product and in such cases the optimum design tends towards deeper extraction and a lower HHV LNG. Also, once LPG extraction is part of the design, there are many more facilities required for the plant, such as LPG storage and loading; so higher extraction tends to be required to justify the extra cost of these new facilities.

If 95% of the propane and 100% of the butane were extracted in the above example, the HHV becomes 40.0 MJ/Sm3 and the Wobbe Index becomes 51.6 MJ/Sm3. If lower values are needed then ethane must also be extracted. Ethane extraction has its own difficulties and is discussed later.

Nitrogen. Another way to reduce heating value is to add nitrogen. In fact, if achieving Wobbe Index is the goal, adding nitrogen can be more effective than extracting LPG. Typical US pipeline specifications allow for a maximum of 3 mol % nitrogen.

At many sites the nitrogen is not 'added'. The nitrogen is present in the natural gas and in some cases the concentration is over 5 mol %. To keep the nitrogen with the LNG it is only necessary to chill the LNG product colder, but this does take power. The power is significant: on a constant tons/hr basis, it takes roughly 4% more power for a 1 mol % increase in nitrogen concentration. Table 2 shows the effect of adding nitrogen compared to extracting LPG which is also shown graphically in Figure 3.

Table 2 – LPG Extraction vs. Nitrogen Injection.

		Unadjusted	**With LPG Extraction**	**With 3 mol% N_2**
LNG HHV	(MJ/Sm^3)	42.4	40.6	41.4
LNG Wobbe Index	(MJ/Sm^3)	52.9	51.9	51.4

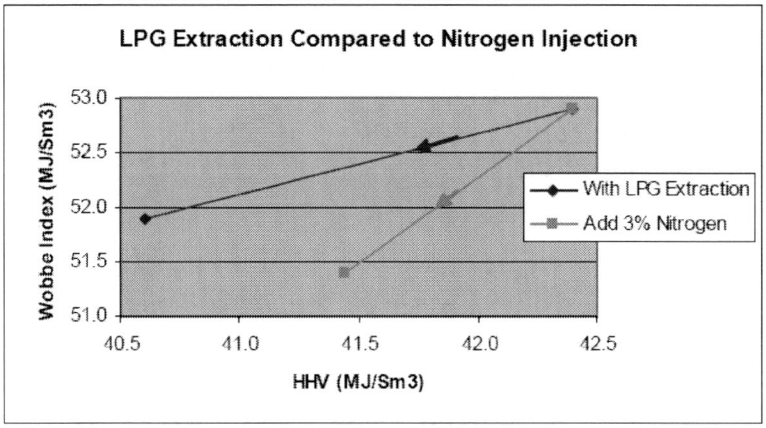

Figure 3 - LPG Extraction vs. Nitrogen Injection

Adding 3% nitrogen (instead of extracting LPG) does not change heating value as much as extraction however the nitrogen has a greater effect on Wobbe Index. Since Wobbe Index is often a more stringent requirement than heating value, increasing nitrogen content may be a good way to meet specifications in some cases.

When considering using nitrogen on the liquefaction end to meet natural gas specs, a design engineer must confirm that the shipping and receiving end can take a high nitrogen LNG because the boil off gas from a 3 mol % nitrogen LNG contains over 50 mol % nitrogen. Also, the receiving terminal might have to send out boil off gas that meets the pipeline specifications prior to recondensing in the LNG send out. In general the nitrogen solution is handled at the receiving end.

Handling Ethane. One of the biggest challenges on the liquefaction end can be the disposition of ethane. The UK specifications (and California specifications in the US) have the effect of limiting the ethane to as little as 6 mol % in the LNG. If the natural gas feed to the LNG plant contains 8-9% ethane then ethane must be used within the plant or exported.

The problem with exporting is that the liquefaction plants tend to be in remote locations where there is little industrial demand for ethane. The value of the ethane usually does not support the cost of the extraction.

Using ethane as fuel within the plant is not a trivial matter; beyond possible problems with excess fuel, there are also limits on how much ethane can be contained in gas turbine fuel while still meeting environmental restrictions.

The amount of fuel an LNG plant requires depends on many factors such as process efficiency, compressor and driver efficiency, and site specific factors such as domestic gas or electric power exports. However for the sake of argument, we can assume a plant requires 10% of the heating value contained in the feed to be consumed as fuel.

Also assume for this example that the plant has a feedstock containing 9 mol % ethane, and the LNG product must be 6 mol % ethane. The resulting fuel composition would be 64% methane, 36% ethane, with a high heating value of 47.90 MJ/Sm3.

Fuel containing 36% ethane may be a problem depending on the type of gas turbines used for compressor drivers and NOx emissions requirements. If NOx emissions up to 160 ppm are allowed then Frame 5 gas turbines are capable of burning this amount of ethane. Frame 6 gas turbines can also be designed to meet this requirement with a high ethane fuel.

Frame 7 gas turbines, on the other hand, require dry low NOx burners and methane must be at least 85% of the reactive material in the fuel gas. This precludes burning a 36% ethane stream in the Frame 7 gas turbines. However, power generators and other process services are often driven with Frame 5 or Frame 6 gas turbines, and segregating the fuel system into high and low ethane systems can help to achieve the amount of ethane "disposal" required while meeting environmental restrictions.

In some cases an ethylene plant may be considered as a solution to the ethane problem. This would be a last resort option because operating an ethylene plant adds such significant scope to the simple objective of transporting natural gas. The numbers work out well on paper at least; in our example if all of the ethane is extracted from the nominal 4.5 MTPY LNG train feed the ethylene plant capacity would be about 1.3 billion lbs. per year, which is a competitive world scale plant.

Transportation of the ethylene is a problem however, and most owners would also build a polymer plant to convert the ethylene into a product more easily shipped. Although ethylene is moved by ships in some cases, the ships are expensive and the ethylene becomes uncompetitive because of transportation costs.

Operations of the ethylene plant would be dependent on the LNG plant for feedstock and this would require a cooperative environment between the upstream LNG plant and downstream chemical plant operations. Ethylene plant operators sometimes run 4-5 years before turn-around for maintenance, while the LNG plants have gas turbine maintenance requirements more frequently (typically annually).

Such inherent differences in technologies are not typically close-coupled together and this would certainly put the "cooperative environment" to the test. The ethylene and LNG parties would both have to face technical realities while responding to production pressures.

Given all the challenges of the ethylene plant option it does not seem likely to be a common solution for future plants. However, in Skikda, Algeria Sonatrach has an ethylene plant with ethane feedstock extracted from LNG trains. The concept has worked at that location at least, and perhaps there are other niches to be discovered in the future.

D. Chretien of Total has proposed another alternative at the World Gas 2006 Conference which converts ethane to methane. The process begins with steam reforming ethane to H_2 and CO, followed by catalytic conversion to methane. For every four molecules of ethane seven molecules of methane are produced, along with water and CO_2: (4 C_2H_6 + 9H_2O □□ 21 H_2 + 7 CO + CO_2 □□ 7 CH_4 + 7 H_2O + CO_2). The conversion process is called CRG for "Catalytic Rich Gas" conversion which is commercially proven and licensed by Davy Process Technology (Chretien, 2006).

As with the ethylene option, the cost is significant and the investment may not pay out on a localized basis. However, if there is a need to remove ethane to make the LNG product marketable this process has the advantage of eliminating by-product streams and the cost associated with purifying and storing them (propane, butane and heavier hydrocarbons can also be converted, but hydrocarbon efficiency drops as the feedstock gets heavier).

Although projects are proceeding with processes modified to extract ethane for export or as fuel within the plant, there are also projects on the receiving end where ethane will be extracted. It is probably a safe bet that ethane extraction will be done on both ends for the foreseeable future.

Increasing HHV

At some locations very lean gas is available for export to Japan or Korea. Historically this trade began with Alaskan LNG with heating value adjusted upwards on the receiving end by way of LPG injection. Currently there are several projects in the Pacific Rim under development with relatively lean feedstocks. If the LNG product does not meet minimum heating value specifications LPG must be injected on the liquefaction or receiving end.

Looking at the problem globally, it would seem the best technical solution is LPG injection at the receiving end. The shipping temperatures of propane and butane are at least as warm as -40 degrees C and thus do not require expensive special materials such as 9% Ni, aluminum or stainless steel. All things being equal, it would cost less to ship and store LPG separately from the LNG, until the LPG is injected into the natural gas send out stream. If the LPG is injected at the liquefaction end then the LPG is transported in special materials suitable for cryogenic service. At the same time, land constraints on the receiving end may still lead to a solution where LPG is injected into the LNG prior to export.

If LPG must be purchased and injected at the liquefaction location the first challenge is finding a local source with sufficient quality and then installing the facilities for unloading and storage. Then more refrigeration is required because the LPG must be chilled from -40 degrees C (in the case of propane) down to -160 degrees C. This requires energy which will be applied as additional boil off gas compression or refrigeration compression. If the LPG is injected at -40 degrees C into the LNG then the LPG chilling is accomplished by vaporizing methane. The vaporized methane must be compressed to high enough pressure to be used as fuel or material for the natural gas circuit of the liquefaction process.

Simply injecting -40 °C LPG into the LNG is equivalent to pure component refrigeration with methane at -160 degrees C in a single stage, which is not very efficient. A better approach is to stage the injection into liquid methane at several pressure levels, or better yet use refrigeration to chill the LPG to LNG temperatures prior to injection. This can be easily accomplished by adding an LPG injection pass to the MCHE, or adding another exchanger in parallel to the MCHE, using liquefaction level refrigerant to provide the necessary cooling duty.

While there are technical problems and solutions for LPG injection, the commercial side is not as easy to overcome. The refrigeration for chilling the LPG to LNG temperatures is expensive and the LPG itself generally sells for more as a separate product than as a component of LNG. Add to this the more expensive transportation when the LPG is shipped at -160 degrees C, and the prospects for purchasing LPG and injecting on the liquefaction end do not look favorable.

Multiple Products

At some facilities there may be a desire to market multiple products, such as a high HHV product and a low HHV product. For a liquefaction plant with a lean feed gas this might be accomplished by importing LPG and injecting the LPG on a cargo by cargo basis. The plant operator runs the LPG injection equipment for a cargo destined for a high HHV market, but the equipment remains idle when loading a ship headed for a low HHV market. In theory it is possible to chill the LPG prior to injection using liquefaction refrigeration as mentioned above, but every loading would significantly change operations within the train. For this reason it is more likely that actual designs would have a stand-alone refrigeration unit, or would rely on greater boil off gas compressor capacity.

An important technical feature of the design is to avoid sudden vaporization of the LNG as the LPG is injected. This can be achieved by pumping the LNG to higher pressure before using the LNG to chill the LPG.

If the opposite situation exists where a plant with significant LPG in the feed is considered for both low and high HHV markets, the LPG can be extracted and stored or exported until a high HHV cargo is needed. At that time the stored LPG is injected during loading, but this concept has the same disadvantages mentioned earlier – expensive LPG transport and diluted LPG product value.

Perhaps the most promising way to accommodate multiple products is based on a large facility where two separate products can be produced in different trains with the two products stored separately. This becomes more convenient if two berths are also present, each with their own loading lines. The disadvantage of this method is the extra storage capacity that would have to be installed compared to a single product facility.

A plant producing multiple products will cost more than a plant marketing a single product. The multiple product plant may still be competitive if it is closer to the consumer or has economy of scale advantages over single product plants in the same market.

RECEIVING END SOLUTIONS

There are certain situations which make adjustment on the receiving end of the LNG chain attractive. As with the liquefaction end, the selected solutions can depend on whether the facility is new or a modification to an existing plant. For example, in existing receiving terminals "grandfathering" is an issue, and minor modifications to lines containing LNG raise the possibility that the entire facility will have to be brought up to current code revisions. Such factors obviously have a large bearing on the process selection.

Reduce HHV

HHV can be reduced either by adding inert nitrogen or by extracting LPG. Typical U.S. pipeline specifications limit the amount of inert material to 3 mol %. Thus, adding inert material is limited to 0.9 to 1.2 MJ/Sm^3 (25-35 Btu/scf) heating value reduction, depending on the amount of nitrogen in the LNG as it is received. However, if this is the only adjustment needed, the process is relatively simple and there are no other products to deal with besides LNG send out. There are also consumers on the U.S. east coast and in the U.K. who can take up to 5 mol % nitrogen. This makes greater HHV adjustment by adding inert material possible.

Add nitrogen. Adding nitrogen to the natural gas product would seem to be a straight forward proposition, and yet there are a myriad of options. The first option is the method for generating the nitrogen, and the following are possibilities:

- Membranes
- Pressure swing absorption
- De-oxygenation of submerged combustion vaporizer flue gas using hydrogen
- Inert gas production by combustion
- Cryogenic Fractionation

The suitability of these processes depends to a great extent on the capacity requirement and pipeline gas specification. Some pipeline specifications limit oxygen concentration to only 1 ppm, which effectively eliminates all but the last option from consideration. Furthermore, if the design send-out rate is one billion standard cubic feet per day (1 bscfd) and the nitrogen addition is 3% of that, the 1,250,000 scf/hr nitrogen plant capacity makes cryogenic air separation the only practical choice.

Once the nitrogen is generated, injection into the natural gas stream can be done in a straightforward manner by compressing to pipeline pressure and then mixing the nitrogen and natural gas streams together. This simple approach has the advantage of not requiring modification to lines containing cryogenic material which may be important in a grandfathering situation. The disadvantage is that there may be another process option requiring less power.

LNG receiving terminals typically take advantage of the LNG cold temperature sink as a way to reduce boil off gas compression. The LNG is pumped out of the storage tank to an intermediate pressure, which puts the LNG in subcooled state. It is then possible to add heat by "recondensing" boil off gas where the boil off gas is mixed with the LNG and the combined mixture is a single liquid phase. The higher pressure allows the boil off gas to be recondensed back into the LNG and the LNG remains in the liquid state (though at a higher temperature).

At many terminals, the amount of LNG cold is more than enough to recondense the boil off gas, and in such cases there is cold available for recondensing at least some of the nitrogen to be added. The recondensed nitrogen is then pumped to high pressure with the LNG rather than being compressed, and thus requires less power.

Separating air requires power. Based on an ambient temperature of 22°C and adding 3% nitrogen to a pure methane send out, air separation and compressing the nitrogen to pipeline pressure requires roughly 0.25-0.3% of the natural gas send out as fuel. This amount of fuel is relatively minor, given that a rule of thumb for terminals with submerged combustion vaporizers is 1.8% of the send out required as fuel.

The LNG cold can also be used to assist the air separation process, and a typical concept is to cool air using a non-flammable intermediate fluid. This concept can reduce the energy consumption somewhat and reduce cost. One thing to remember, though, when exploring gas quality modifications on the receiving end is that receiving terminals are usually expected to operate 365 days per year, and send out flow rates can also vary widely. As a result, instead of one large air separation unit, it may be necessary to have two 50% units running and one identical unit as a spare. Adding a spare unit and losing some economy of scale increases cost.

Extract LPG. Like adding nitrogen, LPG extraction also offers many options, and selecting between them depends on the amount of extraction desired, whether there is a higher than ambient temperature heat source available, and the economics of compression versus pumping. With LPG extraction, LPG product is likely to have a high enough value to support the additional equipment and operations cost.

Once the decision is made to extract LPG, the question is then how much to extract. It is possible to limit the extraction to only what is needed to meet the natural gas pipeline specs, or it is possible to extract more to increase LPG product sales. Generally speaking, if any extraction at all is required, the tendency is to extract as much as possible.

As with the liquefaction end, ethane may also present challenges if heating value or Wobbe Index is too high after extracting all the propane and butane. Hopefully there is a consumer in the vicinity to take the ethane product via pipeline. This scenario is currently more likely on the U.S. Gulf coast than on the east and west coasts; thus no general solutions are available. If no consumer is available, then the owner will have to use other measures such as burning the ethane as gas turbine generator fuel to produce electric power.

If ethane extraction is not required, a process such as the one shown in Figure 4 will extract propane and butane to low concentration levels in the LNG. This process is also capable of extracting up to 80-85% of the ethane. It is possible to extract over 95% of the ethane by adding a condenser to the recovery column. Basically, the processes can be divided into three different types based on the way the residue gas (the demethanizer overhead product stream) is handled: residue gas compression; residue gas compression and condensing; and residue gas condensing. Representative examples of each type are shown in Figures 4, 5, and 6.

The optimal process for a given site is a trade off between cost and process complexity, according to the following summary:

- Residue compression is simple and easy to operate but requires expensive vapor compression.
- Residue condensing followed by liquid LNG pumping and vaporization requires more complex process controls; however, the expensive compression is eliminated.
- Several processes are available which are combinations of the two described above. Some compression is required but not to the extent of the residue compression type. Some compression savings are achieved while still maintaining relatively simple operation and flexibility.

At some locations (such as the U.S. Gulf Coast) it may be possible to tie into existing LPG pipelines to transport the extracted LPG products. If the LPG is transported by truck or ship, then LPG storage will be required, and the storage inventory must be factored into the safety assessment. It may be necessary to acquire more land, depending on the size of the storage and LPG spill criteria.

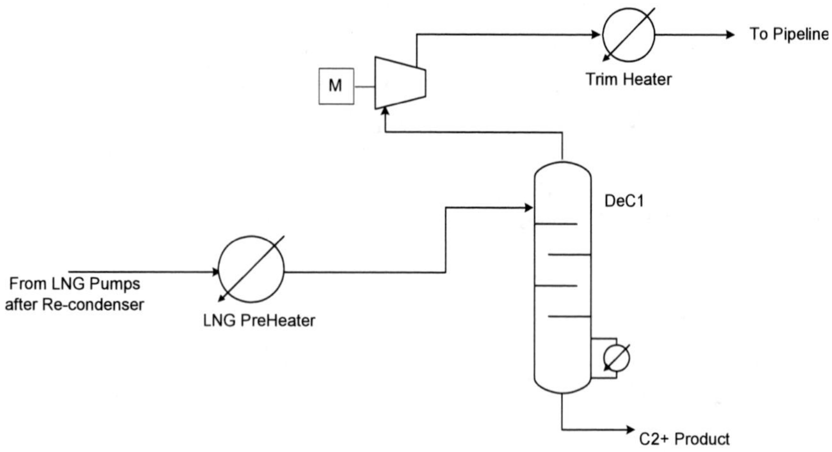

Figure 4. Residue Gas Compression (Rambo *et al.*, 1992).

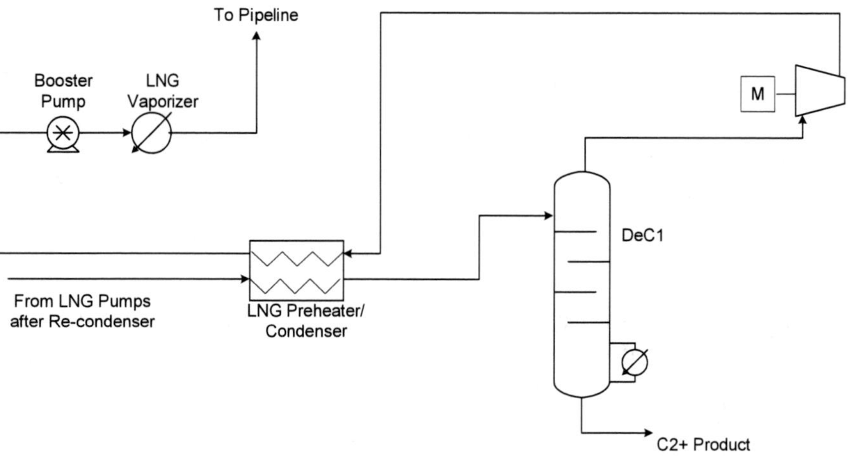

Figure 5. Residue Gas Compression and Condensing (McCartney, 2003).

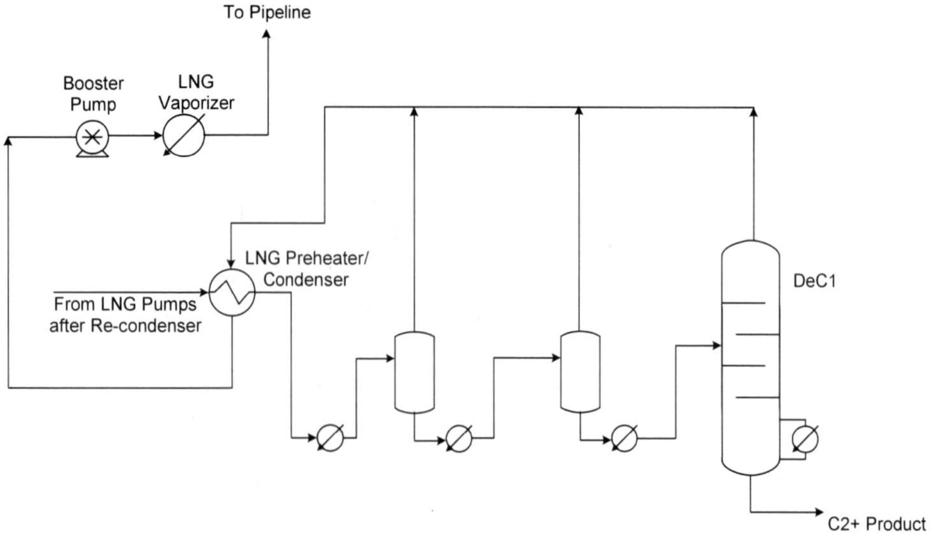

Figure 6. Residue Gas Condensing.

Increase HHV. Adding LPG is a practical way to increase HHV outside of blending with gas from another source. In cases where nitrogen content in the LNG is significant then nitrogen removal could also be considered if the desired specification is not too far away. Adding LPG is relatively simple from a process standpoint since it can be pumped to pipeline pressure as a liquid. At pipeline pressure, propane and butane are dense phase (as is the methane) and mixing the LPG with the natural gas is not difficult.

LPG injection is common in Japan where the heating value specification for natural gas is high to maximize utilization of the infrastructure. One source of LNG to Japan is Kenai, Alaska, where the LNG is nearly pure methane. This LNG must have LPG injection for interchangeability purposes.

If received LNG has a high heating value of 1,020 Btu/scf that must be increased to 1070 Btu/scf, it takes about five propane trucks (each truck carrying 3,000 gallons or 11.4 m^3) per billion scf to achieve the required heating value. The resulting concentration of propane in this case is just over 3 mol %.

It is likely that some amount of LPG storage is needed to accommodate the LPG delivery package size and to handle possible transportation delays. As with the case where LPG is extracted and stored on site, the storage capacity and spill criteria may influence the plot space required.

COMMERCIAL SOLUTIONS

Imagine if a rich LNG (that is, one with a high HHV) is produced in Australia for a market in California where the specification requires a low HHV. At the same time, a very lean LNG produced in Alaska is being shipped to Japan where LPG is injected to raise the heating value. Technically, it would be better for the Australian LNG to go to Japan and the Alaskan LNG to go to California. There would be alignment on LNG quality as well as shorter shipping distances. But with the usual long-term contracts between buyer and seller, it would be difficult to make this happen in spite of the incentives.

Given the growing spot market, perhaps this arrangement could be carried out occasionally based on cargoes that are available beyond the long term contract requirements. Eventually, if LNG really does become a commodity, this sort of deal could be done on a regular basis by selling cargoes FOB to an arbitrageur who would then find a suitable market and price for the delivery.

In August 2006, the Dubai Techno Park and Dubai Multi Commodities Centre announced plans to build a massive storage facility. This facility may be a solution for some interchangeability challenges. Dubai has access to Pacific and Atlantic markets, and the storage hub would provide some flexibility in quality blending and shipping that does not now exist.

CONCLUSIONS

Gas interchangeability is becoming a more important issue as the LNG market becomes more global. At the same time, it is very unlikely that the players in different markets could ever reach agreement on a uniform standard on a global basis. Indeed, the FERC has decided, for the time being at least, that implementing a universal standard within the U.S. would be disruptive to existing trade. This reality will continue to be a factor in plant designs as LNG buyers and sellers alike try to keep their options open.

Here, several points are relevant:

- LPG extraction is already common and will continue to be a preferred method for adjusting LNG HHV downward both on the receiving and liquefaction ends.
- Nitrogen injection is also employed in existing terminals to reduce HHV, and will probably continue to be implemented as a low cost method on the receiving end, especially when the controlling specification is Wobbe Index.
- LPG injection to increase HHV is more likely to be seen on the receiving end instead of liquefaction because of transportation costs (more expensive to ship with the cryogenic LNG).
- Ethane disposal on the liquefaction end will be a real challenge at some locations. If the ethane cannot be burned as fuel within the plant, proven technical solutions, while expensive, do exist.

Finally, where there is enough economic incentive, it may be possible to swap cargoes and save on plant modifications and shipping. This involves more parties each with their own risk and reward needs, but after the nightmares of negotiations all parties may awaken to a bright day of enhanced returns on their investments.

REFERENCES

Haynes, D. and Martin, P., "LNG is Coming Home", Gastech 2005, 14-17 March 2006.

US Energy Information Administration, Annual Energy Outlook 2004.

Chretien, D., "Process for the Adjustment of the HHV in the LNG Plants," 23rd World Gas Conference, June 2006.

THE SPEAKER

Frank Del Nogal graduated as Chemical Engineer from the *Universidad Simón Bolívar* in Venezuela, 1997. He performed as Technical Consultant on process simulation with Aspen Technology for five years. In 2001 he joined the Centre for Process Integration at UMIST in the UK, nowadays The University of Manchester, obtaining an MSc with distinction on Process Integration, a PhD on optimal LNG process design and winning the GPA Europe's Aungier Award 2003 with a previous publication. Frank joined M.W. Kellogg Limited in the UK in 2005 and is currently a Principal Technical Professional with their parent company, KBR, in Houston, TX.

Affinity Law Deviations in High Pressure LNG Pump Performance

Sarah D. Alison-Youel
Research and Development
Ebara International Corporation
salison-youel@ebaraintl.com

AIChE Spring National Meeting, April 2008
8th Topical Conference on Natural Gas Utilization
New Orleans, Louisiana, April 6-10, 2008

Abstract

Liquefied natural gas pumps are subject to the affinity laws and divergence from an affinity law surface fit is common. According to American Petroleum Institute (API) Standard 610, deviations are measured from the tested data point to the surface fit curve only in the vertical direction. However, as suggested by Hjorth et al (2006) and Pauly et al (2002), the true deviations of a tested point to the surface fit curve occur in both the vertical and horizontal directions, thereby the error being measured by the shortest distance, or perpendicular distance between the tested data point and the surface fit curve. The error related to the shortest distance produces smaller deviations between the tested data and the surface fit than the API error. High pressure LNG pump performance data is utilized to investigate the shortest distance error.

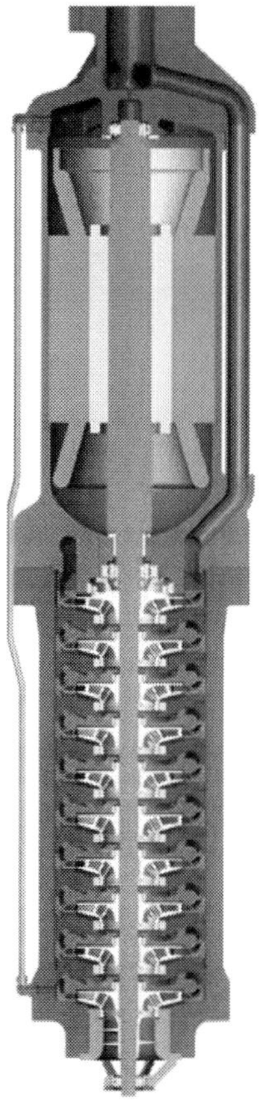

Fig. 1: Multi-Stage High Pressure Pump

High Pressure Pump

The LNG high pressure pump data employed is from an Ebara International Corporation (EIC) multi-stage high pressure pump, model number 8ECC-1516. This particular pump is designed for land-based applications, has an 8 in discharge flange, 15 in impeller diameter, and has 16 stages. Figure 1 is a solid model of a similar Ebara multi-stage high pressure pump.

Deviations

The high pressure pump performance data was applied through a least square error fit to both traditional affinity law performance equation (Kimmel, 1996) as well as general performance equation (Alison-Youel, 2008). The standard deviation for the tested data was found to be 29.41 for the traditional performance equation and 28.07 for the general performance equation (Fig. 2).

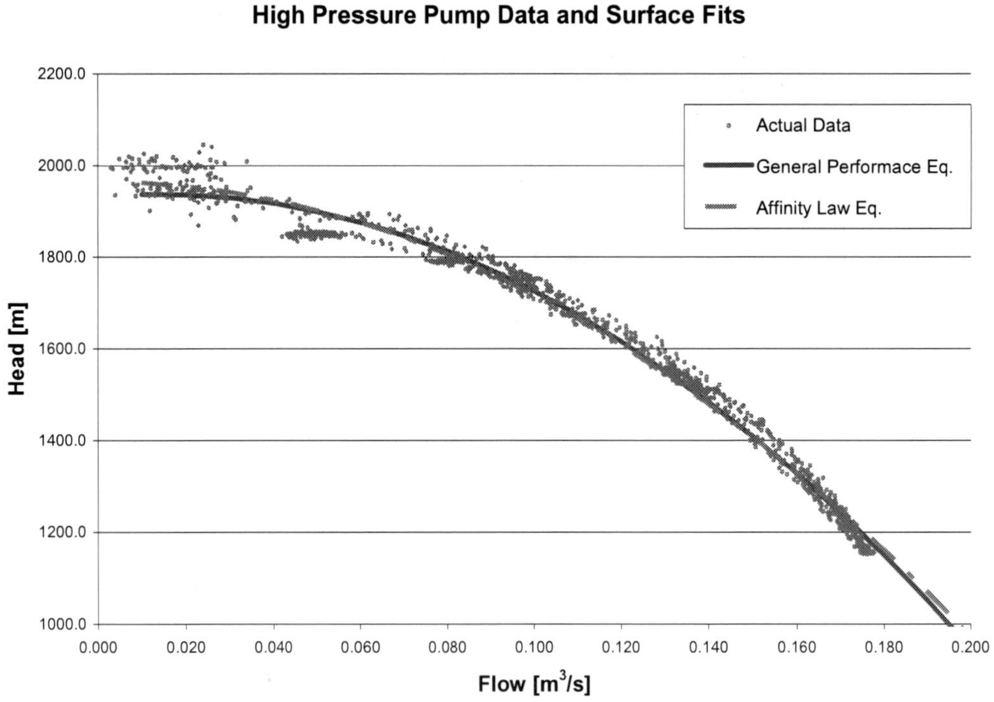

Fig. 2: Typical Performance and General Performance Equation Surface Fit to Tested High Pressure Pump Data

The API error is the vertical distance between the tested data point and the surface fit curve. In the case of high pressure pump performance the API error is strictly measured in terms of the head. The arithmetic mean of the absolute API deviation with respect to the general performance equation surface fit curve was found to be 21.57. The magnitude of the perpendicular distance, or shortest distance between each tested data point and the general performance equation surface fit curve was then calculated, whereby the arithmetic mean of the shortest distance was found to be 5.87. The shortest distance deviation method accounts for the deviations seen in both the head and the flow while more accurately describing the data fit (Fig 3).

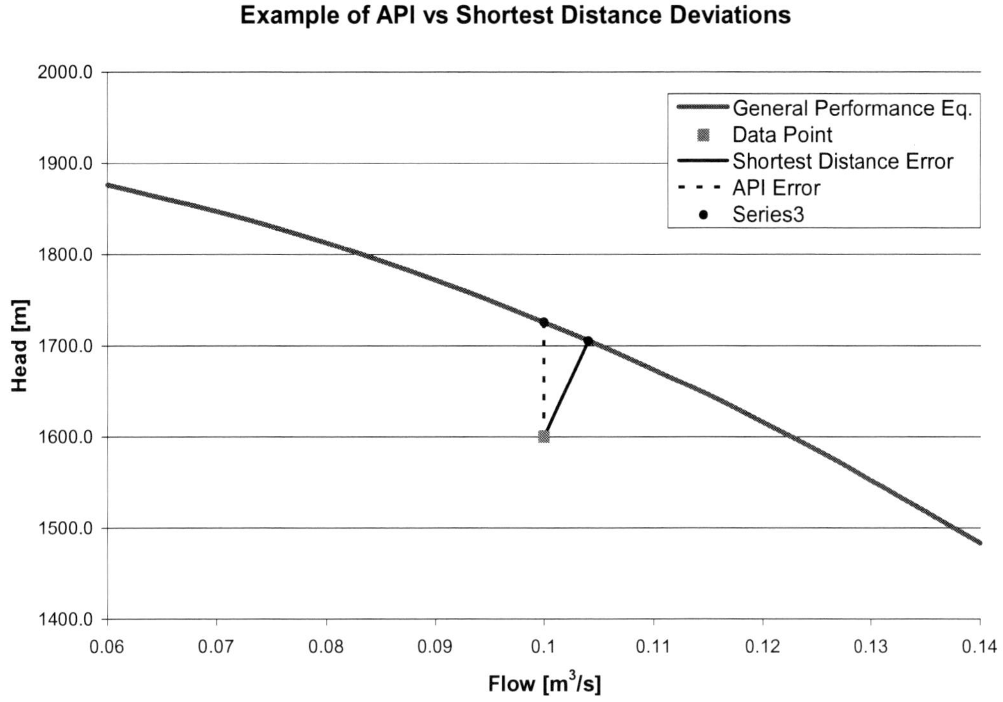

Fig. 3: Sample plot showing difference between API error and shortest distance error

Summary and Conclusions

By calculating the perpendicular distance between the surface fit curve and the tested data point the true deviation is known. The shortest distance method includes error in both the vertical and horizontal directions, in the case of high pressure pumps this is deviations in both the flow and the head. This result no only produces smaller errors; the errors found are the correct measure of the deviation between the tested data and the surface fit curve.

References

1. Hjorth, P.G. et al. "Distance of Sample Measurement Points to Prototype Catalog Curve." Journal of Physics: Conference Series 52 (2006) 27-35.
2. Pauly, M. et al. "Efficient Simplification of Point-Sampled Surfaces." Proceedings of the Conference on Visualization '02. IEEE Computer Society 2002.

3. American Petroleum Institute (API). *Centrifugal pumps for Petroleum, Petrochemical, and Natural Gas Industries, ANSI/API Standard 610, 10th Edition*. October 2004.
2. Kimmel, H. "Variable speed turbine generators in LNG liquefaction plants." GASTECH 96, Vienna, Austria, 3-6 December 1996.
3. Alison-Youel, S.D. "Observation and analysis of affinity law deviations through tested performance of liquefied gas reaction turbines." ISROMAC-12, Honolulu, Hawaii, 17-22 February 2008.

29d Radio Frequency Applications in Industrial Processes

Ben Wilson, PSC. Inc., A Litzler Company, 21761 Tungsten Rd., Cleveland, OH 44117

Radio Frequency (RF) heating is used in many industrial processes to significantly increase production and improve product quality. Any industrial processes requiring heating, curing and drying is a potential application for RF technology. This paper will cover the benefits of RF processing and review how these benefits have been realized in specific industrial applications. By understanding the benefits of RF technology with real world applications, industrial customers can evaluate how their processes might benefit from the use of RF technology.

Paper #1171014

ACSAICHE 99014 – Microwave and RF drying of heat sensitive and difficult to dry materials.

Robert F. Schiffmann, R.F. Schiffmann Associates, Inc.
149 West 88 Street, New York, NY 10024
microwaves@juno.com

Microwaves and radio frequency (RF) heating offer unique means of drying heat sensitive and difficult to dry materials. These closely related energy sources create unique thermal patterns within a material and, as a result, may cause rapid moisture or solvent removal, often under much lower ambient temperatures. However, these energy sources are generated by electricity and, thus, may have high capital cost. Therefore, both microwave and RF drying should only be used where their many unique benefits cannot be achieved by any other means.

Background: The terms microwaves, radio frequency, and dielectric can be confusing and need to be defined. Dielectric heating can be applied to all electromagnetic frequencies up to and including the infrared band. For our purposes, we will use the terms microwave and RF and link these to specific frequencies and wavelength based upon the availability of equipment (Table 1), and the term dielectric to refer to both.

Table 1
Wavelength, Frequency & Penetration Depth of RF and Microwave System Frequency (MH_z)

	5.0*	13.56*	27.12*	40.0*	915**	2450**
Wavelength						
Meters	60.0	22.1	11.1	7.5	0.328	0.122
Feet	196.9	72.6	36.3	24.6	1.07	0.400
Depth of Penetration						
Meters	23.9	8.8	4.4	3.0	0.130	0.049
Feet	78.4	28.9	14.4	9.8	0.425	0.158

*: RF frequency
**: Microwave frequency
Note: The Depth of Penetration is shown for Douglas Fir, and represents the depth at which 63 % of the energy has been attenuated within the material.

Process Advantages of Microwave and RF Systems

Advantages of Microwave and RF Heating:

Heating and drying with microwave and RF energy is distinctly different from conventional means. Whereas conventional methods depend upon the slow march of heat from the surface of the material to the interior as determined by differential in temperature from a hot outside to a cool inside, heating with RF and microwave energy is, in effect, bulk heating in which the electromagnetic field interacts with the material as a whole. The heating occurs on the interior of the material, nearly instantaneously.

Advantages of Microwave and RF Drying

The mechanism for drying with microwaves and RF is different from that for conventional drying. In conventional drying, moisture is initially flashed off from the surface and the remaining water diffuses slowly to the surface. Although the potential of energy transfer for heating is the temperature gradient, which results in energy transfer to the interior of the material, the potential for mass transfer is the mass concentration gradient existing between the wet interior and the drier surface. This is often a slow process, diffusion rate limited, which requires high external temperatures to generate the temperature differences required.

With internal heat generation, in microwave and RF systems, mass transfer is primarily due to the total pressure gradient established because of the rapid vapor generation within the material. Most of the moisture is vaporized before leaving the sample. If the sample is initially very wet and the pressure inside rises very rapidly, liquid may be removed from the sample under the influence of a total pressure gradient. The higher the initial moisture, the greater is the influence of the pressure gradient on the total mass removal. Thus there is in effect a sort of "pumping" action, forcing liquid to the surface, often as a vapor. This leads to very rapid drying without the need to overheat the atmosphere and perhaps cause case hardening or other surface overheating phenomena. Table 2 summarizes the advantages of microwave and RF drying.

Table 2
Advantages of Microwave and RF Drying

Efficiency: in most cases, the energy couples into the solvent, not the substrate

Nondestructive: drying can be done at low ambient temperatures; no need to maintain high surface temperatures, leading to lower thermal profiles and elimination of case hardening.

Reduction of migration: the solvent often mobilized as a vapor; therefore does not transfer other dissolved materials to the surface.

Leveling effects: coupling tends toward the wetter areas.

Speed: drying times can be shorted by 50% or more.

Uniformity of drying: by a combination of more uniform thermal profiles and leveling.

Conveyorized systems: less floor space, reduced handling.

Product improvement in some cases: eliminates case hardening, internal stresses, and other problems.

Of great interest today are the potential energy savings achievable from such a system. This is due to speed of drying, the direct coupling of energy into the solvent, possible lower drying temperatures, far more effective use of conventional heating in combination with the RF methods, and less overall heat loss.

A word of caution must be expressed here. These systems can heat and dry quickly, but too rapid heating can be destructive. Care must be taken not to heat so fast that the material may scorch, burn, or be otherwise damaged or dry so quickly that the steam or other vapors cannot escape quickly enough, leading to internal pressure buildup, which can lead to rupture of the piece or an explosion.

When drying with dielectric heating it is usual to combine hot air with the electromagnetic energy, particularly in microwave systems. This is because it usually improves the efficiency and the economics of the drying process. Hot air is, by itself, relatively efficient at removing free water at or near the surface, whereas the unique pumping action of microwave and RF heating is an efficient way of removing internal free water and even bound water. By combining these properly, it is possible to draw on the benefits of each and maximize efficiency and keep the costs of drying down. Note that drying with microwaves or RF alone can be very expensive in terms of both equipment and operating costs.

There are three ways in which microwave and RF energy may be combined with conventional drying methods, as illustrated in Figure 1.

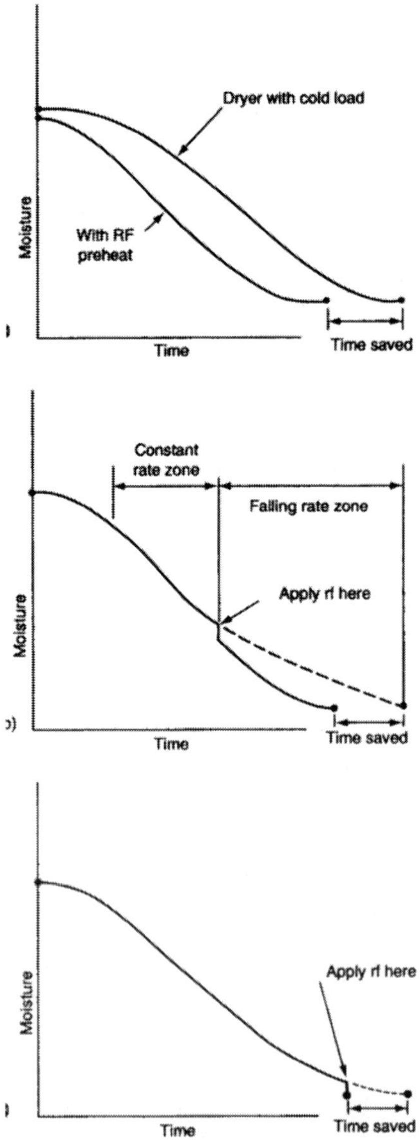

Figure 1 (a, b, c)
Typical drying curves for microwave and RF drying systems: (a) preheating with microwave or RF; (b) booster drying; (c) finish drying.

Preheating (Fig. 1-a)
By applying the microwave or RF energy at or before the entrance to the conventional dryer, the interior of the load is rapidly heated to evaporation temperature, thereby forcing moisture to the surface and immediately permitting the conventional dryer to operate at its most efficient condition, at higher temperatures. The drying curve is steeper, and drying time is shorted.

Booster Drying (Fig. 1-b)
The microwave or RF energy is added to the conventional dryer when the drying rate begins to fall off. At this point, the surface of the material is dry and moisture is concentrated in the center. The added electromagnetic energy, usually at low wattage output, generates internal heat and vapor pressure, forcing the moisture to the surface, where it is readily removed. The drying is sharply increased often with a leverage of 6:1 or 8:1 in terms of increased drying capacity for each unit of electromagnetic energy added. This is most effective on thick, hard to heat materials.

Finish Drying (Fig. 1-c)
The least efficient portion of a conventional drying system is near the end, after which two-thirds of the time may be spent removing the last one-third of the water. By adding a microwave or RF dryer at the exit of the conventional dryer, it replaces the inefficiency of hot air drying with internal heat generation. The conventional drying may also be speeded up, thereby increasing the throughput of the dryer while presenting the dryer with an initially wetter load, thus increasing the efficiency. This method also provides close control of the terminal moisture and moisture leveling while avoiding over drying.

Proper selection and addition of any of the above methods results in an overall increase in drying efficiency and throughput, and can bring about large economic savings in spite of the greater cost for electrical energy than natural gas.

Drying of heat sensitive materials: many materials, such as pharmaceuticals, foods, chemicals, etc. are sensitive to high ambient temperatures, which may alter their composition or otherwise cause damage or deterioration. There are several ways this may be done:

- Reduced ambient temperature: in conventional dryers, the major means of increasing the drying rate is to raise the ambient temperature. However, with heat sensitive or easily case hardened materials this option is limited. On the other hand, microwave and RF dryers use the hot air to remove the moisture driven to the surface by internal heating. Therefore, the ambient temperature may be lower, as its primary function is to transport the evolved moisture out of the dryer.

- Drying in a cold atmosphere: here, the material to be to be dried, usually in powder or granular form, is maintained in an atmosphere considerably below room temperature, while microwaves or RF are applied to couple with the solvent, water or ethyl alcohol for example, to evaporate it. This application may be done in either a batch or continuous applicator has been used in the production of a heat sensitive vitamin.

- The combination of microwaves with vacuum: here, the vacuum lowers the evaporation temperature, while the microwaves couple into and evaporate the solvent. This method has been used for the manufacture of high value pharmaceuticals, such as chemotherapy agents, in a closed bowl system. It has also been used to manufacture fruit flavors and enzymes, both of which are thermo labile.

-

Drying of difficult to dry materials: many materials are difficult or very slow to dry by conventional methods, but may lend themselves to dielectric drying.

- Porous solids: the existence of pores within a material makes conventional heat transfer difficult and slow, leading to long drying times. Some examples are large Styrofoam blocks and catalytic converters.
  - The air in pores and voids is transparent to microwaves and RF, so only the solid portion is visible to the energy. Therefore, heat is easily generated within these materials.
  - The existence of these pores means that these waves can penetrate farther into the material, thereby heating and drying many times faster than the conventional methods, and without the case hardening or damage that accompany high temperature hot air drying.

- Extremely dense materials: such things as ceramic tiles, sanitary ware, gypsum forms and dog bones all have very dense internal structures. Drying must be slow and at reduced temperatures to avoid case hardening or other structural damage. Dog bones, for example, may take as much as 7 days to dry at ambient temperatures. To raise the temperature is to case harden the surface which may crack during drying, creating a very dangerous situation for the dogs. However, successful microwave and RF drying is possible in many instances, by using very low levels of these energy sources combined with warm to hot air at significant velocity.
  - Other examples are: the drying of gypsum forms used for production of airplane parts. Since OSHSA regulations limits the handling of such parts to no more than 140 ° F, the normal drying times of 14 to 26 hours may be reduced to less than or slightly more than one hour by combining low levels of microwave energy continuously with the warm air.
  - Ceramics: these, also, are sensitive to high ambient temperatures that may be significantly shortened by judicious applications of microwaves or RF with hot air. However, the high costs associated with dielectric heating

systems, restricts their use to high value ceramics. In some cases, such heating also provides significant benefits in terms of internal grain structure.

When not to use microwaves or RF: There are some circumstances under which the use of dielectric drying is not practical, unless this drying provides some other significant benefit. Some examples are:

- A successful, efficient, reasonable cost conventional dryer is commonly used: The high cost of the dielectric dryer is likely to make it impractical.
- The intrinsic value of the material is low: If the raw material cost is lower than $ 5.00/lb, or $ 10.00/kilo, a dielectric dryer will probably be too expensive.
- There is a large bulk of material to be dried: The high capitol cost of the dielectric dryer makes it impractical to evaporate large quantities of water or another solvent. However, it may be practical to add before, during or after a conventional dryer, following one of the schemes suggested above, to increase the efficiency or throughput, or shorten the drying time, or ant combination of these.
- Evaporation of flammable or explosive solvents: Because these energy sources have the potential to cause arcs, which could cause catastrophic effects with such solvents, it is best to avoid using dielectric dryers, unless the drying can be done under vacuum or in an inert atmosphere.

Economics of microwave and RF drying: As stated above, the capitol cost of microwave and RF dryers is usually high and this restricts their use to applications to which they are uniquely suited. It is possible to quickly estimate a ball park cost for a system using the following:

- One kilowatt of microwaves or RF will evaporate approximately one kilogram of water per hour. The cost of a complete RF or microwave system can range from $5000 to $9000 per kilowatt. So, after calculating the kilos of water to be evaporated, simply multiply that by the installed costs shown above.
- An in depth method of analyzing cost is provided in a paper by this author: "A Primer for Evaluating the Economics of a Microwave of RF Processing System"

Conclusions: Microwave and RF drying offer unique benefits not achieved any other way, especially for heat sensitive and difficult to dry materials. However, high capital cost restricts their use to those drying challenges that cannot be met any other way. The potential user should work closely with equipment manufacturers or other knowledgeable professionals to determine the practicality and overall costs associated with using such drying systems.

ELSEWare ©
(Electromagnetic Laboratory Simulation Environment Software)
Development of a New Software Package to Simulate Electromagnetic Processes

Brian Warren, Ed Ripley, B&W Technical Services Y-12 National Security Complex

***Abstract*—A feasibility study has been carried out to investigate the possibility of using microwave energy in specially designed industrial furnaces for metal casting applications. The design of such furnaces, however, is a challenging multi-physics problem that would require addressing concurrently electromagnetic, thermal, material and chemical issues. In this abstract the principles of operations of those special furnaces will be demonstrated. The demanding multi-physics modeling of the problem will be presented addressing both the electromagnetic and thermal aspects. The material characterization of the furnace structures will also be briefly visited. Discussions of two furnaces with different sizes will be given stressing the experimental precautions and concerns that need to be taken into account when operating these special furnaces.**

I. Introduction

The heating effect of microwave radiation has been widely used in many applications. Certainly, the microwave ovens are the most popular kitchen appliance to cook or heat food. Microwave radiation has been used also in drying, bonding, curing, pasteurizing and sterilizing applications, however they can also be used for metal casting applications.

The energy savings to U.S. industry in using microwaves to heat metals could be billions of dollars. The casting industry alone is a $20 billion per year industry that spends about $11 billions per year on heating costs for melting metals. Microwave casting could cut that cost by about 30%. That equates to an annual savings of over $3 billions [1]. Microwave furnaces can be used to cast various types of metals including steel, titanium, zirconium, copper, brass, bronze, aluminum and other sorts of metals and alloys. Microwave furnaces operation does not require any water cooling, can fit any specimen size; and can provide a controlled atmosphere for a cleaner casting process with no oxidization.

Like any new technology, microwave metal casting technology has some obstacles associated with its use. For example, the need to avoid the presence of super hot spots, achieving higher operating efficiency, sustaining uniform specimen melting, minimizing microwave power reflection, preventing any plasma formation due to out gassing, and avoiding any arcing due to the presence of sharp edges.

The microwave problems already mentioned can be only solved if we understand the EM phenomena inside the chamber. The material behavior at elevated temperatures is a non-linear dynamic process, in which both the electrical and thermal parameters are time and temperature dependent. This process dependence on both the electrical and thermal properties must be considered for accurate modeling. Therefore, we needed to develop a complete and dynamic EM/thermal model for the heating process of the chamber. Such models should help in improving the casting operation efficiency; provide better control during melting time, and better understanding of various materials behavior especially at elevated temperatures.

However, modeling of such multi-physics process from EM point of view would require simulating a very large structure that is filled with non-linear material parameters, i.e. function of time and temperature. The size of such problem is significantly large and would require extremely long computation run-times and huge computer storage memories.

In this abstract, we present our efforts in developing a simplified multi-physics model showing the electromagnetic and thermal aspects of the problem. The material characterization of the furnace structure will also be briefly presented. The experimental conditions of two different size furnaces will be shown stressing the experimental precautions and concerns that need to be taken into account for operating these special furnaces.

II. Principles of Operations

The microwave oven units studied here are multimode cavities operating around 2.4 GHz, and equipped with vacuum capability, as well as complete control for argon, air, nitrogen and other atmospheres [2]. In general, there are three basic elements that are required to heat and melt bulk metals using microwaves: a very large multimode cavity, a microwave absorbing ceramic crucible, and a thermally insulating casket that is microwave-transparent. This assembly is placed inside the microwave chamber as shown in Fig. 1, and is fed by a high power industrial microwave generator to provide the microwave energy to the cavity in the range of 1-25 kilowatts. Obviously, metals do not directly couple with microwave energy at room temperature because they are electrically conductive, and would reflect all incident electromagnetic power. To avoid this reflection, the desired metal is placed in a uniquely designed crucible as shown in Fig. 1(a); in such an insulating casket is completely covering the crucible material. When the signal generator is turned on, the crucible, being the only lossy material inside the furnace, absorbs most of the energy, and heats up. This hot crucible-in return heats up the metal placed inside it by conduction. This heating process would have different mechanism upon reaching the metal melting point.

Thermal heating results from the interaction between the electromagnetic waves and materials. The governing equation to describe this thermal phenomenon in solid materials is given by

$$\rho_d C_p \frac{\partial T}{\partial t} = \nabla \bullet (k \nabla T) + q_{abs} \quad (1)$$

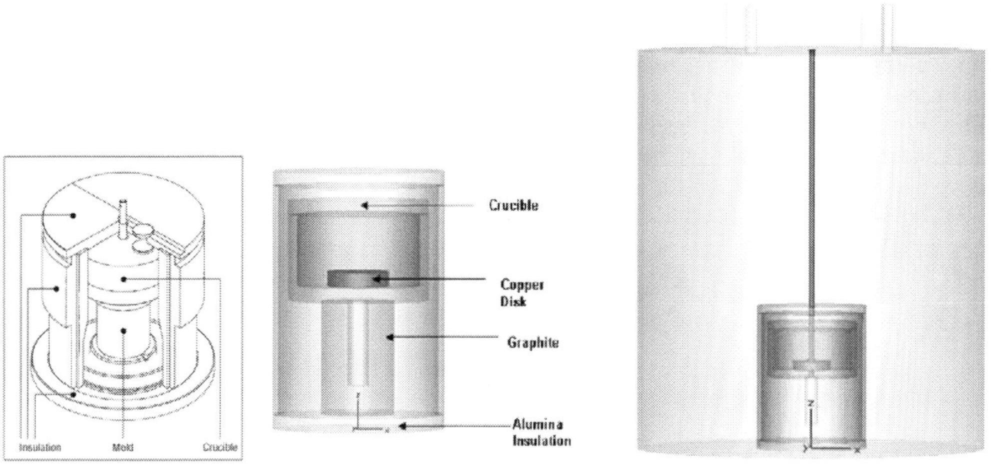

Fig. 1. Typical microwave furnace. (a) Insulation, mold and crucible assembly. (b) Whole structure.

where ρ_d is the density, C_p is the specific heat at constant temperature, k is thermal conductivity and q_{abs} is the power density absorbed by the material, which is related to the electric field by the following equation [3]:

$$q_{abs} = \omega \varepsilon''_{eff} |E|^2 \quad (2)$$

where ω is the angular frequency, |E| is the RMS magnitude of the electric field in the material, and ε''_{eff} is the imaginary part of the complex permittivity of the material.

Thermal properties characterization

For the thermal properties characterization, an instantaneous triangular heat pulse from a laser heat flash system (FL5000) made by Anter Corporation, Pittsburgh, PA was used to obtain the thermo-physical properties in the temperature of range 90°C to 890°C of machined 4 mm-thick 12.7 diameter. The measured specific heat for a Boron Carbide sample is shown in Fig. 3.

The above method has been systematically used to evaluate the various materials used within the microwave furnace. These materials include the alumina gasket and the different ceramic crucible samples.

III. COUPLED ELECTROMAGNETIC THERMAL MODELING

The heating process in the furnace at elevated temperatures is a nonlinear dynamic process, in which both their electrical and thermal parameters are time and temperature dependent. Hence the electrical and thermal problems must be simultaneously considered for accurate process modeling. Therefore, we needed to develop a complete and dynamic EM/thermal model for the heating process of the chamber [5].

EM Modeling

From the electromagnetic point of view, we have an electrically vertical resonant cylindrical cavity of 31cm radius and 62cm height (for the modular chamber). The cavity is fed by a standard rectangular waveguide WG9A of 86 mm $\times$ 43 mm as shown in Fig. 3. The cavity contains a very high loss cylindrical shell (crucible) surrounded by a thermal insulation layer. This cavity should resonate around 2.45 GHz; i.e. within the ISM band.

For analysis, we have utilized the symmetry of this structure to reduce the computational time/storage memory requirements by simulating only quarter of this structure. The boundary conditions are enforced by utilizing perfect electric and magnetic walls as shown in Fig. 2. Early model validation included solving scaled down models and comparing the solution to that of the simplified quarter model results.

It is crucial in the design of the microwave oven to minimize input power reflection and efficiently utilize the available electromagnetic energy applied to the furnace. Fig. 3 shows the reflection coefficient versus frequency obtained using Comsol and has been validated as well by other EM CAD tools such as CST and HFSS. The figure shows very similar calculated return loss performance, i.e. excellent input-match around 2.3 GHz, hence we will operate at 2.3 GHz for

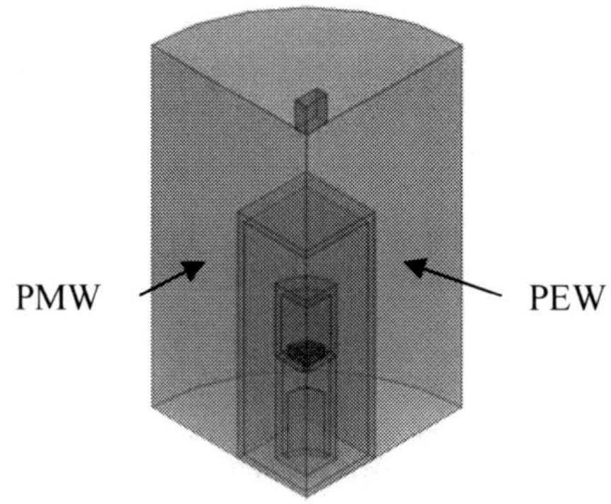

Fig. 2. Quarter of the modular chamber with enforced symmetry boundaries.

this experiment. Although this reflection coefficient at the waveguide input is quite acceptable, it is generally required to add an automatic tuner for dynamic matching of the input port over the wide temperature operating range.

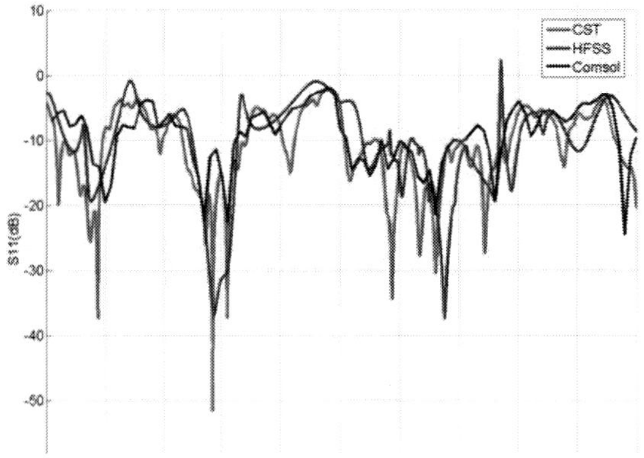

Fig. 3. Reflection Coefficient versus Frequency

Thermal Modeling

Based on simulating the coupled EM-Heat problem using Comsol package and after 80 minutes operation, the calculated results are demonstrated in Fig. 4. The heat flux distribution inside the furnace is shown in Fig. 4(b) and the temperature profile after 80 minutes is shown in Fig. 4(a).

These theoretical results indicate that the crucible will be heated up to 1269° C after 80 minutes, which is close to the measured copper disk temperature as will be shown later. Meanwhile, the theoretical temperature profile shows that the thermal conduction from the crucible to the mold is generated by a temperature gradient on the mold with the top of the mold being at least 200° C higher than its bottom which is necessary for adequate cast solidification.

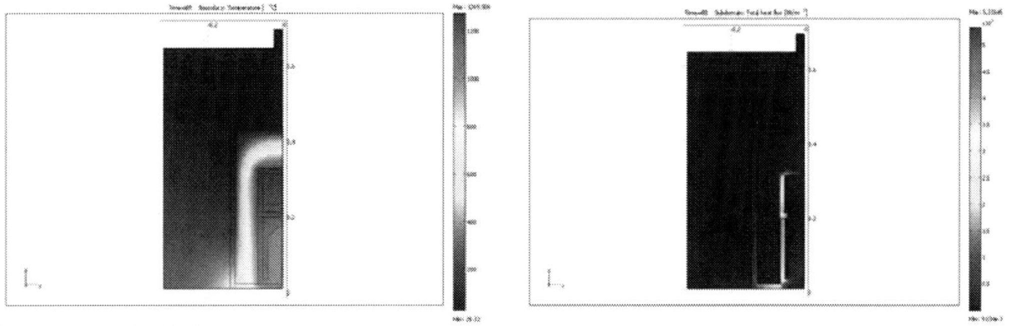

Fig 4. Heat Calculations

(a)

IV. Experimental Results

Modular Chamber

The measured temperatures at two points; one on the metal, and the other on top of the mold are shown in Fig. 5. Basically, during the experiment the temperature of the copper has reached the 1084° C mark. This means that the copper was in the liquid phase at that point. These results are shown in Fig. 5, where the solid lines represent the simulated data from the Comsol model, and the dotted lines are the measured ones using thermocouples that were inserted into the crucible-mold stack. It can be seen that the experimental results have the same trend like the simulated ones, but the data also shows some changes in the temperature vs. time slopes upon getting closer to the metal melting point. Such behavior has been previously related to the drastic change of the metal conductivity upon melting but it has not been included yet in our metal conductivity model as it would require further conductivity evaluation of the metal in its mixed state. Required measurements are currently underway to demonstrate such behavior.

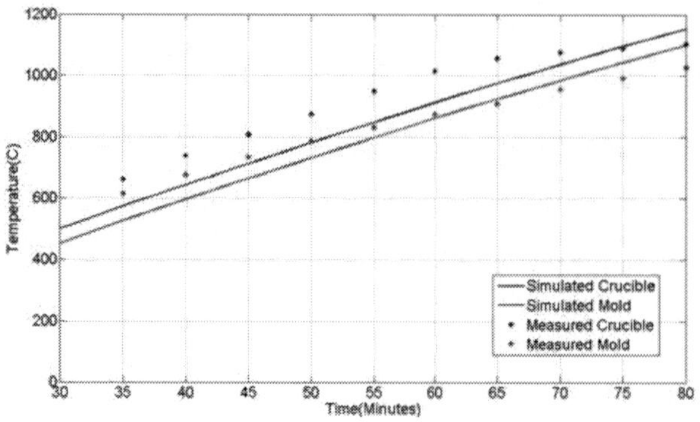

Fig. 5. Modular chamber experimental versus simulation results

A full-size chamber operating also at 2.45 GHz has been built and tested. The chamber utilizes a multimode resonant cavity of 24" radius and 50" height, and has been utilized for melting large metal samples. The developed chamber has a four port input waveguide feed where the power is synchronized by a power divider, and each has an automatic tuner to minimize any reflections. Preliminary experimental results are shown in Fig. 6 and indicate very similar performance to that of the prototype furnace previously modeled in section IV. Furnace models previously developed for the scaled down version, are currently being extended to allow running the process simulation that would require huge memory storage capabilities.

Typical industrial microwave units operate and melt metals based upon the amount of input power being distributed into the chamber. The modular unit has been generally operated up to a maximum input power of six kilowatts, while monitoring the input power ramp rates to avoid plasma formation. Meanwhile, the large scaled unit operates up to twenty five kilowatts and has the capability for a multiphase port excitation. The furnace chamber material configurations are very similar to the modular unit, but significant care should be especially placed on the insulation design and material selection strategy. This is due to the enormous fields that may cause thermal runaway in the metal and thus could blow up the insulation material.

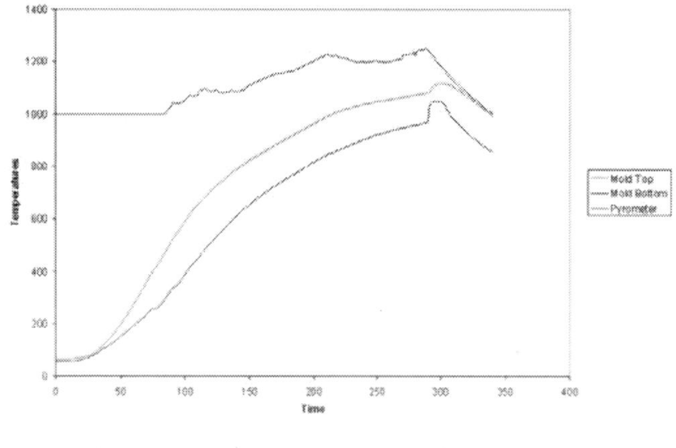

Fig. 6. Full-size chamber experimental results

V. CONCLUSION

The potential advantages for using microwave energy for industrial heating applications include a much cleaner process, and achieving a significant high operating efficiency; which could save a lot of money for the casting industry. A very interesting application is metal casting and has been thoroughly investigated and proved successful. However, it is imperative to provide a multi-physics model to continuously improve its efficiency and avoid possible operation problems such plasma formation, arcs, and out-gassing. The developed models require simultaneous solution of the coupled EM and heating problems that are time-dependent and non-linear processes. Simulations for many input power levels have been carried out and were noted to be representative of the experimental data. In this paper the simulation for the modular furnace provided a feasible estimate of the experimental process, and is currently being extended to model the full-size chamber. Meanwhile, the data from the simulations of the modular furnace provided further validation of our capability to represent the experimental data for the different material properties and process parameters. The model also provided a base line to adding additional modules to study fluid flow and chemical phenomenon within the furnace chamber.

ACKNOWLEDGEMENT

This work was sponsored by B &W Technical Services Y-12 National Security Complex, Oak Ridge, Tennessee and the authors wish to thank Dr. Patrick Mensah from Southern University for providing the thermal measurements and Josh Mosier from University of Tennessee at Knoxville for his assistance in carrying out the dielectric measurements.

REFERENCES

[1] "Microwave and Process Technologies", BWXT Y-12 National Security Complex, Oak Ridge, Tennessee.

[2] Edward B. Ripley and Jason A. Oberhaus, "Melting and Heat Treating Metals Using Microwave heating," BWXT Y-12 National Security Complex, Oak Ridge, Tennessee.

[3] W. Fu and A. Metaxas, "Numerical prediction of three-dimensional power density distribution in a multi-mode cavity," J. Microw. Power Electromagn. Energy, vol. 29, no. 2, pp. 67–75, 1994.

[4] http://cp.literature.agilent.com/litweb/pdf/5968-5330E.pdf

[5] M. Awida, N. Shah, B. Warren, E. Rippley, and A. E. Fathy, "Modeling of Industrial Microwave Furnace for Metal Casting Applications," submitted to IEEE MTT-S Int. Microw. Symp. Dig, Atlanta, GO, Jun. 2008.

Technology and Emerging Applications Using Low Temperature Plasmas

Dr. John Barkanic, Joseph Cresko, Gene Hoffman
Electrotechnology Applications Center, Northampton Community College, Bethlehem, PA

INTRODUCTION

Over the past thirty years, plasma, the fourth state of matter has become increasingly useful in a variety of manufacturing applications. The technology has experienced nearly double-digit growth annually. The plasma state finding use in this regard is not the high temperature collisionless plasma commonly associated with fusion or space physics. The plasma finding industrial application is the low temperature, non-equilibrium plasma operated at sub-atmospheric or atmospheric pressure.

The origin of such growth stems from applications in the manufacturing of semi-conductor integrated circuits. Initial developments in surface cleaning using an oxygen based inductively coupled plasma, have led to applications in etching, surface patterning, activation of surfaces for improved adhesion, surface cleaning, the deposition of thin films and the sterilization of surfaces for medical applications. In this paper, the development of the plasma state will be discussed with respect to physics and chemistry as well as applications in a variety of industry sectors. In addition, equipment for plasma generation and applicator configurations will be presented.

PLASMA CHEMISTRY AND PHYSICS

Plasma is a reactive gaseous state of matter that consists of electrons, anions, cations, excited and neutral state species. A plasma predominately consists of molecular fragments of the compounds used as feed gases. A variety of reactive species are formed in the plasma as a result of electron molecule collisions. In a radio frequency (RF) field, electrons are accelerated and gain momentum as they oscillate. Electron motion in the plasma follows the applied voltage and swings through a sinusoidal cycle.

A variety of gases could be used in low-temperature, non-equilibrium plasma and the selection of the gas often dictates the extent to which the plasma is successful in the application. Commonly, the gases are non-reactive at room temperature. However, when placed in the presence of an oscillating electric field, they dissociate, ionize, and form a reactive state of matter. The applied field can be in the kilohertz (KHz) range, 13.56 megahertz (MHz) for RF applications or 2.2 gigahertz (GHz) range for microwave (MW) plasmas. The electrical properties and voltage distribution of a plasma are important parameters. They control the degree of cation formation and ion bombardment of surfaces exposed to the plasma.

Depending on the frequency of the applied power and the specific application, the applicator or antenna for coupling the power to the load may vary significantly. In some microwave applications, a resonant cavity may be used. In some applications, tuning stubs are used to provide load coupling and reduce reflected power. RF power coupling to a load can be done using capacitive coupling or inductive coupling. In some applications, dielectric barrier discharges are used to develop a plasma at atmospheric pressure.

A typical voltage distribution between two parallel plate electrodes is shown in Figure 1 below for an RF plasma.

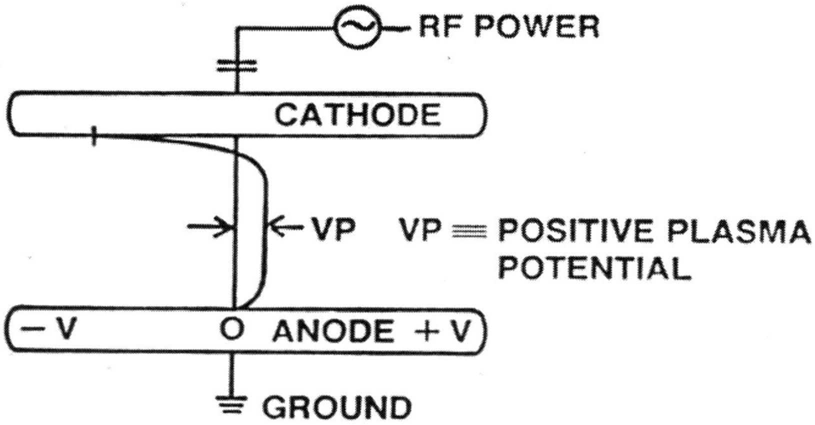

Figure 1. Voltage distribution in an RF parallel plate reactor (1).

In this application, RF voltage applied to the electrode oscillates at 13.56 MHz. The AC power supply is capacitively coupled to load through tunable or variable capacitors. Capacitors are adjusted to maximize load coupling efficiency and minimize reflected power. In this configuration, the cathode develops a negative voltage bias, the potential of the plasma, VP, is shown as positive. The positive value of the plasma potential is indicative of a high relative concentration of positive ions in the plasma state. Electrons generated in the plasma have a greater mobility and move quickly towards grounded surfaces. Positive ions, in general, have a lower mobility in the plasma field and tend to oscillate in the space between the electrodes thereby providing a positive plasma potential.

Electron collisions with halogenated gases result with a very reactive state of matter. The plasma states are often noted for their characteristic colors resulting from photon emissions of excited state species (2). In this regard, visible light emission is a commonly used diagnostic and routinely used to determine the end point of a reaction. For example, a reaction of difluorosilyl with fluorine atoms produces a trifluorosilyl excited state. The excited state emits photonic energy with a maximum intensity at a wavelength of 632 nanometers. The emission can be used to monitor reactions with silicon-based materials.

The plasma also generates cations as a result of electron collisions. A simple molecular ionization reaction can proceed:

$$Ar + e- \rightarrow Ar^{+} + 2e-$$

Molecular ionization with dissociation also called dissociative ionization, can also proceed in a plasma.

$$NF_3 + e- \rightarrow NF_2^{+} + F + 2e-$$

For the dissociative ionization reaction to proceed, electron energy must exceed the nitrogen fluorine bond energy of the NF_3 molecule, 247kJ/mol. Several dissociation energies for neutral species commonly found in radio frequency plasmas are listed below in Table 1.

Table 1.

Dissociation Energies for Neutral Species (3)

Species	Dissociation Energy (kJ/mol)	Rxn. Products
NF_3	247	$^{\cdot}NF_2 + F^{\cdot}$
$^{\cdot}NF_2$	243	$^{\cdot\cdot}NF + F^{\cdot}$
CF_4	544	$^{\cdot}CF_3 + F^{\cdot}$
$^{\cdot}CF_3$	364	$^{\cdot\cdot}CF_2 + F^{\cdot}$

Film deposition and plasma polymerization are also common applications. For example, in plasma polymerization, the plasma generates and sustains a reactive neutral species that tends to forms deposits on a surface. Example of film formation using a dichlorodifluorocarbon (4) is shown below.

$$CF_2Cl_2 + e^{-} \rightarrow (CF_2Cl_2^{-})^{*}$$

$$(CF_2Cl_2^{-})^{*} \rightarrow {}^{\cdot}CClF_2 + Cl^{-}$$

$${}^{\cdot}CClF_2 + e^{-} \rightarrow {}^{\cdot\cdot}CF_2 + Cl^{-}$$

$$n\,{}^{\cdot\cdot}CF_2 \rightarrow (CF_2)_n$$

In this chemistry, the halogen atoms in a compound are electronegative and electron attachment reactions are likely to occur in the plasma. On electron impact, dichlorodifluoromethane is raised to an excited state. For small molecules, stabilization is usually obtained through dissociation. In this regard, the carbon-chlorine bond has a bond energy of only 3.3 electron volts compared to 4.8 electron volts for a carbon-fluorine bond. As a result, the reaction usually proceeds via dissociative electron attachment, breaking the carbon-chlorine bond and forming anionic monatomic chlorine. The chlorodifluoromethyl radical formed is a very reactive intermediate due to the one unpaired electron. The radical can react in the gas phase or deposit on surfaces forming a polymeric material. Chorodifluoromethane can also participate in electron collision reactions forming difluorocarbene and anionic chlorine. Difluorocarbene is a reactive but stable intermediate that can participate in gas phase reactions or deposit on the surfaces forming a polymeric film.

LOW TEMPERATURE PLASMA APPLICATIONS

Semiconductor and Printed Electronics

There are two areas where the use of low-temperature non-equilibrium plasmas, have seen growth in the manufacturing of electronics. The initial application area is associated with the patterning of thin films used to manufacture semiconductor microelectronic devices. Thin films of polysilicon, silicon dioxide, or silicon nitride, are commonly used in the fabrication process and require patterning (5). In this area, the plasma generated typically contains halogen atoms like chlorine or fluorine. The halogen atoms in the plasma react with the thin film of semi-conductor material forming a volatile compound, which is subsequently pumped away. The use of low temperature plasmas in the manufacturing of integrated circuits is one of the largest applications world-wide.

The printed circuit and flexible circuit industry has also found applications for the technology. In multiple layer printed circuit boards, the surfaces are often drilled to provide vias for interconnecting layers. Oxygen based plasmas are often used in desmear applications to clear epoxy resin prior to interconnecting the layers (6). Other applications in the hybrid electronics market are associated with cleaning surfaces. Examples include die attach bonding with an epoxy adhesive and surface preparation prior to wire bonding to a pad. These applications often use an argon plasma to provide surface activation and improve bonding strength.

Thin Film Deposition

Plasma technology has also been used in many industries for the deposition of thin films of materials. Although initial applications focused on electronic materials such as silicon nitride and silicone dioxide dielectric films, the technology has been successfully used to deposit thin films of hard and durable coatings. Examples would be the use of the technology to form films of tungsten carbide on drill bits or turbine blades, the deposition of silicon carbide for radiation hardened devices, and the deposition of single crystal carbon or diamond-like films as a protective barrier against wear.

Surface Cleaning and Activation

A broad-based use of low temperature plasma is associated with the treatment of surfaces. Common applications of surface cleaning are associated with simply removing organic contaminants present on a surface prior to assembly or prior to processing. The removal of organic contaminants on the surface of the material can be accomplished using a plasma-based on oxygen or nitrous oxide.

A growing application is the use of plasmas for surface activation to improve adhesion. This is particularly useful in the area of medical device manufacturing. In many cases, dissimilar polymers need to be joined. In these applications, the plasma provides a surface treatment that modifies surface energy and allows for better surface wetting and improved adhesion.

One rapidly growing application in surface cleaning is the use of plasmas to clean surfaces without ion bombardment. Free radical halogen-based chemistry is used to clean furnace tubes and deposition reactors. One of the main markets using this technology is in manufacturing flat panel displays. This is a market targeted for double-digit growth throughout the next decade. In the process, a halogenated gas in a reactive state is presented to the inside of a furnace tube. The furnace tube is then cleaned as the reactive gas volatilizes the solid deposits. A typical gas used in this application is nitrogen trifluoride.

The technology, however, is in the midst of change as a result of the high global warming potential (GWP) of the feed gases and the large volume of gases used. Hansen and co-workers (7) have brought the subject of global warming to the forefront of debate. They believe that "rapid warming in recent decades has been driven mainly by non-CO_2 greenhouse gases." This is a significant statement from a leading authority on global warming and warrants immediate attention.

Use of semiconductor etchant gas is growing at an environmentally alarming rate, 150% over the last decade (8). On an industry-wide basis, the potential for greenhouse gas emission is estimated at 7.4 teragrams of carbon dioxide equivalents. For the semiconductor industry, discussions regarding non-carbon dioxide based greenhouse gases have been focused on the widespread use of plasmas. Nitrogen trifluoride has a 100-year GWP four orders of magnitude larger than carbon dioxide (9). It is imperative that new chemistries be identified, that are environmentally benign and socially responsible.

Surface Sterilization

Another growing area of interest is the use of plasma technology in surface sterilization. Plasma systems have been developed to expose plastic surfaces to a low temperature plasma environment to efficiently reduce the concentration of surface contaminants. In one technology, a low-temperature hydrogen peroxide plasma was generated to sterilize temperature sensitive equipment that could not be exposed to typical autoclave conditions. Sterilization work has also been conducted on plastics that would be used in exploratory developmental research for the pharmaceutical industry.

References

1. Coburn, J., American Vacuum Society, Educational Committee.
2. Einspruch, N.G., Brown, D. M., VLSI Electronics, Plasma Processing for VLSI, Academic Press, New York, NY, 1984.
3. Barkanic, J.A., Reynolds, D.M., Jaccodine, R.J., Stenger, H.G., Vedage, H., "Plasma Etching Using NF_3: A Review," Solid State Technology, 32(4), 109 (1989).
4. Stoffels E., Stoffels W.W., Vender D., Kroesen G.M.W., deHoog F.J., J. Vac. Sci. Technol., A 13(4), 2051(1995).
5. Sze, S.M., VLSI Technology, McGraw-Hill Company, New York, NY.
6. Barkanic, J.A., Reynolds, D., United States Patent Number 4,787,957.
7. Hansen,J., Makiko, S., Ruedy, R., Lacis, A., Oinas, V., " Global Warming in the Twenty-first Century: An Alternative Scenario," PNAS Early Edition, June 16, 2000.
8. "The U.S. Greenhouse Gas Inventory," Office of Air and Radiation, U.S. Environmental Protection Agency, EPA 430-F-02-008, April, 2002.
9. EPA PFC Emissions Reduction Partnership for the Semiconductor Industry.

CONVENTIONAL AND EMERGING MW/RF PROCESSING METHODS FOR POLYMERS, PLASTICS AND COMPOSITES.

Joseph W. Cresko
Electrotechnology Applications Center, Northampton Community College
3835 Green Pond Road
Bethlehem, PA, 18020, USA

INTRODUCTION

There are great opportunities in the polymer, plastics and composites manufacturing industries to improve energy efficiency, materials utilization and product quality issues. Microwave (MW) and radio frequency (RF) heating technologies are underutilized tools that have the potential to bring competitive advantages to those manufacturers that can exploit the ability of these non-ionizing electromagnetic (EM) waves to volumetrically heat materials.

Traditionally, the use of microwave MW and radio frequency (RF) energy to heat materials has been referred to as dielectric heating. The terminology comes from the class of materials known as dielectric materials that have been historically heated by this method. Dielectric materials have strongly bound electrons that are polarized by these electromagnetic waves. Materials that have a dielectric response due to their molecular structure (i.e., those that have sufficient polarizibility at the applied frequency) are good candidates for MW and RF processing. Until the relatively recent development of conducting polymers, essentially all polymeric macromolecules could be considered dielectric materials when considered in the context of dielectric heating. However, certain polymers show excellent heating response (e.g. polyvinylchloride – PVC), whereas others show virtually zero heating response even when exposed to very high MW or RF field strengths. Indeed, those essentially lossless materials that possess great stability when exposed to high fields (without breaking down) are used as electrical insulators in power applicatons, radomes in communications systems, and often as parts of the structure (conveyor belts, for example) in MW and RF processing equipment.

Although dielectric heating methods have been deployed by industry for more than 50 years[1], many industrialists are still unfamiliar with the equipment and processes. Within academia, the study of MW and RF processing of materials continues to be of great interest to engineers, chemists, physicists, ceramists, environmental scientists, materials scientists, biologists, food scientists and medical doctors. More recently, exciting advances in processing of non-traditional materials has expanded interest to a wider audience of materials scientists. For example, metallurgists are now processing powdered metals with MW energy. As materials evolve, the lines separating classes of materials is blurring. No longer are dielectrics the only materials being processed by these methods.

It is the intent of this paper to review key concepts important from the process development/materials engineering perspective for both traditional and emerging MW and RF applications, and assist the potential user in bridging the gap between science and application. This paper will consider fundamental concepts important from the process development/materials engineering perspective, and will review state-of-the-art of dielectric materials processing in context with its history. This will be followed by an overview of emerging MW and RF methods, such as composites curing, polymer processing, solid phase polymerization, processing of heat sensitive emulsions, processing of conductive materials, curing of carbon fiber reinforced plastics (CFRPs).

BACKGROUND

Early applications of MW & RF technology were adopted food manufacturing (i.e. tempering of frozen foods, pre-cooking of bacon and chicken, pasta drying, finish drying of potato chips and biscuits)[1,2]. Other industries that took early advantage included RF for adhesives curing, textile drying and wood drying in the 1940's. Later, as MW equipment improved, applications included urethane foam heating, rubber curing and foundry sand core drying. Applications such as PVC plastic welding by RF started in the 1960's, and are commonly used today. An entire industry supplies equipment used in the semiconductor industry - MW & RF generated plasmas used to manufacturer semiconductors, as well as diamond and diamond-like coatings. Other specialty industries also exist, such as the use of MW power supplies for ultraviolet curing equipment that was developed in the 1980's. Today, common applications of MW include the curing of rubber extrusions, pre-cooking of bacon and increasingly sintering of ceramics; common RF applications include pre-heating of plastics for thermoforming operations, drying of textiles and curing of laminates. A summary of typical and applications is listed in Table 1.[3,4,5] Emerging applications for MW include powdered metal sintering and accelerated chemistry that started in the 1990's, and more recently the processing of metals, including brazing, grain structure modification, melting.[6]

Table 1. Traditional Industrial Applications of MW and RF Technologies

Radio Frequency	Microwave
Drying (Textiles, Paper, Wood)	Tempering (Meats, Vegetables, Fruits)
Plastics Welding	Curing Rubber
Plastic Pre-form Heating	Vacuum Drying Pharmaceuticals
Post Baking Crackers/Biscuits	Ceramics Sintering
Resins & Adhesives Curing	Bacon Cooking
Plasma Generation	Plasma Generation

The most often cited advantage of MW and RF technology is the effect of *volumetric heating,* which can overcome slow conduction/convection heating rates of poor thermally conducting materials. However, industrial MW and RF heating equipment are more expensive to acquire (often cited is $1,000/kW installed power for MW) than traditional conduction/convection equipment. If the process speed/throughput is substantially improved, the benefits may appear straightforward. However, if a thorough understanding of the process economics is not well understood, the overall operating costs can be higher as well.

Another difficulty encountered when developing MW and RF processes is the complexity of constructing a comprehensive understanding of the physics of the process and system. The physics of EM wave/material interaction is complex, and are compounded by the coupled heat and mass transfer as well as the materials physics and chemistry. Further, because the material to be processed (the load) becomes an integral part of the overall system, the equipment design - especially the applicator design – is far more critical than in traditional heating processes. In recent years, advanced computer modeling techniques have significantly improved the designer's ability to model the process and the system, resulting in improved applicator and systems design.

MW & RF installations that have proven successful can be traced to results that may not be easy to calculate or quantify, or sometimes even anticipate. These can include improved product quality, the ability to process heat sensitive materials, or even the result of new products or new classes of products. Failures can be traced to technological barriers that are most commonly encountered during scale-up, and insufficient analysis of the economics. However, when this technology is applied appropriately, significant efficiency advantages can be realized as value is added, and in many cases the MW/RF process becomes the *enabling technology*.

MATERIALS CONSIDERATIONS

The propagation of an electromagnetic (EM) wave within a material is a function of the material's electrical properties, typically described by complex permittivity and permeability, as well as wavenumber and impedance. As the wave enters a dielectric material, there is some attenuation due to molecular interactions - for instance polar molecules will tend to align themselves with the electric field. A substantial understanding of the physics of this wave/material interaction has been developed,[7,8,9] and is of fundamental importance when developing MW/RF processes and equipment. However, a few key concepts can help describe the expected heating result – especially the values of permittivity and permeability, which are often expressed as a ratio with respect to free space:

$$\varepsilon_r = \varepsilon^* = \frac{\varepsilon}{\varepsilon_o} \qquad \textit{and} \qquad \mu_r = \mu^* = \frac{\mu}{\mu_o} \qquad (1, 2)$$

where ε_o=8.854x10^{-12} Farad/m and μ_o=4πx10^{-7} Henry/m. In the dielectric heating literature, one often sees values of the dielectric constant reported as κ (at a particular frequency). Further, the permittivity and permeability are complex values, so that:

$$\varepsilon = \varepsilon' - j\varepsilon'' \qquad \textit{and} \qquad \mu = \mu' - j\mu'' \qquad (3, 4)$$

The real part of the permittivity ε describes how much of the incident electric field is stored in the material (e.g. the capacitance), and is greater than one for most solids at room temperature. The imaginary part of the permittivity (ε'') is typically referred to as the loss factor and describes how much of the incident field is dissipated. The loss factor (ε'') is greater than zero and is not the same as the loss tangent (although sometimes in the literature it may not be clear which value is being presented). The loss tangent is the ratio of energy lost to energy stored, and mathematically is the vector sum of the real and imaginary components at 90° out of phase:

$$\tan\delta = \frac{\varepsilon''}{\varepsilon'} \qquad (5)$$

An example of a material that is essentially transparent to EM (at 25 °C and 2.45 GHz) is Teflon (polytetrafluoroethylene - PTFE), with typically reported values of ε'=2.04-2.08 and ε''=0.00031-0.0005, which vary little from the values of a vacuum: ε'=1 and ε''=0. Teflon or similar materials with low loss such HDPE are often used as containers, vessels, etc. in MW or RF applications.

Permeability is analogous to permittivity, but describes the interaction of the magnetic field with a material. The real part of the permeability μ is the ratio of the inductance of a material to the inductance of a vacuum. Many materials are nonmagnetic, especially organics, but ferrous materials such as iron have appreciable permeability. It is important to note that the permittivity and permeability affect the interaction of the EM wave with a material. Indeed, such materials are sometimes used as susceptors to improve the heating response of materials in MW and RF heating. When a susceptor is introduced into a matrix, the complex values of the effective permittivity and permeability are altered. There are a number of methods for predicting the effective dielectric properties of certain types of composite materials.[10,11,12]

MEASURING THE DIELECTRIC PROPERTIES OF MATERIALS

Values of the dielectric properties of select materials can be found in the literature, but may only be reported at one or two particular frequencies. Further, the composition of the reported material may not be accurately known, or how the dielectric properties may change during processing (i.e., as a function of temperature). Any comprehensive evaluation of MW and RF heating technology will require accurate values of the material's dielectric properties. Further, since the dielectric properties can depend upon temperature, it is important to determine the values across the temperature range expected in practice.

Electromagnetic energy, at very low power levels, can be used to probe the properties of materials. In the case of MW measurements, transmission and reflection (S_{11} and S_{21}) information is acquired, typically by means of a vector network analyzer (VNA) and test fixtures designed for the particular frequency (i.e. a resonant cavity fixture).[13] The transmission and reflection data are reduced to the complex expressions of the effective permittivity, permeability, wavenumber and impedance via numerical techniques[14,15]. Various methods of measuring the dielectric properties across a frequency range are possible, including transmission line techniques (waveguide & coaxial fixture)[16], coaxial probe[17] and free space techniques.[18,19]

HEATING BY ELECTROMAGNETIC ENERGY

When an EM wave impinges upon a material, some amount of the energy is *reflected* from the interface (the higher the impedance mismatch, the more energy is reflected) or *transmitted* into the material. When that surface is highly conductive (the conductivity σ is σ_m, e.g. the conductivity of a metal) at the given frequency, the energy is principally reflected back from the surface and performs no useful heating. Materials that are very transparent may allow most of the transmitted energy to pass through the material unimpeded, but lossy/conductive dielectrics can convert energy not reflected into heat energy.

The physical/chemical mechanisms by which molecules respond to EM are frequency dependent. A material may have several mechanisms that contribute to its effective permittivity, each of these will exhibit a characteristic resonant frequency. The mechanisms for losses at RF frequencies are generally due to ionic conductivity, and its contribution to the loss factor (ε) falls off as the frequency increases (the ionic conduction contribution to ε is proportional to $1/\omega$). In the MW region, losses are generally due to rotation of induced or permanent dipoles that manifest themselves as a *resonant peak* of ε as ε is *relaxing*. Electronic and atomic polarization can also contribute to the dielectric properties, but the actual resonance response is observed at much higher frequencies (i.e. in the infrared and ultraviolet ranges); for many dry solids electronic and atomic polarization are the dominant dielectric loss mechanisms.

Conveniently, the power density P (W/m^3) inside a material can be described in terms of the angular frequency (ω), the loss factor (ε) and the peak applied electric field ($\hat{E}$)[8]:

$$P = \frac{1}{2}\omega\varepsilon''_{eff}\varepsilon_o\hat{E}^2 \tag{6}$$

When described in terms of the average electric field inside of the dielectric (P_d) (based upon the RMS voltage), the *average* power density is:[9]

$$P_d = \omega\varepsilon''_{eff}\varepsilon_o E_{RMS}{}^2 \tag{7}$$

If there are also magnetic losses within the material (from ferrous iron powder dispersed in the dielectric matrix, for example), an effective permeability (analogous to the effective permittivity) also contributes to the average power density:

$$P_d = \omega\varepsilon''_{eff}\varepsilon_o E_{RMS}{}^2 + \omega\mu''_{eff}\mu_o H_{RMS}{}^2 \tag{8}$$

The average power density does not describe the variation in power density within a material. EM energy entering a lossy material will give up its energy to the material. As a result, the electric field and the power decay. The skin depth (d) is defined as the depth where the electric field is 1/e (about 37%) of its incident value; the power penetration depth (D_p) is defined as the depth where the power density is 1/e (about 37%) of its incident value. Note that $D_p = \delta/2$ and the power at the skin depth is only 14% the incident power. This has important implications in the choice of dielectric heating, for example when considering wavelength with respect to the geometry of the material to be heated.

EMERGING APPLICATIONS

As described in the Background, there are a number of emerging applications of interest to materials scientists, chemists, and chemical engineers. Table 2 lists a number of these emerging applications.

Table 2. Emerging Applications of MW and RF Research of Interest to the CPI

Microwave and Radio Frequency Processing of Polymers
Composites Curing
Polymerization Acceleration
Solid Phase Polymerization
MW Processing of Heat Sensitive Polymer Emulsions
Microwave Synthesis & Reactions
Organic Synthesis – Accelerated Reactions & New Materials
Catalysis
Inorganic Reactions – Hard Coatings, Oxides and Complexes
Synthetic Diamond Production
Microwave Processing of Conductive Materials
Curing of Carbon Fiber Reinforced Plastics (CFRPs)
Sintering of Powdered Metals
Metals and Conductive Materials Processing

A number of the emerging applications are of interest to the chemical process industries. Our applied research center, the Electrotechnology Applications Center (ETAC), has been involved in the development of a number of applications in the polymer, composites, and plastics areas.

Polymerization Acceleration/Solid Phase Polymerization

Typical polymerization reactions are accomplished when reactants are fluid, e.g. as a gaseous phase monomer, or more likely in the liquid phase (polymer solutions and polymer melts). However, condensation polymerization reactions that build molecular weight in polyamides such as polycaprolactam can be advanced at elevated temperatures when the material is in the solid phase, known as solid-phase polymerization (SPP). SPP is typically performed as an independent unit operation subsequent to the initial polymerization reaction since excessive reactor residence times would be required to achieve the high SPP molecular weights. Further, the high number-average molecular weights cause the melt viscosity to rise rapidly, which interferes with efficient stirring and heat transfer within the polymer melt resulting in increased energy consumption and polymer uniformity problems. Hence, SPP by convection heating methods is performed on finished resin, but this operation has its own limitations. Traditional, industrial convection heating methods of SPP are rate limited by poor thermal conductivity coupled with slow moisture diffusion in the polymer. This manifests itself by yielding undesirable variations in molecular weight in the bulk resin as well as in individual resin pellets, which adversely affects the processing and properties of the final polymeric product.

An industrial radio frequency (RF) solid phase polymerization (SPP) process has been developed.[22,23] A unit operation for SPP of polyamides (nylons) has been designed that accelerates the SPP reaction with RF energy, while removing the condensation polymerization reaction by-products (water) to ensure the reaction equilibrium continues toward the desirable higher molecular weight product. Since RF heats volumetrically, the result is a tighter molecular weight distribution resulting from better temperature homogeneity. Further, the continuous application of RF energy can significantly increase the rate of nylon-6 and nylon-6,6 condensation polymerization in the solid phase. Initial studies indicate RF energy gives an 80% reduction in processing time over SPP techniques that utilize convection heating.

Composites Curing/Curing of Conductive Carbon Fiber Reinforced Composites

Fiber reinforced composite materials of an epoxy matrix are traditionally cured in an autoclave under a well-defined heat and pressure cure cycle. These cure cycles are often on the order of hours, especially for massive parts, since heat transfer is conduction limited. In recent years, MW and RF dielectric heating have been investigated as techniques for accelerating the cure cycle of fiberglass/epoxy composites. By this method, select composite parts can be cured by dielectric coupling to the epoxy matrix, where the glass reinforcement is mostly transparent to the electromagnetic energy. But the relatively high conductivity of carbon fiber (CF) in CF/epoxy composite systems precludes the use of dielectric heating as a cure mechanism. An alternative approach is to inductively couple to the conductive component.[24] A modified induction process has been developed to inductively couple to the conductive carbon fiber and induce curing of the epoxy.

MW Processing of Heat Sensitive Polymer Emulsions

A highly efficient separation and recovery process has been developed[25,26] that utilizes electromagnetic (EM) energy as the driving force to significantly increase the efficiency of stripping unreacted monomer from select polymer product streams. During manufacture, direct steam injection is a standard industrial method of heating emulsions of polyvinyl chloride (PVC) to the stripping temperature. After stripping, the product is dried to a final, powdered form. Replacing direct steam injection with a highly controllable MW heating step is advantageous since significant energy savings are realized by reducing the amount of water to be removed in the final drying step. In addition, improved stripping efficiency results due to an increased bulk processing temperature as well as by preferential EM heating of the monomer, yielding lower subsequent volatile emissions.

An applicator was explicitly designed to heat aqueous emulsions of polyvinyl chloride (PVC) in order to increase the efficiency of stripping unreacted monomer from PVC polymer streams. Important to the design process are the dielectric properties of the water/PVC emulsion and the temperature dependence of the emulsion to assess temperature stability criteria. Experimental heating tests of PVC emulsions using the applicator and its associated shielding demonstrated that the heating efficiency was highly dependent upon minor variations in system geometries. The computer modeling was accomplished with *QuickWave-3D*, electromagnetic modeling software well suited for a parametric study of those geometric factors affecting the system efficiency. The simulation allowed a far more effective optimization of the applicator design than if analyzed by empirical methods alone. Further, the setup can be utilized to heat other fluid systems, and the modeling software is an important tool in maximizing the microwave heating efficiency when utilizing the applicator to heat fluids exhibiting different dielectric properties.

CONCLUSIONS

MW & RF technologies have evolved from traditional drying heating applications. Emerging applications incorporate the heating of materials that just several years ago were considered technically unviable. Research performed by interdisciplinary teams has helped advance an integrated concept of materials engineering and process development, especially for composites and new classes of materials. MW and RF methods are not only important tools to achieve process efficiency improvements, but in many cases are the enabling technologies.

REFERENCES

1. Osepchuck, J.M., "A History of Microwave Heating Applications," *IEEE Trans. Microwave Theory Tech.*, Vol. MTT-32, pp. 1200-1223, (Sept. 1984).
2. Decareau, R.V., "Microwaves in the Food Processing Industry," Academic Press, Inc., Orlando, FL (1985). ISBN: 0-12-208430-6
3. "Microwaves: Theory and Application in Materials Processing IV" (Ceramic Transactions Vol. 80 - *Proceedings from the 1st World Congress on MW Processing*), Edited by Clark, D.E., Sutton, W.H. and Lewis, D.A., (1997). ISBN: 1-57498-025-4
4. "Microwaves: Theory and Application in Materials Processing V" (Ceramic Transactions Vol. 111 - *Proceedings from the 2nd World Congress on MW & RF Processing*), Edited by Clark, D.E., Binner, J.G.P. and Lewis, D.A., (2001). ISBN: 1-57498-103-X
5. "Microwave and Radio Frequency Applications (Ceramic Transactions Vol. 111 - *Proceedings from the 3rd World Congress on MW & RF Applications*), Edited by Folz, D.C., Booske, J.H., Clark, D.E. and Gerling, J.F., (2003). ISBN: 1-57498-158-7
6. Kubel, Ed, "Advancements in Microwave Heating Technology,"*Industrial Heating*, January, 2005
7. A.C. Metaxas, "Foundations of Electroheat," John Wiley & Sons, Inc. Chichester, UK, (1996). ISBN: 0-471-95644-9
8. Roussy and J.A. Pearce "Foundations and Industrial Applications of Microwaves and Radio Frequency Fields," John Wiley & Sons, Inc. Chichester, UK, 1995. ISBN: 0-471-93849-1
9. Metaxas, A.C., Heating with Electromagnetic Fields – *A Unified Approach,"Teaching Monograph 4 of the Electroproduction Education Series*, The Electricity Council, London – EC 5055/9.89
10. Maxwell Garnett, J.C., *Philos. Trans. R. Soc.* 203, 385, (1904); 205, 237, (1906).
11. Bruggeman, D.A.G., *Ann. Phys.* (Leipzig) 24, 636, (1935).
12. Looyenga, *Physica*, 31, 401, (1965).
13. ASTM, *Standard Test Methods for Complex Permittivity (Dielectric Constant)*, Designation D 2520-86, Method B (Resonant Cavities), American Society for Testing and Materials.
14. Altshuler, H. M., "Handbook of Microwave Measurements," editors M. Sucher & J. Fox, Wiley & Sons, New York, (1963).
15. A.M. Nicholson and G.F. Ross, "Measurement of the Intrinsic Properties of Materials by Time Domain Techniques," *IEEE vol. IM-TC*, (January, 1990).
16. Baker-Jarvis, J., Vanzura, E.J. and Kissick, W.A., "Improved Technique for Determining Complex Permittivity with the Transmission/Reflection Method," *IEEE Transactions MMT 38 (8)*, p. 1096, (August 1990).
17. Blackham, D., "Use of Network Analyzer and Coaxial Probe to Determine Complex Permittivity," *Mat. Res. Soc. Proc.* vol. 189, pp 123-128, (1991). ISBN: 1-55899-214-6
18. Ghodgoankar, D.K., et al, "Free Space Measurement of Complex Permittivity and Complex Permeability of Magnetic Materials at Microwave Frequencies," *IEEE Transactions on Instrumentation and Measurement, vol. 39 (2)*, (1990).
19. Ghodgoankar, D.K., et al, "Free Space Method for Measurement of Dielectric Constants and Loss Tangents at Microwave Frequencies," *IEEE Transactions on Instrumentation and Measurement, vol. 38 (3)*, (1989).
20. Metaxas, A.C. and Meredith, R.J., "Industrial Microwave Heating," *IEE Power Engineering Series 4*, Peter Peregrinus Ltd., London (1993). ISBN: 0-906048-89-3
21. Cresko, J.W., "Innovative Application of Dielectric Heating Technologies to Concurrently Reduce Energy Usage and Emissions in Polymer Manufacturing Processes," *Proceedings of the 3rd World Congress on Microwave and Radio Frequency Processing,* Sydney, Australia September 22-26, 2002.
22. J.W. Cresko, L.M. Phipps and A.M. Mavretic, "Development of an Industrial Solid Phase Polymerization Process Using Fifty-Ohm Radio Frequency Technology," *Proceedings of the 8th International Conference on Microwave and High Frequency Heating*, Bayreuth, Germany Sept. 3-7, 2001. . Published in: Advances in Microwave and Radio Frequency Processing, Editor M. Willert-Porada, Springer-Verlag, 2006, ISBN: 3-540-43252-3
23. J.W. Cresko, L.M. Phipps and N.L. Brown, "A Vertical Radio Frequency Design as a Unit Operation for the Continuous Processing of Free-Flowing, Bulk Granular or Particulate Products," *Proceedings of the 2nd World Congress on Microwave and Radio Frequency Processing,* Orlando, FL April

2000. Published in: Microwaves: Theory and Application in Materials Processing V 'Bridging Science, Technology and Applications, Ceramic Transactions Vol. 111, April 2001 ISBN: 1-57498-103-X

24. Cresko, J.W. and Roberts, P.L., "Curing Carbon Fiber/Epoxy Composites by Inductively Coupling RF Energy," *Proceedings of the 36th Microwave Power Symposium*, pp. 45-48, San Francisco, CA April 18-20, 2001.
25. J.W. Cresko and V.V. Yakovlev "A Slotted Waveguide Applicator Design for Heating Fluids" *Proceedings of the 9th International Conference on Microwave and High Frequency Heating,* Loughborough University
26. J.W. Cresko, "Innovative Application of Dielectric Heating Technologies to Concurrently Reduce Energy Usage and Emissions in Polymer Manufacturing Processes" *Proceedings of the 3rd World Congress on Microwave and Radio Frequency Processing,* Sydney, Australia September 22-26, 2002. Published in: Microwave and Radio Frequency Applications, Editors D. Folz et al, The American Ceramic Society, 2003 ISBN: 1-57498-158-7

Distillate Hydrotreater Fouling

G. Medine, PhD
B. E. Wright, P.E.

Baker Petrolite Corporation
Sugar Land, TX

Prepared for presentation at the AIChE 2008 Spring National Meeting

Topical Conference on Refinery Processing (16a)
Session 9 – Fouling Mitigation

April 9, 2008

Previously Unpublished

Abstract

Fouling in hydrodesulfurization units is common throughout the refining industry. Hydrotreater capacity utilization and asset availability can be reduced because of the sometimes significant impacts of equipment fouling. Equipment fouling can result in inefficient heat transfer, rate restrictions and equipment downtime. Hydrotreaters are prone to fouling at different stages of the process and due to several different fouling mechanisms.

Low sulfur fuel regulations have resulted in higher severity operations of both naphtha and distillate hydrotreaters. In addition, ever greater utilization and a need to run these units for longer periods of time between turnarounds are common throughout the industry. As distillate hydrotreaters (DHT) are being pushed to greater sulfur removal efficiencies, higher activity catalysts and temperatures are being employed. These changes are resulting in an increased incidence of fouling issues in distillate hydrotreaters. This paper will review the causes of fouling in distillate hydrotreater units along with some of the possible solutions to these problems.

Introduction

During the past decade, fuel regulations have mandated ever decreasing amounts of sulfur. This trend began with gasoline sulfur levels and has rapidly expanded to diesel fuel sulfur limits. The allowable sulfur concentrations in diesel fuel are being stepped down over time based on rules from the various regulating bodies. The Euro V standard specifies a limit of less than 10 ppm by 2009, and the US EPA has set a 15 ppm standard. Many other countries around the world are adopting similar standards. As a result of these regulations, diesel hydrotreaters are being revamped and run at significantly increased conversion rates, including increased pressures, increased temperatures, improved high selectivity catalyst and increased throughput. Virtually all distillate blending stocks must now be desulfurized including catalytically and thermally cracked fluids, along with all straight run feeds, in order to attain the required sulfur concentration for the overall pool.

As a result of the increased demand on hydrotreaters both in terms of throughput and sulfur removal efficiency, an increase in problems and reliability issues have been seen around the world. Fouling in the feed to effluent exchangers and reactor beds have increased, problems in the furnace are becoming more common and hydraulic limits are being reached in some cases well before the projected end of run. In addition, many refiners are processing low cost, and lower quality, crude oils due to the price-advantage these crude oils have over premium crudes. These "opportunity" crude oils usually contain higher quantities of sulfur and/or nitrogen compounds compared to the premium crudes, or they contain significant quantities of naphthenic acids, metals or salts which create numerous operating problems throughout the refinery including issues at the hydrotreaters.

Distillate Hydrodesulfurization Units

A simplified distillate hydrotreater schematic is shown in Figure 1. The most significant differences between distillate and naphtha hydrodesulfurization include operation at slightly higher pressure and temperature, with much lower degrees of feed vaporization through the preheat train and heater. In order to achieve increased sulfur removal efficiencies, refineries have increased the severity of operation through the catalyst bed, including increased temperatures, increased residence times and use of more active catalyst. Some plants have added reactor beds in series in order to achieve the required level of sulfur removal. In cases where an existing unit is being used to achieve significantly increased sulfur removal efficiency, the residence time through the reactor must be increased and as a result, the throughput decreases. This problem coupled with the need to process more of the overall distillate blending stock has produced an increase in the numbers of distillate hydrotreaters operating at each plant and a decrease in the available hydrogen supply. Overall, these factors have come together at one time to yield a situation where diesel hydrotreater unit reliability, run length between turnarounds and operational severity all need to be maximized.

Figure 1 – Typical Hydrotreater Flow Diagram

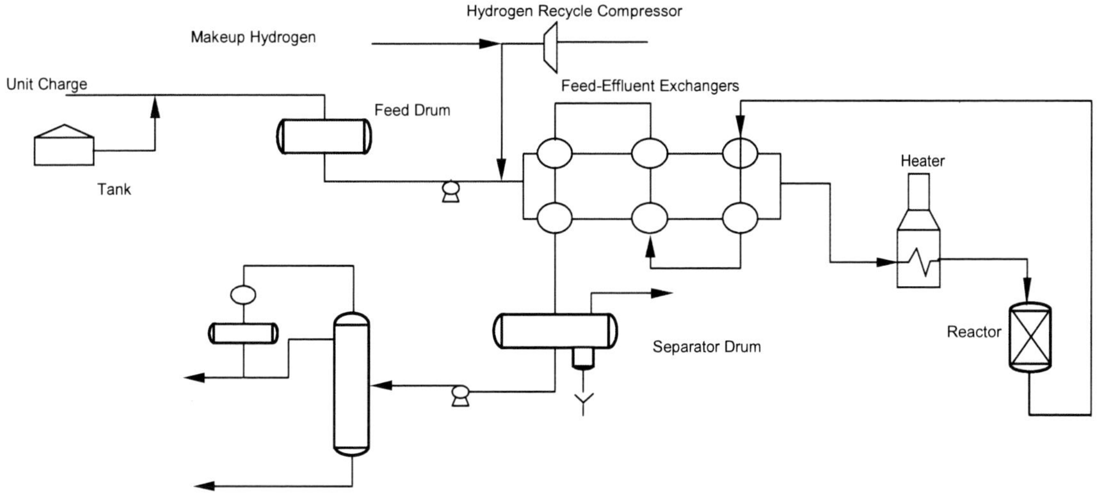

Locations Prone to Fouling

In distillate hydrotreaters, fouling is most common in the feed side of the feed / effluent exchangers. Fouling problems can also be observed in the heater, although this is usually a result of poor flame patterns, the reactor bed, and the compressor.

Types of Fouling Common to Distillate HDS Units

Fouling deposits in distillate HDS units can be inorganic or organic in nature. Fouling can be caused by the deposition of particulate matter (inorganic or organic) and the polymerization of unstable molecules. The predominant fouling material in distillate HDS units is organic. The source of inorganic fouling is usually iron sulfide, other corrosion products, catalyst particles or inorganic salts entrained in the hydrotreater feed stream. Iron sulfide can be present as 'tramp' FeS carried from upstream units or as in situ FeS. For the latter scenario, iron that is dissolved in the feed can react with sulfur in the feed or sulfide films on tube wall surfaces to form crystalline growths that grow in place on hot metal surfaces. The particle size of in situ FeS is typically about one micron near the inlet of the feed-effluent exchangers. At the outlet, the particle size can reach about ten microns and eventually grow to several hundred microns at the inlet to the reactor bed.

Organic fouling in distillate HDS units results from the polymerization of unstable molecules in the feed. Common fouling precursors include olefins, organic acids, mercaptans, ketones, aldehydes, oxygen, organo-nitrogen compounds and organo-sulfur compounds. These compounds have the potential to form polymers with themselves or upon reaction with other unstable compounds.

Oxygen promotes the polymerization of certain compounds, especially olefins. For this reason, oxygen should be excluded from storage tanks of streams known to contain olefins. It is common for feed components which are imported via ship, barge or pipeline to be contaminated with oxygen. Some polymerization reactions can also take place in the absence of oxygen.

Metal ions, especially copper and iron can catalyze certain polymerization reactions. Metals may be present in the feed if there is corrosion occurring upstream of the distillate HDS unit.

Two primary fouling mechanisms dominate polymerization in HDS feed streams: free radical polymerization and condensation polymerization.

Free radical Polymerization

Free radical polymerization is a chain reaction that forms a long chain polymer (commonly referred to as gums). Free radical formation begins with an initiator step that forms an initiator radical. Typical initiators can be heat, oxygen, mercaptans and metals.

Figure 2 Free Radical Polymerization: Initiation Step

Initiation

$$RH \xrightarrow{\text{Initiator} \quad \text{Time}} R^{\cdot} + H^{\cdot}$$

$$ROOH \xrightarrow{\text{Initiator} \quad \text{Time}} ROO^{\cdot} + H^{\cdot}$$

Once an initiator radical is formed, polymers will start to grow as the initiator reacts on various monomers; an olefin is an example of a typical HDS monomer.

Figure 3 Free Radical Polymerization: Propagation Step

Propagation

$$R^{\cdot} + O_2 \longrightarrow ROO^{\cdot}$$

$$ROO^{\cdot} + RH \longrightarrow R^{\cdot} + ROOH$$

$$R^{\cdot} + R' \longrightarrow R\text{-}R'$$

These polymers will continue to grow via additional propagation steps until a termination step is reached that stops the growth of polymer. An efficient way to terminate this chain reaction is to consume the free radicals, or their precursors, using special chemical additives.

Figure 4 Free Radical Polymerization: Termination Step

Inefficient (not highly favorable)

Termination

$$R^{\cdot} + R^{\cdot} \longrightarrow R\text{-}R$$

$$R^{\cdot} + ROO^{\cdot} \longrightarrow ROOR$$

Efficient (more favorable)
Termination with Antifoulant

$$R^{\cdot} + AH \longrightarrow RH + A^{\cdot}$$

$$R^{\cdot} + A^{\cdot} \longrightarrow R\text{-}A$$

Condensation Polymerization

Condensation, or step reaction polymerization, is an example of a non-free radical polymerization. It involves the combination of reactive compounds (ions rather than free radicals) with other molecules to form a larger molecule with the loss of a simple molecule such as water. This larger molecule can continue to react with other reactive compounds in the hydrocarbon stream to make higher molecular weight polymers. Compounds like organic acids, carbonyls, ketones or aldehydes, amines and pyrrole nitrogen will 'condense' with each other to form larger polymeric compounds. An example of this is the formation of amides.

Figure 5 Condensation Polymerization

$$R\text{-}NH_2 + HOOC\text{-}R \xrightarrow{\text{Salt Formation}} R\text{-}NH_3^{+}O_2C\text{-}R^{-}$$

$$R\text{-}NH_3^{+}O_2C\text{-}R^{-} \xrightarrow{\Delta} R\text{-}NHCO\text{-}R + H_2O$$

The concentration of these species in distillate HDS feed determines how significant this form of polymerization is to HDS fouling. Chemical additives are available which prevent this type of polymerization. In some cases, the polymers that form in distillate HDS units are highly cross-linked or highly polar requiring special consideration in the development of an effective antifoulant program.

The reactions discussed above all accelerate with increasing temperatures. Since the distillate hydrotreaters are now being asked to remove ever increasing amounts of sulfur, many are being run at higher temperatures and at full capacity. At the increased severity, the incidence of fouling and reliability issues has increased significantly. The refinery should carefully consider the strategy being employed when changing to ultra low sulfur distillate operation.

Identification of Fouling Precursors – Root Cause Analysis

Baker Petrolite has developed a root cause analysis approach to identify the source of fouling, the mechanism of fouling and solutions (both operational and chemical) to mitigate fouling.

Deposit Analyses

The first step in identifying the source of the fouling is to analyze the deposit that has formed in the HDS unit. Primary areas of concern in distillate HDS that are susceptible to fouling include: feed side of the feed/effluent exchangers, effluent side of the feed/effluent exchangers, heaters, reactor catalyst beds and hydrogen recycle compressors. Analysis of the deposit reveals the most information as to the source of the fouling problem. By using analytical techniques such as XRF, XRD, TGA, FTIR and CHN analysis, it is possible to identify the material as inorganic or organic in nature. These types of analyses can quantify inorganic salts, inorganic corrosion products, or identify the functionality of the organic foulant that has contributed to the deposit.

Feedstock Analyses

Feed characterization is another important tool to help identify the mechanism that is causing the fouling. A typical approach is to characterize the feeds analytically by a specific set of compositional parameters known to indicate fouling potential. Fouling mechanisms are determined by which parameters exceed concern levels.

Identification of the following compounds gives important information on the fouling mechanism and the fouling potential of HDS feeds. Organo-nitrogen and organo-sulfur containing compounds can act as catalysts for free radical formation and for various condensation reactions. Imines, an organo-nitrogen compound, are similar to carbonyl compounds except that the double bond is between a carbon and a nitrogen atom instead of an oxygen atom. Because of this similarity, imines can undergo condensation reactions resulting in a foulant that grows in molecular weight resulting in gums or insoluble solids in the process stream. Enamines, another organo-nitrogen compound, contain an olefin double bond and therefore can react as an olefin in a free radical polymerization as well as undergoing a condensation reaction.

A bromine number test is sometimes run as a simple method to show whether olefins (excluding aromatic olefins) and sulfur compounds reactive to bromine are present. Phenolics, sometimes known as acid oils, tend to increase instability and increase peroxide formation in reactive feed streams.

Mercaptans are acidic compounds. Mercaptans in the presence of oxygen or heat can form a sulfidic radical (contains a free electron) that can initiate free radical polymerization reactions. Mercaptans can also react with metals to form a complex that upon heating will also initiate a free radical formation. Acid numbers are sometimes run to show the acidity of a feed stream. An acidic environment tends to increase the reactivity of olefins and heterocyclics (containing nitrogen, sulfur or oxygen) and accelerate conversion of peroxides to carbonyls.

Oxygen can promote polymerization of various unstable compounds, such as olefins. Therefore, oxygen needs to be excluded from storage tanks, especially if the feed contains cracked, olefinic feedstocks. Typically nitrogen blanketed storage tanks have been used but may be ineffective at controlling fouling for feedstocks exposed to oxygen. This is because nitrogen cannot remove oxygen reaction products like peroxide or hydroperoxides that may have already formed prior to tankage. In addition, oxygen can mix with the feed causing fouling to accelerate in the feed-effluent exchangers if the seals in a transfer pump downstream of a blanketed tank are worn.

Carbonyl formation in hydrocarbons also can be a free radical reaction. An initiator radical can act on oxygen to form an organic peroxide that decomposes to an alcohol radical and a hydroxide radical. The alcohol radical can react with a paraffin molecule shifting the radical to the paraffin and producing a carbonyl compound.

Fouling Tests

In addition to the deposit and feed characterization work, fouling tests can also be completed on the feed streams to the distillate HDS units. Two common tests that are used are the existent gums test and potential gums test. Modifications to these tests are essential as under normal operating conditions distillate feed streams cannot be evaporated. Typically evaporation of these samples requires superheated steam evaporation.

Existent gums are determined by evaporating light hydrocarbons from the field samples and represent the amount of foulant inherent in the sample. The amount of gums (foulant) is reported in units of mg/100 mL. High levels of existent gums typically require dispersant chemistries to mitigate equipment fouling.

If the existent gums are low, then any additional foulant material that forms can be attributed to reactions that occur during processing (*in situ* foulant formation). To investigate this phenomenon, streams can "stressed" under various conditions and the resultant amount of foulant determined. Foulant formed under stress of heat and nitrogen is referred to as *thermal gums*, foulant formed under stress of both heat and air is referred to as *potential gums* and foulant under stress of both heat and oxygen is referred to as *oxidative gums*. These parameters provide reliable indicators of fouling potential.

Once the fouling potential and fouling mechanism have been identified via the root cause analysis approach, the next step is to implement a cost-effective solution to mitigate the fouling. This can be either an operational solution or a chemical solution or a combination of the two.

Fouling Control

There are several operational and fouling control additive program solutions that are available to refiners to control distillate HDS unit heat exchanger fouling problems.

Operational Solutions

Various operational tactics can be used to help reduce fouling in distillate HDS units, including:

- Inert gas blanketing of feedstock storage tanks
- Notching the baffles in heat exchangers
- Increasing the velocity of the feedstock through the exchanger trains
- Installing feed filters to minimize the deposition potential of particulates
- Blending protocols

Baker Petrolite recommends inert gas (for example nitrogen or natural gas) blanketing of storage tanks especially for feedstocks that have been identified as having a high fouling potential. This blanketing eliminates oxygen which can induce fouling tendencies. Tank seals should be in good condition before blanketing to prevent loss of the gas. Gas blanketing is one of the most economical methods to help minimize fouling, but it does not give total protection against fouling. As stated earlier, many polymerization reactions occur at higher temperatures in the absence of oxygen.

Many refiners are notching the baffles in heat exchanger to eliminate areas of stagnation where deposition can buildup. Fouling can begin on or near the baffles and proceed through the exchanger.

The velocity of the feedstock is another important consideration. If the feedstock is not forced through the exchanger at a sufficiently high rate, deposits and polymerization can occur and promote fouling. With sufficient speed, the polymerized products and deposit forming components of the stream do not have adequate residence time to be left on the exchanger tubes.

Installing feed filters can help to minimize fouling problems that can originate from feedstock in tankage or from imported feed. Typically, they are most effective at controlling fouling during periodic upsets. Filter systems often require regular maintenance and only capture particles greater than the filter pore size. This can result in smaller particles continuing to agglomerate and potentially deposit on equipment further downstream, especially in the reactor bed.

The development of a good blending protocol could also help to mitigate fouling problems. The ability to segregate feeds or blend compatible streams can minimize their fouling potential. Feeds that go straight to the HDS unit from the producing unit and bypass storage normally have a lower fouling potential, whereas imported feeds or feeds from tankage can have a higher fouling potential due to possible oxygen exposure. The only drawback to refiners is the reduction in hydrotreating flexibility for poorer quality feed streams.

Fouling Control Additive Solutions

Chemical solutions are also used to help refiners control fouling. Chemical antifoulants include dispersants, antioxidants, oxygen scavengers, mercaptan scavengers and metal deactivators. One of the key components is the dispersant which disrupts polymerization and agglomeration of the foulant.

Typically, dispersants are used to control fouling in a hydrotreater feed system. A properly selected dispersant will act on organic polymers to limit gum formation. Dispersants can minimize the deposition of iron sulfide that forms in situ by keeping particles in a sub micron (< 1 micron) range. A dispersant can act on gums formed by a free radical or condensation polymerization reaction keeping them finely dispersed and suspended in the process stream allowing improved operating conditions for a longer period of time. Therefore, a properly selected dispersant can minimize or eliminate pressure drop concerns in a preheat system and at the reactor bed.

Dispersants reduce deposition by disrupting agglomeration. They can keep the particles small enough to remain suspended in the process stream - sub micron size. A dispersant adds a solubilizing group to improve suspension stability of a potential foulant and limits particle size growth. Polymers agglomerate to form large particles. If left untreated, these polymers will agglomerate and precipitate. One action a dispersant adds is steric stabilization of foulant particles. Basically, this means that a dispersant increases the steric size of potential foulants to keep polar groups apart, thereby limiting growth. As these very tiny particles enter a reactor, they will act as the many other organic compounds present in the feed. They can be desulfurized, denitrified, or hydrogenated. These sterically stabilized particles may decompose somewhat to smaller molecules, but they will not grow to larger ones within a reactor bed.

Antioxidants and antipolymerants are widely used especially in those cases where a free radical polymerization reaction is the source for a foulant. If intermediate storage is provided, an antioxidant should be introduced in the rundown line to tankage. If the formed polymers are highly crossed linked or highly polar, an antioxidant or an antipolymerant will be needed to block the polymer formation. A dispersant cannot act on this form of foulant.

Antioxidants act by disrupting the chain reactions of polymerization by limiting initiation steps and inhibiting propagation or chain extending reactions. Antioxidants will increase the rate of termination. Hence, they terminate formation of free radical chains. Antioxidants can also inhibit peroxide induced carbonyl formation that in turn impacts polymer formation via condensation polymerization.

Fouling Rate Monitoring

Because fouling is seen as a drop in heat transfer efficiency or as a pressure drop limitation, monitoring is needed to show the influence of fouling mitigation actions on the specific limitation. For pressure drop limitations, tracking the specific pressure drop over time will be effective. The increase in pressure drop will appear as an exponential curve rising sharply near the end of a run. If the feed rate is changed significantly, variations in pressure drop can occur because pressure drop is proportional to the square of the mass flow rate.

If the limitation is related to heat transfer efficiency, then a temperature survey across the preheat exchangers is typically performed. In some cases, the reactor effluent temperature to the separator is sufficient to show the effects of fouling or its control. In other cases, calculating and trending duty and "U" value over time for the preheat exchangers can be used to show effects of treatment. In other cases, a rigorous model that calculates actual fouling resistance for the affected exchangers may be needed. Again, the variables selected to show performance should be trended over time. In addition to measuring pressure drop and heat transfer efficiency, measuring temperature and throughput can also be useful parameters to monitor fouling control on the unit. Feed characterization can also be performed on the feedstocks to the unit to monitor the fouling potential of the streams over time. All fouling control programs should incorporate a comprehensive monitoring strategy.

Case History

The following case history is an example of Baker Petrolite's root cause analysis approach to identify the source of the fouling problem and mitigate the fouling via chemical and operational solutions.

<u>Problem</u>

A Gulf Coast refinery was having problems with fouling in the preheat section of the distillate hydrotreater. The fouling was suspected to be occurring on the feed side of the feed/effluent exchangers. The unit was experiencing severe fouling and the reduction of unit feed rates with run lengths on the preheat exchangers of about three months.

The refiner requested a meeting and input from Baker Petrolite to investigate the problem. After meeting with the engineering staff, a plan was put in place to identify the problem, test the feed streams and identify a program to mitigate the problem.

The problem was believed to involve only one stream being fed to the unit. The feeds to the DHT were coming from several units in the refinery as hot rundown, and from storage. The Baker Petrolite technology group determined that the streams should be evaluated individually, and as a blend. In addition to the suspected streams, the refinery knew there were problems with the nitrogen blanketing system on the stored feed in tankage. The feed streams were sampled and evaluated in our laboratories for the existence of fouling precursors, existent gums and potential gums.

Solution

Based on the information and test results from the feed characterization, the Light Cat Gas Oil (LCGO) was identified as the primary fouling stream. The LCGO was stored in a blanketed tank prior to being fed to the DHT. In addition to identifying the specific stream, a likely mechanism was identified. With this information, the Baker Petrolite technology group was able to choose suitable products for further laboratory testing.

The fouling mechanism identified in the laboratory tests was sulfur or H_2S induced polymerization. Based on this laboratory testing, Baker Petrolite identified a proprietary antifoulant as the product of choice for treating the problem stream. Other recommendations offered by Baker Petrolite included an improvement of the blanketing system on the feed tank and a capital improvement project to feed the LCGO directly to the DHT. Each of these solutions would have provided improvement to the system as well.

The recommendation, based on the laboratory testing, was to treat the LCGO stream with best performing antifoulant at a dosage of 20-40 ppm (by volume), prior to storage. The LCGO was about 25% of the total feed to the DHT. Treating only the “problem” stream minimized program costs, resulting in a significant ROI for the refiner. The program was initiated after the DHT was shut down and the preheat exchangers cleaned.

Results and Benefits

Under ideal conditions, the DHT would be shut down, and the preheat exchangers cleaned once a year. The refinery is currently in its 19th month of the run since initiating treatment, with little or no fouling noted in the system. This case history is presented for illustration purpose only as results may vary between applications.

The estimated return on the fouling control chemical program is over 200 percent. This value considers only improvements in throughput and fuel efficiency (less furnace firing) and assumes no mechanical outages.

Other savings not considered in this economic evaluation include deferred capital expenses to modify the system to reroute the LCGO feed stream directly to the DHT unit, and for improvements to the gas blanketing system on the LCGO feed storage tank.

Summary

Fouling in distillate hydrodesulfurization units is caused by inorganic materials from upstream operations and from polymeric compounds which are formed from the feeds being processed. As a result of the increase in HDS operating severity to achieve high sulfur removal efficiency, an increase in the incidence of fouling has been observed in these units around the world. However, various operational changes or fouling control additive programs can be used to control fouling rates. Proper identification of the nature of these materials is a key to developing effective mitigation solutions.

References

1. Jones, L. & Kokayeff, P.; Equilibrium Limitations in Distillate Hydrotreating, AIChE 2005 Spring National Meeting, Paper No. 61d
2. Sanghavi, K. & Torrisi, S.; Getting More With Less: Alon Big Springs Refinery ULSD Project, AIChE 2007 Spring National Meeting, Paper No. 3a
3. McCullough, D.; Edgar, M. & Pistorius, J.; Higher Severity Diesel Hydrotreating, 1987 NPRA Annual Meeting, Paper No. AM-87-58
4. Sayles, S.; Bailor, J.; & Ohmes, R.; ULSD Problems and Solutions, *Petroleum Technology Quarterly*, 2004, Q3
5. Hazlett, R. N.; Schreifels, J. A.; Stalick, W. M.; Morris, R. E. & Mushrush, G. W., *Energy and Fuels,* 1991, 5, 269-273
6. Mushrush, G. W.; Hazlett, R. N.; Pellenberg, R. E. & Hardy, D.R., *Energy and Fuels,* 1991, 5, 258-262
7. Beaver, B.; Gilmore, C.; Veloski, G.; Sharief, V., *Energy and Fuels,* 1991, 5, 274-280
8. Congram, G.E., *The Oil and Gas Journal*, December 9th, 1974, 87
9. Chou, T. S., Causes of Fouling in Hydroprocessing Units, *Petroleum Technology Quarterly*, 2004, Q3, 81-85

203a Tutorial on Refinery Fouling Mitigation

Irwin A. Wiehe, Soluble Solutions, 3 Louise Lane, Gladstone, Gladstone, NJ 07934

Today, when refineries are trying to run at full capacity, shutting refinery units down to remove foulants has severe financial penalties. Fortunately, refinery fouling mitigation is much easier than previously imagined. This is because 90% of fouling has only six common causes and three of these involve insoluble asphaltenes. The analysis of the foulant will usually enable identifying the root cause that is confirmed by finding the precursor in the oil flowing through the fouled unit. By tracing the precursor to the source, one arrives at a number of potential mitigating methods from which a refinery can select the best to implement for the refinery. Case studies will be presented where this strategy was used to mitigate petroleum fouling for different causes.

203b Solvent Ranking for Asphaltenes Stability in Pipeline Transportation

Parviz Rahimi and Simon Cooper. National Centre for Upgrading Technology (NCUT), 1 Oil patch Drive, Suite A202, Devon, AB T9G 1A8, Canada

In this study an optical scanner is used to compare and rank oil stability by observing the compatibility of oils and precipitants. The Turbiscan AGS is capable of sensitively and accurately observing changes in particle concentration and size. The goal of this study is to be able to produce an automated method capable of measuring and comparing the stability of an oil solution. By modifying an existing ASTM method variance in light transmittance through the oil solution is observed and related to flocculation and sedimentation of precipitated insoluble particles.

203e Inhibition of Carbon Deposition from Jet Fuel by Surface Coating

Arun Ram Mohan, Energy and Mineral Engineering, Penn State University, 210 Academic Projects Building, University Park, PA 16802 and Semih Eser, Department of Energy and Mineral Engineering, Penn State University, 114A Hosler Building, University Park, PA 16802.

Solid deposits are formed on metal surfaces from reactions with fuel in a jet engine. The deposits consists of metal sulfides and carbonaceous solids. Coating metal surfaces with inert materials inhibits the formation of these deposits. Alumina films were deposited on the surface of stainless steel by low pressure metal-organic chemical vapor deposition. Alumina-coated surfaces were tested by thermal stressing of jet fuel under autoxdative conditions at 330 - 380 °C. Characterization of the deposits by temperature programmed oxidation and scanning electron microscopy showed that the alumina coatings inhibit the formation of metal suflides and decrease the amount of carbonaceous solid deposits.

Electrospray Ionization FT-ICR Mass Spectrometry of 'ARN' Naphthenic Acids in Sodium and Calcium Naphthenate Deposits, and Crudes: Characterization, Extraction and Quantification

Mmilili M. Mapolelo[1,2], Priyanka Juyal[2], Ryan P. Rodgers[1,2], Andrew T. Yen[3], Justin D. Debord[3], Sam Asomaning[3] and Alan G. Marshall[1,2]

[1]Department of Chemistry & Biochemistry, Florida State Univ., Tallahassee, FL 32306
[2]Ion Cyclotron Resonance Program, National High Magnetic Field Laboratory, Florida State University, 1800 East Paul Dirac Drive, Tallahassee, FL 32310-4005
[3]Baker Petrolite, 12645 W. Airport Blvd, Sugar Land, Texas 77478

Abstract

Many oilfield operators face flow assurance challenges associated with the deposition of naphthenate solids and formation of sodium soaps in production equipment. Calcium and sodium naphthenates are respectively solid deposits and emulsions formed by the interaction of naphthenic acids with divalent (Ca^{2+}, Mg^{2+}) or monovalent (Na^{+}, K^{+}) ions in produced water. Calcium naphthenate formation, an interfacial phenomenon, is thought to depend largely on tetraprotic naphthenic acids known as 'ARN' acids in the crude, whereas the formation of sodium naphthenate originates from less substituted lower molecular weight naphthenic acids. In the current paper, 'ARN' acids extracted from whole crude oils and naphthenate deposits / soaps are characterized by high-resolution FT-ICR mass spectrometry and the results compared. We present our first attempts to quantify ARN-type acids in whole crude oils. The high resolution and mass accuracy of FT-ICR MS provides detailed acidic speciation for all crudes and deposits analyzed.

Keywords: Fourier Transform, Ion Cyclotron Resonance, ICR, FT-ICR, FTMS, Naphthenic acids, "ARN" acids.

Introduction

Many oilfield operators are faced with challenges in flow assurance due to formation of deposits and soaps in the oilfield, generally termed naphthenates.[1, 2] Naphthenate deposition in crude oil production causes major problems for the petroleum producers as recently reviewed.[1, 3-5] The obstruction of pipelines and screens, production irregularities, subsequent unplanned and expensive production shutdowns are encountered due to naphthenate deposition.[3] The problem has recently become more pronounced as more oilfields that contain immature, heavy oils with higher naphthenic acid content begin production.[6-8] A few oilfields associated with formation of deposits are found offshore in Great Britain, West Africa, the North West Shelf in Australia, Bohai Bay in China and the Gulf of Mexico.[1, 9]

Naphthenate deposition occurs mainly during upstream operations, more commonly at the oil/water cutoff point. Calcium and sodium naphthenates are solid and emulsion-like deposits formed by the interaction of naphthenic acids and divalent (Ca^{2+}, Fe^{2+} and Mg^{2+}) or monovalent (Na^{+}, K^{+}) ions present in produced waters. Calcium naphthenates occur as a viscous, sticky solid or sludge commonly found in oil-water separators and electrostatic treaters.[3] Interestingly, calcium naphthenates harden when they are exposed to air and cooling.[3, 10] Crudes with medium to heavy API gravity (typically 18°-27°) rich in naphthenic acids, with a high total acid number (TAN = 0.80-8.0 mg KOH/(g oil)) are at risk to form calcium naphthenates[11] whereas sodium naphthenates occur as a very stable emulsion, soap, or sludge, and are associated with light API (typically > 32° to 41°) crude oils with a low TAN (0.05-0.60) and high concentration of volatile fatty acids.[11]

Naphthenic acids are saturated acyclic or cyclic carboxylic acids in which the carboxylic acid group is attached to the aliphatic side chain or a cycloaliphatic ring (single ring or multiple fused rings).[12-15] The characterization of naphthenic acids has always been a challenge, more especially for naphthenic acids associated with formation of naphthenate deposits. However, naphthenic acids found in Norwegian calcium naphthenate deposits have been well characterized.[5, 7] These naphthenic acids are known by the generic name as "ARN" acids. The acids were shown to be a head to head linked isoprenoid 20-bis-16, 16' biphytane carbon skeleton, with 4-6 cyclopentane rings.[16, 17] The structural resemblance of ARN acids to the tetraethers in archael lipid membrane suggest they are of an archael origin.[16, 17] ARN acids are tetraprotic acids and their molecular weights range from 1227 to 1235 Da (attributed from observations of the deprotonated molecular ions).[3, 18] The homologous series of ARN acids correspond to empirical formulas, $C_{80}H_{138}O_8$, $C_{80}H_{138}O_8$, $C_{80}H_{140}O_8$, $C_{80}H_{142}O_8$, $C_{80}H_{144}O_8$, and $C_{80}H_{146}O_8$ with double bound equivalence (DBE = number of rings plus double bonds to carbon) ranging from 8-12, indicating 4 to 8 rings in the hydrocarbon structure.[3] The presence of ARN acids in crude oil could serve as an indicator for potential calcium naphthenate deposition. However, the presence of ARN acids in crudes does not necessarily imply the certainty of calcium naphthenate deposition,[19] because naphthenate deposition is also influenced by the composition of production water, such as the pH and bicarbonate content.[19] Calcium naphthenate deposition is rarely reported in "dry" oils.

Here, we apply negative-ion electrospray ionization Fourier transform ion cyclotron resonance mass spectrometry (ESI FT-ICR MS) to analyze the ARN-type naphthenic acid extracts from calcium naphthenate deposits. We compare the compositional analysis of calcium and sodium naphthenate deposits from different oilfields with respect to heteratom distribution, aromaticity, and carbon number distribution. Detailed class by class comparisons of the two types of naphthenate deposits are presented. Furthermore, we discuss our first attempts to simply extract and quantify ARN-type acids in whole crude oils.

Materials and Methods

Samples

Two field calcium naphthenate deposits and one sodium naphthenate emulsion sample from the field were analyzed. The calcium naphthenate samples were collected from production separators in which naphthenate deposition has been reported. A sodium naphthenate deposit generated in the laboratory was prepared as follows. An oil sample was received and used without any modification. To form an emulsion, 100 mL was transferred into a 250 mL flask along with 50 mL of synthetic brine consisting of 2000 ppm Na^+ and 0.1 g of $NaHCO_3$. The mixture was agitated briefly by hand (~ 5-10 minutes) and allowed to settle for 2 days. Afterwards, a biphasic mixture of 50% oil and 50% emulsion was noted. The emulsion was sampled at the midpoint of the emulsion layer for subsequent analysis.

Extraction of naphthenic acids from naphthenate deposits

Calcium naphthenate samples were preweighed (~100 mg) and washed with toluene to remove entrapped crude oil. The samples were repeatedly washed until the color of the toluene phase was almost clear. After the toluene wash, half of the samples were subjected to another wash with methylene chloride. The sodium naphthenate samples were not subjected to the wash procedure. After the associated deposits were isolated (sodium naphthenate) and cleaned (calcium naphthenate), they were digested with 1 M hydrochloric acid (HCl) or 5 M glacial acetic acid (CH_3COOH) followed by toluene extraction. The acidic digests were redissolved in 50:50 (vol:vol) toluene: methanol for negative-ion ESI FT-ICR MS analysis.

Quantitation

Three crudes with associated field calcium naphthenate deposits and bitumen (no deposit and serves as a control) were spiked with a "pure" ARN acid standard previously isolated from a solvent-cleaned naphthenate deposit followed by acid digestion. Analysis of the standard by negative-ion ESI MS showed only ARN acid species. The ARN standard was spiked at various concentrations (from ~2 – 200 micromolar) into the whole crudes to determine the broadband MS detection limit. For those concentrations at or just below the broadband detection limit, a modified extraction procedure was performed to preconcentrate the acidic species.[20] Samples were analyzed by negative-ion electrospray ionization (ESI) with a custom-built 9.4 Tesla FT- ICR mass spectrometer and the resulting mass spectral data was processed with an in-house data package (PREDATOR).

Results and Discussion

Analysis of the naphthenate deposits identified acidic species that selectively accumulate and thus are assumed to participate in naphthenate formation. Calcium naphthenate samples were largely composed of a family of tetraprotic acids (ARN acids) whereas the sodium naphthenate deposits contained mostly low DBE saturated naphthenic acids. **Figure 1** shows broadband negative-ion ESI 9.4 Tesla FT-ICR mass spectra of two sodium naphthenate (NaN) samples. Although the mass spectra of the samples are quite similar, the mass spectra of NaN #2 seem to be more complex and extend to a higher m/z relative to NaN #1. In NaN #1 the molecular weight of the extracted naphthenic acids ranged from 250 to 550 Da, whereas in NaN #2 the range is from 300 to 650 Da. The sodium naphthenates consist mainly of low molecular weight naphthenic acids, primarily acyclic saturated fatty acids.

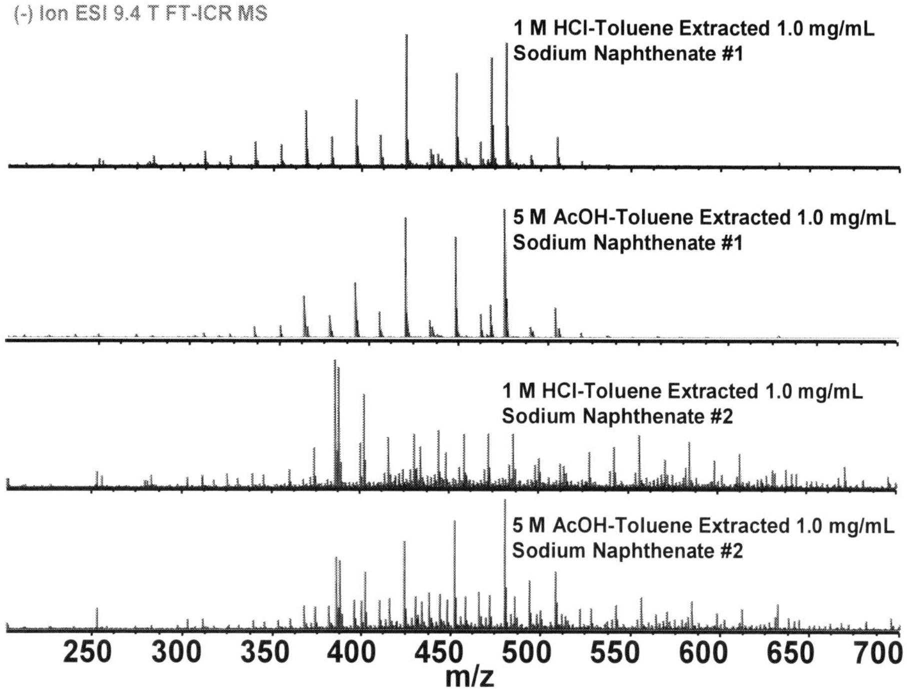

Figure 1. Broadband negative-ion ESI 9.4 T FT-ICR mass spectra for NaN#1 and NaN #2 samples extracted by acidified (AcOH or HCl) toluene.

These low molecular weight naphthenic acids are known to self-associate and are likely to form proton bond multimers, resulting in formation of high molecular weight naphthenic acids. Dimerization is quite common in sodium naphthenate emulsions and is mainly attributed to the presence of (carboxylic) fatty acids. Sodium naphthenates have frequently been reported as emulsions and are sodium salts of *n*-alkyl or slightly branched carboxylic acids (fatty acids), RCOONa.[11] When crudes contact $NaHCO_3$-laden production waters at high pH, saponification takes place and thus produces the sodium salt of fatty acids.[11, 19]

Figure 2 shows negative-ion FT-ICR mass spectra of a calcium naphthenate deposit. The spectra clearly exhibit two distinct peaks from ARN acids with a singly charged cluster at *m/z* 1225-1270 $[M-H]^-$ and a strong doubly charged cluster at *m/z* 615-630 $[M-2H]^{2-}$. The general composition for these species is $[C_{80}H_{142}O_8-2H^+]^{2-}$ that is formed by deprotonation of two of the four carboxylic groups from the parent compound $C_{80}H_{142}O_8$. The base (highest-magnitude) peak in the mass spectrum (singly deprotonated) has m/z 1230.06293, consistent with the molecular formula $[C_{80}H_{141}O_8]^-$. The peaks correspond to the general formula $[C_{80}H_{142}O_8-H^+]^-$ from the parent compound $C_{80}H_{142}O_8$. Tetraprotic acids comprise all of the singly charged cluster peaks and indicate that ARN acids are the major component in the deposits. The range of molecular weights indicates that ARN acids in the calcium naphthenate deposits are not one particular acid, but rather a family of acids with a general formula with varying degrees of unsaturation in the hydrocarbon skeleton.

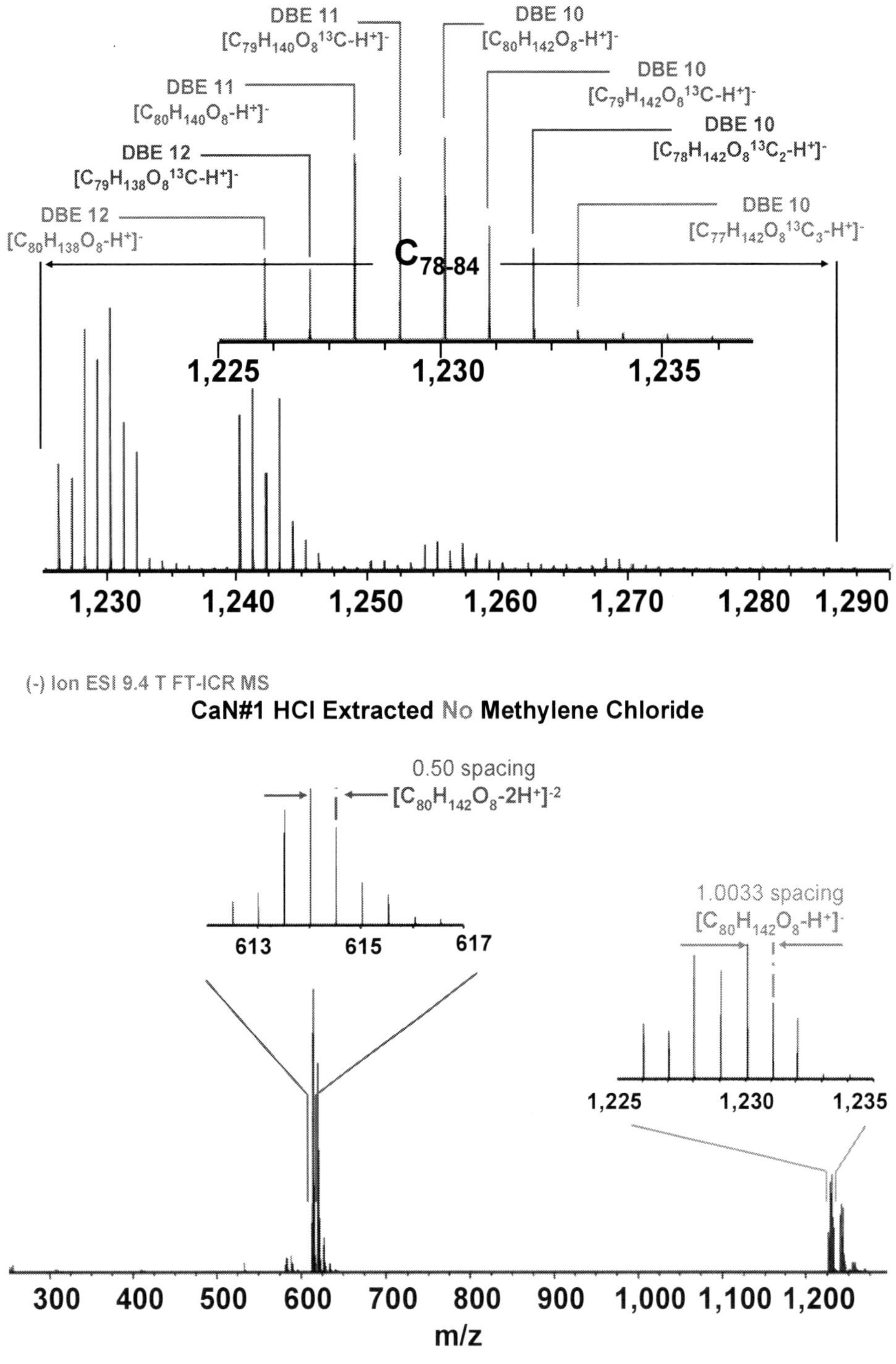

Figure 2. Bottom: Negative-ion ESI FT-ICR mass spectrum of HCl acidified toluene extracted CaN#1 deposit not washed with methylene chloride. Top: Mass scale-expanded segment, 1225-1235 Da, with assigned molecular formulas for the acid species. Insets show identification of singly-charged (right) and doubly charged (left) species based on m/z intervals of 1 or 0.5 between species differing by ^{12}C vs. ^{13}C.

Broadband negative-ion ESI FT-ICR MS analysis of parent crude oils that produce associated calcium naphthenate deposits identifies thousands of acidic species in each crude. Comparison of the acidic species in the parent crude to those isolated in their associated deposits revealed that species characteristic of the deposit are at low relative abundance or unobservable in the broadband mass spectrum. Thus, preconcentration of the acidic species in the parent crude is necessary. Conventional ion exchange extraction of acidic species from crude oil is time consuming and may alter the "true" acid profile of the parent crude through irreversible absorption to the ion exchange resin. Therefore, a modified extraction method was used to isolate acidic species from the parent crude. In order to test the broadband and extraction based detection limits for ARN acids in the parent crudes with associated calcium naphthenate deposits, known amounts of ARN acid were spiked into both the parent crudes and a bitumen sample that does not contain native ARN acids. We shall discuss the results of the quantitation experiments, such as broadband/extraction detection limits and how the acid composition of the parent crudes affects quantitation efforts.

Acknowledgments

Work supported by NSF (DMR-00-84173), Florida State University, and the National High Magnetic Field Laboratory in Tallahassee, FL.

References

(1) Turner, M.; Smith, C. P. *Paper No. SPE 94339* **2005**.
(2) Vinstad, J. E.; Bye, A. S.; Grande, K. V.; Hustad, B. M.; Nergard, B. *Paper No. SPE 80375* **2003**.
(3) Lutnaes, B. F.; Brandal, O.; Sjoblom, J.; Krane, J. *Org. Biomol.Chem.* **2006**, *4*, 616-620.
(4) Clemente, J. S.; Fedorak, P, M *Chemosphere* **2005**, *60*, 585-600.
(5) Baugh, T. D.; Wolf, N. O.; Mediaas, H.; Vinstad, J. E.; Grande, K. V. *Paper No. SPE 93011* **2005**.
(6) Olsen, S. D. *Prepr.-Am. Chem. Soc., Div. Pet. Chem.* **1998**, *43*, 142-145.
(7) Baugh, T. D.; Wolf, N. O.; Mediaas, H.; Vinstad, J. E.; Grande, K. V. *Prepr.-Am. Chem. Soc., Div. Pet. Chem.* **2004**, *49*, 274-276.
(8) Barth, T.; Moen, L. K.; Dyrkorn, C. *Prepr.-Am. Chem. Soc., Div. Pet. Chem.* **1998**, *43*, 134-136.
(9) Smith, P. C. Soaps (Naphthenates) Scales Management from Deepwater Flow Assurance Aspects to Oil Terminal Sludge Processing Workshop. In *Flow Assurance-A Holistic Approach*, December **2004**.
(10) Rousseau, G.; Zhou, H.; Hurtevent, C. *Paper No. SPE 68307* **2001**.
(11) Gallup, D. L.; Curiale, J.A.; Smith, P.C. *Energy & Fuels* **2007**, *21*, 1741-1759.
(12) Barrow, M. P.; Headley, J. V.; Peru, K. M.; Derrick, P. J. *J. Chromatogr., A* **2004**, *1058*, 51-59.
(13) Fan, T.-P. *Energy & Fuels* **1991**, *5*, 371-375.

(14) Seifert, W. K.; Teeter, R. M. *Anal. Chem.* **1969**, *41*, 786-795.
(15) Slavcheva, E.; Shone, B.; Turnbull, A. *Br. Corr. J.* **1999**, *34*, 125-131.
(16) Chappe, B.; Albrecht, P.; Michaelis, W. *Science* **1982**, *217*, 65-66.
(17) Moldowan, J. M.; Seifert, W. K. *Science* **1979**, *204*, 169-171.
(18) Smith, B. E.; Sutton, P. A.; Lewis, C. A.; Dunsmore, B.; Fowler, G.; Krane, J.; Lutnaes, B. F.; Brandal, O.; Sjoblom, J.; Rowland, S. J. *J. Sep.Sci.* **2007**, *30*, 375-380.
(19) Brocart, B.; Bourrel, M.; Hurtevent, C.; Volle, J.-L.; Escoffier, B. *J. Disp. Sci. Technol.* **2007**, *28*, 331-337.
(20) Wang, Y.; Chu, Z.; Qui, B.; Liu, C.; Zhang, Y. *Fuel* **2006**, *85*, 2489-2493.

108a Characterization of Heavy Oil Polar Components by Solubility Separations

John F. Schabron and Joseph F. Rovani Jr. Western Research Institute, 365 North 9th Street, Laramie, WY 82072

An automated separation based on the solubility of precipitated asphaltenes provides a unique approach to the characterization of heavy oil components. The technique employs an inert polytetrafluoroethylene (PTFE)-packed column to conduct on-column asphaltene precipitation and re-dissolution. Quantitative estimation of the distribution of molecular component types by polarity is accomplished using an evaporative light scattering detector (ELSD). The observed separation patterns can be evaluated in terms of the model of asphaltene component molecules consisting of 4-10 pericondensed rings with 4-6 alkyl side chains, with average H/C atomic ratio of 1.2, molecular weights near 750 g/mol, and sulfur and nitrogen functional groups.

Correlation of Atmospheric Pressure Photoionization Nebulization Temperature to Athabasca Bitumen Cut Point Detected by Fourier Transform Ion Cyclotron Resonance Mass Spectrometry

Amy M. McKenna[a,b], Jeremiah M. Purcell[a,b], Parviz Rahimi[c], Ryan P. Rodgers[a,b*] and Alan G. Marshall[a,b*]

[a]*National High Magnetic Field Laboratory, Florida State University, 1800 East Paul Dirac Dr., Tallahassee, FL 32310-4005*

[b]Department of Chemistry, Florida State University, Tallahassee, FL 32306

[c]*National Centre for Upgrading Technology, 1 Oil Patch Drive, Devon, Alberta, AB T9G 1A8 Canada*

(Abstract)

The ultrahigh mass resolving power and sub-ppm mass accuracy of Fourier transform ion cyclotron resonance mass spectrometry (FT-ICR MS) allow the assignment of a unique molecular formula to each peak in the mass spectrum. FT-ICR MS has successfully been coupled to soft ionization techniques such as field desorption, field ionization, electrospray, matrix assisted laser desorption ionization (MALDI) and most recently atmospheric pressure photoionization (APPI). APPI ionizes both nonpolar species (such as PAH, PASH etc.) as well as polar compounds (pyridinic and pyrrolic nitrogen) simultaneously, without the generation of ions from the largely abundant unsaturated hydrocarbons. Here, eight bitumen heavy vacuum gas oil (HVGO) distillation fractions; initial boiling point (IBP)-343 °C, 343-375 °C, 375-400 °C, 400-425 °C, 425-450 °C, 450-475 °C, 475-500 °C and 500-525 °C, as well as a bitumen feedstock were analyzed by APPI FT-ICR MS at each of a series of ion source nebulization temperatures (from 250 - 500°C) to determine the optimal temperature for each boiling point range. Class, type and carbon number distributions reveal both compositional trends among the distillation cuts as well as within a distillation cut at different source temperatures. The detailed compositional information provided by FT-ICR MS for a given distillation cut offers a unique opportunity to investigate the change in molecular weight and class, type and carbon number of species as a function of nebulization temperature. An optimum nebulization temperature maximizes the molecular weight range and limits undesirable thermal degradation of the sample. The systematic analysis of a series of distillation cuts at nebulization temperatures between 250 - 500°C suggests that the optimum desorption temperature falls below the boiling point of the distillation cut. Therefore, analysis of higher boiling species than dictated by the maximum allowable APPI source temperature (500 °C) is possible.

Keywords: APPI, petroleum, Fourier Transform, Ion Cyclotron Resonance, ICR, FT-ICR, FTMS

INTRODUCTION

Ionization methods for petroleum and crude oil successfully coupled to ultrahigh-resolution FT-ICR MS include electrospray,[1-3] field ionization[4], field desorption[5], and most recently, APPI[6, 7]. Electrospray ionizes polar species through protonation/deprotontation but cannot ionize nonpolar hydrocarbons, thiophenes and furans, all of which can be highly abundant in heavy or unconventional crude oil. Field desorption ionizes nonpolar compounds but operates at reduced pressure that complicates the need for high sample throughput [4, 8]. APPI is able to ionize both polar and nonpolar compounds efficiently at atmospheric pressure. A vacuum ultraviolet gas discharge lamp (krypton with 120 nm photons) generates both protonated and radical cations. Direct photoionization occurs if the photon energy is greater than the ionization potential of the analyte. At atmospheric pressure, gases in the source reduce the mean free path of the photon (<1 pm) and direct photoionization is limited. Robb et al.[9] first introduced dopant-assisted APPI to address the issue where a dopant, often toluene, acts as an intermediate between photons and the analyte. Toluene ions facilitate charge exchange and proton transfer reactions to efficiently produce protonated and radical ions. Purcell et al. first applied dopant-assisted APPI to FT-ICR MS analysis of crude oil with assignment of 12,000 unique elemental compositions across a 400 Da mass window [7].

Due to the complexity of crude oil, distillation is used to separate the different types of hydrocarbons to produce useful products based on boiling point. Boduszynski et al.[10-13] correlated boiling point to molecular weight, aromaticity and heteroatom content. Smith et al.[14] used positive- and negative-ion electrospray (ESI) and automated liquid injection field desorption ionization (FDI) coupled to FT-ICR MS to characterize the acidic, basic, and non-polar species in a bitumen HVGO distillation series. APPI uses heated sheath gas to nebulize crude oil dissolved in toluene and create gas-phase ions. Determination of the optimal sheath gas temperature for a given boiling point range can identify (at the level of molecular formula assignment) compounds from various distillation stages. However, because of the heated nebulizer, the source has an inherent thermal desorption limit. The main objectives of this work are to determine the sheath gas temperature necessary to successfully ionize all of the compounds within a distinct boiling point range. Since APPI employs heated sheath gas for photoionization, the additional parameter of heat can be used to correlate boiling point ranges. Each temperature was determined based on the lowest temperature that resulted in the broadest molecular weight distribution without thermal degradation. Surprisingly, the optimal sheath gas temperature is much lower than the boiling point range. Thus the method can be used to successfully ionize higher boiling compounds in crude oil than expected from the maximum allowable source temperature (500°C).

EXPERIMENTAL METHODS

Sample Preparation for APPI MS. A bitumen heavy vacuum gas oil (HVGO) was fractionated by ASTM D-1160 into eight distillate fractions in 25 °C sections from the initial boiling point (IBP) to the final fraction (500 – 525+ °C) which is the residue left in the distillation pot after collection of the 475 – 500 °C fraction. Each distillation cut was prepared in toluene at 0.5mg/mL and analyzed without additional modification.

Instrumentation. Samples were ionized with an Ion Max™ API source (ThermoFisher) in photoionization configuration. Ions are generated at atmospheric pressure as the gas stream containing the vaporized analyte flows orthogonally to the heated metal capillary inlet and the krypton vacuum UV lamp.[7, 9] Dopant-assisted APPI with toluene as both a solvent and dopant increases sensitivity through proton-transfer and charge exchange reactions.[9] Because toluene is an excellent solvent for crude oil, the addition of a separate dopant is not necessary. The solvent flow rate through the APPI source was 50 µL/min and the nebulizer heater was operated at a series of temperatures to determine the optimal temperature for each boiling point range (ie distillation cut). Nitrogen served as an auxiliary and sheath gas and was introduced into the source at 100 psi. Low resolution and high resolution mass spectra were collected with a modified LTQ-FT-ICR (ThermoFisher Corp., Bremen, Germany) mass spectrometer equipped with a 14.5 Tesla superconducting magnet (Magnex Scientific, Oxford, UK).[15] An additional wired octopole directly behind the LTQ allows for a second method of ion accumulation: ions are first mass-selected in the linear trap and passed to the storage octopole for numerous cycles prior to transfer to the ICR cell.[15, 16]

RESULTS AND DISCUSSION

Each sample was analyzed at a series of temperature ranges and the optimal sheath gas temperature determined through analysis of the molecular weight distribution, the absence of thermal degradation (i.e., the temperature at which high mass compounds begin to pyrolyze into lower mass compounds) and ion abundance. **Figure 1** is a low resolution LTQ mass spectrum of a middle distillation cut collected between 450 – 475 °C. At a sheath gas temperature of 250 °C, well below the boiling point range, the higher molecular weight compounds are easily ionized and thermal degradation is minimal.

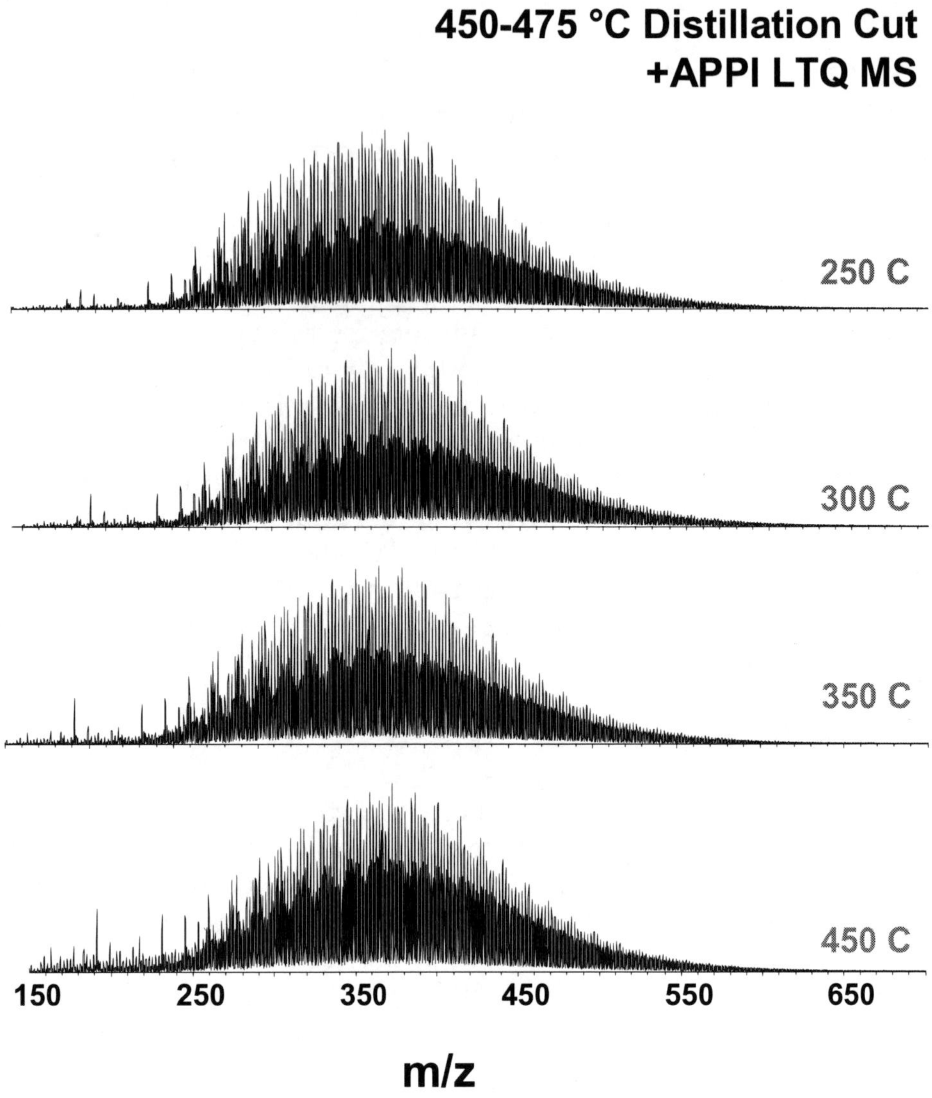

Figure 1. Low-resolution LTQ-MS for the distillation cut collected between 450-475 °C at successively higher sheath gas temperatures. All spectra show essentially the same mass distribution from 200-600 Da. As the temperature is raised from 250 to 450 °C, the low mass compounds increase slightly in abundance and could result from thermal degradation; high-resolution FT-ICR MS can answer that question.

After preliminary analysis with the LTQ mass spectrometer, samples were analyzed with the hybrid FT-ICR mass spectrometer operating at 14.5 Tesla. Because ion guides are necessary to move ions through a high magnetic field, it is necessary to compare the mass distribution from both instruments for independent verification of the molecular mass range. **Figure 2** shows mass spectra (LTQ (top) and FT-ICR (bottom)) of the highest boiling distillation cut (and thus the most complex mixture) with number-average mass of ~525 Da.

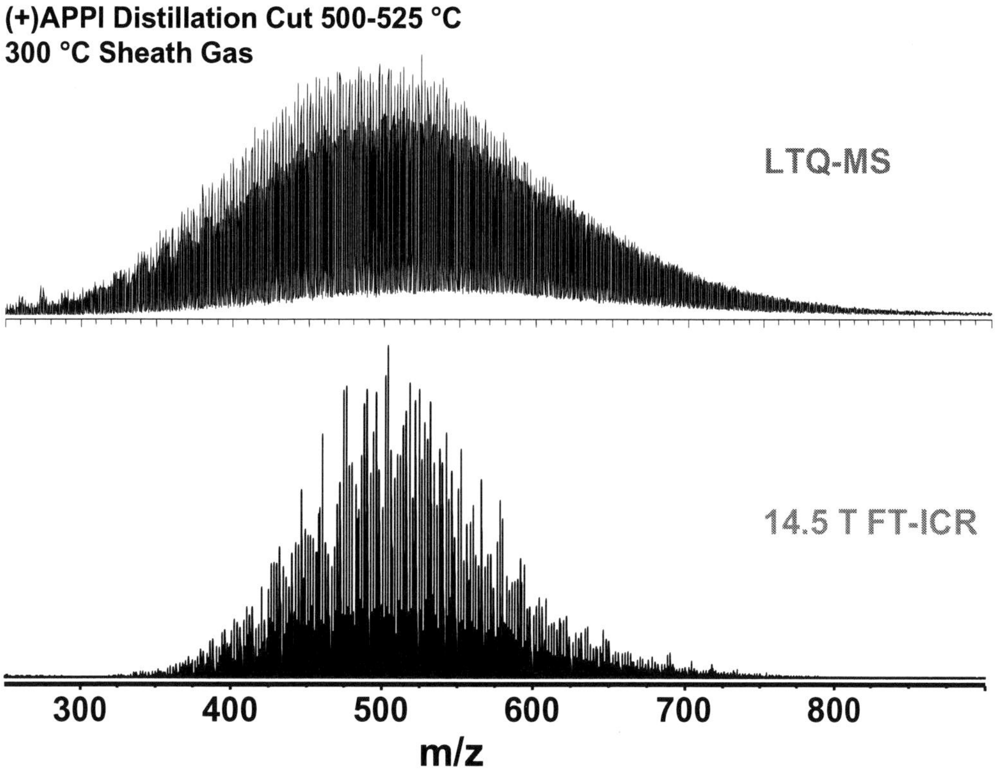

Figure 2. Low-resolution LTQ (top) and ultrahigh-resolution (bottom) FT-ICR mass spectra for the highest distillation cut, to determine whether or not the mass range observed in the FT-ICR spectrum accurately represents the mass range of this sample.

High resolution analysis of each sample is required to determine the optimal sheath gas temperature for each boiling point range. Each peak was assigned a unique elemental composition and sorted into heteroatom class, DBE (double bond equivalents = number of rings plus double bonds to carbon) and homologous alkylation series. Isoabundance color contour plots for each heteroatom class show carbon number and aromaticity (reflected by DBE) for a specific heteroatom class. **Figure 3** shows DBE vs. carbon number plots for the 500 – 525+˚C distillation cut for the pure hydrocarbon compounds at incrementally increased sheath gas temperatures. At 300 ˚C, the carbon number ranges from 25-51 with 12 double bond equivalents. As the sheath gas temperature increases to 325 ˚C, the increased thermal energy facilitates transfer of the heavier, higher carbon number compounds into the gas phase and the carbon number range increases to 30-55 with DBE = 12-14. At 350 ˚C sheath gas temperature, the carbon number range and DBE remains constant. An increase in the sheath gas temperature to 450 ˚C probes the effect of thermal degradation on carbon number and DBE distribution for the hydrocarbon class. The increased heat degrades the heavier compounds and results in an increase in the lower carbon number compounds with 25-50 carbon atoms and DBE = 8-12. The most abundant heteroatom class in each sample at a series of nebulization temperatures was analyzed in the same manner for all eight distillation temperatures.

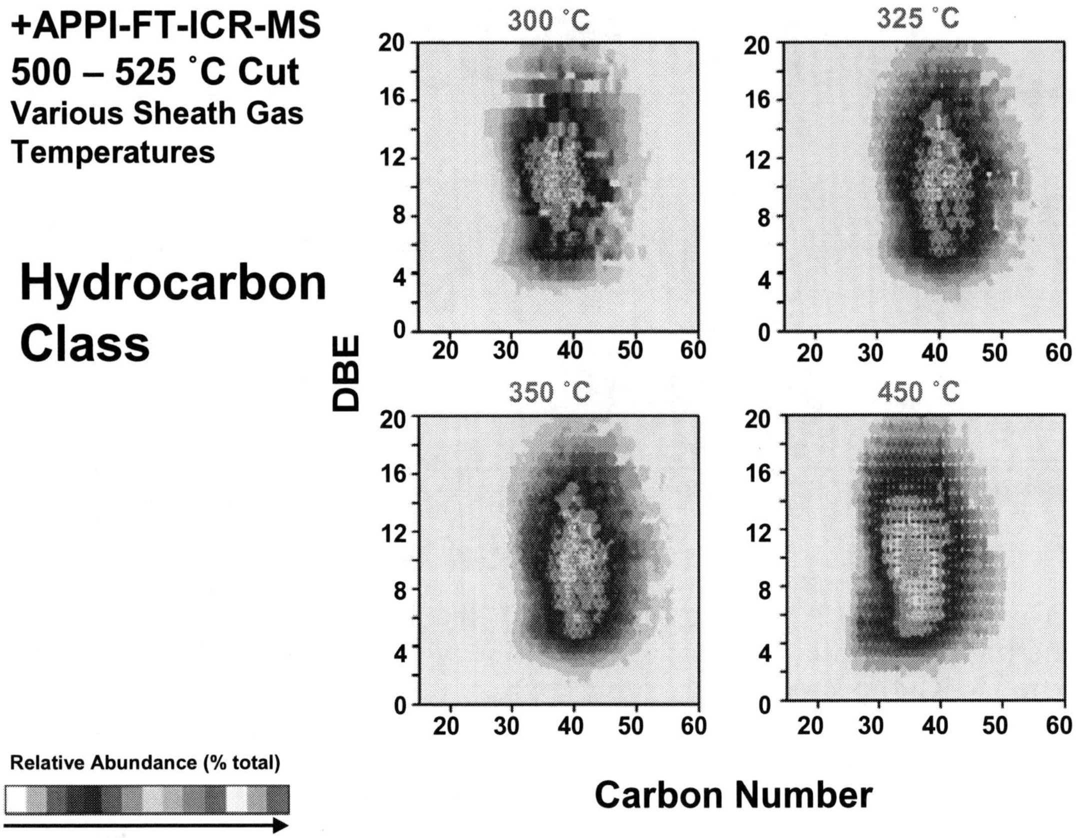

Figure 3. DBE vs. carbon number plots for the hydrocarbon class of a bitumen HVGO vacuum distillation, containing material boiling from 500 – 525 °C as well as any material left in the pot after that fraction was collected.

The broadband FT-ICR mass spectra of the same distillation cut at a series of sheath gas (nebulization) temperatures (**Figure 4**) combined with the results summarized in Figures 3 and 4, determine the optimal sheath gas temperature for a given boiling point range. For example, a nebulization temperature of 350 °C is optimal to analyze compounds boiling between 500-525 °C. Each optimal temperature achieved a wide mass distribution (ionization range) without thermal degradation for a given distillation temperature.

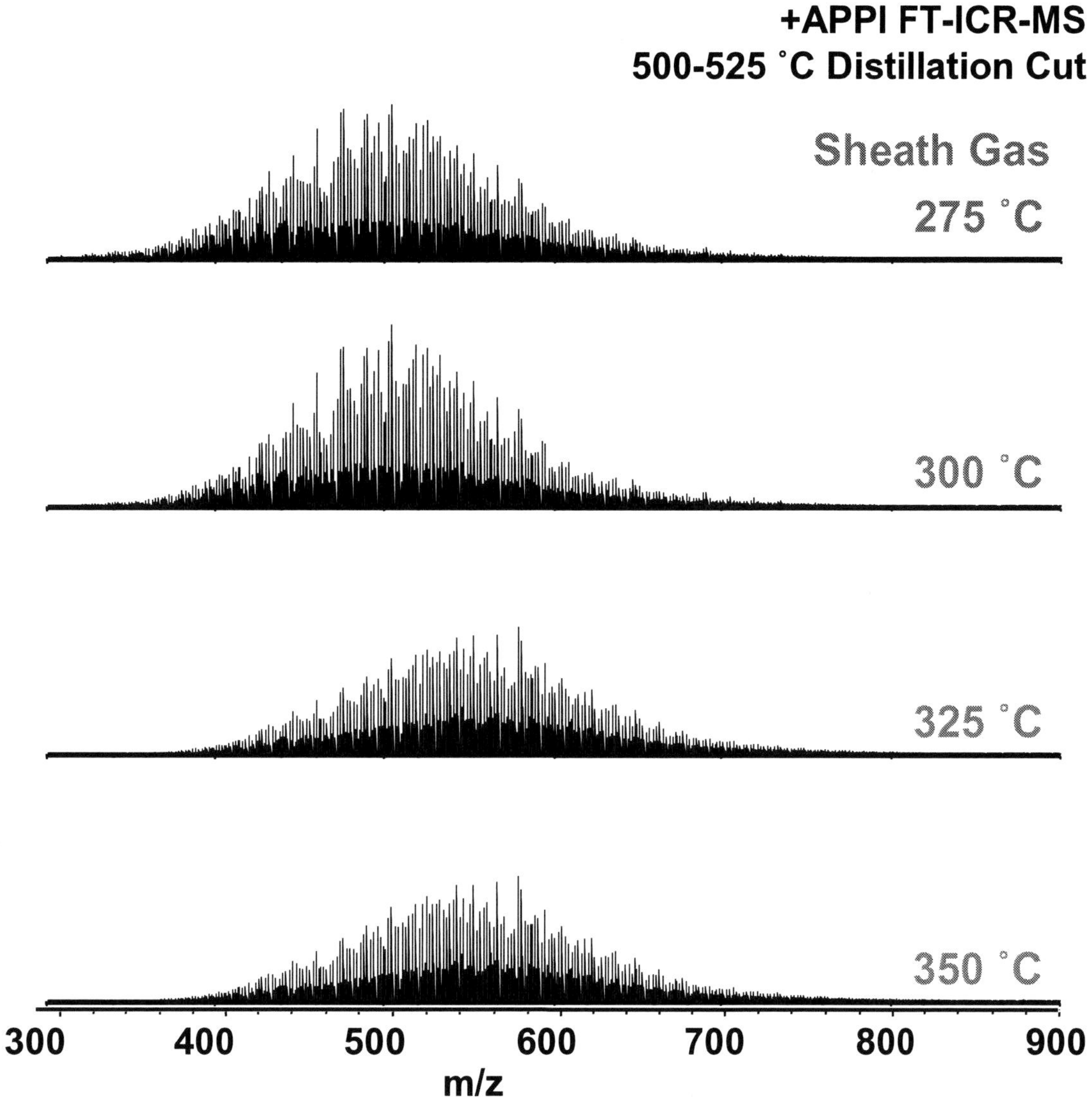

Figure 4. Positive-ion APPI-FT-ICR mass spectra for the highest-boiling distillation cut. Each spectrum is scaled to 100 for the highest-abundant peak among all four spectra to show differences in ion abundance at each temperature. Although the higher sheath gas temperature produces an increased abundance of higher molecular weight compounds, the lower two temperatures produce the broadest mass range.

After each individual distillation was correlated with a sheath gas temperature, the original feedstock was analyzed at each of the optimized temperatures for comparison. Although the HVGO feedstock showed increased complexity as expected, similar compounds were observed at each boiling point range. Through APPI analysis at various nebulization temperatures, the types of compounds that boil within a given temperature range can be extensively characterized. Because the required sheath gas temperature is much lower than the boiling point, the range of crude oil compounds is increased to include the heavier, higher boiling ends that have been problematic for ionization in the past. Future work will continue to analyze higher boiling point compounds in crude oil despite the 500°C temperature constraint of the APPI source.

References

(1) Hughey, C. A.; Rodgers, R. P.; Marshall, A. G. *Anal. Chem.* **2002**, *74*, 4145-4149.
(2) Klein, G. C.; Rodgers, R. P.; Teixeira, M. A. G.; Teixeira, A. M. R. F.; Marshall, A. G. *Abstracts of Papers of the American Chemical Society* **2003**, *225*, U844-U844.
(3) Wu, Z.; Jernström, S.; Hughey, C. A.; Rodgers, R. P.; Marshall, A. G. *Energy & Fuels* **2003**, *17*, 946-953.
(4) Schaub, T. M.; Hendrickson, C. L.; Qian, K. N.; Quinn, J. P.; Marshall, A. G. *Anal. Chem.* **2003**, *75*, 2172-2176.
(5) Schaub, T. M.; Hendrickson, C. L.; Quinn, J. P.; Rodgers, R. P.; Marshall, A. G. *Anal. Chem.* **2005**, *77*, 1317-1324.
(6) Purcell, J. M.; Hendrickson, C. L.; Rodgers, R. P.; Marshall, A. G. *J. Am. Soc. Mass Spectrom.* **2007**, *18*, 1682-1689.
(7) Purcell, J. M.; Hendrickson, C. L.; Rodgers, R. P.; Marshall, A. G. *Anal. Chem.* **2006**, *78*, 5906-5912.
(8) Schaub, T. M.; Linden, H. B.; Hendrickson, C. L.; Marshall, A. G. *Rapid Commun. Mass Spectrom.* **2004**, *18*, 1641-1644.
(9) Robb, D. B.; Covey, T. R.; Bruins, A. P. *Anal. Chem.* **2000**, *72*, 3653-3659.
(10) Altgelt, K. H.; Boduszynski, M. M. *Energy & Fuels* **1992**, *6*, 68-72.
(11) Boduszynski, M. M.; Altgelt, K. H. *Energy & Fuels* **1992**, *6*, 72-76.
(12) Boduszynski, M. M. *Energy & Fuels* **1988**, *2*, 597-613.
(13) Boduszynski, M. M. *Energy & Fuels* **1987**, *1*, 2-11.
(14) Smith, D. F.; Schaub, T. M.; Rodgers, R. P.; Rahimi, P.; Marshall, A. G., *Oilsands 2006*, Edmonton, Alberta, Canada, 22-24 Feb 2006.
(15) Schaub, T. M., Blakney, G.T., Hendrickson, C.L., Quinn, J.P., Senko, M.W., Marshall, A.G., *55th Amer. Soc. for Mass Spectrom. Ann. Conf. on Mass Spectrom. & Allied Topics*, Amer. Soc. for Mass Spectrom., Indianapolis, IN 3-7 Jun 2007.
(16) Marshall, A. G., Hendrickson, Christopher L., Emmet, Mark R., Rodgers, Ryan P., Blakney, Greg A., Nilsson, Carol L., Schaub, Tanner M., *234th Amer. Chem. Soc. Natl. Mtg.*, Boston, MA, 19-23 Aug 2007.

FRACTIONATION OF AN ASPHALTENE FROM A BRAZILIAN VACUUM RESIDUE

Abstract ID: 117129

*Lina Constanza Navarro Quintero [1] *, Peter Rudolf Seidl [1] *, Ljubica Tasic[2], Katia Z. Leal[3], Sonia M. C. Menezes[4]*

[1] Escola de Química, Universidade Federal do Rio de Janeiro. [2]Instituto de Química, Unicamp. [3] Instituto de Química, UFF. [4]CENPES, Petrobras. Brazil.

*linacons2@yahoo.com.ar; pseidl@eq.ufrj.br

Introduction

Solid deposition is one of the most important problems that arise in oil production. Well and oil pipeline clogging can result in very significant economic losses and operational delays. Asphaltenes play an important role in the understanding of this kind of problem because they appear to be mainly responsible for sedimentation that may occur during recovery operations, owing to their tendency to flocculate, a process in which molecular interactions have strong influence. In order to investigate such problems, an understanding of the chemical and physical characteristics of asphaltenes is required. These characteristics remain unclear in spite of the large amount of work that exists in the literature on their origin and composition. Asphaltenes correspond to complex materials and have been studied as averages of large molecules having great differences in properties, which makes it extremely difficult to define their molecular structure. Vacuum residues represent an alternative source of asphaltenes. They correspond to constituents that have been subjected to considerable heating and have probably undergone reactions that would remove more labile functionalities. They must also concentrate the constituents that would tend to associate and form aggregates.

We have used vacuum residues from Brazilian crudes that are commonly processed in local refineries to study their solubility under certain conditions[1]. Modeling and simulation studies require reliable representations of molecular structure thus our work has been concentrated mainly on the constituents that tend to aggregate rather than the types of species that may be present. In a view of the recent success of fractionation methods[2-8] in characterizing different types of asphaltenes that are present in a sample, we tested one of these methods on a sample that we have been more thoroughly investigating [1]. As heptane and toluene are commonly used as solvents in asphaltene separations, we used different proportions of their mixture (Heptol) to verify if we could fractionate our asphaltene into representative constituents.

Experimental

Asphaltenes from a Brazilian vacuum residue were fractionated by the IP-143 modified method, using mixtures of heptane and toluene (Heptol) in different concentrations, as precipitation solvents. The sample may be considered rich in asphaltenes (around 9% by weight).

Precipitation and Fractionation of Asphaltenes. The vacuum residue was heated gently until it became fluid and homogenized before an aliquot was taken. Precipitation was based on the Institute of Petroleum of London (Standard Methods for Analysis and Testing of Petroleum and Related Products-vol 1, IP-143, ASTM-6500-00) procedure.

Initially the percentage of asphaltene for each heptol fraction was quantified. Approximately 1 g of sample was weighed for each analysis. The solvent employed was a mixture of n-heptane reagent grade, 99,5% pure and toluene (reagent grade), "heptol" in proportions than vary between 0% v/v ("pure" asphaltene), and 40% v/v of toluene (asphaltenes are completely soluble in higher concentrations of toluene). The efficiency of fractionation was monitored by the asphaltene content of each fraction while chemical properties of sub-fractions are reflected by their ^{1}H chemical shifts given in Table 1.

Chemical Characterization. ^{1}H Nuclear Magnetic Resonance (NMR) data were collected from each sub-fraction. ^{13}C NMR and elemental analysis were used for the representative fractions.

Elemental analysis was determined on Perkin-Elmer CHN 2400 equipment, according to ASTM 5291-95 methods. Sulfur was determined on a Phillips PW 2400 x-ray fluorescence spectrometer. The sample was pulverized to 325 mesh and 1 mg was analysed on a plate.

^{1}H spectra were run in 5 % solutions of a 1:1 (v/v) $CDCl_3$:C_2Cl_4 mixture at 300 MHz using 45° pulses, 1.0 s intervals between pulses and 128 transients were accumulated. NMR (^{13}C, and DEPT) experiments were performed on a Varian INOVA 500 MHz spectrometer equipped with 5 mm probes (bbsw) at 25°C (298 K). Standard Varian software was used to acquire and process the NMR data. The spectra were referenced using the central $CDCl_3$ signal (77.0 ppm). NMR samples were prepared in $CDCl_3$ (600 μL) with sample concentration of ~100 mg. To obtain quantitative ^{13}C NMR spectra, chromium acetylacetonate ($Cr(AcAc)_3$) was added in the concentration of 0.05 mol.dm-^{3}, as relaxation reagent and the decoupler was adjusted in the gated mode to avoid NOE. DEPT analyses were performed to identify CHn groups. Chemical shifts were integrated according to ranges corresponding to the functional groups of interest[1].

Results and Discussion

The solubility profile of the asphaltenes that were fractionated or precipitated from vacuum residue A is given in Figure 1.

Asphaltenes are completely soluble in neat toluene and Heptol mixtures with a higher aromatic content (> 40% v/v toluene) but their behavior depends on the amount of asphaltene in each sample. For concentrations above 30% v/v of toluene in Heptol, the amount of asphaltene that is precipitated is almost insignificant, thus only concentrations of up to 30% of toluene in heptol were used in preparative fractionations. Figure 1 confirms that the amount of precipitate decreases with the increase in toluene concentration and the decrease in-heptane concentration of the solvent.

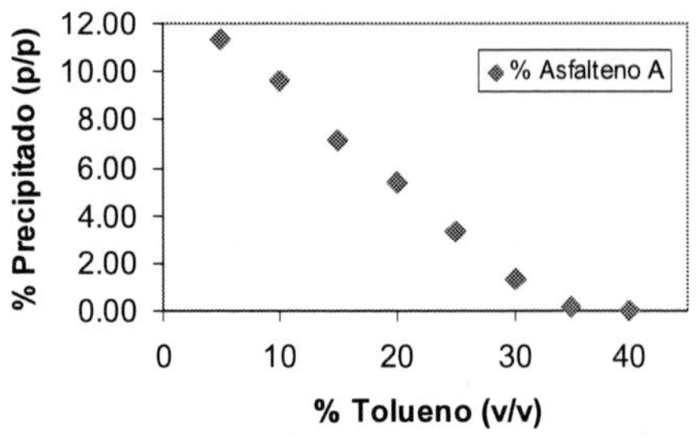

Figure 1. Solubility profile of asphaltenes from vacuum residue A in Heptol (n-heptane/toluene).

According to the literature, adding toluene to n-heptane increases the dispersive and polar component of the solubility parameter[9]. This results in flocculation and precipitation of the asphaltene aggregates, which may change their composition depending on the change in polarity of the solvent. This process of precipitation seems to be controlled, as suggested in some studies[10], principally by dispersion and by π–π interactions between an asphaltene and the solvent or mixture of solvents.

Properties of the fractionated asphaltenes. The first analysis run on the sub-fractions was proton NMR (Table 1). It was used in order to determine whether significant differences among them could be observed and to choose the fractions that had changed the most, particularly regarding the concentration of highly aromatic molecules.

Analyses by proton NMR show a significant variation in nuclei that are part of the aromatic system of each sub-fraction. In the sub-fractions, the aromatic proton content (H_{ar}) does not reveal a regular pattern (Table 1). Most of the sub fractions show values of H_{ar} between 7-10%. The highest value was observed for fraction A–10% (13.1%). Hα remains almost constant ≅ 20%, Hβ show results between 49-57%, and Hγ between 14.4-18.1%. These results show how efficient this fractionation is and the concentration of molecules with higher aromaticity in sub-fraction A–10%.

Preliminary ^{13}C NMR and elemental analyses (Table 2), were run on A-0%, A–10% and A–20% sub-fractions. These data show that the A–10% sub-fraction has the highest percentage of ring junction aromatic carbons 18.4%, against 13.7% w of fraction A-20.

Some structural observations could be made when the measured values of aromatic hydrogen (H_{ar}) and aromatic carbon (C_{ar}) are analyzed. The A–10% sub-fraction corresponds to high values of H_{ar} and C_{ar}, this shows low substitution of peripheral aromatic carbons and strongly suggests extensive aromatic ring fusion[11]. In contrast, A–20% has a low H_{ar} content, showing high substitution of peripheral aromatic carbons and lower aromatic ring fusion. This sub-fraction also show significant increment in the oxygen content that can be related to a decrease in carbon content.

Asphaltene A	Sample		A - 0%	A - 5%	A -10%	A -15%	A - 20%	A - 25%
	% Toluene in Heptol (%)		0	5	10	15	20	25
	% Precipitate (%)		11.3	9.59	7.14	5.36	3.31	1.29
		Chemical Shift (ppm)	Integrated Area (%)					
	H mono aromatic	7-6	3.1	1.7	3.4	1.5	2.1	1.5
	H di aromatic	9-7	6.5	5.3	9.7	5.5	5.3	4.0
	H alfa, Hα	4-2	21.6	22.1	22.0	21.8	20.0	20.8
	H Beta, Hβ	2-1	50.8	56.6	49.3	55.8	57.4	57.2
	H Gamma, Hγ	1-0	18.1	14.4	15.7	15.5	15.2	16.4
	H Saturated, Hsat	4-0	90.4	93.0	86.9	93.1	92.6	94.4
	Har-total, Har	9-6	9.6	7.0	13.1	6.9	7.4	5.6

Table 1. ^{1}H NMR Integrated Areas.

Asphaltene A	Integrated Area (%)					Elemental Analysis (%)					
	C Ins	C Sat	C ar Alquil	C ar H	C ar Junc	C	H	N	S	O*	H/C
A - 0%	48.8	51.2	12.4	12.2	24.2	87.0	7.7	2.3	1.0	2.0	1.07
A -10%	38.7	61.3	5.9	14.4	18.4	87.2	7.4	2.2	1.0	2.1	1.02
A - 20%	30.3	69.7	4.6	12.0	13.7	86.5	7.5	2.4	1.0	2.7	1.04

*By difference

Table 2. ^{13}C NMR Integrated Area and Elemental Analysis.

Elemental analysis was used to determine the proportion of nitrogen and oxygen (by difference) to H/C in each fraction. As is shown in the table 2, the H/C ratio does not vary considerably between the "pure" asphaltene and its sub-fractions, indicating that their respective degrees of unsaturation are similar. Their content is practically the same, but small differences in polarity and aromaticity may make significant contributions to their macroscopic proprieties[12].

Some work has been done on the sub-fractonation of asphaltenes using mixtures of aliphatic and aromatic solvents. Szewczyk e co-workers[13], fractionated asphaltenes from an oil using mixtures of o-xylene e n-heptane. They obtained four complimentary soluble and insoluble fractions (37, 56, 64 e 74 % p/p of insoluble asphaltenes). Their results reveal that insoluble fractions had a low H/C ratio and a high heteroatom (Nitrogen e Oxygen) content, compared to the soluble fractions. Similar results were reported by Andersen et al[12] using Boscan oil. Spiecker and co-workers[7], sub-fractionated asphaltenes from four different oils, obtaining the lowest H/C ratio from the least soluble sub-fractions which also showed a modest increment in the percentage of heteroatoms and metals (i.e., V, Ni e Fe). Avid et al[14] fractionated asfaltenes using mixtures of n-heptane and toluene (Heptol), observing a decrease in the hydrogen content and an increase in the content of heteroatoms (S and N) with the increase in toluene in the mixture (least soluble fraction). However, the observed variations fall practically within the experimental error of the method. Comparing previous results with the ones reported in the present work one can observe that the elemental composition of the asphaltenes and their sub-fractions do not reveal fundamental variations that would allow inferences as to their behavior but they confirm the hypothesis proposed by Speight[9] that asphaltenes correspond to well-defined chemical compositions with regard to their carbon and hydrogen content and this also applies their sub-fractions.

NMR results suggest the presence of rigid and polar aromatic centers, bound to naphthenic rings and side chains. The A-0% fraction (which corresponds to the "pure" asphaltene) shows a large extent of aromatic ring fusion. The A-10% fraction shows less fused aromatic rings compared to the "pure" asphaltene, but with a larger proportion then the A-20% fraction. This type of analysis also reveals that the structures in fraction A-10% are relatively rigid, while the A-20% fraction appears to be more flexible, since the aromatic rings could be linked by aliphatic chains or saturated rings. These characteristics suggest that largest proportion of aromatic rings are concentrated in the A-10% fraction and that they may be responsible for aggregation among thus type of asphaltenes.

Conclusions

The solubility and chemical properties of asphaltenes and their sub-fractions from a vacuum residue were studied. The precipitated asphaltenes were totally soluble in toluene concentrations in Heptol of over 40 % (v/v), but above 30 % (v/v) the amount obtained is practically negligible.

Asphaltenes from these vacuum residues and their Heptol sub-fractions reveal a similar degree of unsaturation and heteroatom content. They must correspond to well-defined chemical structures. Variations in behavior must thus be owing to different hydrogen and carbon content.

The sub-fractionation of asphaltenes using Heptol, showed that the aromatic species with higher content of ring fusion aromatic carbons, can concentrate in a specific sub-fraction, and suggests also that this sub-fraction may be mainly responsible for the association of these complex molecules.

Acknowledgement. This research was supported by CNPq and PETROBRAS. P.R. Seidl has a research fellowship from the PRH ANP/FINEP Program and L.C.N. Quintero a doctoral fellowship from CNPq.

References

(1) Carauta, A.; Seidl, P.; Chrisman, E.; Correia, J.; Menechini, P.; Silva, D.; Leal, K.; Menezes, S. de Souza, W.; Teixeira, M. *Energy Fuels*, **2005**, *19*, 1245.
(2) Rahimi, M. E Gentzis, T. Fuel Processing Technology. v 80, N°1, p.69-79, **2003**
(3) Cyr, N., D. Mcintyre, D., Toth, G., Strausz, O. Fuel. v 66, N° 12, p 1709-1714, **1987**
(4) Dehkissia, S., Larachi, F., Chornet, E. Fuel. v 83, N° 10, p 1323-1331, **2004**.
(5) Östlund, J., Wattana, P., Nydén, M., Fogler, H. Journal of Colloid and Interface Science. v 271, N° 2, p 372-380, **2004**.
(6) Spiecker, P., Gawrys, K., Trail, C., Kilpatrick, P. Colloids and Surfaces A: Physicochemical and Engineering Aspects. v 220, N°1-3, p 9-27, **2003**.
(7) Spiecker, P., Gawrys, K., Kilpatrick, P. Journal of Colloid and Interface Science. v 267, p 178-193, 2003bSpiecker, et al. 2003a; Spiecker, et al. **2003**.
(8) Navarro, L. C.; Seidl, P. R.; Tasic, L.; Leal, K.; Menezes, S.; Nunes, E. The 8th International Conference on Petroleum Phase Behavior and Fouling, Petrophase **2007**, 109.
(9) Speight, J. The chemistry and technology of petroleum. 2nd Ed.. Marcel Dekker Inc, p 1-22, 197-255, 309-473, **1991**.
(10) Buckley, J. Predicting the Onset of Asphaltene Precipitation from Refractive Index Measurements. Energy and Fuels. v 13, N° 2, p 328-332-1797, **1999**.
(11) Acevedo, S.; Escobar, O.; Echevarria, L.; Gutierrez, L. B.; Mendez, B. *Energy Fuels*, **2004**, *18*, 305.
(12) Andersen, S., Keul, A., Stenby, E. Petroleum Science and Technology, v15, N° 7-8, p 611-645, **1997**.
(13) Szewczy, V., Behar, F., Behar, E., Scarsella, M. Journal of IFP, v51, N°4, p 575-579, **1996**.
(14) Avid, B., Sato, S., Takanohashi, T., Saito, I. Characterization of Asphaltenes from Brazilian Vacuum Residue Using Heptane-Toluene Mixtures. Energy and Fuels. v 18, N° 6, p 1792-1797, **2004**.

108e Hydrotreating and Catalytic Cracking of Bitumen-Derived Vacuum and Heavy Coker Gas Oils

Warren M. Ewert, Joe E. Bares, Walter E. Alvarez, and Barbara A. Todd. R&D/Shared Services, ConocoPhillips Company, Bartlesville Technology Center, Bartlesville, OK 74004

Many North American refiners are looking to the Canadian oil sands as a plentiful hydrocarbon resource in the face of dwindling supplies of conventional crude oils. However, the oil sands bitumen presents a number of challenges, even after it has been diluted to mimic traditional crudes for transport from Alberta to the United States. As bitumen-derived crude oils become a larger portion of the refinery feedstock slate, more vacuum gas oil and heavy coker gas oil will be fed to the catalytic cracker to convert these heavy fractions to gasoline and distillate-range materials. Gas oil hydrotreating may be required in order to obtain suitable cat cracker conversions and yields. This paper will describe the results of pilot-scale hydrotreating and cat cracking studies conducted on straight-run vacuum and heavy coker gas oils isolated from Canadian bitumen-derived crude oils.

ESTIMATION OF MICROCARBON RESIDUE FOR HEAVY OILS AND FRACTIONS

Azfar Hassan, Lante Carbognani and Pedro Pereira-Almao, University of Calgary, Calgary, AB, Canada.*

Introduction

Many characterization methods have been used to elucidate the structure, composition and properties of heavy oils and bitumens. Of these methods, determination of correct amount of carbon residue has increasingly become important. Microcarbon residue (MCR) method has attracted more attention because of the small amounts of sample used and simpler experimental set up. Noel in 1984 first described this method, and showed the effect of various parameters (gas flow rate, type of gas used, type of crucible etc) on the amounts of MCR determined. It was reported that small amounts of air increases the amount of residue formed [1].

We herein report that the general assumption that presence of small amounts of air, tends to increase the amounts of carbon residue is not always true. We have found that for one of the samples tested, MCR always decreased with introduction of trace amounts of air. This is in contrast to Noel's findings [1]. For sure, introduction of high amounts of air would lead to complete oxidation of the sample resulting in very low MCR values. Our results have shown that it is critical to have inert atmosphere to get correct values for carbon residue. MCR amount may decrease or increase in presence of small amounts of oxygen depending on the nature of the sample. Noel also reported no effect from carrier gas flow rate upon the value of MCR determined. This also is contrary to our findings. In our set up change in gas flow rate profoundly affects the MCR value determined. An apparatus was designed successfully to analyze a maximum of twenty six samples. We herein report that our new set up allows MCR determination in oil samples with high accuracy and precision.

Experimental

Standard samples with known MCR contents (0.35, 2.0, 8.48 and 24.5 wt.%) were obtained from PCA, Texas, USA. Sample that has 24.5 wt.% of MCR, is a solid while the rest are liquids. The liquid samples were well shaken before use.

Studied Athabasca bitumen samples were provided by Suncor Energy and Jacos. Athabasca vacuum residue was also provided by Suncor Energy. A sample of Cold Lake bitumen was provided by Professor B. Maini and Peace River Bitumen by Shell Canada. Professor C. Scott provided a sample of Hamaca heavy oil (Eastern Venezuela). Mr. F. Mazin provided the sample of Arabian Light vacuum residue.

MCR content was determined using a custom made apparatus placed in a Barnstead Muffle furnace, equipped with programmable temperature controller. Analytical Balance from Mettler with ± 0.01 mg sensitivity was used for weighing the samples before and after heating.

The apparatus (Figure 1) was made of Al and stainless steel in order to make it light but robust. The platform on which sample holders are placed is 4 inches in diameter. A known mass (10-40 mg) of sample was placed in a 2 cc glass sample vial. The sample holders were then placed on the platform. Twenty six purge tubings, 3/4 inch long and 1/8 inch wide extend out from the platform for N_2 purge close to sample. A glass cover 4 inches wide and 2 inches high with one orifice of 1/8 inch is then placed to shield the samples form air atmosphere. The system was purged with N_2 at a flow rate of 900 cc/min for 45 min. The samples were then heated to 500 °C for 15 min at a rate of 10 °C/min. After heating, the sample was allowed to cool under N_2 until the temperature dropped to 200 °C. The samples were then placed in a desiccator before measuring their final mass.

FTIR spectra were run with a Nicolet 6700 FT-IR spectrometer from Thermo Electron Corporation. Smart Diffuse Reflectance Cell (DRIFTS) was used for spectra acquisition of KBr diluted/ground samples.
High temperature simulated distillation (HTSD) for bitumen samples was carried out following the ASTM-D-7160-05 method. Details for experimental set up are reported elsewhere [2].

Bitumen viscosities were determined with a Brookfield model RV DV-II+PROCP viscometer. Set up temperatures were maintained with a recirculating glycol bath (Brookfield model TC-102).

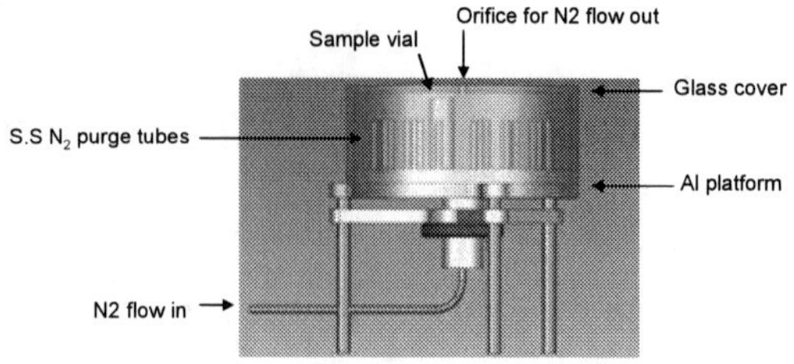

Figure 1. Schematic for MCR multisample set up.

Results and Discussion

Table 1 presents the results obtained for MCR standards. The experiment was repeated at least five times to ensure reproducibility of the measurement. Sample position on the platform was also changed in order to confirm that the values obtained are independent of the position where the sample is kept.

Table 1. MCR values obtained using multi-sample setup at a Nitrogen flow rate of 900 cc/min.

Trial	0.35 wt.%	2.0 wt.%	8.48 wt.%	24.5 wt.%
1	0.39	2.17	7.31	24.32
2	0.30	2.03	8.24	23.74
3	0.36	1.93	8.01	24.15
4	0.47	2.11	8.70	23.56
5	0.40	2.46	8.50	23.75
Average	0.38	2.14	8.15	23.90
Standard Deviation	0.06	0.20	0.54	0.32

It can be seen that very reproducible results were obtained for samples with varying MCR concentration (0.35-24.5 wt.%).

In our system the size of the used vial affects the results. It is crucial for the vials to be longer than the purging tubes. When 0.5 cc vials that were short in length were used, results obtained were neither precise nor accurate. A change in the flow rate affected the MCR value. This was in contrast to Noel's findings [1]. Results presented in Table 2 illustrate these findings.

In order to evaluate the effect of presence of air, experiments were carried out without any initial purging of the system with N_2. The outer glass cover was also removed. However, N_2 was flown into the system during heating. The results presented in Table 3 clearly show that presence of "some" oxygen tends to increase the amount of MCR in most of the samples. This increase in weight is due to the partial oxidation, leading to bond formation of extra oxygen atoms in the sample. Of the four primary standards analyzed, introduction of little oxygen increased the amount of MCR for three samples (0.35, 2.0 and 8.48 wt.%). The 24.5 wt.% standard sample always gave low MCR values. This reflects that there is complete combustion of this sample even in the presence of slight amounts of air. From Table 3 it appears that the nature of oil is critical in deciding whether MCR increases or decreases in presence of slight amounts of air.

Table 2. Effect of N_2 flow rates on MCR values.

Flow cc/min	300		1000	
Run	8.48 wt.%	24.5 wt.%	8.48 wt.%	24.5 wt.%
1	8.31	10.82	7.44	23.67
2	11.45	24.06	9.51	27.71
3	11.99	22.99	10.78	31.41
Average	10.58	19.29	9.24	27.60
Std. Dev.	1.99	7.35	1.68	3.87

However, when the amount of air present is increased by decreasing the N_2 flow, there is a sharp decrease in the amount of microcarbon residue for all samples due to complete combustion of carbonaceous material to form carbon oxides.

Table 3. Effect of initial presence of air on MCR values. N_2 Flow= 900 cc/min

Run	0.35 wt.%	2.0 wt.%	8.48 wt.%	24.5 wt.%
1	0.42	2.65	8.8	0.96
2	1.04	2.77	9.23	3.15
3	1.02	3.4	13.4	2.39
4	0.64	2.56	9.68	0.48

Table 4 shows the effect of presence of trace amounts of air on MCR value obtained for real oil samples. It appears that the amount of sample taken for measurement, affects the MCR value obtained, until a threshold value for the mass taken is achieved. Below this value the final MCR value would always be less than that obtained under inert conditions as the amount of oxygen is high enough to lead to combustion. With increase in mass of the sample taken, a time comes when oxygen present becomes 'limiting', implying that the amount is enough to incorporate oxygen into the sample matrix but not lead to combustion reaction. This reflects upon the fact that "little amount" of oxygen that leads to increase in MCR value is different for different samples. For similar amounts of various samples taken, the MCR values obtained may increase or decrease as compared to the values obtained under inert conditions; depending upon the chemical nature of the sample.

In order to shed light on the possible causes for the former findings got with MCR standards, distillation yields by High Temperature Simulated Distillation (HTSD) and viscosities were determined for these materials. Figure 2 shows their simulated distillation chromatograms. The SimDist elution patterns suggest that the low content MCR samples (0.35 and 2.0%) are gasoils spiked with small amounts of residual fractions that can be observed eluting after about 16 minutes. The 8.48 % standard is a light-paraffinic crude oil with a large proportion of light distillates eluting before 7.5 minutes plus a larger proportion of residual fractions eluting after 16 minutes (the spikes overlapping the hump correspond to paraffins). The heaviest sample, the one possessing 24.5 % MCR, is enriched in heavy hydrocarbons, particularly those comprising the vacuum residue region ranging from about 16-36 min-

Table 4. Effect of trace amounts of air on MCR content for real oil samples

Nature of material	Sample	Mass (g)	% MCR (Trace air)
Whole Bitumen	Athabasca (SUNCOR)	11.40	16.1
	(JACOS)	11.77	12.4
	Cold Lake	5.15	2.7
		6.60	3.5
		7.64	3.3
		11.83	8.9
		14.71	12.6
		17.82	14.0
		19.01	12.8
		66.86	13.1
	Peace River (A)	19.50	15.5
		20.83	13.9
	(B)	23.82	13.8
	Hamaca (Vzla.)	6.78	4.4
		10.84	8.5
		12.68	13.2
		52.29	19.1
Vacuum Residue	Athabasca 23% 525°C-	16.71	20.6
		25.95	23.4
	Arab light 8% 525 °C-	7.89	30.6
		8.12	26.3
		13.3	29.9

utes. The viscosity for the first 3 standards is deemed low while this parameter is extremely large for the 24.5% MCR sample (Table 5). This sudden rise in viscosity for the last sample, which for practical purposes is a solid, can explain its different behavior, discussed before. Solid heavy oil fractions are aggregated structures that might possess porous spaces that liquids can not have. This leads to speculate that probably this is one of the reasons for the differing oxidative properties reported before. Oxygen molecule has easy access to the porous space of the solid, however, it must diffuse through the whole liquid volume of the liquid samples. Other authors have presented evidence that during heating period of oil residua disproportioning of samples and appearance of cracks occur [3].

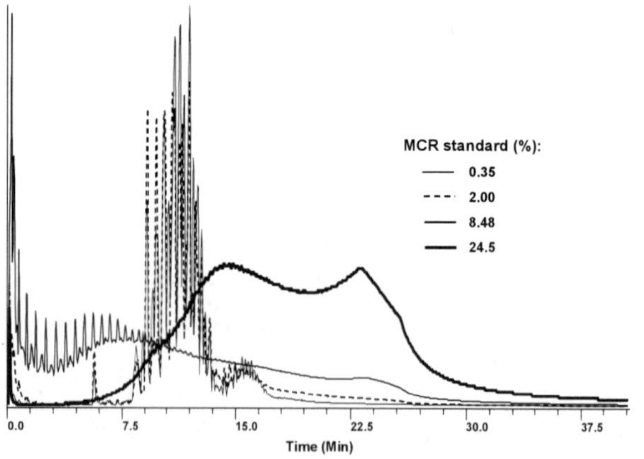

Figure 2. Qualitative Simulated Distillation Chromatographic traces for the four studied MCR primary standards.

Table 5. Viscosity values for the four MCR primary standards

Sample (% MCR)	Viscosity (cP) @ set Temperature (°C)	
	25	145
0.35	46.7	ND*
2.0	78.3	ND
8.48	551.8	4.5
24.5	ND	116,000

*ND: Non determined (not possible under set conditions)

In order to improve our understanding on the reported oxidative phenomenon, IR spectra of MCR standards for 8.48 and 24.5 wt.% were analyzed before and after the carbonization test. Results are presented in Figure 3.The following findings can be deduced from the results presented:

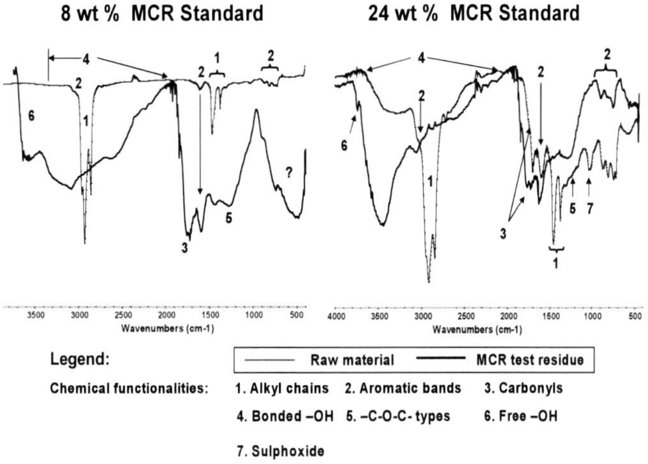

Figure 3. DRIFTS spectra for standards with 8.48 and 24.5 wt. % MCR, before and after heating (in the presence of low amounts of air).

1. The 8.48 wt.% MCR original standard is rich in alkyl moieties (bands #1) whilst the 24.5 wt.% is enriched in aromatics (bands #2).

2. The 8.48 wt.% MCR original sample does not show presence of carbonyl (#3) and/or -OH functionalities (#4). Moreover, only the higher MCR content sample presents the S=O band (#7). These evidences show that the 24.5 wt.% MCR sample is originally oxidized, i.e., enriched in oxygen compounds. The very broad band ranging from ca. 3,600-2,200 cm^{-1} seems to indicate that carboxylic acids are comprised within this sample. Dramatic H-bonding effects from carboxyls can be ascribed for such a wide band.

3. After the carbonization brought by MCR test over the 8.48 wt.% MCR standard, a dramatic increase of oxygen functionalities is evidenced in the IR spectrum. Abundances for the carbonyl (#3), bonded –OH (#4) and C-O-C types (#5) within the MCR residue, are impressive. Aromatics also are observed to increase to a very high level (bands #2) whilst alkyl chains are totally removed (bands #1). The aromatic bending bands seem to increase and shift to lower energies, suggesting that the contribution from these species becomes larger (relative more concentration of these types) and intermolecular interaction/interlayer aggregation is speculated to be the cause of bending vibrations shifted to lower wavelengths (band labeled "?"). Generation of an important new band

located at high frequency (#6), suggest the formation of compounds bearing free –OH bands. With great probability, phenolics are the compounds responsible for these signals.

4. After the carbonization brought by MCR test over the 24.5 wt.% MCR standard, removal of alkyl functionalities (#1) and increase of aromatics (#2) happens following a similar pattern already shown by the previous sample. These changes can be assigned to typical thermal cracking reactions. The product from MCR in this case shows that S=O types (#7) and some of the bonded –OH types (#4) disappeared. Speculation on the COx and SOx production can be proposed in these cases.The most striking features for spectra from this standard sample are the very low increase of carbonyls (#3) plus disappearance of –OH intensity. This later in particular is observed to happen with a decrease in the extension of H-bonding, suggesting the occurrence of decarboxylation reactions during the carbonization process.

Summarizing the findings observed with these experiments, it can be said that different oxidative/cracking mechanisms affect samples submitted to MCR experiments, depending on the chemical nature of the material carbonized. The increase in oxygen types observed with the 8.48 wt.% MCR standard, parallels the results published by Noel [1], showing the feasibility of MCR increase, observed by this author during the initial stages of sample carbonization. However, for 24.5 wt.% MCR sample the mass is reduced because of complete oxidation to CO_2.

This report ends showing an application of the developed MCR methodology that relies on the multisample analyzer. Table 6 lists the carbon residue contents reported for selected bitumen and residua samples and the MCR content for some of these samples determined in our laboratory under N2 and air atmosphere. It is evident that the values determined in our laboratory set up is in good agreement with those reported in literature [1, 4-10]. The samples were analyzed at least five times to determine the average value.

Table 6. Carbon residue contents reported/determined for selected samples.

Nature of material	Sample *	MCR	CCR	RCR	Reference(s)
Whole Bitumen	Athabasca		13.6		4
		12.5	13.2		4
	(Syncrude)		11.8		5
			18.5	14.9	5
	(SUNCOR)	12.6±0.6			This work
	(JACOS)	11.7±0.4			This work
	Cold Lake	12.5	13.2&13.6	11.3	4
			13.6		4
			13.6		5
		13.6			6
		11.7±0.3			This work
	Peace River (Spring)			12.5	4
	(A)	11.3±0.1			This work
	(B)	12.4±0.5			This work
	Hamaca (Vzla.)	14.2			7
	(Same sample)	14.4± 0.6			This work
	Average 305 samples (50 basins)		13.7		8
Vacuum Residue	Athabasca 15% 525°C-	25.4			9
	23 % 525°C-	17.6± 0.4			This work
	Arab light (VR, EXXON)		24.3		10
	8% 525 °C-	20.8±0.3			This work

* (Reported origin and properties provided within referenced document)

Conclusions

The MCR developed set up is simple and allows quick determination of MCR using simple systems available in any laboratory. MCR content for up to 26 samples can be simultaneously determined with high accuracy and precision. MCR determined values for known bitumen samples matched those reported in the open literature.

Presence of small amounts of air may or may not increase the MCR content. This increase or decrease appears to depend upon the extent of oxidation that occurs in the sample, process that seems to depend on both its physical and chemical nature either.

Acknowledgements

The authors would like to acknowledge funding from Alberta Ingenuity Centre for In Situ Energy (AICISE), and the facilities provided by Schulich School of Engineering, University of Calgary, Canada, to carry out this work. Also the financial support provided by AERI is greatly acknowledged. Experimental support provided by Luis Alberto Pineda-Perez and Linda C. Roa-Fuentes is appreciated. Helpful comments and suggestions from our colleagues Clementina Sosa and Dr. Francisco Lopez-Linares are appreciated. The authors are grateful to Suncor Energy, Jacos, Shell Canada, Prof. B. Maini and Prof. C. Scott for providing studied samples.

References

1. Noel, F. (1984), *Fuel*, 63, 931.
2. Carbognani L, Lubkowitz J, Gonzalez MF, Pereira-Almao P. (2007), *Energy Fuels*, 21, 2831.
3. Rahimi P, Gentzis T, Dawson WH, Fairbridge C, Khulbe C, Chung K, Nowlan V, DelBianco A. (1998), *Energy Fuels*, 12,1020.
4. Strausz, O,P, Lown, E,M. (2003), The Chemistry of Alberta Oil Sands, Bitumens and Heavy Oils. AERI, Calgary, AB, Canada,.
5. Speight JG. (1991), The Chemistry and Technology of Petroleum. Marcel Dekker, 2nd ed. NY, USA.
6. Tye, CT, Smith K. (2004), *Catal. Letters*, 95, 20.
7. Delgado O, Bolivar C, Ovalles C, Scott CE. (2005), Hamaca Crude Oil Upgrading Using Formic Acid as Hydrogen Precursor under Steam-Injection Conditions. Chapter 10 in *"Heavy Hydrocarbon Resources. Characterization, Upgrading and Utilization.* Nomura, M., Rahimi, P. M., Koseoglu, O. R. (Eds), ACS Symposium series # 895.
8. Meyer RF, Atanasi E, Freeman PA. (2007), Heavy Oil and Natural Bitumen Resources in Geological Basins of the World. Open File Report 2007-1084.. US Department of Interior, US Geological Survey.
9. Japanwala MS. (2001), Quality of Distillates from Coking of Recycled Residue. MSc Thesis, University of Alberta, Edmonton, AB. Canada.
10. Bearden R, Hou Z, Gorbaty ML, Ferrugehelig DT, Myers D. (2001), Combination Slurry Hydroconversion Plus Solvent Deasphalting Process., EXXON Research and Engineering. Patent WO/2001/027225.

ADSORPTION STUDIES OF HEAVY MOLECULES ON SOLID SORBENTS

Francisco López-Linares[1], Lante Carbognani[1], Clementina Sosa-Stull[1], Azfar Hassan[1], Pedro Pereira-Almao[1] and Ronald J. Spencer[1,2]
Schulich School of Engineering, AICISE[1]; Department of Geosciences, AICISE[2], University of Calgary, Calgary, Alberta, T2N 1N4, Canada

Introduction

Adsorption is a process that play important role in different areas such as surface chemistry,[1] catalysis,[2,3] environmental remediation,[4] biopolymers[5] and separation sciences.[6] Particularly, inside the current research activities performed in AICISE, adsorption is a key step within a bitumen residua upgrading process recently proposed by our research group.[7-9] Development of adsorbents able to capture large molecules, such as the compounds present in oil vacuum residua, is of utmost importance. Further, the ability of this adsorbent to act as a steam gasification catalyst is another key component of our research toward the application of this process.[7-9]

Recently, our research group has paid attention to the carbonaceous materials that remain after a SAGD process. At this point it is not economical to produce this hydrocarbon in place. Typically remaining hydrocarbons after SAGD production vary from 3-6 % w/w.[10] Therefore, the knowledge about the chemical nature of this fraction is of primary importance for proper selection of feasible catalysts for further processing of this material. Adsorption properties of minerals within the reservoir are also very important in this upgrading scheme. How temperature affects the remaining hydrocarbons is believed to be another important parameter to understand, since the adsorptive properties of thermally treated hydrocarbons may change from the original properties of virgin organics in place within the reservoir.

This work present some aspects related to the adsorption kinetics and uptake of virgin and thermally cracked hydrocarbons over natural reservoir materials. To achieve these goals, kinetic studies were carried out with toluene solutions of virgin and thermally cracked Athabasca vacuum residue asphaltenes. Visbroken (mild thermally cracked) asphaltenes were selected as proxies for the expected materials remaining within partially depleted reservoirs exposed to thermal treatments.

Experimental

Materials. Toluene (Spectrophotometric Grade), methylene chloride and n-heptane, (HPLC grade) from Sigma-Aldrich were used as received. A vacuum residue from Athabasca bitumen provided by Suncor was employed in this work. This feedstock contains 26 wt % distillable fractions (545°C-), as determined by gas chromatography simulated distillation (SimDist). An amount of about 0.7 wt % of insoluble materials in methylene chloride was determined to be present within the sample. Visbroken residues (obtained through a mild thermal cracking process) at different conversion levels were prepared according to a previous report.[11] The corresponding visbroken asphaltene (VB-ASP) was obtained by n-heptane precipitation[11] and then re-dissolved in toluene without further washing in the precipitant alkane.[12] Kaolin from Merk-KGA Germany was used as starting material for the preparation of the macroporous kaolin adsorbent. Details of the preparation and characterization of this material could be found elsewhere.[13] Smectite group clay minerals; a Ca-Montmorillonite (from Apache County, Arizona, USA) and a Na-Montmorillonite (from Crook County, Wyoming, USA) were used as received. A representative sample from an Athabasca siltstone (well 88-14-27, 11W4 at a depth of 282.2 m) was also utilized. The siltstone contains about 75% quartz, with the remainder being clay minerals. Kaolinite is the dominant clay mineral present (about 2/3 of the clay fraction), with lesser amounts of illite and smectite.

Adsorption kinetics experiments. Kinetic experiments were carried out by continuously measuring the change of absorbance for asphaltene toluene solutions in contact with adsorbents via UV-Vis spectrophotometry. Details on the methodology have been published previously. [13]

Results and Discussion

Adsorption of asphaltene fractions on different solid surfaces has been reported by different authors.[15-22] However, reports concerning adsorption of thermally cracked asphaltene derivatives from processes such as visbreaking have not been found in the literature. In this sense and considering visbroken asphaltenes as proxies for remaining thermal treated hydrocarbons within spent reservoirs, we studied the adsorption behaviour of such fractions over different solids, including macroporous kaolin, smectite group clays (Ca- and Na-Montmorillonite) and an Athabasca siltstone. The results obtained are presented in the following sections.

Adsorption kinetics of virgin (VR) and visbroken asphaltene (VB-ASP) on macroporous kaolin adsorbent. Initially, macroporous kaolin was selected as adsorbent because we have been studying the adsorption of single molecules and complex hydrocarbon fractions over this solid. Adsorption kinetics was followed for 300 minutes using a toluene solution of virgin (VR) as well as solutions with thermally cracked asphaltenes at different levels of conversion (VB-ASP). Kinetic plots are displayed in Figure 1, where the numbers attached to VB correspond to visbreaking conversions of the Athabasca residue (%w/w).

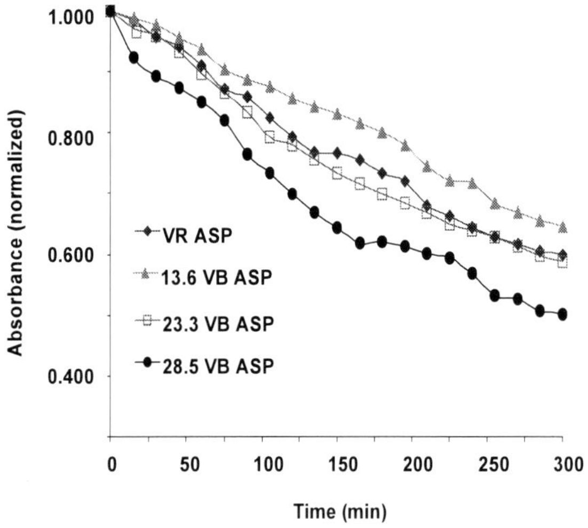

Figure 1. Kinetics of the adsorption of virgin and thermal cracked Athabasca asphaltenes on macroporous kaolin adsorbent @ 295 K. Conditions: $[Adsorbate]_0$=60-80 ppm, adsorbent=0.3 g, toluene solution= 3mL, time 6 h.

As can be seen in Figure 1, the macroporous kaolin adsorbent is able to adsorb the virgin product (VR) and the corresponding thermal cracked asphaltenes obtained from visbreaking at different conversion levels. From this figure, it is observed that different degrees of adsorption are achieved with the degrees of thermal cracking. The kinetic plots from these experiments indicate that at 300 min with the exception of 13.6 VB sample, there is a correlation between adsorption and conversion; higher conversion products are adsorbed to a larger extent over the solid. Kinetic analyses were performed based on the curves displayed on Figure 1, restricted to the linear portion, within the first 120 minutes. The values were fitted to a first order kinetic equation in order to determine kinetic coefficients.[13] The values obtained are presented in Table 1.

Table 1. Kinetic coefficients and regression factors for an initial apparent first order rate fit of the adsorption of virgin (VR) and visbroken Athabasca asphaltene (VB-ASP) on macroporous kaolin

Fraction	$K \times 10^{3}$ $(min)^{-1a}$	R^{2b}
VR	1.4 ± 0.3	0.982
13.3 VB	1.6 ± 0.3	0.952
23.3 VB	2.2 ± 0.4	0.985
28.5 VB	2.8 ± 0.4	0.980

[a] At 295 K. [b] Correlation coefficient for first order fitting

As can be seen from Table 1, at the initial stages of the adsorption, the studied asphaltenes follows a trend such that higher uptakes are obtained from the more converted materials. This behavior might be characterized by fractions with reduced size. Reports on thermal conversion processes propose that most large molecules present in VR are cracked into smaller molecules, these changes being dependent on the severity of the process.[23-27] However, under controlled conditions (before onset of bulk solids precipitation) small amounts of large molecular mass fractions are generated simultaneously, mostly related to coke formation.[24] From the preceding premises, higher severity thermal cracking is expected to afford large proportions of low molecular mass materials. A quantitative estimation of the size of those fractions seems to be imperative. In this sense a methodology regarding to the estimation of a MW of such fractions is currently under development in our lab[14]. Preliminary results confirm the above mentioned; higher severity thermal cracking promote the formation of large proportions of low molecular mass materials. The results presented on Figure 1 indicate that the synthetic kaolin adsorbent does not reach its adsorption capacity during the experiments as absorbance decreases throughout the experiments. The large macropores (>50 nm) of the synthetic kaolin make it suitable for the adsorption of large MM vacuum residual molecules. However, there appears to be a reduction in the apparent rate of adsorption with time.

The total uptake of adsorbate for each fraction at 300 minutes was determined. Results are presented on Figure 2.

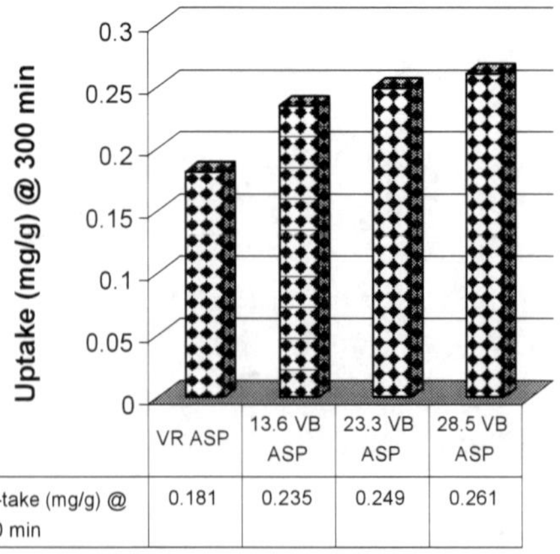

Figure 2. Uptake of virgin and thermal cracked Athabasca asphaltene solution on kaolin adsorbent at 300 min. Conditions: $[Adsorbate]_0$=60-80 ppm, adsorbent=0.3 g, toluene solution= 3ml, T=295 K.

From therein it is clearly showed that higher uptake is achieved with the asphaltene obtained from the more converted products, the 23.3 and 28.5 wt % VB.

In Summary, the results from this section show that mild thermal cracking of the studied residue improved the adsorptive properties of the molecules present within the visbroken residua. The wide available pores on the synthetic adsorbent allow for easy adsorption of such fractions after the thermal cracking process.

Adsorption kinetics of vacuum residual virgin (VR ASP) and visbroken Athabasca asphaltenes (VB-ASP) over reservoir minerals. The adsorption behaviour of the same asphaltenes described above over macroporous kaolin was investigated with another set of natural adsorbents with varying textural properties. In particular, Athabasca siltstone is deemed important to evaluate. The siltstones are present in many hydrocarbon-bearing horizons associated with abundant resources as part of the complex, heterogeneous reservoirs of north eastern Alberta. The important textural properties of the adsorbents studied are displayed in Table 2.

Table 2. Textural properties of solids studied

Sample	S.A.[a] (m^2/g)	Vμ[b] (cm^3/g)	Vs[c] (cm^3/g)
Macroporous kaolin	12	-	0.48
Na-Montmorillonite	21	0.001	0.07
Ca-Montmorillonite	108	0.005	0.22
Athabasca Siltstone	9	0.003	0.03

[a] S.A.=B.E.T. specific surface area
[b] Vμ= micropore volume calculated by *t*-plot
[c] Vs= specific pore volume at P/P0=0.98

As shown in Table 2, a broad range of surface areas are displayed by the solids studied. The micropore volume, determined by *t*-plot, confirms that these solids are mainly microporous. It is worth to mention that that macroporous kaolin was characterized via mercury porosimetry due to its high pore size. For that reason, the microporous volume was not determined by this method. However, the Ca-Montmorillonite and macroporous kaolin possess a higher specific pore volume that may impact the uptake of residual hydrocarbon fractions. By using the experimental methodology employed in the previous section, the kinetics of adsorption of two different adsorbates was carried out. The asphaltene from vacuum residue (VR ASP) and the visbroken product at the highest severity (28.5 VB ASP) were selected. The results obtained during this evaluation are described below.

Adsorption kinetics and uptake of virgin Athabasca Asphaltene over reservoir minerals. The kinetics of adsorption of VR asphaltene over the selected adsorbents was carried out as above, using a toluene solution of the adsorbate and monitoring the absorbance change for a period of 300 minutes. Figure 3 presents the experimental data for VR ASP adsorption over these solids as a function of time. Data for the macroporous kaolin is included as a reference.

Initially it is observed that adsorption of VR asphaltene over macroporous kaolin is nearly constant with time. Nevertheless, as indicated on Figure 3, the other three solids interact with this asphaltene fraction in a very different manner. Initial adsorption rates of the Athabasca siltstone and smectite group clays are all higher than for the macroporous kaolin. However, these initial rates decrease quickly. Adsorption over Athabasca siltstone, while initially quite high, decreases most rapidly. Over the 300 minutes of the experiment it is the lowest. Practically all the asphaltene adsorbed was captured within the first 30 minutes, with a plateau along the remainder of the experiment.

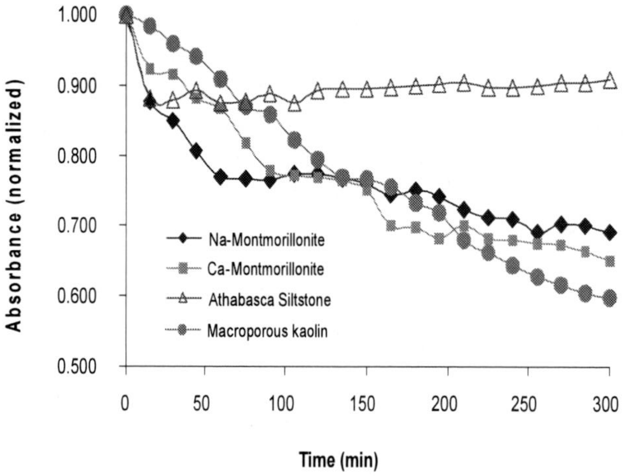

Figure 3. Kinetics of the adsorption of virgin Athabasca (VR) asphaltene over solid adsorbents @ 295 K. Conditions: $[\text{Adsorbate}]_0$=60-80 ppm, adsorbent=0.3 g, toluene solution= 3mL, time 6 h.

The smectite group clay samples display two different adsorption zones. The Na-Montmorillonite adsorbs at a high rate during the first 60 minutes. After this the uptake remains nearly constant up to 200 minutes. This is followed by further, but lesser, adsorption to the end of the experiment. The Ca-Montmorillonite adsorbs at a high rate during the first 90 minutes. This is followed by a short plateau from 90 to 120 minutes. Adsorption resumes at a low rate until the end of the experiment at 300 minutes.

The behaviour displayed by these solids suggests that different adsorption mechanisms occur under the experimental conditions. Based on the curves displayed in Figure 3, kinetic analyses were performed, restricted to the linear part of each curve, that it was observed during the first 60 minutes. Fitting those values to a first order kinetic equations, allows the determination of kinetic coefficients.[13] The results obtained are presented in Table 3.

Table 3. Kinetic coefficients and regression factors for an initial apparent first order rate fit of the adsorption of vacuum residue Athabasca asphaltene (VR-ASP) over different solids

Solid	K x10 3min-1a	R2b
Na-Montmorillonite	4.3 ± 0.3	0.868
Ca-Montmorillonite	2.3 ± 0.4	0.896
Athabasca Siltstone	0.8 ± 0.1	0.904
Macroporous kaolin	1.6 ± 0.3	0.986

As deduced from the results in Table 3, the highest kinetic coefficient was obtained for the Montmorillonite samples, indicating that VR-ASP interacts rapidly with these adsorbents. For macroporous kaolin, it is observed that VR asphaltene interacts to a lesser extent at the beginning but, the adsorption remains constant for a long the time. The Athabasca siltstone displayed the lowest adsorptivity indicating either a poor affinity of the asphaltene on this solid surface or, because this sample is largely quartz silt which has very low adsorption capacity, the available sites for adsorption on the clay fraction may have been saturated.

Adsorption kinetics and uptake of thermally cracked Athabasca Asphaltene over reservoir minerals. As part of our research program, proxy molecules representative of products of the thermal cracking process are being researched. Recently, known biomarker molecules were studied in this regards.[43] Herein, thermally cracked asphaltenes (visbreaking process) were studied as proxies for cracked fractions produced under thermal enhanced oil production processes. The asphaltene obtained under the highest achievable severity (28.5 VB ASP) was selected for this part of the study in order to detect differences more easily. By using the same experimental methodology described in the preceding section, adsorption experiments were conducted and the kinetic profiles are presented on Figure 4.

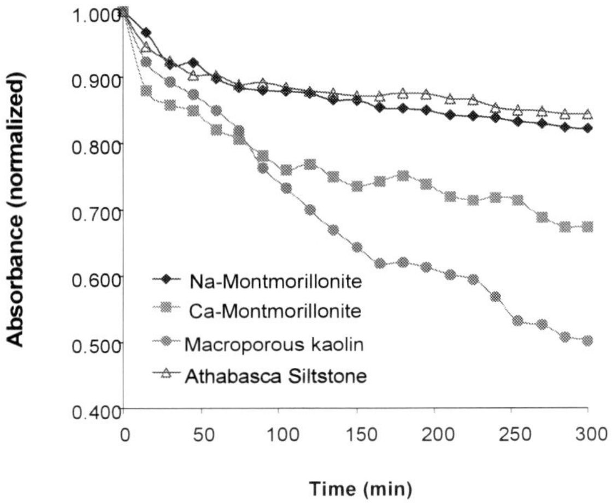

Figure 4. Kinetics of the adsorption of visbroken Athabasca Asphaltene (28.5 VB) over solid adsorbents @ 295 K. Conditions: $[Adsorbate]_0$=60-80 ppm, adsorbent=0.3 g, toluene solution= 3mL, time 6 h.

Thermal cracking of asphaltene at a high conversion level leads to different interactions with the solids evaluated; this is evident from a comparison of Figures 3 and 4. These results indicate induced structural changes in the original asphaltene fraction (VR-ASP). The thermally cracked material adsorbs more continuously over all of the solids as a function of time, indicating that one dominant adsorption mechanism is operating in this case. Plateaus in the adsorption behaviour of the smectite-group clays are much less pronounced, if present at all. It is observed that lower adsorption is obtained for Na-Montmorillonite, comparable in this case to the Athabasca siltstone. Ca-

Montmorillonite again showed a higher adsorption, and, once again, the macroporous kaolin included for comparison displayed the highest adsorption as in the case of the virgin asphaltene (VR ASP).

Based on the curves displayed in Figure 7, kinetic analyses were restricted to the linear part of each curve which is within the first 60 minutes. Fitting those values to a first order kinetics allows the determination of kinetics coefficient constants. The results are displayed in Table 3. As shown in Table 3, the largest kinetic coefficient was obtained for the Ca-Montmorillonite sample, indicating that thermal cracked asphaltene has preference for this solid. However, it is interesting to point out that this type of asphaltene showed the same extent of adsorption as the other studied solids, indicating that the selectivity to the surfaces is diminished after the thermal cracking process.

Table 3. Kinetic coefficients and regression factors for an initial apparent first order rate fit of the adsorption of 28.5 VB Athabasca Asphaltene over different solids

Solid	K x 10 ^{3}min^{-1a}	R^{2b}
Na-Montmorillonite	1.8 ± 0.3	0.914
Ca-Montmorillonite	2.5 ± 0.4	0.898
Athabasca Siltstone	1.7 ± 0.5	0.874
Macroporous kaolin	2.8 ± 0.5	0.986

[a] At 295 K. [b] Correlation coefficient for first order fitting

The study of others minerals presents in a Athabasca type reservoir as well as designed synthetic materials is currently investigated where details regarding to the effect of the textural properties of such solids on the adsorption are covered and the results will be publish elsewhere.

Conclusions

The present study shows that Athabasca vacuum residue virgin asphaltene and derived thermally cracked products interact to different extents with solid surfaces. Evidence suggests that the molecular size of such fractions does not limit their adsorption over synthetic macroporous adsorbents. The interaction of VR Asphaltene over the evaluated solids indicates that different adsorption mechanism is operating. However after visbreaking at the highest conversion, the thermally cracked material adsorbs more continuously over all of the solids as a function of time, indicating that one dominant adsorption mechanism is operating in this case. Finally, first insights into possible organo-mineral interactions within spent reservoirs were achieved in this work relying on mild thermally cracked (visbroken) asphaltenes as proxy molecules for phenomena occurring within the reservoir.

Acknowledgements

This work is being supported by the Alberta Ingenuity Fund through the Ingenuity Scholar support provided to Professor Pedro R. Pereira-Almao. The authors would like to acknowledge funding from Alberta Ingenuity Centre for In Situ Energy (AICISE), and the facilities provided by Schulich School of Engineering, University of Calgary, Canada, to carry out this work. Finally, we would like to thank Lina Diaz from Catalysts for Bitumen Upgrading and Hydrogen Production group (CBUHP) from U of C, for providing the SEC analyses.

References

1. McCash, E. M. *Surface Chemistry*. ACS, 2001.
2. Bowker, M. *The Basics and Applications of Heterogeneous Catalysis*. ACS, 1998.
3. Ocelli, M. L. *Fluid Catalytic Cracking III*. ACS, 1998.
4. Harrison, R.G.; Todd, P.W.; Rutge, S.R.; Petrides, D. *Bioseparations in Science and Engineering*. ACS, 2002.
5. Vandegrift, G. *Environmental Remediation*. ACS, 1998.
6. Giddings, J. C. *Dynamics of Chromatography: Principles and Theory*. Chromatographic Science Series, Vol. 1. Giddings J.C., Keller, R.F (eds). 1965.
7. Gonzalez, M. F.; Carbognani, L.; Pereira-Almao, P. Selective Adsorption of Thermal Cracked Heavy Molecules. 19th. Canadian Symposium on Catalysis, Saskatoon, SA. Canada May 14-17, 2006.
8. Sosa, C.; Gonzalez, M. F.; Carbognani, L.; Perez-Zurita, M. J.; Lopez-Linares, F.; Moore, R.G., Husein, M.; Pereira, P. Visbreaking Based Integrated Process for Bitumen Upgrading and Hydrogen production. *Paper 2006-074. CIPC 2006* (Canada International Petroleum Conference), Calgary, July 2006.
9. Carbognani, L.; González, M.F.; Lopez-Linares, F.; Sosa-Stull, C.; Pereira Almao, P. (in press, *Energy & Fuels*)
10. Maini, B. Personal communication. Univ. of Calgary, Dec. 2007.
11. Carbognani, L.; Gonzalez, M.F.; Pereira-Almao, P. *Energy & Fuels* 2007**,** 21, 1631
12. Alboudwarej, H.; Beck, J.; Svrcek, W. Y.; Yarranton, H. W.; Akbarzedeh, K. *Energy & Fuels* 2002, *16*, 462-469
13. González, M. F.; Sosa Stull, C.; López-Linares, F.; Pereira-Almao, P. *Energy & Fuels*. 2007, 21, 234
14. The asphaltene fractions were analyzed by Size Exclusion Chromatography. Details of the methodology developed will be published elsewhere.
15. Dean, K.R; McAtee, J.M. Jr. *Applied Clay Science*, 1986, 313.
16. Dubey, S. T.; Waxman, M. H. *SPE Reservoir Eng.* 1991, 389
17. Piro, G.; Canonico, L. B.; Galbariggi, G.; Bertero, L.; Carniani, C. *SPE Prod. Facil.* 1996, *August*, 156
18. Pernyeszi, T.; Patzko, A.; Berkesi, O.; Dekany, I. *Colloids Surf., A* 1998, *137*, 373.
19. González, G.; Moreira, M.B.C.: "Adsorption of Asphaltene and Resins on Various Minerals," *Asphaltenes and Asphalts 1*, T.F. Yen (ed.) Elsevier 1994, 219.
20 Acevedo, S.; Castillo, J.; Fernandez, A.; Goncalves, S.; Ranaudo, M. A. *Energy & Fuels* 1998, *12*, 386-390.
21. Acevedo, S.; Ranaudo, M. A.; Escobar, G.; Gutierrez, L.; Ortega, P. *Fuel,* 1995, *74*, 595. 30.
22. Acevedo, S.; Ranaudo, M. A.; Garcia, C.; Castillo, J.; Fernandez, A.; Caetano, M.; Goncalves, S. *Colloids Surf., A.* 2000, *166*, 145
23. Wiehe. I.A. *Ind. Chem. Res.* 1993, 32, 2447.
24. Wiehe, I. A. *Energy & Fuels,* 1994, 8, 536-544.
25. Thomas, M.; Fixari, B.; LePerchec, P.; Princic, Y.; Lena, L. *Fuel,* 1989, 68, 318-322.
26. Zhang, L.; Yang, G.; Que, G.; Zhang, Q.; Yang. P. *Energy & Fuels,* 2006, 20, 208-2012.
27. Dettman, H.; Inman, A.; Salmon, S.; Scott, K.; Fuhr, B. *Energy & Fuels* 2005, 19, 1309-1404

.

Liquid-phase Catalytic Decarboxylation of Naphthenic Acids

Ling Yang[1], Alan R. Nelson[1,2], and Murray R. Gray[1]
[1]Department of Chemical & Materials Engineering, University of Alberta
Edmonton, Alberta, T6G 2G6, Canada
[2]Current address: The Dow Chemical Company, Corporate Research and Development
Midland, MI, 48674, USA

Abstract

Naphthenic acids in crude oil cause serious corrosion in oil production and refining processes, especially at high temperature (230-400°C). Hydrotreating is effective in removing these components, but a non-hydrogenative method for treating gas oils or whole crude oil would be much cheaper. In this study, catalytic decarboxylation of naphthenic acids in liquid-phase was studied experimentally in the presence of alkali metal and alkaline earth metal oxides and carbonates. A commercial mixture of naphthenic acids in a carrier oil was used as a model system. The reagent powders of Li_2O, CaO, $CaCO_3$, BaO and MgO were tested at 385°C in a batch reactor, and their effectiveness in the decarboxylation reaction were determined based on acid reduction. TAN measurement and FTIR analysis showed the trend in decrease of acids using different catalyst (Li_2O>CaO>BaO>MgO>$CaCO_3$). The naphthenate was detected in the post-reacted catalysts by FTIR and TGA. The removal rate of TAN and the production rate of CO_2 were independent of the concentration of NA, and the reaction temperature had a great influence on the production rate of CO_2 but a minor effect on the removal rate of NA.

Key words: Catalytic decarboxylation, naphthenic acid;

1. Introduction

Corrosion problems in petroleum refining operations due to naphthenic acids (NAs) in crude oils have been recognized since early 20th century[1]. The presence of relatively high levels of petroleum acids in crude oils tends to cause serious equipment corrosion, and leads to high maintenance costs. Therefore, reduction in naphthenic acid content would alleviate corrosion-associated problems.

NAs are a complex mixture of acids, and the structure is primarily represented by alky-substituted cycloaliphatic carboxylic acids with the general formula $R(CH_2)_nCOOH$, where R is a cyclopentane/cyclohexane ring. The cycloaliphatic carboxylic acids may also include single ring and fused multiple ring structure. The carboxyl group is usually bonded or attached to a side chain rather than directly to the cycloaliphatic ring. Because of the inherent structural complexity of NAs, the identification of the exact structure of NAs is impossible. Therefore, the corrosion of oil is described by total acid number (TAN). The TAN is defined as the amount of potassium hydroxide required to neutralize one gram of oil sample.

Several methods have been proposed to remove NAs from crude oils, such as thermal decomposition[2], adsorbents[3] and chemical conversion with various catalysts[4, 5]. One of the attractive methods to reduce NAs content is to use heterogeneous catalysts to decompose the acids into neutral compounds. With catalysts active for decarboxylation reactions, NAs can be catalytically converted into neutral ketones and carbon dioxide at moderate temperatures.

The basic reaction for the catalytic decomposition of naphthenic acids is illustrated by Equation (1).

$$2R(CH_2)_nCOOH \leftrightarrow (R(CH_2)_n)_2CO + CO_2 + H_2O \qquad (1)$$

The gas-phase catalytic decarboxylation of model compound of NAs has been studied by Guarino[6]. Guarino found that formation of butyrate is an intermediate step in the carboxylation reaction to produce ketone. The objective of this work is to extend this catalytic approach to the liquid-phase, and to study the kinetics of decarboxylation of NA in liquid phase. In our study, a commercial mixture of naphthenic acids in a carrier oil was used as a model system. The reagent powders of Li_2O, CaO, BaO and MgO, $CaCO_3$ were tested at 385°C in a batch reactor, and their activities in the decarboxylation reaction were determined based on the acid reduction.

2. Experimental

2.1 Catalysts

It has been reported that the alkali and alkaline earth metal oxides and carbonates exhibit high activity in the vapour-phase decarboxylation of NAs[6]. Moreover, Zhang, et al[7] found that MgO has a high reactivity in the decarboxylation reaction of carboxylic acid. Therefore, the commercial products of CaO (A.C.S. reagent, 99%, Sigma), $CaCO_3$ (reagent grade, 98%, Sigma), Li_2O (97%, Sigma), MgO (99+%, Sigma) and BaO(99.99%, Sigma) were applied as catalysts in this work.

2.2 Reactants

A commercial mixture of naphthenic acids, NA-2, in a carrier oil was used as a model system. NA-2 is a complex mixture of alkyl-substituted acyclic and cycloaliphatic carboxylic acids. Paraflex HT oil, a pure saturated hydrocarbon mixture, was used as the solvent due to its high thermal stability.

Paraflex HT oil and NA-2 were purchased from Petro-Canada and OMG, respectively. The density of paraflex HT oil is within the range of 0.85 and 0.8703 kg/L at 15 °C, and the density of NA-2 is in the range of 0.923 to 0.983 kg/L(ASTM D 1963), TAN is between 200 and 215 mg KOH/g sample (ASTM D 1980).

2.3 Reactions

Batch microreactors with a capacity of 30 ml were used for the study of the catalytic decarboxylation reaction. The mixture of paraflex HT oil and NA with specific TAN was charged to a microreactor, and nitrogen as an inert gas was applied to purge air out. Then the reactor was placed in a fluidized sand bath whose temperature was controlled to remain at a reaction temperature. The reactor was agitated within the sand bath so that a good contact between the reactants and the catalyst was achieved (Figure 1).

2.3.1 Analysis methods

After a given reaction time, the reactor was taken out of the sand bath and rapidly quenched in a cool water bath. After about 15 minutes, the reaction residues were separated by the centrifuge at 3000 rpm for 30 minutes. The liquid phase was analyzed by a Fourier transform infrared spectroscopy (FTIR, Nicole 8700 FT with OMNIC software by Thermo Electron Corporation) and TAN measurement (ASTM 664). The solid phase was washed with toluene, and filtered with the membrane filter, then kept in the fume hood for dryness, finally analyzed by FTIR and a thermal gravimetric analysis (TGA, STA409 PC by NETZSCH Thermal Analysis).

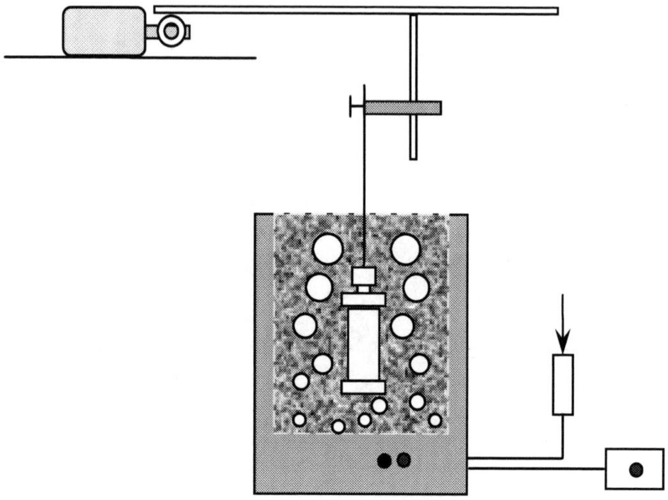

Figure 1. The batch reactor and fluidized sand bath

3. Results and discussion

Guarino[6] studied the gas-phase decarboxylation of model compound of NA at 385°C. In this work, the same temperature was also used for catalyst screening, and the initial TAN of the reactant mixture was as low as 3.

3.1 Analysis of liquid samples

The liquid samples were analyzed by FTIR analysis and TAN measurement. The results are showed in Figure 2.

As shown in Figure 2, the NA content in the mixture decreased by 17% without addition of the catalyst, which was similar to that in presence of $CaCO_3$. However, with addition of the tested metal oxides, NA in the mixture decreased significantly, about 70% of acid was removed in presence of Li_2O. TAN measurement and FTIR analysis showed the same trend in decrease of acids with addition of different catalysts. The effectiveness of the tested catalysts decreased in the order: Li_2O>CaO>BaO>MgO>$CaCO_3$. Li_2O showed the most active for removal of NA, and $CaCO_3$ was the least effective, which is different from the observations of

Guarino[6] that the catalyst of $CaCO_3$ was highly active in the gas-phase decarboxylation reaction under the experimental conditions. Considering the prices of the reagents and their activities shown in Figure 2, CaO was chosen as catalyst for the further study.

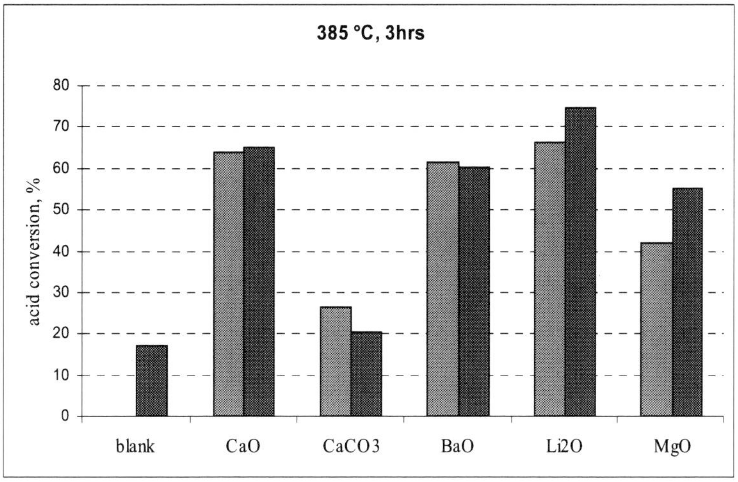

Figure 2. TAN measurement and FTIR analysis of liquid samples

3.2 Analysis of solid samples

3.2.1. FTIR analysis

The catalysts of CaO before and after the reaction were analyzed using FTIR to detect the existence of naphthenate in solid samples (Figure 3). The typical IR adsorption of C=O in naphthenate has a very strong, characteristic band in the region 1610 and 1550 cm^{-1} due to the asymmetric stretching vibration of CO_2^-[8], which is different from C=O in naphthenic acids (around 1700 cm^{-1}).

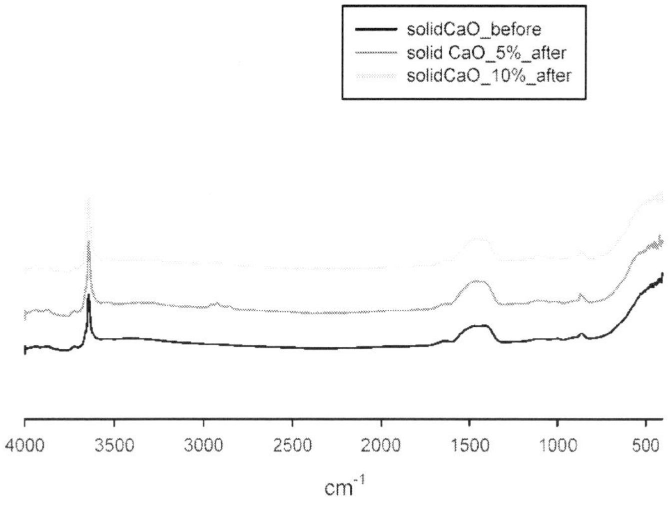

Figure 3. IR Spectra of solid CaO before and after reactions

The data in Figure 3 shows that the IR spectra of CaO before and after the reaction were similar, and the separated peak corresponding to C=O in calcium naphthenate was not found in the spectra. There were two possibilities for this behavior. One was that no calcium naphthenate was indeed formed in the reaction, and the other was that the produced calcium naphthenate couldn't be detected because of its low concentration.

Therefore, the other reaction was run in which the initial TAN of the reactant mixture was increased from 3 to 45, and the molar ratio of CaO to NAs was as low as 0.76. The post-reacted CaO was analyzed by FTIR (Figure 4). The commercial calcium naphthenate solution was also analyzed for comparison.

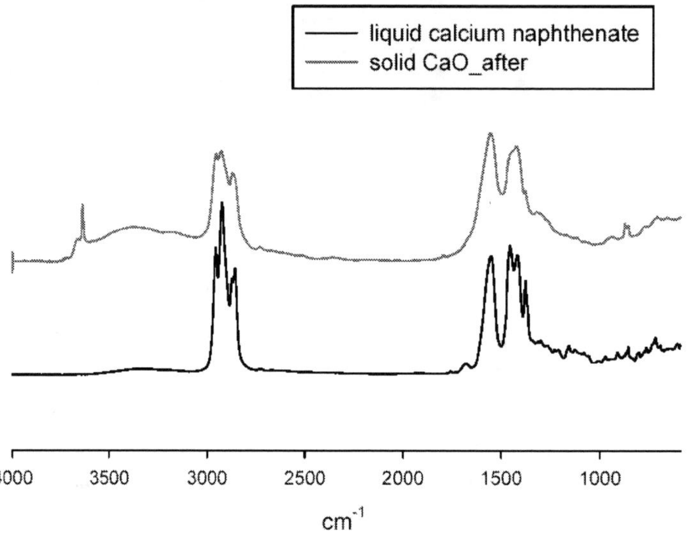

Figure 4. IR Spectra of the post-reacted CaO and liquid calcium naphthenate

It can be seen in Figure 4 that both IR spectra show the peaks at around 1560 cm^{-1}. As described previously, CO_2^- in calcium naphthenate exhibits a sharp peak in the range 1610-1550 cm^{-1}. Therefore, the peak at around 1560 cm^{-1} was attributed to IR adsorption of CO_2^- in calcium naphthenate. The conclusion was drawn that the post-reacted CaO contained calcium naphthenate, in other words, calcium naphthenate was formed in the reaction. At the same time, the second possibility was verified that the produced calcium naphthenate could not be detected just because of its low concentration.

3.2.2. TGA analysis

Figure 5 shows the TGA results of the post-reacted CaO, and the commercial calcium naphthenate solution was also analyzed for comparison. It was found that the mass loss between 400 and 500 °C in the curve c was due to the decomposition of calcium naphthenate. Comparing curves (a) and (b), the mass loss in this temperature range was also observed in the curve (b). The reason for no calcium naphthenate found in the curve (a) was that the initial TAN of the reactant mixture was too low to be detected by TGA.

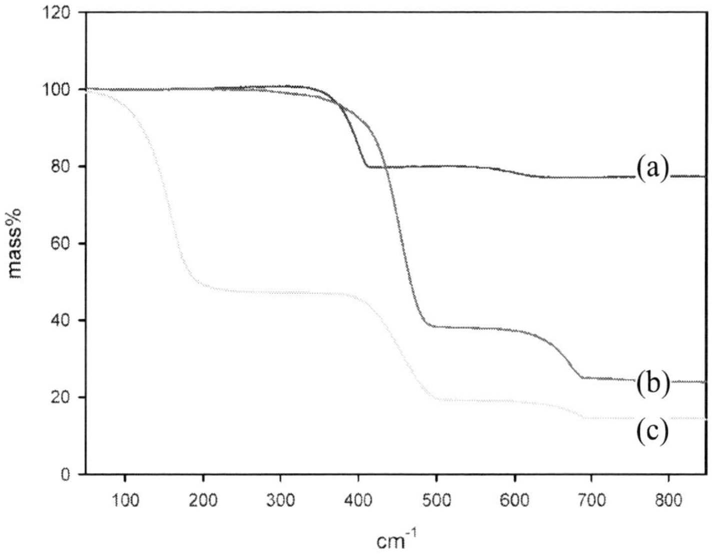

Figure 5. TGA results of post-reacted CaO and liquid calcium naphthenate, (a) TAN=3; (b) TAN=45; (c) liquid calcium naphthenate

3.3 Kinetics

The kinetics of decarboxylation was also studied experimentally using a microbatch reactor. Three mixtures with different TAN of 5.06, 8.48, and 10.82 were prepared for the study. The experiments were carried out at 365, 375, and 385°C, respectively. For each run, 20 wt% of CaO was loaded. The tightened reactor was pressurized with N_2 at 400 kPa. Each series of reactions with the same initial TAN lasted for 0, 10, 25, 40, 55, and 85 minutes, respectively. A series of blank reactions with the initial TAN of 8.48 was performed for comparison.

After the reaction, the gas sample was collected with a syringe directly and analyzed by GC(GC, Varian 3400X with the star chromatography workstation). After that, the supernatant liquid was taken out of the reactor and titrated to measure TAN immediately. Figure 6 shows the TAN decrease and CO_2 production at 365°C against time for the reaction with the initial TAN of 8.48.

The data shown in Figure 6 (curves 1 and 2) demonstrate that the TAN of the mixture decreased against time with and without addition of CaO, respectively, but NA removal was much more effective in the presence of CaO. Curves 3 and 4 represent CO_2 production against time with and without addition of CaO, respectively. The more NA was removed, the more CO_2 was produced.

Figure 7 shows TAN reduction and CO_2 production at 365°C against time for three mixtures with different initial TAN in presence of 20 wt% of CaO, respectively. It can be seen that the TAN decreased in the same trend with time in all three series of reactions with different initial TAN, which means that the concentration of NA had no influence on the removal rate of NA, therefore, the removal rate of NA was zero order in the concentration of NA. In addition, no conclusion can be arrived that CO_2 production was dependent on the NA concentration because all three curves for the released CO_2 show the same trend against time with the different initial TAN.

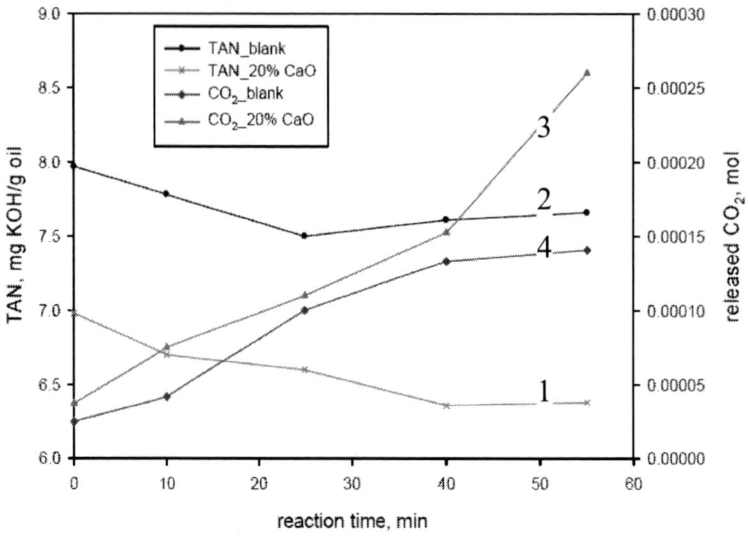

Figure 6. Reduction of TAN and production of CO_2 versus reaction time at 365°C

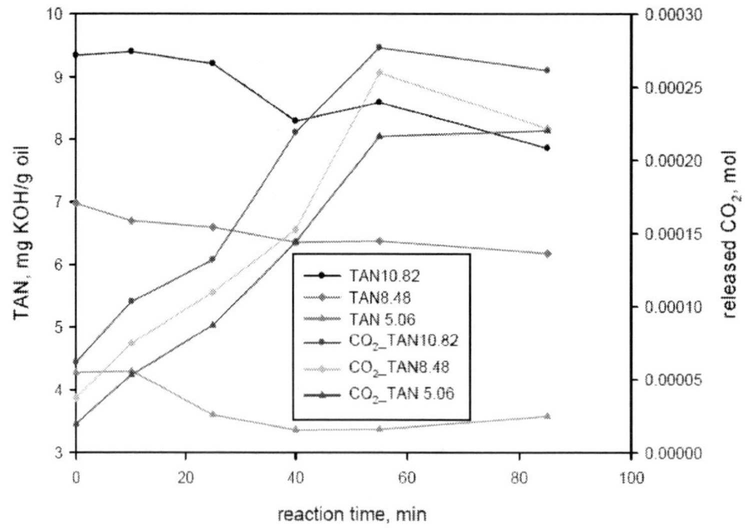

Figure 7. Reduction of TAN and production of CO_2 versus reaction time at 365°C

In order to investigate the influence of temperature on conversion, three series of reactions with the initial TAN of 8.48 were taken place at three different temperatures (365, 375, and 385°C), respectively. Figure 8 shows TAN decrease and CO_2 production at different temperatures.

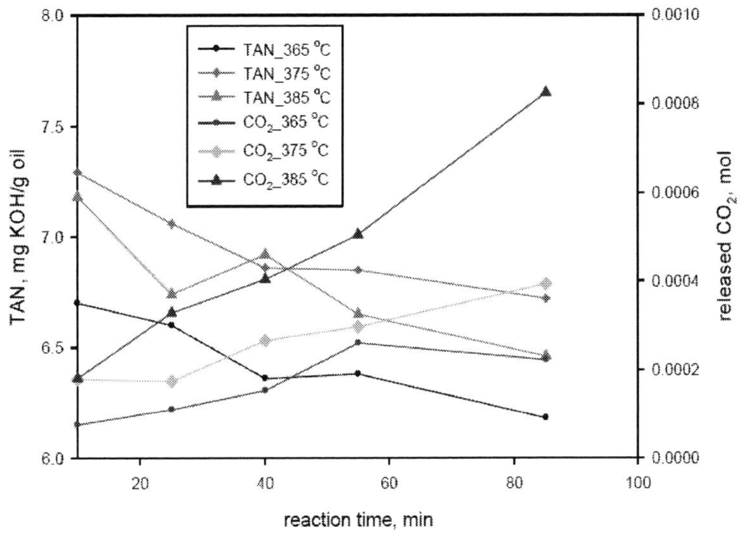

Figure 8. Reduction of TAN and production of CO_2 versus reaction time

The data of Figure 8 show that the higher temperature of reaction, the less NA was removed, but the more CO_2 was released, The reasons for such reverse behaviors were likely that the initial reaction of NA on the surface of catalyst was retarded when the temperature increased because of an exothermic adsorption, but higher temperature was favorable for the decomposition of naphthenate to produce CO_2, therefore, more CO_2 was released when the temperature increased from 365 to 385°C. It was also observed that temperature had a minor effect on the removal rate of NA, but had a great influence on the production rate of CO_2.

Based on the experimental results, a hypothetical mechanism of decarboxylation was proposed: NA firstly was reacted on the surface to form naphthenate; Naphthenate, as an intermediate of the reaction, was then decomposed to produce CO_2. Finally CO_2 was released in the gas phase. Among the above three steps, the decomposition of the naphthenate on the surface was much slower than the others, which is the rate-controlling step.

4. Conclusion

Alkali and alkaline earth metal oxides exhibit highly active in the liquid-phase decarboxylation reaction of naphthenate acid in oil, while the reactivity of carbonates is much lower than the oxides. The reactivity of the tested catalysts decreases in the order, Li_2O>CaO>BaO>MgO>$CaCO_3$. Through analysis of gas samples after the reaction, CO_2 was produced in the reaction, which indicates that the decarboxylation reaction takes place. The further analysis of solid samples before and after the reaction shows that the naphthenate was formed on the surface of the catalyst. The kinetic study shows that the removal rate of NA and CO_2 production rate were independent of the NA concentration, and the temperature had a strong influence on the decomposition of naphthenate on the surface of the catalyst, but only a minor effect on the removal rate of NA.

5. References

[1]W.A. Derungs, Corrosion-national association of corrosion engineers 12 (1956) 41-46.

[2]S.C. Blum, W.N. Olmstead, R. Bearden, US Patent No. 5820750 (1998).
[3]R. Varadaraj, US Patent No. 6454936 (2002).
[4]F.R. Moser, US Patent No. 2186425 (1940).
[5]R. Bearden, S.C. Blum, W.N. Olmstead, W. Robbins, US Patent No. 5891325 (1999).
[6]C. Guarino, Master Thesis at University of Alberta (2006).
[7]A. Zhang, Q. Ma, K. Wang, X. Liu, P. Shuler, Y. Tang, Applied Catalysis A: General 303 (2006) 103-109.
[8]G. Socrates, John Wiley & Sons, Ltd. (1980) P65.

137f Kinetic Study of Extra Heavy Crude Oil Upgrading Using Hydrogen Donors Under Cyclic Steam Injection Conditions

Cesar Ovalles, Jorge Martinis, Edgar Cotte, and Jose Luis Benitez. Centro de Catálisis, Petróleo y Petroquímica, Universidad Central de Venezuela, Caracas, Venezuela

Extra-Heavy Crude Oil Upgrading using a tetralin as hydrogen donor under steam injection conditions (methane and the mineral formation) led to an increase of at least three degrees in API gravity, three-fold viscosity reduction and, approximately, 8% decrease in the asphaltene content with respect to the original crude. Two different kinetic studies were carried out in order to determine the kinetic model for the upgrading process. Firstly, a reaction model involving four pseudo-components (light, medium, heavy and asphaltene fractions) was used and the Arrhenius pre-exponential factors and activation energies were determined. Using a continuous bench scale plant, the reaction between Hamaca pseudo-components under steam injection conditions was evaluated at different temperatures (280-315°C) and residence times (24-64 h). The results showed a good match between the calculated and experimental API gravities of the upgraded crude oil with an average relative error of 4%. Secondly, different experiments at various crude oil to tetralin weight ratios were carried out to determine the tetralin reaction order (n). The results indicate that n = 0.95 with a correlation factor of 0.9966 for several batch experiments at different residency times (12-48h). Finally, combining the four pseudo-component model and the tetralin reaction order (n_=_1), a general reaction scheme was obtained to describe the kinetic model for the upgrading process.

137b Water-Extractable Components from Heavy Crude

Scott Taylor, Analytical Services Group / R&D-Shared Services, ConocoPhillips, 232 Petroleum Laboratory, Bartlesville Technology Center, Bartlesville, OK 74004 and **Bruce B Randolph**, Heavy Oil Group / R&D-Shared Services, ConocoPhillips, 242 RF, Bartlesville Technology Center, Bartlesville, OK 74004.

A new laboratory method, using a microwave oven to extract water soluble chloride from crude oils, has demonstrated excellent correlation with the ASTM D6470 method for measuring chloride in crude oils. This effort was undertaken to offer improved and quicker methods for determining refining desalter effectiveness. In an expansion of this initial work, further examination on the nature of the aqueous phase has been conducted. The pH, elemental content, weak acids content, and the total acid number of the crude oil (pre-and post-extraction) have been determined on selected crudes that are geographically distinct and cover a range of API gravities. This paper will present the results on the expanded study of the aqueous phase after the microwave extraction.

Emulsion Characterization by Ultrahigh Resolution FT-ICR Mass Spectrometry

Brandie M. Ehrmann[1], *Priyanka Juyal*[2], *Ryan P. Rodgers*[1, 2], *Alan G. Marshall*[1, 2]

[1] *Department of Chemistry and Biochemistry, Florida State University, Tallahassee, FL 32306 USA*
[2] *National High Magnetic Field Laboratory, Florida State University, 1800 East Paul Dirac Drive, Tallahassee, FL 32310-4005 USA*

Introduction

Crude oil refineries often introduce water into refinery streams in order to increase product recovery. For example, water addition removes water-soluble salts and facilitates transport through pipelines by lowering viscosity. Emulsion formation (water-in-oil, oil-in-water, and/or both) can occur throughout the refinery process and further downstream water must be removed prior to petroleum product recovery. Production losses and added cost and effort associated with chemicals used to break water and oil emulsions, therefore, make emulsion formation a large problem for the petroleum industry.

Surface-active components, such as resins, asphaltenes, and organometallics, contribute to the stability of crude oil emulsion largely in part due to their hydrogen-bonding abilities.[1, 2] Resins and asphaltenes are solubility defined fractions of crude oil obtained from SARA fractionation.[1] Resins are the fraction of crude oil or bitumen that is soluble in 80:20 toluene:methanol. Asphaltenes, the *n*-heptane insoluble but toluene soluble fraction of crude oil/bitumen, are typically comprised of condensed aromatic rings and heteroatoms, such as nitrogen, oxygen, and sulfur. Organometallics, for example nickel and vanadium porphyrins, have also been identified in petroleum.[3] The hydrogen bonding capabilities of these compounds enables self-association, which can promote the formation of a third (stable) layer between the organic and aqueous phases.[1, 2, 4-6] These compounds contained within this interfacial layer contribute to emulsion stability. Therefore, it is advantageous to characterize and identify the species that exist at the interface. If successful, better chemistry can be targeted at the removal of these compounds, which in turn, may aid in petroleum recovery and water purification.

Previous research primarily suggests that asphaltenes absorb and accumulate at the emulsion water-oil interface and contribute to emulsion stability.[7, 8] However, solvent drop methods used to extract asphaltenes directly from the parent oil have been shown to include a small amount of maltene material, or co-precipitate. Moreover, high resolution mass spectrometric results suggest that the acidic portion of interfacial material is enriched in specific O_x and SO_x species relative to the parent crude.[9] Here we characterize the interfacial material and a crude oil known to cause emulsions in the field at an unprecedented level of detail. Positive/negative electrospray ionization and atmospheric pressure photoionization high resolution FT-ICR mass spectrometry highlight the basic, acidic and nonpolar species in the isolated interfacial material, parent crude and interfacial material-depleted crude.

Methods

Samples

Asphaltenes were precipitated from a crude oil by use of a simplified SARA fractionation. Raw asphaltenes were isolated by simple filtration and separated into pure asphaltene and coprecipitant fractions by Soxhlet extraction in clean *n*-heptane. Subsequent Soxhlet extraction of the pure asphaltenes in fresh toluene reconstituted the purified asphaltenes. The interfacial material (if present) was isolated from emulsions generated from the parent crude, raw asphaltene and coprecipitant fractions.

Sample Preparation

Isolated samples were dissolved to 2 mg/mL in toluene for electrospray analysis. Dilution to a final concentration of 1 mg/mL in 1:1 toluene:methanol allowed for spiking with 1% formic acid or ammonium hydroxide for positive and negative electrospray (ESI), respectively. This procedure allows for efficient protonation/deprotonation. All solvents were HPLC-grade from VWR International. Samples used for atmospheric pressure photoionization (APPI) were diluted from 2 mg/mL to 0.25 mg/mL in toluene and analyzed as is. Samples were delivered to the ionization source via a syringe pump at a rate of 0.5 µL/min for ESI and 25 µL/min for APPI.

Emulsion interfacial material was isolated by experimental procedures outlined by Wu.[10] Emulsions were first prepared by dissolving 10% (w/w) of the crude oil or isolated fraction into a 1:1 solution of *n*-heptane:toluene, commonly termed "heptol." Centrifugation with heavy water (D_2O) isolated the interfacial layer and samples were evaporated to solid with nitrogen gas at room temperature. Interfacial material was then re-dissolved in 1:1 toluene:methanol at 1 mg/mL for analysis by ESI FT-ICR mass spectrometry or to a concentration of 0.25 mg/mL for APPI analysis.

Mass Spectrometry

Samples were analyzed with either a custom-built 9.4 T 22 cm diameter bore (Oxford Corp., Oxford Mead, UK) ESI FT-ICR mass spectrometer or a hybrid LTQ FT-ICR 14.5 T 104 mm bore (ThermoFisher, Bremen, Germany) mass spectrometer. A modular ICR data acquisition system (PREDATOR) was used to collect and analyze the data.[11, 12] Low-resolution mass spectrometry (linear ion trap) provided independent verification of molecular weight distribution. A total of 600 time-domain data sets was co-added and Hanning-apodized, followed by a single zero-fill, fast Fourier transformation and magnitude calculation. Frequency was converted to mass-to charge ratio by the quadrupolar electric potential approximation.[13, 14] All observed species were singly charged, as verified by unit *m/z* separation between mass spectral peaks corresponding to the ^{12}C and $^{13}C^{12}C_{c-1}$ isotope for each elemental composition.

Results and Discussion

Figure 1 shows a flow chart for sample preparation. Simplified SARA fractionation is depicted in the branching to the left. Crude oil, maltene, raw asphaltene, and coprecipitate fractions are selected to test their ability to form an emulsion. If present, interfacial material is then isolated and analyzed by FT-ICR mass spectrometry. Smith *et. al.* showed that aggregation and multimerization of petroleum compounds is prevalent at concentration above 0.5 mg/mL.[15] Therefore, prior to FT-ICR mass spectrometry analysis, analysis by low-resolution mass spectra independently verifies monomer molecular weight distributions. Tandem MS in the linear trap establishes a true monomer molecular weight distribution for samples with multimodal distribution. Low-resolution mass spectra of a SARA fractionated crude oil, isolated crude oil and interfacial material are shown in **Figures 2 and 3**.

Low mass resolution limits the information obtainable from complex crude oil samples. Thus, high-resolution mass spectrometry methods are employed to allow alkylation, heteroatom content and aromaticity speciation. **Figure 4** compares the spectra obtained from the ion trap mass spectrometer (a) and FT-ICR (b) in negative ESI. The center of the molecular weight distribution for both spectra is ~375 Da with a width of 250-550 Da. Ultrahigh-resolution FT-ICR MS allows for the assignment of unique elemental compositions for each *m/z* value.

The most abundant heteroatom classes for the acidic compounds in the emulsion interfacial material are summarized in **Figure 5**. Relative abundances of the monoisotopic (^{12}C, ^{1}H, ^{16}O, ^{14}N, etc) species are plotted on the y-axis. The O_2 species are the most abundant in the interfacial layer at ~43% of all monoisotopic compounds, consistent with Stanford *et al.*,[9] who showed that O_x and SO_x species preferentially accumulate at the interface of emulsions relative to their parent crude oil.[9]

Figure 6 shows the isoabundance contour plots of double bond equivalents vs. carbon number for the acidic O_2 class from Figure 5. Double bond equivalents (DBE) is a measure of aromaticity, and is defined as number of rings plus double bonds to carbon for each elemental composition; each additional double bond or ring results in loss of two hydrogen atoms. The O_2 compounds are comprised largely of non-aromatic (DBE < 4) compounds with 15 – 35 carbon atoms.

Future work will focus on emulsion interfacial material generated from other SARA fractions. Analysis of the isolated interfacial material confirms the minimum components necessary to form a stable emulsion. Examination of compounds that preferentially adsorb at interfaces provides information about interfacial components and can be applied to develop agents that destabilize emulsions and increase product recovery.

Work supported by NSF DMR-00-84173, Florida State University, and the National High Magnetic Field Laboratory in Tallahassee, FL.

References

(01) Acevedo, S.; Escobar, G.; Ranaudo, M. A.; Kahazen, J.; Borges, B.; Pereira, J. C.; Mendez, B. *Energy and Fuels* **1999**, *13*, 333-335.

(02) Eley, D. D.; Hey, M. J.; Lee, M. A. *Colloids Surf.* **1987**, *23*, 173-182.

(03) Rodgers, R. P.; Hendrickson, C. L.; Emmett, M. R.; Marshall, A. G.; Greaney, M. A.; Qian, K. *Can. J. Chem.* **2001**, *79*, 546-551.

(04) Goual, L.; Harvath-Szabo, G.; Masliyah, J. H.; Xu, Z. H. *Langmuir* **2005**, *21*, 8278-8289.

(05) Ignasiak, T. M.; Strausz, O. P. *Fuel* **1978**, *57*, 617-621.

(06) Moschopedis, S. E.; Speight, J. G. *Fuel* **1976**, *55*, 187-192.

(07) McLean, J. D.; Kilpatrick, P. K. *J. Colloid Interface Sci.* **1997**, *196*, 23-34.

(08) Yarrington, H. W.; Hussein, H.; Masliyah, J. H. *J. Colloid Interface Sci.* **2000**, *228*, 52-63.

(09) Stanford, L. A.; Rodgers, R. P.; Marshall, A. G.; Czarnecki, J.; A., W. X.; Taylor, S. *Energy & Fuels* **2007**, *21*, 963-972.

(10) Wu, X. *Energy and Fuels* **2003**, *17*, 179-190.

(11) Senko, M. W.; Hendrickson, C. L.; Pasa-Tolic, L.; Marto, J. A.; White, F. M.; Guan, S.; Marshall, A. G. *Rapid Commun. Mass Spectrom.* **1996**, *10*, 1824-1828.

(12) Blakney, G. T.; van der Rest, G.; Johnson, J. R.; Freitas, M. A.; Drader, J. J.; Shi, S. D.-H.; Hendrickson, C. L.; Kelleher, N. L.; Marshall, A. G., Chicago, IL, May, 2001 2001; Amer. Soc. Mass Spectrom.; WPM265.

(13) Hendrickson, C. L.; Laude, D. A., Jr. *Anal. Chem.* **1995**, *67*, 1717-1721.

(14) Ledford, E. B., Jr.; Rempel, D. L.; Gross, M. L. *Anal. Chem.* **1984**, *56*, 2744-2748.

(15) Smith, D. F.; Schaub, T. M.; Rahimi, P.; Teclemariam, A.; Rodger, R. P.; Marshall, A. G. *Energy and Fuels* **2007**, *21*.

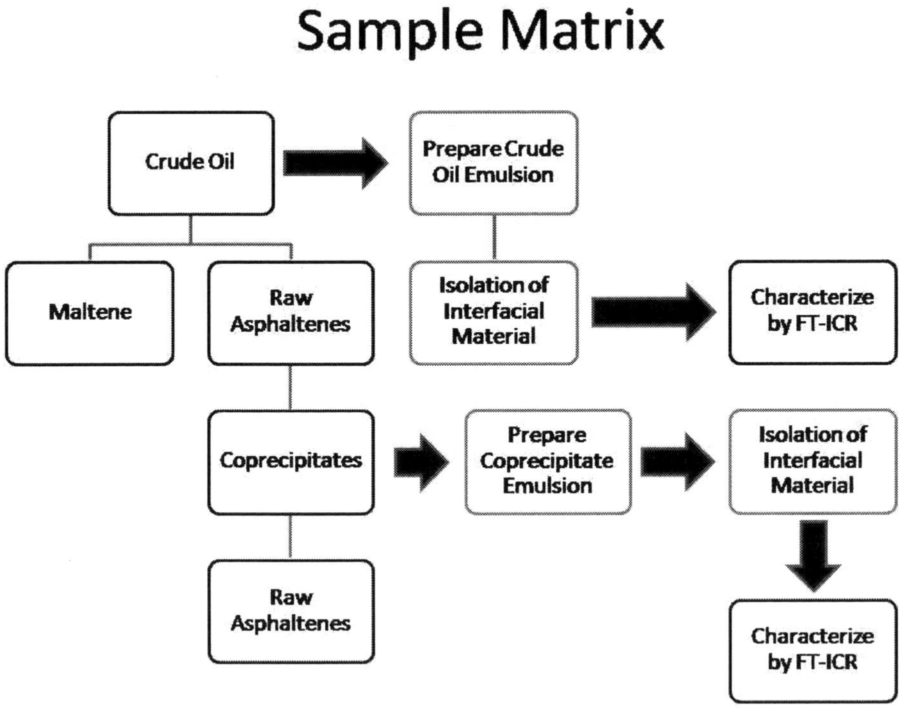

Figure 1. Simplified sample matrix for crude oil fractions isolated from the SARA method. Selected fractions were then tested for their role in emulsion formation. If interfacial material was present, the interface was isolated and examined by FT-ICR mass spectrometry.

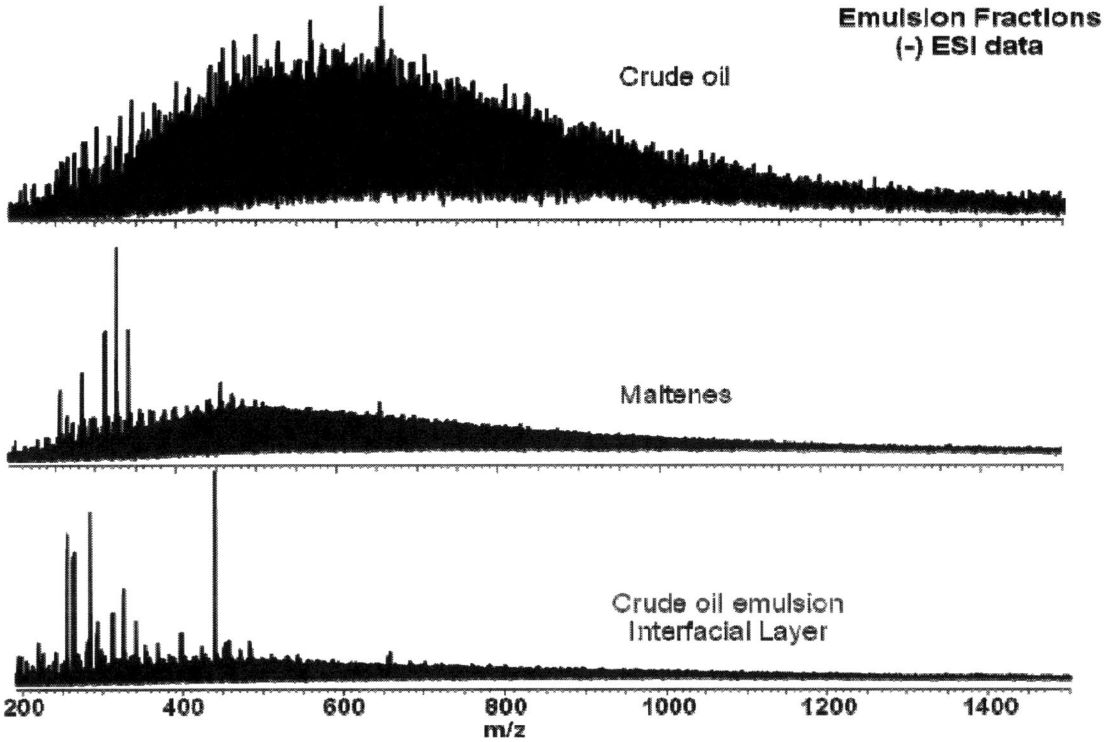

Figure 2. Low Resolution LTQ (ion trap) mass spectra of crude oil, fractionated maltenes, and isolated emulsion interfacial material.

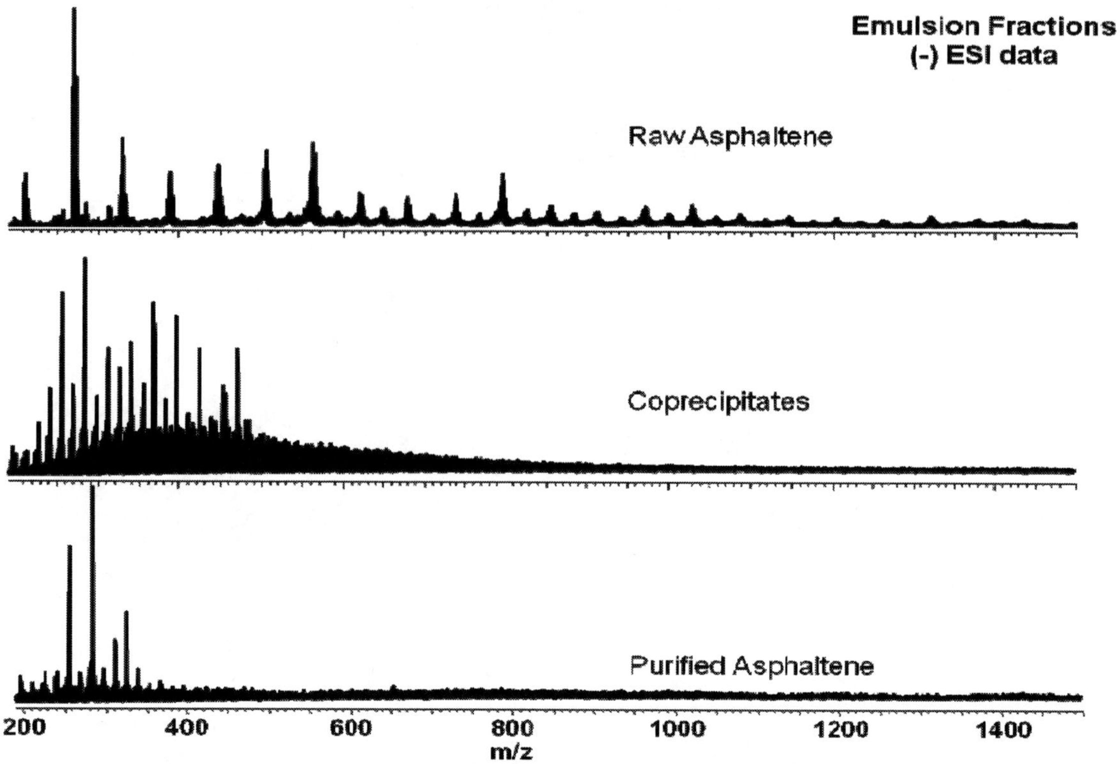

Figure 3. LTQ (ion trap) mass spectra of isolated raw asphaltenes, co-precipitates, and purified asphaltenes.

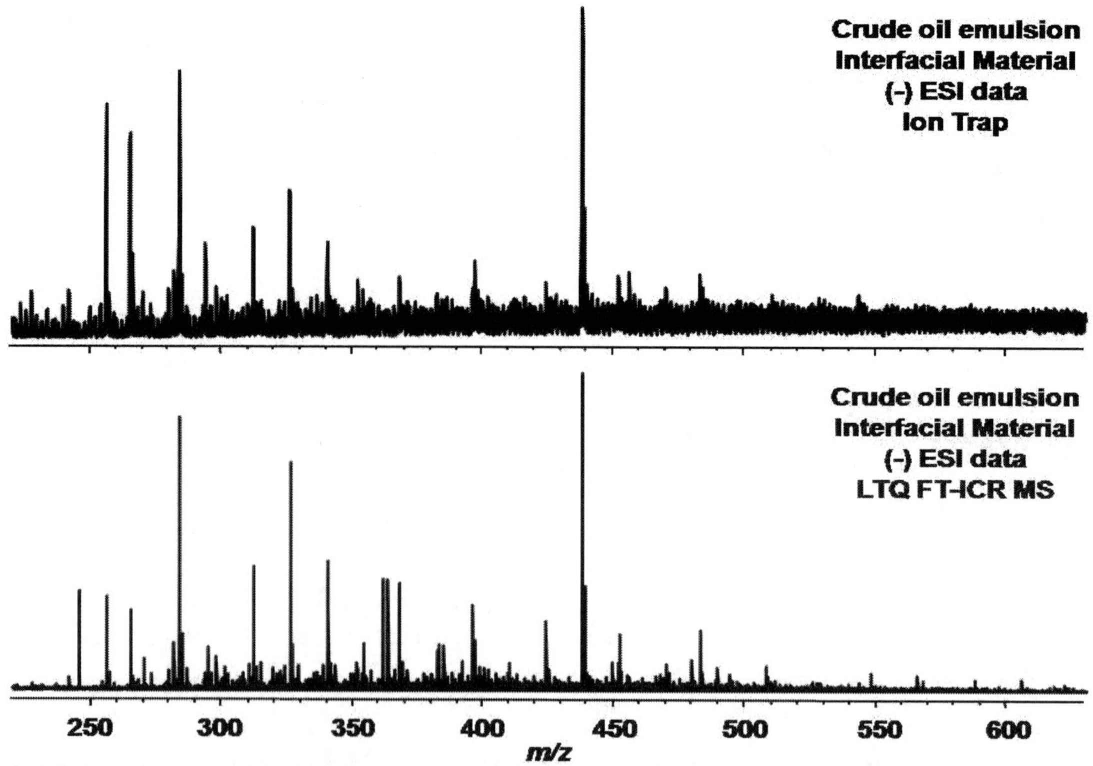

Figure 4. Molecular weight distributions collected by (-)ESI from (top) low-resolution (ion trap) mass spectrometer and (bottom) 9.4 T FT-ICR mass spectrometer.

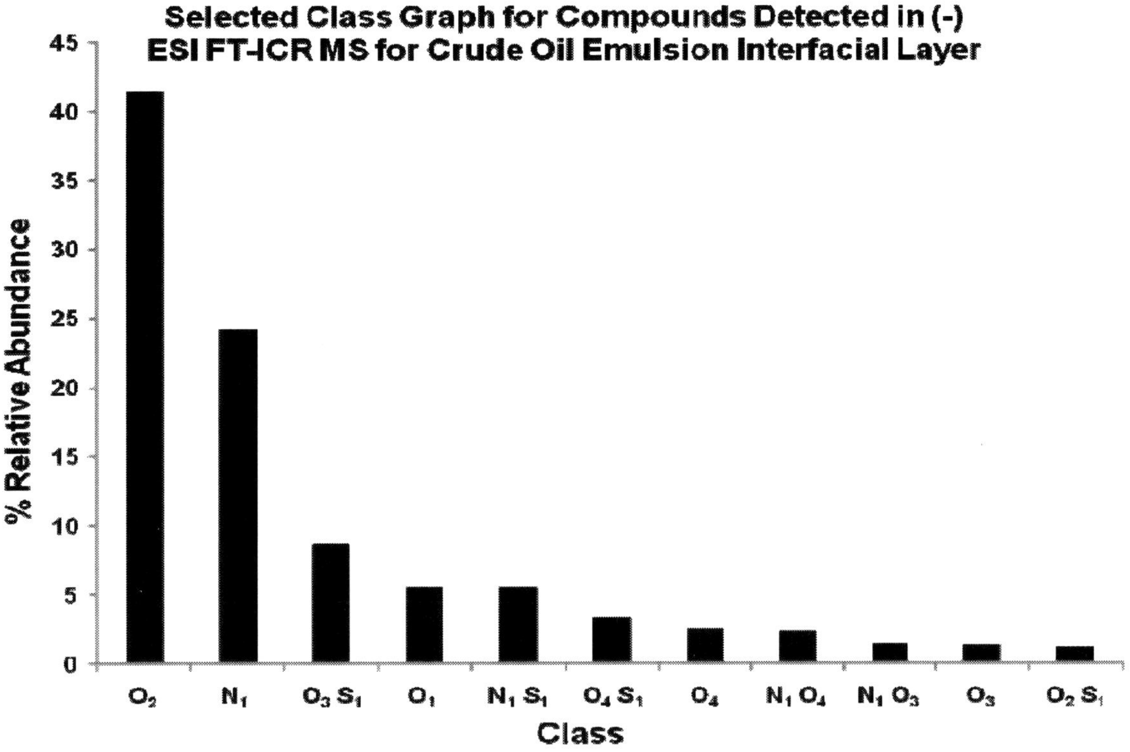

Figure 5. Acidic heteroatom class relative abundances (> 1%) for ions detected by FT-ICR MS in the emulsion interfacial material of Figure 3.

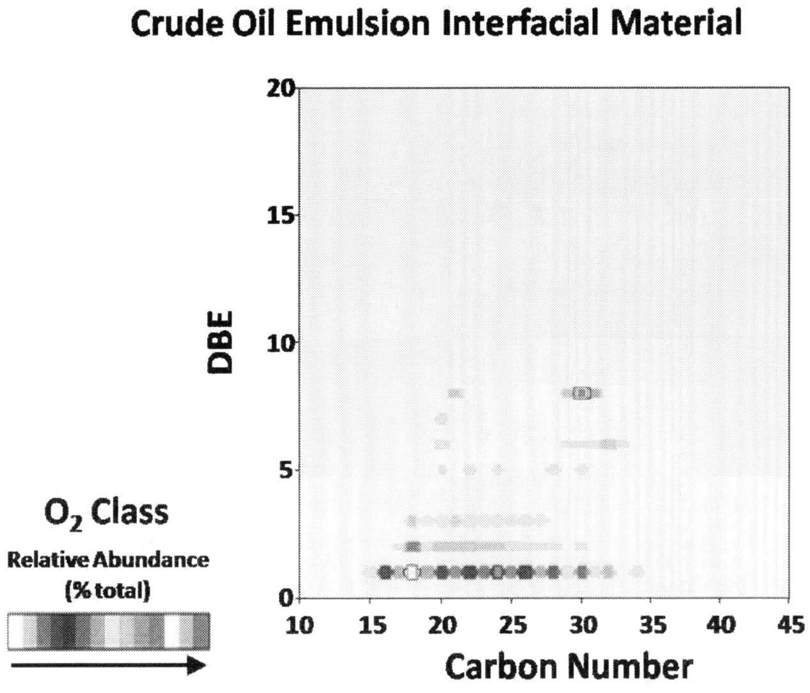

Figure 6. Relative isoabundance-contour plots of double bond equivalents vs. carbon number for the acidic O_2 compounds from crude oil emulsion interfacial material.

ADSORPTION OF VIRGIN AND THERMAL CRACKED ATHABASCA ASPHALTENES FOR FURTHER GASIFICATION

Clementina Sosa-Stull, Francisco Lopez-Linares, Lante Carbognani and Pedro Pereira-Almao,*
University of Calgary, Calgary, AB, Canada.

*To whom correspondence should be addressed. Tel.: +1 (403) 220-4799 Fax: +1 (403) 284-4852.
E-mail: ppereira@ucalgary.ca.

Introduction

To date, the most common process for hydrogen generation used in upgrading of heavy hydrocarbons is Steam Reforming of natural gas and/or naphtha [1]. Nevertheless, natural gas and naphtha have much better alternative uses. Moreover, bitumen upgrading using hydrogen generated from Steam Reforming of natural gas and/or naphtha increases upgrading cost significantly. An alternative and cost effective process for hydrogen production can be gasification of the heaviest hydrocarbon molecules, preferably the ones closest to instability, at low temperature after selectively removing them from the heavy oil stream. The objective of this work was to design inexpensive, thermally stable macro-porous selective adsorbents with catalytic active phases capable of steam reforming adsorbed heavy hydrocarbons by attacking them from multiple surface sites with the oxygen coming from water splitting. Ideally, the oxidation product would preferably be CO_2 in order to ensure maximum H_2 production. The macroporous adsorbent should be prepared at low cost, tolerant to steam and to the severity of gasification conditions as well as capable of accumulating metals present in the feedstock with minor negative impact on the performance. Another objective of this work was to accelerate the development of the process by using commercially available model molecules with similar defining asphaltenes characteristics to test their adsorption on the designed solid. This work demonstrated, at a conceptual level and laboratory scale, the viability of the proposed upgrading scheme comprising the integrated converted asphaltenes retention and hydrogen production processes at temperatures as low as 600°C.

.

Experimental

C7-Asphaltene from Athabasca bitumen (AB-C7-A) was obtained by n-heptane precipitation [2,3] and then re-dissolved in toluene without further washing in the alkane. Thermally treated samples, or visbroken residues (VB), starting from Athabasca vacuum residue at different level of conversions (13.6%, 23.3% and 28.5%) were obtained in our laboratory [4]. Toluene from Sigma-Aldrich (Spectrophotometric Grade) was used to dilute the asphaltenes used for the adsorption experiments while n-heptane from Sigma-Aldrich (Spectrophotometric Grade) was used for the asphaltenes precipitation studies. The adsorbents were prepared using kaolin and their characteristics were designed for the solid to have an important macro- pore's proportion with an average diameter higher than 50 nm in order to allow the penetration throughout these pores of the asphaltenes and to accommodate them in the macro-porous surface of the solid adsorbent. The details of the preparation are proprietary and will not be unveiled in this paper [3]. The adsorbents prepared were impregnated with active metals to produce steam gasification catalysts. For this study, potassium nitrate and calcium nitrate salts, from Sigma-Aldrich, were used. After the preparation of the solids, their characterization was performed (pore volume, surface area, pore diameter, etc) by BET method using a CHEMBET-3000 equipment from Quantachrome Instruments and mercury porosimetry.

For the adsorption experiment, several solutions of adsorbates (60 ppm in toluene) were prepared. The kinetics of the adsorption experiments were carried out by continuously measuring the change of UV-Visible absorbance in the solution of the different adsorbates when in contact with the macro-porous solid. The procedure was as follows: 3mL of toluene solution of the adsorbate were placed in a screw-cap spectrosil 3.5 m cuvette from Sigma-Aldrich with the following characteristics: rectangular cell quartz, open-top cap, light path: 10 mm, and the initial absorbance was measured (Ao). Then, 0.3 g of the macroporous solid were added, the system was closed and the cuvette cell was placed in the Spectrophotometer where every 15 minute a spectra was recorded for a period of 12 h in a static mode at room temperature. The kinetics plots were obtained by transforming the absorbance A(t) in relative or normalized absorbance RA(t) in order to make it independent of the initial concentration, valid within the large dilution range used as function of time.

The catalytic steam gasification experiments were carried out in a thermo balance (TGA) from TA Instruments, and the evolution of the gases was followed using a Quadrupole mass spectrometer (MS) from Ametek Process Instruments. This study was carried out at atmospheric pressure under a steam environment where the partial pressure of steam was equal to 29.28 kPa. The steam was generated inside the equipment after pumping distilled water from a Cole-Parmer model 74900 syringe pump attached to the TGA. Helium was used as carrier gas for all the experiments. The TGA study was used to evaluate the kinetics of the catalytic steam gasification of adsorbed carbonaceous materials like cracked heavy molecules or asphaltenes to prove at small scale the concept of the whole process. TGA analyzed samples consisted of the catalyst developed plus adsorbed heavy hydrocarbons. Samples were crushed in an Agate mortar in order to produce homogeneous materials. After this procedure, each sample was placed on a small sample holder. The holder was placed over a horizontal balance arm inside the TGA system. Once the sample was introduced into the TGA, helium (He) (flow, 100 cc/min) was admitted as carrier gas for the whole run. The first step was to dry the sample in He at 200 °C. After that, at the same temperature, the water was injected into the TGA to produce the steam. This is an isothermal step and was kept for 15 minutes to guarantee that steady state conditions were attained, confirmed with continuous monitoring via the Quadrupole Mass Spectrometer. Then, the temperature was linearly increased at a rate of 200 °C /min up to the steam gasification temperatures set to 550, 575, 600, 625 and 650 °C and held for 30 min. The initial gasification rate for each case was determined as a function of time and also as a function of the weight of reacting carbon. The interval to calculate this term was 5 minutes after reaching each reaction temperature and a relative conversion was calculated assuming that the maximum weight loss or total gasification for each sample was achieved at 650 ºC.

Results and Discussion

In figure 1, the normalized or Relative Adsorbance (RA(t)) versus time for the different samples was plot. It can be observed, at the initial concentration of 60 mg/L, all samples display different levels of adsorption during the evolution of the experiment. At 120 minutes, it is observed that the 23.3 Visbroken asphaltenes in contact with the adsorbent-catalyst with 2 wt % K-Ca as metal loading showed the highest uptake of all the samples. An interesting trend between conversion level and adsorption uptake on Kaolin with no catalyst is observed where the adsorption order is as follows: 28.5 VB ASP> 23.3 VB ASP> VR ASP. The above results confirm that the adsorption over this solid is not limited by the conversion during visbreaking; it seems that even larger molecules such as Vacuum Residue asphaltenes are able to adsorb on this solid as originally designed by using a large pore diameter. Also it seems that the 13.3 VB asphaltenes sample may have different characteristics in comparison with the others, like nitrogen content. Finally, the 23.3 Visbroken Asphaltenes contacting the adsorbent-catalyst at a higher metal loading of 8 wt K-Ca % showed the poorest adsorption. This result may be linked to the modification of the surface with the presence of a higher content of a basic oxide such as KCaOx. Using the data reported in Figure 1, it was possible to calculate the kinetics constants and fit those values to a first order reaction, Table 1.

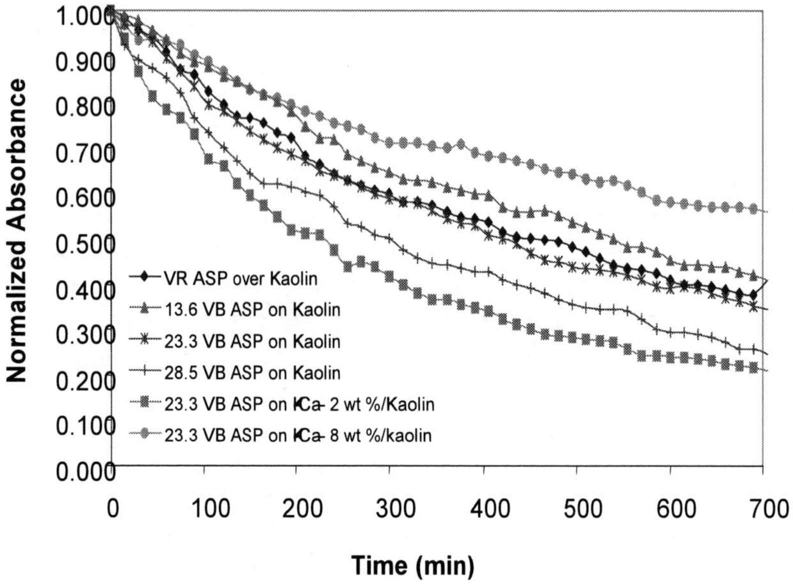

Figure 1.- Kinetics of the adsorption of asphaltenes from VR and VB products in toluene solution on macroporous kaolin and catalysts at 295 K. Conditions: Concentration = 60 mg/L, adsorbent: 0.3 g, solvent/adsorbent=10

Table 1. Apparent first order rate constant for the adsorption of the asphaltenes obtained from Virgin and visbroken Athabasca products in toluene solution over macroporous kaolin after 120 min at 295 K. Correlation coefficient for first order fitting.

Asphaltene fraction	**k x 10^3 (min^{-1})**	**R^{2b}**
AB-VR	1.4 $\pm$ 0.5	0.938
13.6 VB	1.4 $\pm$ 0.6	0.952
23.3 VB	2.2 $\pm$ 0.5	0.985
28.5 VB	2.8 $\pm$ 0.5	0.980
23.3 VB on K-Ca/Kaolin 2 wt%	3.4 $\pm$ 0.5	0.983
23.3 VB on K-Ca/Kaolin 8 wt%	1.0 $\pm$ 0.4	0.941

Literature on thermal conversion processes, suggest that large molecules present in Vacuum Residua would be transformed into small molecules depending of the severity of the Visbreaking process [5,6]. Thus, fractions with lower molecular sizes are expected with increased conversion. This could cause products from higher conversion to display higher adsorption up-takes. As can be seen in Table 1, increasing Visbreaking conversion concomitantly increases the adsorption of Visbroken (VB) asphaltenes as reflected by the kinetic constants. The 13.3 VB does not follow the expected trend, indicating that this fraction may display some particular characteristics. Molecular characterizations of all fractions are in progress in our research group with the aim to identify possible differences between those fractions that may affect their adsorption. Finally, the characteristics of the adsorbent-catalyst play an important role in the adsorption of converted asphaltenes, exemplified with the sample 23.3 VB adsorbed in the catalyst with the highest metal loading.

Moreover, the adsorption results presented before demonstrate that the phenomenon is not limited by the sizes of the molecules present in such fractions. Aiming to find what factor can be controlling the adsorption of such heavy fractions, the elemental analysis of the converted and unconverted asphaltenes used in our adsorption experiments was determined and the results are shown in Table 2.

Table 2. Elemental analysis of the Athabasca asphaltenes obtained from VR and VB products.

	Elemental analysis (wt%)				**Atomic ratio**		
Sample	**C**	**H**	**N**	**S**	**H/C**	**N/C**	**S/C**
VR	78.350	8.580	0.806	9.740	1.305	0.010	0.124
13.6 VB	80.690	7.686	0.988	9.670	1.135	0.012	0.120
23.3 VB	81.260	7.315	1.054	7.650	1.073	0.013	0.094
28.5 VB	81.830	7.548	0.993	6.850	1.099	0.012	0.084

From Table 2 it is observed that the atomic H/C ratio decreases in parallel with visbreaking conversion. Decreases in H/C reportedly correlate with increases in aromaticity [7-9]. It was found that visbroken products with increased aromaticity were adsorbed to higher levels over the developed macro-porous solid. Further, an increase in Visbreaking severity corresponds to an increase in the nitrogen content. Previously, it was shown that this macroporous kaolin has affinity for nitrogen, which enhances the adsorption uptake [10]. This fact explains why 23.3 VB and 28.5 VB have kinetic constants very close, associated with the higher values of the N/C ratio. However, when the macroporous kaolin surface is modified with K and Ca oxides at different loading, the uptake has an optimum with respect to the metal loading. It is suspected that a synergism effect between metal oxides content, N/C and H/C ratio are controlling this adsorption

It has been demonstrated that the asphaltenes fraction isolated from VB products could adsorb over the adsorbent and/or an adsorbent-steam gasification catalyst. Consequently, initial catalytic steam gasification experiments using Visbroken residue at 23.3% conversion adsorbed over the macro-porous kaolin adsorbent/catalyst were conducted. The reason for choosing 23.3% conversion sample versus 28.5% conversion sample was related to the availability. The results of gasification will be discussed in the following section. In TGA, the reactivity of carbonaceous materials can be correlated with the weight loss during the reaction [11]. Figure 2 shows initial rate curves at different reaction temperatures for different metal loadings. The raw data was taken from TGA software and was exported to Excel files to be analyzed. From Figure 2 it can be seen that there are roughly three different reaction regions in the whole process of catalytic steam gasification of heavy molecules within a TGA setup. a) Water vaporization comprised from room temperature up to 200°C. b) Hydrocarbon devolatilization and thermal cracking zone from 200°C up to reaction temperature.c) Catalytic steam gasification carried out under varying reaction temperatures kept for 30 minutes (550, 575, 600, 625 and 650°C). Thus, the weight loss for each adsorbate/catalyst system after the devolatilization period is due to catalytic steam gasification of the remaining heavy molecules adsorbed. It is important to mention that from the beginning of the gasification reaction the rate decays due to catalyst/carbon contact loss and by exhaustion of the carbon in the catalyst surface. Applying the initial reaction rate method [12], the kinetics data is obtained. For each reaction temperature, only the data collected up to 5 minutes after reaching the reaction temperature was used to analyze the kinetics reaction parameters. A zero order reaction was found after analyzing the data obtained by TGA. In Table 3 the reaction rate coefficients are shown for each catalyst at different reaction temperature.

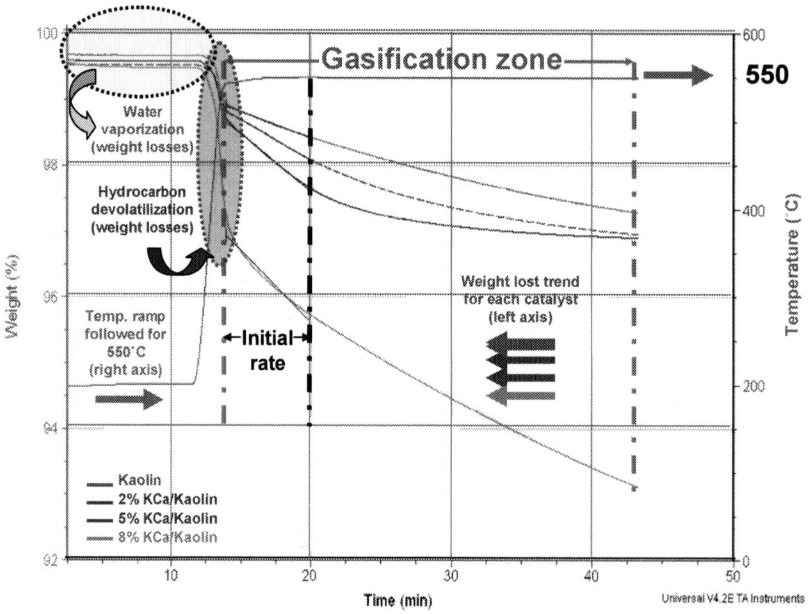

Figure 2. A typical TGA data plot of catalytic steam gasification reactions. The presented example was carried out at a reaction temperature of 550 °C.

Table 3 shows that at relatively low reaction temperatures from 550°C up to 600°C, the initial rates seem to follow a typical trend where the reaction rate increases with temperature. Once the reaction temperature is above 600°C, the trend is not the same particularly for the catalyst KCa8 wt%/kaolin, because its initial rate decreases with increasing reaction temperature. This fact could be related to the fact that this catalyst is the most active and besides has the lowest amount of adsorbed heavy molecules. Because of this, the carbonaceous material is consumed within the time selected to study the initial reaction rate.

Table 3. Reaction rate coefficient for each catalyst at different reaction temperature

Reaction temp., (°C)	k (mg/g min) for 2 wt% KCa,	k (mg/g min) for 5 wt% KCa,	k (mg/g min) for 8 wt% KCa,
550	0.1326 ± 0.0037	0.1830 ±0.0045	0.1948 ± 0.0067
575	0.1666 ± 0.0028	0.1916 ±0.0053	0.2216 ± 0.0075
600	0.1825 ± 0.0069	0.2083 ±0.0058	0.2312 ± 0.0012
625	0.2073 ± 0.0032	0.2432 ±0.0076	0.2075 ± 0.0086
650	0.2201 ± 0.0092	0.2661 ±0.0089	0.1822 ± 0.0098

The evolution of gases during gasification was measured using a quadrupole mass spectrometer (QDP-MS) but due to the low amount of adsorbed carbonaceous material in each sample and the high carrier gas flow (100 mL/min) the determination of product ratios was not possible. The QDP-MS was only used to determine the steady-state condition of the system before starting the reaction and to detect the presence of hydrogen once the reaction temperature was reached.

Conclusions

The kinetics of the adsorption of virgin and thermal cracked Athabasca asphaltenes was found to be first order, showing the reaction rate dependence on the initial concentration of the adsorbate. Also, the rate seems to be controlled by the chemical nature and specific content of heteroatoms present on these molecules.

This study, conducted at laboratory scale, clearly proved the viability of catalytic steam gasification at low temperature (550 to 650 °C), of virgin and thermal cracked asphaltenes over tailor designed catalysts. Initial gasification rates were affected by temperature and metal loading. It was found that a metal loading of 8 wt% shows the highest rate of steam gasification at 600°C. This temperature range is lower than typically published in the open literature [13]. A zero-order reaction model based on the carbon concentration fitted the determined rates adequately and permitted a comparison between the effects of the variables studied using this model.

Acknowledgements

The authors would like to acknowledge funding from Alberta Ingenuity Centre for In Situ Energy (AICISE), and the facilities provided by Schulich School of Engineering, University of Calgary, Canada, to carry out this work. Also the financial support provided by AERI is greatly acknowledged. The authors are grateful to Suncor Energy, for providing the Athabasca bitumen and its vacuum residue.

References

1. Institut Francais du Petrole, I.F.P., *Conversion Processes*. Petroleum Refining, ed. P. Leprince. Vol. 3. **2001**, Paris, France: Editions Technip.
2. González, M., *Stability Improvement of Thermal Treated Oil Residua by Adsorption of Heavy Molecules over Solid Surfaces*, in *Chemical and Petroleum*. **2006**, University of Calgary: Calgary. p. 130.
3. Sosa, C., *Adsorption of Heavy Hydrocarbons for the Purpose of Hydrogen Production in Chemical and Petroleum*. **2006**, University of Calgary: Calgary. p. 130.
4. Carbognani, L, Gonzalez, M. F., Pereira-Alamao, P. *Energy Fuels*, **2007**, 21(3), 1631.
5. Speight, J.G. and Dekker, M., *The Chemistry and Technology of Petroleum*. **1999**, New York.
6. Strausz, O,P, Lown, E,M. (**2003**), The Chemistry of Alberta Oil Sands, Bitumens and Heavy Oils. AERI, Calgary, AB, Canada.
7. Dias, J.R. *Acc. Chem. Res.* **1985**, 18, 241-248.
8. Ruiz-Morales, Y. Molecular Orbital Calculations and Optical Transitions of PAH and Asphaltenes. Chapter 4 in "Asphaltenes, Heavy Oils, and Petroleomics", Mullins, O.C.; Sheu, E.Y.; Hammami, A.; Marshall, A. G. (eds), Springer, NY **2007**, pp.95-137.
9. Carbognani, L. *Energy Fuels* **2001**, 15, 1013-1020.
10. González, M. F.; Sosa Stull, C.; López-Linares, F.; Pereira-Almao, P. *Energy Fuels*. **2007**, 21, 234.
11. Wigmans, T.,van Dam, F., and Moulijn, A., *Comparison of a block-flow reactor and thermogravimetric analysis in the steam gasification of different types of carbon. Carbon,* ***1981****. 19(4): p. 309-320.*
12. *Bartholomew, C.H. and Farrauto, R.J., Introduction and Fundamental Catalytic Phenomena, in Fundamentals of Industrial Catalytic Processes.* ***2006****, Wiley-Interscience: Hoboken, New Jersey.*
13. Navarro, R. M., Pena, M. A., Fierro, J. L. G. *Chem Rev*. **2007**, 107, 3952.

USER CONTROLLED KINETIC NETWORK GENERATION WITH INGEN

Craig A. Bennett[1], Zhen Hou[1], and Michael T. Klein[1]

[1] Department of Chemical & Biochemical Engineering, Rutgers University, Piscataway, NJ 08854 US

1. Introduction

The increasingly greater competition for dwindling conventional energy resources has driven the need for more efficient upgrading and conversion processes where the selectivity to desirable species is paramount. Meanwhile, governmental regulation has increased the accountability of companies with regard to environmentally harmful by-product species. To this end, detailed, molecule-based kinetic models are being used both in the selection of feeds and in the management of their conversion to valuable end products[1].

The goal of molecule-based models (MBM) is to represent the chemistry and kinetics of a chemical process with a species for every molecule. The challenge is that as the carbon number increases, the number of possible molecules in the kinetic model's *footprint* explodes exponentially, and it is sometimes important to group together isomers in order to limit the number of simultaneous balance equations. Various techniques can be employed, but it is efficient to make the lumping decisions at the model building step of the creation of the reaction network[1-5].

MBM's can be especially tedious to build "by hand" for complex feeds with many molecules (often hundreds or thousands), which has motivated the development of automated model building tools. Reaction network generators such as NetGen[2] were limited in scope to specific chemistries and allowed only expert programmers to change the specific rules that govern the network generation. A new tool, INGen (Interactive Network Generator), bridges the gap between modeler and programmer, allowing chemical process experts the ease of creating, changing, and adapting their own kinetic networks.

INGen is a kinetic network modeling environment that allows for the quick generation of large networks via intuitive chemistry, rules, and seed selection. The overall goal of INGen is to seamlessly integrate the creation of the footprint, network, and species properties with other programs that: handle the feed and product composition determination (CompGen); create the C code based executable file for the ODE (ODEGen[2]); and provide easy editing, running, and analysis capabilities for the model (KME[6]).

2. User Controlled Network Generation

2.1 Important Concept: Model Footprint

The footprint of the model is the full fixed set of molecules that occur within the network. This will be determined at the completion of the network generation, but it is an important concept to understand up front. When modeling a system, the entirety of the process should share a common footprint. The reaction network generation should supply a set of molecules that can be accurately represented by property-to-composition algorithms (e.g. CompGen). In addition, the set of balance equations used by the model should also be able to handle the size and scope of the footprint. A footprint too large may take too long to tune, whereas a model too small might not be able to accurately represent the chemical process.

The integrated, Excel-based nature of INGen and its companion tools (CompGen and KME) allows the user to quickly and easily change the footprint of the model using specific chemical rules (e.g. no side chains longer than five carbons). By making these decisions once, the user's choices are reflected in all aspects of the modeling process, thereby maintaining a constant footprint.

The common footprint layout will be a matrix where the columns represent molecule types (PIONA-like break-downs), the rows represent carbon number, and the elements themselves will become the concentrations.

2.2 Important Concept: Model Feed

The term feed refers to the physical composition entering the chemical process. Often the feed is specified as a set of properties and must be converted into molar flows or fractions for each of the molecules in the model. Ultimately, the feed is the non-zero subset of species found in the footprint before the chemical process begins. It is important to note that this implies that the footprint has already been established by the time the feed is constructed, which in turn entails that network is not created based on the composition of the feed.

This is not to say that considerations for available properties have not been taken into account at network generation time. As stated in the footprint section above, the model needs to have a common set of molecules. If the feed analysis can provide in depth characterization, network generation rules need to be selected such that the footprint will ultimately create all known species that exist not only in the feed, but the model itself.

2.3 Important Concept: Model Seed

Instead of providing "feed" molecules to the network generation, the user must select "seed" molecules. The seed to INGen is the starting set of molecules that are only used for the creation and shaping of the footprint and reaction network. The seed molecules are allowed to react given the various rules supplied to the network generator. In some cases, a single molecule could create an entire, complex network (e.g. n-C_{10} is allowed to undergo unlimited cracking and ring closure reactions). In others, the size of the network may need to be limited by allowing only a certain number of product generations away (i.e., DelPlot[7] rank) from the initial seed molecules. If an n-C_{80} was allowed as much freedom as the stated n-C_{10} example, the footprint would quickly ~~blow up~~expand into millions of possible molecules. This exponential footprint growth can be controlled by placing limits such as any molecule over C_{10} can only crack in the middle of the chain, all C_{15+} isomers should be collapsed into representative molecules, or the maximum product generation be set to three.

If these rules are implemented in the n-C_{80} model, it is obvious that the footprint would be controlled, but the resulting network might not be representative of the chemistry. The C_{80} would crack to C_{40} which would crack to C_{20} which would then crack to C_{10}. The C10s would no longer be allowed to react based on the generation rule, but what if the light products are important in the chemistry. Instead of simply seeding the model with C_{80}, the user can seed it with C_{80}, C_{10}, C_9, C_8, etc. The same can be done if more detail is needed with the molecules between C_{80} and C_{10} if those species are desirable in the model.

Again, it is important to state that the user must have an idea of the chemistry of the situation and the general shape of the footprint and reaction network. NetGen limited some of these decisions to those with programming backgrounds, thereby taking total control away from the chemists and reaction engineers.

Because such network creation artistry requires that the user is able to quickly and repeatedly generate kinetic networks, INGen allows the user simple point and click access to a database of predefined species in order to select which species should seed the model. With INGen, network generation time is on the scale of minutes, thereby allowing the user the opportunity to quickly find a reasonable network with which to model.

2.4 Chemistry Rules

As stated previously, along with the selection of seed molecules, the user defines the rules that will govern the generation of the network. There are two main types of rules regarding network creation, the first of which is a set of chemistry-based rules. Chemistry-based rules are the physically inherent limitations that govern the network. They determine what various types of reactions can occur to which species. INGen allows the user to specify which types of reactions should be allowable for each type of species. For allowed species type-reaction selections, the program determines whether each species can actually be physically reacted by analyzing the structure of the molecule via graph theory.

Each molecule is represented by an adjacency matrix, and by using depth-first-search and cycle-counting algorithms, the structure and type of each molecule can be identified. The type of molecule triggers a list of which possible reactions can occur. Each reaction type has a set of site-selection criterion based on the chemistry of the reaction. Algorithms read through the structure of the molecule looking for site matches. Once found, the molecule is allowed to react. In Figure 1 below, the labeled C1 and C2 atoms have been identified as reaction sties. Reaction takes place via matrix augmentation (the two reactants are combined into one matrix), permuation (the reaction sites are placed in a specific order), reduction (non-reacting atoms are disregarded), and addition (a reaction matrix is added to the augmented matrix, thereby transforming the reactant sites). Once the reaction is complete, algorithms that identify the products begin.

INGen incorporates many chemistry reaction types, rules, and site selection criteria. The user is afforded the opportunity to limit the possible reaction types for each molecule type from the front end. In addition, if multiple sites are possible for a certain type of reaction, the user has the ability to disable them (e.g. for steric reasons). By coding as many types of chemistries as possible and allowing the user the ability to turn them on and off, the need for multiple network generation tools is avoided.

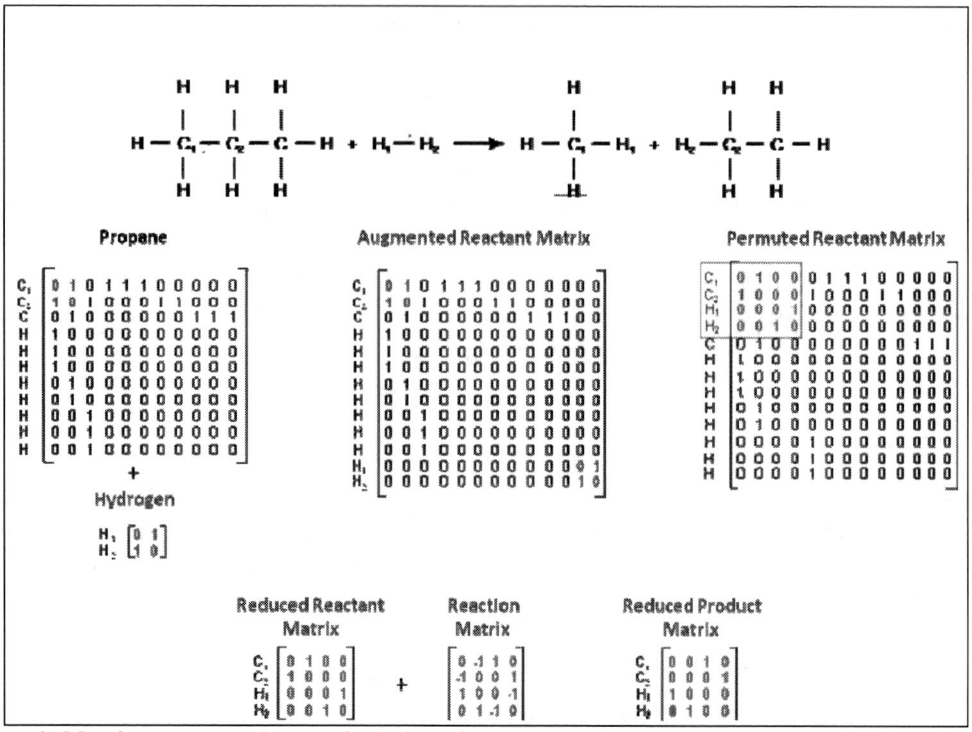

Figure 1- Matrix ~~representation~~ manipulation of a representative chemistry

2.5 Generation Rules

The second type of network building rules are known as generation rules in that they control the generation of the network. Such rules fall more into the art of model building, and it is therefore clearly important that they are under the control of the chemist and not the programmer. As mentioned in the Footprint section above, the generation rules will create the look and feel of the network. If the user wishes to have fine-grain characterization on the light ends, coarse analysis on the middle products, and outright lumping on the heavy ends, the generation rules need to be able to accommodate these issues.

INGen accomplishes this goal by allowing the user to specify various constraints on overall network generation (e.g., the rank limitation mentioned above). In addition, each of the species type-reaction categories may have user specified constraints (e.g., allow paraffin isomerization only on species with carbon numbers between 4 and 10). This level of runtime user customization, unique to INGen, is essential for the rapid creation of large scale kinetic networks.

2.6 Species Properties and INGen Output

Once the user has configured the network generator, the execution takes only a few seconds. Upon completion, the user will be supplied with lists of the reactions and species occurring within the model. The reaction file is immediately importable into the KME environment. It also includes the reaction type distinction so that the user can make of KME's LFER[5] group kinetic simplification.

Each species in the footprint has a corresponding adjacency matrix (two-dimensional representation of connected atoms), string code (unique linear representation), CML designation (Chemical Markup Language), thermodynamic property list (from MNDO type analysis), and physical property list (name, carbon count, etc.). These properties will be used by the CompGen program to find the composition of a sample mixture.

By maintaining a constant footprint and environment, common I/O sheets can be used thereby making transition from network generation, to feed characterization, to model tuning simple.

2.7 Overall INGen Benefit

INGen was created as a piece of a larger suite of kinetic modeling applications that allows for the seamless integration of all aspects of modeling. It is an important and sometimes overlooked concept that each step in building a model must work within the confines of the footprint. INGen allows the modeler to quickly and easily change the network, and because of its integrated nature, so too will the composition and analysis be easily changed.

3. Acknowledgements

The authors would like to acknowledge discussions held with Gang Hou, Steve Jaffe, and Wei Wei.

4. References

1. Quann, Richard J. and Steve B. Jaffe (1992), "Structure Oriented Lumping. Describing the Chemistry of Complex Hydrocarbon Mixtures", Industrial & Engineering Chemical Research, Volume 31, pp. 2483–2497.
2. Broadbelt, Linda J., Scott M. Stark, and Michael T. Klein (1994), "Computer-generated Pyrolysis Modeling — On-the-fly Generation of Species, Reactions, and Rates", Industrial & Engineering Chemical Research, Volume 33, pp. 790–799.
3. Hou, Gang and Michael T. Klein (1999), "Molecular Modeling of Gas Oil Hydrodesulfurization", Abstracts of Papers of the American Chemical Society, pp. U610–U611.
4. Joshi, Prasanna V., Sadasivan D. Iyer, and Michael T. Klein (1997), "Computer Assisted Modeling of Gas Oil Fluid Catalytic Cracking (FCC)", Abstracts of Papers of the American Chemical Society, pp. 80–Petr.
5. Korre, Styliani C. and Michael T. Klein (1996), "Development of Temperature-independent Quantitative Structure/reactivity Relationships for Metal-and-acid-catalyzed Reactions", Catalysis Today, Volume 31, Issues 1-2, pp. 79-91
6. Wei, Wei, Craig A. Bennett, Ryuzo Tanaka, Gang Hou, Michael T. Klein Jr., and Michael T. Klein (2007), "Computer Aided Kinetic Modeling with KMT and KME", Fuel Processing Technology, In Press.
7. Bhore, Nazeer A., Michael T. Klein and Kenneth B. Bischoff (1990), "Species Rank in Reaction Pathways: Application of Delplot Analysis", Chemical Engineering Science, Volume 45, Issue 8, pp. 2109-2116.

Developing New Wide Band Gap Semiconductors for Water Photolysis

Sarif Masud
Materials Science and Engineering, University of Texas at El Paso, 1127 Los Angeles Dr, Apt. 3, El Paso, TX 79902 and Geoffrey B Saupe, Chemistry, University of Texas at El Paso, 500 West University Ave, Chemistry, El Paso, TX 79968.

We are developing dye-sensitized metal oxides for photochemical energy conversion. We would like to understand the basic structural and chemical characteristics that control electron conduction and reactivity in particle-based porous photocatalysts. These issues are fundamental to the design of efficient high surface area heterogeneous photocatalysts. With the information gained from these experiments, we can increase the yields of sun driven direct water photolysis to provide a cost-effective source of hydrogen gas. We have synthesized highly porous metal oxides made of wide bandgap semiconductors. These new materials work well as electron-hole producing catalysts under UV light. However, we are interested in using them in conjunction with visible light sensitizers to develop visible light photocatalysts for a direct water splitting process. Here we have tested the un-sensitized catalyst for its ability to reduce gold and platinum ions in water to their metallic state. The metals are deposited on the surface of the catalyst at locations that are most catalytically active. This provides some insight on the best catalyst structure.

Steam Reforming of Methane in a Novel Microreactor

Pankaj V. Mathure, Anand V. Patwardhan and Ranajit K. Saha
Department of Chemical Engineering, Indian Institute of Technology, Kharagpur - India

AIChE Spring Meeting, April 2008
8th Topical Conference on Natural Gas Utilization
New Orleans, LA, April 6-10, 2008

Abstract

Almost 90% of industrial hydrogen manufactured today is produced by steam reforming of natural gas as this is by far the most efficient process available. The objective of this article is to review recent trends in steam reforming of methane and natural gas and identify possible strategies for use of novel small dimension reactors for the same process. The focus of our work is on the development of novel micro channeled monolithic reactors which can be coated with suitable catalysts and used for steam reforming of methane. The paper presents the performance characteristics of such a novel microreactor. The structured catalyst called PS–CAT is a cordierite based monolithic structure coated with nickel oxide. The testing was carried out by varying the process parameters such as temperature (673 to 1048 K), molar ratio of steam to methane in the feed (from 2.5:1 to 7:1), space time (W/F_{CH4} = 10 to 55.0 gm–cat min/mol). The catalyst exhibited good conversion (above 90%) of methane for a molar ratio of steam to methane from 2.5 to 7. Maximum hydrogen yield of 3.8 mol of H_2/mol of CH_4 reacted was observed at a W/F_{CH4} of 55.1 gm–cat min/mol at a temperature of 1048 K and molar ratio of steam to methane of 3.0. The performance of this developed catalyst was compared with the performance of commercial nickel–based catalyst in small diameter (3 mm and 6 mm) fixed bed reactors. The catalyst exhibited a minimum increase of 10% and a maximum increase of 22% in the conversion of methane over the commercial catalyst.

Introduction

Catalytic steam reforming of hydrocarbons, alcohols and light oil fractions involves the reaction of steam with methane, ethane, natural gas, LPG, naphtha, gasoline, alcohols like methanol, ethanol, and propanol, over catalysts at elevated temperatures (473 to 1173 K) and pressures (1 to 30 atm) to produce a mixture of hydrogen, carbon monoxide and carbon dioxide [1]. The mixture of hydrogen and carbon monoxide is known commonly as syngas. The technology has matured over the last 55 years. However the endothermic nature of the reactions makes the process energy–intensive. Coke formation which results in catalyst deactivation is also another major concern. Work is still in progress to use milder operating parameters, increase reactor through–put and minimise coke formation increasing catalyst life and thus reduce reformer down time. It is reported that 90% hydrogen generated globally, is produced by the steam reforming of natural gas and light oil fractions [2]. The major reason for this is the commercial viability of such plants by which hydrogen can be produced at $2.11/kg of H_2 [3,4]. This is by far the most efficient process available in comparison to other processes like, carbon dioxide reforming, coal gasification, pyrolysis, water electrolysis and photobiochemical techniques.

A novel way of developing micro or meso−scaled reactors which can be readily integrated in existing steam reformer plants for bulk manufacture of hydrogen is the use of wash–coated monoliths which can be fitted inside reformer tubes. The size of the monolith and the base material (ceramic or metal) can be selected as per the process requirements. The basic support monolith can be purchased from a number of suppliers in the market and subsequently the wall coating can be modified to suit the feedstock. Monoliths or honeycomb reactors have been used as a support for catalytic converters for treatment of automobile exhaust gases and as filters for processing exhaust gas streams from industries [5–7]. These structures are either made of ceramic material or aluminium containing metals. Monolith reactors act as multifunctional reactor systems having the advantages of low pressure, efficient contact of the reactants with (most frequently) solid catalyst and hence avoiding the problems related to partial wetting of the catalyst with liquid/gas phase; the possible choice of combining reaction and separation processes; the controlled loading of reactants, product removal, etc. Thus in monolith reactors one can combine the advantages of conventional multiphase reactors (e.g. fixed bed, slurry and trickle bed reactors) [5]. This increases process efficiency and its cost effectiveness. One disadvantage of monolith reactors is the poor distribution of liquid and/or gas along the reactor's cross-section. This makes their use difficult, especially under higher flow rates.

Concept of the novel, channel reactor

Conventional catalysts for steam reforming are available in the form of cylindrical pellets, tablets and spheres and are simply dumped in the reformer tube. This is called as a fixed bed or packed bed reformer. The catalyst pellets are randomly placed in the reformer tube and do not contribute to a defined structure to the bed. There is no structure to the bed as a whole and it can be imagined as a bed containing a large number of convoluted or tortuous paths through which the feed gas flows. As the gas flows through the path, there is an increased pressure drop (relative to the use of structured catalysts). Monoliths are uniform extruded structures composed of parallel flow–through channels. The basic monolith structure varies in dimensions and shapes depending on the purpose of use. They consist of a number of channels or cells of hexagonal, rectangular or circular shapes. Usually exhaust gas purifier systems consist of a closely packed array of small square channels. The active catalytic material is coated on this structure. The structure is then placed within a suitable housing (such as the reformer tube) and can be used as a catalytic reactor. Thus by using such a monolithic structure in the form of a micro or meso–scale, channeled catalyst (coated with a special formulation) which exactly fits the reformer tube inner diameter, one can replace the dumped (randomly packed) catalysts in the steam reforming unit. This will not only decrease the pressure drop (relative to packed beds), but also avoid "short circuiting" or "channeling" through the reformer bed. This will improve the performance of the reformer bed due to the improved contact between the reactants and the active catalytic ingredient. Ceramic monoliths have been used successfully in the treatment of automotive exhaust gases. Their success mainly is due to their ability to withstand extreme process conditions like high temperature and mechanical stress due to vehicle vibrations. Such ceramic monoliths are usually made of cordierite which has a composition of $2MgO.5SiO_2.2Al_2O_3$. There are hardly any references (in literature) to the use of cordierite, monolith channels for steam reforming of methane (heat provided external to the reformer tube). It was therefore decided to use ceramic based (cordierite) monoliths having rectangular channels as the base material for the monolithic catalyst for steam reforming of methane.

The catalyst block (named as PS–CAT)[8] consists of a cordierite channeled, monolithic structure which is wash coated (a specified number of times to achieve the desired catalyst loading) with a nickel–based salt solution (2M concentration), dried, calcined, and reduced with hydrogen. The structured catalyst consists of a number of thin-walled channels which exactly fit the reformer tube inner diameter. The wall thickness (average) was found to be around 270 μm (micron). The cell/channel dimension was found to be 1500 × 1500 μm. A number of such blocks can be stacked one above the other to fill the entire length of the reformer tube. Also, the diameter of such blocks can be as per the process requirements. In this study, experiments were conducted with 4 channels (figure 1) each part of a larger monolith consisting of 100 cpsi (cells per square inch). This reactor concept can be used in the existing reformer infrastructure without any major change in process equipment to significantly increase hydrocarbon feed throughput. With higher throughput, the same steam reformer unit can produce larger quantities of syngas. The main advantages are increased throughput, simplicity of the arrangement, and minimum plant modification. Figure 1 shows a 4-channel PS-CAT. Figure 2 show the schematic diagram of a monolith and the monolith when placed in the tubular reactor.

Figure 1: PS-CAT (4-channel): A novel, channel, structured catalyst.

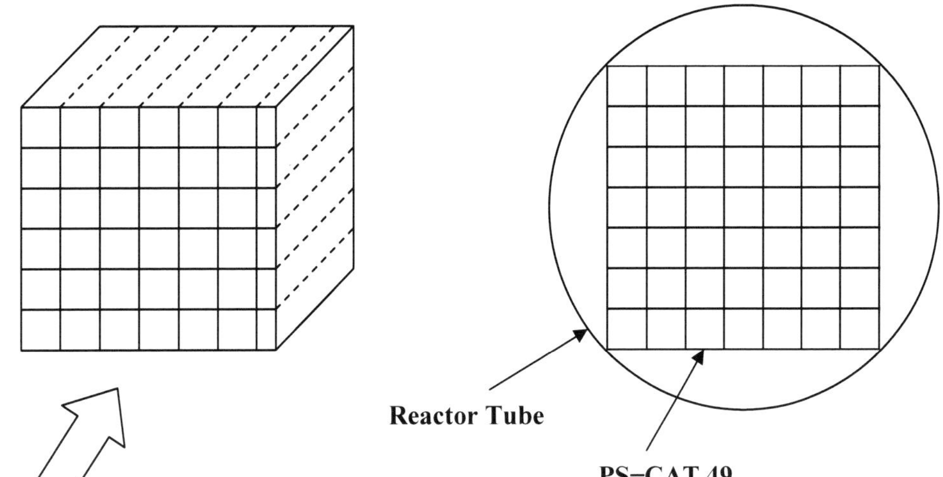

Figure 2: Typical schematic of a meso-scale, channeled monolith. The figure on right shows the block placed in a reactor tube.

Literature review

The main focus of this section is review work on the use of monolithic, channel reactors for steam reforming of methane for production of hydrogen. Tomasic and Jovic [2006] have published a state of the art review on the use of monolithic catalysts [5]. According to the authors development of monolithic catalysts and/or reactors has been one of major achievements in the field of heterogeneous catalysis and chemical reaction engineering of the recent years. The work highlights the advantages of monolithic catalysts and/or reactors with respect to the conventional ones, with particular focus on the integral approach to the catalyst and reactor design. The paper is detailed and gives basic definitions and classification of monolithic catalysts, including basic features of the monoliths and factors that have proceeded to the development and application of the monolith structures. The paper also discusses the preparation of monolithic catalysts and their commercial application with particular emphasis on the less known applications, and those which are still under development. These applications include the use of monolithic catalysts for Fischer–Tropsch synthesis, selective hydrogenation, oxidation, catalytic fuel combustion units, and gas cleaning devices for industries and automobiles. The authors have concluded that further research should be directed towards the development of new constructional designs by changing geometries to resolve the distribution problems at the entry of monolithic structures.

Hessel et al. [2005] have reviewed work on micro–structured devices (both etched channels and monoliths) developed for reforming of methane and other hydrocarbons [9]. They have reported a number of devices which can have been used mainly for partial oxidation or autothermal reforming of methane, propane, isooctane and gasoline.

Meille [2006] has discussed the methods used to deposit a catalyst on structured surfaces [6]. The review outlines both physical methods such as physical vapour deposition and chemical methods (sol–gel, chemical vapour deposition, direct synthesis, etc.) The coating of catalysts based on oxide, zeolite or carbon support is detailed on various surfaces such as silicon or steel microstructured reactors, cordierite monoliths or foams, fibres, tubes, etc.

Zhou et al. [2006] have investigated alumina supported Pt group metal monolithic catalysts for selective oxidation of CO in hydrogen–rich methanol reforming gas for proton exchange membrane fuel cell (PEMFC) applications [10]. The results show that Pt/Al_2O_3 was the most promising candidate to selectively oxidize CO from an amount of about 1 vol% to less than 100 ppm. They have investigated the effect of the O_2 to CO feed ratio, the feed concentration of CO, the presence of H_2O and/or CO_2, and the space velocity on the activity, selectivity and stability of Pt/Al_2O_3 monolithic catalysts. The Pt/Al_2O_3 catalyst was scaled up and applied in 5 kW hydrogen producing systems based on methanol steam reforming and autothermal reforming.

Giroux et al. [2005] have published a review paper for the technical and market potential of hydrogen for use in varied applications namely, stationary fuel cell power generation units, on-site generators of hydrogen for industrial uses, and hydrogen fueling stations for fuel powered automobiles [11]. They have proposed the use of monolithic structures as alternatives to particulate catalysts for the reforming of hydrocarbons for hydrogen generation. They have further proposed that new process designs for fuel processing can take advantage of the successful experience of environmental catalysis in automotive and stationary applications using monolithic catalysts and other substrates such as heat exchangers.

Brobova et al. [2005] have studied syngas formation by selective catalytic oxidation of liquid hydrocarbons in a short contact time adiabatic reactor [12]. Their research involved pilot plant scale exploration of syngas formation from liquid fuels like isooctane and gasoline by selective catalytic oxidation at short contact times at nearly adiabatic reactor conditions. Monolithic catalysts with different supports (micro-channel ceramics and metallic honeycomb structure) have been used in the experiments. Their results demonstrated that over the range of the operational parameter $O_2/C = 0.50–0.53$ required for syngas generation, equilibrium syngas was produced over the catalysts employed, thus proving the evidence of their high activity and selectivity.

Qia et al. [2005] have studied the performance of La–Ce–Ni–O monolithic perovskite catalysts for gasoline autothermal reforming system [13]. Autothermal reforming of gasoline or its surrogates, n-octane with or without thiophene additive, was carried out on either bulk perovskite pellet or monolithic perovskite catalyst. During the 220-hour long-term test, the pellet catalyst exhibited high thermal stability and activity with hydrogen yield approaching to the theoretical maximum value and only minor amount of CH_4 slipping through. It possessed fairly good sulfur tolerance although it could still be seriously poisoned when subjected to high concentration of sulfur. Their raw cordierite monolith could be an effective catalyst for autothermal reforming of gasoline, exhibiting superior performance to the catalyst of 0.3 weight % $Rh/CeO_2–ZrO_2$/cordierite at a temperature range of 923 to 1073 K.

Pasel et al. [2004] have studied different water-gas-shift catalysts for combined autothermal reforming with water-gas-shift (WGS) reaction [14]. The autothermal reforming of liquid hydrocarbons using different conditions have been studied. In the first step, the dry reformate from autothermal reforming and a separate stream of steam were used to conduct the WGS reaction. Strong differences concerning the catalytic activity between the three investigated commercial monolithic catalysts could be observed. The most active one showed a very promising catalytic behaviour. At a gas hourly space velocity (GHSV) of 12,250 h^{-1}, CO conversion amounted to 85% at 553 K. CO concentration in the reformate reduced from 6.1 to about 0.9 vol. %. In another step, unconverted water from autothermal reforming was fed to the reactor for the WGS reaction together with additional components of the reformate. No catalyst deactivation was observed within almost 90 h under autothermal reforming conditions generating only traces of carbon dissolved in the water.

Lindstrom et al., [2003] have carried out an experimental investigation on hydrogen generation from methanol using monolithic catalysts [15]. They have evaluated the activity and carbon dioxide selectivity for the reforming of methanol over various binary copper-based materials, Cu/Cr, Cu/Zn and Cu/Zr. The methanol reforming was performed using steam reforming and combined reforming (a combination of steam reforming and partial oxidation). This combined reforming process was carried out at two modes of operation: near autothermal and at slightly exothermal conditions. Their results show that the choice of catalytic material has a great influence on the methanol conversion and carbon dioxide selectivity of the reforming reaction. The zinc-containing catalyst showed the highest activity for the steam reforming process, whereas the copper/chromium catalyst had the highest activity for the combined reforming process. The copper/zirconium catalyst had the highest CO_2 selectivity for all the investigated process alternatives.

Apart from these there are a few patents which refer to the use of ceria–coated zirconia monoliths for catalytic partial oxidation process of hydrocarbons for syngas production and the use of platinum and palladium monoliths for autothermal reforming of hydrocarbons [16–17].

Vergunst [1999] has published a detailed thesis on the preparation aspects of carbon-coated monolithic catalysts and three phase hydrogenation of cinnamaldehyde using the same [7]. The author has covered preparation of monolithic catalysts, testing of monolithic catalysts in the (selective) hydrogenation of benzaldehyde, cinnamaldehyde, methylstyrene, benzene, and γ−butyrolactone, testing of monolithic catalysts in solid acid catalyzed reactions (acylation and esterification) and in the selective oxidation of cyclohexanol into adipic acid, development of internally finned monolithic structures, hydrodynamics, and scale up. The work is extremely important for any work related to monolithic catalysts. However the manufacture of these catalysts and their subsequent application is totally different from the current application (steam reforming) being discussed.

As can be observed from the above work, wash−coated monoliths with meso or micro−scaled channel sizes are a promising substitute for conventional large and small scale fixed bed reactors. Once the coating procedure and coating material is finalized, these monoliths provide cost-effective reactor geometries to retrofit existing reformer units. To the best of our knowledge, there are hardly any references (in literature) to the use of meso−scale, ceramic, monolith channels for steam reforming of methane (heat provided external to the reformer tube). It was therefore decided to use ceramic based (cordierite) monoliths having meso−scale rectangular channels as the base material for the monolithic catalyst for steam reforming of methane. As nickel possesses hydrogenation activity and is a low-cost but effective catalyst for cleavage of O—H, —CH_2—, —C—C— and —CH_3 bonds, it was decided to exploit these properties and use nickel nitrate solution for the coating of the monoliths [1,18].

Experimental study

The performance of the developed catalyst was tested extensively using the experimental set up shown in figures 3 and 4 [19]. The testing was carried out by varying the process parameters such as temperature (673 to 1048 K), molar ratio of steam to methane in the feed (from 2.5:1 to 7:1), space time (W/F_{CH4} = 10 to 55.0 gm–cat min/mol). Figure 3 shows the actual photograph of the experimental rig whereas figure 4 gives the schematic flow diagram of the same. The length of each block was 0.025 m. In case of experiments with the 4 channel PS−CAT, the tubular reactor housing with inner diameter 6 mm was used. 5 blocks each having 4 channels were placed end to end within the tubular reactor housing such that the centre of the catalyst bed corresponded to the central heating zone of the furnace. Rest of the tubular reactor housing (both above and below the catalyst blocks) was filled up with inert particles. As 5 blocks were used in series, the total length of the channel reactor was 0.125 m. The reactor was mounted in the furnace assembly. Fresh catalyst was used for each experimental run. The catalyst was activated in a hydrogen flow of 40 N ml/min with a steady increase of temperature. The catalyst was maintained at the desired reaction temperature for two hours. Methane gas (make: BOC, XL grade, purity: 99.9 %, specific gravity: 0.553, calorific value: 55530 kJ/kg) was fed to the reactor at the desired flow rate using a Bronkhorst make mass flow meter. The preheater temperature was maintained at 443 K to ensure complete vaporisation of the feed water even at high liquid flow rate. A HPLC (high pressure liquid chromatography Make: SSI, USA) pump was used to feed water to the reactor at the desired flow rate (between 0 to 5 ml/min). The product stream was condensed using the double pipe condenser at the outlet of the reactor. The liquid product (mostly water) was collected in a separator while the outlet gas stream flow rate was measured using a wet gas flow meter. The gas was periodically sent to the GC (gas chromatograph) for analysis using a three-way valve. Gas analysis was done using a thermal conductivity detector (TCD) with Spherocarb column (1/8" diameter and length 8') to find the composition of the outlet gas stream. The product gas was found to contain methane, carbon monoxide, carbon dioxide and hydrogen. No other hydrocarbon species was detected. The results of the gas analysis of the product gases indicated whether steady state conditions had been achieved. The measurements for the study were done once similar gas compositions were obtained after the start of the experiment. As this is a fast reaction in general steady state conditions were observed within 15 to 20 minutes from the start of the experiment. The pressure for all experimental runs was maintained at 1 atmosphere. The liquid product was weighed and analysed periodically in the GC using the TCD in a Porapak Q column (1/8" diameter and length 6'). The liquid product was found to be only water. The reaction was stopped after 3 to 6 hours of steady state operation. Reproducibility of the experiments was verified by conducting repeat runs for a few random reaction conditions. The results showed a maximum error of ±4.0%.

Figure 3: Experimental rig used for steam reforming of methane.

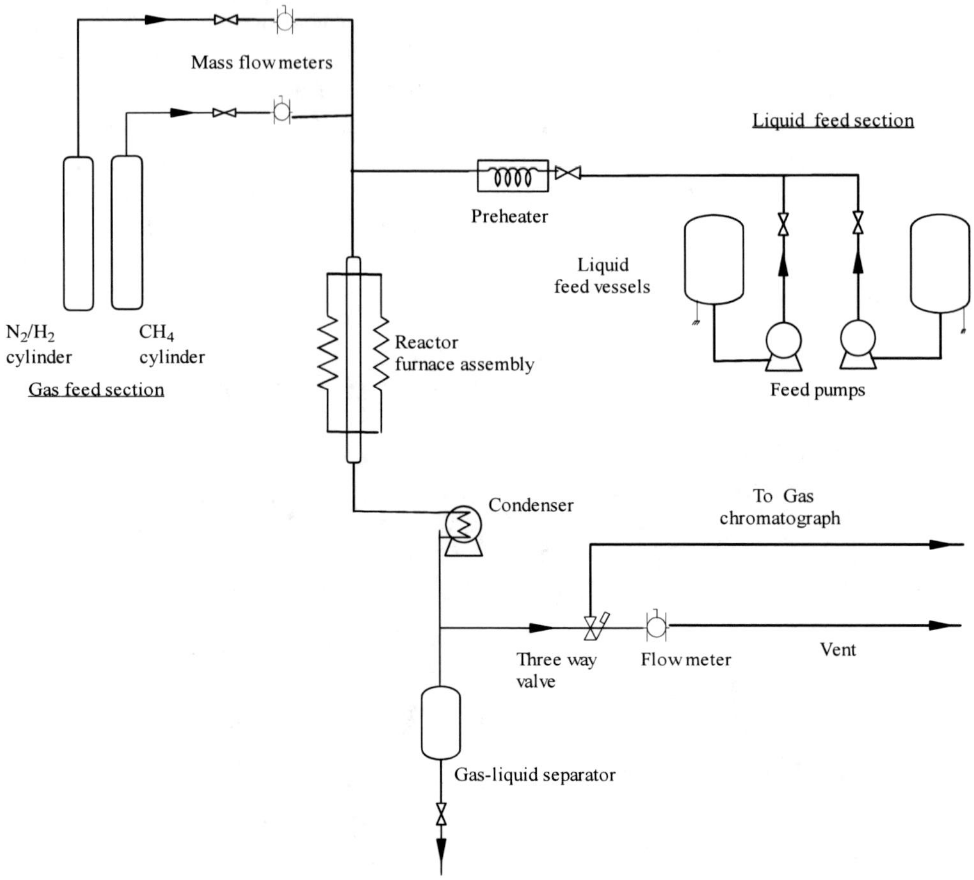

Figure 4: Schematic of the experimental rig used for steam reforming of methane.

Results and discussion

Effect of temperature

Steam reforming of methane was carried using PS–CAT at different temperature 673 K to 1048 K keeping the molar ratio of feed to steam (MR) and the W/F_{CH4} constant. Figure 5 shows the effect of temperature on the conversion of methane. As temperature increases the conversion of methane gradually increases from 30% to nearly 100%. The catalyst exhibited a methane conversion of more than 90% for a temperature above 973 K. The catalyst showed nearly complete conversion (above 98 %) of methane at a temperature above 1048 K at a W/F_{CH4} close to 55.0 gm–cat min/mol and a molar ratio (CH_4:H_2O) close to 3.0. With increase in temperature the carbon monoxide content in the product gas stream increased from as low as 0.21 % to 22.43% (v/v). The carbon dioxide content decreases with increase in temperature. Thus the catalyst promotes the endothermic steam reforming reaction but does not promote the exothermic water–gas–shift reaction which is favoured at low temperature. This result is in accordance with the presence of nickel as the active ingredient of the catalyst which promotes bond cleavage in the steam reforming reaction but has limited water–gas–shift activity [1,18].

Figure 6 shows the effect of temperature on the yield of hydrogen at constant molar ratio of steam to methane. Maximum hydrogen yield of 3.8 mol of H_2/ mol of CH_4 reacted was observed at a W/F_{CH4} of 55.1 gm–cat min/mol at a temperature of 1048 K and molar ratio of steam to methane of 3.0.

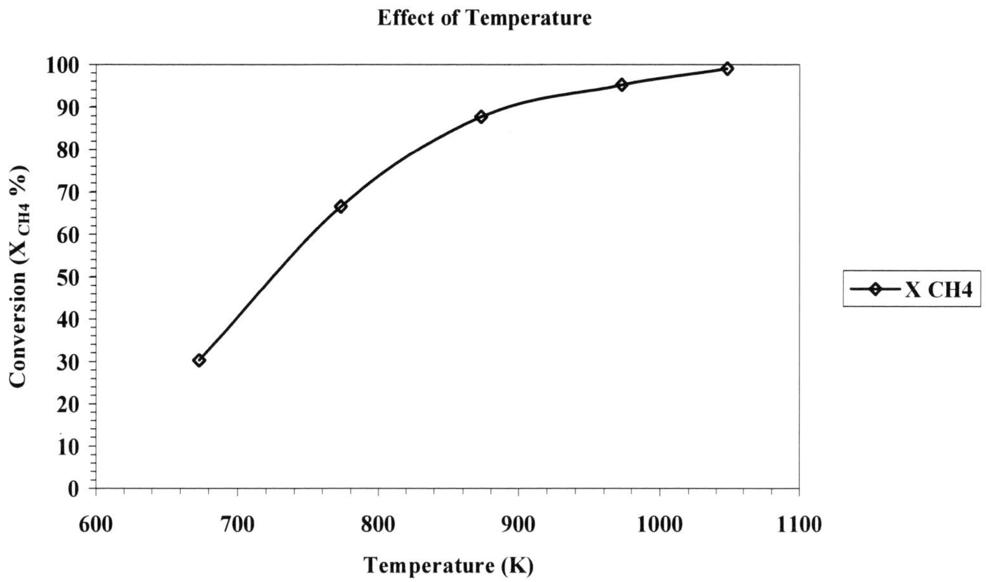

Figure 5: Effect of temperature on methane conversion. Other conditions: PS−CAT 4 channel block, 5 blocks in series, W/F_{CH4} = 56 gm–cat min/mol, steam to methane molar ratio in feed = 3:1

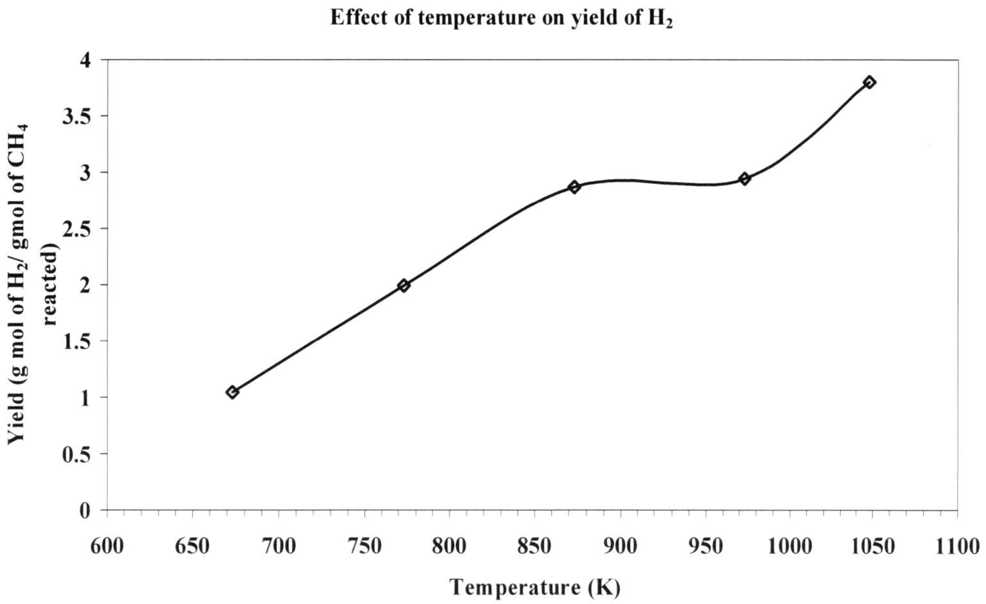

Figure 6: Effect of temperature on yield of hydrogen. Other conditions: PS−CAT 4 channel block, 5 blocks in series, W/F_{CH4} = 56 gm–cat min/mol, steam to methane molar ratio in feed = 3:1

Effect of molar ratio of steam to hydrocarbon

The effect of molar ratio of steam to hydrocarbon in feed on methane conversion was studied for molar ratio a range of molar ratio from 2.5 to 7 at a constant temperature of

1048 K and W/F_{CH4} = 56 gm–cat min/mol. The catalyst exhibited good conversion (above 90%) of methane for a molar ratio of steam to methane from 2.5 to 7. Best results (methane conversion above 97%) were obtained for a molar ratio of steam to methane between 2.5 and 4. Figure 7 shows the effect of molar ratio of steam to hydrocarbon in feed on methane conversion. Figure 8 shows the effect of molar ratio of steam to hydrocarbon in feed on the yield of hydrogen. Maximum hydrogen yield of 3.8 mol of H_2/ mol of CH_4 reacted was observed at a W/F_{CH4} of 55.1 gm–cat min/mol at a temperature of 1048 K and molar ratio of steam to methane of 3.0. The result corresponds with the value of optimum molar ratio of steam to hydrocarbon in other published literature for steam reforming of methane using commercial nickel catalysts [1].

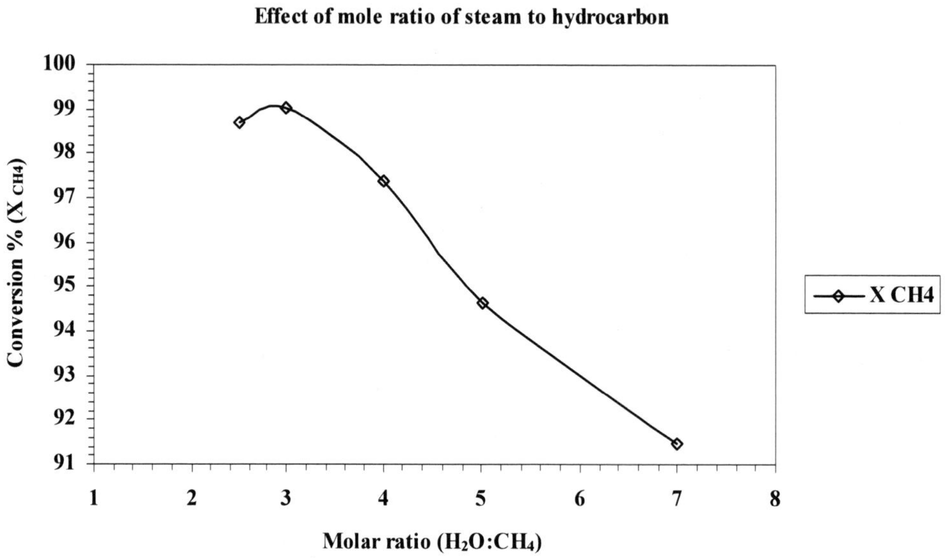

Figure 7: Effect of molar ratio of steam to hydrocarbon in feed on methane conversion. Other conditions: PS–CAT 4 channel block, 5 blocks in series W/F_{CH4} = 56 gm–cat min/mol, T = 1048 K.

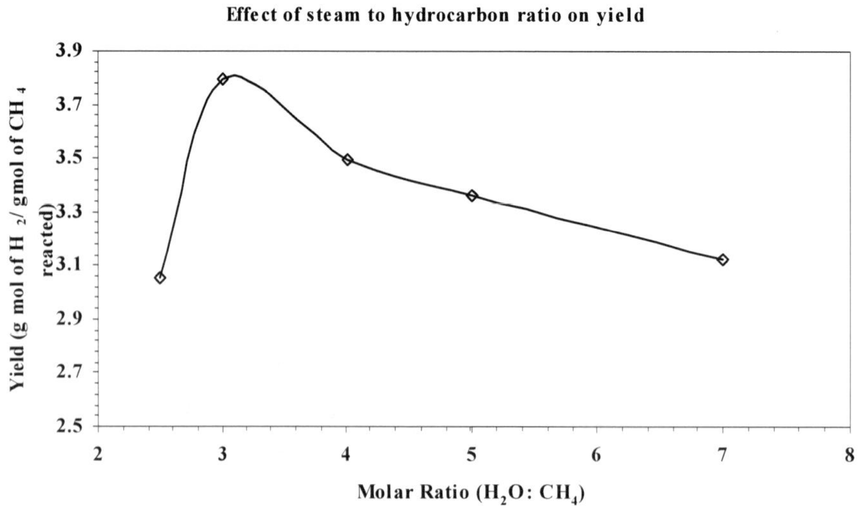

Figure 8: Effect of molar ratio of steam to hydrocarbon in feed on hydrogen yield. Other conditions: PS–CAT 4 channel block, 5 blocks in series W/F_{CH4} = 56 gm–cat min/mol, T = 1048 K.

Effect of space time

The effect of space time was studied by varying the W/F_{CH4} ratio from 10 to 55.0 gm–cat min/mol for three temperatures namely 873, 973 and 1048 K at a molar ratio of steam to methane of 3.0. Figure 9 shows the effect of space time on methane conversion. The conversion of methane increases with increase in W/F_{CH4} ratio. The catalyst exhibits a conversion of more than 90% for a W/F_{CH4} ratio above 37.3 gm–cat min/mol for temperature above 923 K.

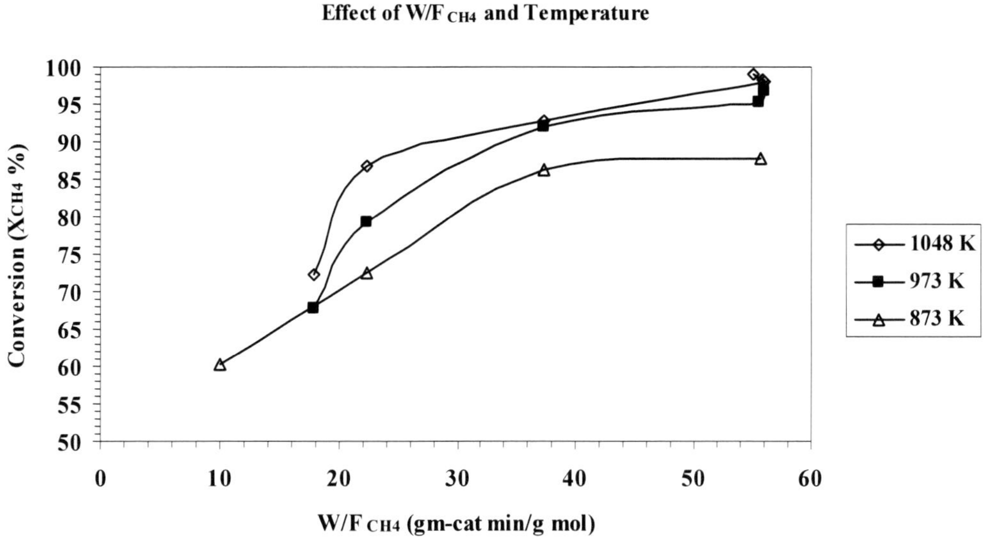

Figure 9: Effect of space time on methane conversion. Other conditions: PS–CAT 4 channel block, 5 blocks in series, steam to hydrocarbon molar ratio in feed = 3.0

Comparison of performance with a commercial catalyst

As mentioned in the literature review section, to the best of our knowledge there are hardly any references (in literature) to the use of ceramic, monolith channels specifically for steam reforming of methane (heat provided external to the reformer tube). Thus direct comparison with other data was not possible. In order to compare the performance of the developed catalyst, experimental runs were conducted using a commercial nickel containing catalyst in a tubular fixed bed reactor of inner diameter 6 mm and 3 mm. In case of the 3 mm and 6 mm reactors the experiment was conducted maintaining the ratio of the length of the catalyst bed to the diameter of the reactor to enable a common basis of comparison between the two cases [20]. This ratio was maintained at 10. The composition of the commercial catalyst was determined using EDS–X-ray analysis. The catalyst consisted of 25 % (elemental, weight basis) nickel, 15% alumina and 55% magnesia (weight basis). Swamy [2007] has reported a detailed kinetic study of steam reforming of methane using the same catalyst [21]. The commercial catalyst was crushed, sieved to obtain the required size of particles (approx 1.0 mm). The required weight of catalyst (0.1 gm) was filled in the tubular packed bed reactor of diameter 3 mm or 6 mm such that the centre of the catalyst bed corresponded to the central heating zone of the furnace. The remainder of the reactor was filled with inert ceramic material. The reactor was mounted in the furnace assembly. The catalyst was activated in a hydrogen flow of 40 N ml/min with a steady increase of temperature. The catalyst was maintained at the desired reaction temperature for two hours or till steady state conditions were achieved which can be determined from the temperature of the bed and the reactor skin temperature (outer wall of the reactor). At identical conditions (feed to steam molar ratio of 1:3 at a W/F = 56 gm−cat min/mol , pressure = 1 atm and temperature 1048 K), when a commercial nickel−based steam reforming catalyst was tested for steam reforming of methane in a 6 mm packed bed reactor the corresponding conversion obtained was 79.7 %. The conversion of methane using PS−CAT at identical conditions was found to be 98.27 % exhibiting a massive increase in the conversion of methane. Figure 10 shows the effect of temperature on methane conversion while comparing the performance of PS−CAT with commercial catalyst at identical process conditions. As can be observed, PS−CAT outperforms the commercial catalyst. The reactor exhibited a minimum increase of 10% and a maximum increase of 22% in the conversion of methane that directly translates into increased production of syngas and hydrogen with existing plant infrastructure. Thus PS–CAT is a very promising substitute for packed bed reactors for production of syngas and hydrogen which can be used for manufacture of fertilisers. The better contact between the reactants and the active catalytic ingredient in case of the use of the structured catalyst shows a substantial increase in the performance in comparison to a fixed bed catalyst consisting of randomly packed catalyst. In the case of the 3 mm and 6 mm diameter fixed bed reactors, the conversion of methane in the 3 mm reactor was higher than the 6 mm reactor. This is in agreement with principles of use micro or meso–scale devices [22]. Decreasing the linear dimensions, for a given difference in a physical property (like temperature or concentration) increases the respective gradient. Consequently the driving forces for heat transfer, mass transport or diffusion flux per unit volume or unit area increase. Literature reports the effect of changing diameter of a membrane reactor system on methane conversion for the partial oxidation process [20]. The authors have reported an increase in methane conversion with decrease in reactor diameter (at the same length to diameter ratio). Even though the system reported is different from this study the reasoning for the increase in methane conversion for smaller reactor diameters has a similar basis and can explain the increase in conversion for the 3 mm reactor.

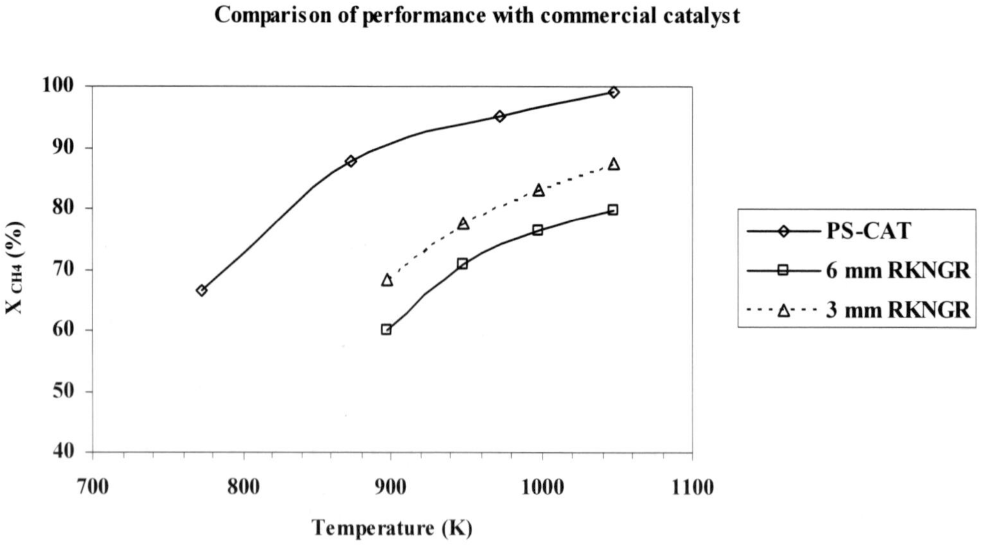

Figure 10: Comparison of performance of PS−CAT with commercial catalyst, effect of temperature on methane conversion. Other conditions: PS−CAT 4 channel block, 5 blocks in series, W/F_{CH4} = 56 gm–cat min/mol, steam to hydrocarbon molar ratio in feed = 3.0

Conclusions

A novel, micro–scale, structured, channel reactor was developed and successfully tested for steam reforming of methane. The structured catalyst called PS–CAT is a cordierite based monolithic structure coated with nickel oxide. The testing was carried out by varying the process parameters such temperature (673 to 1048 K), molar ratio of steam to methane in the feed (from 2.5:1 to 7:1), space time (W/F_{CH4} = 10 to 55.0 gm–cat min/mol). The catalyst exhibited good conversion (above 90%) of methane for a molar ratio of steam to methane from 2.5 to 7. The catalyst showed a conversion of more than 90% for a W/F_{CH4} ratio above 37.3 gm–cat min/mol for temperature above 923 K. Maximum hydrogen yield of 3.8 mol of H_2/mol of CH_4 reacted was observed at a W/F_{CH4} of 55.1 gm–cat min/mol at a temperature of 1048 K and molar ratio of steam to methane of 3.0. Finally the performance of this developed catalyst was compared with the performance of commercial nickel–based catalyst in small diameter (3 mm and 6 mm) fixed bed reactors. The catalyst exhibited a minimum increase of 10% and a maximum increase of 22% in the conversion of methane that directly translates into increased production of syngas and hydrogen with existing plant infrastructure. The main advantages of use of PS–CAT are increased throughput, simplicity of the arrangement, and minimum plant modification. Thus PS–CAT is a very promising substitute for packed bed reactors for production of syngas and hydrogen which can be used for manufacture of fertilisers.

Nomenclature

F_{CH4}	=	molar flow rate of methane, mol/min
Q_{CH4}	=	volumetric flow rate of methane, N ml/min
Q_{H2O}	=	volumetric flow rate of water, ml/min
T	=	temperature, K
v/v	=	composition of gas species expressed as volume percent of the total product gas %
W	=	weight of catalyst, gm
W/F	=	ratio of weight of catalyst to molar flow rate of feed (space time), gm−cat min/mol
X_{CH4}	=	fractional conversion of methane

References

1. Rostrup−Neilsen, J.R., Catalytic Steam Reforming in *Catalysis, Science and Technology*, *vol.5,* ed. Anderson, J. R. and Boudart, M., Springer−Verlag, Berlin **(1984)**.
2. Das, D., Veziroglu, T. N., Hydrogen production by biological processes: A survey of literature, *International Journal of Hydrogen Energy* **2001**, *26*, 13–28.
3. Momirlan, M., Veziroglu, T.N., Current status of hydrogen energy, *Renewable and Sustainable Energy Reviews* **2002,** *6*, 141–179.
4. Harayanto, A., Fernando, S., Murali, N., Adhikari, S., Current status of hydrogen production techniques by steam reforming of ethanol: A review, *Energy and Fuels* **2005,** *19,* 2098–2106.
5. Tomasic, V., Jovic, F., State-of-the-art in the monolithic catalysts/reactors, *Applied Catalysis A: General* **2006,** *311*, 112–121.
6. Meille, V., Review on methods to deposit catalysts on structured surfaces, *Applied Catalysis A: General*, **2006**, *315*, 1–17.
7. Vergunst, T., *Carbon coated monolithic catalysts preparation aspects and three phase hydrogenation of cinnamaldehyde*, PhD Thesis, Technical University, Delft, The Nethelands **(1999)**.

8. Mathure P.V., Swamy B., Patwardhan A.V., Saha R.K., Ganguly S., A structured catalyst for steam reforming of methane for production of syngas *Application filed in the name of I.I.T., Kharagpur –Indian Patent Application No. 1515/KOL/2007* **dated 02/11/2007**.
9. Hessel, V., Lowe, H., Muller, A. and Kolb, G., *Chemical Micro Process Engineering – Processing and Plants,* Wiley–VCH, Germany **(2005).**
10. Zhou, S., Yuan, Z. and Wang, S., Selective CO oxidation with real methanol reformate over monolithic Pt group catalysts: PEMFC applications, *International Journal of Hydrogen Energy,* **2006**, *31*, 924 – 933.
11. Giroux, T., Hwang, S., Liu, Y., Ruettinger, W. and Shore, L., Monolithic structures as alternatives to particulate catalysts for the reforming of hydrocarbons for hydrogen generation, *Applied Catalysis B: Environmental,* **2005**, *56*, 95–110.
12. Bobrova, L., Zolotarskii, I., Sadykov, V., Pavlova, S., Snegurenko, O., Tikhov, S., Korotkich, V., Kuznetsova, T., Sobyanin, V. and Parmon, V., Syngas formation by selective catalytic oxidation of liquid hydrocarbons in a short contact time adiabatic reactor, *Chemical Engineering Journal*, **2005**, *107*, 171–179.
13. Qia, A., Wang, S., Fub, G., Nib C. and Wub, D., La–Ce–Ni–O monolithic perovskite catalysts potential for gasoline autothermal reforming system", *Applied Catalysis A: General*, **2005**, *281*,233–246 (2005).
14. Pasel, J., Cremer, P., Wegner, B., Peters, R. and Stolten, D., Combination of autothermal reforming with water–gas–shift reaction – small scale testing of different water–gas–shift catalysts, *Journal of Power Sources*, **2004**, *126*, 112–118.
15. Lindström, B., Agrell, J., Pettersson, L.J., Combined methanol reforming for hydrogen generation over monolithic catalysts, *Chemical Engineering Journal*, **2003,** *93* ,91–101
16. Jiang, W., Park, S., Tomczak, M. S., Acharya, D. R., Tamhankar, S. S., Ramachandran R., Monolith based catalytic partial oxidation process for syngas production *United States Patent 7,090,826*, **2006**
17. Hwang, H., Heck, R.M., Yarrington, R. M., Fuel cell electric power production *United States Patent 4,522,894*, **1985**
18. Vaidya, P.D., Rodrigues, A.E., Insight into steam reforming of ethanol to produce hydrogen for fuel cells, *Chemical Engineering Journal* **2006**, *117*, 39–49.
19. Mathure, P.V., *Studies on steam reforming for the production of hydrogen*, Ph.D. Thesis, I.I.T., Kharagpur, India **(2008)**.
20. Hoang D.L., Chan S.H., Effect of reactor dimensions on the performance of an O_2 pump integrated partial oxidation reformer–a modeling approach, *International Journal of Hydrogen Energy,* **2006**, *31*, 1–12.
21. Swamy, B., *Studies on methane steam reforming for production of hydrogen*, Master's Thesis, I.I.T., Kharagpur, India **(2007)**.
22. Lowe H. V., *Microreactors New technology for modern chemistry*, Wiley–VCH, Germany (**2000**).

Development of Dense Membranes for Hydrogen Separation from Coal Gasification Streams

U. Balachandran
T. H. Lee, C. Y. Park, and S. E. Dorris. Energy Systems Division - Ceramics Section, Argonne National Laboratory, 9700 South Cass Avenue, Bldg. 212, Argonne, IL 60439

We are developing dense cermet (i.e., ceramic-metal composite) membranes for separating hydrogen from product streams generated during coal gasification. Hydrogen separation with these membranes is nongalvanic (i.e. requires no electrical power), and the hydrogen separated from the feed stream is of high purity, making post-separation purification steps unnecessary. Using various feed gas mixtures, we measured the hydrogen permeation rate, or flux, for the membranes in the temperature range of 500-900°C. This rate varied linearly with the inverse of membrane thickness and reached ≈33 cm^3[STP]/min-cm^2 at 900°C for an ≈15-μm-thick membrane when 100% H_2 at ambient pressure was used as the feed gas. When membranes were tested in a gas mixture that contained CH_4, CO, and CO_2 for times approaching ≈500 h, performance did not degrade. Because the membranes will encounter H2S, we also tested the membranes in gas mixtures that contained H_2, CH_4, CO_2, CO, and H_2S. Hydrogen flux measurements showed that at 900°C, the membranes are stable for up to 1200 h in gases that contain 400 ppm H_2S. The cermet membranes are found to be stable in the temperature range of 500-800°C in simulated synthesis gas mixtures containing ≈21% steam. The present status of membrane development will be presented in this talk.

Work supported by the U.S. Department of Energy, Office of Fossil Energy, National Energy Technology Laboratory's Hydrogen and Syngas Technology Program, under Contract DE-AC02-06CH11357.

Autothermal Reforming of Propane Over Alkali-Doping on the Hydrotalcite Catalysts

Sun Young Park[1], Jong Ho Kim[2], Jae Soon Shin[2], Nam Cook Park[2], and Young Chul Kim[2]

(1) Advanced Chemicals, Chonnam National University, Gwangju, South Korea,
(2) School ofApplied Chemical Engineering, Chonnam National University, Gwangju, South Korea

The performance of hydrotalcite catalyst in propane autothermal reforming(ATR) for hydrogen production was investigated in fix-bed flow reactor. Reactions were conducted with a feed stream of H2O/C/O2=3/1/0.37 in the temperature range from 300 to 700 °C. Catalysts were characterized by XRD, TGA and TEM methods. During the manufacturing of alkali metal modified Ni/MgAl of hydrotalcite-like catalysts. The effect of the added of alkali promoter were much smaller size of nickel particles by depressing Ni sintering. And coke formation rates lower than Ni/MgAl catalysts. When added the alkali metal of small quantity, the Ni particle size has an effect on degree of dispersion. It is expected that added of alkali metal enhanced propane conversion and hydrogen selectivity. Also, carbon deposit reduction caused that reduce catalysts deactivation.

109a Industrial Microwave Systems: Important Tools for Increased Productivity

Martin Yonnone, Cober Electronics, 151 Woodward Ave., South Norwalk, CT 06854

Industrial microwave systems offer speed, precision, efficiency and high quality production in difficult heat processing applications. This paper will discuss the critical factors for deciding when to use a microwave process and will show examples of resulting installations.

109b Optimizing Performance and Minimizing Cost in Microwave Heating Systems

John Gerling, Gerling Applied Engineering, Inc, P.O. Box 580816, Modesto, CA 95358

Microwave heating systems range in size and complexity from small consumer microwave ovens to large multi-chamber conveyor systems. Each system is made up of three basic elements: one or more microwave power generators, a means of microwave power delivery, and a cavity or applicator within which material is heated. These basic elements can be designed and configured in many ways, typically depending on process requirements and cost constraints. In most cases, the need to minimize cost by reducing system complexity conflicts with process requirements for performance. This paper discusses important and fundamental considerations for system design to achieve an optimal balance between cost and performance. Examples such as simple, low cost microwave ovens and highly complex and expensive semiconductor fab equipment highlight the overriding factors that influence design decisions.

109c Application of Microwave Energy for Processing of a Variety of Materials: An Overview

Dinesh Agrawal, Prof. of Materials Science and Eng.; Prof. Eng. Science & Mechanics; Dir. of Microwave Processing and Eng. Center, The Pennsylvania State University, 107 Materials Research Laboratory, University Park, PA 16802

Microwave energy has been in use for over six decades in diverse fields such as communication, medicine, chemistry, polymers, food processing, rubber vulcanization, drying, and materials processing. Microwave processing can provide distinct and substantial advantages over conventional thermal processes, including: energy efficiency, quality improvement, simplicity, speed, and eco-friendliness.

Most of the research over the last 20 years has focused on ceramics processing. However, only modestly successful lab samples emerged from such studies. Today, due to specific new developments, microwave processing is commercially viable in many new areas. Successes in microwave heating/sintering of oxide ceramics have been extended to cemented carbides for cutting and drilling applications that demonstrate improved performance. More recently, microwave processing has been shown to efficiently sinter, melt and braze metallic materials, often exhibiting better mechanical properties. This presentation will give an overview on these findings and discuss the commercial prospects and implications of the developed technology.

MICROWAVE PROCESSING OF METALS

Edward B. Ripley
Brian C. Warren
Jason A. Oberhaus
Y-12 National Security Complex
Oak Ridge, TN

INTRODUCTION

The Oak Ridge Y-12 National Security Complex has developed and applied microwave metal processing techniques to a wide variety of metals and alloys. This paper is an abbreviated discussion of several emerging metal working technologies.

Y-12, has been melting and casting metals using microwave energy for over a decade. This process has developed from a lab-scale curiosity to a production scale process. The technology has recently been licensed to several companies for commercial use. The metals that have been melted and cast to date include steels, titanium, zirconium, uranium, copper, brass, bronze, aluminum, and many other metals and alloy systems. Melts range in size from a few pounds to over 750 pounds. We have only begun to explore the huge potential of this technology, and additional scale-up is possible.

Bulk metals do not readily couple directly with microwave energy at room temperatures. Since metals are electrically conductive they readily reflect the incident energy. However, the electromagnetic field generated by the microwave field does allow for the loosely bound electrons to move and concentrate at surfaces, edges and points. This results in discharge of the energy in the form of arcing[1]. However in the case of powder metals, microwaves couple very efficiently at room temperatures[2] and sintering and melting can be achieved very rapidly.

The topics that will be covered are:

- Microwave melting, and casting of metal and alloy systems
- Heat-treating and stress relief of metals

THE EQUIPMENT AND SET UP

The microwave unit that was used for this work is a relatively standard 2.45 GHz multi-mode cavity. This cavity has been equipped with vacuum capability as well as complete control for inert, air, nitrogen and other atmospheres. The chamber is equipped with a mode stirrer to break up any standing waves and create a multi mode, 2.45 GHz, field within the cavity. A pair of 6 kW COBER S6F Industrial Microwave Generators are used to provide the microwaves to the cavity. The waveguides are equipped with dual couplers and a pair of Agilent Power Meters that supply a signal to an Agilent E44198B EPM Power Meter. A set of quarter wave tuning stubs has been placed in each wave-guide to help tune the cavity and reduce the reflected power. In addition, wave matching features are included at the windows where the wave-guide enters the chamber to prevent heating of the windows. One waveguide is directed into the cavity in transverse magnetic (TM) mode, the second is directed into the cavity in transverse electric (TE) mode.

MICROWAVE METAL MELTING AND CASTING

Microwave metal melting is counter-intuitive given common misconceptions about the behavior of metals in microwave fields. There are a few fundamental principals that make metal melting in the microwave possible. These are outlined below.

In order to heat and melt bulk metals using microwaves, three basic ingredients are required. These are a multimode microwave cavity, a microwave-absorbing ceramic crucible, and a thermally insulating casket that is microwave-transparent. The metal charge is placed in an open ceramic crucible (no lid) and the insulating casket is positioned to completely cover the open crucible. The casket and crucible assembly are then placed into a high-power multimode microwave cavity capable of uniformly heating the crucible to the desired temperature as shown in Fig. 1. When microwave energy is applied to the cavity, the energy is strongly absorbed by the crucible. The metal charge in the crucible is quickly heated through radiation, conduction, and convection with the heated crucible walls. The thermally insulating casket increases the energy efficiency of the microwave system by trapping the heat generated in the crucible. In this way, metal objects that could not be directly heated by microwave energy can be melted easily and efficiently.

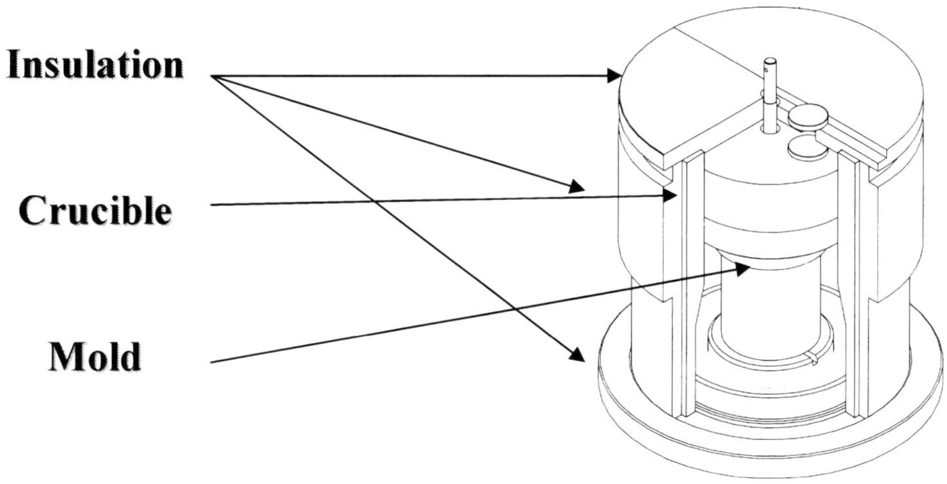

Figure 1. Microwave energy is easily transferred to metal shapes using a suscepting crucible.

HEAT-TREATING AND STRESS RELIEF OF METALS

There are three distinct methods employed to heat treat metals using a microwave furnace. They are:

- Molten Salt Bath Processing
- Granular Suscepting Media
- Fluidized Bed Processing

Molten Salt Bath Processing

Molten salt processing has traditionally been used for such processes as hardening, annealing, carborizing and nitro-carborizing. The composition of traditional eutectic salts for heat-treating include nitrate salts, nitrite salts, carbonate salts, cyanides, chloride salts and caustic salts. The work piece to be heat-treated is immersed in the molten salt bath. The type of heat treatment is determined by careful control of the time and temperature. Quenching of the metal part is often required to lock in the desired phases. One advantage of traditional salt baths it that they quickly and uniformly heat the work piece by conduction. By controlling the composition of the salt a variety of surface properties can be imparted to the work piece[5].

Traditionally the method of heating the salt to the required processing temperature has been by submerged electrode, internally heated, externally heated, electrical resistance, or gas. Each of these has their limitations.

Using microwaves as a heat source for molten salt processing has several advantages over other methods, these include:

1. The salt bath can be controlled at a wide variety of processing temperatures with a single salt bath and power supply.
2. The salt bath can be allowed to cool and solidify when not needed, and easily restarted when needed.
3. A variety of salt compositions can be used.
4. Can be operated a s a batch or continuous process.
5. The same microwave power supplies can be used for other heating processes when the salt bath is not in use.

This method has been used to melt and re-melt traditional hat treating salt in a laboratory scale. Scale-up to a full size production is being investigated.

Granular Suscepting Media

By placing a part in a granular material that directly absorbs microwave energy, the part can be quickly and efficiently heated. When metal items are placed into a microwave field, the energy tends to concentrate at edges and points. This could provide areas where stresses could initiate a crack and provide a failure mechanism. The use of a granular suscepting media helps to normalize and diminish any non-uniformity in the microwave field, and provide an evenly heated part. A variety of processing atmospheres can also be used. This method provides good thermal conduction and results in a simple and easily controlled process. This method has several distinct features:

1. It is completely scaleable, a small simple unit based on a "home" microwave has been demonstrated, and the data at the end of this report is from a 12 kW production scale microwave system.
2. A variety of suscepting powders can be used, depending on the metal being processed.
3. A wide range of temperatures can be achieved and maintained.
4. Can be operated as a batch or continuous process.
5. This process can be performed in a standard industrial microwave chamber, and does not require the purchase of a new system.

Fluidized Bed Processing

A alternative method of heat treating is the fluidized bed process. In this process a suspension of particles are suspended in a gaseous stream. The suspended particles behave like a liquid. A variety of processing gases, such as LPG, natural gas, ammonia and nitrogen, can be used to produce the desired surface properties to the metal and provide the fluidization. The atmosphere within the furnace can also be varied easily and quickly, according to the heat treatment required. Quenching is also commonly employed lock in the desired phases[5].

This processing method has advantages in energy usage, material characteristics, and processing flexibility.

1. It is scaleable, a small simple unit based on a "home" microwave has been demonstrated
2. The thermal conductivity of this method is comparable to molten salt methods.
3. Does not cause hydrogen embrittlement.
4. Can be heated quickly, used for small production runs and shut off when not in use.
5. A variety of suscepting powders can be used, depending on the metal being processed.
6. A wide range of processing gasses can be used depending on processing requirements.
7. A wide range of temperatures can be achieved and maintained.
8. Can be operated as a batch or continuous process.

Case Study

Heating metals using microwave energy is also being pursued as a means of heat treating different metals. The following is a case study of using microwaves to heat treat metal using the granular suscepting method previously described. The goal was to use microwave energy for annealing. It is important to achieve uniform heating throughout the work piece, without having point, edge or skin effects. As previously described, these effects cause non-uniform heating of a material which could lead to unwanted changes in material properties. Experiments were performed to compare conventionally-annealed cartridge brass to microwave-annealed cartridge brass. Cartridge brass was selected based on material properties and available data for comparison. Work was performed in three phases. In Phase I, cartridge brass coupons were heated to 426°C. In Phase II cartridge brass coupons were heated to 537°C, and in Phase III a cartridge brass gear was heated to 649°C. In each of these phases, the microstructure and hardness of the microwave heat-treated samples were compared to conventional heat-treated samples (Table 1). In all three phases, the microstructure of the microwave samples duplicated the microstructure of the conventional samples. The hardness values of the microwave samples were 5-15% higher than the conventionally treated samples in all the phases. It is believed that this is a result of slightly longer times at temperature for the conventionally processed samples than microwave processed samples.

During this task, homogeneity of the work piece was achieved and no negative effects were observed from the use of the microwave. There were no edge effects, surface effects, or arcing of the metal. The microwave successfully duplicated conventional methods of heat treating a metal.

	Temperature (°C)	Avg. Hardness (DHP)	Grain size (ASTM)
As-received	N/A	135	7.5
Conventional	426	107	7
Microwave	426	110	7
Conventional	537	73	6
Microwave	537	84	6
Conventional	649	69	2
Microwave	649	78	2

Table 1 Hardness and grain size of annealed samples[3]

Performing the heat treatment at the higher temperatures resulted in a significant change in microstructure from the as-received samples. As shown in Figure 2, the 649°C heat treatment had significant grain growth. The grain growth of the microwave is identical to the grain growth of the conventional heat treat.

The final test was to compare the heat treatment of a representative shape. A gear that included rounded, sharpened, flat, and typical teeth was used. Shapes cut into the body of the gear to make it more complex. Pictures of the gears can be seen in Figure 3, which shows the sample locations, where metallography and hardness testing were performed. This design was chosen because it represents a broad range of angles and curvatures a component may have that would be desired to heat-treat. If there were any negative effects caused by the use of microwaves as a heat source, it would be shown in a component of this design.

Figure 2 Microstructure of as-received cartridge brass (left), conventional heat-treated (middle), and microwave heat-treated (right) at 649°C. (Equivalent magnification of 100X).[3]

Figure 3 Sample locations of the gears used for analysis, with the gear used for conventional heat treatment (left) and the microwave heat treatment (right).[3]

The gears were heated to 649°C and held for 1 h in both the conventional furnace and the microwave. After the heat treatment, the gears went through the same evaluation as in the previous phases (see Figure 4).

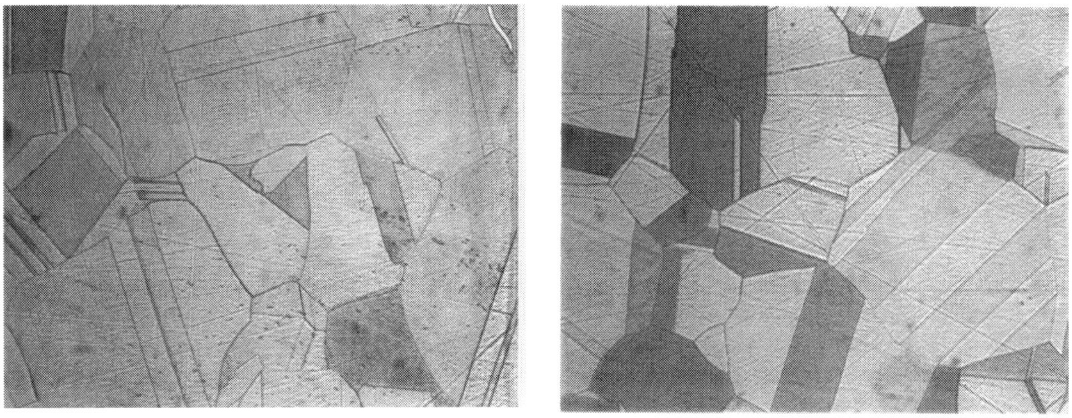

Figure 4 Microstructure of the gear at location 4T at an equivalent magnification of 100X, which has the conventional heat treatment (left) and the microwave heat treatment (right).[3]

No negative surface, edge, or point effects were observed by using the microwaves as a heat source for annealing. This shape incorporated several different challenge shapes and curvatures. This did not inhibit the ability of the microwave to successfully heat treat any of the teeth or the base of the gear. The surface finish of the microwave-annealed gear was comparable to the conventional heated gear. The microstructure of the gear heated in the microwave showed homogeneity throughout the entire structure. Arcing is most likely to occur at sharp points, and no arcing was observed during the heat treatment in the microwave.

Based on the results of these experiments, the microwave can be used to successfully heat treat metal components without any negative effects on the metal. In every phase of the project, the hardness data and microstructures were all similar to that of the conventional heat treatment. Microstructure, hardness and homogeneity were achieved in the work pieces.

ACKNOWLEGEMENTS

We would like to thank the following people for their efforts related to microwave processing technology at the Y-12 National Security Complex:

- Cressie Holcombe and Stan Morrow, former employee of Y-12, for thier work in ceramic applications using microwave technology and microwave innovation
- Don E. Schechter, former employee of Y-12, for his dedication and development of microwave applications
- Alan F. Moore, Y-12, for his work in metallurgical applications
- Terry White, Patrice Eggleston, Hal Clift, Hugo Huey and Sam Lariviere for their invaluable help on various aspects of this field work at Y-12
- References 6, 7 and 8 as listed below.

REFERENCES

1) A. C. Metaxas and R. J. Meredith, "*Industrial Microwave Heating,*" Peter Peregrinus Ltd. 1988
2) R. Roy, D. Agrawal, J. Cheng, and S. Gedevanishvili, "Unexpected sintering of powdered metals parts in microwaves", Nature, 399, 664 (June 17, 1999)
3) J. A. Oberhaus, "*Microwave Annealing,*" Y/DV-1832 Y-12 Internal Document., (2003).
4) Morrow et al, "US Patent 6,554,924, B2," *Metallic Diffusion Process and Improved Article Produced Thereby.*
5) Report by the Australian Government, Department of the Environment and Heritage "Cleaner Production - Replacement of Salt Baths with Fluidised Bed Furnaces" - Quality Heat Technologies Pty Ltd; January 1997
6) L.H. Van Vlack, "*Elements of Materials Science and Engineering,*" Third Edition, Addison-Wesley, pp 140-145, Reading Mass., (1975).
7) D. E. Clark and W.H. Sutton, D.A. Lewis, "*Microwaves: Theory and application in Materials Processing IV,*" American Ceramic Society, Westerville, OH.
8) R.F. Schiffmann, "*Principals of Industrial Microwave and RF Heating,*" First World Congress on Microwave Processing, Jan 5-7, 1997.

29f Microwave Assist Technology for Chemical Processing

Shawn Allan, Morgana Fall, Inessa Baranova, and Holly Shulman. Ceralink, Inc., 105 Jordan Road, Troy, NY 12180

Microwaves are being used to assist chemical processes, with surprising results, including super-fast reactions and new compounds. Many discoveries have been limited to laboratory scale work, waiting on appropriate industrial microwave equipment. The need for equipment and process development to occur in concert is now being addressed. This talk focuses on the scale-up and commercialization of microwave assisted processes and equipment. New commercially available equipment will be described, including microwave autoclaves, and microwave assist kilns. A review of important applications relevant to diverse industries will be given. These include catalysts, pharmaceuticals, proppants, additives, electroceramics, and nanomaterials. Several examples will be discussed in detail, including 1) calcination reactions using Microwave Assist Technology (MAT) and 2) precious metal reclamation with microwave autoclaves.

AUTHOR INDEX

AUTHOR INDEX

AUTHOR INDEX

AUTHOR INDEX

AUTHOR INDEX

American Institute of Chemical Engineers
3 Park Avenue
New York, New York 10016-5991

ISBN: 978-1-60560-226-4